Commercial Flowers

— Volume 5 —

THE EDITORS

Professor T.K. Bose obtained his M.Sc. (Ag.) and Ph.D. degree from Calcutta University, West Bengal, India. Trained at the Royal Botanic Garden, Kew, UK. Former Secretary of the Agri-Horticultural Society of India; Project Coordinator, All India Coordinated Floriculture Improvement Project (ICAR); Professor and Head, Department of Horticulture; Dean, Faculty of Agriculture; Director of Research, Bidhan Chandra Krishi Viswavidyalaya, West Bengal; Member, State Agriculture Commission, West Bengal; Member of Scientific Advisory Committees of ICAR, CSIR (Govt. of India) and Government of West Bengal. Awarded (D.Sc.) Honoris Causa from OUAT, Bhubaneshwar and UBKV, Coochbehar, India. Guided 45 Ph.D. students, published more than 250 research papers and authored 30 books on horticultural crops.

Dr. Chenna Reddy Aswath obtained his B.Sc. (Hort), M.Sc. (Hort.) in Floriculture from University of Agricultural Sciences, Bengaluru and Ph.D. degree in Biotechnology from Guwhati University, Assam. Serving as Principal Scientist in ICAR-IIHR, Bengaluru and has made significant contributions in the field of Floriculture and Biotechnology. Recipient of "Dr BP Pal Best Scientist Award for Agriculture" from NABS, Chennai, became FELLOW of the NABS, Chennai, ISOH, New Delhi and SAB, Dharwad, CHAI, New Delhi. Received the 2012 SAB Award of Excellence in Horticultural Biotechnology. Recipient of Brain pool Post-Doctoral Fellowship from South Korea Science Federation for the outstanding research contributions in Horticultural Biotechnology. His short term fellowship in Japan under INSA has recognized him among the Orchid tissue culture group. Published more than 110 research papers besides several books. Guided more than 10 post-graduate students.

Dr. Tarak Nath Saha graduated in Agriculture from Banaras Hindu University, Varanasi (Uttar Pradesh), obtained his Masters and Doctoral degree in Horticulture with specialization in Floriculture from India Agricultural Research Institute (IARI), New Delhi. Presently working as Senior Scientist at ICAR-Directorate of Floricultural Research, Pune. He has an experience of more than 12 year in research, teaching and extension, besides coordinating work of All India Coordinated Research Project (AICRP) on Floriculture. He has contributed 25 research papers, 41 technical articles, 5 book chapters, 20 technical bulletins, 6 folders, besides in the editorial team of 30 Annual Reports and 10 AICRP workshop proceedings. Developed one variety and registered one genetic stocks of tuberose with ICAR-NBPGR, New Delhi. He is actively involved in PG teaching; recognized as faculty of MPKV, Rahuri (also at IARI, New Delhi 2009-11) and guided more than 5 students in Floriculture as advisory.

Dr. Reeta Bhatia is working as senior scientist at the Division of Floriculture and Landscaping, ICAR- Indian Agricultural Research Institute, New Delhi. She has more than 13 years of research and teaching experience in the field of floriculture. She has obtained her Master's degree from Punjab Agricultural University, Ludhiana and doctoral degree from Indian Agricultural Research Institute, New Delhi. She has been awarded with the prestigious Endeavor Research Fellowship by the Government of Australia and pursued her postdoctoral research at the University of Western Australia. She has an excellent academic and research career and awarded with several medals. She has published more than 65 research papers in the reputed high impact scientific journals and written more than 30 review/popular/scientific articles. She is also serving as editor for the Journal of Ornamental Horticulture.

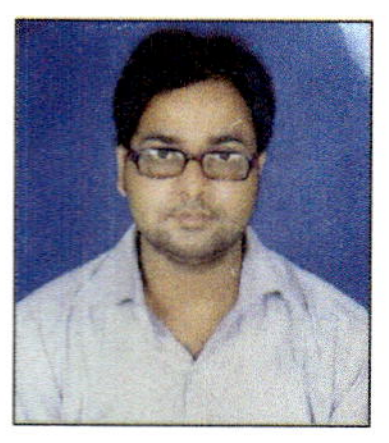

Dr. K. Dutta obtained his M.Sc. (Hort.) from CCS Haryana Agricultural University, Hisar, Haryana and Ph.D. from Uttar Banga Krishi Viswavidyalaya, West Bengal. Acting as Assistant Secretary of The Agri-Horticultural Society of India, Kolkata, West Bengal. Recipient of Inspire Fellowship from Department of Science and Technology, New Delhi. Published several scientific papers in national journals on different aspect of horticulture particularly in tissue culture and written book chapters on nursery management and jasmine. Associated with micropropagation of different of horticultural crops during last 7 years.

Debmala Mukherjee is persuing Ph.d. in Department of Vegetable Science from Bidhan Chandra Krishi Viswavidyalaya (BCKV), West Bengal, India. She has completed her B.Sc. (Horticulture) and M.Sc. (Vegetable Science) from BCKV and now continuing her research under All India Coordinated Vegetable Improvement Project (AICVIP) at Kalyani, West Bengal centre. She was recipient of the Bayer Merit Scholarship during her M.Sc. programme and awarded with ICAR- National Talent Scholarship in the discipline of Horticulture and also qualified NET-2019 (National Eligibility Test) conducted by Indian Council of Agricultural Research (ICAR) in the discipline of Vegetable Science. She has published more than 5 research publications in several national and international journals of repute and has written three book chapters mainly focused on seed production of vegetables and also has published a number of articles in several journals.

Commercial Flowers

— *Volume 5* —

— *Senior Editor* —

C. Aswath
Division of Flower and Medicinal Crops
ICAR-Indian Institute of Horticultural Research
Bengaluru 560089, Karnataka, India

— *Editors* —

T.K. Bose
Faculty of Horticulture
Bidhan Chandra Krishi Viswavidyalaya
West Bengal 741 252, India

Tarak Nath Saha
ICAR-Directorate of
Floricultural Research,
Pune 411005
Maharashtra, India

Reeta Bhatia
Division of Floriculture and
Landscaping,
ICAR- Indian Agricultural
Research Institute
New Delhi 110012

Koushik Dutta
The Agri-Horticultural Society of India
1, Alipore Road,
Kolkata 700027
West Bengal, India

Debmala Mukherjee
Faculty of Horticulture,
Bidhan Chandra Krishi Viswavidyalaya
West Bengal 741252, India

2022

Daya Publishing House®
A Division of
Astral International Pvt. Ltd.
New Delhi – 110002

Third Revised & Illustrated edition 2022

Cataloging in Publication Data--DK

Courtesy: D.K. Agencies (P) Ltd. <docinfo@dkagencies.com>

Commercial flowers / senior editor, C. Aswath ; editors, T.K. Bose [and four others]. -- Third revised edition.

Volumes cm

Contributed articles.

Includes index.

ISBN 978-93-54610-21-9 (HB)

1. Floriculture. 2. Flowers--Pictorial works. I. Aswath, C. (Principal scientist at the Indian Institute of Horticulture Research), editor. II. Bose, T. K., editor.

LCC SB405.C66 2021 | DDC 635.9 23

Published by : **Daya Publishing House**®
A Division of
Astral International Pvt. Ltd.
– ISO 9001:2015 Certified Company –
4736/23, Ansari Road, Darya Ganj
New Delhi-110 002
Ph. 011-43549197, 23278134
E-mail: info@astralint.com
Website: www.astralint.com

Digitally Printed at : Replika Press Pvt. Ltd.

From the Desk of Prof. T.K. Bose

The undersigned had initiated the publications entitled 'Fruits: Tropical and Sub-tropical', 'Vegetable Crops' and 'Commercial Flowers', one volume each in the 1980's in collaboration with co-editors, humbly expresses from the core of his heart that the updated revised editions of the above mentioned publications with colour illustrations in thirteen volumes which was planned in June, 2020, are being printed from January this year, a rare event in Horticultural Science has been possible by the blessing of Almighty God, with the cooperation of dedicated horticulturists as editors and contributors.

T.K. Bose
Editor

Preface to the Third Edition

The first edition of the book entitled **'Commercial Flowers'** published in 1989, reprinted in 1993 and 1998 was greatly appreciated and proved valuable scientific review of commercially important flowers, to the students, teachers, researchers and extension specialists in horticulture in general and floriculture in particular, in different parts of the world. The book was revised in 2002 with the same title in two volumes due to its increasing demand, which contained updated scientific information of commercially grown flower crops and considered as valuable contribution on the subject. After about 18 years from the last revised publication in 2002, voluminous information pertaining to crop improvement, production, protection, postharvest management, biotechnology has been generated necessitating further revision of the existing chapters and inclusion of more ornamentals plants of commercial importance in four volumes, with relevant colour illustrations.

Fifth volume in this edition includes chapters on Digitalis, Delphinium, Calceolaria, Pansy, Statice, *Impatiens, Catharanthus*, Gazania, Pelargonium, Iris, Hemerocallis, Spathiphyllum, Haemanthus, Eucharis, Pancratium, Habranthus, Shrubs- beautiful and perennial ornamental plants, Hydrangea, Hibiscus, Bougainvillea, Camellia, Azalea (Rhododrendron), Poinsettia, Gardenia, Fuchsia, Ixora, Mussaenda, Adenium, Calliandra, Crossandra and Solidago with relevant colour illustrations, covering more than 850 pages.

Apart from own collection of photographs, editors have also taken some photographs which are in public domain from different websites, as far as possible. Therefore, we acknowledge with enormous thanks to all State Agricultural Universities and Research Institutes under Indian Council of Agricultural Research, New Delhi, India; Public and Private International Agricultural Institutes across the world for using photographs from their websites to make this book a valuable source of knowledge for the readers. We think images can be powerful teaching tools, as illustrations to in-class lectures, or for studying concepts outside of the classroom. We also think that identification of flower crops, their importance and uses, species and cultivars, methods of propagation, cultivation practices including greenhouse cultivation, diseases and pests, harvesting and post harvest technology and achievement of biotechnology, with the help of suitable illustrations in a consolidated manner will make a dent for the students, researchers and teachers of floriculture and ornamental horticulture.

We sincerely thank the contributors for preparing updated revised manuscripts. Our thanks are due to Shri Prateek Mittal, Director, Astral International (P) Ltd., New Delhi, for his keen interest in bringing out this publication.

Kolkata

T.N. Saha
Reeta Bhatia
K. Dutta
Debmala Mukherjee

Contents

List of Contributors

Anand, P.

Division of Floriculture and Landscaping, ICAR-Indian Agricultural Research Institute, Pusa Campus, New Delhi 110012, India

Aswath, C.

Division of Flower and Medicinal Crops, ICAR-Indian Institute of Horticultural Research, Bengaluru 560089, Karnataka, India

Arora, J.S.

Department of Vegetable Crops and Landscaping, Punjab Agricultural University, Ludhiana 141004, Punjab, India

Bal, S.

Department of Vegetable Science, Faculty of Horticulture, Bidhan Chandra Krishi Viswavidyalaya, Mohanpur 741252, Nadia, West Bengal, India

Barman, K.

Department of Horticulture, Institute of Agriculture Science, Banaras Hindu University, Varanasi 221005, Uttar Pradesh, India

Bhatia, R.

Division of Floriculture and Landscaping, Indian Agricultural Research Institute, New Delhi 110012, India

Bhunia, A.

Department of Vegetable Science, Faculty of Horticulture, Bidhan Chandra Krishi Viswavidyalaya, Mohanpur 741252, Nadia, West Bengal, India

Biswas, B.

Department of Horticulture, Bidhan Chandra Krishi Vishwavidyalaya (State Agricultural University), Mohanpur 741 252, West Bengal, India

Bose, T.K.

Department of Horticulture, Bidhan Chandra Krishi Vishwavidyalaya (State Agricultural University), Mohanpur 741 252, West Bengal, India

Das, P.

Regional Plant Resource Centre, Bhubaneswar 751015, Odisha, India

Dhatt, K.K.

Department of Floriculture and Landscaping, Punjab Agricultural University, Ludhiana 141004, Punjab, India

Dhiman, M.R.

ICAR-Indian Agricultural Research Institute, Regional Station, Katrain, Kullu-Valley 175129, Himachal Pradesh, India

Dutta, K.

The Agri-Horticultural Society of India, 1, Alipore Road, Kolkata 700027, West Bengal, India

Ganga, M.

Department of Floriculture and Landscape Architecture, Horticultural College and Research Institute, Tamil Nadu Agricultural University, Coimbatore, Tamil Nadu, 641003, India

Gurung, A.

Division of Floriculture and Medicinal Crops, ICAR-IIHR Bengaluru 560089, Karnataka, India

Islam, S.M.

Faculty of Horticulture, Bidhan Chandra Krishi Vishwavidyalaya (State Agricultural University), Mohanpur 741 252, West Bengal, India

Jadhav, P.R.

ICAR-Directorate of Floricultural Research, College of Agriculture Campus, Shivajinagar, Pune 411005, Maharashtra, India

Jain, R.

Division of Floriculture and Landscaping, ICAR-Indian Agricultural Research Institute, New Delhi 110012, India

Kad, S.

ICAR-Directorate of Floricultural Research, College of Agriculture Campus, Shivajinagar, Pune 411005, Maharashtra, India

Kadam, G.B.

ICAR-Directorate of Floricultural Research, College of Agriculture Campus, Shivajinagar, Pune 411005, Maharashtra, India

Kakote, A.S

Department of Horticulture, Mahatma Phule Krishi Vidyapeeth, Rahuri 413722, Maharashtra

Kumar, K.R.

Dr.YSRHU, Horticultural Research Station, Kovvur, Andhra Pradesh, India

Kumar, P.N.

ICAR-Directorate of Floricultural Research, College of Agriculture Campus, Shivajinagar, Pune 411005, Maharashtra, India

Kumar, R.

ICAR-Indian Agricultural Research Institute, Regional Station, Katrain, Kullu-Valley 175129 , Himachal Pradesh, India

Kumar, R.

Division of Flower and Medicinal Crops, ICAR-Indian Institute of Horticultural Research, Bengaluru 560089, Karnataka

Kundu, S.

Department of Vegetable Science, Faculty of Horticulture, Bidhan Chandra Krishi Viswavidyalaya, Mohanpur 741252, Nadia, West Bengal, India

Lalramhlimi, B.

Department of Vegetable Science, Bidhan Chandra Krishi Vishwavidyalaya (State Agricultural University), Mohanpur 741 252, West Bengal, India

Madhubala

Department of Floriculture and Landscaping, Punjab Agricultural University, Ludhiana 141004, Punjab, India

Mitra, M.

Department of Horticulture, Bidhan Chandra Krishi Vishwavidyalaya (State Agricultural University), Mohanpur 741 252, West Bengal, India

Mukherjee, D.

Department of Vegetable Science, Bidhan Chandra Krishi Vishwavidyalaya (State Agricultural University), Mohanpur 741 252, West Bengal, India

Parthasarathy, V.A.

ICAR – Indian Institute of Spices Research, 32/482 C Narmada Nilayam, Bharathan Bazar, Chelavoor, Calicut 673571, Kerala, India

Pal, P.

Department of Horticulture, Bidhan Chandra Krishi Vishwavidyalaya (State Agricultural University), Mohanpur 741 252, West Bengal, India

Patel, D.A.

Department of Floriculture and Landscape and Architecture, S.D.Agricultural University, Jagudan, Mehsana, 384460, Gujarat, India

Prasad, K.V.

ICAR-Directorate of Floricultural Research, College of Agriculture Campus, Shivajinagar, Pune 411005, Maharashtra, India

Raju, D.V.S.

ICAR-Directorate of Floricultural Research, Regional Station, Vemagiri, Andhra Pradesh, India

Saha, S.

ICAR-Indian Agricultural Research Institute, Regional Station, Pune 411008, Maharashtra, India

Saha, T.N.

ICAR-Directorate of Floricultural Research, College of Agriculture Campus, Shivajinagar, Pune 411005, Maharashtra, India

Sambandamurthi, S.

Division of Flower and Medicinal Crops, ICAR-Indian Institute of Horticultural Research, Bengaluru 560 089, Karnataka, India

Sangeetha Priya, S.

Division of Flower and Medicinal Crops, ICAR-Indian Institute of Horticultural Research, Bengaluru 560 089, Karnataka, India

Sanyal, D.

Department of Crop Improvement, Horticulture and Agricultural Botany, Institute of Agriculture, Viswa-Bharathi, Sriniketan 731 236, West Bengal, India

Shreya

Department of Basic Science, College of Horticulture, S.D. Agricultural University, Jagudan, Mehsana, 384460 Gujarat, India

Singh, B.

Division of Floriculture and Landscaping, ICAR-Indian Agricultural Research Institute, New Delhi 110012, India

Singh, S.

Deparment of Floriculture and Landscaping, Punjab Agricultural University, Ludhiana 141004, Punjab, India

Tiwari, A.K.

Division of Floriculture and Landscaping, ICAR-Indian Agricultural Research Institute, New Delhi 110012, India

Vanlalruati

Division of Floriculture and Landscaping, ICAR-Indian Agricultural Research Institute, Pusa Campus, New Delhi 110012, India

Varun, M.H.

Division of Floriculture and Landscaping, ICAR-Indian Agricultural Research Institute, New Delhi 110012, India

Visalakshi, M.

Department of Floriculture and Landscape Architecture, Horticultural College and Research Institute, Tamil Nadu Agricultural University, Coimbatore 641003 Tamil Nadu, India

Digitalis

P. Naveen Kumar, Tarak Nath Saha, Ganesh B. Kadam, D.V.S. Raju and K.V. Prasad

1.0 Introduction

Foxglove, also known as Digitalis, Fairy's Gloves, Witches' Fingers and Fairy Thimbles is one of the most popular garden flowers. The name digitalis from latin word 'Digit' means finger, which decribes finger-shape flowers the plant bears. *Digitalis* is a genus of some species of flowers that grow wild in the eastern hemisphere, and are widely planted as ornamentals by gardeners, despite being poisonous. Several species have been used medically for centuries and are source for digoxin, a drug still used to treat cardiac arrhythmia. The medical use of digitalis was popularized by a British physician, William Withering, whose book, *An Account of the Foxglove*, was first published in 1785 (Withering, 1785 and 2018). The term digitalis is also used for drug preparations that contain cardiac glycosides, particularly one called digoxin, extracted from various plants of this genus. Foxglove has medicinal uses but can also be toxic to humans and other animals. The plant flowers for a shorter duration of time. The flowers are tubular in shape, produced on a tall spike and vary in colour with species, from purple to pink, white, and yellow.

2.0 Importance and Uses

2.1.0 Aesthetic Uses

This biennial is often grown as an ornamental plant due to its vivid flowers which range in colour from various purple tints through pink and purely white. The flowers also possess various marks and spottings. They are excellent cut flowers that last longer in a vase. It is suitable for beds, border and also used in the shrubbery with fine effect, as its tall, spire-like spikes, crowned with its large thimble, bell-shaped purple or white flower, finely contrast with the green foliage of the shrubs in any garden.

2.2.0 Therapeutic and Other Uses

Digitalis is an example of a drug derived from a plant that was formerly used by herbalists. It was largely abandoned because of its narrow therapeutic index and the difficulty of determining the amount of active drug in herbal preparations. Once the usefulness of digitalis in regulating the human pulse was understood, it was employed for a variety of purposes, including the treatment of epilepsy and other seizure disorders, which are now considered to be inappropriate treatments. Digoxigenin (DIG) is a steroid found in the flowers and leaves of *Digitalis* species. Digoxigenin can be used as a molecular probe to detect mRNA *in situ* and label DNA, RNA and oligonucleotides.

Although, the parts of the plant that grow above the ground can be used for medicine, foxglove is unsafe for self-medication. All parts of the plant are poisonous. Chemicals taken from foxglove are used to make a prescription drug called digoxin. *Digitalis lanata* is the major source of digoxin (https://www.webmd.com). Symptoms of digitalis poisoning include nausea, vomiting, severe headache, dilated pupils, problems with eyesight and convulsions in highest level of toxicity.

Uses of Digitalis

As Cut Flower

In Rockgarden

In Pots

As Borders

Along Pathways

Surrounding Lawn

In Landscaping

In Gardens

Uses of Digitalis

In Bouquets **In Home Garden** **Floral Arrangements**

Cut Bunch **Border**

In Garden Pathways **In Beds**

Source: www.cdn8.dissolve.com, www.hort.extension.wisc.edu, www.images.fineartamerica.com, www.reneesgarden.com, www.betweennapsontheporch.net, www.a4.pbase.com, www.live.staticflickr.com, www.in.pinterest.com, www.i.pinimg.com/originals, www.rayagarden.com, www.in.pinterest.com, www.gardenerspath.com, www.lovenfreshflowers.com, www.ccfc.org

Only two species, *viz. Digitalis lanata* and *Digitalis purpurea* are of economic interest. The Grecian foxglove (*D. lanata*) is preferred over *D. purpurea* as a source of glycosides for pharmaceutical industry.

3.0 Origin and History

The discovery of digitalis is credited to the Scottish Doctor William Withering (https://www.ch.ic.ac.uk). The plant is a biennial, native to Europe, North Africa and Central Asia. The common name, Foxglove, refers to the fact that the spire of blossoms resembles clusters of gloves and the areas where foxgloves grew naturally were thought to be inhabited by fairies. Thus the plants were thought to be fairies' gloves. The plant had been known as far back as 1000AD. It has been cultivated since the 1400's in England, but was not grown in American gardens until the 1700's. Joseph Breck in his 1851 book, The 'Flower Garden' (Breck, 1851), describes five varieties with the most popular being *Digitalis purpurea*, the purple foxglove. Breck writes, 'The plant is a violent poison, but invaluable in medicine'. By the late 1700's, the plant's value as a heart stimulant was well known and it had become a valued medicinal plant as well as a garden flower. The drug discovery had been made by a British, Dr. William Withering in 1785 when he had tried, but failed to relieve a woman who appeared to be dying from dropsy. Weeks later, he was informed that the woman had been cured by drinking an herbal tea. Withering found that the ingredient in the tea was foxglove and the active ingredient that had cured the woman was digitalis. That same year he published, 'An Account of the Foxglove'. This book propelled digitalis to the forefront of treatments for the heart (https://harvesting-history.com).

4.0 Taxonomy

Luckner and Wichtl (2000) divided the genus *Digitalis* into five sections, namely *Frutescentes, Digitalis, Grandiflorae, Tubiflorae*, and *Globiflorae*, comprising 19 species. They included the genus in the Scrophulariaceae family, which has previously been put in the order of the Lamiales (Dahlgren, 1980). Chemo-taxonomic considerations (Luckner and Wichtl, 2000) and approaches were not fully convincing and a few studies focusing on chemotaxonomic markers have been done over the past 20 years (Taskova *et al.*, 2005). Rigorous approaches of assessing phylogenetic relationships using DNA sequences finally made it necessary to disperse the order Scrophulariales (Douglas *et al.*, 1998). The classification of the genus Digitalis still followed the morphological analyses of Werner (Luckner and Wichtl, 2000; Werner, 1961, 1965). More recently, morphological, bio-geographical and molecular data were combined to challenge Werner's phylogentic tree (Carvalho and Culham 1997, 1998 and Nebauer, 2000). Moreover, the Scrophulariaceae were regarded as a collection of unrelated lineages and were dismembered. Olmstead *et al.* (2005) used three plastid genes and revealed the existence of at least five distinct monophyletic groups and placed the tribus Digitaleae into the Veronicaceae. Albach and Chase (2004) still found several incongruencies and proposed the "new" Plantaginaceae integrating the genus *Digitalis* and other members of the former Scrophulariaceae (Albach *et al.*, 2005). Currently, the genus *Digitalis* comprises 23 species including the four species of the former genus Isoplexis based on the molecular phylogeny of Bräuchler *et al.* (2004) and Herl *et al.* (2008).

5.0 Morphology

The genus Digitalis consists of herbaceous perennial plants, shrubs and biennials. Several species are cultivated for their attractive flower spikes. Foxgloves typically grow to a height of 45-150 cm. They produce alternating, ovate to oblong leaves towards the lower part of the stem, which is capped by a tall, one-sided cluster of pendulous, bell-shaped flowers, each of which may be up to 6.5 cm long.

The flowers are be purple, yellow, or white and are often marked with spots within. Flowers are zygomorphic and arranged in terminal, bracteate racemes. The calyx is equally five-lobed and shorter than the corolla tube. The corolla, with a cylindrical-tubular to globose tube, is often constricted at the base and the limb is more or less two-lipped. The upper lip is usually shorter than the lower, which is spotted or veined inside (Brauchler *et al.,* 2004).

The fruits are dry capsules with numerous small seeds. The common, or purple, foxglove (*Digitalis purpurea*) is cultivated commercially as the source of the heart-stimulating drug digitalis. The drug is obtained from the dried leaves (https:// www.britannica.com).

6.0 Species and Cultivars

6.1.0 Species

The digitalis comprises of more than 20 species and the description of some of the important species are given hereunder.

Digitalis canariensis

This species is part of section Isoplexis, which was temporarily accepted as an own genus. The synonym *Isoplexis canariensis* also continues to be used. In general, as of 2017, opinions concerning the taxonomic status of *Isoplexis* species differ depending on the source. Individuals of these species are small, evergreen plants growing into rounded shrubs up to 150 cm tall. The plant has lanceolate-ovoid leaves with toothed margins. The leaves are spirally arranged. The inflorescence is a cluster of orange-reddish flowers (3 cm) with short petals and noticeable upper lip and lower lip. *Digitalis canariensis* contains cardenolids (cardiac glycosides), which are toxic in the case of an overdose. Its medical use is well documented. *Digitalis canariensis* is endemic to the Canary Islands where it occurs on the islands of Tenerife, La Gomera and La Palma. It grows in sub-tropical humid forest areas.

D. ciliate

It is commonly called Hairy Foxglove. It has thimble-shaped, yellow to cream coloured flowers produced on perennial plants with evergreen foliage. It is native to the Caucasus and is grown as an ornamental. The species name is derived from the fine hairs that cover the plants stems and flowers. Individuals of these species are herbs that grows between 30-60 cm tall. The seeds of *D. ciliata* are rich in the steroid glycoside digitonin, along with other lipids that are important in cardiac treatments.

D. ferruginea

It is also known as the rusty foxglove. It is native to Hungary, Romania, Turkey and the Caucasus. The Latin specific epithet *ferruginea* means "rusty coloured", referring to the flowers. It is a biennial or short-lived perennial plant growing to 1.2 m, which forms a rosette, oblong dark green leaves and carries spikes of brown, tubular flowers in summer. All parts of the plant are poisonous and contact with the leaves may also cause an allergic reaction. This plant has gained the Royal Horticultural Society's Award of Garden Merit.

D. grandiflora

The yellow foxglove, big-flowered foxglove, or large yellow foxglove, is a species of flowering plant in the genus *Digitalis*. It is native to southern Europe and Asia. In mountains it grows on warm, bushy slopes or areas left after logging. The Latin specific epithet *grandiflora* means "large flowered". It is a herbaceous perennial growing from a short rootstock with fibrous roots. It has glossy green, veined leaves, whose flowering stem can reach a height of 70–120 cm. The pale yellow bell-shaped flowers are spaced out on the stem, 3–4 cm (1–2 in) long and show a netted brown marking in their interior. In the wild, plants bloom in June and July. *Digitalis grandiflora* is long lived perennial, it has gained the Royal Horticultural Society's Award of Garden Merit.

D. laevigata

The common name iss Grecian foxglove or giraffe foxglove. This species is native to southern Europe. It grows wild in the Balkans. It grows to about 70–90 centimetres in height. This perennial herbaceous plant has erect stems with lance-shaped leaves, while basal leaves are oblong to ovate. It produces spires of orange or yellow-brown bell-shaped flowers with a large whitish lower lip and purple veined, speckled interiors.

D. lanata

It is often called as woolly foxglove or Grecian foxglove. It gets its name due to the woolly indumentum of the leaves and is toxic in all parts of the plant. The plant is also harmful to other animals. It contains a powerful cardiac glycoside that may be used by patients with heart conditions. Digoxin (digitalin) is a drug that is extracted from this species.

D. mariana

It is a perennial foxglove with evergreen foliage and rose-red coloured flowers produced in summer. It is native to Portugal and Spain. The calyx has sepals which are patent or subpatent and the corolla tube abruptly tightens towards its base. The capsule is equal or larger in length than the calyx, and the lower lip of the corolla has lateral lobes which are little developed, rounded, not auricular, nor split to the mouth of the tube.

Species of Digitalis

Digitalis purpurea ***Digitalis lutea*** ***Digitalis canariensis***

Digitalis ciliate ***Digitalis ferruginea*** ***Digitalis grandiflora***

Digitalis laevigata ***Digitalis lamarckii*** ***Digitalis lanata***

Digitalis mariana ***Digitalis minor*** ***Digitalis obscura***

Source: **www.en.wikipedia.org, www.species.wikimedia.org**

D. minor

It is also known as dwarf Spanish foxglove. It is a biennial or short-lived perennial species of foxglove which is endemic to the Balearic islands with large, pendulous, pink or purple flowers. Closely related to the common purple foxglove, it is best distinguished by its small fruits. It is one of the only foxgloves to grow in calciferous, alkaline soils.

D. obscura

It is commonly called willow-leaved foxglove or dusty foxglove or spanish rusty foxglove. It is also grown as an ornamental flower. This foxglove is a woody perennial plant. It is similar to many of the foxglove species in its high toxicity and medicinal use as a source for the heart-regulating drug digoxin. Its strikingly distinctive amber- to copper-coloured flowers give the species its name and help distinguish it from other members of the genus.

6.2.0 Cultivars

Foxglove is very variable throughout its natural range, and several subspecies, varieties and forms have been named. Gardeners have produced several hybrids, and many cultivars have been selected. 'Alba' has white flowers. 'Giant Shirley' has larger bell-shaped pink flowers with crimson spots. 'Gloxinioides' has fringed flowers that can be pink, yellow or purple and are blotched on the inside of the corolla. 'Excelsior Hybrids' have pink, white or yellow flowers that stand out horizontally and are arranged all around the spike rather than just on one side. 'Foxy' is smaller and is the only foxglove that flowers in its first season from seed. *D.* x *mertonensis* (strawberry foxglove) is a very robust tetraploid hybrid created by crossing *D. purpurea* with the perennial, yellow foxglove (*D. grandiflora*); it has coppery-pink flowers that are larger than either parent, it lives and blooms for several years, and it comes true from seed. Strawberry foxglove was created in 1925 by gardeners at the John Innes Horticultural Institute in England (www.floridata.com). Description of some of the cultivars is as follows.

Polkadot Polly

'Polkadot Polly' foxgloves are covered with bell-shaped flowers in bright apricot-pink. The stems are longer and therefore suitable for vases and bouquets. As it is a hybrid, the plants do not produce seeds.

Digiplexis 'Illumination Flame'

This beautiful hybrid is a cross between a foxglove (*Digitalis*) and an *Isoplexis*, a related plant from the Canary Islands. Flame-coloured flowers appear from spring to late summer, and the plants are hardy in zones 8-11. The plants are sterile, so they donot set seeds.

Mountains Mixed

It has strong stems and unique, upward-facing flowers that attract bees. These foxgloves grow in full sun to part shade; the seeds germinate in 14-30 days. The plants bloom in June and are hardy in zones 4-9.

Cultivars of Digitalis

Polkadot Polly

Digiplexis 'Illumination Flame'

Mountains Mixed

Rose Shades

Pam's Choice

Excelsior Hybrid Mix

Waldigone

Camelot Series

Foxlight Ruby Glow

Strawberry Foxglove

Dusky Maid Sunset

Milk Chocolate

***Source*: www.hgtv.com, www.homestratosphere.com, www.hgtv.com**

Rose Shades

This is nearly carefree foxglove, produces pale to deep pink flowers in late spring and grows up to 75 cm.

Pam's Choice

It bears spikes of pure white flowers with splashes of burgundy in their throats. They are suitable in the border or in a cutting garden. They are deer resistant and flower through the summer.

Excelsior Hybrid Mix

'Excelsior Hybrid Mix' foxgloves are suitable for early season flowering. These speckled blossoms found in pink, cream, rose, purple, white and primrose, and bloom in the second season after sowing. It is suitable for the edge of garden.

Digitalis 'Waldigone' (Goldcrest Foxglove)

Introduced by British plant breeder David Tristam, 'Waldigone' is a hybrid that bears spikes of big, apricot-coloured flowers speckled with brown in the throats. The plants form clumps with shiny, evergreen foliage.

Camelot Series

Digitalis purpurea, 'Camelot Series', features large vibrant blooms. This old-fashioned favourite foxglove blooms in late spring to early summer, followed by repeat later in summer if faded flowers are removed.

Foxlight Ruby Glow

It is suitable for container garden and has a potential for cut flower.

Strawberry Foxglove

It prefers full sun to part shade and behave as a biennial plant.

Dalmatian Crème Hybrid

Flowers are cream-coloured with bell-shaped petals open up to reveal purple specks inside and it goes from seed to bloom in less than four months. The 'Dalmatian Crème Hybrid' is winter-hardy to zone 5. They are suitable for containers and garden beds.

Dalmatian Hybrid (*Digitalis purpurea*)

They are very early and their blooms are well-packed and compact. The flowers are slightly bell-shaped and are little upturned. They are tall (180 cm) suitable for borders, alongside fences and walking paths.

Dalmatian Peach Hybrid

It is early and plant bears peach coloured, trumpet shaped flowers. They grow up to 180 cm in height and produces bloom within four months. The 'Dalmatian Peach Hybrid' is winter-hardy to zone 5 and blooms very well even in the first year.

Dalmatian Purple Hybrid

The flowers are having deep purple-lavender petals and maroon-and-white speckled insides. They are planted one week earlier to other Dalmatians. The plants grow up to 90 cm in height and thrive well in small pots as well as in large containers. It is winter-hardy to zone 5.

Description of some of the hybrids developed in the past is mentioned below:

Dobbie's hybrids

These are the result of crossing *D.* ×*fucata* Lutz hybrids with *D. purpurea* 'Giant Shirley'. They are vigorous with large, white flowers spotted in chocolate. They were developed by Dobbie & Co, the then a Scottish seed company.

John Innes Tetra (*D. grandiflora* × *D. lanata*)

This is a robust, fertile, tetraploid form of a hybrid raised at the John Innes Horticultural Institution in 1926. It combines the flower shape of *D. grandiflora* with the markings of *D. lanata* to provide white-lipped flowers, which are coppery yellow outside and spotted and flushed with bronze inside, and with reddish tints on top of the flowers.

Knee High Series

These are dwarf sterile cultivars from Charles Valin at Thompson & Morgan, initially in three colours. They are shorter (50-60 cm) than the similar 'Lucas' and 'Martina' and are well-branched with many side shoots with long flowering season.

Polkadot Series

This a series of six seed-raised, sterile, long-flowering hybrids developed by Charles Valin at Thompson & Morgan from selections of *D. purpurea, D.* × *mertonensis* and 'John Innes Tetra'. They have rectangular mouths to their flowers and reach about 90 cm in height (Graham, 2017)

7.0 Genetics and Breeding

The basic chromosome number, x = 7. There exists strong polyploidy in the genus *Digitalis*. Among the 20-25 species of this genus, *D. lutea* is the only one that presents 2n = 56, 112 (Chène *et al.*, 2016)

7.1.0 Pollination

The bumble bee (*Bombus hortorum*) is the main pollinator in *Digitalis purpurea* (common foxglove) (Arthur *et al.*, 2012). Though the flowers are self-compatible, the spatial arrangement of the anthers and stigma prevents self-fertilization. Flower lifespan was terminated by corolla abscission, 5–6 days after stigma opening in the unpollinated flower. Increased pollen load produced increased seed set and reduced flower longevity progressively to a minimum of one day after pollination with pure pollen. Weakening of the abscission zone was detectable 8 h after pollination, while the pollen tubes were still within the stigmatic zone, suggesting that a stimulus, moving at minimum 4 mm/h, was transmitted through the style and ovary. Soon after pollination removal of the stigma prevented the pollination-induced corolla abscission. The progressive induction of the stigma and style took place at a rate of 1.5 mm/h, in advance of the pollen tubes which grew at 0.75 mm/h. It was not possible to reproduce the pollination effects by application of indoleacetic acid (IAA) to the stigma or the style (Stead and Moore, 1979).

In another study by Romaisa *et al.* (2008) using three species (*Digitalis purpurea, D. grandiflora* and *D. lanata*) revealed that floral bud opening in an inflorescence in all three species occurs sequentially from the bottom to upwards in an acropetal succession. The male and female phases are temporally separated with basal flowers entering the male phase first followed by the upper flowers in a regular sequence. Only after shedding the pollen grains completely, flowers shift to the female phase. Resource allocation in all three species is more towards maleness than femaleness during anthesis. In general, cross pollination is observed across the species. The species are principally insect pollinated and selfing is denied because of protandry and spatial separation of the stigma and anthers. Copious amounts of viable pollen grains produced by these species aid in successful pollination, fertilization and seed production. Studies revealed that none of these species operates autogamy because hand-pollinated flowers with self-pollen or bagged flowers (foreign pollen excluded) failed to set seed. High seed output in open-pollinated flowers and high pollen–ovule ratios also indicate the outbreeding nature of these species. Thus, any reduction in the number of individuals constituting a population would significantly reduce the floral display and consequent rewards to the pollinators, with a subsequent influence on seed production.

7.2.0 Breeding

7.2.1 Selection

Mass selection programs conducted in the Netherlands (Mastenbroek, 1985) and France (Brugidou and Jacques, 1987) resulted in varieties with increased leaf yield, and also with improvement in other traits such as upright habit of leaves and resistance to diseases and bolting. There are only a few varieties of yellow foxglove through selection. 'Carillon' is a dwarf selection that grows only 30 cm tall and has smaller yellow flowers. It often blooms in the first year, but may act more as an annual or biennial than perennial. 'Temple Bells' is a larger flowering selection (https://hort.extension.wisc.edu)

7.2.2 Mutation Breeding

The terminal flower mutant, sometimes called *Digitalis purpurea monstrosa*, has been described by botanists from the mid-nineteenth century onwards. Breeding experiments have shown that the terminal flower mutation in digitalis is inherited as a simple Mendelian recessive, and can be reproduced from seed *via* either the peloric or normal flowers of the same plant, which are all fertile. Mutations that can be inherited and reproduce by seed could theoretically be capable of establishing new plant lineages, and would therefore be "hopeful monsters" (Goldschmidt, 1940) or "prospecies" (Bateman and DiMichele, 2002).

'Terminal Peloria' is the particular mutant of Digitalis where only the terminal flower is different, so the location of the flower on the plant is significant. Normally, foxgloves have indeterminate inflorescences, which means that the apex of the flower spike continues to produce flowers until it becomes exhausted, sometimes achieving a considerable height in a single flowering season. However, in the mutant plant with terminal peloria, the flower stalk is terminated by the strange abnormal "bell-flower", which means that no further flowers could be produced (www.kew.org). Pelorism is the term used for the formation of 'peloric flowers' which botanically is the abnormal production of actinomorphic flowers in a species that usually produces zygomorphic flowers

The peloric flower frequently shows an increase in the number of parts. The peloric flowers are perfectly fertile with their own pollen, the inheritance being the same through peloric or normal flowers of the plant.

7.2.3 Polyploidy

Digitalis purpurea × *D. ambigua* (syn *D. grandifora*) is an example of a constant intermediate hybrid resulting from chromosome doubling. The tetraploid giant hybrid between *Digitalis purpurea* and *D. ambigua* has bred true for five generations. It can be crossed back to the parental species only with difficulty and the triploid progeny are almost entirely sterile. It has the essential characteristics of a new species and is named *Digitalis mertonensis* (2n = 112). It has yielded a diploid parthenogenetic seedling which is sterile and resembles the original hybrid morphologically and in its chromosome behaviour. The triploid back-cross to *D. purpurea* gives a variety of new types with unbalanced chromosome complements. Their form and variation is a proof of the qualitative differentiation of the chromosomes. The morphological characteristics of the hybrid and its derivatives are described. In the balanced forms the influence of each species is as usual proportionate to the number of its chromosome sets present. *D. mertonensis* has yielded a hybrid with *D. obscura*, having 112 chromosomes. The fertility and sterility of tetraploid and diploid forms is shown to agree with the rule, earlier suggested, that there is an inverse correlation between the fertility of diploids and that of the tetraploids to which they give rise (Buxton and Darlington, 1932).

Spontaneous allotetraploids were formed in 90 per cent of F_2 progenies of interspecific crosses between *Digitalis ambigua* and *Digitalis purpurea*, which are common ornamental plants (Ramsey and Schemske, 1998).

7.2.4 Hybridization

Digitalis hybrids were created at the John Innes Horticultural Institution back in the 1920s and 1930s. The first true Digitalis cross, between *D. purpurea* and D. *grandiflora*, was by CF Gartner in 1849, but this and most subsequent efforts produced sterile seedlings. The first breakthrough came with the creation of two tetraploid hybrids at the John Innes Horticultural Institution. These were *D.* × *mertonensis* and *D.* 'John Innes Tetra' and have remained popular. Genetic incompatibilities between species have limited the success of hybrid pollinations and difficulties in propagation have often prevented successful crosses becoming widely grown.

Digitalis x *mertonensis*, known as strawberry foxglove, is an early hybrid dating from 1924. Digitalis Goldcrest ('Waldigone') was created by British breeder David Tristram in 2001.

Digitalis are relatively appealing for non-professional breeders, and many hybrids have been developed by gardeners or small nurseries such as the short-lived, but very active, Europa Nursery. The flowers are large and easy to pollinate and each pollination produces a significant amount of seed – more than 300 per capsule in some cases. Also, the flowers are protandrous and, as the lowest flowers on the raceme open first, flowers producing ripe pollen and those with a receptive stigma can easily be distinguished.

8.0 Growth and Flowering

The digitalis is basically a summer plant and bloom twice a year. Foxglove plants are very easy to cultivate due to the fact that they are adopted to different conditions. They prefer sunny location but can be grown in partial shady areas with dry light soil that are rich in nitrogen.

8.1.0 Temperature

Foxglove can be grown in a wide range of temperatures, ranging as low as -25°C. Due do the cold tolerance ability, the foxglove easily colonize wild habitats like the temperate forests, open forest, river banks, sea cliffs, dry hilly pastures and rocky mountain slopes. However, they are commonly grown in gardens as well in normal conditions. Foxgloves perform better in cooler temperatures and may wilt in temperatures over 32°C. The ideal temperature for seed germination is 18-21°C and 20- 30°C for growth and development.

8.2.0 Photoperiod

Digitalis is a facultative long-day plant and has a critical day length of approximately 14 hours. Foxglove plant grows better in sunny location to partial shaded areas, although it perform best in partial sunny condition. During maturity the plant can tolerate well to dry shade but not full shade. Under controlled environment the plant can tolerates light upto 53,800 Lux.

8.3.0 Growth Regulators

Digitalis respond well to growth regulators like B-nine/Alar (daminozide) and Bonzi. B-nine/Alar (daminozide) at 2000 ppm (3.0 g/l of 64 % formulation or 2.5 g/l of 85 % formulation) can be applied at 2-3 weeks after sowing. If necessary, repeated spray can be undertaken after a week. In warmer conditions, spraying of Bonzi (paclobutrazol) at 5 ppm (1.25 ml/l 0.4 % formulation) or Sumagic (uniconazole) at 3ppm (5.5 ml/l of 0.055 % formulation) is recommended. In Northwest Europe, weekly spraying of Tilt (propiconazole) at 0.3 ml/l is good for control of plant height. Application of 'Topflor' is very effective in digitalis, as higher doses leads to stunting and delayed flowering.

9.0 Propagation

9.1.0 Seed Propagation

They are hardy biennial or perennial plants. Seed germinates well at 15-18°C in presence of light. Seeds are sown during spring and are transplanted to a permanent location in fall.

9.2.0 Vegetative Propagation

Perennial species can be propagated by division of clump. Lateral buds, with and without roots, were carefully removed (Miller, 1914). Root division and propagation of lateral buds have both been tried with several varieties of Digitalis. It was expected that clonal varieties could be obtained in this manner.

9.3.0 Micropropagation

Studies on micropropgation indicated that leaf tissue appeared to have the best callus formation (Bosila *et al.*, 2003). Leaf explants from 2 week (*in vitro*) or 12 week (*in vivo*) seedlings can be cultured on MS solid medium supplemented with 5.0 mg/l 2,4 D + 0.5 mg/l BA and incubated for 45 days at 26°C and 16 hr/day (Bosila *et al.*, 2003).

Shoot tips and nodes of *D. purpurea* produced *in vitro* from seeds cultured on MS medium free from hormones were cultured on MS medium supplemented with different concentrations of BA or Kinetin for multiplication. The shoots produced from multiplication stage cultured on MS medium supplemented with different concentrations of IBA for rooting. Data revealed that multiplication of nodes was better as compared with shoots tips, and BA was better in multiplication compared with Jinetin. Nodes cultured on MS medium supplemented with 0.75 ml/l BA gave the highest shoots number (30.2 shoot/explant) compared to 4.0 ml/l (13.4 shoot/explant). Shoots produced *in vitro* gave 100 per cent rooting when cultured on MS medium free from hormones or on MS medium supplemented with different concentrations of IBA. Highest root number and root length were obtained from cultured shoots on MS medium supplemented with 0.1 ml/l IBA after four weeks (Alabasi and Bashi, 2013).

10.0 Soil and Climate

It requires a well drained sandy soil rich in organic matter. A soil with poor drainage is detrimental. Optimum pH range of soil is 5.5-6.5.

The crop prefers a temperature range of 20-30°C for better growth and development. At lower temperature, vegetative growth of the plants as well as the glycoside synthesis is retarded. They do not have any specific humidity requirements. But too low and high humidity adversely affect the crop. The plants grow to a elevation of 1200-1800 m above mean sea level.

Propagation and Growing of Digitalis

Seeds

Seedling Germination

Seedling

Offshoot from the Main Stock

In Plug Trays

Transplanting in Pots

In Pots

In Open Field

In Greenhouse

Propagation and Growing of Digitalis

In vitro **propagation of** ***Digitalis purpurea*****, (a) seeds as explant source, disinfected with 5 per cent NaOCl, (b) seedlings after 15 days growing on MMS culture medium +1.0 mg/l 6- BAP + 0.1 mg/l IAA, (c)** ***in vitro*** **plants ready for subculturing, (d) individual long shoots, (e) shoot multiplication, (f) hyperhydric shoots, (g) plantlets growing on TIS, (h) harvested biomass from TIS (www.semanticscholar.org)**

Source**: www.alamy.com, www.woodlandtrust.org.uk, www.seasonalpreferences.com, www.cautiouslyadventurous.com, www.seasonalpreferences.com, www.victoriawade.co.uk, www.hort.extension.wisc.edu**

11.0 Cultivation

11.1.0 Sowing and Transplanting

Digitalis is raised from direct seeding or by raising in a nursery and then transplanting to the main field. The land should be thoroughly ploughed, levelled and weeded for direct seeding. Before planting, the land should be ploughed 2-3 times and leveled properly. Then the field is divided into plots of convenient size in which the pits are made at a spacing of 60 cm x 60 cm. The seeds may be broadcast or sown in lines. Deep sowing should be avoided. The nursery beds are raised in well prepared places and the seeds are sown in lines at 10 cm intervals and lightly covered with about 1cm of fine sand. The seedlings will be ready for transplanting 40-45 days after sowing. Good air circulation and adequate spacing is essential for proper growth and development. Spacing recommended is 45 cm between rows and 30 cm between plants in row. Planting spreads from May – June for direct sowing and February for nursery sowing. The optimum seed rate is 8 kg/ha for direct sowing and 2 kg/ha for transplanting. In order to reduce the impact of drought and high temperature, addition of mulches around the plant is recommended.

11.2.0 Manuring and Fertilization

Application of farmyard manure is beneficial to the crop. Nitrogen @ 30kg and phosphorus (P_2O_5) 8 kg may be applied every year. In the first year the first dose may be given 2 months after germination or during February-March in the second year. It also responds to foliar nutrition in the form of urea (1.0 per cent).

11.3.0 Irrigation

Irrigation is given immediately after planting and fertilizers are applied. Later on, irrigation once in 5-6 days is sufficient. During summer, depending on the weather condition irrigation frequency needs to be increased. Irrigation during summer is very important. Critical stages for watering are flower development stage and after fertilizer application.

11.4.0 Weed Management

Since, weeds compete with the plants for water and nutrients; an area of one metre around the plant needs to be kept clean by hand weeding at periodical intervals. The plot requires 1 – 2 weedings in the initial stages.

12.0 Diseases and Pests

12.1.0 Diseases

12.1.1 Downy Mildew (*Peronospora digitalidis*)

Cool wet weather favours this disease. Conidia of this fungus are dispersed by wind or rain-splashes. Initial symptoms on leaves consist of light green, rectangular areas that are vein-delimited. These spots later become chlorotic, may coalesce and turn necrotic. The purple-gray sporulation of the pathogen can be seen primarily on lower leaf surfaces. However, in some cases, extensive fungal growth may develop on the upper surface of lower leaves. Rotation of fungicides like Fosphite (2-3 ml/l), Heritage or Mefanoxam from different groups is recommended for resistance management.

12.1.2 Leaf Spot (*Ramularia variabilis/Colletotrichum fuscum/Septoria digitalis*)

The disease is favoured by wet, rainy conditions. Septoria may be seed borne. Leaf spots are typically favoured by warm wet conditions. Symptoms appear as round to irregular, white to brown spots up to 0.63 cm in diameter with red to purple borders. Lower and older leaves are most severely affected and may shrivel and die. Ramularia infections are characterized by tufts of white conidia growing out of stomata in the leaves. Septoria infections exhibit tiny black pycnidia, which extrude tendrils of spore masses under wet conditions. Spray of any foliar fungicide like Mancozeb (2g/l), chlorothalonil or iprodione are recommended for control of this disease.

12.1.3 Leaf Blight (*Rhizoctonia solani*)

It is characterized by semicircular, water-soaked lesions on leaves just above the soil line at the blade-petiole junction and later along the leaf margins. Lesions expanded for several days along the mid-vein until the entire leaf was affected. Blighted leaves turned brown, withered, clung to the shoots, and matted on the surrounding foliage (https://pubmed.ncbi.nlm.nih.gov).

12.1.4 Powdery Mildew

Mildew is called fair-weather fungus. The plants are weakened by long dryness or hot summer days, then the mildew-pathogens have the best conditions to attacks the plant. It shows a dusty, mealy-grey layer on the surface of the shoots and leaves. The dusty layer can be wiped away.

12.1.5 Root Rot

The root rot is a problem of the foxglove. It is triggered by fungal pathogens. This disease is often a result of compressed soil and water-logging condition. The root is attacked by pathogens and rotting continues until its complete decomposition.

12.1.6 Crown Rot

This is a common problem with foxgloves as a result of overwatering, poor air circulation, or poor drainage. Signs of crown rot include leaves turning yellow and wilting, a spongy brown rotten base, and white fungal spores around the base of the plant. There is no cure for crown rot, and uprooting the infected plants can prevent the disease from spreading. The disease can be prevented by proper spacing (50 cm), good ventilation and growing them in well-drained soil and with judicious use of water. (www.trees.com)

12.2.0 Pests

12.2.1 Leaf-Miner

The larvae regularly leave a mine to restart elsewhere. Pupation occurs outside the mine. Neither larva nor mine can be distinguished from that of related species.

12.2.2 Foxglove Pug (*Eupithecia pulchellata*)

The larvae feed on the inside of Foxglove (*Digitalis purpurea*) flowers, especially on the stamens and developing seeds, protected by closing off an individual floret. This moth is a brightly coloured member of the Pug family with alternating bands of grey-brown and buff-orange across the forewings. The hindwings have similar markings but are generally paler. Flies during the day in May and June and is attracted to light. The caterpillars, feed inside the flowers of Foxglove and can be seen from late June to mid-August before overwintering as pupae in loose earth (https://butterfly-conservation.org).

12.2.3 Aphids

The permanent foxglove aphid does not host alternate. *Aphis armata* only feeds on foxglove (*Digitalis purpurea*). The adult apterae of *Aphis armata* are black, rarely

Diseases and Pests of Digitalis

Downy Mildew

Leaf Spots

Japanese Beetle

Snail Inside Flower

Foxglove Pug

Powdery Mildew

Spider Mites

Aphids

Slugs

Source: www.walterreeves.com, www.butterfly-conservation.org, www.picuki.com, www.in.pinterest.com, www.gardenerspath.com, www.upload.wikimedia.org, www.pnwhandbooks.org, www.seasonalpreferences.com

with any white wax spots. The middle abdominal tergites in apterae are usually without dark sclerotic bands, but they may have a few scattered sclerites, and the largest apterae may have segmental bands. The longest hairs on the hind femora and the tibiae are about twice as long as the least width of the tibiae. Immatures often have discrete wax spots. The aphids attacks the young leaves (https:// influentialpoints.com).

12.2.4 Mealybugs

They feed on plant sap in clusters on foxglove stems and leaves. It favours sheltered areas such as between a stem and leaf. In small numbers, the damage caused is insignificant, but large colonies slow plant growth and damage leaves (https://homeguides.sfgate.com).

12.2.5 Japanese Beetles

The Japanese beetle is a serious pest in Foxglove. They have a metallic green and bronze red shell with tufts of white hair protruding from the sides of the abdomen. The adults emerge from the soil in early to mid-summer and fly onto ornamental plants like foxgloves to eat the leaves. Japanese beetle damage gives leaves a lacy appearance where the insects have eaten the soft tissue between the leaf veins. Chemical pesticides suitable for controlling Japanese beetles include malathion, carbaryl or orthene (https://homeguides.sfgate.com).

13.0 Harvesting and Postharvest Management

Digitalis are popular equally both for aesthetic as well as medicinal purposes. They are planted in beds or borders of a garden. They spikes of cut flowers are harvested when half of the florets are open.

Digoxin, a cardenolide, is produced from dried leaves of *Digitalis lanata* (Clemente *et al.*, 2011). For medicinal purpose, leaves are the economic parts. 2-3 harvests of leaves can be made in the first year, while in the second year 2 harvests besides a seed crop can be taken. It is recommended to keep the plantations as annuals with 2-3 harvests in the first year and in the second year, some portion of the first year crop should be earmarked for seed production and from this crop, leaves have to be harvested just before flowering. Harvesting consists of plucking 8-10 cm long leaves without petioles.

14.0 Biotechnology

Digitalis has been studied extensively using various biotechnological tools mainly intended for enhancing the active principles which have medicinal properties. These include micropropagation (through cell/shoot/hairy root culture; somatic embryogenesis, proptoplast derived plants; haploid plants) as well as transgenics

14.1.0 Molecular Markers

RAPD markers were used to study inter-specific variation among six species of the genus, Digitalis. *D. obscura, D. lanata, D. grandiflora, D. purpurea, D. thapsi* and *D. dubia*, and the hybrid D. excelsior (*D. purpurea* × *D. grandiflora*). A total of 91 highly reproducible bands amplified with four arbitrarily chosen decamer primers were obtained. Homology of the coemigrating RAPD markers was tested by blot hybridisation and sequencing of selected bands. The application of a range of statistical approaches for RAPD data analysis, including distance and parsimony methods, family clustering and the analysis of molecular variance (AMOVA), indicated that these molecular markers were taxonomically informative in Digitalis. The species relationships revealed were fully consistent with those previously obtained using morphological affinities. The hybrid *D. excelsior* seems to have stronger affinity to the section Digitalis than to Grandiflorae. This is the first known report of the application of RAPD markers for the study of genetic relationships among species of the genus Digitalis (Nebauer *et al.*, 2000)

RAPD markers also proved to be a powerful tool for the detection of spatial genetic variation in Spanish wild populations of *D. obscura* and *D. minor* and for the identification of individuals from the different populations (Nebauer *et al.*, 1999; Sales *et al.*, 2001). The RAPD-based analyses of molecular variance (AMOVA) revealed that most genetic variation in *D. obscura* and *D. minor* was recorded within populations (84.8 per cent and 92.9 per cent, respectively), a result consistent with those from most other outcrossing plant species (Hamrick and Godt, 1996). More recently, *D. grandiflora* populations growing in Russia were also analyzed by RAPD and ISSR (intersimple sequence repeats) markers (Boronnikova *et al.*, 2007); results from both markers revealed a weak population structure, with most of the genetic variation accounting for within population variability.

14.2.0 Suspension-Culture

Suspension-cultured Digitalis cells, which do not synthesize cardenolides *de novo* (Reinhardt *et al.*, 1975; Kreis *et al.*, 1993) as well as roots or shoots cultivated *in vitro* (Theurer *et al.*, 1998), are able to take up exogenous cardenolides and modify them. Cell suspension cultures established from different plants producing cardiac glycosides did not produce cardenolides (Luckner and Diettrich, 1985; Seidel and Reinhardt, 1987; Stuhlemmer *et al.*, 1993).

14.3.0 Somatic Embryogenesis

Somatic embryos, green shoot cultures, as well as plants obtained by organogenesis or somatic embryogenesis were found to produce cardenolides (Diettrich *et al.*, 1991). Palazo *et al.* (1995) evaluated the effect of phenobarbital on organogenesis from *D. purpurea* callus; the interaction of this xenobiotic with IAA not only reduced the production of biomass but also increased the volume of mitochondria per cell and the formation of shoot buds in callus tissues. Somatic embryogenesis can also be easily achieved from primary explants or callus cultured on media with auxin alone or with several auxin–cytokinin combinations; at least in *D. obscura*, gibberellic acid favoured both differentiation and normal embryo development. It is worth noting that a cell strain line of *D. lanata* was used to characterize somatic embryogenesis of the species by means of two-dimensional gel electrophoresis of *in vivo* and *in vitro* synthesized polypeptides (Reinbothe *et al.*, 1992).

Cell, tissue, and organ cultures showed morphogenetic capacity for regeneration but this capacity decreased in long-term cultures. Somatic embryo cultures derived from embryogenic cell lines of *D. lanata* were reported (Garve *et al.*, 1980; Tewes *et al.*, 1982; Diettrich *et al.*, 1986a, b, 1991). Embryogenic cell lines were also established from other species, *e.g.*, *D. obscura* (Arrillaga *et al.*, 1986, 1987). Root cultures are another system in which several aspects of growth and development can be studied (Pe´rez-Bermu´ dez *et al.*, 1987). Haploid cell cultures from anthers, pollen, non-fertilized egg cells, and pistils have also been generated (Ernst *et al.*, 1990; Pe´rez-Bermudez *et al.*, 1990; Diettrich *et al.*, 2000).

14.4.0 Genetic Transformation

The first transgenic Digitalis plants were obtained by Lehmann *et al.* (1995), who reported an *Agrobacterium tumefaciens*-mediated transformation of *D. lanata* by using protoplasts derived from an embryogenic line. Pradel *et al.* (1997) also reported the regeneration of D. *lanata* plants from hairy root cultures established after infection with several wild type strains of *Agrobacterium rhizogenes*. A more efficient transformation protocol was developed by Sales *et al.* (2003) for *D. minor*. About 8.4 per cent of the leaf explants infected with *A. tumefaciens* formed at least one transformed plant. This transformation efficiency allowed the genetic engineering of this plant species using an Agrobacterium strain containing a Ti plasmid with the catalytic domain of the 3-hydroxy-3-methyl-glutaryl coA reductase gene (HMG1) isolated from *A. thaliana* (Sales *et al.*, 2007). Constitutive expression of this transgene resulted in an increased sterol and cardenolide production in both *in vitro* and greenhouse-grown plants. Although a clear correlation between HMG1 expression and cardenolide accumulation in transgenic plants could not be established, since HMGR up-regulation in *D. minor* seems to be more effective in modifying sterol than cardenolide metabolism, lines with the higher HMG1 also showed higher cardenolide contents. These results suggest that additional critical steps exist in the cardenolide biosynthesis pathway. However, further progress in this field is limited due to the lack of sufficient information concerning the genes involved in the biosynthesis of these secondary metabolites.

15.0 References

Alabasi, I.N. and Bashi, B.Z.K. (2013) *Mesopotamia J. of Agric.*, **41**(4): 85-95.

Albach, D. and Chase, M. (2004) *Mol Phylogenet Evol.*, **32**: 183–197.

Albach, D.C., Meudt, M. and Oxelman, B. (2005) *Am. J. Bot.*, **92**: 297–315.

Arrillaga, I., Brisa, M.C. and Segura, J. (1986) *J. Plant. Physiol.*, **124**: 425–430.

Arrillaga, I., Brisa, M.C. and Segura, J. (1987) *Plant Cell Rep.*, **6**: 223–226.

Arthur, A.D., Broadbent and Andrew, F.G. Bourke. (2012) *Journal of Pollination Ecology*, **8**(7): 48-51.

Bateman, R.M. and DiMichele, W.A. (2002) *Generating and filtering major phenotypic novelties: neoGoldschmidtian saltation revisited*. Pp. 109–159 in: Cronk, Q.C.B. *et al.*, eds. Developmental Genetics and Plant Evolution. Taylor & Francis.

Boronnikova, S.V., Kokaeva, Z.G., Gostimsky, S.A., Dribnokhodova, O.P. and Tikhomirova, N.N. (2007) *Russ. J. Genet.*, **43**: 530–535.

Bosila, H.A., Mohamed, S., El-Gamal, S. and Bekhit, M. (2003) *Acta Horticulturae*, **597**: 289-298.

Bräuchler, C., Meimberg, H. and Heubl, G. (2004) *Plant Syst Evol.*, **248**: 111–128.

Breck, J. (1851) *The Flower Garden or Brecks Book of Flowers*, John P. Jewett & Company, Boston.

Brugidou C. and Jacques, M. (1987) *Agronomie*, **7**: 685–694.

Buxton, B. H. and Darlington, C. D. (1932) *The New Phytologist*, Vol. XXXI, **4**: 225-240.

Carvalho, J.A. and Culham, A. (1997) *Amer. J. Bot.*, **84**: 180.

Carvalho, J.A. and Culham, A. (1998) *Bol. Mus. Mun. Funchal.*, **5**: 109–127.

Chène, P., Bourge, M. and Verlaque, R. (2016) *Flora Mediterranea*, **26**: 11-18.

Clemente, E.S., Müller-Uri, F., Nebauer, S.G., Segura, J., Kreis, W. and Arrillaga, I. (2011) *Digitalis*, Chapter 5. In: Kole C, ed. Wild Crop Relatives: genomic and breeding Resources, Plantation and ornamental Crops. Berlin, Heidelberg: Springer; 73–112

Dahlgren, R.M.T. (1980) *Bot. J. Linn. Soc.*, **80**: 91–124.

Diettrich, B., Steup, C., Neumann, D., Scheibner, H., Reinbothe, C. and Luckner, M. (1986a) *J Plant Physiol.*, **124**: 441–453.

Diettrich, B., Ernst, S. and Luckner, M. (2000) *Plant Med.*, **66**: 237–240.

Diettrich, B., Haack, U. and Luckner, M. (1986b) *J. Plant Physiol.*, **126**: 63–73.

Diettrich, B., Schneider, V. and Luckner, M. (1991) *J. Plant Physiol.*, **139**: 199–204.

Douglas, E., Soltis, P. and Doyle, J.J. (1998) *Molecular Systematics of Plants II. DNA Sequencing*. Boston: Kluwer Academic Publishers, pp. 574.

Ernst, S., Scheibner, K., Diettrich, B. and Luckner, M. (1990) *J Plant Physiol.*, **137**: 129–134.

Garve, R., Luckner, M., Vogel, E., Tewes, A. and Nover, L. (1980) *Planta Med.*, **40**: 92–103.

Goldschmidt, R. (1940) *The Material Basis of Evolution*. Yale University Press.

Graham Rice. 2017. *Plant breeding: A survey of Digitalis hybrids* -The Plantsman Magazine March Issue pp. 28-36.

Hamrick, J.L. and Godt, M.J.W. (1996) *Effects of life history traits on genetic diversity in plant species*. Philos Trans R Soc Lond Ser B, 351: 1291–1298.

Henkel, A. (1912) *Drug. Circ.*, **56**: 128-135.

Herl, V., Albach, D.C., Müller-Uri, F., Bräuchler, C., Heubl, G. and Kreis, W. (2008) *Plant Syst Evol.*, **271**: 65–78.

https: / / www. americangardener.net

https: / /www.aaas.org

https: //www.aggie-horticulture.tamu.edu.

https: //www.britannica.com

https: //www.butterfly-conservation.org

https: //www.ch.ic.ac.uk

https: //www.en.wikipedia.org

https: //www.floridata.com

https: //www.harvesting-history.com

https: //www.hgtv.com

https: //www.homeguides.sfgate.com

https: //www.homestratosphere.com

https: //www.hort.extension.wise.edu

https: //www.indiaagronet.com

https: //www.influentialpoints.com

https: //www.innovativeagriculture.in

https: //www.kew.org

https: //www.plantopedia.com

https: //www.pnwhandbooks.org

https: //www.pubmed.ncbi.nlm.nih.gov

https: //www.semanticscholar.org

https: //www.ukflymines.co.uk

https: //www.webmd.com

Kraemer, H. (1903) *Am. Jour. Phar.*, **75**: 553-569.

Kreis, W., Hoelz, H., May, U. and Reinhardt, E. (1993) *Plant Cell Tiss Org Cult.*, **20**: 191–199.

Laci, H. (2012) *Digitalis lantana* - New Crop Summary & Recommendations

Lehmann, U., Moldenhauer, D., Thomar, S., Diettrich, B. and Luckner, M. (1995) *J. Plant Physiol.*, **147**: 53–57.

Luckner, M. and Wichtl, M. (2000) *Digitalis – Geschichte, Biologie, Chemie, Physiologie, Molekularbiologie, Pharmakologie, medizinische Anwendung*. Stuttgart: Wissenschaftliche Verlagsgesellschaft.

Luckner, M. and Diettrich, B. (1985) *Formation of cardenolides in cell and organ cultures of Digitalis lanata*. In: Neumann KH, Barz W, Reinhard E (eds) Primary and secondary meta- bolism of plant cell cultures. Springer, Berlin, Germany, pp. 154–163.

Mastenbroek. (1985) *Br. Heart J.*, **54**: 262–268.

Miller, F.A. (1914) *The Propagation of Medicinal Plants*. Bulletin of the Torrey Botanical Club, Published by: Torrey Botanical Society, **41**(2): 105-129.

Nebauer, S.G., Del Castillo-Agudo L. and Segura, J. (2000) *Theor Appl Genet.*, **100**: 1209–1216.

Nebauer, S.G., Del, Castillo-Agudo L. and Segura, J. (1999) *Theor Appl Genet.*, **98**: 985–994.

Olmstead, R.G., de Pamphilis, C.W., Wolfe, A.D., Young, N.D., Elisons, W.J. and Reeves, P.D. (2005) *Amer. J. Bot.*, **88**: 348–361.

Pe´rez-Bermu´ dez, P., Falco´, J.M. and Segura, J. (1987) *J. Plant Physiol.*, **130**: 87–91.

Pe´rez-Bermudez, P., Cornejo, M.J. and Segura, J. (1990) *Digitalis spp.: In Vitro production of haploids*. In: Bajaj YPS (ed) Biotechnology in agriculture and forestry, vol 12: Haploids in crop improvement I. Springer, Berlin, Germany, pp. 277–289.

Pradel, H., Lehmann, U., Diettrich, B. and Luckner, M. (1997) *J. Plant Physiol.*, **151**: 209–215.

Ramsey, J., and Schemske, D.W. (1998) *Annual Review of Ecology and Systematics.*, **29**: 467-501.

Reinbothe, C., Tewes, A., Luckner, M. and Reinbothe, S. (1992) *Plant J.*, **2**: 917–926.

Reinhardt, E., Boy, M. and Kaiser, F. (1975) *Planta Med Sup.*, **27**: 163–168.

Romaisa, N., Zafar, R. and Bashir, A.W. (2008) *Plant Species Biology*, **23**: 59–70.

Sales, E., Mun˜ oz-Bertomeu, J., Ros, R., Arrillaga, I. and Segura, J. (2007) *Planta Med.*, **73**: 605–610.

Sales, E., Nebauer, S.G., Mus, M. and Segura, J. (2001) *Am. J. Bot.*, **88**: 1750–1759.

Sales, E., Segura, J. and Arrillaga, I. (2003) *Planta Med.*, **69**: 143–147.

Seidel, S. and Reinhardt, E. (1987) *Planta Med.*, **53**: 308–309.

Stead, A.D. and Moore, K.G. (1979) *Planta*, **146**: 409–414.

Stuhlemmer, U., Kreis, W., Eisenbeiss, M. and Reinhard, E. (1993) *Planta Med.*, **59**: 539–545.

Tagheiei, M., Talebi, B. and Tabatabaei, S. (2018) *Journal of Crop Production and Processing*, **8**(3): 51-62.

Taskova, R.M., Gotfredsen, C.H. and Jensen, S.R. (2005) *Phytochemistry*, **66**: 1440–1447.

Tewes, A., Wappler, A., Peschke, E.M., Garve, R. and Nover, L. (1982) *Z Pflanzenphyiosl.*, **106**: 311–324.

Theurer, C., Kreis, W. and Reinhardt, E. (1998) *Planta Med.*, **64**: 705–710.

True, R.H. (1906) *Progress in drug-plant cultivation*. Yearbook Dept. Agr. 1905: 533-540.

Werner, K. (1961) *Wuchsform und Verbreitung als Grundlagen der taxonomischen Gliederung von Digitalis* L. [Dissertation]. Halle: Martin-Luther-Universität Halle-Wittenberg.

Werner, K. (1965) *Feddes Repert.*, **70**: 109–135.

Withering, W. (2018) *An account of the foxglove*. Reproduction of the original, An Account of the Foxglove and some its medical uses with practical remarks on dropsy and other diseases by Outlook Verlag GmbH, Deutchland.

2

Delphinium

Ganesh B. Kadam, Tarak Nath Saha, P. Naveen Kumar, D.V.S. Raju, Sanjay Kad and K.V. Prasad

1.0 Introduction

Delphinium is an important perennial flowering plant belongs to Ranunculaceae family. It is mainly grown in cooler climatic conditions around the world for showy blue coloured flowers. It is also known as Larkspur, refers to the horn-shaped nectary of the flower which resembles the spur of a lark's claw. The other names are larks heel, larks toes, larks clawe and monkshood. The word *Delphinium* (Larkspur) is originated from Greek and derivative of *delphis* (stem delphin-) dolphin; so called from the shape of the nectary of flower. In India, delphiniums are popular by generic name Jadwar and commonly named as *Narbasi/Nirbisi* due to its antidotal properties (Dymock *et al.*, 1890). *Nir* means to oppose or to remove and *Bisi* means *Bis* or *Vish* (poison). In regional languages it has names *viz.* Nirbisi (Hindi), Neelamabari or Jadwar (Kannada), Nirvasi (Malayalam), Nirvasi (Sanskrit), Nirvisha (Tamil), Nirbasi (Telugu), Nirvisi (Marathi), *etc.*

In India, it is popularly grown in northern plains and hilly regions. The commercial cut flower and pot plant production of delphinium has also started in some northern Indian states. Some species are adapted to subtropical and others to sub-temperate and temperate climatic conditions (Polunin and Stainton, 1984). It is commercially cultivated as cut flower as well as potted ornamental plants in USA, New Zealand and Europe in cottage-style gardens and cutting gardens.

2.0 Importance and Uses

2.1.0 Aesthetic Uses

Delphiniums are widely grown for home gardens, beds, borders, pots, rock garden, and hanging baskets. Flowers of these plants/species are available in the shades of blue, purple, white, pink, lilac and sometimes even yellow or red, although this is rare. The dwarf and medium-sized as well as tall (growing up to 180 cm or more) delphiniums can be effectively used in back of a mixed border or as a centerpiece to an island flower bed, in groups or in mass effect or along with fencing. It is also commercially cultivated as a cut flower.

2.2.0 Therapeutic and Other Uses

All parts of the delphinium plant are toxic. The plants are most toxic when it is young, but seed oil is used for external applications in some of the medicines. Very few species like *Delphinium caeruleum, Delphinium cashmerianum, Delphinium denudatum, Delphinium elatum, Delphinium menziesii, Delphinium semibarbatum, Delphinium staphisagria, etc.* have been reported for medicinal purpose with limited known uses. *Delphinium staphisagria* is most commonly reported for use in antiparasitic ointment. Seeds are the major source of alkaloids Delphinine, and Staphisagrine found in the shells. A parasiticide is obtained from the leaves of *Delphinium brunonianum* which is highly toxic and is applied for external use only. In the Himalayan region it is only used to destroy ticks on animals (Polunin and

Uses of Delphinium

In Cut Flower In Beds In Pot

In Garden Border In Cottage Garden Hanging Arrangement

In Rock Garden Wedding Arrangements Window Box

Interior Scaping In Hedge Screening of Wall

Source: www.delphinium-planten.nl, www.pinterest.co.uk, www.flowerpatchfarmhouse.com, www.ebay.co.uk, www.growingcolours.com, www.flowerona.com, www.plantsoftheworldonline.org, www.pinterest.com, www.plantedwell.com

Stainton, 1984). The juice of the flowers, particularly *D. consolida*, mixed with alum, gives a blue ink. *D. nudicaule* known as sleep root has narcotic properties. The growing plants and flowers are poisonous to grazing animals and can cause severe illness or even death due to poisonous alkaloids (aconitine). The roots of (*Delphinium denudatum*) the plant have long been used in the Unani system of medicine for the treatment of aconite poisoning, numbness, palpitation, opium addiction, scorpion sting, snakebite, paralysis, epilepsy and tonic for brain and nerves (Husain, 1897).

Since delphiniums are rich in pigments therefore several earlier studies have attempted to isolate anthocyanin pigments from Delphinium flowers and the first anthocyanin (delphinin) identified was from *D. consolida* by Willstatter and Mieg in 1915. Delphinium flowers are also an important source of natural dyes. The yellow dye for silk has been extracted from flowers of *D. zalil* for a long period of time in Iran and India (Fang and Benn, 1992).

3.0 Origin and History

The genus *Delphinium* includes about 370 species (Blanché, 1991) comprising of annuals, biennials, and perennials (Orellana *et al.*, 2009b). They are mostly native to the northern temperate regions of the world and higher mountains of South Africa. The Asia is having centre of diversity in South-West China and Eastern Himalaya, where Delphinium has 150 species (Wang and Warnock, 2001). In India *Delphinium* has documented 27 species mainly confined to the temperate and alpine zones of the Himalayas (Rau, 1993) and one species, *D. malabaricum* is endemic to Western Ghats of South India. Therefore, Himalayas has been considered as primary centre for speciation in the genus *Delphinium* (Blanche, 1990).

4.0 Taxonomy

Delphinium is perennial flowering plant with few biennial species and rarely annual types are also reported. It belongs to the family Ranunculaceae, Order Ranunculales, Superorder Ranunculanae, Class Magnoliopsida and Tribe Delphinieae. The first reference to the name 'Delphinium' was found in the work of Dioscorides De Materia Medica. The first valid publication of the name Delphinium was made by Linnaeus (1753: 530). He placed the genus into the group called "Polyandria trigynia", which is characterized by an androecium with numerous stamens and a gynoecium with three carpels, although this group also included species with a single carpel in Delphinium.

Based on seed morphology, Malyutin (1987) divided the genus Delphinium L. into subg. Staphisagria (J. Hill) Peterm. and subg. Delphinium, and further divided the latter one in sect. Anthriscifolium Wang and sect. Delphinium (confirming Wang's (1979) classification for this subgenus). This classification was supported by a combination of morphological and cytological characters (Blanché, 1990). Later, based on breeding systems, pollination ecology, cytology and isozyme variations, he proposed the division of the sect. *Delphinium* into four series *Cossoniana* C. Blanché, Molero and P. Simon, *Balansae* C. Blanché, Molero and P. Simon, *Macropetala* C.

Blanché, Molero and P. Simon and *Halterata* B. Pawl (Blanché *et al.*, 1997). Integrating the results of molecular data, Jensen *et al.* (1995) placed *Delphinium* L. together with *Consolida* (DC.) Gray and *Aconitum* L. in the subtribe Delphiniinae Benth. (Delphinieae Warm., Ranunculoideae Hutch). Using 65 morphological characters combined with four plastid and nuclear DNA sequence data, Wang *et al.* (2009) placed the genera *Delphinium* L., *Consolida* (DC.) Gray and *Aconitum* L. within the tribe Delphinieae. Through a morphological analysis, Trifonova (1990) proposed to consider *Consolida* (DC.) Gray and *Aconitella* Spach as different genera based on morphological characters. However, this was challenged by Jabbour and Renner (2011a) who found that both genera were nested within *Delphinium* L. using molecular data and a large taxon sampling, leading them to proposed an extended genus *Delphinium* L. including both *Consolida* (DC.) Gray and *Aconitella* Spach. Based on the results of a molecular phylogenetic analysis, they decided to resurrect the genus *Staphisagria* J. Hill, including *D. staphisagria* L. (Jabbour and Renner, 2011b). They recognized three genera in the tribe Delphinieae: *Delphinium* L., *Aconitum* L. and *Staphisagria* J. Hill. (Jabbour and Renner, 2012a). The genus *Aconitum* L. was further split into *Aconitum* L. *sensu stricto* and the monotypic *Gymnaconitum* (Stapf.) Rapaics (Wang *et al.*, 2013).

Typical *Delphinium* flowers are composed of four categories of organs: sepals and petals composing the perianth (the diversity of terms used in the bibliography of these organs is reviewed in Jabbour and Renner, 2012b), and the two types of sexual organs, stamens and carpels (Blanché, 1990). From outside in, there are five free petaloid sepals quincuncially arranged: two ventral, two lateral and a spurred dorsal one, four free petals located in the dorsal half of the flower: two dorsal organs forming nectariferous spurs inserted into the spur of the dorsal sepal and with an exserted limb, and two lateral organs with a wide limb and a narrow claw. The other petals (corresponding to four ventral primordia) stop developing shortly after organogenesis (Payer, 1857; Jabbour and Renner, 2012b). The stamens are arranged in eight spiral series, and the gynoecium is composed of 3(-5) free carpels (Pawowski, 1964) turning into follicles after fertilization. In *Delphinium*, bilateral symmetry is established through two phenomena: 1) the development of dorsal spurs; and 2) the arrested development of the ventral petals (Jabbour *et al.*, 2009).

5.0 Species and Cultivars

5.1.0 Species

The Delphinium genus has 370 species (Blanché, 1991) distributed in North temperate regions of the world. Munz (1967 and 1968) documented 244 species from Asia with the centre of diversity in South-West China and Eastern Himalaya where Delphinium has 150 species (Wang and Warnock, 2001). Some of the important species are listed below.

Delphinium andersonii

It is a perennial species of larkspur known as Anderson's larkspur, native to North America. The plants are erect reaching about 50 cm in height. It has small leaves on long petioles with the leaf blades divided into long finger like lobes. The top of the slender stem is occupied by a cylindrical inflorescence of flowers, each flower 2-4 cm wide with a spur measuring nearly 2 cm in length.

D. californicum

It is endemic to California and also known as California larkspur. The stem is tall (100-200 cm), leaves arise on long petioles divided into finger like pointed lobes. The top of the stem is occupied with a very large inflorescence. Each flower rises on a pedicel with tubular florets. The sepals point forward to make a cup. The flower is generally white to greenish white to light lavender.

Consolida regalis

This is an annual herbaceous plant popularly known as forking larkspur, rocket-larkspur, and field larkspur. The stem is erect, hairy and branched at the top reaching a height of 30–80 cm. The inflorescence is a cluster with 5-8 hermaphrodite flowers. The flowers are dark blue or purple with 5 sepals.

D. elatum

It is known as alpine delphinium or candle larkspur and is a source of several present day hybrids. Its flower ranges from blue and purple to pink, cream, and white which are popular ornamental plants.

D. parishi

It is a perennial herbaceous plant, known as desert larkspur. The plants are tall (17–60 cm) with palmately lobed leaves which often shrivel by the time of flowering. The flower colour vary from dark blue to purplish, sky blue to pink. Each flower can be multi-hued with upper true petals white, nectar spur darker blue or purple, and sepals light blue with darker tips.

D. polycladon

It is a species of larkspur known as mountain marsh larkspur. It is native to the mountains of eastern California. It is a perennial herb producing one or more erect stems which easily exceed a meter in height but often remain dwarfed in high-elevation. The leaves are small and mostly located near the base of the plant. The inflorescence is a raceme and flowers are often arranged along one side of the stem. The sepals are deep blue to purple, 1-2 cm long, and with a spur upto 2 cm in length.

D. malabaricum

It is the only species of the genus found in Western Ghats of India. It is an endemic, vulnerable, perennial herb with very attractive racemes. The plants are tall (50-100 cm) with hollow stems. The leaves are palmately lobed, arranged alternately. Lower leaves are large, reniform with 5- 7 lobed whereas stem leaves are with narrower segments and hairy upper surface. Flowers are blue in colour and occur in long racemes at the end of branches. Flowers have five sepals with a white spot near the top and two small petals which are bi-lobed.

Species of Delphinium

Delphinium andersonii  ***Delphinium bakeri***  ***Delphinium californicum***

Delphinium consolida  ***Delphinium elatium***  ***Delphinium leucophaeum***

Delphinium luteum  ***Delphinium parishii***  ***Delphinium malabaricum***

Delphinium pyramidale  ***Delphinium tricorne***  ***Delphinium variegatum***

***Source*: www.en.wikipedia.org, www.efloraofindia.com**

D. tricorne

It is native to the eastern United States and commonly known as dwarf larkspur. It is perennial and found in rich mesic forests, often over calcareous rocks. The plant is tall, stringy thin stems with few leaves and bears attractive flowers in shades of blue.

D. uliginosum

It is also known as swamp larkspur and bog larkspur and endemic to California. This is a perennial herb producing a hairless, erect stem up to 70 cm tall. It can be identified by its leaves, which are fan-shaped, a characteristic unique among the larkspurs. The inflorescence bears up to 45 flowers, each on an upright pedicel with blue flower.

D. umbraculorum

The common name of this species is umbrella larkspur due to its growing habitat, usually shady and cool places. It is a perennial herb endemic to California. It is often confused with *D. parryii* which has similar flowers and *D. patens* which has similar stems and leaves. The plant is tall (40-80 cm) and the hairless leaves are located at the base of the plant and along the stem as well. The inflorescence bears several flowers with reflexed dark blue sepals and a good spur long.

D. viridescens

This rhizomatous perennial herb popularly known as Wenatchee larkspur. The stem is tall (1.5-2.0 m), hollow and the upper part of the plant is coated in yellow glands. The mid-stem leaf blades are each divided into three wedge-shaped lobes, each of which is divided and subdivided into narrow, pointed lobes. Leaves higher on the stem are linear in shape. The inflorescence is a dense raceme of up to 80 flowers, each borne on a pedicel up to 2 cm long. The flower sepals have a purplish, yellow-green-tinged colouration and the smaller petals are purple to yellow. It is pollinated by bumblebees only.

5.2.0 Cultivars

The present day *Delphiniums* grown in gardens are mostly hybrids, involving species *Delphinium elatum, Delphinium grandiflorum, Delphinium exaltatum* and *Delphinium formosum*. The development of these hybrids began in the late 1800s and since then these hybridisation resulted into many hybrids which are grown in various gardens till today. The parentage of today's horticultural varieties contain characteristics of several species due to hybridization happened over the period. Based on the growth pattern and height, the delphiniums are classified into different groups.

5.2.1 Belladonna Group

The plants of this group are upright, loose and branching perennials with single flowers that grow 7-10 cm tall. This group is thought to originate from hybrids between cultivated garden delphiniums and cultivated forms of

D. grandiflorum. Initially belladonna delphiniums were sterile but this is not always true with addition of new cultivars in this group. Belladonna delphiniums have a highly branched inflorescence, with the side spikes comparable in importance to the primary raceme. This structure results in floriferous plants suitable for garden display. 'Blue Bees' is a Belladonna producing clear blue flowers with white centers. 'Belladonna' delphiniums include 'Bellamosum' (blue), 'Blue Donna','Blue Donna Imperial', 'Blue Shadow', 'Cliveden Beauty' (blue), 'Oriental Blue', 'West End Blue'; 'Casablanca' (white), 'Connecticut Yankees' (blue/white), 'Delft Blue' (blue/white) and 'Stelchen Strain' (blue/white). Series such as 'Arrow', 'Dasante', 'New Millennium' and 'Plagu' predominate in blue cultivars with some whites and pinks.

5.2.2 Elatum Group

These are clump-forming perennials, bloom with numerous single, semi-double, or double flowers in early to mid-summer. Flowers, 5-7 cm across fill stalks that are commonly 150-180 cm tall. These are the tallest spiked hybrids growing to 180 cm or more. 'Blue Nile' is a medium plant bearing semi-double, bright, and mid-blue flowers with white centers (called bees). 'Bruce' is a tall Elatum bearing semi-double, violet-purple flowers, paler towards the center, with brown bees. The series grown include 'Aurora', 'Centurion', 'Clear Springs', 'Crown', 'Guardian', 'Highlander', 'Magic Fountain', 'Mistral', 'New Century hybrids', 'Prism', 'Raider', 'Scent' and 'Summer', which usually include shades of blue, white and pink. Less usual colours include creamy white/yellow (such as 'Sungleam') and salmon-pink ('Princess Caroline' selections such as 'Red Caroline').

5.2.3 Pacific Hybrids

This is similar to Elatum Group, but with lesser height. This hybrid is short-lived and often grown as annuals or biennials and having semi-double florets. The Pacific hybrids, also known as Pacific Giants, can grow up to 120-180 cm tall. Pacific Hybrids' have been subject to much hybridisation and selection and may appear under various series names, such as 'Giant Pacific Court', 'Pacific Giant', 'Pacific Premium' or 'Round Table'; they include numerous blue, white and pink shades and are recognisable by their Camelot-themed cultivar names ('Black Knight', 'Guinevere', 'King Arthur' 'Galahad', 'Blue Jay' and 'Blue Bird').

5.2.4 Grandiflorum Group

Also called Chinese or Siberian delphinium, this variety has a compact, bushy habit and grows 30-60 cm tall and nearly as wide. Grandiflorums bloom blue or white flowers mid-summer and tend to have a higher heat tolerance than other *delphinium* species.

5.2.5 New Zealand Hybrids

The New Zealand hybrids are medium-sized, and include such favorites as 'Pagan Purples', 'Dasante Blue' and the New Millennium series.

Every year many cultivars are being developed and released in various part of world under different groups. Some of the cultivars tested and selected for reference at Royal Horticulture Society (Wesley and Bassett, 2004) are *D.* 'Constance Rivett',

Cultivars of Delphinium

Blue Indulgence, King Arthur, Blue Bird

Centurion Sky Blue, Atholl, Blue Dawn

Blue Nile, Bruce, Can-Can

Cherub, Clifford Sky, Conspicuous

Source: Source: www.mikepeel.net, www.in.pinterest.com

Cultivars of Delphinium

Galileo	Holly Cookland Wilkins	Jill Curley
Kennington Classic	Kestrel	Langdon's Blue Lagoon
Langdon's Pandora	Lilian Bassett	Lord Butler
Lucia Sahin	Margaret	Michael Ayres

Source: www.delphinium-society.co.uk

Cultivars of Delphinium

Min | Oliver | Our Deb

Purple Velvet | Rosemary Brock | Rosy Future

Spindrift | Sungleam | Sunkissed

Sweethearts | Tiddles | Walton Gemstone

Source: https://delphinium-society.co.uk/delphiniums1/delphinium-awards

D. 'Elizabeth Cook', D. 'Olive Poppleton', D. 'Sandpiper', D. 'Atholl', D. 'Lilian Bassett', D. 'Celebration', D. 'Sungleam', D. 'Sunkissed', D. 'Claire', D.'Summerfield Miranda', D. 'Our Deb', D. 'Cherub', D. 'Langdon's Royal Flush', D. 'Rosemary Brock', D. 'Lucia Sahin', D. 'Conspicuous', D. 'Walton Gemstone', D. 'Emily Hawkins', D. 'Gillian Dallas', D. 'Min', D. 'Oliver', D. 'Tiddles', D. 'Spindrift', D. 'Blue Dawn', D. 'Loch Leven', D. 'Lord Butler', D. 'David Mannion', D. 'Fenella', D. 'Langdon's Pandora', D. 'Faust', D. 'Galileo', D. 'Blue Nile', D. 'Völkerfrieden', D. 'Atlantis', D. 'Can-can', D. 'Cassius', D. 'Giotto', D. 'Bruce', D. 'Michael Ayres', D. 'Coral Sunset', D. 'Darwin's Blue Indulgence', D. 'Darwin's Pink Indulgence', D. 'Delft Blue', D. 'La Bohème', D. 'Merel', D. 'Purple Ruffles', D. 'Sarita', D. 'West End Blue' and D. 'Wishful Thinking'.

6.0 Genetics and Breeding

6.1.0 Cytology

The basic chromosome number of the genus Delphinium was reported to be x=8 (Darlington and Janaki-Ammal, 1945). The diploid chromosome number 2n=16 has been observed in most of the Delphinium species such as *D. ajacis* (Langlet, 1927), *D. cardinale* (Mehlquist *et al.*, 1943), *D. grandiflorum* (Propach, 1940), *D. altissimum* (Mehra and Kaur, 1963), *D. nudicaule* (Gregory, 1941), *D. zalil* (Gage, 1953) and *D. malabaricum* var. malabaricum (Pai *et al.*, 2007). In the genus Delphinium diploid species are by far the most numerous, tetraploids are occasional and hexaploids are known only in cultivated forms (Gage, 1953). The existence of chromosome counts for the studied species of the genus *Delphinium* has revealed the presence of diploid (2n=16), triploid (2n=24), tetraploid (2n=32) and hexaploid (2n=48) (Langlet, 1927; Al-Kelidar and Richards, 1981 and Subramanian, 1985). Cytological data is of great importance for systematic and evolutionary studies of plants and have been widely used to elucidate intraspecific and/or interspecific relationships and delimit specific or generic circumscriptions (Stebbins, 1971; Raven, 1975; Hong and Zhang, 1990).

6.2.0 Pollination

Inflorescence of delphinium is raceme and flowers are arranged aesthetically all around the unbranched, indeterminate type of its axis. The *delphinium* genus is cross pollinated because of the presence of a built in system preventing self-pollination. Delphiniums attract hummingbirds, butterflies and bees as pollinators. The large circumboreal genus *Delphinium* is characterized by short-spurred blue-purple flowers pollinated primarily by bees. Two exceptions to this pattern are the long-spurred red flowers of two western North American species known to be hummingbird-pollinated, and the long-spurred evening-scented white flowers of an east African species, *Delphinium leroyi* (Johnson, 2008).

6.3.0 Interspecific Hybridization

Natural hybridization regularly occurs among members of the genus Delphinium, although hybrids are usually limited to a small percentage of the parent species population (Dodd and Helenurm, 2002). Hybridization has been reported between *D. robustum* and *D. sapellonis*, as well as between other pairs of Delphinium species *e.g.*, *D. ramosum* × *D. barbeyi*, *D. ramosum* × *D. glaucum* (Warnock, 1993; Warnock, 1997). Interspecific hybrid plants of *D. cardinale* × *D. grandiflorum* and *D. grandiflorum* × *D. nudicaule* were successfully obtained through ovule culture (Honda and Tsutsui, 1997). Interspecific crosses was carried out between *D. elatum* and *D. nudicaule* or *D. cardinale* to produce the red flowering *Delphinium* varieties (Katoh *et al.*, 2004). Honda (*et al.*, 2003) attempted ovule culture by making reciprocal crosses between the diploid species *D. grandiflorum* L. and the tetraploid cultivar *D. hybridum* 'Galahad' and an interspecific hybrid plant was obtained after confirming the hybridity by chromosome counting. Interspecific hybrid plants of *D. cardinale* × *D. grandiflorum* and *D. grandiflorum* × *D. nudicaule* were successfully obtained by Honda and Tsutsui (1997) and their hybridity was confirmed by isozyme analyses. Legro (1961) succeeded in producing red flowered delphiniums by interspecific hybridisation of red flowering *D. nudicaule* and *D. cardinale* with *D. elatum*. Cheng *et al.*, 2020 observed strong cross-incompatibility in interspecific cross between *D. yunnanense* × *D. elatum* which resulted into hybrid sterility.

7.0 Growth and Flowering

Delphinium prefers moist, cool summers and do not perform well in hot and dry weather. Most hybrids flower from early to mid-summer. Immediately after flowering, the flower stalks are removed and it encourages additional flower spikes in late summer or early fall in northern hemisphere. Under Indian conditions it starts flowering late winter to early summer season. *Delphinium malabaricum*, the only species native and grown in southern part of India flowers from last week of September with the appearance of buds, and full blooming occurred by the last week of November. However, flowering continued until the end of December. The duration from flower-bud initiation to full bloom is 25–35 days under western Indian conditions.

7.1.0 Temperature

Temperature plays a significant role in all the developmental stages of delphinium plants. Larkspur plants require vernalization to trigger flower development. The ideal temperature during the vernalization period is 10-13°C and in the absence of cool period, the plants will not flower well. In a study on *Delphinium nuttallianum*, artificial warming resulted in fewer flowering plants than those under control (unheated) plots. The study also found that individual plants in heated plots on an average had fewer flowers than control plots (Saavedra *et al.*, 2003). The most rapid and uniform germination generally occurred at constant 20°C (Carpenter and Boucher, 1992). Under greenhouse production it requires a day temperature between 18-21°C and a night temperature of 13-17°C. Delphiniums are naturally compact, but a variety of cultural techniques can be used to control plant height. When grown at cooler temperatures, the plants tend to be more compact and additional height control may not be necessary. DIF (temperature differential) plays an important role in its development and flower yield and quality in delphinium are enhanced by cool conditions (Armitage, 1993).

7.2.0 Photoperiod

Delphinium plants require 6-8 hours of light and exposure to early morning sun is better for growth and development. For germination to occur, seeds are independent of light. Seeds require vernalization for germination. Long day conditions are required for better growth, longer stems with quality flowering. Delphiniums grow best in full sun (8 or more hours of direct sunlight). They can also be grown in partial shade (4-8 hours of direct sunlight). Supplemental lighting is recommended under low light conditions. High pressure sodium lighting (500 to 600 foot candles) improved the quality of 'Summer Blues' plants by increasing the flower number. Supplemental lighting hastened flowering of 'Summer Blues' and 'Volkerfrieden' by about seven days, probably because the lights increased the temperature of the plants. Runkle (1996) found that *Delphinium elatum* is day-neutral plants (DNP) and is insensitive to photoperiod with respect to flowering; flowering percentage, time to flower, and flower count are similar under all photoperiods. *Delphinium grandiflorum* thrives well under high light, up to 7,500-foot candles/80,000 lux, as long as optimum temperatures are maintained. Long-day photoperiod promotes stem elongation and hastens flowering in delphinium (Wilkins, 1985). Garner *et al.* (1997) found that night break lighting to provide long days also appears to be an effective component to increases yield and reduces time to harvest in delphinium.

7.3.0 Growth Regulators

Plant growth regulators can be used to enhance flowering in delphinium. Gibberellic acid is used to promote growth and increase stem length of cut flowers in Delphinium (Latimer and Whipker, 2012). Growth regulation may be necessary to control stem elongation and should be applied as the stems begin to elongate above the basal foliage. If necessary, the same may be applied after 7 days. A-Rest (ancymidol, SePRO Corporation) is a more active compound than B-Nine or Cycocel. It is labelled for use on containerized plants in greenhouses and nurseries in hybrid Delphinium @ 35-132 ppm (Latimer, 2001). It is recommended to provide adequate spacing to control height. It is also suggested to apply Bonzi @ 10-20 ppm approximately 3 weeks after transplant (once roots have reached bottom of pot). If additional control is needed, applying multiple sprays of Bonzi @ 10-20 ppm is effective.

8.0 Propagation

Delphiniums are hardy annuals/perennials, usually propagated by seeds as well as cuttings.

8.1.0 Seed Propagation

In northern hemisphere, delphinium seeds sown in mid-winter flowers during first year itself, whereas late sown seeds like early summer or spring will flower in following year. Most hybrids come true from seed with a minimum of variability. Germination is most successful with fresh seeds at 18-23°C. The germination improves when they are chilled before planting for about 7 days at 2°C in sealed

Propagation and Growing of Delphinium

Cuttings | Preparation of Cuttings | Treatment with IBA

Seeds | Seedlings in Plug Tray | Growing on Bed

Growing Plant | Beds of with Wire Support | Net Support

Delphinium in Full Bloom | Greenhouse Production | Harvesting Stage

Source: www.florapodium.com, www.gardenerspath.com, www.floraldaily.com, www.line.17qq.com, www.florapodium.com, www.facebook.com

containers. Plants have tap root system and do not transplant easily, therefore it better to sow them directly. Since, delphiniums require darkness for germination, the seeds need to be covered completely beneath the soil. Germination is usually high if the soil temperature is appropriate, but seeds will not germinate if soil temperature is above 13°C. The soil should be consistently kept moist during germination and usually takes 14-21 days for germination. For better germination stratification and other plant growth regulators can be used. Some species requires 2-3 months cold/ moist stratification for satisfactory germination. Stratification (5°C) and application of gibberellic acid both improved germination in *D. elatum* cultivars. Chilling the dry seed at 3°C for 1 week prior to sowing improved germination of seeds (Carpenter and Boucher, 1992). The seeds of *Delphinium malabaricum* are very sensitive to high temperature. Therefore, no germination takes place it sown in the month of March-May. Seeds remain dormant until temperatures and moisture are sufficient to trigger germination (Kolar *et al.*, 2017).

8.2.0 Vegetative Propagation

Delphiniums can also be propagated by softwood cuttings or root cuttings. Softwood basal cuttings can be taken from young shoots. The clear white and solid shoots of about 7-10 cm long can be taken from the crown portion of base. The hollow, blackened shoots should be avoided as they may rot in rooting media itself. The shoot should be cleaned and lower leaves are removed, retaining the tip and one leaf. They are treated with fungicides to avoid microbial growth. For better rooting, the cuttings should be kept in a shade maintaining a temperature around 14°C. It takes 4-6 weeks for rooting.

8.3.0 Micropropagation

Agrawal *et al.* (1991) developed a micropropagation technique for *Delphinium malabaricum* using nodes from inflorescence stalks. Maximum shoot proliferation was obtained on Murashige and Skoog's medium supplemented with 2-1P (10 mg/l) and inositol (100 mg/l). Samantha *et al.* (1993) studied the uptake of macronutrients from MS medium and shoot growth rates. With increased concentration of total nutrients significantly increased both fresh weight gain and multiplication rate at the high plant density. Bott (1980) produced the rapid callus from plants developed through crosses between tetraploid *D.nudicaule* × *D. cardinale* and *D. elatum* on basic Murashige and Skoog (MS) agar medium + 1 mg/l 2,4-D and 0.2 mg kinetin/l by using apical meristems with young (1-5 mm long) leaves. Morphological and physiological characteristics of micropropagated plants of Delphinium *cv.* Princess Caroline were studied by Santamaría *et al.* (1993). Leifert *et al.* (1994) found that bacterial contaminants in Delphinium tissue cultures can be prevented by acidification of the tissue culture medium. Leifert *et al.* (1992) examined the phytotoxic effects of antibiotic treatments on micropropagated Delphiniums and found that antibiotic toxicity to plant growth was maximum in streptomycin and least in carbenicillin. Murphy *et al.* (1998) revealed the advantages of using closures with filters or vented vessels, which allow gas exchange, increasing the photosynthetic capacity, the multiplication rate and the survival of plants after transfer to *ex vitro* conditions.

9.0 Soil and Climate

9.1.0 Soil

Delphiniums prefers a rich soils with high organic matter and grow well in well-drained but moist soil (Phillips and Rix, 1992; Huxley, 1992). They are traditionally grown in basic soils with pH levels of 6.5-7.0. As these are susceptible to root rot, use of a well-drained medium is critical and plants should be grown on sunny locations with optimum soil moisture content for better performance (Brickell, 1990).

9.2.0 Climate

Delphiniums should be planted in areas receiving 6-8 hours of sunlight, preferably in the morning. It needs to be protected from strong winds and heavy rainfall to avoid damage to the tall flowering stalks. Flowering is induced when the temperature is higher than 15-18°C. Minimum night temperature should be 13°-15°C for better flowering.

10.0 Cultivation

10.1.0 Seedling Raising

Seedlings can be raised in a nursery beds. The dimension can be kept as per the requirements. The site should receive full sunlight away from any shade. The soil should be well drained and rich in organic matter. The top soil should be free from weeds and well rotten manures/leaf mold should be mixed and levelled. Seeds should be sown evenly and thinly in lines and cover with fine soil for germination. Use fresh seeds, as germination can be poor after a year. Pre-soaking is recommended to improve the germination rate and seeds germinates in three weeks. Delphiniums are very sensitive to root-rot diseases. Fertilizer can be applied as necessary if leaves look deficient, but avoid excessive fertilizer. Young seedlings are very sensitive to high temperature and therefore it is recommended to maintain average temperature of 25°C (day) and 15°C (night) until 1-2 true leaf. Thereafter, the temperature is reduced to 20°C (day) and 10°C (night) to prevent premature flower initiation under high-temperatures. Overgrown root bound plug seedlings should not be planted.

10.2.0 Planting and Spacing

The field may be prepared well in advance and soil may be tilled to loosen the soil to a depth of 30-40 cm before planting. Well decomposed compost may be used and mixed properly before planting. The planting distance may vary based on delphinium types and species. The plants can be planted at distance of 30×40 cm when growing for more than one year and 23×23 cm to 30×30 cm for annual production, equivalent to 8, 11 or 19 plants/m^2. Elatum types of delphinium can perform well when spaced at 36 plants/m^2 (Hanks and Mason, 2016).

10.3.0 Mulching

Weeds compete with the plants for water, space and nutrients during the establishment phase. Therefore, use of different types of mulches can prevent weeds and maintain optimum level of moisture which is required for delphiniums to grow faster. Mulches also help retain soil temperatures. For perennial delphiniums organic mulch (bark, shredded leaves, grasses, *etc.*) can be used on beds which also helps in improving soil properties after decomposition. Since delphiniums are susceptible to rotting, the organic mulches should be avoided near crown and stem portion.

10.4.0 Manuring and Fertilization

The crop prefers rich soil with adequate organic matter. Therefore addition of compost and other manures is recommende at 5-8 kg/m^2 during soil preparation. A basal dose of fertiliser can be applied while field preparations. One week after transplanting, a liquid fertiliser dose of 100–150 ppm nitrogen should be applied weekly as part of a complete balanced fertiliser schedule. Controlled or slow release fertiliser (@ 60–80g/m^2) may be applied to avoid higher ammonium and nitrogen levels as the plants' roots are sensitive to high salt levels. When grown in pots, the plants are feed with NPK 12:10:10 every 2 weeks for better growth.

10.5.0 Irrigation

Excessive soil wetness and drought should both be avoided for delphiniums cultivation. Watering should be done thoroughly once a week to help new roots to grow deep. The roots and crown part are highly susceptible to rotting and therefore excess watering or water stagnation should be avoided. After harvesting of flower stems, overhead irrigation can result in rotting of the hollow ends of the cut stems. Irrigation should be given at the base and foliage should not be sprinkled as wet foliage might attract diseases.

10.6.0 Staking

Staking is recommended for tall types of delphiniums. The taller and hallow stems make them susceptible to wind damage. Therefore, each stem can be individually staked or wire support can be provided on bed from both the sides.

10.7.0 Pot Culture

Delphiniums can be grown successfully in different types of pots and containers for terrace gardening, balcony, home gardens, *etc.* The magnificent flowering spikes are attractive focal points when grown in pots and it can be relocated as per the need. Earthen or plastic pots (4.5-8 litre) can be used for cultivation of delphiniums. They can also be planted in smaller pot and as the plant grows to a certain height they can be transferred to a larger pot. Older pots can also be used provided they are well scrubber and disinfected before planting. Different types of planters or grow bag are also used for growing delphiniums. The ideal pot mixture for delphinium should contain plenty of organic matter. It is recommended to use slow release fertilizer in the soil compost mixture before planting. The pot should be fortnightly watered with liquid fertiliser (NPK 12:10:10) until flower bud initiation. A gap of 2 - 3 cm should be kept between the soil surface and the top of the pot which will help in retaining the water in the pot. Staking is also an important activity when plants are grown in pots. Stakes can be placed at three corners around the edge of the pot during early growing stage to prevent root ball damage. Dwarf type of delphiniums are highly suitable for pot cultivation. Mini Delphiniums (formerly known as Dwarf Stars) have been specifically bred for pot growing.

11.0 Diseases and Pests

11.1.0 Diseases

11.1.1 Crown and Stem Rots

It is caused by soil borne fungi like *Pyhytophthora* spp., *Pythium* spp., *Rhizoctonia solani,* and *Fusarium* spp. The leaves of the affected plant turn yellow and plants remain stunted. Foul smelling ooze forms under wet conditions and stems may topple. It is recommended to remove and destroy infected plants to avoid further spread. Drenching of soil with fungicide or removal of all plant parts from the soil which touches ground is recommended for effective management of disease complex. Some fungi can live in the soil for years, so planting of delphiniums in the same area should be avoided.

11.1.2 Powdery Mildew (*Erysiphe polygoni*)

The disease occurs on the top of the leaves in humid weather conditions particularly in dry spells after rainy conditions. First symptom can be seen on upper side of leaves when white spots of mildew growth appears. The leaves appear to have a whitish or greyish surface and may curl. It is recommended to have good plant spacing and pruning. The disease can be managed by removal of infected leaves and spraying of suitable fungicides.

Virus

11.1.3 Cucumber Mosaic Virus

It causes misshapen shoots, stunted growth and necrotic yellow leaf spots. It can only be managed through control of the vectors (aphids and/or thrips).

11.1.4 Aster Yellows

The aster yellow shows symptoms on economical part *i.e.* flowers. The petals turn green, deformed and leaf like tissues are formed instead of flower parts. This virus like disease is spread by leafhoppers. Therefore, leafhopper management is important for control of aster yellow in delphiniums along with removing infected plants and weeds which serve as alternate hosts to the disease.

11.2.0 Pests

11.2.1 Aphids

Aphids are greenish, red, black or peach coloured sucking insects feed on the undersides of leaves. They leave a sticky residue on foliage that attracts ants. It is recommended to use natural predators like lady beetles and wasps who feed on aphids. In case of severe infestation suitable insecticides can be used.

Diseases and Pests of Delphinium

Aster Yellow

Bacterial Stem Rot

Bacterial Crown Rot

Powdery Mildew

Black Leaf Spot

Broad Mite

Delphinium Worm

Aphids

Slug

Source: www.images.app.goo.gl, www.gardening.usask.ca, www.gardenerspath.com, www.pnwhandbooks.org

11.2.2 Mites

The mite damage plants by sucking juice from stems and leaves which resulted into distorted and stunted growth and sometimes plant may not flower. Flowers looks distorted, streaked and blotched. Leaves become cupped, curled, dwarfed and thickened. These are very tiny, but multiply very rapidly in hot, dry weather. The infected plants can be discarded and spray of acaricides can be undertaken for its control.

11.2.3 Leaf Miners

It bore under the leaf surface causing irregular serpentine lines. The larvae are yellow cylindrical maggots and the adults are small black and yellow flies. They do not usually kill plants, but disfigure the foliage.

11.2.4 Borer

The insect feed inside the plant stem causing the plants to wilt. By the time the plant wilts, it is too late to save it. The larva is 1.5 inches long, greyish brown with one dorsal stripe and two lateral stripes on each side. The lateral stripes on the front half are interrupted and the lower brown stripe extends forward onto the side of the head. The eggs hatch in May to early June, after the moth lays them the previous September or October. It can be controlled by spraying suitable insecticides.

11.2.5 Slugs and Snails

It can also become a major problem for delphinium growers especially when new shoots emerge and develop. Clean cultivation practices should be followed for its management. In case of severe infestation suitable metaldehyde be used.

12.0 Harvesting and Postharvest Management

12.1.0 Harvesting

Delphiniums are normally harvested when lower 1 or 2 flowers are opened on the spike. After harvesting, stems should be placed immediately in water solution containing suitable anti-ethylene substance for at least four hours at ambient temperatures (but not in a cold store).

12.1.0 Postharvest Treatements

The stem can also be placed in water with a postharvest conditioner for storing for one to two days at 3–5°C. Delphiniums are very sensitive to ethylene, which causes rapid loss of all the flowers on the spike. Flower spikes can also be pre-treated with 1-MCP or STS to extend their vase life and protect them from exposure to ethylene. Ichimura *et al.* (2003) studied the effect of soluble carbohydrates on sepal abscission. The result suggests that the treatment with mannitol, a major carbohydrate in Delphinium, delayed the abscission of sepals by reducing the sensitivity to ethylene. Mannitol further acted, not merely as an osmolyte, but as an apparent source for carbohydrate metabolism in the flower. Kikuchi *et al.* (2003) studied the activities of cell wall and soluble acid invertase, sucrose synthase (SS) and mannitol dehydrogenase (MDH) in relation to sink mechanism in sucrose- and mannitol-translocating plants. The result found that SS activity that is known to be important in sink metabolism also decreased during sepal wilting, suggesting that SS was related to the wilting. The delphinium requires carbohydrates in the vase solution to continue their metabolic activities; therefore, cut stems should be kept continuously in a sucrose solution after harvest (Dole *et al.*, 2013; Ichimura *et al.*, 2000).

12.3.0 Packaging and Storage

These flowers are often packed with or without water, but may also be packed in standard horizontal boxes. They are sensitive to gravity, so horizontally-packed flowers should be pre-cooled and maintained at proper holding temperatures. Delphinium should be stored at 0-1°C. For longer term storage, they should be wrapped in perforated polyethylene to reduce water loss.

12.4.0 Seeds Production

For seed production the plants are directly cultivated in raised beds planted at 30x30 cm. The flowers are arranged vertically on long stems, which later on produce small legumes enclosing seeds. The crop reaches maturity after mid March till mid April. Flowering branches are harvested manually when the pods show sign of drying and slight splitting. The yield varies 30-40 kg/acre depending on the health of crop and prevailing temperature.

13.0 Biotechnology

13.1.0 Genetic Engineering

Development of blue colour in flower crops like roses, carnations, chrysanthemums has started by accumulation of delphinidin (Dp)-type anthocyanins in their petals *via* the heterologous expression of a flavonoid 3′,5′-hydroxylase gene which is responsible for bluish colours in these crops. Although the floral colours of these transgenic plants still lack a true blue colour, the basis for the future molecular breeding of truly blue flowers is *via* the engineering of anthocyanin pathways. The 7-polyacylation process during the biosynthesis of purple-blue anthocyanins in delphinium (*Delphinium grandiflorum*) was found to occur in vacuoles using acyl-glucose as both the glucosyl and acyl donor. The 7-polyacylation proceeds through the alternate glucosylation and p-hydroxybenzoylation catalyzed by these enzymes. p-Hydroxybenzoyl-glucose serves as the p-hydroxybenzoyl and glucosyl donor to produce anthocyanins modified with a p-hydroxybenzoyl-glucose concatemer at the 7-position. This novel finding has provided a potential breakthrough for the genetic engineering of truly blue flowers, where polyacylated Dp-type anthocyanins are accumulated exclusively in the petals (Sasaki and Nakayama, 2015).

13.2.0 Genetic Transformation

Hirosei *et al.* (2002) successfully developed transformation method on *Delphinium* spp. using *Agrobacterium tumefaciens*. They have used explants like elongated hypocotyl and cotyledon petiole segments which was infected with

Agrobacterium tumefaciens strain LBA4404, which harbored a binary vector plasmid, pIG121-Hm, which included the β-glucuronidase (GUS) gene (with an intron) as a reporter gene, and the neomycin phosphotransferase II gene and the hygromycin phosphotransferase gene as selection markers. In Delphinium, genes encoding ethylene receptors have also been isolated by Kuroda *et al.* (2003) and Tanase and Ichimura (2006).

14.0 References

Agnihotri, P., Jena, S.N., Husain, D. and Husain, T. (2014) *Pleione*, **8**(2): 344 - 352.

Agrawal, D.C., Pawar, S.S., Morwal, G.C. and Mascarenhas, A.F. (1991) *Annals of Botany*, **68** (3): 243–245.

Al-Kelidar, R.K. and Richards, A.J. (1981) *Cytologia*, **46**: 623–633.

Blanche, C. (1990) *Delphinium* L. Subgen. *Delphinium*: Origin and evolutionary trends *Collectamea Boanica.* (Barcelona). **19**: 75-95.

Blanche, C. (1991) Revisio biosistemàtica del gènere *Delphinium* L. a la Peninsula Ibèrica i a les illes Balears. Arxius de la Secció de Ciències, Institut d'Estudis Catalans, Barcelona. pp. 1–288.

Blanche, C., Molero, J., Rovira, A.M., Simon, J. and Bosch, M. (1997) *Lagascalia* **19**: 59-82. https: //idus.us.es/handle/11441/62784

Bott, J. C. (1980) *Plantsman*, **2**(3): 169-171.

Brickell, C. (1994) The RHS gardener's encyclopedia of plants and flowers, Dorling Kindersley Ltd., London, *pp*. 453-454.ISBN 0-86318-386-7

Carpenter, W.J. and Boucher, J.F. (1992) *Hort Science*, **27**: 989-992.

Darlington, C.D. and Janaki-Ammal, E.K. (1945) *Chromosome atlas of cultivated plants*. Allen and Unwin, London.

Diaz, J., Juan, G.R. and Gabriel, F. (2000) *J.Natural Prod.*, **63**: 1136 - 1139.

Dole, J.M., Carlson, A.S., Crawford, B.D. and McCall, I.F. (2013) *Acta Hort.*, **1000**: 63-70.

Dymock, W., Warden, C.J.H. and Hooper, D. (1890) *Pharmacographia Indica* Vol-I (Bishen Singh, Mahendra Pal Singh, Dehra Dun and Periodical Expert, Delhi): 20-23.

Espinosa, F., Deroin, T., Malécot, V., Wang, W., Pinedo, M., Nadot, S. and Jabbour, F. 2021. *Historical note on the taxonomy of the genus Delphinium L.* (*Ranunculaceae*) *with an amended description of its floral morphology. Adansonia, sér.* 3, **43** (2): 9-18. https: //doi.org/10.5252/adansonia2021v43a2. http: //adansonia.com/43/2

Fang, S. and Benn, M. (1992) *Phytochemistry*, **31**: 3247–3250.

Gage, M.A. (1953) *Ann. Mo. Bot. Gard.*, **40**: 113–183.

Garner. J.M., Jones, S.A. and Armitage, A.M. (1997) *Hort Science*, **32**(1): 61-63.

Gregory, W.C. (1941) *Trans Am Phil Soc.*, **31**: 443–520.

Hanks, G. and Mason, L. (2016) *Delphinium as a cut flower crop grown in Spanish tunnels*. Agriculture and Horticulture Development Board. Information Sheet 7.

Honda, K. and Tsutsui, K. (1997) *Euphytica*, **96**: 331–337. https: //doi.org/10.1023/A: 1003058923669

Honda, K., Watanabe, H. and Tsutsui, K. (2003) *Euphytica*, **129**: 275. https: //doi.org/10.1023/A: 1022206731894

Hong, D.Y. and Zhang, S.Z. (1990) *Cathaya*, **2**: 191–197.

https: //www.en.wikipedia.org

https: //www.flowersofindia.net

https: //www.ngb.org/

Husain, S.M. (1897) *Qarabadeen-e-Kabir*: Vol I. Lucknow, India Matba Munshi Nawal Kishore, pp.561-565.

Huxley. A. (1992) *The New RHS Dictionary of Gardening*. MacMillan Press 1992 ISBN 0-333-47494-5

Ichimura, K., Kohata, K. and Goto, R. (2003) *Physiol. Plant.*, **108**(3): 307 - 313

Jabbour, F. and Renner, S.S. (2011a) *Taxon*, **60**: 1029-1040. https: //doi.org/10.1002/tax.604007

Jabbour, F. and Renner, S.S. (2011b) *PhytoKeys*, **7**: 21-26. https: //doi.org/10.3897/phytokeys.7.2010

Jabbour, F. and Renner, S.S. (2012a) *Molecular Phylogenetics and Evolution*, **62**: 928-942. https: //doi.org/10.1016/j.ympev.2011.12.005

Jabbour, F. and Renner, S.S. (2012b) *International Journal of Plant Sciences*, **173**: 1036-1054. https: //doi.org/10.1086/667613

Jabbour, F., Ronse De Craene L. P., Nadot, S. and Damerval, C. (2009) *Annals of Botany*, **104**: 809-822. https: //doi.org/10.1093/aob/mcp162

Jensen U., Hoot S.B., Johansson J.T. and Kosuge K. (1995) *Systematics and phylogeny of the Ranunculaceae* – a revised family concept on the basis of molecular data. Plant Systematics.

Katoh, N., Tokuhiro, K. and Nakabayashi, T. (2004) *Breeding Science*, **54**(2): 99–103.

Kikuchi, K., Kanahama, K. and Kanayama, Y. (2003) *Engei Gakkai zasshi*, **72**(1): 37-42.

Kolar, F., Pawar, N. and Dixit, G. (2011) *J. Appl. Hort.*, **13**(1): 18-24.

Kuroda, S., Hakata, M., Hirose, Y., Shiraishi, M. and Abe, S. (2003) *Plant Physiol. Biochem.*, **41**: 812–820.

Langlet, O.F. (1927) *Svensk Bot. Tidskr.*, **21**: 1–17.

Latimer, J. and Whipker, B. (2012) *Selecting and Using Plant Growth Regulators on Floricultural Crops* Extension Bulletin, Floriculture Crops, North Carolina State University.

Latimer, J.G., Scoggins, H.L. and Banko T.J. (2001) *Extension Specialist*, Virginia Tech, Hampton Roads AREC, Virginia Beach.

Legro, R.A.H. (1961) *Euphytica*, **10**: 1–23.

Leifert, C., Camotta, H. and Waites, W.M. (1992) *Plant Cell Tissue Organ Cult.*, **29**: 153-160.

Leifert, C., Waites, B. and Keetley, J.W. (1994) *Plant Cell Tiss Organ Cult.*, **36**: 149–155. https: //doi.org.https.jxnydx.proxy.chaoxing.com/10.1007/BF00037713

Malyutin, N.I. (1987) *Bot. Zhurn.*, **72**: 683–693.

Mehlquist, G.A.L., Blodgett, O. and Bruscia, L. (1943) *J. Hered.*, **34**: 187–192.

Mehra and Ramanandan, P. (1972) *Cytologia*, **37**: 281–296.

Mehra, P.N. and Kaur, B. (1963) *Cytological study of some Himalayan Ranunculaceae. Proceedings of the Indian Science Congress*, **50**B: 453–454.

Munz, P. A. (1968) *J. Arn. Arb.*, **49** (1 & 2): 73 - 166 and 233 - 259.

Munz, P.A. (1967) *J. Arn. Arb.*, **48** (3 & 4): 249 - 303 and 476 - 545.

Murphy, K.P., J.M. Santamaria, W.J. Davies and P.J. Lumsden, (1998) *Journal of Horticultural Science and Biotechnology*, **73**: 725-729.

Orellana, M.R., Lypez-Pujol, J., Blanché, C., Rovira, A.M. and Bosch, M. (2009b) *Genetica*, **135**: 221–232.

Pai, S.R., Kamble, M.Y., Yadav, S.R., Dixit, G.B., Pawar, N.V., Chavan, P.D. and Yadav, U.S. (2007) *Cytologia*, **72**: 319–322.

Paul, S.R. (1975) *Geobios.*, **2**(1): 27 - 28.

Pawowski, B. (1964) *Delphinium*, In: Tutin T. G., Heywood V. H., Burges N. A., Valentine D. H., Walters S. M., Webb D. A. (eds), Flora Europaea. The University Press, Cambridge.pp. 213-216.

Payer, J. (1857) *Traité d'organogénie comparée de la fleur*. Masson, Paris, 748 p. https: //doi.org/10.11588/diglit.9757

Phillips. R. and Rix. M. (1991) *Perennials Volumes 1 and 2*. Pan Books 1991 ISBN 0-330-30936-9

Polunin, O. and Stainton, A. (1984) *Flowers of the Himalaya*. Oxford University Press, Delhi. pp. 6 - 8.

Prajapati, N.D. and Kumar, U. (2003) *Argo's Dictionary of Medicinal plants*. Agrobios (India), Jodhpur, India. pp. 107.

Propach, H. (1939) *Gartenschonh*, **20**: 283–284.

Rahman, A., Nasreen, A., Akhtar, F., Shekhani, M. S., Clardy, J., Parvez, M. and Choudhary, M.I. (1997) *J.Natural Food Prod.*, **60**: 472 – 474.

Rau, M.A. (1993) *Ranunculaceae*. In: *Flora of India*. vol. I. Eds. Sharma, B.D.; Balkrishnan.

Raven, P.H. (1975) *Ann. Missouri Bot. Gard.*, **62**: 724–764.

Runkle, E.S. (1996) The effects of photoperiod and cold treatment on flowering of twenty-five species of herbaceous perennials. MS Thesis, Michigan State Univ., EastLansing.

Saavedra, F., Inouye, D.W., Price, M.V. and Harte, J. (2003) *Global Change Biology.*, **9**(6): 885-894.

Said, H.M. (1970) *Hamdard Pharmacopoeia of Eastern Medicine.* Hamdard National Foundation, Times Press, Karachi, Pakistan. pp. 49.

Samantha, P., Lumsden, P.J., Berger, F. and Leifert, C. (1993) *Journal of Horticultural Science*, **68**: 5, 807-813. DOI: 10.1080/00221589.1993.11516417

Sasaki, N. and Nakayama, T. (2015) *Plant Cell Physiol.*, **56**(1): 28-40. doi: 10.1093/pcp/pcu097.

Stebbins, G.L. (1971) *Chromosomal evolution in higher plants.* Edward Arnold Ltd., London.

Subramanian, D. (1985) *Cytologia*, **50**: 759–768.

Wang, W., Liu, Y., Yu, S.X., Gao, T.G. and Chen, Z.D. (2013) *Taxon*, **62**: 713-722. https: //doi.org/10.12705/624.10

Wang, W., Lu, A.M., Ren, Y., Endress, M.E. and Chen Z.D. (2009) *Perspectives in Plant Ecology, Evolution and Systematics*, **11**: 81-110. https: //doi. org/10.1016/j. ppees.2009.01.001

Wang, W.T. and Warnock, M.J. (2001) *Delphinium* L. In: Wu, Z.Y., Raven P.H. (Eds.), *Flora of China*, Vol. 6. Science Press/Missouri Botanical Garden, Beijing/St. Louis. pp. 223 - 237.

Warnock, M.J. (1993) *Delphinium.* In: Flora of North America North of Mexico, Volume 2. Flora of North America Editorial Committee, Oxford University Press.

Warnock, M.J. (1995) *Physiologia*, **78**: 73–101.

Warnock, M.J. (1995) *Phytologia*, **67**(6): 468-493.

Warnock, M.J. (1997) *Delphinium.* In: Flora of North America North of Mexico, Volume 3. Flora of North America Association, Oxford University Press: New York, NY. Online version available at http: //www.fna.org.

Wilkins, H.F. (1984) *Delphinium*, In: A.H. Halevy (ed.). Handbook of flowering, supp. CRC Press, Boca Raton, Fla. pp. 89-91.

Willstatter, R. and Mieg, W. (1915) *Ann. Chem.*, **408**: 61.

3

Calceolaria

P. Naveen Kumar, D.V.S. Raju, Tarak Nath Saha, Ganesh B. Kadam and K.V. Prasad

1.0 Introduction

The genus *Calceolaria* comprises of showy, unique and brightly coloured house plants. The genus name came from the Latin word 'calceolus' meaning a 'slipper', a reference to the appearance of the flower. They are commonly known as pocket-book plants, slipper flowers or slipperwort or lady's purse, as these stunning, decorative species usually have pouches at the bottom of the flower that looks like purses or slippers. They are sometimes mistaken as an unusual type of orchids (Bailey, 1976).

There are over 250-300 different species in this genus, but only a few (a couple of them) are in cultivation. Generally, the flowers are attractive, show stopping and in bright yellow and orange colour sometimes with red spots. Flower size varies with origin and cultivars. Calceolaria flower usually last for one season and need to be replaced as they rarely rebloom (Johnstone, 2021).

2.0 Importance and Uses

2.1.0 Aesthetic Uses

Calceolaria is a delightful annual flower that adds value to any garden (mostly in pots). As it is a small plant, it is ideal for decorating table-tops on decks, patios, and balconies in early spring (late winter) before other common flowers are available. It grows well with pansies, violas, sweet alyssum and dianthus, hence, can be grown in the garden (flower beds or borders) or in a variety of containers for a container garden (https://www.costafarms.com/plants/calceolaria).

These cool-weather-loving plants show off unique flowers mimicking orchids are very popular in US, UK and Europe. The delightful flowers are held on stems above the dark green, arrow-shaped leaves, making calceolaria unique. Potted plants of calceolaria are a popular option for gifting in late winter and early spring. They are produced and marketed typically during Easter and/or Mother's Day in the United States (Erwin, 1994). They are also suitable for windowsill in winter deu to their bright colours and unusual flowers. Although they can sometimes be used as bedding plants, calceolaria are usually grown in pots, both for outdoor and indoors. They prefer cool place with diffused sunlight. The plants should be watered regularly and the potting medium should be light and fertile, if grown in pots.

2.2.0 Therapeutic and Other Uses

Calceolaria species, in addition to the above are having some medicinal uses and some of them are toxic to insects and are found to be resistant to microbial attack (Cespedes *et al.*, 2014). These properties are probably associated with the presence of terpenes, iridoids, flavonoids, naphthoquinones and phenyl-propanoids previously demonstrated to have interesting biological activities.

The aerial parts of *Calceolaria* species possesses medicinal activities that are used for digestive, diuretic and antimicrobial treatments (Harborne and Baxter, 2001; and Woldemichael *et al.*, 2003). This genus is characterized by the presence

Uses of Calceolaria

In Containers

In Home Decor

In Flower Vase

In Rock Garden

In Pots

In Garden

In Basket as a Gift

In Landscaping

As Border Plants

Uses of Calceolaria

In Flower Beds

In Bouquet

Near Window

As Cut Flower

In Window Sill

In Exhibitions

Source: www.budivel.ru, www.live.staticflickr.com, www.newplantsandflowers.com, www.parks.ox.ac.uk, www.searcharchives.vancouver.ca, www.s.ecrater.com, www.portraitsofalpineplants.com, www.greenhousemag.com, www.gapphotos.com, www.thespruce.com, www.in.pinterest.com, www.cdn.shopify.com, www.worldoffloweringplants.com

of substances with important agrochemical and pharmacological potential such as insecticide (Khambay *et al.*, 1999) and antimicrobial properties (Falcao *et al.*, 2006; and Woldemichael *et al.*, 2003). Bio-pesticides from *Calceolaria integrifolia* exhibited anti-feedant, antifungal, insecticidal and antibacterial activities (Muñoz *et al.*, 2013a, 2013b, 2013c; Céspedes *et al.*, 2013a, 2013b). The n-hexane extract of *C. talcana* exhibited a good anti-feedant index activity against *Spodoptera frugiperda* and *Drosophila melanogaster* with values of 55.5 per cent and 87.5 per cent at 50.0 mg/ml, respectively (Muñoz *et al.*, 2013b).

3.0 Origin and Distribution

3.1.0 Origin

Calceolaria is the largest genus of the Calceolariaceae, comprising about 250 species distributed from Central Mexico to southern Argentina. It is perhaps one of the widest genera in the Andean region, distributed from Patagonia to central Mexico (Molau, 1988 and Cespedes *et al.*, 2013c). Calceolaria are native to the mountainous regions of Central and South America ranging from Mexico to Chile. These are easily recognized by its yellow (also available in orange and red) bilabiate flowers with saccate lower lip and two stamens (Molau and Sanchez, 1986).

3.2.0 Distribution

Calceolaria species ranges from Mexico to Tierra del Fuego, and is distributed mainly at high altitudes, between 2000 and 4000 m in tropical and subtropical regions. In temperate regions of southern Argentina and Chile certain species reach the sea level at both the Atlantic and Pacific coasts (Cosacov *et al.*, 2009).

Two main centers of diversification have been recognized in Calceolaria: at middle latitudes of Chile and at the south of the Huancabamba Deflection in northern Peru (Molau, 1988; and Ehrhart, 1997, 2000). This deflection, located at 6°S, is the deepest and widest depression present in the high Andes, representing a major geographical break and a significant barrier for plants and animals (*e.g.*, Simpson, 1975; Fjeldså, 1995; Weigend, 2002). Jovellana, with six species, is distributed in Chile (two species) and New Zealand (four species), and the monotypic Porodittia has a narrow distribution in the Peruvian Andes. The three genera differ markedly in flower morphology; *Porodittia* is characterized by a three-lobed corolla and three stamens, while *Jovellana* and *Calceolaria* have a bilabiate corolla and two stamens. The distinctive floral trait of *Calceolaria* is an inflated lower lip with an infolded median lobe, bearing a patch of oil-secreting trichomes, the elaiophore (Vogel, 1974 and Molau, 1988).

The majority of *Calceolaria* species are shrubby, but there are also perennial and annual herbs, and the genus is characterized by a noteworthy floral variation (Molau, 1988; and Ehrhart, 2000). Most of this variation is defined by the shape and size of the corolla lobes, the position of the elaiophore or lack thereof, stamen morphology, and sometimes, by the shape of the calyx.

4.0 Taxonomy

Calceolaria has been subdivided into three subgenera with 24 sections. The vast majority of the sections are in subgenus *Calceolaria,* mainly distributed in tropical regions, while subgenera *Cheiloncos* and Rosula are mainly distributed in temperate South America. Subgenus *Calceolaria* includes shrubs, subshrubs, and herbs that have stamens with filaments shorter than anthers. Subgenus *Cheiloncos* also includes herbs and subshrubs, but their stamens have filaments that are several times longer than their anthers. Subgenus *Rosula* includes rosulate or subrosulate herbs that have stamen filaments about as long as the anthers and includes the two strictly extratropical sections in the genus: Corymbosae and Kremastocheilos (Molau, 1988). Many sections as described by Molau (1988) have been shown to be polyphyletic (Anderson, 2006; Cosacov *et al.,* 2009).

Subgenus *Calceolaria* contains most of the sections previously placed in subgenus *Cheiloncos* by Pennell (1951), and includes 19 sections and 184 species mainly from the tropical Andes.

Subgenus *Cheiloncos* is restricted to two sections and 30-35 species and probably builds up a monophyletic group distributed almost exclusively in temperate zones of Chile and Argentina (Molau, 1988).

Subgenus *Rosula* contains three sections and 35-40 species of temperate and cold regions of Chile and Argentina.

Calceolaria herbeohybrida Voss (Scrophulariaceae) is a group of cultivars generally grown as flowering pot plants in the greenhouse and was probably derived from *Calceolaria crenatiflora* Cav. Recent molecular studies (Judd *et al.,* 1999; Olmstead *et al.,* 2001; and Oxelman *et al.,* 2005) proved their polyphyly and Scrophulariaceae were split into several families. One of these newly established groups was the family Calceolariaceae, comprising the three genera *Calceolaria, Jovellana* and *Stemotria* (=Porodittia) (Olmstead *et al.,* 2001). Discontinuation from Scrophulariaceae and elevation as a family of Calceolariaceae has been supported by the findings of Mayr and Weber (2006) based on the floral morphology and development in Calceolaria.

5.0 Morphology

From the many species of this genus, the only *Calceolaria* generally grown indoors is the *Calceolaria herbeohybrida.* These have very decorative foliage and flowers, but can be kept only for a single season. Its indoor life span is about a month or several weeks of flowering time before being discarded. The plants are about 10-15 cm in height and width. Leaves are large – up to 20 cm across – roughly heart-shaped and grouped around the base of a single central stem. Above the leaves rise several branching stalks 30-60cm tall, each carrying loose clusters of pouch or slipper-shaped flowers, which vary in size (according to seed stain) from 1-5cm across, in any one of various shades of red, orange, yellow or reddish brown. Most of the flowers of *Calceolaria herbeohybrida* are spotted or blotched with contrasting colour (https://www.plantsrescue.com/tag/calceolaria-crenatiflora/).

Flowers are produced in yellow, orange or red in solid, spotted or bicolor; 2.5cm across with swollen lower lip resembling a slipper, pocket-book or pouch; held in dense corymbs often covering foliage. The calyx is in 4 parts: the corolla is 2-lipped and pouched – the upper lip is smaller and the lower lip is much lager and is inflated. There are 2 stamens per flower. Flowers are zygomorphic and bilabiate composed by two lobes: one superior and reduced and the other inferior and inflate. The latter lobe includes a gland or elaiophore that secrete oil (Vogel, 1974). Sérsic (2004) gave a broader account on the diversity of the flower biology in Calceolaria based on observations of pollination mechanisms in other species, the study of functional morphology, considering also aspects of oil chemistry and breeding systems.

6.0 Species and Cultivars

6.1.0 Species

Calceolaria is one of the most diverse genera in Central and South America, including 250 species distributed from Mexico to Patagonia (Molau, 1988). *Calceolaria corymbosa* Ruiz et Pav is one of the most widely distributed in a variety of habitat types. Along its distribution, six subspecies have been described that are distinguished mainly by their red spot pattern on the corolla lobes and the altitude that they inhabit (Ehrhart, 2000).

Calceolaria herbeohybrida (syn. *C. crenatiflora* or *C. Multiflora*)

It is commonly called pocket-book plants which are known for their showy flowers. These hybrids are typically crosses between several species including *C. crenatiflora*, *C. corymbosa* and *C. cana*. They are primarily grown in commercial greenhouses for the floral. *C. herbeohybrida*, *C.* × *profusa* (*C. clibranii*) and *C. multifora nana* are suitable for growing in higher elevations (temperate region) only (Randhawa and Mukhopadhyay, 2001).

C. mexicana

The plants are about 30 cm high and can be grown only in the northern higher plains of Mexico and at medium elevations. They are excellent as pot plants.

C. integrifolia

It is a popular species and has seeds that are easier to grow. This is one of the taller and hardier varieties sometimes sold as a bedding plant.

C. uniflora (*C. darwinii*)

It is known as Darwins' slipper flower and was discovered by Charles Darwin in his voyage around South America. These unique flowers almost look like little orange penguins marching over the rocks. The Darwin slipper flower is winter hardy and prefers cool climate. The best germination temperature is between 10-23°C and the germination takes around 3-5 weeks.

Species of Calceolaria

Calceolaria crenatiflora  — *Calceolaria uniflora*  — *Calceolaria integrifolia*

Calceolaria bilfora  — *Calceolaria martinezii*  — *Calceolaria bilatata*

Calceolaria corymbosa  — *Calceolaria tripartite*  — *Calceolaria cana*

Calceolaria lepida  — *Calceolaria paralia*  — *Calceolaria williamsii*

Source: www.in.pinterest.com, www.ebay.co.uk, www.patagoniawildflowers.org, www.commons.wikimedia.org, www.pinterest.se,www.chileflora.com, www.worldoffloweringplants.com, www.en.wikipedia.org

C. × hybrida

All the modern F_1 hybrids and some of the improved varieties are designated as *C.* × *hybrida*.

6.2.0 Cultivars

The Herbeohybrida Group of calceolarias (*Calceolaria herbeohybrida* syn., *C. crenatiflora* or *C. multiflora*), commonly called pocket-book plants, are known for their showy flowers. They are primarily grown in commercial greenhouses for the floral industry (https://www.britannica.com/plant/slipper-flower). The flowers (1-5cm) have an enlarged lower petal and come in a range of colours like yellow, orange or red and are solid, spotted or bicolor. The blooms are held in dense corymbs often covering the basal foliage. Plants are primarily sold in late winter or early spring and will bloom continuously for several weeks. All the commercial varieties/cultivars belong to this group. Some more important species from commercial front include *C. andina, C. bicolor, C. biflora, C. corymbosa, C. falklandica, C. filicaulis, C. integrifolia, C. mexicana* and *C. uniflora*.

Moe (1977b) has divided Calceolaria cultivars into four groups according to flower and plant size (Table 1). Many cultivars are available, with continued improvements being made. The F_1 hybrids are more uniform in size and colour, bloom 4-5 weeks earlier and are more widely grown than open pollinated selections.

Table 1: Grouping of Calceolaria Cultivars Based on Flower Size

Sl.No.	Class	Group	Flower Size (cm)	Plant Height (cm)
1	Grandiflora	Tall	3.5-5	30-40
2	Grandiflora primula compacta	Medium	4.5-5	20
3	Multiflora	Medium	3-4	25-30
4	Multiflora nana	Small to medium	2-3	15-30

The members of *C. integrifolia* are mostly shrubs. Important varieties include Stewarti, Pink Beauty, Talisman, Rosacker's Brilliance and Theodore Wirth. These are propagated by cuttings (Poesch, 1937).

Development of new varieties/cultivars is mostly undertaken by private sector. Some of the important varieties include Aida, Sunset and Golden Rain. The cultivars of *C. herbeobyrida* recommended by the University of Vermont, Burlington, US are presented in Table 2. Leaves measuring (10-15cm long and 7.5-10cm wide) are opposite, pubescent, obovate to ovate and toothed (http://pss.uvm.edu/pss123/fpcalce.html).

The Plant Worlds Seeds, UK has good number of varieties in Calceolaria. These include Calceolaria-'John Innes', Calceolaria Mexicana 'Gold Purse', Calceolaria Bicolor 'Goldcrest', Calceolaria-'Andina', Calceolaria-'Biflora', Calceolaria-'Falklandica', Calceolaria-'Filicaulis', Calceolaria-'Corymbosa' and Calceolaria-'Integrifolia'.

Varieties of Calceolaria

Calceolaria hybrida
'Red'

Calceolaria hybrida
'Red & Yellow'

Calceolaria hybrida
'Bronze'

Calceolaria hybrida
'Yellow with Spots'

Calceolaria calynopsis
'Dark Red'

Calceolaria calynopsis
'Yellow'

Calceolaria hybrida
'Calynopsis Red'

Calceolaria hybrida
'Calynopsis Yellow'

Calceolaria hybrida
'Calynopsis Red+Yellow Eye'

Calceolaria Sunset
'Orange'

Calceolaria integrifolia
'Kentish Hero'

Calceolaria Sunset
'Yellow'

Source: www.ballseed.com, www.darcyeverest.co.uk, www.shireplants.co.uk, www.costafarms.com, Sakata Ornamentals

Varieties of Calceolaria

Balloon Orange Tiger

Calynopsis™ Orange

Calynopsis Cerise

Calceolaria Purple

Calceolaria Mexican

Calceolaria Gentle

Calceolaria Wrinkled (Solidistic)

Calynopsis Purple

Sunset Mixed

Calceolaria mexicana 'Gold Purse'

Calceolaria 'John Innes'

Redoak Calceolaria

Source: www.plant-world-seeds.com, www.thompson-morgan.com, www.budivel.ru, www.gpnmag.com, www.flipkart.com, www.plant-world-seeds.com, www.whitehallgardencentre.co.uk, www.images-na.ssl-images-amazon.com, www.selectanorthamerica.com

Table 2: Cultivars of *C. herbeohybrida* Recommended by the University of Vermont, Burlignton, US

Cultivar	*Flowers*	*Other*
Anytime Mix	Bicolor, spotted, solid	Very early, compact 15-20cm tall
Brite 'n Early	Bright, solid	18-22cm tall, large heads and florets
Glorious Mix	Bicolor, spotted, solid	20-25cm tall, large heads
Gold Fever	Golden yellow, solid	15-20cm tall, 20-25cm wide, early

Note: Sometimes 'Gold Fever' is listed under *C. integrifolia; Source:* (http://pss.uvm.edu/pss123/fpcalce.html)

Calceolaria rugosa F_1 Golden Bunch (SAKATA Seeds) - Golden Bunch is bred to have smaller leaves and a compact habit so it is an attractive bedding plant in areas with moderate Summer climates. It is also very suitable for mixed containers. Plants are covered with bright, golden-yellow flowers. Compact habit with small leaves for high-density production (www.sakataornamentals.eu).

Calceolaria Calynopsis Dark Red (Early Blossoms) - Calynopsis Dark Red calceolaria bears intriguing dark red flowers over rich green foliage. It grows 30×30 cm tall and wide (https://www.costafarms.com/plants/calceolaria).

Calceolaria Calynopsis Yellow (Early Blossoms) - Calynopsis Yellow calceolaria shows off dark green foliage and bright yellow flowers. It grows 30cm tall and wide. (https://www.costafarms.com/plants/calceolaria).

7.0 Genetics and Breeding

7.1.0 Chromosome Number

The basic number in sect. Calceolaria is x = 8. The chromosome numbers of *C. crenata* ssp. australis, *C. adenanthera*, *C. gossypina*, *C. semiconnata*, *C. stricta*, *C. sericea*, and *C. comosa* (all 2n = 36) have not previously been reported. The other species *C. mexicana*, *C. tenuis*, *C. tripartita*, and *C. chelidonioides*, have their chromosome numbers (2n) of 32, 60, 64 and 60, respectively (Molau, 1981).

7.2.0 Pollination

Calceolaria, a highly diversified Andean and Patagonian genus, represents one of the most specialized plant-pollinator systems because flowers produce non-volatile oils, a very unusual floral reward that attracts particular solitary oil-collecting bees (Vogel, 1974, 1988). Calceolaria species with an elaiophore are pollinated by specialized oil-gathering bees (Vogel, 1987). While species lacking this structure are either autogamous or pollinated by common bees or bumble bees (Molau, 1988; Sersic, 2004). A unique association with birds has been reported for two *Calceolaria* species of southern Patagonia (Sérsic and Cocucci, 1996; Roitman *et al.*, 2002) where the elaiophore-bearing lobe of the flowers has functionally differentiated trichomes modified into a juicy fruiting body.

Most *Calceolaria* species have nototribic flowers (*i.e.*, pollen is deposited on the bee's dorsum or bird's front), but a few have sternotribic flowers (*i.e.*, pollen is deposited on the bee's sternum). The number of species in Calceolaria as well as the variation in floral reward, shape, and size; in the relative position of the elaiophore bearing lobe; in the filament and style length; and in the branching pattern of flowering shoots (*i.e.*, syn.florescence) make the genus an interesting model of evolution for understanding the role that pollinators have played in diversification processes (Sérsic, 1991; Ehrhart, 2000; Johnson and Steiner, 2000).

7.3.0 Pollen Biology

Calceolaria species, although, are one of the most diverse genus (Fuentes *et al.*, 1995), their biology, systematics and evolution have been recently the subject of detailed research (Molau, 1988; Sérsic, 2004; Cosacov *et al.*, 2009). The common *Calceolaria herbeohybrida* Voss is believed to be derived from *C. arachnoidea*, *C. purpurea* and *C. crenatifora* Cav.

The pollen biology in terms of pollen production, pollen-ovule ratio estimation, pollen viability and pollen fertility was studied by Ashoke (2017) to determine the impact of plant and flower density on pollen viability and fruit set of *Calceolaria tripartita* Ruiz & Pav in Darjeeling Himalaya. He found that higher the density of the plants and flowers, the productivity of pollens and ovules and the viability and fertility of pollen grains become higher showing a positive correlation between plant-flower density *vs.* pollen biology. Fruit set is higher in dense flowering site than low ones which might be due to higher pollen fertility rate, low sterility and greater number of pollen load upon stigma.

7.4.0 Breeding System

Murua *et al.* (2014) conducted a pollination-controlled experiment to determine the breeding system of Calceolaria in two species, *Calceolaria filicaulis* and *Calceolaria arachnoidea* (120 buds per species). Buds were bagged with a mesh and each of the following treatments were randomly assigned: (1) control (C), intact flowers exposed to natural pollination; (2) hand cross-pollination (HC), emasculated flowers pollinated with pollen from another plant situated at least 1 m apart; (3) automatic pollination (AP), bagged buds without manipulation; and (4) hand self-pollination (HS), emasculated flowers pollinated by its own pollen. Pollination success was estimated as seed set per plant. Pollinator assemblages of both *Calceolaria* species were totally composed of hymenopterans. However, it differ in terms of the number of species responsible for pollination (Table 3).

The results of the pollination experiment showed that no seeds were produced in the absence of pollinators. Tukey's post hoc comparison showed that for both species, HS produced less seed compared to HC. In *C. filicaulis*, neither HS nor HC showed significant differences with control conditions, while in *C. arachnoidea*, natural pollination produced more seeds per fruit than HS and HC treatments. The results revealed that *Calceolaria* species are highly dependent on pollinator for reproduction; however, they are able to self-fertilize and showed differences in their reproductive strategies. *C. filicaulis* is allogamous and self-incompatible, whereas *C. arachnoidea* is partially autogamous and present an incomplete self-compatibility. In addition, both Calceolaria species were mainly visited by oil-collecting bees, but they differ in their principal pollinator. This finding suggest that Calceolaria and their oil-collecting bees could represent a highly specialized system, where plant species develop different reproductive strategies possibly in response to their ecological context.

Table 3: Floral Visitors in *Calceolaria* Species

Plant Species	*Pollinators*	*Visitation Rate (Visits/flower/h) of Visits*	*Proportion*
C. filicaulis	*Chalepogenus caeruleus*	0.29 ± 0.07	0.831
	Bombus terrestris	0.06 ± 0.03	0.169
C. arachnoidea	*Centris nigerrima*	0.30 ± 0.04	0.872
	Bombus terrestris	0.04 ± 0.20	0.121
	Chalepogenus sp	0.002 ± 0.002	0.006
	Megachile semirufa	0.001 ± 0.001	0.001

Pollinator identity, mean visitation rate (±S.E.), and proportion of visits per insect species.

The traits related to pollination attraction (*i.e.*, corolla lobes) were found to be very similar among *C. corymbosa* subspecies, however those directly related to plant reproduction (*i.e.*, anthers and stigma length), specially the distance between the reproductive structures (*i.e.*, hercogamy), may be good candidates to explain subspecies differences. The reproductive system of many Andean species is still unknown, especially of those species with specialized pollination system. In particular, it has been observed that *Calceolaria* species show a wide spectrum of reproductive strategies from complete to partial autogamy (Murúa *et al.*, 2014), which seems to be closely related to the availability of their specific pollinator species (*i.e.*, oil-collecting bees) (Cisterna and Murua, 2018).

7.5.0 Interspecific Variation

The variation in flower colours was evaluated in 26 populations of *Calceolaria uniflora* Lam. in Southern Patagonia, Argentina. Computerized analysis of high-resolution photo-images was used to estimate the proportions of red, orange and yellow in surfaces of two corolla parts, 'instep' and 'throat', in field samples of 20–35 flowers per population. The between-populations component accounted for 48 per cent of variance for instep colours and 24 per cent for throat colours. Geographic differentiation was found between populations with a uniform red instep in the Andes in the west, and populations with a maculate yellow-and-red instep in the Magellanic steppe to the east. Mixed populations occurred in a transition zone. Throat colours showed a different, north-south geographic trend (Mascó *et al.*, 2004).

7.6.0 Breeding for Neutrality

Efforts were made to breed day-neutrality into the crop. In other words, flowering would occur under a variety of environmental conditions and is controlled by the level of maturity of the crop rather than solely the environmental conditions the crop is being grown under, *i.e.* temperature and/or photoperiod. These were found successful as there were 'Anytime Series' and others developed which are day-neutral in their flowering response.

7.7.0 Polyploidy

The species *C. clibrani* (2n=18), *C. dentata* (2n=18) and *C. Banksii* (2n=18) were found to be diploids whereas the aneuploid species, *C. mexicana*, has 2n=60 (Srinath, 1940). According to Molau (1988) who has described the Calceolarieae, erstwhile sub-family of Scrophulariaceae there are two centers of diversification in Calceolaria, one in central Chile, and the other one in northern Peru. Cytological studies (Ehrhart, 1997 and 2000) also distinguished two diversity centers; one includes predominantly tetraploid tropical species and the other contains mostly diploid temperate species of Argentina and Chile. There is almost no overlap between both groups of species.

7.8.0 Hybridization

Most of the Calceolaria species are allogamic, pollinated by oil collecting bees (Molau, 1988). About 49 species are autogamous or visited by pollen collecting bees. Only one species, *C. uniflora* Lam is known to be pollinated by birds which obtain a food tissue as reward rather than pollen or oils (Sercic *et al.*, 2001).

Calceolaria herbeohybrida syn., *C. crenatiflora* or *C. multiflora* hybrids are typically crosses (intra-specific and inter-specific) between several species including *C. crenatiflora*, *C. corymbosa*, and *C. cana* (Dole and Wilkins, 2005). It is also believed that most of the commercial cultivars are bred through hybridization. Information on crossing programme/hybridization, however, is either not published or not available in public domain. The feasibility of artificial hybridization between species of *Calceolaria* was already mentioned by Darwin (1859) in "The origin of species". He reported that hybrids obtained from *C. integrifolia* × *C. plantaginea*, two species with very dissimilar habits, produced fertile seeds. Presently, there are many Calceolaria cultivars bred for ornamental purposes, known by names such as C. "anytime", C. Bikini series, C. Monarch Series, C. "Sun- shine", C. "John Innes", C. "Ualter Shrimpton" (Bailey, 1949 and Brickel, 1991). Many of them were apparently obtained by artificial selection and hybridization, such as the commercially important cultivars of C. "herbeohybrida".

Murua *et al.* (2017) in a hand cross-pollination experiment involving *C. filicaulis* and *C. arachnoidea* found that intra-specific crossings produced significantly more seeds than inter-specific ones in an Andean ecosystem of Chile. They also reported that pollinator species are responsible for most of the reproductive isolation in these *Calceolaria* species.

Breeding and development of new varieties including F_1 hybrids is extensively taken up by private companies in western countries including US and UK are John Innes Seeds, Etsy, Rare Exotic Seeds, Exotic Plants, Sakata Seeds, Thompson and Morgan, Chiltern Seeds and Seedsnpots.

8.0 Growth and Flowering

8.1.0 Temperature

Calceolaria prefers cool temperatures (18°C) to produce good vegetative growth prior to flower bud initiation. Early work by Post (1937) determined that calceolaria flowering was hastened by temperatures from 10-17°C and inhibited by temperatures from 17-21°C. Optimal flowering occurred when plants were held at 7.5-14.5°C. In general, flower induction of most calceolaria will not occur unless temperatures are held below 15.5°C for a minimum of 6 weeks.

8.2.0 Photoperiod

The flower induction is also affected by photoperiod and irradiance. Calceolaria is a long-day-plant particularly at high irradiance (Runger, 1975). Runger (1975) found Calceolaria 'Zwerg Meisterstuck' to be a long-day plant with a critical daylength of 14-15 hours when grown at high irradiance. At low irradiance (winter), flowering in long days must be preceded by treatment in cool temperatures of 10°C or short days at 15.5-18°C. Day-length had little effect during the cool temperature treatment. Johansson (1976) observed flowering during long days, but not during short days, at 15.5°C. Temperature and day-length affected plant quality as characterized by number of flowers and plant height.

In other words, flower induction is promoted when the length of the day exceeds the length of the night. The critical day-length is 14-15 hours (night length of 9-10 hours). A minimum of 10 long days are required for flower initiation. Although less 'clear-cut' than photoperiod, higher irradiance appears to hasten flowering of calceolaria. Therefore, calceolaria flower induction is stimulated by both cool temperatures and long-day conditions (Table 4).

Table 4: Flower Induction Requirements for Calceolaria

Temperature	*Long Day (LD)*	*Short Day (SD)*
Cool temperature (10°C)	Flowering	Flowering
Warm temperature (20°C)	Flowering	No flowering

These two flower promoting processes appear to be additive in their effect on calceolaria flower induction. In other words, the shorter the chilling period, the greater the need for continued growth under long day conditions to maximize flower induction. If plants receive a 30-40 day cool temperature period, plants will require at least a 2 week long-day treatment. Delivering long-days by night interruption lighting is less effective in promoting calceolaria flowering than delivering long-days by day extension lighting. In addition, day extension lighting using incandescent lamps is more effective than day extension using fluorescent lamps but fluorescents are most often used to provide long-days since incandescent lamps cause excessive stem elongation. A minimum of 10 footcandles (2 umol/m^2/s) is required for the plant to perceive long days.

The limited inductive photoperiod, photoperiod sensitivity at different developmental stages, and the effect of photoperiod on visual quality of potted plants were studied in the long-day plant, *Calceolaria* × *herbeohybrida* Voss 'Dainty' by transferring plants between short day and long day conditions. It was found that long days enhanced plant height, main stem length, flowering uniformity and visual quality of potted plants. When plants were transferred from long day to short day conditions, the later the plant were transferred, the heavier the upper ground part, the root, and the total plant were. The reverse results were observed when plants were transferred from short day to long day conditions. The juvenile stage ended when the plant had 3 pairs of leaves under experimental conditions. In practice, reasonable management of photoperiod can be employed to control flowering time with acceptable quality. If seeds of 'Dainty' are sown in early September and treated with at least 21 long days at 3-leaf-pair stage, they can flower in early January to meet the market demand of Spring Festival (Sun *et al.*, 2011).

8.2.1 Source of Light

The source of artificial light was also found to be significantly influencing the plant response. To provide long days, artificial light from Mazda lamps was found to be superior to either neon or mercury lamps in producing earliness of bloom, greater flower development, and general growth of plants. Additional light (LDs) with warm temperature (15.5°C) produced slightly earlier blooming. However, plants grown in the 10°C were better. They were more stocky and had more flowers with much stiffer stems. Foregoing the small difference in time of flowering (earliness), for the quality of plants and bloom it is advisable to maintain cooler temperature when going for providing artificial LDs (Poech, 1937).

In calceolaria, the breeding programme resulted in the development of day-neutral cultivars (Anytime Series and others). These cultivars do not appear to require a cold treatment or long days to promote flowering. These plants are grown at constant 15.5°C and typically flower 3-6 weeks earlier than the traditional cultivars. However, it was recently noticed that environmental, water or nutritional stress in initial developmental stages may promote excessively early flower initiation (Erwin, 1994).

8.3.0 Flower Scheduling

Flower scheduling can be done in calceolaria as per the occasion (marketing date). Obviously, the larger plants are allowed to grow prior to flower induction, the larger the plant at flower and the more flowers on the plant. In general, calceolaria production requires approximately 4 months when inducing plants to flower using long days. In contrast, calceolaria induced to flower using temperature only requires 6 months. Newer day-neutral cultivars require less than 4 months to flower. Expected size of the plant also influences the time for flowering. Larger (7.5-15 cm) pinched plants require up to 30-45 days longer to flower than a 10 cm size potted crop. The varieties/cultivars vary in their induction requirements. In other words, schedules are to be adjusted for each cultivar, especially when some of the 'day-neutral' cultivars are used.

Several schedules were proposed for the production of blooming calceolaria plants without the traditional lengthy exposure to cool temperature. However, few points are to be considered while scheduling calceolarias, (1) incandescent lighting (12 W/m^2) should be used to extend the daylength (Runger, 1975); (2) plants must have developed 4-5 pairs of leaves before long-day exposure; and (3) selection of cultivars. Whatever schedule one uses, long days should be provided during flower initiation and early development to increase floriferousness.

Kho and Franken (1980) at the Institute of Horticultural Plant Breeding in Germany could select strains of pot-plant calceolaria [*Calceolaria* × *herbeohybrida*] that require only 12-16 weeks from sowing to flowering, compared with 6 months for the existing *cvs*, and that have no cold requirement for flowering so that year-round production is possible. Supplementary lighting during the darkest months advanced flowering by 2 weeks.

8.4.0 Growth Regulators

Chlormequat has been used as an effective height control agent for Calceolaria. Two foliar sprays of 400 ppm, with the first applied when the flower buds were about 1-1.5 cm diameter and the second 2 weeks later, reduced plant height by 18 per cent. Single sprays at 800 ppm gave the same height control but resulted in some foliar phytotoxicity (Johansson, 1976). White (1976) reported that a chlormequat drench at 3000 ppm at the time of visible flower buds reduced plant height by 50 per cent without injury. Chlormequat did not affect flowering time.

9.0 Propagation

Calceolaria is mainly propagated by seeds. Vegetative propagation (stem tip cuttings) is practiced for the members of those species (*e.g. C. integrifolia*) which are shrubs (Ehrhart, 2005). Micropropagation for rapid multiplication and for the production of true to type plants unlike those from crossed seed (without following adequate isolation distance) which fail to produce same flowers/plants that of the parent (due to crossing/cross pollination), but is seldom practiced on commercial scale.

9.1.0 Seed Propagation

Calceolaria is seed propagated. The seed are small, with 600-1400 seeds per gram. They germinate in 8-10 days at 17-20°C. Seed are sown on the surface of the germination medium and are not covered. These are suitable for raising in winter season and sowing is done during July-August to September-October so that the seedlings would be ready for transplanting during October-November. While sowing early, care should be taken that there is sufficient shade and they are not exposed to direct sun. Damping-off can be a serious problem, so proper sanitation and watering is required.

Seeds can be sown in plug trays filled with fine cocopeat or on raised beds. It takes about 30-35 days for the seedlings to get ready for transplanting. The germinating seeds need warm temperatures (not from direct sun), plenty of light, and they should not be covered (or covered lightly). A light cover of medium with

Propagation and Growing of Calceolaria

Seeds  Germinating Seeds  Cutting

In Plug Tray  In Pot  Under Open Field

In Natural Habitat

Diseases and Pests of Calceolaria

Calceolaria Gray Mold Calceolaria Tomato Spotted Wilt Mealy Bugs

Source: www.gene.affrc.go.jp, www.pnwhandbooks.org, www.researchgate.net, www.ourbreathingplanet.com, www.live.staticflickr.com, www.i2.wp.com, www.budivel.ru, www.rassadacvetov.com, www.gardenchronicle.com, www.joom.com

vermiculite helps to maintain optimum humidity. After germination, the seedlings need consistent, very light moisture; otherwise, they can suffer from damping-off. However, the soil should not be allowed to dry out. Minimum temperature of 18-20°C may be maintained throughout the entire plug stage. This is especially critical for larger plug cells (128) as the plants are receptive to flower bud initiation at 3-4 leaf pairs. After germination, plug trays may be placed in a bright greenhouse with good air movement and a day temperature of 21-22°C and a night temperature of 18-20°C. Supplemental lighting can help hasten development under low light periods. A light feed of 75-100 ppm Nitrogen from a well-balanced calcium nitrate-based fertilizer is recommended. Fertilize as needed to maintain strong growth and allow the media to dry slightly between irrigations. Calceolaria is sensitive to sunburn at all production phases so it is not recommended to expose the plants to light intensities above 5,000-foot candles (54,000 lux). After 4-5 weeks, seedlings become ready for transplanting. Transplant when there are 4-5 true leaves into pots using a rich and well-drained soil/media having a pH of 5.5 to 6.0.

It is a common practice to purchase potted calceolarias rather than raising them at home. This is because they last for one season (flowering in pots may last for 4-6 weeks or sometimes upto 8 weeks). Seeds collected from home garden or from the potted plants by the individuals may not be producing plants of true to type as they are cross pollinated.

9.2.0 Vegetative Propagation

The members of *C. integrifolia* are mostly shrubs and these are propagated by cuttings. The stock plants should be selected in March of the previous year and should not be allowed to flower. Stock plants should be kept in the coolest house (shade-net house) available, as the summer heat is destructive. Calceolarias, in general, are intolerant to high temperatures. Hence, care should be taken to avoid exposure to high temperature either it is maintenance of stock plants, raising of seedlings from seeds or its cultivation. Cuttings should be taken during August. The cuttings root within a month and should be potted. Deep potting is recommended to prevent the toppling over of stems in pots.

9.3.0 Micropropagation

In order to produce true to type off-springs in *Calceolaria crenatiflora* a rapid propagation by tissue culture technology was attempted by Xu (2002) and found that the germination rate of explant being its stem tip is higher than that of the other parts of the plant. The contamination rate was reduced notable by having the explants sterilized directly and it is very effective for reducing the secondary contamination during the subculture period.

Ying (2007) standardized the micropropagation techniques from seeds of *Calceolaria crenatiflora* so as to solve the problem of low sowing propagation rate. Seeds were used as explants and found that sterilizing explants (seeds) with 70 per cent alcohol for 30s resulted in lowest contamination (18.7 per cent) and highest germination (62.5 per cent). Optimum medium for *C. crenatiflora* adventitious buds was found to be MS + BA 1.0 ml/l + IAA 0.1 ml/l.

In order to isolate and multiply the desirable variation in *Calceolaria mexicana* through tissue culture, it was found that MS+ 0.5 mg/l 6-BA + 3.0 mg/l 2,4-D was the optimum medium for inducing the callus from the tender stems; MS+1.0 mg/l $AgNO_3$ +0.8 mg/l 6-BA +0.2 mg/l NAA was the optimum medium for callus differentiation; $^1/_3$ MS+ 15 g/l sucrose + 2.0 mg/l ABT was the optimum medium for adventitious buds, rooting and rooting of the subculture. The survival rate of transplanted tube seedlings was 91.2 per cent and colonization survival rate was as high as 96.8 per cent. The colonization of plantlets maintained not only all biological traits but also the desirable variation (such as orange spots on pale yellow petals) (Song *et al.*, 2012). Earlier to this, Yuping (2007) reported that that MS + 6-BA 2 mg/l+NAA 0.1 mg/l was suited for bud induction; MS+6-BA 1 mg/l+NAA 0.1 mg/l for bud multiplication; and ½ MS+NAA 0.05 mg/l+IBA 0.5 mg/l or MS+IBA 1.0 mg/l for root regeneration.

10.0 Soil and Climate

10.1.0 Soil

Calceolaria grows best in a well-drained and loose potting medium. Adequate drainage is needed when grown in pots. If they are planted in flower beds, enriching the soil with compost or peat moss help to extend the bloom period. Soil (or potting media) with a pH range of 6.0-7.0 (slightly acidic) is desirable for calceolaria. A potting media consisting of 3 parts loam, 2 parts manure (FYM) and 1 part sand is best. The ideal cultivation range for the species with respect to EC is 0.12-0.35 mS/cm (Cavins *et al.*, 2000).

10.2.0 Climate

These are winter season plants and sowing is possible from around June or July to October so that flower production happens in winters to early spring. The pots for indoors should have bright light, cool temperatures and medium humidity. While using in the windowsill, they should be placed in windows receiving intense light/heat. Also, excess shade may result in a shortened bloom period and the petals may wilt.

The pocket-book plants are mostly grown in pots (container/indoor gardening) (Lorenzi, 2008) but to some extent in flower beds also (NAU, 2011). These are cultivated in greenhouses in northern climates and outdoors in its native range. The plant should never be put in too warm environment which may result in incidence of aphids. These are purchased from greenhouses and florists and used as an annual due to its high cultural demands.

10.2.1 Humidity

Calceolaria prefers a medium to high humid atmosphere, but the soil must not be too damp. They are mostly grown for indoors, if they are in outdoors requires partial shade. Pots can be moved to shade as and when temperatures rise because excessive heat in outdoors is harmful for plants. The plants are sensitive to frost.

The plant prefers good ventilation. But misting the plants while indoors may be beneficial in order to increase the humidity, but care should be taken not to spray directly on the plant (Johnstone, 2021). It result in gray rot forming on the foliage.

11.0 Cultivation

11.1.0 Planting

Calceolarias are planted when the seeldings attain 4-5 true leaves. Planting can be taken in pots as well as in beds or windosills. The potting media should be rich with a provision of good drainage.

11.2.0 Irrigation

Regular watering is essential for ensuring sufficient moisture in the media. Drying of media in between two waterings does not harm growth. On the other hand, overwatering can be detrimental to plant's roots as it is sensitive to water-logging. Prolonged dry condition (lack of moisture) may result in wilting and petal drop. Water should be given directly into the medium as wetting of leaves resulted in grey rot of leaves. Yellowing of the foliage may be due to overwatering, too acid or too alkaline soil, or a lack of nitrogen. Due care should be taken for the re-potted plants until new roots are visible (Poesch, 1937).

11.3.0 Inter-culture Operation

Pruning in calceolaria is not necessary but old flower stems can be removed to encourage re-blooming. The plants may be pinched back to maintain shape and encourage blooming. The flower spike sometimes require support of wire hooks. The wire hook method is a much faster means of supporting the plants and are inserted into the soil as close as possible to it (Poech, 1937).

11.4.0 Manuring and Fertilization

Calceolaria does not require any fertilizer. However, any general-purpose garden fertilizer can be used. Winhelmann *et al.* (2017) found that for potted *Calceolaria* × *herbeohybrida*, application of 1.2-1.5g/l nitrogen is good for growth.

Through fertigation, about 75-100 ppm of Nitrogen is recommended for better growth and flowering of calceolaria. To maximize the length of the bloom period, application of weak solution of a balanced fertilizer once a week is beneficial.

White *et al.* (1984) found that with the increase in dose of ammonium sulphate in *Calceolaria hybrida*, there is a decrease in dry weight.

12.0 Diseases and Pests

Calceolaria is severely affected by stem rot (*Sclerotinia sclerotiorum*). The infected plant shows irregular, necrotic areas extending from the petiole into the lamina. Cottony mycelium and black sclerotia of the pathogen are found in pith resulting in dry, crumble stem. Apothecia were formed on sclerotia in the soil and cinder substratum upon which the pots were placed. At least 75 per cent of the grower's stock of several thousand plants suffered from the disease (Whetzel and Dimock, 1940). It can be controlled by application of suitable fungicide.

The other diseases includes botrytis, root and crown rots and Tomato Spotted Wilt Virus (TSWV).

Major insect pests affecting calceolarias are whiteflies, aphids and mites which can be easily controlled by spraying of a suitable insecticide.

13.0 Biotechnology

13.1.0 Molecular Markers

Calceolaria is a diversified American genus whose species are mainly visited by oil-collecting bees, representing a highly specialized plant-pollinator system. However, this is a unique plant system, different aspects of its ecology and evolution remain unexplored. Characterization using fifteen polymorphic microsatellites for four Andean Calceolaria species was done using next generation sequencing. Thirty-three of the 40 primers tested on fifteen samples per species amplified, of which thirteen were monomorphic, five showed unreliable amplification and fifteen were polymorphic. A set of fifteen polymorphic microsatellites able to cross-amplify among the studied Calceolaria species. The number of alleles per locus (Na) for the study species ranged from 2 to 11, and the observed (Ho) and expected (He) heterozygosity ranged from 0 to 0.85 and 0.3 to 0.87, respectively, while FIS ranged from 0.03 to 1.0. Among the four species, *C. cana* showed the greatest genetic diversity followed by *C. filicaulis. C. arachnoidea* and *C. lanigera* showed similar, lower genetic diversity. These markers are a useful tool to conduct future studies on population structure and gene flow. These markers will be helpful to estimate selfing rates as a first approximation to the understanding of the role of specialization in plant mating system evolution (Murúa *et al.*, 2018)

14.0 References

Anderson, S. (2006) *Taxon*, **55**(1): 125–137.

Ashoke, B. (2017) *Trends in Biosciences*, **10**(31): 6572-6577.

Bailey, L.H. (1949) *Manual of cultivated plants*. 2nd edn. The MacMillan Company, New York, pp. 1–1116.

Bailey, L.H. (1976) *Hortus* Third. MacMillan Publishing Company, Inc., New York, pp.1290

Beaujard, F. (1982) *Problems posed by the mycrohizal infection of heathers in the nursery*. Comptes Acad. Agric. France. (Hort. Abst., 53: 4405).

Brickel C. (1991) *Plants and flowers. The Royal Horticultural Society Gardener's Encyclopaedia*. (ed.) Dorling Hindersley, London, pp. 1–608.

Cavins, T.J., Whipker, B.E., Fonteno, W.C., Harden, B., Mccall, I. and Gibson, J.L. (2000) *Monitoring and managing pH and EC using the Pour-Thru Extraction Method. Horticulture Information Leaflet 590*, New 7/2000. North Caroline State University, pp. 17.

Céspedes, C.L., Molina, S.C., Muñoz, E., Lamilla, C., Alarcón, J., Palacios, S.M.,Carpinella, M.C. and Ávila, J.G. (2013a) *Ind. Crop Prod.*, **42**: 78–86.

Céspedes, C.L., Muñoz, E., Salazar, J.R., Yamaguchi, L., Werner, E., Alarcón, J. and Kubo, I. (2013b) *Food Chem. Toxicol.*, **62**: 919–926.

Céspedes, C.L., Salazar, J.R. and Alarcón, J. (2013c) *Phytochemistry Reviews*, **12**(4): 733–749.

Céspedes, C.L., Salazar, J.R., Castolo, A.A., Yamaguchi, L., Avila, J.G., Aqueveque, P., Kobo, I., and Alarcón, J. (2014) *Environmental Research*, **132**: 391–406.

Cisterna, J. and Murua, M. (2018) *Gayana Bot.*, **75**(1): 544-548.

Cosacov, A., Sérsic, A., Aosa, V., De-Nova, A., Nylinder, S. and Coccuci, A.A. (2009) *American Journal of Botany*, **96** (2): 2240-2255.

Darwin C. (1859) *The Origin of species*. Reprint of Oxford University Press, Oxford, pp. 437

Dole, J.M. and Wilkins, H.F. (2005) *Floriculture: principles and species*. Upper Sadle River: Prentice Hall, 1023p.

Ehrhart, C. (1997) *Sendtnera*, **4**: 41-59.

Ehrhart, C. (2000) *Bibliotheca Botanica H.*, **153**: 1–283.

Ehrhart, C. (2005) *Syst Bot*, **30**: 383–411.

Erwin, J. (1994) *Calceolaria production, Minnesota Commercial Flower Growers Association Bulletin*, **43**(5): 1-6.

Falcao, D.Q., Costa, E.R., Alviano, D.S., Alviano, C.S., Kuster, R.M. and Menezes, F.S. (2006) *J. Pharmacog*, **16**(1): 73–76.

Fjeldså, J. (1995) *Geographical patterns of neoendemic and older endemic species of Andean forest birds: The significance of ecologically stable areas. In* S. P. Churchill, H. Balslev, E. Forero, and J. L. Luteyn [eds.], Biodiversity and conservation of neotropical montane forests, 89 – 102. New York Botanical Garden, Bronx, New York, USA.

Fuentes, E.R., Montenegro, G., Rundel, P.W., Arroyo, M.T.K., Ginacchio, R. and Jaksic, F.M. (1995) *Functional approaches to biodiversity in the Mediterranean-type ecosystems of central Chile. In*: Davis GW, Richardson DM (eds) Mediterranean-type ecosystems: the function of biodiversity. Springer-Verlag, New York.

Harborne, J. and Baxter, H. (2001) *Chemical dictionary of economic plants*. Wiley, Chichester, pp. 236.

https: / /www.chilternseeds.co.uk

https: / /www.costafarms.com/plants/calceolaria

https: / /www.etsy.com

https: / /www.exotic-plants.de

http: / /www.flowersofindia.net/catalog/slides/Slipper per cent 20Flower.html

https: / /www.plantsrescue.com/tag/calceolaria-crenatiflora/

https: / /www.plant-world-seeds.com

https: / /www.plant-world-seeds.com/store/view_seed_item/3180

http: //www.pss.uvm.edu/pss123/fpcalce.html

https: //www.rarexoticseeds.com

http: //www.sakataornamentals.com/plantname/dainty

https: //www.sakataornamentals.eu

http: //www.seedsnpots.com/product/calceolaria-seeds

https: //www.thompson-morgan.com

Johansson, J. (1976) *The regulation of growth and flowering in Calceolaria X speciosa Lilja. In* Symposium on Production of Potted Plants and Cut Flowers **64**: 239-244.

Johnson, S.D. and Steiner, K.E. (2000) *Trends in Ecology and Evolution,* **15**: 140 – 143.

Johnstone, G. (2021) *Calceolaria Plant Profile* accessed on May 8, 2021 (https: //www.thespruce.com/calceolaria-plant-profile-5073219).

Judd, W., Campbell, C, Kellog, E. and P.Stevens (1999) *Plant Systematics.* A Phylogenetic approach. – Sinauer Associates, Inc. Sunderland, pp.464.

Khambay, B.P.S., Batty, D., Cahill, M., Denholm, I., Mead-Briggs, M., Vinall, S., Niemeyer, H.M. and Simmonds, M.S.J. (1999) *J. Agric. Food. Chem.,* **47**(2): 770–775.

Kho, Y.O. and Franken, J. (1980) *Vakblad voor de Bloemisterij,* **35**(4): 102-104.

Lorenzi, H. (2008) *Plantas ornamentais no Brasil: arbustivas, herbáceas e trepadeiras.* Nova Odessa: Instituto Plantarum, pp.1088.

Mascó, M., Noy-Meir, I. and Sérsic, A. (2004) *Plant Syst. Evol.,* 244: 77–91. https: //doi.org/10.1007/s00606-003-0083-1

Mayr, E. M. and Weber A. (2006) *Am. J. Bot.,* **93**: 327– 343. (doi: 10.3732/ajb.93.3.327)

Molau, U. (1981) *Nordic J. Botany,* **1**(5): 595-615. (doi.10.1111/j.1756-1051.1981.tb01419.x)

Molau, U. (1988) *Scrophulariaceae–Part I.* Calceolarieae. Flora Neotropica, 47: 1–326

Molau, U. and Sánchez, I. (1986) *Calceolaria (Scrophulariaceae): Species from the Department of Cajamarca,* Peru. Bull. Lima **43**: 837–851.

Muñoz, E., Avila, J.G., Alarcón, J., Kubo, I. and Céspedes, C.L. (2013c) *J. Agric. Food Chem.,* **61**: 4336–4343.

Muñoz, E., Escalona, D., Salazar, J.R., Alarcón, J. and Céspedes, C.L. (2013b) *Ind. Crop Prod.,* **45**: 283–292.

Muñoz, E., Lamilla, C., Marín, J.C., Alarcón, J. and Céspedes, C.L. (2013a) *Ind. Crop Prod.,* **42**: 137–144.

Murúa, M., Cisterna, J. and Rosende, B. (2014) *Revista Chilena de Historia Natural,* **87**: 7.

Murúa, M., Espíndola, A., González, A. and Medel, R. (2017) *Evolutionary Ecology,* **31**(4): 421-434.

Murúa, M., Peralta, G., Carrasco, L. and Pérez, F. 2018. *Genetics and Molecular Research,* **17** (1): gmr16039863

NAU (2011) *J. Ball Redbook: Crop Production*. Chicago: Ball Publishing, pp.785.

Olmstead, R.G., Depamphilis, C.W., Wolfe, A.D., Young, N.D., Elisons, W.J. and Reeves, P.A. (2001) *American Journal of Botany*, **88**: 348–361.

Oxelman, B., Kornhall, P., Olmstead, R.G. and Bremer, B. (2005) *Taxon*, **54**: 411–425.

Pennell, F. (1951) *The genus Calceolaria in Ecuador, Columbia and Venezuela. – Proc.* Acad. Nat. Sci. Philadelphia. **103**: 85-196.

Poesch, G.H. (1937) *Greenhouse Potted Plants*, Ohio Agricultural Experiment Station, Wooster, Ohio, Bulletin No.586: 24-27.

Post, K. (1937) *Proc. Amer. Soc. Hort. Sci.*, **34**: 627-629.

Randhawa, G.S. and Mukhopadhyay, A. (2001) *Floriculture in India*. Allied Publishers.

Rasmussen, C. (1999) *M.Sc. Thesis*, University of Aarhus, Aarhus, pp.IV87.

Roitman, G.G., Sérsic, A.N., Cocucci, A.A. and Montaldo, N.H. (2002) *Bolet í n de la Sociedad Argentina de Bot á nica*, **37**: 71 – 77.

Runger, W. (1975) *Sci. Hort.*, **3**: 45-64.

Sérsic, A.N. and Cocucci, A.A. (1996) *Botanica Acta*, **109**: 172 – 176.

Sérsic, A. N. (2004) *Pollination biology in the genus Calceolaria L.(Calceolariaceae)* (Vol. 82). Biology Centre of the Upper Austrian Museums.

Sérsic, A.N. (1991) *Kurtzian*, **21** : 153 – 164.

Sérsic, A.N., Mascó, M. and Noy-Meir, I. (2001) *Plant Systematics and Evolution*, **230**(1): 111-124.

Simpson, B.B. (1975) *Paleobiology*, **1**: 273-294.

Song, Y., Zhang, Y.F., Li, Q.R., Ning, S.X. and Jiang, C.Y. (2012) *Hunan Agricultural Sciences*, 11.

Srinath, K.V. (1940) *Cytologia*, **10**(4): 467-491.

Sun, Z., Zhang, S., Li, M. and Li, L. (2011) *Acta Horticulturae Sinica*, **38**(6): 1121-1128.

Vogel, S. (1987) History of the Malpighiaceae in the light of pollination ecology. – Abstr. XIV Int. Bot. Congr., Berlin, pp. 308.

Vogel, S. (1988) *Zeitschrifi fir Zoologische Systematik und Evolutionsforschung*, **26**: 341-362.

Vogel, S.P. (1974) *Akademie der Wissenschaften und der Literatur, Tropische und Subtropische Pflanzenwelt*, **7**: 1-267.

Weigend, M. (2002) *Botanical Review*, **68**: 38 – 54.

Whetzel, H.H. and Dimock, A.W. (1940) *Sclerotinia sclerotiorum on Calceolaria*. Plant Disease Reporter, **24**(14).

White, J.W. (1976) *Year-around flower crop in the pocket-book plant*. Science in agriculture.

White, J.W., Bierrnbaum, J.A. and Holcomb, E.J. (1984) *Orn. Hort.*, **11**: 512.

Winhelmann, M.C., Grzeça, G.T., Paris, P., Tedesco, M., Palolazzi, J., Avrella, E.D., Fior, C.S. and Schafer, G. (2017) *Ornamental Horticulture,* **23** (2): 144-151. Available at: <https: //ornamentalhorticulture.emnuvens.com.br/rbho/article/view/1010>. Date accessed: 22 may 2021. doi: https: //doi.org/10.14295/oh.v23i2.1010.

Woldemichael, G.M., Waechter, G., Singh, M.P., Maiese, W.M. and Timmermann, B.N. (2003) *J. Nat. Prod.,* **66**(2): 242–246

Xu, W. (2002) *Journal of Shanxi University* (Natural Science Edition), **25**(3): 254-257.

Ying, X. (2007) *Journal of Anhui Agricultural Sciences,* 32.

Yuping, Z. (2007) *Journal of Anhui Agricultural Sciences,* **35**(15): 4440.

Pansy

K.K. Dhatt

1.0 Introduction

Viola is a large genus of flowering plants comprising around 500 species distributed around the world. Pansy (*Viola wittrockiana*) is an important perennial trailing herb in the temperate region and behaves as an annual in the plains and biennial in hills. The name 'pansy' originated from a French word 'pensee' meaning 'thought', had been derived from the wildflower called the Heartsease or Johnny Jump Up (*Viola tricolour*), and is sometimes given the subspecies name *Viola tricolour hortensis*. However, many garden varieties are hybrids are referred to as *Viola* × *wittrockiana*. The modern pansy, *Viola* × *wittrockiana*, is thought to have derived from hybridization among *Viola tricolour* ($2n = 26$) with *V. lutea* ($2n = 48$), *V. cornuta* ($2n = 22$) and *V. altaica* (Beckett, 1983).

2.0 Importance and Uses

2.1.0 Aesthetic Uses

Viola× *wittrockiana* Gams. (garden pansy) is most common among annual bedding flowering plant worldwide due to unusual flower shape, long flowering duration and adaptability to wide climatic conditions (Fernandes *et al.*, 2017). The flowers have a wide range of colours, including different shades of blue, purple, red, bronze, yellow, cream, orange, pink, salmon, black, white and beautifully marked stripes or penciled blotches. The large flowered types like 'Swiss Giant, 'Masterpiece' or 'butterfly' hybrids with ruffled petals produce less flowers and hence, are suitable for pot culture and exhibition purposes (Pearson *et al.*, 1995). The small flowered types *Viola tricolour* and *Viola cornuta* group flowers profusely and hence, are suitable for hanging baskets, window boxes and rockeries,walkways and bedding purpose (Bailey, 1998).

2.2.0 Therapeutic and Other Uses

Due to the presence of flavonoids and other phenolic compounds, pansies represent a promising source for natural antioxidants (Evans *et al.*, 1997). Pansy is used in folk medicine and herbalism for epilepsy, eczema, skin diseases, urinary and respiratory ailments such as asthma, bronchitis, cold and cough. The properties of pansy are suitable for treating bronchitis as its mucilaginous fiber helps to remove secretions, reduces inflammation and helps to reduce coughing and soothes irritated throat. Infusion of a teaspoonful of flowers in a cup of water three times a day is highly beneficial for cough and associated symptoms of asthma, cold, flu, *etc.* It is diuretic plant having mucilage, potassium, myrecitin and gaultherin that favours elimination of body fluids, very appropriate in case of obesity and heart diseases. It has anti-inflammatory properties which are beneficial for treating gout, pneumonia and rheumatism.

The flower buds or young leaves are consumed raw or cooked or added to soups to thicken it. Herbal tea is made from leaves and small attractive flowers are used as garnishes in fruit salads, green salad, desserts or in soups. Cookies, cakes, tartlets and beverages are decorated with pansy petals.

Uses of Pansy

In Garden Pot | In Cut Flower | In Pot

In Hanging Basket | In Window Box | Designer Garden Beds

In Garden Centre | Along Path | In Border

In Edible Dessert | In Salad Decoration | In Cake Decoration

Source: www.mygardenlife.com, www.hgtv.com, www.in.pinterest.com, www.marthastewart.com, www.theviewfromgreatisland.com, www.shop.floretflowers.com

3.0 Origin and History

Pansy (*Viola tricolour* L.), commonly known as Garden Pansy, Heart's Ease is a member of family Violaceae. It is important bedding annual and is a native of central Europe. Most species are found in the temperate Northern Hemisphere, however violets, pansies or heartsease are also found in widely divergent areas such as Hawaii, Australasia and the Andes in South America (Chattopadhyay, 2007). The genus comprises of annual, biennial and herbaceous evergreen, semi-evergreen perennials and a few deciduous sub-shrubs found in varied habitat in temperate region worldwide. *Viola tricolour* is a native of cooler part of Europe and was first grown in the gardens of England.

The history of pansy is closely linked with its ancestor the violas. Violas are thought to be endemic to the cooler regions of Europe and were cultivated by the Greeks during the fourth century B.C primarily for medicinal purposes. Some while later, a watchful gardener noticed a particular viola that flourished in alpine meadows with abundant sunshine. This strain of viola was given the common name of wild pansy.

European gardeners prized both violas and wild pansies and grew them in abundance. It was in the 1800's that an English nobleman Lord Gambier and his gardener, William Thompson, began crossing various species of the genus *Viola* (*Viola tricolour*, *Viola lutea* and *Viola altacia*). These crosses resulted in the origin of 'face' type pansies with huge block of colours on the lower petal. The flowers are not only unusual in colours, colour combinations but with large size as well. History credits William Thompson with making the cross that resulted in the new (and now familiar) species *Viola × Wittrockiana*. The flowers of many previous crosses possessed only lines of contrasting colour (Beckett, 1987).

By the mid 1800's pansy breeding was being carried out in many European countries including England, Scotland and Switzerland. These breeding efforts resulted in the development of hybrids with greater plant vigor and flowers that had no dark blocks or lines. Thus, 'clear' pansies got their start in the horticultural world.

The first pansy hybrids with dark central blotch appeared in 1814 and by 1835 there were 400 named varieties in the market. The other series namely 'Medora' was discovered in 1839 and by 1850 many new strains of pansies became popular with gardeners and breeders throughout Europe. Hybridization was used to breed more plant vigour and flowers that had no dark blocks or lines in Scotland, England and Switzerland. All British strains raised by 1861 from the imported continental types were classified as 'Fancy pansies' which were perfectly circular in outline (7 cm across), petals flat, smooth and velvety, lower petals covered lower half of flower, lower three petals with large blotches. These clear pansies without a face were bred about the turn of the 20th century by Scottish grower, Dr. Charles Stewart.

In the past 50 years, modern pansy breeding is largely concentrated in Germany, the United States, and Japan with intensive efforts to improve quality such as vigor, heat tolerance and free flowering. These efforts have produced pansies with new colours, including new shades of pink, rose or orange, as well as flowers with unusual bicolor designs.

4.0 Taxonomy

The modern pansy probably represents crosses among the *Viola tricolour* and other members of the *Viola* family. The *Viola tricolour,* so named because it is generally a combination of three colours - white, yellow and either blue or purple. Pansy (*Viola × wittrockiana*) is an important perennial plant in the temperate climate and being popular as an winter annual in the plains and biennial in hills. The leaves of garden pansy are ovate to lanceolate in shape, have crenate leaf margin and pubescent texture. *Viola* species typically have heart-shaped, scalloped or palmate leaves with stipules that are often leaf-like. The flowers of pansy have pentamerous structure and consist of two superior petals, two lateral petals and one inferior petal. The inferior petal is approximately symmetrical with respect to its central axis, but the two lateral petals and the two superior petals are symmetrical with respect to the central axis of the flower. The morphologies of these three petal types are used in varietal registration because they directly affect flower shape and thus flower appearance (Yoshioka *et al.,* 2006). The flowers possess sporophytic self-incompatibility and set seed through hand cross pollination.

The flowers are hermaphrodite, pedicellate, bracteates, zygomorphic, hypogynous and pentamerous structure *i.e.,* five petals, two overlapping, side-by-side upper petals. The lower half of the flower consists of three petals - a large center spurred petal and two side petals. Each petal is covered with tiny hairs that get longer toward the center of the flower and these hairs give a velvety feel to petals. The lower petal is folded to hold the nectar and the internal two lower petals have blotch of contrasting colour. The female parts of the flower are in the pistil in the center of the flower where the petals meet. The pistil contains the stigma, style and ovary. It is shaped somewhat like a vase, with the stigma the top of the vase, the style the neck of the vase and the ovary the wide vase bottom. The male parts of the flower surround the stigma like a cone in the center of the flower. The stamen, consisting of a filament topped by an anther, produces the pollen. There are five anthers in each flower. Each anther is covered with a dusting of yellow pollen, which germinates when transferred to the stigma, fertilizing the ovule to create seeds. The seeds of the pansy develop in an ovule inside the ovary. At the top of the flower stem, or peduncle, are five green and leafy sepals. The flowers possess saprophytic self incompatibility; hence bud pollination is done to get the seed. The fruit is capsule bearing obovate seeds, capsules when ripe splitting into three boat shaped valves with thick rigid keels.

4.1.0 Classification According to Flower Colour and Size

Pansies can be divided into pure-colour flowers and multicoloured flowers. Pure-colour varieties have a single colour on the flower and are called 'clear.' Multicoloured flowers that have a very dark blue/black centers are called 'blotched' or 'faced.' Some blotched pansies may have a different colour blotch than the usual dark blue/black face. Other multicoloured pansies have white or light coloured edges or petals of differing colours. Pansies are classified into three categories as Large: 9 to 11.5 cm, Medium: 6 to 9 cm and Multiflora: 4 to 6 cm according to flower size.

4.1.1 Large

4.1.1.1 Accord Series

Introduced in 1992, these pansies have large blooms on short sturdy stems. There are eight single colours, both faced and clear, and a mix. Abundant blooms are produced on strong plants.

4.1.1.2 Lyric Series

It was bred in 1985 and characterized by weather tolerance and free flowering. It is available in five colours, all faced, or in a mix.

4.1.1.3 Majestic Giant Series

The series has a free-flowering habit and stands up to heat and cold particularly well. There are six colours and a mix, all with faces. This series Majestic Giant Mix and Majestic Giant White Face were All-America Selections Winners during 1966.

4.1.1.4 Medallion Series

The pansies in the Medallion series have extra large flowers in six colours, and the pansy face.

4.1.1.5 Swiss Giants

It is an old-fashioned European pansy bred in Switzerland. The large flowers on long stems are good for cut flower bouquets. Many of today's hybrid varieties were developed from 'Swiss Giants'.

4.1.2 Medium

4.1.2.1 Crown Series

All clear colours, no faces adorn the nine colours and mixture of the Crown series. All are early flowering and bright spring colours.

4.1.2.2 Imperial Series

A series prized for its non-fading colours and vigorous growth. Most of the blooms are faced. A longstanding favorite of gardeners is imperial Blue, an All-American Selections Winner in 1975. There are unique colours in this series such as pink shades.

4.1.2.3 Joker Series

The striking colour combination of Jolly Joker garnered it an A.A.S. Award in 1990. This pansy is a velvety purple with orange face. The plant has a compact habit and may grow 20 cm wide.

4.1.2.4 Roc Series

This unusual name has two meanings - Roc refers to *V.* × *wittrockiana*, the pansy botanical species. Roc also refers to the rock solid garden performance. Introduced in 1985, the Roc series contains nine colours with or without a face.

4.1.2.5 Springtime Series

This series has a wide rainbow of 17 colours. The plants are heat and cold tolerant, resulting in a reliable performance.

4.1.3 Multiflora

4.1.3.1 Crystal Bowl Series

All are clear colours without a face. There are eleven colours in this series. The compact plants do not sprawl in the garden.

4.1.3.2 Maxim Series

The fourteen Maxim colours have faced flowers that bloom prolifically on compact plants. One of the series, Maxim Marina, is an A.A.S. 1991 Bedding Plant Winner. Its unusual flower colour is light blue with a dark blue face that is outlined in white. Padparadja is a 1991 A.A.S. Winner named for the rare brilliant orange sapphire. The blooms do not fade under heat stress.

4.1.3.3 Universal Series

These pansies have masses of early blooming flowers that may be clear colours or faced. There are 13 colours and a mixture. The plants have proven to be cold tolerant during southern winter conditions and heat tolerant in northern summers.

5.0 Species and Cultivars

5.1.0 Species

The pansies are classified under Section Melanium of genus *Viola* and the important species are described here.

Viola tricolour

It is is a native of cooler part of Europe and was first grown in gardens of England. It is hardy, herbaceous perennial with long trailing stems up to 12-15 cm in length, rhizomatous with spreading stems basal leaves cordate or round cordate, glabrous or nearly so, stalked and crenate dentate, stipules large, pinnately parted towards the base stems becoming long and branched. The flowers are large and of three colours, the spur usually twice as long as the appendage of calyx. Many hybrid forms were derived from a long series of crosses among *Viola tricolour*, *Viola lutea* and *V. altacia.*

V. lutea sub sp *V. elegans*

It is native of Britain, wild, perennial plant having yellow flowers, which was the first grown commercially in England, Scotland and Europe. The plants are cylinder, creeping, rhizomatous, evergreen, bearing ovate lower stem leaves and ovate to lance shaped or entire upper leaves, bright yellow, blue violet and red violet flowers of 3 cm across.

Species of Pansy

Viola tricolour — *Viola x wittrockiana* — *Viola pedunculata*

Viola biflora — *Viola pedata* — *Viola cornuta*

Viola altaica — *Viola odorata* — *Viola lutea*

Viola arborescens — *Viola guestphalica* — *Viola orientalis*

Source: www.plants.ces.ncsu.edu, www.en.wikipedia.org, www.alamy.com

V. wittrockiana

It is native to Europe, erect, bushy evergreen perennials, grown as annual or biennial, derived from cross breeding *V. altaica, V. lutea* and *V. tricolour*. The plants are larger than their parents, spreading stems, ovate to heart shaped shallowly lobed, shiny mid to deep green leaves. Flowers are 6-10 cm across, with the lateral petals overlapping the lower and upper petals and with short spurs. Flowers are self coloured, usually in blue, violet, yellow, orange, pink, red or purple, bi-coloured with face like markings.

V. odorata

It is native to Europe and commonly called sweet violet. Rhizomatous, semi evergreen, perennial with slender stolons, short, erect stems that bear tufts of heart shaped or rounded, toothed, bright green leaves. Flowers are sweetly scented, blue or white 2 cm across with spurs to 5 cm long. Self seed freely so excellent for wild or woodland garden. The hybrid types are also cultivated commercially for the extraction of scent, the flower colour includes shades of white, pink, lavender, violet, purple and cream yellow.

V. cornuta

It is native to Pyreness in northern Spain. Foliage small, oval, pointed, dentate forming a compact tuft of vegetation. The plants are spreading, rhizomatous, evergreen perennial with flowers violet to lilac blue peatls, the lower ones with white markings and slender spurs to 1.5 cm long.

V. aetolica (syn. *V. saxatillis aetolica*)

This species is native to East Europe especially Balkan Peninsula. It is tufted species resembling *V. tricolour* but is more prostrate, growing to 10 cm in height. It has 2 cm long, greyish-green, ciliate and ovate to lanceolate leaves with wedge shaped base and bright yellow pansy like flowers about 1.8-2.5 cm across.

V. biflora

This species is native to Northern Hemisphere. The plants grow hardly up to 8 cm height and 30 cm wide. The leaves are bright green, reniform and flowers are vivid yellow shaped and 1.2 – 2 cm across.

V. bosniaca (syn. *V. elegantula*)

This species is native to Southern Western Europe. It is a semi-trailing species growing up to 13 cm in height and bearing pointed oval leaves with undulated margins. Flowers are small and pinkish purple with a yellow mark at the lower petal's base.

V. canina

It is a shade loving hardy species native to Europe. It has all the characteristics of *V. odorata* but without scent. It grows up to 23 cm height and bears dark green, dentate and pointed leaves along with large, violet blue or white flowers.

V. elatior

This species is native to Europe and Western Asia. It is deciduous, tufted to clump forming species that grows up to 50 cm in height with erect stem. It bears hairy, lanceolate to narrowly ovate and 4.0-7.5 cm long leaves. The flowers are pale blue and 2.0-2.5 cm across.

V. gracilis

A mat forming species is native to Balkan Peninsula and Asia Minor. It grows 15 cm tall with decumbent stems and mid green, broadly ovate to oblong with rounded teeth and 2-3 cm long leaves. Flowers are pansy shaped, prolific, 2-3 cm across, yellow or violet blue in colour.

V. hederacea

This species is native to South-Eastern Australia. It is stoloniferous and mat forming tender species that grows 10 cm tall and spreads to 30-60 cm. It bears pale green, kidney shaped, undulated leaves and violet purple with white tipped flowers that grow well above the leaves throughout the year making it an excellent pot plant.

V. riviniana

A tufted and colonizing species from Europe including Iceland, Northern Africa and Madeira grows hardly 15 cm tall and bears ovate –orbicular and cordate leaves. The flowers are blue violet and 1.5-2.0 cm across. The flower resembles *Viola odorata* but is scentless and non-stoloniferous.

V. rupestris

This species is native to Northern Hemisphere and appears as a small version of *Viola riviniana*. It grows up to 4 cm, bears 5-10 mm long leaves and pale purple flowers 1.0-1.5 cm across.

V. pedata Torr & Gray

It is native to eastern and mid-western United States. A very small herbaceous perennial of tufted habit. Leaves basal, divided into segments like a bird's foot. The lower three flowers petals are lilac and upper ones dark violet.

5.2.0 Cultivars

The following four varieties of pansy were developed by Punjab Agricultural University, Ludhiana, Punjab, India.

Punjab Purple Wave: The flowers are purple with dark purple blotch and one plant produces about 71 flowers. The plant height is 30.00 cm, takes 64 days to flowering and flowering duration is of 76 days. It is suitable for pot culture and bedding.

Punjab Choco-Gold: Flowers are yellow with prominent chocolate blotch and one plant produces about 83 flowers. Its leaves are light green, plants are 30 cm tall, takes 66 days to flower and flowering duration is of 80 days. This variety is suitable for pot culture and bedding.

Cultivars of Pansy

Delta Orange Blotch | Matrix Series | Mammoth Series

Halloween II | Sunrise | Yellow Blotch

Amber Kiss | Accord Black Beauty | Jolly Joker

Swiss Giants Rhinegold | Swiss Giants Ullswater | Swiss Giants Silver Bride

***Source*: www.gardenerspath.com, www.in.pinterest.com**

Varieties Developed by Punjab Agricultural University, Ludhiana, Punjab, India (www.pau.edu)

Punjab Purple Wave

Punjab Choco Gold

Punjab Neelma

Punjab Sunaina

Punjab Sunaiana: Its plants are dwarf, leave light green and flowers purple and yellow colour. It is an early variety and bears small sized flowers for four months. One plant produces about 480 flowers and this variety is suitable for pot culture and bedding.

Punjab Neelma: Its plants are dwarf, leave dark green and flowers purple colour. It is an early variety and bears small sized flowers for four and half month. One plant produces about 500 flowers and this variety is suitable for pot culture and bedding.

5.2.1 Other Cultivars/Series of Pansy

Delta Orange Blotch: The plants are 10-15 cm tall and spread upto 15-20 cm. The flowers are orange with black blotched center and has heat tolerance. These are highly suitable for window boxes or planting along a walkway due to plant vigour and thrive well in moist well drained soils.

Matrix series: These are available in a variety of colours from scarlet to light blue to orange. The blooms are solid coloured or bi or tri-colour blooms with 7.5-9 cm diameter. The plant spread ranges between 15-20 cm.

Mammoth series: The plants are 15-20 cm tall with equal wide. The plants produces large, flat, tricoloured blooms growing on stems. The plants are a little less cold hardy. With three- to four-inch flowers, the Mammoth series (*V.* x *wittrockiana*) truly earns its name.

Halloween II: The blooms are, inky-black flowering during October in temperate zones. The size of bloom is large around 7.5 cm, flowers are purple-tinged yellow eyes, and therefore 'Halloween II' might just be the most thrilling pansy of all. This cold-hardy cultivar grows 15-20 cm tall and loves moist, well-drained soil.

Sunrise: It has pink and gold petals and it changes petal colour as they grow older or a slice of sunshine at any time of the day. The plants are 22-30 cm wide.

Yellow Blotch: The blooms are yellow with brown blotch in the center. This annual is especially well-suited to container growing, walkway or window box with its 20-30 cm spread.

Swiss Giants Rhinegold: The plants grow 10-15 cm tall, with a spread of 20-30 cm. The blooms are in bright yellow have pretty dark brown patches on three large petals.

Swiss Giants Ullswater: The plants are biennial grow 10-15 cm tall with spread of 20-30 cm and bloom in a ravishing electric blue with dark patches.

Swiss Giants Silver bride: The plants grow 10-15 cm tall, with a spread of 20-30 cm. This silver-white bloom features deep purple patches that resemble a bride adorned with her flowers and lace.

Amber Kiss : The golden-orange blooms of 'Amber Kiss' (*V. cornuta*) are true beauties, with tidy 2.5 cm blossoms and a delightfully sweet fragrance and bloom for around 10 weeks. They grow 20-25 cm tall with a spread of 15cm. This cold-hardy cultivar grows 15-20 cm tall and loves moist, well-drained soil.

Accord Black Beauty: The petals are deep-black in colour with bright yellow center and give any garden an elegant look.

Jolly Joker: The blooms are in medium size with deep purple upper petals and purple margined lower petals.

6.0 Genetics and Breeding

6.1.0 Cytology and Pollination

The basic chromosome number in *Viola tricolour* L n=13, *Viola lutea* n=24, *V. cornuta* (n = 11) *Viola wittrockiana* octoploid (n = 6, 8x = 48) which is thought to be derived from cross combinations among *V. tricolour* (n = 13), *V. lutea* (2n = 48), and *V. altaica*. Utilization of polyploidy, either autopolyploidy from induced mutagenesis or allo-polyploidy from spontaneous or artificial interspecific hybridization, is a useful breeding method for creating new ornamental genotypes with desirable phenotypic traits such as plant vigour, compactness, larger flowers, novel flower colours, tolerance to environmental stress and diseases. The cultivated pansies have diverse pollination conditions and the flowers may exhibit cleistogamy. *V.* × *wittrockiana* is reported to be highly fertile but self-incompatible hence hand pollination is done at $3/4^{th}$ open stage of flower to get seed. In pansy, sporophytic self incompatibility system is used for F_1 hybrid seed production. The highest pod set (65 per cent) was observed at the 3-4th opened stage followed by fully opened stage (50 per cent) and at half opened stage (12 per cent) (Chandran, 1982; Singh (1987). The maximum capsule set (81.94 per cent) with or without hand pollinations and minimum set (4.16 per cent) with natural selfing in pansy (Jhon and Saini 1987) Hand pollination improved the pod set from 86.66-93.33 per cent, while cross pollination was limited to only 0 to 13 per cent (Singh,1987).

Observations on insects visiting pansy fields grown for seed production indicates that bumble bees (*Bombus* spp.) and moths of *Plusia gamma* were the only efficient pollinators, as pansy pollen could be found on the proboscis immediately after visiting the flowers. Tongues of the pollinating bumble bees are long enough to reach the nectar in the spur. At 40 per cent of the visits self pollination may occur between two flowers of the same plant, if no pollen of former visits is still present on the proboscis. Pollinating insects cover rather short distances when undisturbed and thus makes it improbable that contamination by pollination between different varieties frequently occurs (Veerman, 1965).

6.2.0 Hybridization

The open pollinated varieties of large flowered pansy were uniform for height, flower size, colour and seed set. F_1 hybrid production was practiced commercially in large glass houses by emasculation, followed by pollen application of desired male inbred (Watt, 1980). Flower appearance is an important target for improvement in floriculture because it is directly related to the commercial value of the product. In many floricultural crops, aspects of flower appearance such as shape and colour have long been genetically improved to meet different consumer demands. The breeders and breeding companies have developed many inbred lines as varieties, or as parents, and have chosen the best combinations of parents of new F_1 hybrids from these lines based on outstanding flower appearance. An F_1 hybrid first became commercially available in the 1960s, and breeders have since bred dozens of F_1 hybrids. Although phenological and physiological characteristics such as earliness and plant vigor were the main targets during the early stages of F_1 hybrid breeding, morphological characteristics like round flowers are currently popular. The productivity in terms of number of flowers and fruits in the parents varied in hybrids due to incomplete dominance of the more productive parents (Novotna, 1984). The plant height in 21 F_1 crosses in *Viola tricolour* was highly positive correlations with plant width, fresh weight and peduncle length whereas plant height, days to flowering, number of flowers and number of leaves showed high effects in narrow sense heritability (Song, 2002). In a study on the evaluation of F_1 hybrids of pansy, Accessions A-1 in yellow and A-11 in yellow with chocolate blotch were found suitable for pot culture and window boxes due to more number of flowers and longer flowering duration (Kumar, 2011).

The highest coefficient of variation was observed in pod setting (8.82 per cent) and lowest for flowering duration (2.08 per cent). High PCV and GCV were observed for flower number followed by branch count, days from bud initiation to flowering, days to flowering and final plant height. High estimates of heritability were acquired for flower number, branch count and days from bud initiation to flowering with high genetic advance as per cent mean (>60 per cent) indicating additive gene action in expression of these traits (Kaur and Dhatt, 2020). Genetic analysis for plant growth and floral parmeters in pansy indicated that inbred line Pa-32 was found as the best general combiner for plant height at first flowering, plant height at peak flowering, plant spread, days to first flower opening, duration of flowering and flowers per plant. Higher genotypic coefficient (18.97 per cent) was found with number of flowers per plant and heritability ranged between 49.01 per cent and 98.71 per cent for different parameters. The estimates of genetic advance (per cent) varied from 4.61 to 38.30 (Kaur, 2018). Kroon (2008) described the method of breeding diploid garden pansies and occurrence of fertile aneuploids in the various generations is related to the composite genome.

An artificial crossing was made between four *Viola* × *wittrockiana* (2n = 48) lines and one *Viola cornuta* inbred (2n = 26) line as well as among *V. wittrockiana* lines to investigate the cytologic characteristics, fertility, and heterosis of hybrids. The sizes of flowers, stomata and the leaf areas of the *V. wittrockiana* plants were larger than those of the *V. cornuta* plants. The karyologic stability of *V. cornuta* was greater than that of *V. wittrockiana* and 20 per cent of individuals in the latter deviated from the given 2n = 48. Capsule setting after interspecies hand pollination ranged from 28 per cent to 94 per cent, and F_1 seed germination ranged from 16 per cent to 88 per cent, revealing no obvious pre- or post-zygotic selection mechanisms. The chromosome numbers in *V. wittrockiana, V. cornuta* F_1 seedlings varied (2n = 34–38), with about half the cells exhibiting the intermediate (2n = 36) value. These interspecific hybrids exhibited positive heterosis in flower size (range, 11 per cent to 66 per cent) and negative heterosis in plant height (range, –12 per cent to –57 per cent). Both of these traits are desirable in pansy breeding programs (Du *et al.*, 2018).

6.3.0 Mutation Breeding

Effects of polyploidy on phenotype and four pollination methods on capsule and seed set of *Viola × wittrockiana* were studied using ten octoploid (2n = 8x) and two hexadecaploid (2n = 16x) genotypes, originated from induced mutagenesis. Principal component analysis, using 19 phenotypic, phenological and physiological characteristics, revealed that hexadecaploids showed larger differences to the corresponding standard cultivars than octoploids. Number of seed per capsule was similar among genotypes. Capsule set with open pollination was 32-64 per cent, with self-pollination by hand 18-49 per cent and with cross-pollination by hand 14-43 per cent, while no plant successfully set capsules with seeds under isolators if not pollinated manually. Thus, is self-compatible but requires an agent-mediation for successful pollination. The induced phenotypes were found stable over four generations. Hexadecaploids had more attractive phenotypes but fewer seeds than octoploids. However, variation in seed set enabled selection of plants with high fertility, and average seed set increased over generations. Thus, new varieties, fulfilling aesthetic criteria as well as economic and agronomic traits, can be bred from induced mutagenesis (Dalbato *et al.*, 2013).

7.0 Growth and Flowering

Flowering in Viola is often profuse, and may last for much of the spring and summer. *Viola* is most often spring-blooming with chasogamous flowers that have well developed petals pollinated by insects. Many species also produce self-pollinated cleistogamous flowers in summer and autumn that do not open and lack petals (Walter *et al.*, 1996). In some species the showy chasmogamous flowers are infertile (*e.g.*, *Viola sororia*). *Viola odorata* and *Viola cornata* plants grow vigorously during summer when there are long days with high temperature and produce cleistogamous flowers. They produces chasmogamous flowers throughout autumn, winter and early spring along with few runners when they get short days (Post, 1949; Allard and Garner, 1940; Cooper and Watson, 1952 and Mayers, 1986). The temperature and light environment delivered to a pansy crop has a great impact on crop timing and quality. In general, the highest-quality pansies are grown under high light levels and cool temperatures, while the poorest quality occurs under low light and warm temperatures. Understanding how temperature and light impact pansy growth and development is useful to design growing environments, identify production problems and schedule crops properly (Faust and Lewis, 2001). For summer/fall pansy production, the outdoor environment is often superior to the greenhouse environment. Greenhouse temperatures are frequently hotter than outdoor temperatures, and the light levels are much lower inside the greenhouse. High temperature combined with high shade is the worst condition to produce high-quality pansies.

7.1.0 Temperature

Pansies can flower throughout the year if temperatures remain cool, as high temperature limits the growth and development of plants. Temperature, significantly affected the time to flowering, as reported for many other species. Both low and high temperature affects the growth and flowering.

Pansy and viola plants can be grow best at 4-13°C temperatures and above this temperature, the growth becomes weak with increased lengths of internodes (Holcomb and Mastalerz, 1985). Higher temperatures promote rapid growth and internode elongation in Viola (Pearson *et al.,* 1995). Faster growing rates affect both vegetative and reproductive growth. The high rate of development after bud initiation causes the flower to be smaller. The decline in flower size was directly proportional to increase in temperature and its exposure duration *i.e.* prolonged exposures to higher temperatures led to progressively smaller flowers (Pearson *et al.,* 2015).

7.1.1 Low Temperatures

Cold temperatures can create problems in the landscape or inside unheated green houses or cold frames. Specifically, frozen soil or media, which occurs when soil temperatures drop below -2°C, can prevent water uptake by the roots. If the air temperature is sufficiently high while the soil is frozen, the shoots gradually lose water. Since no water is available from the roots, the shoots and flowers dry out, or desiccate. Desiccation can result in flower death and even plant death, depending on the duration the plants remain in these conditions. Desiccation damage occurs much more frequently than actual freeze damage: the two can be differentiated by the appearance of the damaged tissues. Desiccation damage results in dry, shriveled flowers; freeze damage results in mushy, dark brown or black flowers.

7.1.2 High Temperatures

Pansies experience heat stress when average daily temperatures rise above 26°C. Pansies continue to photosynthesize and grow when day temperature is greater than 32°C indicating that they can tolerate high day temperatures provided that night temperatures are sufficiently cool so that average daily temperature falls below 26°C. The time from transplanting a plug to flower take 2-3 weeks when the average daily temperatures are around 24°C. High temperature leads to water stress too. Watering poses a real challenge during high temperatures. The root growth gets restricted at high temperature and becomes prone to root rot attack. If plants get too little water, drought stress is created, whereas over-watering creates root rot problems. High temperature enhances the rate of flower development but flower size is reduced. Long days accompanied with high light levels and warm temperatures lead to rapid flowering but flowers have thin and stretched appearance. These plants quickly bolt or produce flower on primary shoot before lateral shoots develop.

Temperature prior to visible bud stage had little effect on final flower size, but all subsequent stages were equally sensitive to increased temperatures. The photo-thermal model can be used to develop improved sowing date schedules. However, by using the photo-thermal model, schedules can be tailored for individual growers, since the model considers environmental factors which vary between different locations; *i.e.* photoperiod, irradiance and temperature (Adam *et al.,* 1997a). Furthermore, the flowering model could be used 'on-line' by growers to advise how maturity dates can be manipulated to meet defined targets. Such an approach would be most powerful if combined with models, such as those proposed by Pearson *et al.* (1995) and Adams *et al.* (1997b), for predicting the effects of environment on plant quality.

Fastest rate of flower development from visible bud was found at a temperature of 25°C.When the pansies are grown at cooler temperatures, a slower growth rate is achieved allowing for a longer duration for flower development creating a larger flower. When plants of pansy and viola are grown within temperature range of 4-13°C and high light, no increase in height is recorded whereas temperature of 16°C or above increases intermodal length that makes plants taller. The time from transplant to flower is influenced by both temperature and light quantity. Rapid flowering occurs at 24°C and at high-light levels (greater than 50 percent of full sunlight or 5,000 foot candles) whereas low light levels with shading or natural low-light levels from late fall to early spring delay flowering and number of flowers. Temperatures below 21°C or above 26°C increase the time to flower.

The size of flower increases as temperatures decreases, so flowers developing at 13°C are larger than those at 18°C or 24°C. The flowers that open in early spring are often large as these flowers grow slowly for longer duration but flowers under warmer conditions expand quickly and grow for shorter duration. Thus, pansy flowers have greater opportunity to expand under cool temperatures.

Adjusting fertility programs according to peak demand periods and production temperatures helps to prevent periodic nutrient disorders during the life of pansy. At 22°C, pansies flowered earlier on lateral branches than at 12°C, but these plants also suffered a loss in quality due to unfavorable growth characteristics and the development of nutritional disorders (Hamlin and Mills, 2001). If maximum growth is to be obtained in warm temperatures, the use of NH_4^+ containing fertilizers should be reduced or eliminated. However, the choice of NO_3^- : NH_4^+ ratio for nutrition may be less important under cool growing conditions. (Hamlin *et al.*, 1999).

7.2.0 Photoperiod

Pansies are facultative long-day plants, thus flowering occurs fastest under long days, but most varieties flower under short days. Winter flowering improves, in terms of flower number and flower time, when night-interruption lighting is provided.

They appears to be day neutral with respect to flower initiation and a quantitative LDP with respect to flower development. The promotion of reproductive development was linearly related to the promotion of extension growth. Therefore, it appears that in LDPS such as pansy, light duration and quality concomitantly promote extension growth and flowering, and cannot readily be separated with lighting strategies. Pansy has the ability to remain in flower under moderate climates and both quantity and quality of light is critical for its successful production (Runkle and Heins, 2004). In pansy flower development and stem extension are promoted by quality and duration of night interruption (Runkle and Heins, 2003). Both the amount of light, or light quantity, and the day length impact pansy quality.

7.2.1 Light Quantity and Quality

Pansies are a relatively high-light-requiring crop. Whereas many floriculture crops perform well receiving 2,000-4,000 foot candles or 10 moles per day (20 per cent of outside summer light levels), pansies perform best when receiving greater than 4,000 foot candles or 20 moles of light per day (greater than 40 per cent outside summer light levels). Time to flower is reduced, while flower number and branching increase, under high light quantities.

High light levels can cause leaf curling, leaf cupping or leaf discolouration, the margins of the leaves bend upwards relative to the mid-vein, forming a V-shaped leaf. The outer edge of the leaf curl upwards, creating a spoon-like appearance. These symptoms are visible on small plants but as plant grows, leaf curling disappears. Low light levels result in floppy plants with poor branching and flowering. Leaves are also thinner and more pliable than high-light leaves. Under high-light conditions, the leaves are lighter green, slightly duller and thicker. Petiole length increases and the leaves are held upright under low-light conditions whereas plants display fuller appearance with increased branching, shorter petioles and horizontally arranged leaves under high-light conditions.

Viola× wittrockiana is a quantitative LDP although its response to photoperiod is comparatively less as compared to *Viola tricolour*. Due to its weak response to photoperiod, it is often referred as 'winter flowering pansy' that flower over the winter months. The effect of light integral on flowering could be accounted to increased assimilate availability and growth. The information on regulation of flowering with light integral could enable growers to control flowering, and therefore grow to defined schedules, by manipulating the light environment, either by shading or through the use of supplementary lighting. The flowers of good quality appears 10 weeks after start of the 8h SD though spring propagated stock starts producing flowers in following autumn (Cooper and Watson, 1952).

Most of the species can tolerate bright sun light in garden/field but best growth is under shaded conditions. Even under polyhouses during summer heavy shading is must for optimum growth. Pansies and violas may not require shading for plug production during winter but shading is must to reduce temperature during summer (Post, 1949, Seeley, 1985 and Nau, 1998). The application of LD through High Intensity Discharge (HID) supplementary lighting during 2-6 weeks after germination proves to be effective for flowering and increases flower number and intermodal lengths (Erwin *et al.*, 1997) and under short days short and sturdy stems were recorded by Seeley (1985).

The quality and quantity of light also influence plant weight, height, internode length, branch number and leaf size (Fan *et al.*, 2013). Red-orange (600-700 nm wavelengths) and blue light have been found to stimulate hypocotyl length and stem elongation (Chia and Kubota, 2010; Johkan *et al.*, 2010; Yang *et al.*, 2012). Red orange light also promotes leaf area index and activate phytochrome (Johkan *et al.*, 2012).

Artificial lighting during winter months through LED increase the plant growth when there is insufficient light (Sirtautas *et al.*, 2014). The use of red-orange LED lights creates a significant difference in terms of plant weight, length of branches, number of leaves, number of flowers and length of pedicel of pansy plants (Koksal *et al.*, 2015). The additional far-red or ultra-violet light has potential benefits to the production of edible pansy. The light wavelength partly affected the flower quality and UV-band affected petal colour or anthocyanin contents. The positive effect of FR would be derived from high R: FR ratio (2.3) which was 1.1 in natural sun lighting system (Randall and Lopez, 2013).

Time of supplementing pansy plants with light is crucial factor that need to be focused to enhance growth and flowering. Supplemental lighting during last two-third or entire plug stage increased shoot dry weight, number of leaves and accelerated flowering but supplemental lighting at initial stage or photoperiodic lighting decreased the shoot dry mass and flower bud number. Thus, light supplemented at later in plug stage had greater effect on transplant quality and flowering compared to that provided earlier (Hurt *et al.*, 2019). The light duration and quality concomitantly promote extension growth and flowering in pansy and cannot readily be separated with lighting strategies. Pansy appears to be day-neutral with respect to flower initiation and a quantitative LDP with respect to flower development. Under Far Red deficient environment, 2 or 4 hours of night interruption with light having moderate or low R:FR ratio, respectively, increased the per cent flowering, but 4-hour night interruption with high R:FR ratio failed to promote flowering (Runkle and Heins, 2003).

If pansy plants are provided with lighting for longer durations than recommended could cause undesired stem elongation instead of improving flower timing or uniformity. Pansies flower rapidly and uniformly when photoperiods are at least 14 hours long or when long nights are interrupted with 4 hours of light. Promotory effect of supplemental light is when natural photoperiods are less than 14 hours, only then turning lights on at sunset until a total of at least 14 hours of light has been delivered, or interrupt nights for 4 hours (Erik *et al.*, 1998).

7.3.0 Growth Regulators

The foliar spraying of pansy plants with different growth regulators such as gibberellic acid (GA) and Benzyl adenine (BA) had profound effect on its vegetative growth and flowering (Lateef and Hade, 2018). The foliar application of GA had profound effect on number of branches and flowering diameter whereas paclobutrazol application reduced the plant height (Biswas *et al.*, 2018). The increase in height under high temperature could be controlled by spraying 2-15 ppm of paclobutrazol (bonzi) or 3-6ppm uniconazole (sumagic) or 2500 ppm daminozide (B-Nine) or 3-15 ppm ancymidol (A-rest) or 2.5-7.5 ppm flurprimidol (Topflor). As pansies are highly sensitive to doses, the precise application of retarding chemicals should be done at particular growth stage. In case of pansy plugs, 10 ppm spraying of ancymidol (A-rest) twice is quite effective (Wieland *et al.*, 1997). Further, excessive growth could be checked through lighting and controlled fertilization of nitrogen and phosphorous in the medium.

Flowering in viola and pansy can be delayed with spray of florel. The rate of spray depends upon the ambient light, temperature and magnitude of desired response. Runkle and Heins (2004) found that florel sprayed at the concentration of 500 ppm after 4-5 weeks of seed sowing can delay flowering of viola and pansy. When spray is repeated, delay in flowering can be prolonged. The flowering time in pansy was not affected with the application of daminozide and chlormequat. Although these plant growth regulators retarded vegetative growth but paclobutrazol and uniconazole were found to be more effective in pansy (Lewis *et al.*, 2004). Regular applications of plant growth regulators are required to keep plants compact and maintain plant quality but spring pansies do not require PGRs to produce compact flowers. Bonzi, Sumagic, B-Nine/Cycocel tank mixes and A-Rest are the most effective PGRs on pansies when applied through appropriate method and dose. Florel application consistently delayed flowering, but did not improve branching or reduce stem elongation.

8.0 Propagation

8.1.0 Seed Propagation

The most common species namely *Viola × wittrockiana*, *Viola tricolour* and *Viola cornuta* are grown by seed. The seeds are sown in pots, wooden boxes or nursery beds in mixture comprising of equal quantity of farm yard manure, sand and garden soil. The basic idea behind this is that the substrates should be composed of finer smaller particle size components by repeated diggings to ensure adequate aeration and good moisture retention for pansy seed germination. The seeds are sown in lines and very lightly covered with that mixture and start germinating within 10 days. Seed sowing is done by mid September to end September under north Indian conditions and are transplanted by end October to first week of November. The plug trays are widely used by the nurseries due to sure survival ease in transportation and 100 per cent survival and better establishment.

8.1.1 Seed Priming

Primed seed of pansy (*Viola × wittrockiana*) had faster, higher and more uniform germination in growing medium in greenhouses or plant growth chambers. The optimum conditions were in aerated osmotic solutions of polyethylene glycol 8000 (PEG 8000) at – 1.0 M Pa for 7 days at 15°C. Priming at – 0.8 M Pa and 15°C caused 8 per cent to 26 per cent of germination of seeds during 4 to 13 days of priming (Carpenter and Boucher, 1991). Priming regimes were evaluated on pansy seeds in high temperature germination tests. The seed priming of *Viola × wittrockiana* with GA_3 @ 20 mg/1 was found to be the best for improving the emergence, growth, flowering and seed production when planting was done in November (Rishu *et al.*, 2016 and 2018).

Different priming regimes *i.e.*, $CaCl_2$, K_2HPO_4, NaCl, KCl, KNO_3, Na_2SO_4, and polyethylene glycol (PEG_{15000}) at 1.0 and 2.0 MPa, for 3, 6 or 9 d at 23°C were evaluated for high temperature germination tests. Primed and non-primed seeds were then germinated at 25°C, 30°C and 35°C. Priming with $CaCl_2$ at -1.0MPa for 3 days at 23°C had recorded higher germination (Han *et al.*, 1997). Among all the priming treatments, GA_3 @20 mg/l had recorded the highest germination percentage (93.12 per cent), seedling length (6.55 cm), seedling vigour index (609.6), seedling fresh weight (15.80 mg) and seedling dry weight (1.50 mg) (Rishu *et al.*, 2012).

Propagation and Growing of Pansy

Seed Pods **Seeds Sowing** **Seedlings in Plug Tray**

Seedlings Ready for Transplanting

Open Field Planting

Protected Cultivation

Micropropagation in Pansy

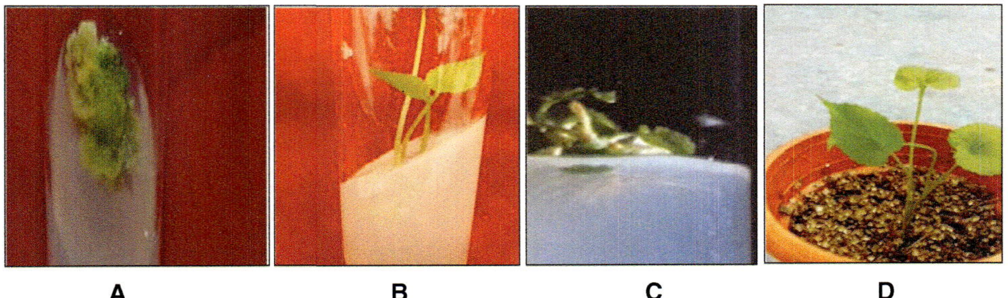

A **B** **C** **D**

(A) Callus induction (Viola serpens) on MS medium supplemented with BAP (2.0 mg/l) + 2, 4-D (0.13 mg/l); (B) induction of the shoot on MS medium augmented with NAA (0.50 mg/l) + KIN (2.25 mg/l); (C) root induction on MS medium with IBA (2.0 mg/l); and (D) hardening of plantlets.

Source: www.sakataornamentals.com, www.in.pinterest.com, www.floraldaily.com, www.newgarden.com, Jha *et al.*, 2020 doi:10.3390/plants9020246.

Three seed priming techniques (hydropriming, halopriming and osmopriming) were compared for their effects on germination and vigour of pansy (*Viola* × *wittrockiana* Gams.) seeds at 20°C, 30°C and 35°C. Seeds were hydro primed in the restricted volumes of water (600 and 700 µl H_2Oâg seed- 1, 2, 3 or 4 days), haloprimed in KNO_3 solution (-1.5 MPa, 5 days), and osmoprimed in polyethylene glycol solutions (-1.0, -1.25 or -1.5 MPa PEG 8000, 7 days) at 15°C or 20°C. It was observed that hydropriming negatively affected the speed of germination, the percentage of germinating seeds and germination capacity. Halopriming accelerated seed germination at 20°C, 30°C and 35°C but did not influence the percentage of germinating seeds and germination capacity. Osmopriming of seeds in PEG solution of osmotic potential –1.0 MPa at 20°C not only improved germination rates at 20°C, 30°C, and 35°C to the highest extent, but also increased percentage of germinating seeds at 30°C and 35°C most effectively and positively affected seed germination capacity at 20°C and 30°C (Dorna *et al.*, 2014).

The effects of a pre-sowing seed treatment with a temperature drop (*i.e.*, seed thermo-priming) on the growth and development of *Viola wittrockiana* Gams.(pansy), were examined. A temperature drop from 22°C to 10°C was applied for 2 h each day during seed germination, over the 6 d before sowing. This seed priming increased plant dry mass, hastened flowering suggesting involvement of certain phytochromes in these temperature-dependent effects on time to flowering (Sysoeva *et al.*, 2010)

8.2.0 Micropropagation

Tissue culture can be used as a tool for the propagation, conservation and commercialization of *Viola odorata* because of its wide popularity and demand. A protocol for *in vitro* plant regeneration can be useful for mass propagation, genetic transformation and long-term storage of horned pansy (*Viola cornuta*) has been developed (Trajkoviæ *et al.*, 2019). Calli formation and shoot induction were obtained in petiole and hypocotyl culture on half-strength MS mineral salts with full concentration of Na–Fe EDTA and vitamins (½MS medium) with 2,4-dichlorophenoxyacetic acid (2,4-D, 0.1 mg/l) and 6-benzylaminopurine (BAP, 2.0 mg/l) and leaf culture on ½MS medium with thidiazuron (TDZ,1.0 mg/l). Subsequent shoot multiplication was achieved on ½MS medium with α-naphthalene acetic acid (NAA, 0.1 or 0.5 mg/l) and BAP (1.0 mg/l). Rooting of shoots was obtained on ½MS medium with low concentration (0.1 mg/l) of auxins. In vitro-derived plantlets were acclimatized under greenhouse conditions. All plants developed normally, bloomed and set seeds. Shoot tips were cryopreserved successfully using modified plant vitrification 3 (PVS3-based vitrification procedure). Cold acclimation for 2 weeks significantly improved shoot regrowth (64 per cent) after thawing in comparison to non-acclimated shoots (39 per cent).

Other reports of successful callus formation are: root hypocotyls and cotyledonary segments of *Viola tricolour* (Babber and Kulbhushan, 1991), petiole callus of wild viola (Sato *et al.* (1995) and from unfertilized ovule in *Viola odorata* (Wijowska *et al.*, 1999).

9.0 Soil and Climate

9.1.0 Soil

The two species namely *Viola tricolour* and *Viola × wittrockiana* are mainly grown for ornamental purposes. Therefore the sandy loam soil textured with good water holding capacity and medium in fertility are suitable for these species. The plants does not thrive well in heavy tenured clayed soils as high moisture causes root rot and plant mortality Pansies grow well in soils with pH range of 6.5-7.5. The ideal temperature range for growing pansies is from about 4.5-15°C day and night.

9.2.0 Climate

Pansy requires mild climate for luxuriant growth and flowering. In high temperature plant growth is ceased and thus affects the flowering adversely. The *Viola tricolour* is transplanted in end October to mid November for seed production under sub tropical climatic conditions of India. The crop is in full bloom from January to April and matures for seed in March. However the *Viola wittrockina* is little temperature sensitive and remains in flower till mid April. The seeds are sown in September-October in the plains and a little earlier in the hills to enable the plants to become established before the temperature cool down. Pansy thrive well under temperate climate of USA or other countries but will require protection in North, where summers are relatively short and cool. The seeds are sown early in July and this is also the time to start seedlings that are to bloom under glass in the winter. In hot weather like Southern parts, seeds are sown in July, or in August it usually takes 6 weeks for seeds to become pansy seedlings of a size suitable for transplanting to their permanent locations. In this way the plants will winter well and will be ready for early spring bloom.

10.0 Cultivation

10.1.0 Transplanting

Under North Indian plains, pansy seedlings like other biennials or perennial plants of temperate region are transplanted in the end of October and plants are in full bloom from February-April and set seed in March-April. This is one of the most promising flowering annual for pot and containers for hanging baskets. They should be filled with equal quantity of fertile garden soil and leaf manure. The seedlings at four leaf stage should be planted in the pots as per the size of pot. These days the plugs raised in propagation trays ensure better success and early flowering for a beginner also. Irrigation should be given after 7-10 days depending on the temperature. Over watering should be avoided especially at the time of flowering. Dhatt (2015) reported 20th October as optimum transplanting time for *V. tricolour* for vigorous plant growth and maximum seed yield as compared to late planting by mid December. The optimum spacing of 30 cm × 30 cm resulted in maximum flower and capsule production in pansy (Kaur and Kumar, 1998).

Viola tricolour is commercially grown for seed production in Punjab, Haryana, Uttar Pradesh on contractual basis for further export to European countries. For production of seed from small flowered pansies, slightly raised beds are prepared with mechanical bed makes (117 cm wide) and the seedlings are transplanted at 40 × 30 cm in the first week of November.

10.2.0 Manuring and Fertilization

10.2.1 Manuring

The beds for display, either for exhibition purpose or mass effect should be thoroughly prepared incorporated with farm yard manure @ 8-10 kg/m^2 or leaf manure @4-5 kg/m^2. The field for seed production is thoroughly prepared by ploughing and farm yard manure is applied a@15-20 tonnes per acre at least two months before transplanting. After the application of well rotten farm yard manure at the rate of 4 kg/m^2, N, P and K were applied in the soil at the rate of 31.97, 4.6 and 67.05 kg per acre, respectively.

10.2.2 Fertilization

The plants require specific fertilizer for good growth and flowering. More of ammoniacal nitrogen can cause stems to stretch and become succulent during warm weather. It weakens the plant and makes them more susceptible to winter injury. The dose @ 30 g N and 20 g P_2O_s/m^2 was found to be adequate for the pansy (Swiss Giant) for better plant growth and flower production, when grown on a soil of average fertility (Kaur and Kumar, 1998) gave the highest seed yield at 20x20 cm (Singh,1979).

10.2.2.1 Fertilization vs. Spacing

The plant spacing of 20x30 cm with 20g nitrogen and 20 g phosphorus resulted in better plant growth and flower production (Jhon and Saini,1984). Similarly another experiment on seed production of *Viola tricolour* to examine the effect of different levels of nitrogen and plant density was conducted under sub tropical climatic conditions. The optimum plant spacing was 15x25 cm and application of 20g N resulted in maximum plant spread, number of leaves, number of flowers per plant, flower duration, flower diameter, number of capsules and seed yield (Dogra and Sirohi,2020).

10.2.2.2 Fertilization vs. Temperature

Pansy (*Viola* x *wittrockiana* Gams.) producers often observed nutrient disorders among plants grown during warm periods (>18°C) of the growing season. These disorders typically are not seen when production temperatures are optimal (≥18°C) even though fertility regimes may remain the same. Hamlin *et al.* (1999) assessed the effects of temperature and nitrogen (N) fertility on growth and nutrition of pansy *cv.* 'Crown White'. Treatments consisted of two temperatures (12 and 22°C) and three NO_3^-:NH_4^+ molar per cent ratios (100:0, 62:38, and 25:75) with a total concentration of 100 mg N/l. A modified Hoagland's solution was used with NO_3^- N supplied as $Ca(NO_3)_2$ and KNO_3 and with NH_4^+ N as $(NH_4)_2SO_4$. The plants were grown until lateral branches had open flowers. Cumulative nutrient absorption and foliar nutrient content were determined when plant lateral branches flowered. Root and shoot growth were limited when NH_4^+ was present in solutions at high ambient air temperature (22°C), but not at low temperature (12°C). Individual absorption and accumulation of plant nutrients varied with N regimes and temperatures. Overall, pansies absorbed more total N, NH_4^+, NO_3^-, calcium (Ca), potassium (K), magnesium (Mg), phosphorus (P), zinc (Zn), and less iron (Fe) and manganese (Mn) at 12°C than at 22°C. In addition, absorption of NO_3^- by pansy was negligible if any NH_4^+ was present in solutions at 22°C. Results suggest that pansy growers should adjust fertility programs according to production temperatures to avoid possible nutritional disorders and maximize plant growth. If maximum growth is to be obtained in warm temperatures, the use of NH_4^+ containing fertilizers should be reduced or eliminated. However, the choice of $NO_3^-:NH_4^+$ ratio for nutrition may be less important under cool growing conditions.

10.2.2.3 Fertilization *vs.* Electrical Conductivity

The effects of fertilizer concentration on growth of subirrigated pansy (*Viola* × *wittrockiana*) and the leachate electrical conductivity (EC) and pH were quantified, using two growing media. Leachate EC increased as the EC of the fertilizer solution increased from 0.6 to 3.6 dS/m (70 to 530 mg/l N). The leachate EC was fairly constant over time when the EC of the fertilizer solution was 0.6 dS/m, while it increased throughout the experiment at higher fertilizer concentrations. MetroMix 300 leachate consistently had a higher EC than did MetroMix 500. Leachate pH of both growing media was similar throughout the growing season. The pH decreased over time and was lower with higher fertilizer concentrations. Optimal plant growth occurred with a fertilizer EC of 1.2 or 1.8 dS/m, and a leachate EC between 1.5 and 4 dS/m. Increasing the concentration of the fertilizer solution resulted in increased shoot tissue levels of P and Mn and decreased tissue levels of K, Mg, and Na. The results of this study indicate that pansy is not very sensitive to the EC of the growing medium and can be grown successfully in a closed sub irrigation system (van Iersel, 1999).

10.3.0 Phytoremediation

Sychta *et al.* (2020) conducted a study to determine heavy metal tolerance level, accumulation efficiency and sexual reproduction in *Viola* × *wittrockiana*, a non-metallophytic ornamental cultivar in comparison to *V. tricolour*, a metallophyte, after zinc (Zn) or lead (Pb) treatment (0, 10, 100 and 1 000 ppm) in pot experiments. The seed germination frequency that was not reduced in comparison to the control, the effective Zn absorption from the soil and exclusion strategy for Pb, as well as the regular sexual reproduction of *V.* × *wittrockiana* treated with heavy metals all indicate the tolerance of this plant to heavy metals. The lack of a seed set under experimental conditions of *V.* × *wittrockiana* was due to the absence of pollinators, rather than the negative impact of heavy metals, as pollen viability and ovule development were normal under the treatments. The results indicate that *V.* × *wittrockiana* represents similar tolerance to *Viola metallophytes* and could be considered as a good material for the reclamation of polluted areas. The exceptional tolerance to heavy metals, the ability to initiate new generations in heavy-metal-burdened soil, which are additionally coupled with the unique beauty, make the garden pansy a good candidate to be potentially used in the future for phytoremediation purposes.

11.0 Diseases and Pests

11.1.0 Diseases

11.1.1 Cercospora Leaf Spot (*Cercospora violae*)

It is the most common leaf spot disease in pansy. The symptoms include large, spreading purple spots with a feathered margin start on the lower leaves and tiny purple spots and flecks appear on the upper leaves. Over time, irregular leaf spots develop tan centers with purple borders. Lower leaves turn yellow and fall off, reducing the plant vigor and appearance. All plant debris should be discarded or destroyed at the end of the production cycle to reduce pathogen survival and remove the severely infected leaves. Application of suitable fungicide such as chlorothalonil copper oxichloride controls the incidence.

11.1.2 Anthracnose (*Colletotrichum violae-tricolouris*)

The leaf spots are typically pale or grey with thick, black margins that sometimes appear to have a concentric ring pattern. Sunken, elongated brown to tan lesions often develop on the petioles, peduncles, or stems of infected plants. The lesions eventually girdle the stem and kill the plant. Anthracnose also may cause brown spotting on infected flowers. The disease is most severe during periods of warm, wet conditions. The fungus survives from year-to year in infected plant residue. Removing and destroying infected plants, as well as plant residue, at the end of the production season can reduce disease survival and spread.

11.1.3 Black Root Rot (*Thielaviopsis basicola*)

It is caused by *Pythium*, *Rhizoctonia*, and *Fusarium*. Infected plants show stunted growth, wilting and leaf yellowing with dark spots or bands on the roots due to fungal spores. Root rot typically occur in wet substrates due to over-watering, poor drainage, or containers standing in water. As it is soil borne, they usually develop in the production area due to poor sanitation practices such as re-using potting substrate and unclean containers, as well as from plant and soil debris. Good sanitation, proper plant care, and use of suitable fungicides can reduce the incidence.

Viruses (Horst, 1990) have also been recorded infecting pansy and violas by causing streaking, distortion, mosaic and certain other symptoms.

11.2.0 Pests

11.2.1 Aphids

Aphids appear in winter months on leaves and flower buds. The aphids suck cell sap and discolour the leaves and buds fail to open. These can be controlled by spraying suitable insecticides Under North Indian condition infestation generally in February and March.

12.2.2 White Ants

They attack the plants and damage the root system and slowly plants die completely. White ants can be controlled by applying 5 per cent BHC.

Diseases and Pest of Pansy

Cercospora Leaf Spot | Anthracnose | Black Root Rot

Root and Crown Rot | Powdery Mildew | Gray Mould/Botrytis Blight

Pansy Worms | Aphids | White Bugs

Spodoptera litura | Tomato Spotted Wilt Virus | Cucumber Mosaic Virus

Source: www.ag.umass.edu, www.pnwhandbooks.org, www.ugaurbanag.com, www.content.ces.ncsu.edu, www.walterreeves.com, www.houzz.com, www.sakataornamentals.com

12.0 Harvesting and Postharvest Management

1.1.0 Harvesting

Pansy is usually grown for beds and pots. The pots are displayed when the flowering buds just opened. Violas and pansies are easy to harvest as there stems are quite fragile but at the same time great care should be done. The stem is cut down at the base of the plant (along with stalk) so that average stem length is 20-30.5 cm long. For cut flower, the optimum stage of harvesting *Viola odorata* and *Viola* × *wittrockiana* is when these are almost fully open and these can be stored at low temperature (1-5°C) for a week (Nowak and Rudnicki, 1990). They remain fresh up to 4-5 days in ordinary tap water and addition of floral preservative add 2-3 days more to its vase life. Another quality of pansy that adds to their longevity is that they do not shed their petals and slowly start curling inwards.

12.2.0 Postharvest Management

There are different reports pertaining to sensitivity of flowers to ethylene as Rogers (1985) reported pansies are sensitive to ethylene gas whereas, Silva *et al.* (2018) reported that pansy flowers are not sensitive to ethylene as longevity of flowers remain unaffected with exogenous ethylene or abscisic acid. The spray of 1-methyl cyclopropene was found to be effective in maintaining cut flower quality. Heins *et al.* (1994) reported that pansy plugs can be stored up to 16 weeks at 5°C with 5fc lighting or at 0-2°C under darkness.

Brightly coloured flowers can be dried by different methods for different purposes. The flowers can be press dried or embedded for three weeks in sand with face up position for short stemmed and horizontal for long stem (Sharavani and Sree, 2018).

12.3.0 Seed Yield

Hand pollination and use of growth regulator (NAA) may be necessary to obtain good seed yields. Seeds should be harvested before shattering begins when the pods show browning but before splitting. Since plants flower continuously, handpicking of seedpods may be necessary. The capsules are collected in boxes and dried in a air tight containers under well-ventilated place. Seeds are usually extracted manually by rubbing or crushing. The clean seeds should be stored in a dry, cool place till next planting season (Dhatt, *et al.*, 2017).

The maximum flower number, seed pods and seed yield was recorded under October planting of pansy at 20x 20 cm (Kumar, 1996) The maximum germination and seed viability of *Viola* × *wittrockiana* was recorded in cold store (0-4°C and 90+2.7 per cent RH) after 18 months of storage and desiccator or poly set bin was the best packaging irrespective of storage temperature (Dhatt, 2016).

13.0 Biotechnology

13.1.0 Molecular Markers

Genetic relationships between eighteen inbred lines of pansy (*Viola wittrockiana*) were estimated on the basis of morphological characters and RAPD data. Thirty-one morphological characters were analysed with an unweighted pair-group method using an arithmetic average (UPGMA). For RAPD analysis, twenty-one 10-mer random primers were selected and 127 polymorphic bands, ranging from 0.2 kbp to 2.0 kbp, were obtained. UPGMA cluster analysis using Nei and Li's distance coefficient, and parsimony analysis using Wagner method, were then performed. In addition, principal component analysis (PCA) was carried out based on the RAPD data. Inbred lines 1 and 14 were isolated in all analyses, and inbred lines 15, 16 and 17 formed a monophyletic group in the RAPD cluster analyses. Inbred lines 6 and 7, and inbred lines 10 and 11 also showed close relationships. The results of cluster analyses indicated that flower colour patterns were not suitable for the classification of inbred lines of pansy (Vemmos, 2005).

13.2.0 Genetic Transformation

The petals of pansy (*Viola* × *wittrockiana* Gams.) 'Mengdie' exhibit a cyanic blotched pigmentation pattern. The accumulation of anthocyanins, cyanidin and delphinidin, was detected in the upper epidermal cells of the cyanic blotches. In order to elucidate the mechanism by which cyanic blotches are formed in pansy petal, the expression level of genes involved in anthocyanin synthesis was measured and compared between cyanic blotches and acyanic areas of the flower. The use of primers in conserved regions allowed the successful isolation of six cDNA clones encoding putative anthocyanin enzymes from pansy petals. The clones isolated encoded chalcone synthase (CHS), chalcone isomerase (CHI), flavanone 3-hydroxylase (F3H), flavonoid 3′-hydroxylase (F3′H), dihydroflavonol 4-reductase (DFR) and anthocyanidin synthase (ANS). The transcription patterns of seven genes (VwCHS, VwCHI, VwF3H, VwF3′H, VwDFR, VwF3′5′H, and VwANS) in cyanic blotches and acyanic areas of the petals at seven stages of flower (Qin *et al.*, 2014).

13.3.0 *In vitro* Callus Culture

Efficient callus induction was observed in MS media with 15 g/l sucrose with BA (2.5mg/l) and 2, 4-D (0.15 mg/l) hormonal concentration in *Viola odorata* after 40 days of incubation. Best shooting occurred on media having TDZ and $AgNO_3$ along with plant hormones GA_3 and NAA. The best shoot regeneration (having 4-5 cm length and 2-3 branches) was recorded on medium supplemented with 1-naphthaleneacetic acid (NAA) 0.5 mg/l, gibberellic acid (GA_3) 1.5 mg/l, $AgNO_3$ 0.42 mg/l and thidiazuron (TDZ) 2.5 mg/l. An assay of the antioxidant potential of the in vitro grown callus and the wild plant extract was determined by DPPH (alpha, alpha-diphenyl-beta-picrylhydrazyl) method shown that the antioxidant activity of in vitro formed callus is higher than that of wild plant (Naeem *et al.*, 2013).

14.0 References

Adam, S.R., Pearson, S. and Hadley, P. (1997a) *Ann Bot.*, **79**: 107-112.

Adam, S.R., Pearson, S. and Hadley, P. (1997b) *Ann Bot.*, **80**: 219-225.

Allard, H.A. and Garner, W.W. (1940) *USDA Tech. Bull.* 727, pp. 1-64.

Ashok, A.D., and Velmurugan, S. (2020) *J. Pharmacogn. Phytochem.*, **9**(5): 190-198.

Babber, S. and Kulbhushan, S. (1991) *Ann. Biol.*, **7**(1): 93-95.

Bailey, D.A. (1998) *Hort. Inf. Leaflet.*, **521**: 1-8.

Beckett, K.A. (1985) *The Concise Encyclopedia of Garden Plants,* Orbis Publishing Ltd., London.

Beckett, K.A. (1987) *The RHS Encyclopaedia of House Plants including Greenhouse Plants,* Salem House Publishers, Massachusetts, USA, pp. 130-131.

Biswas, A., Mandal, T.,Das, S. and Thakur, B. (2018) *Int. J. Curr. Microbiol. App. Sci.* **7**(1): 2125-2130.

Carpenter, W.J. and Boucher, J.F. (1991) *Hort Sci.,* **26**(5): 541-544.

Chandran, K.V. (1982) *M. Sc. thesis,* Punjab Agricultural University, Ludhiana.

Chattopadhyay, P.K. (2007) *Commercial Floriculture.* Gene-Tech Books, New Delhi, India, pp. 236-257.

Chia, P.L. and Kubota, C. (2010) *Hort Science,* **45**: 1501-1506.

Cooper, C.C. and Watson, D.P. (1952) *Proc. Amer. Soc. Hort. Sci.,* **59**: 549-553.

da Silva, T.P., de Araujo, F.F. and Finger, F.L. (2018) *Ornam. Hortic.,* **24**(2): 103-108.

Dalbato, A.L., Kobza, F. and Karlsson, L.M. (2013) *Hort Science,* **40**(1): 22-30.

Dhatt K.K. (2016) *J. Orn. Hort.,* **19** (3-4): 97-105.

Dhatt, K. K., Misra, S., Misra, R.L., Raju, D.V.S. and Singh, M.K. (2017) *Commonly used ornamental plants.* Kurger Bentt Publishers UK Ltd. pp.755-763.

Dhatt, K.K. (2015) *Agric Res, J.,* **52**(4): 52-55.

Dogra, S. and Sirohi, H.S. (2020) *J. Pharm. Innov.,* **9** (7): 596-599.

Dorna, H., Li, W. and Szopiñska, D. (2014) *Acta. Sci. Pol. Hortoru.* **13**(6): 15-29.

Du, X., Wang, M., Slomka, A. and Liu, H. (2018) *Hort Science,* **53**(9): 1300–1305.

Erwin, J.E., Warner, R., Smith, T. and Wagner, R. (1997) *Hort Science,* **32**: 466.

Fan, X.X., Xu, Z.G., Liu, X.Y., Tang, C.M., Wang L.W. and Han, X.L. (2013) *Sci. Hortic.,* **153**: 50-55.

Faust, J.E. and Lewis, K.P. (2001) *Big Grower.* December issue

Fernandes, L., Casal, S., Pereira, J.A., Saraiva, J.A. and Ramalhosa, E. (2017) *J Food Compos Anal.,* **60**: 38-5.

Hamlin, R.L. and Mills, H.A. (2001) *J. Plant Nutr.,* **24** (12): 1975-1985.

Hamlin, R.L., Mills, H.A. and Randle, W.M. (1999) *J. Plant Nutr.,* **22** (10): 1637-1650.

Han, Y., Harvery, L. J. and Greg, C. B. (1997) *Hort Science,* **32(2)**: 248-250.

Heins, R., Lange, N., Wallace T.F. and Carlson, W. (1994) *Plug Storage, Cold Storage of Plug Seedlings.* Meister Publishing Willoughby, Ohio, USA.

Holcomb, E. J., and Mastalerz, J. W. (1985) *Bedding Plants III* (3rd edn.,eds Matalerz, J. W. and Holcomb, E. J.) pp. 87-125.

Horst, R. K. (1990) *Westcott's Plant Disease Handbook* (5th edn) p. 753.

https: //www.gardenerspath.com/plants/flowers/best-violet-varieties

https: //www.gardenloversclub.com

https: //www.healthbenefitstimes.com/wildpansy

https: //www.marthastewart.com/1525112/edible-pansies desserts

https: //www.pau.edu

https: //www.whatscookingamerica.net/EdibleFlowers/EdibleFlowers

Hurt, A., Lopez, R.G. and Craver, J.K. (2019) *Hort Science,* **54**(2): 289-296.

Jain, R., Hiremath, V.M. and Misra, S. (2017) *In*: Misra, R.L. and Misra, S. (eds.), *Commercial Ornamental Crops Cut Flowers,* Kruger Brentt Publisher, U.K., pp. 85-92

Jha, S.R., Naz, R., Asif, A., Okla, M.K., Soufan, W., Al-Ghamdi, A.A. and Ahmed, A. (2020) *Plants,* **9**(2): 246. Doi:10.3390/plants9020246.

Johkan, M., Shoji, K., Goto, F., Hahida, S. and Yoshihara, T. (2010) *Hort Science,* **45**: 1809-1814.

Johkan, M., Shoji, K., Goto, F., Hahida, S. and Yoshihara, T. (2012) *Environ. Exp. Bot.,* **75**: 128-133.

Kaur, R. and Dhatt, K.K.(2020) *Indian J. Agric. Sci.,* **90**(4): 835-837.

Kaur, R. and Kumar, R. (1998) *Indian J. Hortic.,* **55**(4): 337-339.

Koksal, N., Incesu, M. and Teke, A. (2015) *Hortic. Bras.,* **33**: 428-433.

Kroon, G.H. (2008) *Pl. Biotech.,* **21**: 165-70.

Krug, B.A., Whipker, B.E., Frantz, J. and McCall, I. (2009) *Hort Science,* **44(6)** : 1566-1572.

Kumar, A. (2010) *M.Sc. Thesis,* Punjab Agricultural University, Ludhiana.

Lateef, S. A. A. and Hade, S. (2018) *IOSR-JAVS.* **11(2)**: 12-17.

Lewis, K. P., Faust, J. E., Sparkman, J. D. and Grimes, L. W. (2004) *Hort Science,* **39 (6)**: 1315-1318.

Mayers, A.M. (1986) *Handbook of Flowering.* (vol. V, ed. Havley, A. H.) pp. 372-384.

Naeem, M., Naveed, I., Naqvi, S.M.S. and Mahmood, T. (2013) *Pak. J. Bot.,* **45**(1): 197-202.

Nau, J. (1998) *Ball Red Book* (16th edn, ed. Ball, V.) pp. 775-776.

Novotna, I. (1984) *Sbornik UVTIZ, Zaharadnictvi,* **11**: 315-321.

Nowak, J. and Rudnicki, R.M. (1990) *Postharvest Handling and Storage of Cut Flowers, Florist Greens and Potted Plants.* Timber Press, Portland, Oregon, USA.

Pearson, S., Parker, A., Adams, S.R., Hadley, P., and May, D.R. (1995) *J. Hort Science,* **70**(2): 183-190.

Post, K. (1949) *Florist Crop Production and Marketing,* pp. 839-845.

Qin, L., Wang, J., Sun,H.Y. and Shang, X. (2014) *Plant Physiol Biochem,* **84**C: 134-141.

Rogers, M. (1985) *Bedding Plants III* (3rd edn., eds Matalerz, J. W. and Holcomb, E. J.) pp. 274-314.

Runkle, E. S. and Heins, R. D. (2003) *J. Amer. Soc. Hort. Sci.* **128**(4): 479–485.

Runkle, E. S. and Heins, R. D. (2004) *Crop Cultivation*, p. 36-38.

Runkle, E.S. and Heins, R.D. (1998) *Greenhouse,* **16** (10) : 60-66

Sato, T., Kwon, O., Miyake, H., Tangiguchi,T. and Maeda, E. (1995) *Plant Cell Rep.,* **14**: 768-772.

Seeley, J. G. (1985) *Bedding Plants III* (3rd edn., eds Matalerz, J. W. and Holcomb, E. J.)

Setia, M., Bala, M. and Singh, S. (2020) *Int J Radiat Biol.,* **96**(10): 1309-1316.

Sharavani, S.R. and Sree, G.D. (2018) *J. Postharvest Technol.,* **6**(3): 97-108

Sharma, R., Dilta, B.S., Baweja, H.S. and Sharma, B.P. (2012) *Seed Res.,* **40**(1): 67-72.

Sharma, R., Dilta, B.S., Messar, Y., Pal, S. and Kannan D. (2016) *Eco. Env. and Cons.* **22**(2) : 213-217.

Sharma, R., Dilta, B.S., Baweja, H.S., Poonam, K. and Sahare, H.A. (2018) *Ann. Biol.,* **34**(2): 176-178.

Singh, M. (1987) *M.Sc. thesis,* Punjab Agricultural University, Ludhiana.

Sirtautas, S., Virsile, A., Samuoline, G., Brazaityte, A., Miliauskiene, J., Sakalauskiene, S. and Duchovskis, P. (2014) *Zemdirbyste,* **101**: 75-78.

Song, C. Y. (2002) *J. Kor. Soc. Hort. Sci.,* **43**: 239-243.

Sychta, K., S³omka, A. and Kuta, E. (2020) *Plant Soil Environ.,* **66**(6): 272-280.

Sysoeva, M.I., Markovskaya, E.F., Sherudilo, *e.g.* and Shibaeva, T.G. (2010) *J. Hortic. Sci. Biotechnol.,* **85**(3): 238-240.

Trajkoviæ, M., Antoniæ, D., Cingel, A., Ghalawenji, N., Subotic, A. and Jevremovic, S. (2019) *Biotech.,* **9**: 17.

van Iersel, M. (1999) *Hort Science,* **34**(4) : 660-663.

Veerman, A. and Van Zon, J.C.J. (1965) *Entomol. Exp. Appl.,* **8**(2): 123-134.

Vemmos, S. N. (2005) *J. Hortic. Sci. Biotech.,* **80**: 529-536.

Walter, D. R. and Keil, D. J. (1996) *Vascular Plant Taxonomy* (4th ed.). Kendall/Hunt Publishing Company. ISBN 978-0-7872-2108-9.)

Wang, J. and Bao, M. (2005) *J. Horti. Sci. Biotech.,* **80**(5): 537-542.

Warner, R.M. and Erwin, J.E. (2006) *Scientia Hort.* **108**(3): 295-302.

Watt, L. (1980) *Flower and vegetable plant breeding,* Grower books, London, pp. 60-83.

Wieland, C.E., Barrett, J.E., Bartucka, D.G., Clark, D.G. and Nell, T.A. (1997) *Hort Science,* **32**: 508-509.

Wijowska, M., Kuta, E. and Przywara, L. (1999) *Acta Biol Cracov Ser Bot.,***17**: 1-11.

Yang, Z.C., Kubota, C., Chia, P.L. and Kacira, M. (2012) *Scientia Hort.,* **136**: 81-86.

Yoshioka, Y., Iwata, H., Hase, N., Matsuura, S., Ohsawa, R. and Ninomiya, S. (2006) *Euphytica,* **151**: 311-319.

Statice

Anamika Gurung, V.A. Parthasarathy,
Rajiv Kumar, Kalyan Barman,
J.S. Arora and P. Pal

1.0 Introduction

Statice is commonly called as sea lavender. The generic name *Limonium* has been derived from the Greek word *leimon* which means 'a meadow' because certain species are found growing in salt marshes (McCormick, 1981). They also have grown in coastal sites and deserts. The annual prefers cool climate for their growth and flowering, need protection against wet winter. Besides, some species like *Limonium cossonianum* Kuntze is an interesting endemic plant in the southeastern Iberian Peninsula and the Balearic Islands that can be used in wetland restoration and as an ornamental species in saline areas (Luque *et al.*, 2013).

Limonium is a genus of 120 flowering plant species. Members are also known as sea-lavender, statice, caspia or marsh-rosemary. Despite their common names, species are not related to the lavenders or to rosemary. They belong to family Plumbaginaceae, the plumbago or leadwort family. The generic name is from the Latin *lîmônion*, used by Pliny for a wild plant and is ultimately derived from the Ancient Greek *leimon* ('meadow') (https://en.wikipedia.org/wiki/Limonium).

2.0 Importance and Uses

2.1.0 Aesthetic Uses

Statice is a popular flower for garden decoration and cut flower. The dry flowers are also used for indoor decoration. Statice is suitable for borders, rock gardens, pot plants and for an alpine house. Van-Roozerndaal (1994) and Haines (1996) reported that Kenya is a major producer and cut flower exporter of *Limonium sinuatum*. Limonium is grown in several regions of the world for use as a cut flower for both fresh and dry-flower arrangements (Bruna *et al.*, 2004).

2.2.0 Therapeutic and Other Uses

Historically, the flower of *Limonium sinuatum* was often used as a form of tea for the prevention of ageing. In studies by Li *et al.* (2014), the flower of *Limonium sinuatum* has been found to have strong antioxidant capacity, therefore, could potentially be a rich resource of natural antioxidants. The antioxidants from the flower of *Limonium sinuatum* could be used as dietary supplement or food additive. The Ultrasound-assisted extraction (UAE) was observed as an efficient and feasible method in the extraction of desired natural antioxidants from *Limonium sinuatum* flower (Xu *et al.*, 2017). The roots of the *Limonium profusum* are used to extract medicine. The crude ethanolic extracts of stems of *L. stocksii* showed antifungal and antibacterial activities. The pharmacological research showed that *Limonium* species have antipyretic, hemostatic, and depurative, antiviral and antitumoral activities (Ali *et al.*, 2013). *L. gmelinii* roots are used in folk medicine as an astringent and for diseases of the upper respiratory and gastrointestinal tracts (Tuohongerbieke *et al.*, 2019).

Uses of Statice

In Bed · In Border · In Garden

Pot Plant · Cut Flower · Bouquet

In Flower Arrangement · In Dry Flower · Dried Flower Arrangement

Pressed Flower · In Flora Wreath · Floral Crown

Source: www.nz.carousell.com, www.pinterest.com, www.123rf.com, www.shopee.ph, www.afloral.com, www.journals.ashs.org, www.evergreennursery.com, www.flowerexplosion.com, www.flowermeaning.com, www.plantmaster.com, www.mr-fothergills.co.uk

3.0 Origin and Morphology

The genus *Limonium* is native to Europe, the Mediterranean regions, Asia and the Canary Islands. *Limonium tataricum* is a native to Caucasus mountains whereas *L. bonduelli* is native to Africa. Most botanical species of the genus *Limonium* are endemics in the Mediterranean region, but many species have their centre of origin in Caucaso, Turkestan, Caspian Sea, Russia, Iran, China, and South Africa (Bruna *et al.*, 2004).

Limonium spp. are annuals, biennials, and deciduous and evergreen perennials. Plants 60-65 cm tall; leaves simple, entire or pinnatified, tapering, almost stalkless, rosettes; flowers papery, bracts in more or less one-sided; panicles corymb-like; corollas 5 lobed, joined at the base; calyces tubular and persistant.

4.0 Species and Cultivars

4.1.0 Species

Statice is a member of the Plumbaginaceae family and originally was assigned to the genus *Statice*, but was later classified as Limonium Mill. Most species are native to the Mediterranean region and the Canary Islands and commonly are called Sea Lavenders (Wilfret *et al.*, 1973). Several perennial types exist but are not very commonly cultivated. There are about 150 species under this genus (Brickell, 1996). Important annual species are *L. sinuatum* and *L. suworowii* while *L. latifolium* and *L. vulgare* are grown as perennial. However, Burchi *et al.* (2006) reported that the genus *Limonium* (fam. *Plumbaginaceae*) consists of about 300 species, well adapted to the Mediterranean environmental conditions, particularly to the Southern regions of Italy where several wild species are endemics. In a subsequent report, Bruna *et al.* (2004) and Teani *et al.* (2012) confirmed this number of 300. According to Teani *et al.* (2012), the genus *Limonium* includes over 300 species, about 200 of which originate in the Mediterranean basin.

Several studies were carried to clarify the phylogenetic relationships between species. Many of these studies, however, were not clear and exhaustive as they were often limited to restricted geographical areas (Teani *et al.*, 2012). Both sexual (diploid and tetraploid) and apomictic (spanning triploid to hexaploid cytotypes) species have been reported in section *Limonium*. Nevertheless, diploid species are few, and polyploid agamic species constitute the largest portion of the diversity currently known in this section. Several competing hypotheses, based mainly on karyological data, regarding the origin of polyploid *Limonium* species have been postulated (Palacios *et al.*, 2000).

The genus *Limonium* is notably of difficult taxonomic interpretation: the problem arises from the fact that the morphological groups (phenon) that can be recognised in nature are more often determined by the diversity of habitat conditions rather than by genetic differences (Arrigoni and Diana, 1999 cited by Bruna *et al.*, 2004). Moreover, the evolution of the genus has differentiated a high number of species in different areas (Mediterranean area, Central Asia, Siberia, Far-East, and South Africa), some of them have a large distribution others are extremely endemics. This fact is due to the reproductive biology characteristics of the genus, such as the natural trend to hybridisation, the pollen-stigma dimorphism and the apomixis (that occurs when cross-pollination is prevented). These factors make the *Limonium* a genus in continuous evolution (Bruna *et al.*, 2004).

Limoniums can be distinguished in two groups according on flowering period and its extension: seasonable and continuous blooming. The species belonging to the second one are the most diffusely cultivated and differ in yield and growing techniques (Barbieri *et al.*, 2000).

Limonium aureum

Dwarf, 30 cm tall, erect, branched perennial, grown as an annual; leaves narrowly spoon to lance-shaped, 1-5 cm long, tapering gradually to stalks; panicles small, flowers tiny, 5 mm long, funnel-shaped, orange-yellow; calyces papery. Good for cut flowers.

L. latifolium* syn. *L. platyphyllum

Plant perennial, branched, 60 cm or more high; leaves elliptic to spoon-shaped, 30 cm long; panicles spikelets on wiry stem; flowers 6 mm long short tubular, deep lavender-blue; calyces white. Cvs. Blue Coud : flowers 7 mm across, mauve; Violetta : flowers deep violet.

L. sinuatum

Plants erect, perennial, branched, 45-60 cm tall, hairy, grown as an annual; leaves 15 cm long, oblong to lance-shaped, deeply lobed, wavy margined, basal rosettes; stem bears panicles of clustered spikelets; flowers tiny, funnel-shaped, pink, white or blue; calyces hairy, white or pale violet. Good for cut flowers. Var. Art Shades : flowers orange, salmon-pink, yellow, rose-pink, red, caramine-red, blue and creamy-white. California series : flowers ranging from rich, deep purple to pure white; Forever series : spikes large, flowers a mixture of 6 or 7 colours, including blue, pink and yellow; Forever Gold : plant 60 cm tall, flowers yellow; Fortress series : plants branched; flowers in about 6 vivid shades including bright blues, pastels and apricot-yellow; Pacific series : flowers deep rose-pink, apricot, yellow blue and white; Bouquet series : very dwarf, tightly bunched spikelets, flowers blue, purple, salmon-pink, white, lemon-yellow or golden yellow; Sunburst series : tall (70 cm), flowers orange-peach, apricot yellow and rose-red (Brickell, 1996). It is native to the countries of the eastern Mediterranean region. It is commercially grown as an annual cut-flower crop in many parts of the world and is used for both fresh and dry flower arrangements (Cohen *et al.*, 1995).

4.2.0 Cultivars

According to Wilfret *et al.* (1973) seed companies have selected colour forms of *L. sinuatum* Mill, and market these under such names as Iceberg, American Beauty, Midnight Blue, Kampf's Blue Improved, and others. The yellow annual statice is a different species, *L. bonduellii* Kuntze and selections are marketed as Bonduellii or Gold Coast. *L. sinuatum* has flattened or angular stems, while, *L. bonduelli* has round stems. Plants vary widely in leaf shape, plant form, disease susceptibility, bract colour, and most importantly from a cultural stand point, yields and time to flower.

Species of Statice

Limonium sinuatum *Limonium suworowii* *Limonium latifolium*

Limonium vulgare *Limonium aureum* *Limonium perezii*

Limonium arborescens *Limonium californicum* *Limonium carolinianum*

Limonium gerberi *Limonium otolepsis* *Limonium ramosissimum*

Source: www.biodiversidadvirtual.org, www.ukrbin.org, www.fotomontaro.com, www.calphotosberkeley.edu, www.wildflower.org, www.plantsystematics.org, www.uksouthwest.net, www.itsfun.com.tw, www.diacos.com.au, www.gardenia.net, www.pinterest.com.

Cultivars of Statice

Iceberg | Midnight Blue | American Beauty

Gold Coast | Rose Light | Tainan No.1

Beltalard | Pretty Wings | Navy Sun Birds

Caldo Sun Birds | Ara Lavender | Raspberry Wings

Source: www.trianglenursery.co.uk, www.colouriginz.com, www.royalvanzanten.com, www.colouriginz.com, www.journals.ashs.org, www.shop.excitations.com.au, www.chilternseeds.co.uk, www.australianseed.com.

'Gold Coast' and 'Roselight' are the earliest varieties to flower, averaging 166 and 168 days to flower after seeding, respectively, although some plants of the former variety may flower within 88 days. The blue varieties flower the latest, while, the white and lavenders are intermediate in flowering time. 'Gold Coast' produces the most inflorescences but they are the lightest in weight of all the varieties. Midnight Blue produces the fewest flowers but the individual inflorescences are the heaviest in weight of the varieties. Optimum yields range from 25-60 thousand kilogram per hectare depending upon variety and cultural conditions (Wilfret *et al.*, 1973).

In India, under Dharwad conditions of Karnataka, the cultivar 'Turbo White' was good in quality parameters by recording maximum panicle length, girth and number of branches per panicle. It was also vigorous in vegetative growth. Therefore, the cultivar could produce better quality panicles and found to be suitable for semi-arid regions (Nataraj *et al.*, 2009).

5.0 Genetics and Breeding

5.1.0 Cytogenetics

Erben (1979) cited by Palacios *et al.* (2000) noted that within section *Limonium*, diploid species show two basic chromosome numbers (*n* 58 and *n* 59) and postulated that the complement of the polyploids arose through several combinations, involving reduced and unreduced gametes, of the *n* 58 and *n* 59 genomes. Hence, triploid (2*n* 5 24, 25, 26, and 27) and tetraploid (2*n* 5 34, 35, and 36) taxa combine genomes of the two basic cytotypes (Palacios *et al.*, 2000).

A detailed study on male sporogenesis and gametogenesis to determine male reproductive output in seed production in diploid *L. ovalifolium* and polyploid *L. multiflorum* was carried out by Rois *et al.* (2012). They showed that *Limonium ovalifolium* is mostly composed of diploid cytotypes with 2n = 16 chromosomes and *L. multiflorum* species by tetraploid cytotypes with 2n = 32, 34, 35, 36 chromosomes and had a genome roughly twice as big as the former one. In both species, euploid and aneuploid cytotypes with large metacentric chromosomes having decondensed interstitial sites were found within and among populations, possibly involved in chromosomal reconstructions. *L. ovalifolium* diploids showed regular meiosis resulting in normal tetrads, while diverse chromosome pairing and segregation irregularities leading to the formation of abnormal meiotic products are found in balanced and non-balanced *multiflorum* tetraploids.

Before anther dehiscence, the characteristic unicellular, bicellular, or tricellular pollen grains showing the typical Limonium micro- or macro-reticulate exine ornamentation patterns were observed in *L. ovalifolium* using scanning electron microscopy. Most of these grains were viable and able to produce pollen tubes *in vitro*.

In both balanced and unbalanced *L. multiflorum* tetraploids, microspores only developed until the "ring-vacuolate stage" with a collapsed morphology without the typical exine patterns, pointing to a sporophytic defect. These microspores were unviable and therefore never germinated *in vitro*. *L. ovalifolium* individuals presented larger pollen grains than those of *L. multiflorum*, indicating that pollen size and ploidy levels are not correlated in the Limonium system. Cytohistological studies in mature seeds from both species revealed that an embryo and a residual endosperm were present in each seed. Flow cytometric seed screens using such mature seeds showed quantitative variations in seeds ploidy level. It is concluded that male function seems to play an important role in the reproduction modes of Limonium diploids and tetraploids (Rois *et al.*, 2012).

5.2.0 Breeding

Phenotypic homogeneity was evaluated by Pedroza *et al.* (1997a) in *L. sinuatum* cv. Midnight Blue plants propagated by seed or micropropagated. Statice seedlings exposed to 11°C for a week responded to the treatment by early flowering. They were selected and allowed to cross pollinate with the entire population of selected individuals. Seeds were collected and planted as half-sib families and recurrent selection was applied within and between the families for 5 generations. Non-vernalized plants of the selected strain flowered 3 months before non-vernalized plants of the original population. Yield curves of the non-vernalized selected plants were similar to those of the non-selected population that had been exposed to 6 weeks of vernalization (Cohen *et al.*, 1995).

In Italy, the interest on Statice is increasing for many reasons: i) it can be easily cultivated in Mediterranean climates, in particular in marginal areas where environmental conditions are hard for other ornamentals; ii) in nature there is a large genetic variability not yet utilized for ornamental purposes; iii) it is relatively easy to produce further genetic variability crossing different species; iv) these species can be easily manipulated *in vitro* (Burchi *et al.*, 2006).

5.2.1 Objectives

i. To produce *Limonium* as a flowering potted plant. *Limonium* has the advantage of maintaining flower colour for a very long time because the coloured calyces remain open after the flowers have senesced. Thus, even though only one or a few inflorescences may be produced, the plants retain their colour for a very long time. Breeding potted plants of *Limonium* using *L. perigrinum* is being attempted (Morgan *et al.*, 2001).
ii. Subsequent breeding has had two aims – to increase the flower colour range, but maintain the form of 'Chorus Magenta', and to generate a range of *Limonium* hybrids for the potted plant market (Morgan *et al.*, 2001).

5.2.2 Breeding for Specific Traits

5.2.2.1 Earliness

In its natural habitat, the seeds of *Limonium sinuatum* Mill germinate in the fall following the early rain, producing flat rosettes. The plants remain in the vegetative rosette stage until their vernalization requirement is fulfilled and bolting is induced. Thus, plants which are naturally vernalized produce their flowers in the spring and throughout the summer. The cultivated statice, grown mainly as a summer crop, flowers following a similar course of events to that in a natural population (Cohen *et al.*, 1995).

Israeli growers export flowers to the European markets during the winter months. In order to induce flowering in the winter, seedlings of statice have to be vernalized for 8 weeks prior to the planting in the field in mid-August. *Limonium sinuatum* is a mandatory cross-pollinator and, therefore, a genetic variation exists in both commercial and natural populations with respect to both the level of vernalization needed for flowering and the rate of flower production. This variation served as the basis for a breeding program which aimed to select statice plants requiring lower levels of vernalization or no vernalization at all for flowering (Cohen *et al.*, 1995).

The procedure followed the following steps: 1) Seedling populations of blue statice were exposed to 11°C for 1 week: 2) individual plants which responded to the treatment by flowering early and producing many flowers were selected and allowed to cross-pollinate with the entire population of selected individuals: 3) seeds were collected from the selected seed-producing parents and planted as half-sib families in the following season: 4) recurrent selection was applied between and within families for five generations. Finally, seed samples of all selected families were lumped to produce a synthetic cultivar (Cohen *et al.*, 1995).

A low vernalization-requiring, yellow colour Limonium cultivar 'Tainan No. 2', was developed by Chang *et al.* (2010) by crossing an accession of *Limonium sinense* as pollen parent which flowers any time of the year if temperature, light intensity, and duration are appropriate and Hybrid Limonium 'Yellow Diamond', a commercial cultivar that have an absolute vernalization requirement and flowering occurs only once during late spring or summer, as seed parent. 'Tainan No. 2' is selected for cut flower production in sub-tropical climates. It can flower year-round with good cut flower quality except in the summer. The cut inflorescences can be used in both fresh and dried forms, because the papery calyx remains open and retains its colour.

5.2.3 Inter-specific Hybrids

There are opportunities to develop new *Limonium* cultivars through interspecific hybridization (Burge *et al.*, 1995). Pollination barriers include lack of pollen tube penetration of the stigma and inhibition of pollen tube growth down the style. Post-pollination barriers (embryo abortion) also exist.

Several new *Limonium* hybrids have been produced using ovule-embryo rescue techniques. Fifteen *L. perigrinum* x *L. purpuratum* embryos were successfully rescued when *L. perigrinum* was the female parent. No embryos were produced by the reverse cross. Later, *L. pereziihas* been hybridised with *L. sinutum* (Burge *et al.*, 1995). *Limonium perigrinum* was hybridized with *L. purpuratum* and plant with stems more than 60 cm selected from this cross for further development. The cultivar 'Chorus Magenta' was selected from the hybrids which is relatively resistant to many pests and diseases (Seelye *et al.*, 2000).

Further, Morgan *et al.* (2001) crossed plants of 'Chorus Magenta' and one of its F_1 siblings with *L. perigrinum*. There was small number of plants with a *L. perigrinum* like inflorescence, a small attractive plant that when grown from cuttings produced flowering plants with a total height of approximately 30 cm, which may be suitable as potted plants.

A breeding programme on *Limonium* is carried out in Sanremo and Pescia in Italy since 1998. An incomplete diallelic crossing was done by Burchi *et al.* (2006) between wild species and commercial varieties were with the aim to obtain new varieties suitable for cultivation in Mediterranean conditions with low energy requirements. A first group of selected progenies derived from crosses among *L. latifolium, L. gmelinii, L. caspia, L. bellidifolium, L. otolepis* and *L. serotinum*. The hybrids were cultivated with minimum tillage and low input of fertilisers and pesticides, productivity and commercial quality were found at par or also higher than the commercial control cultivars. A second group derived from the cross *L. bonduelli* x *L. sinuatum*. Selected hybrids were evaluated in comparison with commercial cultivars of Statice. They showed a significantly higher production, significantly shorter and harder stems than *L. bonduelli*, a number of branches per stem significantly higher than *L. sinuatum* and commercial varieties, and different colour combinations of calyx and corolla. A third group derived from selected progenies of *L. tataricum* obtained from free pollination. These genotypes were characterized by good production, high number of flowers per stem and very attractive architecture. The last group derived from crosses among *L. aureum, L. sinensis, L. tetragonum, L. fortunei* and commercial cultivars. The new hybrids showed the valuable agronomical and ornamental traits of the parental cultivars and the stress tolerance of the botanical species (Table 1).

Table 1: Productivity, Plant Biometrical Traits and Susceptibility to Parasites of the New I.S.F. Varieties in Comparison with Commercial Cultivars (Mean data of 2 years x 3 replicates x 24 plants per replicate)*

Name	*Product. (No. of stems/ plant)*	*Stem Height (cm)*	*No. of Branches/ Stem*	*Stem Weight (g)*	*Susceptibility[1]*	
					Oidium	*Peronospora*
Hybrids among *L. gmelinii, L. altaica, L. latifolium* and *L. serotinum*						
I.S.F.Silvia	28.1 a	68.4 ab	7.6 a	27.3 ab	4	0
'Emille'	12.9 b	56.1 cd	5.6 ab	19.0 bc	3	0
I.S.F.Elitop 1	11.1 b	73.5 a	7.7 a	33.1 a	1	0
I.S.F.Violet Maury	9.2 bc	62.6 bc	6.6 ab	32.0 a	3	0
'Lavand Emille'	8.4 bc	59.8 cd	6.3 ab	24.9 ac	3	1
I.S.F.Elitop 2	3.1 c	53.3 d	5.3 b	16.3 c	2	1
'Tall Emille' (1 repl. only)	7.0	63.0	6.3	21.4	4	0
'Topian Blue' (2 repl. only)	1.8	82.2	8.7	50.9	0	3
Hybrids of *L. latifolium* x *L. otolepis* or *L. latifolium* x *L. caspia*						
I.S.F.Eliemi	6.7 a	107.6 a	12.7 a	63.6 a	0	2
'Avignon'	6.2 a	101.6 ab	9.9 b	44.7 b	1	3
I.S.F.Eligil	5.4 a	88.9 b	10.3 ab	47.1 b	0	2
'Ocean Blue'	2.1 b	97.0 ab	10.4 ab	44.6 b	0	4
'Misty Pink'	1.6 b	96.4 ab	10.0 b	58.2 ab	2	4
'Beltlaard'	1.6 b	109.3 a	12.8 a	56.7 ab	0	5
'Mysty White' (1 rep. only)	2.4	104.8	10.7	59.9	0	5

Name	Product. (No. of stems/ plant)	Stem Height (cm)	No. of Branches/ Stem	Stem Weight (g)	Susceptibility[1]	
					Oidium	Peronospora
Progenies deriving from free pollination of *L. latifolium* or *L. gmelinii*						
'Blue Rock'	5.3 a	64.5 ab	10.1 a	104.4 a	0	1
I.S.F.Annalisa	5.0 a	66.7 ab	8.6 ab	105.9 a	0	0
I.S.F.Jola	4.6 a	68.6 ab	8.3 ab	59.1 b	1	0
I.S.F.Elifabi	4.6 a	73.8 a	7.2 b	45.8 b	2	0
I.S.F.Eliblu	2.2 b	59.3 b	8.4 ab	58.3 b	0	0
I.S.F.L125 (1 rep. only)	7.4	69.0	6.7	31.9	2	0
'Drovandi' (1 rep. only)	6.8	67.8	5.9	32.7	2	0
I.S.F.L36.6 (1 year only)	6.8	118.4	8.9	52.4	4	0
'Riva Ligure' (1 rep. only)	3.9	58.7	7.7	82.2	0	2
I.S.F.Domenica (1 year)	3.9	56.6	6.8	55.4	1	0
I.S.F.Sofia (1 rep. only)	3.6	58.5	6.6	42.0	2	0
I.S.F.Ada (1 rep. only)	3.1	54.5	6.0	44.5	0	1
I.S.F.Blue Maury (1 rep.)	1.8	70.4	7.9	30.1	0	2
I.S.F.L104.1 (1 year only)	1.7	63.4	7.2	32.1	1	2

Mean values showing the same letter are not statistically different at $P \leq 0.05$ (according to Duncan's test). Mean data of the varieties evaluated for only 1 year or in only 1 or 2 replicates are shown but were not considered for statistical analysis.

1 Susceptibility to *Oidium* (*Erisiphe poligoni*) and *Peronospora* (*Peronospora statices*): score 5 = very high; score 0 = very low.

*Burchi *et al.* (2006).

Chang *et al.* (2010) developed 'Tainan No.1', an early flowering *Limonium sinuatum* cultivar through eight continuous generations of mass selection from the parental cultivar Fortress Mix. The calyx is funnel-shaped and purple-coloured and corolla is white. It is selected for cut flower production in sub-tropical climates. The cut inflorescences can be used in both fresh and dried forms, because the papery calyx remains open and retains its colour.

5.2.4 Polyploid Breeding

Limonium bellidifolium (2n = 2x = 18), a perennial statice, is an important breeding material for the production of hybrid cultivars in the genus. Chromosome doubling in *L. bellidifolium* was attempted to increase the variability in horticultural traits (Mori *et al.*, 2016). Seeds of this species were treated with an antimitotic agent, colchicine, at different concentrations and exposure periods. More than 50 per cent of the seedlings treated with colchicine for 24 or 48 h, irrespective of the concentration, survived for 4 months after treatment. Most of the seedlings treated for 72 h at any concentration showed very poor growth and abnormal thickening of the hypocotyl, and ultimately died. Flow cytometry analysis of leaf tissues showed that 2.5-5 per cent of plants that received 0.05 per cent colchicine for 72 h, 0.25 per cent colchicine for 24 h, 0.25 per cent colchicine for 48 h, or 0.5 per cent colchicine for 48 h were tetraploid (4x) or mixoploid (2x + 4x). The highest frequencies of tetraploids and mixoploids occurred following treatment with 0.05 per cent colchicine for 72 h. These results showed that colchicine treatment of seeds is effective for chromosome doubling in *L. bellidifolium*.

After 3 years of cultivation, the morphological characteristics of diploid and tetraploid *L. bellidifolium* plants at the flowering stage were investigated. The stomatal density was lower in all tetraploid and mixoploid plants than in the control diploid plant. The stomatal length was 1.1- to 1.5-fold higher in all tetraploid and mixoploid plants than in the control. Tetraploid plants tended to have wider and thicker leaves than the control and also produced larger flowers (Mori *et al.*, 2016).

6.0 Growth and Flowering

6.1.0 Temperature

Semenuik and Krizek (1973) studied the influence of temperature on seed germination, growth and flowering of six cultivars of annual statice. The influence of three germination temperatures, 16/13°, 21/18° and 27/24°C (day and night) and two growing temperatures, 24/16° and 26/24°C (day and night) under long days was studied. The cultivars showed variation in growth and flowering due to the growing temperature. Except for Gold Coast, plant grown at 24/16°C generally had the earliest and largest number of mature flower stalks and the highest rate of flower development. Low temperature seed treatment of *cv.* 'Sophia' reduced the period from sowing to flowering and increased the yield of cut flowers, particularly between November and February (Yoo *et al.*, 1998). Katsutani *et al.* (1998) studied the effect of low temperature treatment of *L. sinuatum* seedlings with illumination and found that chilling combined with illumination at 2-3°C for 50 days produced a greater number of cut flowers.

6.2.0 Photoperiod

Yoo and Lee (2000) investigated the effects of artificial light sources for light period extension on growth and flowering of statice 'Sophia' and 'Early Blue'. High-pressure sodium lamps (HPS), incandescent lamps (Il), and fluorescent lamps (Fl) for day length extension (16-h photoperiod) as compared with short day (8-h photoperiod) were tested. HPS gave the greatest photosynthetically active radiation (PAR). HPS or Fl as compared with Il showed high ratio of red/far-red light. HPS promoted photosynthesis as much as light period extension, while more respiration than photosynthesis occurred under Fl affected by low PAR. Long day condition as light period extension hastened flowering of statice, and HPS or Il were more effective than Fl on flowering among artificial light sources tested. Photosynthesis hastened at high temperature, but amount of photosynthesis at vegetative stage showed much higher than flowering stage under the condition below 20°C. These results indicated that day length extension with HPS increased productivity and quality for cut flower of statice.

6.3.0 Growth Regulators

Single application of 500 ppm GA_3 on *cvs*. 'Iceberg' and 'Midnight Blue' made at 87, 101 and 115 days after seed sowing resulted in early flowering (Wilfret and Raulston, 1975). They also reported that gibberellin substituted for cold under long day conditions but was less effective under short-day conditions. Flowering in *Limonium* is promoted by long days and low temperatures (Semeniuk and Krizek, 1972), and response to gibberellin applications also seems to be related to cumulative exposure to cold days and photoperiod (Shilo and Zamski,1985). Additionally, GA_3 was ineffective on plants that had not been exposed to low temperatures or in which vernalization had been saturated (Shilo, 1977). Proper timing of GA_3 applications under cool, long-day conditions may optimize the flowering response (Garner and Armitage, 1996). Gibberellin partially replaces vernalization requirements, thereby reducing pre-cooling requirements (Shilo and Zamski,1985).

Huh *et al.* (1995) applied GA_3 at 500 ppm, 1-8 weeks after planting of *L. sinuatum* cultivars 'Early Blue' and 'Sophia' between June and March. The number of flower stems/plant increased in October and November following GA_3 treatment 3 weeks after planting. GA_3 treatment did not affect flower quality.

Statice cultivars 'Iceberg' and 'Midnight Blue' were sprayed with different concentrations gibberellic acid (GA_3) 110 days from seeding for earliest flowering and maximum yield. Initial flowering of 'Iceberg' was not affected by GA_3 but 200-300 ppm and 400-2000 ppm GA_3 accelerated flowering time of 'Midnight Blue' approximately 30 and 55 days, respectively. Yields of both *cvs*. treated with 500 ppm GA_3 were increased approximately 4000 kg/ha to December 26 and 8000 kg/ha to February 12, as compared to untreated plants. Although $GA_3$1000-2000 ppm increased yields as compared to untreated plants, yields were less than at 400-500 ppm GA_3, which was the optimum concentration range for earliest flowering and maximum yields (Wilfret and Green, 1975). Treatment at 4 weeks after planting with GA_3 400mg/l resulted in the greatest acceleration of flowering and increase in stem yield (Garner and Armitage, 1996). Under unheated greenhouse, during September, twice spray of gibberellic acid at 500 mg/l on one-year-old plants of *Limonium gmelinii* increased late production flush and total flower yield (Farina *et al.*, 2017).

Salicylic acid (SA) at 150 mg/l on Limonium var. Misty Blue recorded higher plant height, early flowering, maximum inflorescence length, whereas, gibberellic acid (GA_3) at 200 mg/l produced maximum number of shoots/plant, flowering duration, weight and number of inflorescence/plant and vase-life (Patel *et al.*, 2018).

Liu *et al.* (2019) investigated the effects of exogenous salicylic acid (SA) application on seed germination in the *Limonium bicolor* under salt stress. Germination was significantly suppressed by a 200 mM NaCl treatment but was significantly improved when 0.08 mM SA was concurrently applied. During germination, the seeds treated with SA had high levels of gibberellic acid (GA) and high levels of amylase and α-amylase activity, but low abscisic acid (ABA) contents. The SA treatment upregulated the expression of key genes involved in GA biosynthesis while downregulating those involved in ABA biosynthesis, thereby triggering a favourable hormonal balance between GA and ABA that enhanced seed germination under salt stress.

7.0 Propagation

7.1.0 Seed Propagation

The best method of propagation is from seed. Seeds are small in size, 10000 seeds/28 g, and take 5-9 days for germination at 13°-18°C. The optimum germination temperature range for *Limonium aureum, L. bicolor, L. gmelinii* and *L. otolepis* are 15-20°C, 20-25°C, 25-30°C and 20-25°C, respectively (Wang and Li, 2012). Seedlings should be fertilized with low concentration on N fertilizer. Statice is easily grown from seed. Seed should be sown in a sterile artificial medium such as vermiculite, peat, perlite or mix tures of these and should be covered lightly. Statice seed germinates well over a wide temperature range in either darkness or light with optimum germination between 18-21°C. When seedlings begin to show first true leaves (ca. 1/8-1/4" in length), they can be transplanted into individual packs such as pots, cell-packs, Jiffy-7's, or Seedling trays. Any of the light-weight media commonly used for growing bedding plants are satisfactory for the containers (Wilfret *et al.*, 1973).

Ayala-Garay *et al.* (2011) compared growth of statice seedlings during the winter under three production conditions: a polyethylene-covered tunnel, shade netting that cuts light transmission by 50 per cent, and in the open to find the effect of temperature on seedling emergence and growth. Compared to the outdoor treatment (control), the polyethylene tunnel and shade netting produced higher emergence percentage values; 95 per cent and 93 per cent, respectively. The polythene tunnel produced higher dry weight biomass than the other conditions with an average temperature of 19.5°C and high thermal oscillation, but without frost. In the polyethylene tunnel, maximum dry weight, leaf area and absolute growth rate (88 mg/seedling, 15.1 cm^2/seedling, and 7.4 mg/d, respectively) were observed at 35 days after planting. The polythene cover prevented temperatures below 0°C, which explains the greater seedling growth in this environment.

Maximum seed germination in *Limonium cossonianum* was recorded in the lowest saline media (100 mmol/l NaCl) and distilled water (0 mmol/l NaCl) at 20/10°C. No seeds germinated in concentrations above 200 mmol/l NaCl at the higher temperatures. The increase in salinity delayed the beginning and end of germination, reduced final germination percentage, and increased mean time to germination. The rate of germination decreased with an increase in salinity and temperature (Giménez Luque *et al.*, 2013).

Seed storage techniques have greatly improved and recently a technique of separating and cleaning seed from the dried flower heads (decortication) has provided seed which can be sown easily and germinates quickly and uniformly. Ultra drying of *L. aureum* seeds is beneficial for maintaining the vigor (Li *et al.*, 2007). Dried flower heads contain ca. 350 seed per ounce (28g) and require 15-20 days from seeding to transplanting. Three and a half kg of this type of seed has been recommended to plant a hectare (Wilfret *et al.*, 1973).

Propagation and Growing of Statice

Seeds  | Sowing of Seeds  | Germination of Seed

Micropropagation of Statice

***Source*: Khadka *et al.* (2019).**

Seedling for Planting | Transplanting | Statice in Greenhouse

***Source*: www.devasgarden.com, www.floraldaily.com, www.fleurametz.com, www.floretflowers.com, www.budscape.net, www.youtube.com**

7.2.0 Vegetative Propagation

Perennials are propagated through division of clumps in spring. Each portion should contain few roots for quick establishment. Root cutting is a useful method for *L. latifolium* (Brickell, 1992).

7.3.0 Micropropagation

Pedroza *et al.* (1999) successfully propagated *cv.* Midnight Blue *in vitro* and the colour stability among the *in vitro* derived plants was evaluated. Vitrification was solved by increasing the concentration of sucrose to 4 per cent, agar to 0.7 per cent and calcium pantothenate to 2 mg/l. Once micropropagated, they were rooted and adapted to field conditions. Vitrification of plantlets was controlled in media supplemented with 2.0 mg calcium pantothenate/l and 4 per cent sucrose. The highest number of non-vitrified shoots was observed in the medium supplemented with 0.5 mg NAA/l, 2.0 mg calcium pantathenate/l and 4 per cent sucrose, with 0.7 per cent agar (Pedroza *et al.*, 1997b).

The shoot tips of a hybrid statice (*L. altaica* × *L. caspium*) were cultured on 25 ml of liquid half-strength MS medium supplemented with 0.4 µM BA and 0.054 µM NAA and one month after inoculation of shoot tips, numerous small globular structures were formed and propagated vegetatively at a high rate following subculture (Matsumoto *et al.*, 1997).

In axillary buds of the floral stage, BA induced a reversion from floral to vegetative growth. Proliferation of the cultures was initially accelerated by 1.2 mg/l BA but the treatment induced malformed growth with repeated subcultures. The percentage of rooted explants was low (43 per cent) and tended to decrease with the number of subcultures (Oviedo and Guevara, 1988). Lateral buds from flower stalks of annual statice (*L. sinuatum*) *cv.* 'Early Blue' was cultured *in vitro* at 20 or 27°C on a half strength MS medium containing 0.01 mg NAA/l, 1.0 mg BA/l, 30 g sucrose/l and 8 g agar/l (Miyama *et al.*, 1998). The emerging shoots were excised from clumps of multiple shoots and were sub-cultured for 4 weeks at 20°C on a hormone-free rooting medium. The rooted plantlets were transplanted into pots and grown at 20°C. *In vitro* micro-propagation of *L. sinuatum cv.* 'Citron Mountain' was successfully achieved using inflorescence stems as explants. The rate of shoot proliferation was increased when concentrations of both BA and kinetin were increased. BA performed better than kinetin in shoot proliferation, as expressed by number of shoots/explant (Hosni *et al.*, 2000).

Pedroza *et al.* (1997a) observed that physiological performance was similar for sexually and asexually propagated plants; however, the micropropagated plants showed total homogeneity in the expression of the violet blue colour 90A in the sepals up to flowering, irrespective of the origin of the initial explant, whereas the sexually multiplied plants expressed colour differences in this floral structure.

Leaf segments of *Limonium altaica cv.* 'Emille' was cultured on MS medium. The best response in terms of frequency of shoot regeneration (99.5 per cent) and number of shoots per explant was observed on medium supplemented with 2.85 µM IAA and 1.14 µM TDZ (Jeong *et al.*, 2001). Igawa *et al.* (2002) reported establishment plant regeneration system from cell suspension cultures in *L. sinuatum* cv. 'Early Rose'. The regeneration ability of the calluses induced was tested on medium containing 0.25 per cent gellan gum-solidified ½ MS medium containing 1.0mg/l zeatin, which showed high proliferation ability with about 80 times increase in fresh weight during the 2-week interval of subculture.

Jin *et al.* (2007) reported that ½ Linsmaier and Skoog Medium (LS)+NAA 0.5 mg/l+IBA 1.0 mg/l is the best medium for rooting of *Limonium bicolor*, and the plantlets transplanted to the substrate with vermiculite and peat (36″1) gained the highest survival rate and the best condition of growth.

Vegetative and reproductive buds of a selected deep blue clone, from the cultivar Midnight blue were used to establish an *in vitro* culture to stablise the colour. Micropropagated plants showed a complete homogeneity at expression of flower colour (Deep Blue Colour 90A of the RHS Chart), as opposed to the seed derived ones (Pedroza *et al.*, 1999).

Bose *et al.* (2017) optimized an efficient micropropagation protocol for direct organogenesis from root explants. 61.43 per cent of root explants directly formed shoot buds on their surface after 4-weeks of culture in media containing ½ MS, 43.82 mM sucrose 2.22 µM BA and 1.07 µM NAA.

The behaviour of four blue cultivars of statice (*Limonium sinuatum*) with respect to four distances (18.0, 12.5, 8.0, or 2.0 cm) between culture tubes and the light source during their *in vitro* multiplication stage was studied. The light treatments assayed resulted in no significant differences in *in vitro* propagation ratio and *ex vitro* survival and crop production and quality, however, *in vitro* growth and levels of pigments and antioxidants showed variations among cultivars (Sánchez-Sánchez *et al.*, 2018).

7.3.1 Somatic Embryogenesis

The first report of somatic embryogenesis in the Plumbaginaceae was made by Aly *et al.* (2002). They cultured cotyledon explants from perennial statice *Limonium bellidifolium* (*Statice caspia* Willd.) on MS basal medium supplemented with various levels of 2,4-D, kinetin and sucrose. Embryogenic calluses developed over a period of 10 days with the highest response at 4.5 µM (1 mg/l) 2,4-D, 0.5 µM (0.1 mg/l) kinetin and 117 mM sucrose. Somatic embryos at the globular, heart-shaped, torpedo, and cotyledonary stages developed. Induction, maturation and germination of somatic embryos on the optimized media were equally effective using cotyledon, hypocotyl and root explants (Aly *et al.*, 2002).

7.3.2 Synthetic Seeds

Encapsulations of juvenile shoots were carried out by 3 per cent sodium alginate and 100 mM $CaCl_2$ which were again successfully stored at 4°C for 30 days along with 56.79 per cent of plant recovery in MS + 0.44 µM BA + 4.5 µM IBA + 87.64 mM sucrose containing medium. 150 synthetic seed derived full grown plants were successfully acclimatized in green house, where a total of 101 plants survived after secondary hardening. The ISSR analysis revealed genetic homogeneity of synthetic seed derived hardened plants (Bose *et al.*, 2017).

8.0 Soil and Climate

8.1.0 Soil

Any garden soil is suitable for growing statice but preferably fertile and drained loamy soil. It is fairly drought resistant and therefore, can be grown in dry and stony soil. Growth and development of Limonium *cv.* 'Misty Blue' plantlets, cultured in tosilee mix, vermiculite or perlite contained in plug cells, were evaluated and compared with growth in an agar medium (Jeong *et al.*, 2000). Fresh weight, height, roots length and chlorophyll concentration of Limonium plantlets increased in tosilee mix. The agar medium gave the poorest results. Statice loves sunny location and it flowers under a wide range of temperature, but prefers cool environment for early and better flowering.

8.1.1 Medium

The effect of different substrates on plant growth, yield and quality for 'Compindi White' and 'Compindi Deep Blue' varieties of *Limonium sinuatum* growth in greenhouse conditions was studied. Two cultivars of statice were grown in peat, perlite, tuff and equal amounts mixed media (peat, perlite and tuff 1:1:1), soil (control application) were tested for suitability for soilless cultivation. In the first production period, the highest yield was obtained in the peat, in the second production period the highest yield was obtained in the mixed media (peat, perlite and tuff 1:1:1). For *Limonium sinuatum* 'Compindi Deep Blue', the highest yield values were obtained in the peat, in both production periods. The highest values for flower stem length, flower stem weight, leaf number, weight of root and the upper part of the plant, root length were observed in peat and mixed media (Saraçoðlu *et al.*, 2017).

Three cultivars of hybrid Limonium of the Misty series ('Misty White', 'Misty Blue' and 'Misty Pink') were grown in pumice with the aim of evaluating the suitability for soilless cultivation. Plants in pumice of all tested cultivars gave better quality stems than ones from plants in normal soil. In both growing conditions cultivar 'Misty White' produced the highest amount of flower (12 and 11.5 stems/plant for soilless and traditional culture, respectively), while *cv.* Misty Blue gave the longest (106.6 and 90 cm) and heaviest (73.5 and 46.4 g) stems (Zizzo *et al.*, 2002).

8.1.2 Salinity Tolerance

Since most species of *Limonium* are reported to be salt and drought tolerant (Rathinasbapathi, 2000), they can be easily grown on coastal areas, salt marshes, saline soil and water deficient areas where other flowering plants are not conducive to grow. *Limonium cossonianum* Kuntze is an interesting endemic plant in the southeastern Iberian Peninsula and the Balearic Islands that can be used in wetland restoration and as an ornamental species in saline areas. Maximum germination was recorded in the lowest saline media (100 mmol/l NaCl) and distilled water (0 mmol/l NaCl) at 20/10°C. No seeds germinated in concentrations above 200 mmol/l NaCl at the higher temperatures. The increase in salinity delayed the beginning and end of germination, reduced final germination percentage, and increased mean time to germination. The rate of germination decreased with an increase in salinity and temperature (Luque *et al.*, 2013).

8.2.0 Climate

8.2.1 Temperature

Low temperature (11–13°C) is a prerequisite for flower initiation of blue *cvs.* Statice, 'Dunkelblau' and 'Midnight Blue'. Further development of flowers is favoured by higher temperatures (12–16°C at night and 22–27°C during the day). Day length does not affect flower initiation but LD promotes the development of flowers (Shillo, 1976).

Temperature and photoperiod influences time of flowering and percentage of plants flowering (Wilfret *et al.,* 1973). The blue varieties flower earlier with more inflorescences under long day (LD) than under short day (SD) conditions. Vernalization of the seedlings by cool growing conditions promotes earlier and more consistent flowering. Midnight Blue grown at 16/13°C day/night temperatures for 42 days before movement to a warm greenhouse flowers 100 per cent in 130 days whereas none of the plants grown at 27/24°C flower in the same period (Wilfret *et al.,* 1973).

8.2.2 Light

Limonium sinuatum has been reported as a facultative, long-day plant with a critical photoperiod of 13 h (Semeniuk and Krizek, 1972; Shillo and Zamski, 1985). Long days are also likely to promote flowering of *L. perezii* (Armitage, 2003). Observations of leaf number accumulation rate (LNAR) and light integrals (DLI) were used to develop a predictive model for time to flower for a novel hybrid of *L. sinuatum* (L.) Mill. × *L. perezii* (Stapf) Hubb. (Chen *et al.,* 2010). Plants were established in a temperature-controlled greenhouse at seven planting times from fall to late spring. Long days were maintained using day length extension lighting. Two light regimes, full sun or 50 per cent shade, were also used. DLI was highly correlated with the time to appearance of the first visible flower bud, explaining in excess of 80 per cent of the variation. When combined with plant growth variables describing either LNAR or rates of increase in groundcover index, a second model was able to predict the date of first visible flowers and accounted for more variation than DLI alone. Daily average temperature (DAT) did not significantly contribute to variation in time to first visible flower because temperatures were uniform between successive plantings at 18-21.78°C. However, DAT was significant for the period from visible flower through to flower harvest maturity. Growers of these hybrids for cut flowers can therefore use historical records of DLI to determine planting dates to schedule flowering. Once planting has occurred, by measuring actual DLI, DAT, and leaf number per plant, growers can use the second model to more accurately predict the dates for visible flowers and flower harvest (Chen *et al.,* 2010).

9.0 Cultivation

9.1.0 Selection of Cultivars

Trials were conducted by Cocozza-Talia *et al.* (1999) on four *L. sinuatum* cultivars. The yellow cultivar gave the highest cut flower yields (average 192.0 stems/m^2). Its quality made it suitable for use also as a dried flower. The blue cultivar gave the lowest cut flower yield (87.0 stems/m^2). Whipker and Hammer (1994) studied the potentiality of *L. sinuatum* as a cut flower crop. Yellow cultivars had more stems harvested than rose, apricot and blue cultivars, but individual stems of the yellow cultivars weighed less.

9.2.0 Planting

The seedlings are transplanted when they attain a height of 4-5 cm and for this 3-4 weeks are required. Plants should be about 10 cm in diameter and ready for setting in the field 4-5 weeks after transplanting (Wilfret *et al.*, 1973). The effects of sowing time on growth and flowering were compared by Yoo *et al.* (1994) using two cultivars of statice. The first flowering dates for both cultivars was early July when planted on 20 February, the middle of June when planted on 10 March and late July for the last planting date (30 March). Cut flower production was much higher for the earliest 2 plantings (20 February and 10 March) than the last one (30 March). Flower quality (number of branches/cut flower, number of spikes/cut flower, number of spikelets/spike, cut flower, length and weight and percentage of first flowers) peaked from mid-July to early August for sowing on 20 February and 10 March and in late August for sowing on 30 March. Farina *et al.* (1993) extended the production period of a blue and a white flowered cultivar of *L. sinuatum* from the traditional period of April-May. They produced earlier crops (end of February) by vernalizing young plants at 6°C for 5 weeks before planting on 22 November without sacrificing quality or yield. When planted on 5 August only the white-flowered cultivar gave a reasonable yield (7-8 flowering stems/plant) from 15 October to 30 November.

Kumar and Kaur (1997) recorded highest plant height (55.33 cm) and spread (41.16 cm) in cv. 'Forever Purple' at a spacing of 60 × 30 cm. Flower fresh weight was higher when planted in November and the closest plant spacing (60 × 30 cm) gave the highest flower fresh weight and seedlings were transplanted in November.

In a field trial on statice (*L. sinuatum*) *cv.* 'Forever White' produced the highest flowering stems (30.88 cm) at 60 × 30 cm with November planting. Plant spacing of 60 × 45 cm produced the highest number of stems (23.05) with December planting. 'Forever White' took longest to bolt (104.66 days), whereas 'Forever Purple' bolted in the shortest time (87.44 days) in November and (88.33 days) in December (Kumar *et al.*, 1998). Planting distance between rows and plants should be 25-30 cm for border or rock garden.

A wide range of spacing has been used in Statice field production. Paparozzi and Hatterman (1988) used 30 cm x 69 cm gives planting density of 4.8 plants/m^2, while, Whipker and Hammer (1994) used 30 cm x 30 cm spacing, which is equivalent to 11.111 plants/m^2 or 111,111 plants/hectare. Wilfret *et al.* (1973) used two rows of plants with 30 cm between plants and 30 cm between rows. Plants were grown on raised beds 76-90 cm wide and 135 cm between bed centers, resulting in approximately 50,000 plants per ha.

Widest spacing of 60 cm x 45 cm effectively improved plant growth and flower quality but had shorter stalk length and lower yield as compared to closest plant density 30 cm x 15 cm, which had significantly higher yield but unsatisfactory quality. The medium plant density 45 cm x 30 cm achieved moderately good yield with desirable quality of flowers (Deshpande *et al.*, 2001).

A trend of rise in the number of marketable and unmarketable flowering stems per plot was observed as the plant density increased from 4 to 10 plants/m^2. From the tested planting densities, 10 plants/m^2 had shown better results as compared to the other densities in terms of marketability, premium fresh weight and stem length (Mellesse *et al.*, 2013). The spacing of 45 cm × 45 cm and nitrogen dose of 20g/m^2 is optimum for growth and flowering of statice (Jain *et al.*, 2018).

9.3.0 Irrigation

Statice is a herbaceous plant and vegetative growth is too fast at early stage. It takes about 60-65 days to complete vegetative growth and enter into reproductive stage. At all stages of growth, adequate amount of moisture in the soil is essential. Moisture stress at any stage of development adversely affects normal growth and flowering.

9.4.0 Manuring and Fertilization

Statice requires balanced fertilizer to meet its demand for nutrients. Field should be fertilized with well rotten farmyard manure @ 50 tonnes/ha and incorporated 10-15 days before transplanting. Application of 100 kg each of N, P and K per hectare will be adequate for the proper growth and flowering. Half of the N and K, full P and K should be incorporated in the soil 3-4 days before transplanting and remaining half of the N should be applied one month after first application. Raulston and Geraldson (1971) observed that statice was successfully grown under saran cloth-shaded structures in beds receiving a single preplant application of 6-8-8 fertilizer or osmocote (14-14-14) at 300-900 lb/acre or liquid injection of 20-20-20 fertilizer into irrigation water at 200 ppm nitrogen are satisfactory (Wilfret *et al.*, 1973). Mulching with polythene or paper did not make much difference to crop quality.

Application of 75 per cent NP + 100 per cent K + VC + Azotobacter + PSB recorded significant increase in plant height, plant spread, number of leaves and suckers. However, 50 per cent NP + 100 per cent K + VC + Azotobacter + PSB favourably influenced spike emergence, initiation of flower, flower harvesting, spike length, spread, number of branches/spike, flower yield (4.93 spikes/plant and 31.4 spikes/m^2) in statice *cv.* Blue Diamond (Gayithri *et al.*, 2004). Humic acid as foliar spray has a potential effect and can be used for increasing nutrient uptake and availability, thus, stimulating growth and flowering of statice *cv.* Sunday Lavander (Ibrahim *et al.*, 2016).

9.5.0 Weeding

Two to three weeding and hoeing are required to raise a healthy crop. In a trial with 6 herbicides for pre-emergence and 7 for post-emergence treatment in *Limonium tartaricum*, no herbicide was found phytotoxic and pronamide, nitrofen, diphenamid, pyrazon plus dalapon (post-emergence only) and oxadiazon proved effective (Agmalian *et al.*, 1975). The effect of herbicides on growth and yield of flowers of *L. sinuatum* and *L. tartaricum* grown for drying was reported by Hoffmann and

Kneissl (1976). Kerb 50 W (propyzamide), tenoran (chloroxuron), sencor (metribuzin) and gesatop 50 (simazine) damaged the plants unless the herbicides were applied to fully grown plants. Gilreath (1989) applied pre-emergence herbicides in field-grown statice (*L. sinuatum*) cv. Midnight Blue, before transplanting, with a second application over the top before flowering. DCPA (chlorthal-dimethyl), oxadiazon and oxyfluorfen generally provided safe and effective control of weeds. Ivanova (1999) evaluated the efficiency and selectivity of 6 herbicide systems applied to *L. sinuatum i.e.* (i) pendimethalin (1320 g/ha) + chloroxuron (4500 g/ha)/oxadiazon (1000 g/ha); (ii) metolachlor (200 g/ha) + oxadiazon (100 g/ha)/fluazifop-butyl (375 g/ha); (iii) metolachlor (2000 g/ha) + chloroxuron (4500 g/ha)/oxadiazon (1000 g/ha); (iv) pendimethalin (1320 g/ha) + oxadiazon (100 g/ha)/fluazifop-butyl (375 g/ha); (v) trifluralin (960 g/ha) + oxadiazon (1000 g/ha)/fluazifop-butyl (375 g/ha); (vi) trifluralin (960 g/ha) + chloroxuron (4500 g/ha)/fluazifop-butyl (375 g/ha) + oxadiazon (1000 g/ha) and concluded that all 6 herbicide systems were very effective even under very high weed density, including annual grasses and dicotyledonous weeds.

9.6.0 Greenhouse Cultivation

In temperate climate, statice is also grown under greenhouse condition for cut flowers. Magnani *et al.* (1997) evaluated the agronomic performance of 2 greenhouse films made of low density polyethylene (LDPE) using new stabilizing systems, tinuvin 494 and tinuvin 492, based on tertiary HALS (hindered amine light stabilizers). The film using tinuvin 494 allowed earlier harvesting of *L. sinuatum* than with traditional HALS films and gave superior production results both in terms of quantity and quality.

Microclimatic conditions in protected culture are related to the characteristics of the greenhouse (size, orientation, shape, covering materials, shading, cooling, *etc.*) and influence crops yield and quality. Study on five two-span greenhouses with different roof opening systems, with natural or forced ventilation, in the presence of a Limonium, Misty series, (Misty White and Misty Blue) cultivation was conducted by Fascella and Zizzo (2004) to evaluate the effect of vent typology on internal microclimate and crop response. Roof vent typology moderately affected air temperature and relative humidity. More sensible thermic reductions during the warmest days were recorded both in the greenhouse with gull-wing vents and equipped with 2 head-ventilators and in the structure with total roof opening by complete removal of the plastic film cover. Roof opening system did not significantly influence yield but only the qualitative parameters of production. The two above-mentioned environments had the longest (96 and 86 cm, respectively) and the heaviest (34 and 35 g) stems. Tested cultivars did not differ in stem quality. Misty White produced a higher amount of flowers than 'Misty Blue' (26, 4 and 16, 7 stems/plant, respectively).

9.7.0 Soilless Cultivation

Statice is cultivated in normal soil, in open air and greenhouse, as annual, biennal and perennial crop for cut flower production. Three cultivars of hybrid Limonium of the Misty series (Misty White, Misty Blue and Misty Pink) were grown in pumice with the aim of evaluating the suitability for soilless cultivation. Plants in pumice of all tested cultivars gave better quality stems than ones from plants in normal soil. Stem yield, length and weight depended also on the cultivar: *cv.* Misty White produced, in both growing conditions, the highest amount of flower (12 and 11.5 stems/plant for soilless and traditional culture, respectively), while *cv.* Misty Blue gave the longest (106.6 and 90 cm) and heaviest (73.5 and 46.4 g) stems (Zizzo *et al.*, 2002).

10.0 Diseases and Pests

10.1.0 Diseases

Statice is attacked by several diseases.

Fungal Diseases

10.1.1 Damping Off

Damping off mainly caused by *Botrytis cinerea* is generally observed in nursery beds. Soaking the seed in hot water (52°C) for 30 minutes or in 0.5 per cent sodium hypochloride for 2 minutes or in a benomyl solution (1 lb/100 gal) at planting eliminated subsequent damping off without reducing germination (Striden, 1973). Diaz *et al.* (1999) concluded that the integrated control (iprodine and captan + *Tricoderma hamatum* + cultural management) was the most effective method in controlling *B. cinerea* of cv. Midnight Blue, providing the highest number of exportable bunches/plant.

10.1.2 Stem Rot

Stem rot of perennial statice (*Limonium sinuatum*) caused by *Rhizoctonia solani* was observed by Kang *et al.* (2009). Stem and leaf of statice at soil line were dried and turned brown, initially. As the disease became severe, other stem parts and crowns were turned dark brown and then sunken.

10.1.3 Stem Canker

A stem canker disease of a Limonium hybrid 'Chorus Magenta' was identified at several sites throughout the North Island of New Zealand (Harvey *et al.*, 2000). The causal agent was identified as *Phomopsis limonii* sp. nov. First symptoms included leaves of diseased plants turning yellow, often on one side of the leaf blades. Cankers developed blue-black discolouration, and dark fruiting bodies of a fungus were located below the epidermis of cankers on dying plants.

10.1.4 Anthracnose

Anthracnose is caused by *Colletotrichum gloeosporioides.* This fungus also produces crown rot and lesions on leaves, stems and flower parts (Sohi, 1984). Cultivar like Gold coast is more susceptible (Engelhard *et al.*, 1972). Spraying the crop with bavistin (0.1 per cent) or dithane Z-78 (0.2 per cent) at weekly interval can control the disease effectively. Healthy seeds should be used while raising the nursery. The diseased plant debris should be destroyed.

Diseases and Pests of Statice

Damping Off

Fire Blight

Cercospora Infection

Downy Mildew

Gray Mold Rot

Phytoplasma

Anthracnose

Aphid

Lanternfly

Source: www.pnwhandbooks.org, www.forestryimages.org, www.influentialpoints.com, www.line.17qq.com, www.accuweather.com, www.sites.science.oregonstate.edu, www.ucanr.edu, www.link.springer.com/article/10.1007/s10327-016-0657-8, www.semanticscholar.org

Bobev *et al.* (2009) first reported the occurrence of anthracnose in statice (*Limonium tataricum*) in Plovdiv Region of Bulgaria. The causal agent was identified as *Colletotrichum dematium*. Rehman *et al.* (2012) reported the incidence of anthracnose caused by *Colletotrichum dematium* in Pakistan for the first time. The characteristic symptoms were first, brownish spots on leaves and branches, and then drying of branches starting from plant tips down to the collar region, causing collar rots and leading to the death of the whole plant.

10.1.5 Powdery Mildew

In autumn, powdery mildew of *Limonium* was observed in Mie and Oita prefectures, Japan (Hagiwara *et al.*, 1998). First report of powdery mildew caused by *Erysiphe limonii* on perennial hybrid statice (*Limonium latifolium x bellidifolium*) in Korea was done by Choi *et al.* (2016). Symptoms first appeared as thin white patches, which progressed to abundant hyphal growth on both side of the leaves and stems. However, first report of powdery mildew disease associated with an *Oidium* sp. was recorded on *L. sinuatum* in Korea by Kwon *et al.* (2001).

10.1.6 Downy Mildew

Koike *et al.* (1998) first reported the occurrence of downy mildew in statice (*L. bellidifolium*× *L. latifolium*) in San Diego, Monterey and Santa Cruz counties, California, USA and by Shirai *et al.* (2016) in Japan. The causal agent was identified as *Peronospora statices*. Symptoms include yellowish leaves with gray to white downy growth on the surface, mainly on the abaxial side (Shirai *et al.*, 2016). Bluish gray sporulation occurs on undersides of leaves. On upper surface of leaves, infections appear as light green areas that turn yellow, then coalesce, and eventually the leaf dies. In California it is currently found on cultivars in the "Misty" series; in Europe it occurs on many Limonium species. Unlike powdery mildew fungi, this fungus requires very wet conditions to flourish. Favored by cool temperature. High relative humidity (90 per cent or greater) and free moisture required for spores to germinate and infect plants. Spores require a minimum of 8 hours of wetness before infection occurs. Skrzypczak (1999) reported that azoxystrobin when applied as a spray at doses from 125 to 375 $\mu M/cm^3$ suppressed production and germination of *P. statices* conidia. Spraying over 15 weeks significantly decreased downy mildew severity on leaves and the number of infested plants. In greenhouse, provide good air circulation and keep relative humidity below 85 per cent. Avoid wetting foliage, using drip instead of overhead irrigation if possible. Remove infected plants immediately. Protect foliage with fungicides if necessary.

10.1.7 Leaf Spot

Alternaria sp., *Cercospora* sp., *Fusicladium staticis* and *Phyllosticta* spp. cause leaf spot on statice. Nicolettia *et al.* (2003) observed occurrence of *Cercospora insulana* on statice (*Limonium sinuatum*) in Italy. Infections first occurred in June and spread during the summer. Leaf spots were circular, brown with a darker edge, 3–6 mm in diameter, and surrounded by an orange or reddish halo. Lesions were also observed on the wings of the flower scapes, while scapes proper were not involved. Old lesions enlarged and merged, causing early yellowing and senescence of leaves;

heavy infections resulted in severe defoliation and retarded growth or death of panicles. Liu (1991) reported that clorothalonil (as daconil W.P.) at 75 per cent gave the most effective control of cercospora leaf spot disease followed by propiconazole (as tilt E.C.).

10.1.8 Gray Mold (*Botrytis cinerea*)

Rot of leaves, stems, and flowers. Rot may enter crowns and kill plants. Woolly gray fungus sporulation develops on decayed tissues. Flower stubs remaining after flower harvest is particularly susceptible. The fungus also may kill seedlings. Fungus has a wide host range and develops on dead plant parts. Spores are airborne. The disease is prevalent in California in cool (below 77°F) rainy weather. Clean up plant debris. Incorporate crop residue into soil as soon after harvest as feasible. Avoid overhead irrigation especially when flowers are present. Protect plants with iprodione or fenhexamid (Koike and Wilen, 2020).

10.1.9 Verticillium Wilt (*Verticillium dahliae*)

Plants are stunted. Lower leaves yellow, wilt, and dry. The fungus, which has a wide host range, survives in soil as microsclerotia. Fumigate soil with methyl bromide to chloropicrin combination. Heat, fumigate, or solarize soil used in production of transplants (Koike and Wilen, 2020).

Other fungal diseases like southern blight and seedling blight have also been reported. Broad bean wilt is also known to damage the crop (Hein *et al.*, 1977). Kasuyama and Inoue (2009) reported leaf brown spot of statice (*Limonium hybrid*) caused by *Phomopsis* sp.

Bacterial Diseases

Anderson and Tisserat (1994) first reported the occurrence of bacterial leaf spot of statice caused by *Pseudomonas andropogonis*, in USA.

Viral Diseases

Aster yellow is caused by the California strain of the virus *Chlorogenus callistephi* var. *californicus* (Pirone *et al.*, 1960). Latham and Jones (1997) reported the incidence (57 per cent) of tomato spotted wilt tospovirus (TSWV) in statice. Vovlas and Cillo (1966) observed cucumber mosaic cucumovirus (CMV) on statice cultivated under plastic tunnels in Spulia, southern Italy. The infected plants showed stunting, mosaic on leaves and the absence of flowers. In *L. sinuatum*, CMV was not transmitted via the seeds of infected plants.

Fujinaga *et al.* (2009) first reported of a Grapevine Algerian latent virus (GALV) disease on statice plants (*Limonium sinuatum*) in Japan. The chlorotic leaf spot, necrotic stunt, and dwarfing of plant are the symptoms of GALV.

Phytoplasma

The presence of phytoplasmas in *L. sinuatum* plants showed leaf and flower discolouration and malformation, shoot dieback and seed sterility was detected by Kaminska *et al.* (1999) for the first time using polymerase chain reaction assays with primers rU3/fU5 amplifying phytoplasms 16S rDNA regions. Shiomi *et al.* (1999) also reported the presence of numerous phytoplasma particles in the sieve tubes of diseased *L. sinuatum*.

Symptoms include stunted growth, multiple production of short stalks, malformation and bunching of young leaves. Leaves on more mature plants are reddened in the basal rosette. Flowers are abnormal: reduced size, abnormal shape and colour (frequently green), and may fail to open. Phytoplasma are transmitted by leafhoppers (Koike and Wilen, 2020).

Gera *et al.* (2004) observed yellow disease on *Limonium latifolium* hybrids. Symptoms included small and or deformed/discoloured flowers; small, narrow basal leaves, often yellow in colour; excessive leaf growth (witches' broom or asparagus fern) and eventual plant death. Analysis of PCR products, associated with infected Limonium, clustered within two major group of phytoplasmas (16SrV and 16SrIX), elm yellows and almond witches' broom.

Weintraub *et al.* (2004) observed the presence of one known phytoplasma vector, *Orosius orientalis* in very large numbers, ˜10-fold greater than any other species; three other known vectors were present in low to moderate numbers (*Circulifer haematoceps, C. tenellus* and *Exitianus capicola*).

The study by Chung *et al.* (2005) provided the first evidence of association of aster yellows (AY) phytoplasma with ststice wittchea's broom in Korea by transmission electron microscopy observation, and by determination of 16S rRNA gene sequences of phytoplasma.

Mosaic (Turnip Mosaic Virus)

Plants are stunted and have a mosaic pattern of light and dark green in leaves. Plants infected when young often die. Leaves may be distorted. Virus is transmitted by several different aphids and is common in many weedy plants in the mustard family (Cruciferae). Control nearby cruciferous weeds. Control aphids (Koike and Wilen, (2020).

10.2.0 Pests

Cutworms, armyworms, aphids, mites, mealy bugs, thrips, spider and nematodes are the pests causing damage to statice.

11.0 Harvesting and Postharvest Management

11.1.0 Harvesting

Statice flower panicles are harvested as cut flower as the heads start to open. A slightly angled cut is given as near as possible to the base of the stem (Brickell, 1992). Gao *et al.* (1996) reported that statice flowers are harvested when calyces of individual flowers have mostly opened and are showing colour.

The peak harvest season extends from January through May with approximately a 3 month harvest period from a given planting. Statice is harvested when an individual inflorescence has most of the calyxes open and coloured. Stems are cut by a knife as low to the plant as possible but without removing leaves (Wilfret *et al.*, 1973). An individual planting can be harvested 8-12 times at 5-7 day intervals.

11.2.0 Yield

The yield of cut flower varies with cultivars and time of planting. Farina *et al.* (1993) obtained 7-8 cut flowers/plant from 5 August planting. Cocozza-Talia *et al.* (1999) obtained highest cut flower yield (192/stems/m^2) from yellow cultivar whereas blue cultivar gave 87/stem/m^2.

Data collected from different cultivars of statics under field conditions indicated that total fresh stem weight, stem count, and average stem weight differed significantly among cultivars. Yellow cultivars had more stems harvested than the rose, apricot, and blue cultivars, but stems of the yellow cultivars weighed less. The number of stems harvested over time tended to be concentrated in the first 8 weeks after flowering begins. The average stem fresh weight was significantly different among the apricot, blue, and rose cultivars, but the number of stems harvested was significantly different only between the blue and rose cultivars (Whipker and Hammer, 1994).

11.3.0 Postharvest Management

Standard statice is harvested when fully open, but before the true flowers (small white flowers within the brightly coloured bracts) senesce. Hybrid statice cultivars should be harvested when the first flower on each spike is open. Avoid bunches where flowers have yellowing stems and leaves, as well as bunches with obvious fungal growth or wilted flowers. Remove from shipping boxes and re-cool flowers immediately. Also remove stem ties and separate stems to improve air circulation and reduce the likelihood of Botrytis infection.

Quality statice is of the correct maturity, and is free of defects and disease. Standard statice is normally bunched in 25's; other cultivars are bunched by bunch size. Sleeves help to prevent tangling of the delicate branchlets. Statice flowers are sensitive to ethylene, which causes accelerated wilting. In standard statice this effect is not apparent, since the 'flowers' are papery bracts surrounding the true flowers. In the hybrid statice cultivars, effective treatments to inhibit the action of ethylene greatly improve flower quality and vase life.

Cut flower can be stored for 2-3 weeks at 2°C. Statice (*L. sinuatum*) was used to investigate the rate of cooling during vacuum cooling. Although the cooling rate could be increased with both water spray and stem water absorption, the stem water absorption achieved excellent cooling effects of flowers in a short time period (Gao *et al.*, 1996). The vase life of flowers is 1-2 weeks in fresh arrangement.

In Limonium hybrid 'Beltlaard' cut flowers the combined use of 1-MCP and STS resulted in highest quality (Philosoph-Hadas *et al.*, 2003). This combined treatment also enabled storage under sea transport conditions (8 days at 2°C).

Conditioning of the leaves of in gibberellic acid or placing without conditioning them in benzyladenine at concentrations of 25 and 50 mg/dm^3 extended the postharvest longevity of leaves of the *Limonium latifolium*. Conditioning of the

Postharvest Management of Statice

Manual Harvesting

Mechanical Harvesting

Grading

Stem Cutting

Sorting

Bunching

Packaging of Bunches

Packaging in Boxes

Drying

Source: www.floraldaily.com, www.ozexport.nl, www.dreamstime.com, www.alamy.com, www.germanstatice.com

leaves in gibberellic acid at the concentration of 25 and 50 mg/dm^3 influences the inhibiting effect on chlorophyll degradation (Janowska and Schroeter-Zakrzewska, 2010). Gibberellic acid, benzyladenine and topolins applied to leaf conditioning had a favourable effect on the index of leaf greenness (Janowska *et al.*, 2013).

Pulsing *Limonium sinuatum* cv. Girlie Wings in STS at 500 mg/l for 1/2 h followed by transfer to 8-hydroxyqunoline sulphate (200 mg/l) + sucrose (20 g/l) prolonged vase life and delayed senescence (Khella *et al.*, 2018).

11.3.1 Conditioning

Conditioning is a simple and effective method of post-harvest longevity of both flowers and florists' greens. Janowska *et al.* (2013) assessed the influence of four growth regulators from the gibberellin and cytokinin groups applied as a 4-h conditioning treatment of leaf blades and to determine any effects on post-harvest longevity of leaves of *Limonium latifolium*. Gibberellic acid, benzyladenine, *meta*-methoxytopolin and its riboside at the concentrations of 25, 50 and 75 mg/dm^3 were taken up in a solution for 4-h leaf-conditioning of *L. latifolium* at the temperature of 18–20°C. Gibberellic acid, benzyladenine and topolins applied as leaf conditioning treatment extended the post-harvest longevity of leaves of *Limonium latifolium*. Examined growth regulators, with the exception of riboside of *meta*-methoxytopolin, also had a favourable effect on the index of leaf greenness.

Conditioning of the leaves in gibberellic acid or placing without conditioning them in benzyladenine at concentrations of 25 and 50 mg/dm^3 extended the postharvest longevity of leaves of the *Limonium latifolium*. Conditioning of the leaves in gibberellic acid at the concentration of 25 and 50 mg/dm^3 influences the inhibiting effect on chlorophyll degradation (Janowska and Zakrezewska, 2010).

11.3.2 Vase Life

Regardless of their maturity at harvest, the vase life of cut inflorescences of the hybrid *Limonium* 'Fantasia' placed in deionized water was 4 to 5 days. A vase solution containing Physan (a quaternary ammonium disinfectant solution) at 200 ml/l and 20 g sucrose/l not only prolonged the longevity of individual florets but also promoted bud opening so that the vase life of cut inflorescences extended to 17 days. Pulse treatment with 100 g sucrose/l in combination with Physan at 200 ml/l for 12 hours partially substituted for a continuous supply of sucrose. Including 30 mg gibberellic acid/l in the vase solution was without benefit (Doi and Reid, 1995).

L. perigrinum has a vase life of 30-40 days because the sepals remain open and are of decorative value long after the petals have deteriorated. Preservative vase-solutions or 24-h pulse treatments extended the period over which flower buds continued to open, but did not increase overall display life because stems decayed (Lewis and Borst, 1993). The cause of this stem decay was not determined. Preliminary investigations identified two factors that detracted from the appearance of harvested stems of 'Chorus Magenta'. These were the cessation of flower bud opening after harvest and shrivelling of the inflorescence stem. The shrivelling of the stem appears to be caused by desiccation of the inflorescence.

The postharvest display life of the Limonium cultivar 'Fantasia' is also reduced by the short length of time flowers are open after harvest (Doi and Reid, 1995). The water uptake of these stems declined rapidly over the first 5 days from harvest and was only slightly increased by sucrose vase solution treatments. However, the sucrose treatments increased the length of time that flowers were open from 4 to 17 days, and the longevity of individual flowers from 2.7 to 4.1 days. In another Limonium species, *L. sinuatum*, flower bud opening also ceases soon after harvest. However, sucrose vase-solution treatments did not extend the period over which flower buds opened (Steinitz and Cohen, 1982).

Hybrid statice benefits enormously from pre-treatment with STS or 1-MCP, and pulse pre-treatment with a sugar solution containing 10 per cent sugar and 200 ppm Physan-20 for 12 hours. The sucrose present in the preservative extends the life of the individual florets, as well as promotes flower opening resulting in up to 3-fold increases in the longevity of the inflorescences (Burge *et al.*, 1998). Water uptake and stem weight of harvested inflorescences of Limonium 'Chorus Magenta' declined rapidly when the stems were placed in water or a biocide solution, and florets remained open for only 5 days after harvest (Burge *et al.*, 1998). They observed that 24 h wetting-agent pulse treatments effectively reduced the rapid decline in water uptake and, when combined with sucrose (50 or 100 g/l) vase-solution treatments, extended the period that flower buds continued to open to over 16 days.

Statice should be stored, dry, at 0-1°C. Statice intended for long-term storage should be packed in polyethylene-lined cartons and pre-cooled. They may be held for up to 2 weeks in dry storage if the temperature is maintained steady and close to the freezing point. Statice is normally packed in horizontal fiberboard boxes. Many statice species can be dried and used for years in permanent flower arrangements. Yet, when used as a fresh flower, they may last only a few days before leaf yellowing (on statice only, not German statice) or Botrytis infestation occurs. Storing statice under lights can retard leaf yellowing, but this may be hard to do in commercial practice.

11.3.3 Drying of Flowers

Dried flowers of statice remain attractive for one year or more. For drying, the flowers are harvested at about half open stage, to maintain the shape. Harvested flowers are hanged upside-down in a warm, well-ventilated area or positioning them erect in containers for 1-2 weeks (Bhutani, 1990). Dried flowers are hygroscopic in nature and fragile, require careful handling.

Ciotta and Nunes (2011) quantified the time of dehydration of flower stalks of *Limonium sinuatum*, and assessed the durability of flower stalks after harvest, in different concentrations of sucrose solutions. The results show that total greenhouse dehydration occurred at 36 hours and on the open air after the 11th day of drying. In both forms, the flowers kept their characteristics especially colour, while the stems are dim. The longevity of the flowers was longer in that was kept in solutions containing sucrose and stored under room temperature than in cold storage. The durability of these conditions was equal to or greater than 8 or 4 days, respectively.

12.0 Biotechnology

12.1.0 Tissue Culture

12.1.1 Suspension Culture

A plant regeneration system from cell suspension cultures was reported in *Limonium sinuatum* Mill. cv. 'Early Rose' (Igwa *et al.*, 2002). Friable callus was initially induced from leaf segments of *in vitro*-cultured seedlings on 0.25 per cent gellan gum-solidified half-strength Murashige and Skoog [½MS] medium containing 1.0 mg/l (4.14 μ*M*) picloram. These calluses were maintained as cell suspension cultures, which showed high proliferation ability with about 80 times increase in fresh weight during the 2-wk interval of subculture. Shoot formation was observed at different frequencies on four of five cell lines whereas one cell line showed no shoot differentiation. Regenerated shoots detached from callus readily rooted 1 mo. after the transfer onto 0.5 per cent gellan gum-solidified ½MS medium lacking plant growth regulators. The plantets were successfully transferred to the greenhouse after acclimatization. No ploidy changes were observed in the callus induced or in the regenerated plantlets. The regenerated plantlets that were transferred to the greenhouse after acclimatization grew normally and did not any morphological signs of somaclonal variation (Igwa *et al.*, 2002).

12.2.0 Molecular Markers

12.2.1 Evolution

The genus *Limonium,* due to the patchiness of the natural habitats of its species as well as the high frequency of hybridization and polyploidy and the possibility of reproduction by apomixis, provides an example of all the principal mechanisms of rapid speciation of plants (Palacios *et al.*, 2000). As an initial study of evolution in this genus, Palacios *et al.* (2000) have analyzed intra- and interspecific variability in 17 species from section *Limonium,* the largest in the genus, based on RFLPs of cpDNA and nuclear rDNAITS sequences. In the cpDNA analysis, 21 restriction enzymes were used, resulting in 779 fragments, 490 of which were variable and 339 parsimony informative. *L. furfuraceum* exhibited two relatively divergent cpDNA haplotypes. The relationships found among the species based on cpDNA restriction fragments were coincident using different methods of phylogenetic analysis. Due to the presumed reticulate evolution in the genus *Limonium,* the comparison of these results with data from the nuclear DNA was necessary; ITS sequences were analyzed. The final alignment contained 488 characters, of which 198 were variable and 156 parsimony informative. Two relatively divergent ITS types were present at the intra-individual level in *L. delicatulum,* a triploid species.

Each type was related to ITS from different groups of diploid *Limonium* species, one with a base haploid chromosome number *n*58 (represented by *L. cossonianum*) and the other with *n* 5 9 (represented by *L. minutum*). The different phylogenetic inference methods used for the analysis of ITS sequences rendered very similar topologies. In general, the relationships among the species studied were coincident with those obtained with the chloroplast genome. Both nuclear and cytoplasmic markers support the polyphyly of section *Limonium*, with at least two species, *L. narbonense* and *L. vulgare*, clearly divergent from the rest. Moreover, the remaining subsections into which section *Limonium* is currently divided seem to be artificial (Palacios *et al.*, 2000).

12.3.0 Taxonomy and Phylogeny

RAPD analyses were used for the study of genetic relationships in *Limonium* (Bruna *et al.*, 2004). Thirteen wild species were tested with 10 primers. The dendrogram obtained from cluster analysis showed high similarity among three species. In order to clarify the genetic relationships, further analyses were carried out on several genotypes. The new dendrogram, obtained scoring 151 RAPD bands, showed that the genotypes did not group in clear clusters. Analysis of molecular variance (AMOVA) confirmed this trend: the highest genetic variation resulted among genotypes and only 6,58 per cent of the total variation resulted among the species. The use of RAPD markers was useful for clarifying the highly probable identity of the three *Limonium* species, that is notably of difficult interpretation (Bruna *et al.*, 2004).

Phylogenetic relationships within *Limonium* (Plumbaginaceae) are evaluated using sequence data from three plastid regions (*rbcL*, the *trnL* intron, and the *trnL-trnF* intergenic spacer). Sixty-six species representing the major genera of Staticoideae, have been analyzed using four species of Plumbaginoideae as an outgroup *Afrolimon* is embedded in *Limonium* and related to *L. vulgare*, the type of *Limonium*. *Limonium* is split into two major clades corresponding to sub-genera, but otherwise the current infrageneric classification proved to be artificial. The presence of an isolated taxon in the Canary Islands is used as a calibration point for age estimates of the major events in the genus, including migrations to the Southern Hemisphere, the Canary Islands, and Asia. The rapid radiation of *Limonium* in the Mediterranean basin appears to coincide with the desiccation of the Mediterranean Sea in the Messinian (late Miocene) (Lledó *et al.*, 2005).

Teani *et al.* (2012) listed and characterized the collection using both morphometric markers, provided by CPVO, and AFLP molecular markers. The use of morphometric parameters, which is essential for characterization of new cultivars, has been adapted for interspecific evaluation through the identification of species-specific parameters, *i.e.* type of inflorescence, calyx shape and colour, *etc.* Both morphometric and molecular analyses were effective in clarifying most of the phylogenetic relationships among the species of the *Limonium* collection.

12.4.0 Genetic Transformation

Very few reports are available on genetic transformation of Statice. An efficient indirect somatic embryogenesis and *Agrobacterium*-mediated transformation protocol for *Limonium sinense* has been established, wherein neomycin phosphotransferase II (npt II) and β-glucuronidase (GUS) genes were used as selectable and screenable markers, respectively. The efficiency of plantlet regeneration from transformed tissue was compared between direct embryogenesis from leaf and indirect embryogenesis from callus. *Agrobacterium tumefaciens* strain LBA 4404 containing the vector pBI121 was used for the transformation. Transient GUS expression frequency was evaluated and putative transgenic plants were successfully grown on culture medium in presence of kanamycin (80–100 mg/l). PCR analysis of putative transgenic plants confirmed the presence of GUS and *nptII* genes. The transformation efficiency obtained through indirect embryogenesis from calluses (4 per cent) was much higher than through direct embryogenesis from leaf explants (0.9 per cent) (Dam *et al.*, 2017).

12.5.0 Cryopreservation

Cold-hardened, pre-cultured small segments of shoot primordia were successfully cryopreserved in liquid N by vitrification. Vitrified and warmed segments plated on solidified half-strength MS medium produced shoots about 21 days after planting (Matsumoto *et al.*, 1997). *In vitro*-grown apical meristems of hybrid statice (Limonium) *cv.* Blue Symphonet were cryopreserved using 3 different procedures : vitrification with encapsulation; vitrification without encapsulation; and a revised encapsulation/dehydration technique (Matsumoto *et al.*, 1998).

Barraco *et al.* (2011) successfully cryopreserved shoot tips of *Limonium serotinum* using the droplet-vitrification technique. Growth recovery of cryopreserved shoot tips was possible only when samples were pre-treated for 16 h in liquid medium with 0.3 M sucrose, then for 5 h in liquid medium with 0.7 M sucrose before performing the cryopreservation protocol. Re-growth of cryopreserved samples was slow but always direct, without callus formation.

13.0 References

Agamalian, H. S., Elmore, C. L. and Earnham, D. S. (1975) *Weed Flower and Nursery Report*, July-August, pp. 4-5.

Ali, R.A., Inam, S. and Perveen, S. (2013) *J. Basic Appl. Sci.*, **9**: 116-119.

Aly, M.A.M., Rathinasabapathi, B. and Kelley, K. (2002) *Plant Cell Tiss. Organ Cult.*, **68** : 127–135.

Anderson, D. and Tisserat, N. (1994) *Plant Dis.*, **78**: 1218.

Armitage, A.M. (1984) *J. Amer. Soc. Hort. Sci.* **109**: 233–236.

Ayala-Garay, O.J., Carrillo-Salazar, J.A., Hernández-García, E., Díaz-Martínez, E., Livera-Muñoz, M. and Almaguer-Vargas, G. (2011) *Rev.Chapingo Ser. Hortic.*, **17**(2): 129-140.

Barraco, G., Sylvestre, I., Iapichino, G. and Engelmann, F. (2011) *Sci. Hortic.*, **130**(1): 309-313.

Bobev, S.G., Jelev, Z.J., Zveibil, A., Maymon, M. and Freeman, S. (2009) *Plant Dis.*, **93**(5) : 552-552.

Bose, S., Karmakar, J., Fulzele, D.P., Basu, U. and Bandyopadhyay, T.K. (2017) *Plant Cell Tiss. Organ Cult.*, **129**: 313–324.

Brickell, C. (1992) *Encyclopedia of Gardening Dorling Kindersely*, pp. 187.

Brickell, C. (1996) *A-Z Encyclopedia of Garden Plants*, Dorling Kindersely, pp. 618-619.

Bruna, S., De Benedetti, L., Mercuri, A., Schiva, T., Burchi, G., Pecchioni, N. and Agrimonti, C. (2004) *Acta Hortic.* **651**: 155-160.

Burchi, G., Mercatelli, E., Maletta, M., Mercuri, A., Bianchini, C. and Schiva, T. (2006) *Acta Hortic.*, **714** : 43-51.

Burge, G.K., Morgan, E. R., Konczak, I. and Seelye, J. F. (1998) *New Zealand J. Crop Hort Science*, **26**(2): 135-142.

Burge, G.K., Morgan, E.R., Seelye, J.F., Grant, J.E., Zhang, C. and Hopping, M.E. (1995) *Acta Hortic.*, **420**: 78-80.

Chang, Y.T., Yeh, D.M. and Yang, W.J. (2010) *Hort Science*, **45**(12): 1897-1898.

Chen, J., Funnel, K. A. and Morgan, E.R. (2010) *Hort Science*, **45**: 1441-1446.

Choi, I.Y., Choi, C.H., Cho, S.E., Park, J.H. and Shin, H.D. (2016) *Plant Dis.*, **100**(3): 652-652.

Chung, B.N., Huh, K.Y. and Jeong, M.I. (2005) *Plant Pathol. J.*, **21**(4): 383-386.

Ciotta, M.N. and Nunes, E.D.C. (2011) *Rev. Bras. Hortic. Ornam.*, **17**(2): 159-166.

Cocozza-Talia, M.A., Cristiano, G. and Lucia, B. de (1999) *Culture Protette*, **28**: 71-74.

Cohen, A., Harazy, A., Rabinowitch, H.D., Stav, R. and Weiss, D. (1995) *Acta Horti.*, **420**: 118-124.

Dam, A., Bhattacharya, C. and Bandyopadhyay, T.K. (2017) *The J. Hortic. Sci. Biotechnol.*, **92**(4): 411-423.

Deshpande, G.M., Sonawane, P.C. and Dutt, M. (2001) *J. Appl. Hort*, **3**(2): 98-99.

Diaz, N.C., Barrera, M.J., de Granada, *e.g.* and Angarita, A. (1999) *Acta Hortic.*, **482**: 235-238.

Doi, M. and Reid, M.S. (1995) *Hort Science*, **30**: 1058-1060.

Engelhard, A. W., Howard, C. M. and Wilfret, C. J. (1972) *Plant Dis. Reptr.*, **56**: 894-895.

Erben, M. (1979) *Webbin*, **34**: 409–417.

Farina, E., Assenza, M. and Donzella, G. (1993) *Colture Protette*, **22**: 69-72.

Farina, E., Dalla Guda, C., Scordo, E., Castello, S., Paterniani, T. and Palagi, M. (2003) *Acta Hortic.*, **614**: 109-113.

Fascella, G. and Zizzo, G.V. (2004) *Acta Hortic.*, **659**: 171-176.

Fujinaga, M., Ogiso, H., Wakabayashi, H., Morikawa, T. and Natsuaki, T. (2009) *Gen. Plant Pathol.*, **75**(2): 157-159.

Gao, J.P., Sun, Z.R., Guo, K. and Xu, Y.Y. (1996) *Transactions of the Chinese Soci. Agri. Eng.*, **12**: 194-198.

Garner, J.M. and Armitage, A.M. (1996) *Hort Science*, **31**(2): 247–248.

Gayithri, H.N., Jayaprasad, K.V. and Narayanaswamy, P. (2004) *J. Ornam. Hortic.*, **7**(1): 70-74.

Gera, A., Maslenin, L., Rosner, A., Zeidan, M., Pivonia, S. and Weintraub, P.G. (2004) *Hort Science*, **39**(5): 1056-1059.

Gilreath, J.P. (1989) *Hort Science*, **24**: 794-796.

Giménez Luque, E., Delgado Fernández, I.C. and Gómez Mercado, F. (2013) *Botany*, **91**(1) : 12-16.

Hagiwara, H., Abiko, K., Izutsu, S., Tomikawa, A., Kuroda, K. and Okamoto, J. (1998) *Ann. Phytopath. Soc. Japan*, **64**: 506-509.

Haines, B. (1996) *Flora Culture International*, **6** : 16-19.

Harvey, I.C., Morgan, E.R. and Burge, G.K. (2000) *N. Z. J. Crop Hort Science*, **28**(1): 73-77.

Hein, A., Lesemann, D. E. and Chuerfurth, G. (1977) *Phytopath Z.*, **89**: 340-346.

Hoffmann, K. and Kneissl, P. (1976) *Gartenwelt*, **76**: 285-286.

Hosni, A.M., Hosni, Y.A. and Ebrahim, M.A. (2000) *Ann. Agri. Sci. Cairo*, **45**: 327-339.

Huh, K.Y., Kim, W.S., Jang, S.J. and Lee, H.K. (1995) *RDA J. Agri. Sci. Hort.*, **37**: 427-431.

Ibrahim, H.E., El-Fadaly, H.G. and El-Naggar, A.A. (2016) *Alex Sci. Exch. J.*, **37**(3): 515-528.

Igawa, T., Hoshino, Y. and Mii, M. (2002) *In Vitro Cell. Dev. Biol. Plant*, **38**(2): 157-162.

Ivanova, I. (1999) *Rasteniev dni Nauki*, **36**: 212-215.

Jain, R., Singh, M.K., Swaroop, K., Reddy, M., Janakiram, T., Kumar, P. and Pinder, R. (2018) *Indian J. Agric. Sci.*, **88**(7): 1108-1114.

Janowska, B. and Zakrzewska, A. S. (2010) *Nauka Przyr. Technol.*, **4**(1): 3.

Janowska, B., Grabowska, R. and Ratajczak, E. (2013) *Hort Science*, **40**(4): 172-176.

Jeong, B.R., Lee, E.J. and Klerk, G.J. de (2000) *Acta Hortic.*, **520**: 157-162.

Jeong, J.H., Murthy, H.N. and Paek, K.Y. (2001) *Plant Cell, Tiss. Organ Cult.*, **65**(2): 123-128.

Jin, D., Yu, H.Q. and Yi, M.F. (2007) *J. Agric. Univ. Hebei*, pp. 6.

Kaminska, M., Korbin, M. and Rudzinska, L. (1999) *Phytopathologia Polonica*, **18**: 37-45.

Kang, M.H., Cheong, D.C., Choi, C.H., Lim, H.C., Song, Y.J., Noh, T.H., Lee, D.K. and Kim, H.M. (2009) *Res. Plant. Dis.*, **15**(1): 54-56.

Kasuyama, S. and Inoue, K. (2009) Bulletin of the Agricultural Experiment Station, Okayama Prefectural General Agriculture Center (Japan).

Katsutani, N., Kajihara, S., Kawamoto, Y. and Hara, H. (1998) *Bull. Hiroshima Prefectural Agri. Res. Centre*, No. 66, pp. 53-59.

Khadka, S., Rana, N. and Rajbahak, S. (2019) *J. Pl. Res.*, **17**(1): 147-154.

Khella, E.A., Abd-ElMoniem, A.M. and Mohamed, A.M.B. (2018) *Sci. J. Flowers Ornam. Plants*, **5**(1): 89-103.

Koike, S.T. and Wilen, C. A. (2020) http: //ipm.ucanr.edu/PMG/r280112911. html#ipmpagetop.

Koike, S.T., Nolan, P.A., Tjosvold, S.A. and Robb, K.L. (1998) *Plant Dis.*, **82**: 591.

Kumar, R. and Kaur, K. (1997) *J. Oranam. Hortic.*, **5**: 20-25.

Kumar, R., Kaur, K. and Singh, M. (1998) *Indian J. Hort.*, **55**: 172-176.

Latham, L.J. and Jones, R.A.C. (1997) *Australian J. Agri Res.*, **48**: 359-369.

Lewis, D.H. and Borst, N.K. (1993) *New Zealand Journal of Crop and Horticultural Science*, **21**: 359-365.

Li, A.N., Li, S., Li, H.B., Xu, D.P., Xu, X.R. and Chen, F. (2014) *J. Funct. Foods*, **6**: 319-330.

Li, Y., Feng, H.Y., Chen, T., Yang, X.M. and An, L.Z. (2007) *J. Integr. Plant Biol.*, **49**(5): 569-575.

Liu, H.L. (1991) *Bull. Taichung District Agricultural Improvement Station*, **33**: 25-35.

Liu, J., Li, L., Yuan, F. and Chen, M. (2019) *Plant Signal. Behav.*, **14**(10): 644-595.

Lledó, M.D., Crespo, M.B., Fay, M.F. and Chase, M.W. (2005) *Amer. J. Bot.*, **92**: 1189 -1198.

Luque, *e.g.*, Fernández, I.C.D., and Mercado, F.G. (2013) *Botany*, **91**: 12-16.

Magnani, G., Bonora, M. and Lelli, N. (1997) *Colture Protette*, **26**: 85-92.

Matsumoto, T., Nako, Y., Takahashi, C. and Sakai, A. (1997) *Hort Science*, **32**: 309-311.

Matsumoto, T., Takahashi, C., Sakai, A. and Nako, Y. (1998) *Scientia Hort.*, **76**: 105-114.

McCormick, M. (1981) *Annuals and Perennials*, Marshal Cavendish Ltd., London, pp. 52.

Mellesse, B., Kassa, N. and Mohammed, A. (2013) *Afr. J. Plant Sci.*, **7**(11): 528-537.

Miyama, T., Inamoto, K., Doi, M. and Imanishi, H. (1998) *J. Japanese Hort. Sci.*, **67**: 632-634.

Morgan, E.R., Burge, G.K. and Seeley, J.F. (2001) In *XX International Eucarpia Symposium, Section Ornamentals, Strategies for New Ornamentals-Part I*, **552**: 39-42.

Mori, S., Yamane, T., Yahata, M., Shinoda, K. and Murata, N. (2016) *Hortic. J.*, **85**(4): 366-371.

Nataraj, S.K., Gangadharappa, P.M., Reddy, B.S., Naik, K.B., Prashanth, S.J. and Prakash, D.P. (2009) *J. Hort Science*, **4**(2): 184-186.

Nicoletti, R., Raimo, F., Pasini, C. and D'Aquila, F. (2003) *Plant Pathol.*, **52**(3): 418-418.

Oviedo, Y. del C. and Guevara, E. (1988) *Agronomia Costarricense*, **12**: 113-122.

Paparozzi ET, Hatterman HM (1988) *Hort Science*, **23**: 157-160.

Patel, M.G., Patel, R.B., Chawla, S.L., Patil, S. and Patel, D.K. (2018) *Curr. Hortic.*, **6**(2): 45-47.

Pedroza, J., Angarita, A. and Corchuelo, G. (1997b) *Agronomia Colombiana*, **14**: 22-27.

Pedroza, J., Corchuelo, G. and Angarita, A. (1997a) *Agronomia Colombiana*, **14**: 13-21.

Pedroza, J., Corchuelo, G. and Angarita, A. (1999) *Acta Hortic.*, **482**: 321-327.

Philosoph-Hadas, S., Golan, O., Rosenberger, I., Salim, S., Kochanek, B. and Meir, S. (2003) *In VIII International Symposium on Postharvest Physiology of Ornamental Plants*, **669**: 321-328.

Pirone, P.P., Dodge, B.O. and Rickett, H.W. (1960) *Disease and Pests of Ornamental Plants*, Constable and Co. Ltd., London.

Rathinasbapathi, B. (2000) *Ann Bot* **86**(4): 709-716.

Raulston, J.C. and Geraldson, C.M. (1971) *Proc. Fla. St. Hort. Soc.*, **84**: 403-407.

Rehman, A., Mehboob, S., Akem, C., Batool, A. and Idrees, M. (2012) *Afr. J. Biotechnol.*, **11**(42): 9950-9954.

Rois, A.S., G. Teixeira, T. F. Sharbel *et al.* (2012) *Sex Plant Reprod* **25**: 305–318.

Sánchez-Sánchez, J., López-Orenes, A., Ferrer, M.A. and Calderon, A.A. (2018) *Biotech.*, **8**: 335. https: //doi.org/10.1007/s13205-018-1367-z

Saraçoðlu, Ö.A., Akat, H., Güne', A., Çakar, H. and Kýlýç, C.C. (2017) *Ege Üniv. Ziraat Fak. derg.*, **54**(2): 187-195.

Seelye, J.F., Burge, G.K. and Morgan, E.R. (2000) *Hort Science*, **35**(6): 1179.

Semeniuk, P. and Krizek, D.T. (1972) *Hort Science*, **7**: 293.

Semenuik, P. and Krizek, D. T. (1973) *J. Amer. Soc. Hort. Sci.*, **98**: 140-142.

Shillo, R. (1976) *Acta Hortic.*, **64**: 197-204.

Shillo, R. and Zamski, E. (1985) *Limonium sinuatum*, pp. 293–301. In: Halevy, A.H. (ed.) CRC handbook of flowering. CRC Press, Boca Raton, FL.

Shiomi, T., Tanaka, M., Uematsu, S., Wakibe, H. and Nakamura, H. (1999) *Ann. Phytopath. Soc. Japan*, **65**: 87-90.

Shirai, K., Nishiwaki, Y., Kobayashi, S. and Satou, M. (2016) *J. Gen. Plant Pathol.*, **82**(4): 212-215.

Skrzypczak, C. (1999) *Phytopathologia Polonica*, **18**: 47-55.

Sohi, H. S. (1984) *Diseases of Ornamental Plants and Their Control*, I.C.A.R. Publication, New Delhi.

Steinitz, B. and Cohen, A. (1982) *Hort Science*, **17**: 903-904.

Strider, D. L. (1973) *Plant Dis. Reptr.*, **57**: 969-971.

Teani, A., Antonetti, M. and Burchi, G. (2012) *Acta Hortic.*, **953**: 91-96.

Tuohongerbieke, A., Kaiyrkulova, A., Guan, Y.Q., Zhu, J., Shi, L.L., Sabir, G. and Aisa, H.A. (2019) *Chem. Nat. Compd.*, **55**(4): 734-735.

Van-Roozendaal, G. (1994) *Biotech. Development Monitor*, **20**: 6-7.

Vovlas, C. and Cillo, F. (1996) *Informatore Fitopatologico*, **46**: 7-8.

Wang, J.N. and Li, D.L. (2012) *Pratacultural Science*, 2.

Weintraub, P.G., Pivonia, S., Rosner, A. and Gera, A. (2004) *Hort Science*, **39**(5): 1060-1061.

Whipker, B.E. and Hammer, P.A. (1994) *Hort Science,* **29**: 638-640.

Wilfret, G. J. and Raulston, J. C. (1975) *Hort Science,* **10**: 37-38.

Wilfret, G.J. and Green, J.L. (1975) *Proc. Fla. Hort. Soc.,* **88** : 527-530.

Wilfret, G.J., J. C. Raulston1, S. L. Poe and A. W. Engelhard. (1973) *Proc. Fla. Hort. Soc.,* **87**: 399-405.

Xu, D.P., Zheng, J., Zhou, Y., Li, Y., Li, S. and Li, H.B. (2017) *Food Chem.,* **217**: 552-559.

Yoo, D.L. and Lee, S.W. (2000) *Hort Science,* **35**(3): 507A-507.

Yoo, D.L., Kim, W.B., Nam, C.W., Hahn, B.H., Kim, B.H., Kim, J.K. and Choi, K.S. (1994) *RDA J. Agri. Sci. Hort.,* **36**: 435-443.

Yoo, D.L., Ryu, S.Y., Kim, W.B., Nam, C.W., Choi, Y.E., Kim, B.H. and Kim, J.K. (1998) *RDA J. Hort. Sci.,* **40**: 150-154.

Zizzo, G.V., Fascella, G., Costantino, G. and Agnello, S. (2002) In *VI International Symposium on Protected Cultivation in Mild Winter Climate: Product and Process Innovation,* **614**: 235-239.

Impatiens walleriana

T. N. Saha, P. Naveen Kumar, Swati Saha, Ganesh B. Kadam, D.V.S. Raju and K.V. Prasad

1.0 Introduction

Impatiens walleriana is one of the important and most widely grown bedding plants in the world. It is native to eastern Africa from Kenya to Mozambique. The name *'walleriana'* was given in honour of Horace Waller, a British missionary and member of Royal Geographic Society. The name of genus *'Impatiens'* comes from Latin word meaning 'violent method of seed dispersal from ripen pod'. It was previously known as *I. sultani* or *I. holstii* and popularly known as 'Busy Lizzie' or 'Patience Plant'. *Impatiens walleriana* is one of only a few bedding plants available in almost all colours. Its preference for the shade and adaptability to containers, has contributed to a sustained high level of worldwide popularity for this plant. New breeding advances continue to add interest to the class. It is easy to culture and the wide availability of seed or cuttings makes this plant a favourite among both growers and consumers (Uchneat, 2007).

2.0 Importance and Uses

2.1.0 Aesthetic Uses

It is a half-hardy annual in temperate regions, though pot-grown plants can be successfully overwintered indoors. It is one of the most popular of all bedding plants for parks and gardens worldwide, typically grown in containers but also in bedding schemes. It is ideal for both indoor and outdoor cultivation. For indoor cultivation it can be grown in pots, hanging baskets as well as in window boxes. For outdoor cultivation it can be used for landscaping as beds (in shade), mixed borders, along walks, woodland gardens, ground cover, rock garden, *etc.* The plant remains a source of great interest among gardeners and breeders, and new hybrids and varieties are developed every year, facilitating the introduction and spread of this species into new areas and thus increasing the likelihood that further invasions takes place (www.cabi.org).

2.2.0 Therapeutic and Other Uses

In North America, impatiens has been used as herbal remedies for the treatment of bee stings, insect bites, and stinging nettle (*Urtica dioica*) rashes. They are also used to prevent rash development after contact with poison ivy (*Toxicodendron radicans*). Its rich mucilage forms a soothing and demulcent jelly that is used in the treatment of irritations of the gastro-intestinal canal. The Potawatomi and tribes of the Appalachian area uses the impatiens to prevent allergic reaction, athlete's foot and dandruff.

3.0 Origin and Distribution

Impatiens walleriana is native to eastern Africa particularly areas adjoining south-eastern Kenya, Tanzania, southern Malawi, western Mozambique and eastern Zimbabwe. In Africa, it is widely naturalized and also cultivated as ornamental. The species was unknown to Europe until 1896, when Dr. John Kirk brought it to England. In Puerto Rico, this species was first reported as naturalized in 1923

Uses of *Impatiens walleriana*

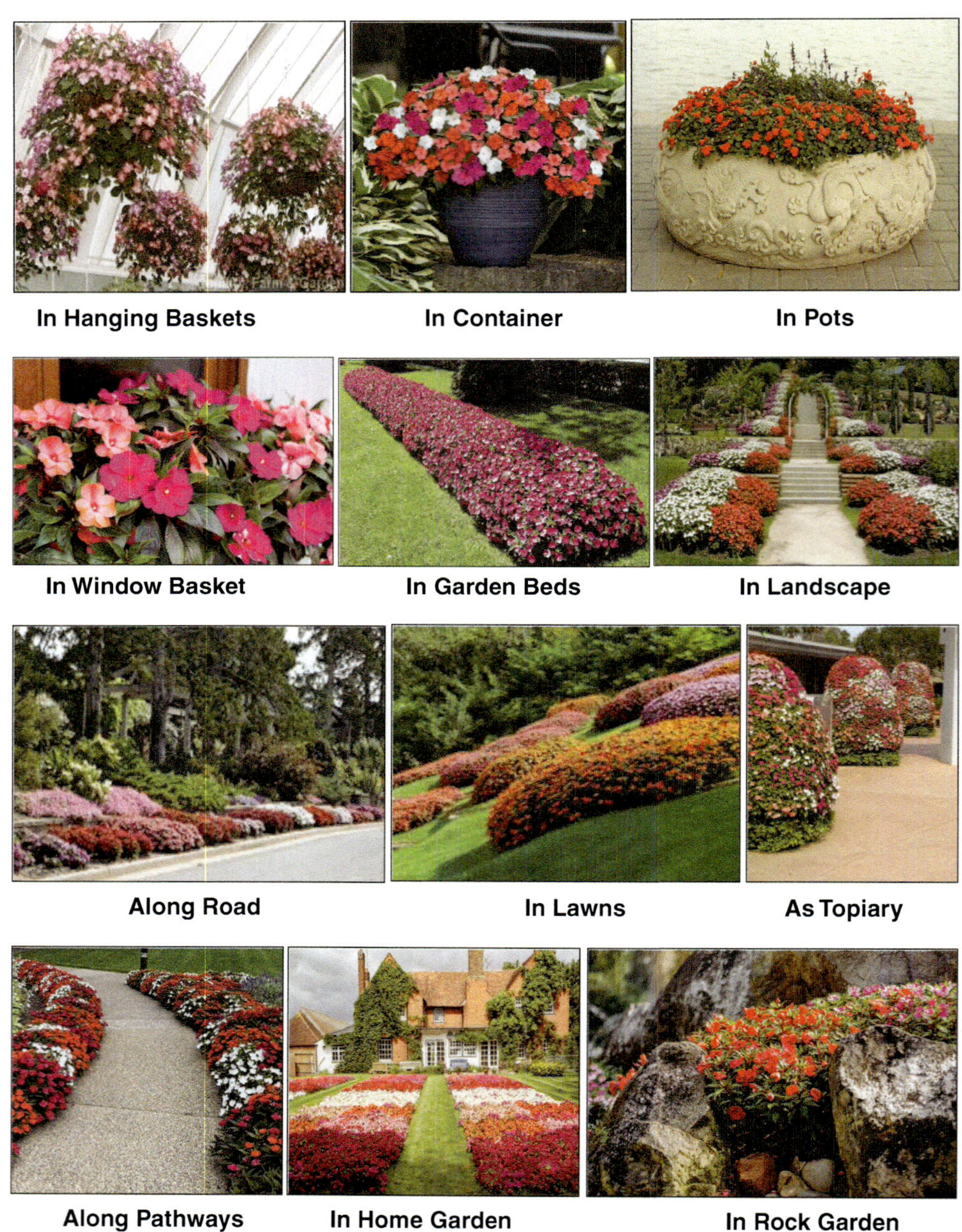

In Hanging Baskets | In Container | In Pots

In Window Basket | In Garden Beds | In Landscape

Along Road | In Lawns | As Topiary

Along Pathways | In Home Garden | In Rock Garden

Source: www.i.pinimg.com, www.theenglishgarden.co.uk, www.encryptedtbn0.gstatic.com, www.s3.euwest2.amazonaws.com, www.michigangardener.com, www.gpnmag.com, www.hillierliveblob.blob.core.windows.net, www.123rf.com, www.plantcaretoday.com

(Rojas-Sandoval and Acevedo-Rodríguez, 2015). It is widely cultivated in warm and temperate zones of whole world including Pacific islands. Generally *wallerianas* are found in damp or moist shady locations in forests and forest margins or along river margins. Significant variation in leaf shape and flower colour has been observed. As many as ten species are contained within the *I. walleriana* aggregate as described by Grey-Wilson (1980a).

4.0 Species and Cultivars

4.1.0 Species

Impatiens walleriana is very similar to *Impatiens sodenii* and *I. balsamina*. These species can be distinguished by the vegetative and floral traits (Christenhusz, 2009). A list of important species are described hereunder.

***Impatiens sodenii* (Syn. *Impatiens sultanii; Impatiens holstii; Impatiens petersiana*)**

The plants are larger upto 1.5 m tall with sessile leaves, all in dense whorls. Flowers are white or pale pink. Older plants have woody base with herbaceous branches.

I. cinnabarina

The plants are found in rainforest of Tanzania, in east Africa. Plants are erect, glabrous, perennial with stems arising from a crown which surmounts a fleshy tuberous rootstock, the tubers long-fusiform and 1–1.5 cm in diameter; stems moderately branched or occasionally unbranched, fleshy, up to 70 cm tall with good foliage. The flowers appear almost constantly round the year.

I. hamata

It is spreading, much-branched, glabrous, perennial and grows upto 1.5 m tall; stems sometimes rooting at the lower nodes. It is habitat of wet shaded places in upland rain-forest and at forest margins.

I. messumbaensis

The plants are sparingly branched perennial herb grows upto 1.5 m tall; stems finely rusty-pubescent when young, becoming sub-glabrescent below. This species is native range is Tanzania.

I. pseudohamata

The plants are perennial tall (upto 1.75 m), glabrous; stems erect, sparsely branched, becoming rather woody and leafless below. It is found in moist places rather than open forest.

I. pseudoviola

It is commonly known as the 'Jewel weed' or 'touch me not' native to Kenya and Tanzania. The plant grows between 6-12 cm height. It flowers bright pink during summer and autumn. This perennial species has ovate, alternate toothed leaves. This species prefers semi-shade environment. It can be propagated with

Species of *Impatiens*

Impatiens niamniamensis — *Impatiens munronii* — *Impatiens walleriana*

Impatiens tinctoria — *Impatiens paucidentata* — *Impatiens rosulata*

Impatiens pseudoviola — *Impatiens balfourii* — *Impatiens xanthina*

Impatiens zombensis — *Impatiens glandulifera* — *Impatiens marianae*

Impatiens walongensis — *Impatiens zironiana* — *Impatiens lohitensis*

Source: www.en.wikipedia.org, www.species.wikimedia.org, www.commons.wikimedia.org, www.researchgate.net, www.novataxa.blogspot.com, www.google.com

Species of *Impatiens*

Impatiens paramjitiana

Impatiens pathakiana

Impatiens rugosipetala

Impatiens tatoensis

Impatiens arunachalensis

Impatiens haridasanii

Impatiens nilalohitae

Impatiens pseudocitrina

Impatiens roingensis

Impatiens walongensis

Impatiens scapiflora

Impatiens pallida

Impatiens burtonii

Impatiens tricornis

Impatiens suksathanii

Source: www.flowersofindia.net, www.en.wikipedia.org, www.researchgate.net, www.novataxa.blogspot.com, www.dekochi.com, www.indpaedia.com

Species of *Impatiens*

Impatiens cinnabarina

Impatiens pseudohamata

Impatiens hamata

Impatiens serpens

Impatiens sodenii

Impatiens usambarensis

Impatiens hawkeri

Impatiens balsamina

Source: www.en.wikipedia.org, www.strangewonderfulthings.com, www.plantsrescue.com, www.imago.indiana.edu, www.plantsoftheworldonline.org, www.flowershots.net

stem cuttings during the spring or summer months. However, it is not hardy and is recommended to keep indoors during the winter months.

I. serpens

This species is low creeping perennial; prostrate stem, rooting at the lower nodes with procumbent to erect flowering shoots 10–15 cm. tall and glabrous. It is found in dense upland rain-forest, especially in deep forest, generally amongst moss and low herbs on the forest floor.

I. usambarensis

It is a vigorous perennial species with an upright, bush-like growth habit. The branches and leaf stems are found in attractive shades of red. The plant flowers all the year round. It is a rare species from east Africa with pink flowers and survives in mild climate with cool nights.

I. hawkeri

It is native to Papua New Guinea and the Solomon Islands. It has been bred and hybridized in cultivation to produce a line of garden plants. It was first collected in the Territory of Papua in 1884 and soon became popular as a greenhouse plant. After its discovery, fifteen other similar New Guinea taxa were collected, all of which were later determined to be forms of *I. hawkeri*. The plants of this species are found with a diverse flower and leaf colours. The species has been crossed with *Impatiens aurantiaca* and *I. platypetala* to improve characteristics such as drought resistance.

I. balsamina

This species is native to India and Myanmar region of Indian sub-continent. It is an annual plant growing upto 20–75 cm with thick soft stem. The leaves are spirally-arranged, 2.5–9 cm long and 1–2.5 cm broad, with a deeply toothed margin. The flowers are pink, red, mauve, lilac, or white, and 2.5–5 cm diameter usually solitary in leaf axils. Lateral petals completely united, shallowly incised, unevenly bi-lobed and fruits are hairy. They are pollinated by bees and other insects and also by nectar-feeding birds.

***Impatiens niamniamensis* (Congo cock, Congo cockatoo, parrot impatiens)**

It is a species of flowering plant in the family Balsaminaceae and is widely cultivated. In temperate zones it requires protection from temperatures below 10°C. It needs a sheltered position in a partially shaded spot.

I. munronii

It is mainly distributed in Western Ghats of India.

***I. walleriana* (Busy Lizzie, balsam, sultana, or simply impatiens)**

It is native to eastern Africa from Kenya to Mozambique.

I. paucidentata

It is an epiphytic species of Impatiens native to Uganda and the eastern Democratic Republic of the Congo.

I. balfourii (Kashmir balsam)

It is commonly known as Balfour's touch-me-not, Kashmir balsam and poor man's orchid.

I. glandulifera (Himalayan balsam)

It is a large annual plant native to the Himalayas. Through human introduction it is now present across much of the Northern Hemisphere and is considered an invasive species in many areas. Uprooting or cutting the plants is an effective means of control. Its aggressive seed dispersal, coupled with high nectar production which attracts pollinators, often allow it to outcompete native plants. Himalayan balsam also promotes river bank erosion due to the plant dying back over winter, leaving the bank unprotected from flooding. Invasive Himalayan balsam can also adversely affect indigenous species by attracting pollinators (*e.g.* insects) at the expense of indigenous species. It is considered a "prohibited noxious weed" under the Alberta Weed Control Act 2010.

I. walongensis

This is a new species of balsam. The species was discovered from Anjaw district of Arunachal Pradesh, India. About a meter tall with ovate elliptical leaves and light pink flowers, the plant was named after Walong, the locality where it was found.

I. zironiana

It was discovered from the Lower Subansiri district of Arunachal Pradesh. The plant bears pale yellow lanceolate floral buds. The flowering and fruiting occurs during rainy season from July to September.

4.2.0 Cultivars

Accent: The flowers are large with high-quality blooms. It is ideal for larger retail-ready containers, but also adaptable to production in small pots or packs. The plant excels in climates with short growing seasons, where strong plant vigour makes flower beds vibrant and colourful. It can also withstand cooler growing conditions and other environmental stresses.

Accent™ Red Star: It is slightly more vigorous than Accent™ Premium with unique star pattern colours and solid hues.

Double Impatiens: These varieties are available with flower colours from bright to pastel to stripe - and all have lush green foliage. They mix well with annuals, perennials and shrubs. It is ideal for beds, borders, baskets, containers, window boxes and for combination plantings.

Tempo Series: They are bushy perennials grows upto 25 cm tall, but grown as annuals. The plant have simple, elliptic leaves and flowers 3-5cm wide in shades of purple, mauve, pink, red and orange, as well as picotee or bicolored.

Super Elfin Series: They are mounding annuals with green leaves and single flowers in shades of pink, orange, red and purple. Super Elfin series was bred by Claude Hope in Costa Rica. Hope developed this species from its native wild form into one of the most popular bedding plants in the world.

Cultivars of *Impatiens walleriana*

Accent™ 'Premium Mix' Tempo 'Lavender' Accent™ 'Red Star'

Accent™ 'Deep Pink' Accent™ 'Rose Star' Accent™ 'Star Mix'

Super Elfin 'Blue Pearl' Super Elfin 'Bright Orange' Super Elfin 'Lipstick'

Super Elfin 'Pink' Super Elfin 'Red' Super Elfin 'Salmon'

Source: www.longfellowsgarden.com, www.syngentaflowers-us.com, www.patlis.com

Cultivars of *Impatiens walleriana*

Super Elfin 'Violet'

Super Elfin 'White'

Rockapulco® 'Rose Double Impatiens'

Fiesta Ole Peppermint

Rockapulco® 'Wisteria' (Double Impatiens)

Rockapulco® Coral Reef (Double Impatiens)

Dwarf Baby Scarlet

Silhouette® Red Double Impatiens

Silhouette® 'Purple'

Silhouette® 'Appleblossom'

Sun Harmony 'Orange'

Sun Harmony 'Purple'

Source: www.syngentaflowers-us.com, www.gpnmag.com, www.in.pinterest.com, www.lucasgreenhouses.com, www.flowerseedsuk.co.uk, www.provenwinners.com, www.davesgarden.com, www.longfellowsgarden.com

Fiesta Olé™ Peppermint: The plant produces rose like blooms and does not requires removing of faded flowers or pruning. These plants have large flowers, dark green to purple foliage and a variety of colour choices. They are suitable in shade combination planters, pots, baskets and the landscape.

Dwarf Baby Scarlet: It is a fantastic dwarf variety of Busy Lizzie that grows upto 25 cm. It produces a wonderful display of single red flowers. It is a very popular hanging basket variety as well as an excellent bedding plant.

Silhouette® Purple: The plants have continuous flowering with unique double blooms. It is ideal for shade, containers and in ground plantings.

The Sun Harmony series: These are variety with highly branched, compact and perform well in full sunlight. They are available in four colours *viz.* Magenta, Orange, Purple and Salmon. The Sun Harmony series blooms early and consistently from spring into fall. Bright, sizzling blooms rest against medium green leaves. The plants form tight, attractive mounds with good height (40-50 cm) and equal spread.

The other popular cultivars are: 'Accent Series', 'Blackberry Ice', 'Confection', 'Eclipse', 'Elfin White', 'Extra Dwarf', 'Fiesta Ole', 'Lipstick, Red Star', 'Super Elfin Series', 'Tempo Series', 'Wink and Blink', 'Divine Series', 'Divine Violet', 'Magnum Series', 'Magnum Wild Salmon', 'Sun Harmony Series', 'Sun Harmony Deep Orange', 'Sun Patiens Series', 'Compact Blush Pink', 'Compact Electric Orange', 'Compact Orange', 'Spreading Variegated Salmon', 'Spreading Variegated White', 'Bonfire Orange', 'Celebrate Peach', 'Celebrate Salmonv, 'Celebration Apricot', 'Celebration Bonfire Orange', 'Celebration Light Lavender', 'Celebration Light Salmon', 'Light Salmon', 'Raul', 'Fanfare Orchid', 'Fiesta Double Impatiens', 'Fiesta Ole Peppermint', 'Bounce White', 'Fiesta Ole Purple Stripe', 'Bounce Pink Flame', 'Divine Mix', *etc.*

5.0 Genetics and Breeding

5.1.0 Cytology

The genus *Impatiens* contains over 1000 documented species. *Impatiens* is a large genus with annual or perennial species distributed primarily in the highlands and mountains of the Old-World tropical and subtropical regions. Grey-Wilson (1980a) estimated conservatively the total number of species to be around 850, but it perhaps has more, as many new species are still being described from different regions such as south-western China and Madagascar. Considering all the available chromosomal data, x=7, 8, 9, 10 are the most frequent basic numbers of the family. Zinovea- Stahevitch and Grant (1984) determined that n=8, and 2n=2x=16 for *I. walleriana*. It is suggested that x=8, x=9, or x=10 are all possible candidates of the ancestral basic numbers in Impatiens. Geographic distributions of the most frequent basic numbers show interesting patterns: x=7 and 8 occur in Africa, x=7, 8, 10 in Southern India and Sri Lanka, x=7, 9, 10 in the Himalayas, x=7, 8, 9, 10 in Southeast Asia, and x=10 in northern Asia, Europe and North America (Song *et al.*, 2003).

5.2.0 Pollination

The flowers of *I. walleriana* are hermaphroditic and come in a wide variety of colours including pink, red, orange, white and purple. These flowers are usually visited and pollinated by bees, beetles, butterflies, moths and hummingbirds (Christenhusz, 2009).

Impatiens are protandrous with the anthers generally maturing one to three days prior to the stigma being receptive (Grey-Wilson, 1980a). Though flower size and shapes vary considerably but are always zygomorphic. Stamens are fused and anthers project downward acting as a brush on the pollinator. The fused anther cone blocks access to the stigma and self pollen generally does not come in contact. Anthers are shed as a single unit prior to the stigma becoming receptive. Ovary growth tends to push the anther cone off. Self pollination (self compatibility) does occur in some species *I. walleriana*, and certain species are cleistogamous (Clevenger, 1971); within a species these characters may be genotype specific. Pollinator access to the spur and nectar is gained from below the anthers.

5.3.0 Breeding

Breeding in impatiens began in the late 1940's, with Bill Marchant and Bob Rieman among the first to work on this crop. Much of their germplasm was passed to Claude Hope who did his breeding work in Costa Rica. Claude is widely regarded as the father of modern impatiens, conducting his first breeding work on *Impatiens walleriana* in 1961.

Many *Impatiens* species are available for commercialization as separate products (*e.g.* New Guinea impatiens, *I. hawkeri*) or for use as parents for interspecific hybrids. The challenges of breeding a crop within the Balsaminaceae are with explosive seed dispersal mechanisms. Important traits for continued crop improvement include seed germination, flower colour, earliness, flower patterns, flower size, foliage colour, pack performance, garden performance, double flowers, habit and branching. Insect and disease resistances, sun tolerance, and new flower colours are also important traits for new or improved series.

Van Went (1981) studied aspects of cytoplasmic male sterility in *I. walleriana*. The information available suggests that microspore development in the sterile anther is blocked by a malfunctioning tapetum. In general, cytoplasmic male sterility is widely used by breeders to produce hybrid seed. CMS in impatiens would provide additional assurances that harvested seeds were not contaminated with seeds resulting from self pollination.

Impatiens walleriana found virtually in all colours except blue and yellow. In wild populations both white and purple were not generally observed. Inheritance for flower colour appears to have dominant and recessive relationship. With the availability in variable colours, it is but obvious that some minor genes or modifiers play a substantial role. Clevenger (1971) studied about the biochemistry for flower initiation in *Impatiens wallerana*. Flavonoid pigments of *I. wallerana* are known to include leucocyanidin, leucodelphinidin, cyanidin, malvidin, peonidin, kaempferol and quercitin. It is also reported that there were different flavanol compositions among plants with the same flower. Cyanidin is the basic pigment and disappears with the appearance of other colours.

The 'stardust gene' which confers a star pattern in the middle of the flower is controlled by a single recessive allele. The novel stardust pattern, however, exhibits an exceptionally high degree of stability from flower to flower. This simple inheritance allows the production of *I. walleriana* plants having the stardust trait with a variety of additional desirable characteristics. Other bicolour flower patterns including the 'star' and 'picotee' patterns are thought to be polygenic in nature (Leue, 1999)

Significant general (GCA) and specific (SCA) combining abilities for floral longevity were detected. GCA mean squares were 37 times larger than SCA mean squares, revealing that additive genetic effects play a more important role in the inheritance of floral longevity in impatiens. This information, coupled with the significant amount of variation for floral longevity among inbreds, indicates that there is good potential for breeding for floral longevity in impatiens to improve the floral display of hybrids (Howard *et al.*, 2012).

5.4.0 Polyploidy

Treating *I. walleriana* with oryzalin can produce tetraploid plants. Compared to their diploid counterparts, the tetraploid plants have significant differences in their vegetative and reproductive characteristics. Changes in the form and colour of the flowers can be attributed to the changes in the size of the conical cells on the surface of the epidermis, along with the increase in the total anthocyanin content in petals. *I. walleriana* tetraploid plants had a larger stomatal size with lower densities and higher number of chloroplasts in their guard cells in comparison with diploid plants (Mohammad *et al.*, 2019).

5.5.0 Interspecific Hybridization

The ploidy levels of related African and Asian species can be used for interspecific hybridization with *walleriana*. Within the genus *Imaptiens*, the ability for interspecific crossing varies widely among the various species. Members of the New Guinea group appear to easily make interspecific among related taxa, but within the *I. walleriana* group, interspecific crossing has been achieved. Grey-Wilson (1980) indicated that *I. walleriana* and *I usambrensis* appear to be evolutionary isolated from other species based on pollen morphology and they share the floral morphology with species such as *I. hamata, I. walleriana* and *I. usambrensis* are known to have naturally hybridised in the wild (Grey-Wilson 1980). Beck *et al.* (1974) attempted crossing *I. walleriana* with several New Guinea and Java *impatiens* species, including those with a chromosome number 2n=2x=16. It is also reported that *I. walleriana* is never used as a female parent for *I. cinnabarina*.

Arisumi (1980) attempted interspecific crosses between *I. walleriana* and a number of species. Embryo development is limited and ovule culture is recommended for attempted recovery of a number of interspecific crosses. *I. walleriana* crossed with *I. auricoma, I. flaccid* and *I. hookeriana* all developed to the globular stage. Cross between *I. walleriana* and *I. thomassetii* as being successful without embryo rescue. Gager (1987) reported inter-specific incompatibility when inter-specific crosses were made between *I. walleriana, I. auricoma, I. repens* and

I. capensis. Hybrids between *I. walleriana* and *I. auricoma* have been commercialized (Parker, 1999 and Anonymous 1998). Commercially *Impatiens species* are available as separate products (*e.g*. New Guinea impatiens, *I. hawkeri;* bush balsam, *I. balsamina*) or for use as parents for inter-specific hybrids.

6.0 Growth and Flowering

Walleriana is a day neutral plant and therefore it flowers regardless of day length. The flowers come in a wide variety of colours (i.e. pink, red, orange, white, purple, etc.), but pink is most common. It is basically annual and occasionally perennial herbaceous plant. The succulent perennial herbs grow between 15-60 cm tall, mounding with lower leaves touching the ground. The plant tolerates full sun, but wilts during intense heat (https://plants.ces.ncsu.edu).

6.1.0 Temperature and Photoperiod

Mattson and Erwin, (2003) quantified the impact of temperature regimes on flower initiation and development rate in Impatiens. The seedlings of each cultivar were placed in four controlled environment chambers maintained at constant temperature viz.12, 16, 20, 24°C with 18 hr photoperiod (inductive; 350 µmol/m2/s). The sensitivity of days to first open flower (DTF) to average daily temperature (ADT) was most obvious when assessing the impact of decreasing temperature. The decrease in 1°C temperature in *Impatiens* cultivar 'Super Elfin Lipstick' increased 3 days more to DTF. The DTF decreased from 72±2 to 47±1 days for *Impatiens* 'Super Elfin Lipstick' as temperature increased from 16 to 24°C. The plants grown at 12°C did not survive. The leaf number below the first open flower (LN) decreased for all plants as temperature increased.

Roberto and Erik (2008) quantified the influence of the mean daily light integral (DLI) on growth and development in Impatiens. A mean DLI of 2 and 8 mol/m^2/d are predicted to flower in 83 and 69 days, respectively, if subsequently forced at 20°C. In New Guinea Impatiens 'Harmony White', root and shoot dry weight increased by 1038 % and 82 %, respectively, and subsequent time to flower decreased from 85 to 70 days as the propagation DLI increased from 1.2 to 10.7 mol/m^2/d.

6.2.0 Growth Regulators

Schroeter-Zakrzewska *et al.* (2007) observed the effect of gibberellic acid on growth and flowering of *Impatiens walleriana*. Young plants of the *Impatiens walleriana* 'Spellbound Lilac' and 'Spellbound Pink Imtrarepu' were sprayed with GA_3 at the concentrations of 0, 10, 30, 50, 100 and 150 mg/cubic dm. Measurements of the plants length of main and lateral shoots and the number of buds and flowers were carried out two weeks after application of gibberellic acid. Gibberellic acid forced plant development and resulted in longer main and lateral shoots with abundant flowering. Number of buds and flowers were also increased by 57.4-87.8 percent in comparison to control plants. However, it resulted in adverse effect on the appearance of plants and low chlorophyll content in leaves.

Safari *et al.* (2021) found that exogenous application of 2 mM Salicylic Acid increased growth and physiological indices in drought-stressed impatiens in mitigating the detrimental effects of water deficiency.

Under adverse weather conditions, use of plant growth regulators may tone-up the crop. Impatiens is responsive to B-Nine (daminozide), Bonzi (paclobutrazol) or Sumagic (uniconazol). However, proper application of water, light and fertilizer is needed to keep the growth and flowering under control.

7.0 Propagation

7.1.0 Seed Propagation

Impatiens walleriana is mainly grown from seed. The seeds can be sown in shallow trays of suitable media mixture during early spring. The pots should be kept in partial shade maintaining a temperature of 18-21°C. When plantlets are grown about 4 cm, they are ready for transplanting in beds or pots. They can also be grown in plug trays. The seedlings do not normally suffer from transplant shock, so they can be held for longer periods before transplanting. Plugs need to be transplanted before the seedlings begin to crowd each other and stretch. Depending on growing conditions, crowding usually begins about 6-8 weeks after sowing in a plug tray. The larger the plug size when the seeds are sown, the longer the seedlings can be held before transplanting.

7.2.0 Cuttings

I. walleriana can also however be propagated by tip cuttings. The cuttings, 5-8 cm are rooted in suitable rooting media during summer. Upon rooting they can be transferred in suitable place and grown as mature plant.

7.3.0 Micropropagation

Subotic *et al.* (2008) developed an efficient protocol for *in vitro* plant regeneration of *Impatiens walleriana* L. using urea-type cytokinins, forchlorofenuron (CPPU) and thidiazuron (TDZ) on the axillary shoot induction. The shoot meristem explants produced multiple axillary shoots when cultured on Murashige and Skoog's (MS) medium containing different concentration of CPPU and TDZ (0.01-1.0 µM). Explant activity varied significantly depending on the cytokinin type as well as concentration and found that the 0.1 µM CPPU induced the largest number of axillary shoots (4.89). The cytokinin was more effective for the induction and development of axillary shoots than TDZ in the same concentration. The separated shoots are rooted and elongated on MS medium without hormone. Shoot meristem explants and urea-type cytokinins are useful for improvement of regeneration procedure of this important ornamental species.

8.0 Soil and Climate

8.1.0 Soil

The plants can be grown on a wide range of soil texture from heavy to light and soil types such as clay, sandy and loamy soils with pH in the range 5.6-7.5. It performs best in organically rich, well drained but moist soil. The combined total EC of water and fertilizer should not exceed 1 to avoid leaf malformations. If media EC is >1.25, shoot tip abortion may occur. In problematic soil, suitable amendments with cocopeat and other mix can be used.

Propagation of *Impatiens walleriana*

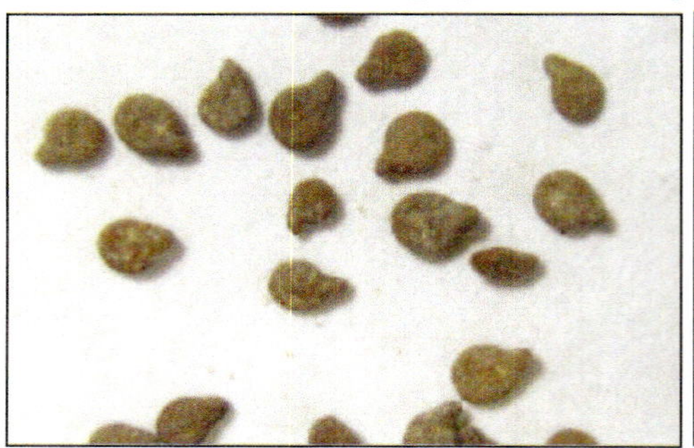

Seeds

Germination in Pot

Seedlings in Plugtray

Seedlings for Transplantation

In Open Field

In Greenhouse

By Cuttings

Propagation of *Impatiens walleriana*

In vitro Propagation of Virus-Free *I. walleriana*

(a) shoot formation on MS media after 7 days, (b) multiplication of shoots on MS supplemented with 0.01 μM CPPU, (c) development of axillary shoot on MS with 1 μM TDZ, (d) axillary shoots on MS with 0.1 μM CPPU, (e) shoot elongation on MS cytokinins-free medium, (f) plants in acclimatization stage (Milosevic *et al.,* 2012)

Source: www.mountfolly.co.uk, www.st3.depositphotos.com, www.previews.123rf.com, www.healthbenefitstimes.com, www.shutterstock.com, www.c8.alamy.com, www.budscape.net, www.tandfonline.com

8.2.0 Climate

8.2.1 Temperature

Impatiens walleriana is adapted to grow in a wide range of environmental conditions, tolerating mean annual temperatures ranging from 10-28°C and mean annual precipitations ranging from 800 mm to >2500 mm. The optimum temperature required for germination is about 20-25°C (Christenhusz, 2009). Temperatures exceeding 25°C may induce thermodormancy, which inhibits germination of Impatiens. Temperature below 21°C decreases the speed and uniformity of germination. Also, the temperatures below 18°C may cause tip abortion and leaf malformation. For optimum growth and development day temperature of 21-24°C and night 16-19°C is essential.

8.2.2 Light

Light plays an important role in the seed germination and radicle emergence in impatiens. Providing a light source of 10-100 foot candles (100-1,000 lux) improves germination and reduces lankiness of seedlings when grown in a chamber. Supplemental lighting at 350-450 foot candles (3,500-4,500 lux) for two weeks produces vigorous, strong-stemmed seedlings in regions where light levels are low. Lighting for more than two weeks during early plug stages can result in bleaching or yellowing of leaves (photo-oxidation). Providing a light shade during growing period is essential.

Carpenter *et al.* (1994) found that seeds having a photodormancy require light for only part of the germination period. Germination ≥85 per cent was achieved after 3, 2, or 1 day of irradiance at 1.5, 15, or 75 µmol/s/m, respectively. Keeping imbibed seeds in darkness for ≥2 days before light exposure caused reduced total germination percentages (G), delayed achieving 50 per cent of the final germination percentage (T_{50}) and increased the days between 10 per cent and 90 per cent germination ($T_{90} - T_{10}$). Light for 6 hours daily at 1.5, 15, or 150 µmol/s/m^2 promoted high G and rapid and uniform germination, but daily 12 to 24 hours of irradiance decreased G and increased T_{50} and $T_{90} - T_{10}$. Estimated rates of decline (increase) in G, T_{50}, or $T_{90} - T_{10}$ with each added day of light (darkness) or increasing daily hours of light were measured by fitting regression equations. Impatiens seed germination was promoted by the initial 1-3 days of light, but light inhibited radicle extension in the latter germination stages. It is recommended that commercial impatiens producers light the seeds at 15 or 75 µmol/s/m for 2 or 1 day, respectively, initially after moistening the germination substrate. Then place the seeds in darkness until emergence of the epicotyls.

9.0 Cultivation

9.1.0 Growth Habit

Impatiens walleriana is commonly cultivated as an ornamental in gardens and greenhouses. It can also be found naturalized in clearings, roadsides, secondary forests, coastal forests, riverine thickets, river margins, wet grasslands, coffee and banana plantations, forest edges, humid places in the understory of tropical and subtropical forests, *etc.*

9.2.0 Spacing

The plants are generally planted at a distance of 20-30 cm. Increasing the planting distance makes them to grow bushy and decreasing it makes them to grow tall. Overcrowded plants do not produce good blooms. For pot culture, a single plant in pot is recommended. For containers on balconies or terraces, the pot should be large enough to give the plant plenty of space.

9.3.0 Manuring and Fertilization

Impatiens requires moderate fertilization. Application of Farm Yard Manure (FYM) is beneficial for plant growth and development. Plants in containers should be fertilized approximately every 2 weeks with a water-soluble fertilizer. In flower beds, incorporating a slow-release fertilizer into the soil prior to planting should be done. Initially nitrogen 25-50 ppm with a calcium-based fertilizer (14-0-14) should be given. As the plant grows slowly increase the dose to 75-100 ppm nitrogen every 2-3 watering with a calcium-based fertilizer (13-2-13 or 14-4-14). Minimal fertilizer is needed to keep the plants compact and promote flowering. Tall, lush-green plants with flowers beneath the foliage, or late flowering indicates excessive fertilizer application.

9.4.0 Irrigation

Impatiens need moderate watering. Plants growing in garden beds usually need to be watered once a week during dry weather. Plants growing in containers or hanging baskets should be watered when the soil surface is dry. Watering should be adjusted as per weather condition. The water schedule is more during hot, dry summer months and less during rainfall. It is also not recommended to overwater the beds and also not allow the pot to stand in water. If the amount of light reaching the plants is reduced, the plant may enter a rest period and they should be watered more sparingly. The top third of the potting mixture should be dry in between watering. But the pot mix should not be dried out completely.

10.0 Diseases and Pests

10.1.0 Diseases

10.1.1 Alternaria Leaf Spot (*Alternaria* sp.)

Brown spots and rings are form on the infected leaves. It is caused by high humidity and poor air circulation in growing environment. Though there are no registered fungicides for this crop. But application of Iprodione and Chlorothalonil may give good control.

10.1.2 Botrytis Blight (*Botrytis cinerea*)

Botrytis is a problem under cool and humid conditions. It usually appears as a gray mold on the soil or leaf surface. The flowers are spotted and sometime rotting of stems took place. Botrytis is especially prevalent when plants need to be held before sale and flowers drop onto the foliage, providing a source of inoculums. It can be managed by proper plant spacing and maintaining good air circulation. Spraying of Tebuconazole or mancozeb @ 1-2 mg/l may give control.

10.1.3 Bacterial Fasciation (*Corynebacterium fascians*)

The infected plant becomes stunted and many short shoots appears at the crown. The affected plant should be immediately discarded.

10.1.4 Damping-off (*Pythium* or *Rhizoctonia*)

It is severely affecting the emerging seedlings. The growing stems at the soil line rot and the plants collapse. *Rhizoctonia* and *pythium* are primarily responsible for damping-off. In matured plants, the roots turn black, causing stunted plants that will eventually die. It can be controlled by using sterilized media for seed sowing. Drenching with suitable fungicide such as Captan 2-3 mg/l is beneficial.

10.1.5 Pythium Root Rot (*Pythium* spp.)

The lower leaves of the affected plant wilts and fall down. In case of severe infection the plant will die. The growing media should be sterilized before planting to minimize the incidence. Drenching with aliette may reduce the incidence.

10.1.6 Thielaviopsis Root Rot (*Thielaviopsis basicola*)

The roots of the infected plant become dark brown and rot. Sunken lesions may appear on the lower stems during infection. The growing media should be sterilized before planting to minimize the incidence.

10.1.7 Verticillium Wilt (*Verticillium dahliae*)

The lower leaves become yellow and falls down. In case of less severity the infected plants may recover. Plants should be grown in a sterilized media to prevent the infection. It is also not advisable to take cuttings from infected plants.

10.1.8 Impatiens Necrotic Spot Virus

Ring spots appear on the leaves and causes death of growing tips. In severe case it causes stunting. Discarding the infected plants and control of thrips population is the only way to manage it.

10.1.9 Downy Mildew (*Plasmopara obducens*)

The disease first occurs on leaves near the tips of branches. The top of leaves develop an irregular yellow-green or purple discolouration. The affected leaves often curl downwards. Infected plants are often stunted. Leaves and flowers eventually drop off, leaving a bare stem. The pathogen sporulates profusely in wet and cool conditions producing a fuzzy white mass on the under sides of leaves, stems and flower buds. Sporulations on undersides of leaves and buds can lead to severe defoliation. Application of ridomil or subdue is beneficial.

Diseases and Pests of *Impatiens walleriana*

Plants Infected with TSWV – (a) necrotic spots, (b) necrotic flecks, line pattern and deformation of leaf, (c) necrosis of petals and sepals, (d) necrosis of stem.

Downy Mildew  **Alternaria Leaf Spot**  **Powdery Mildew**

Leaf Spot  **Botrytis Blight**  **Damping Off**

Tobacco Ringspot Virus  **Impatiens Necrotic Spot Virus**  **Cucumber Mosaic Virus**

Western Flower Thrips **Caterpillar** **Snails**

***Source*: www.researchgate.net, www.forestryimages.org, www.pdfs.semanticscholar.org, www.alamy.com, www.c8.alamy.com, www.biocontrol.ucr.edu, www.pnwhandbooks.org, www.ag.umass.edu, www.extension.psu.edu, www.mcdonaldgardencenter.com, www.researchgate.net**

Diseases and Pests of *Impatiens walleriana*

Aphids

Cucumber Beetle

Moth

Damage by Broad Mites

Spider Mites

Whitefly

Source: www.missouribotanicalgarden.org, www.greenhouses.com, www.candidegardening.com, www.hunker.com, www.theplantaide.com

10.2.0 Pests

10.2.1 Aphids

Melon aphids are one of the most common insect problems on impatiens. They are small, soft bodied insects, usually green and feeds on plant sap. The insects are usually found at the growing tips of the plant or on the undersides of the leaves. Plant symptoms include distorted or stunted growth. The most obvious sign of aphids is the sticky honeydew they excrete on the leaves that can provide a medium for black sooty mold. There are no label claim inseciticides for this crop. However, spraying of neem oil or imidocloprid or dimethoate give some control

10.2.2 Western Flower Thrips

They are difficult to see because they are extremely small and hide within the flowers or the growing tips of the plants. To determine the presence of thrips, shake the plant over a white sheet of paper. The dark-bodied insects can then be easily seen. Thrips cause distorted growth of both the flowers and leaves. It can be controlled with the spray of horticultural oils or pyrethrins or insecticidal soaps to the underside of leaves.

10.2.3 Spider Mites

Two-spotted spider mite occasionally infests impatiens. Spider mites are normally found on the undersides of the leaves, and their activity and growth increase during warmer temperatures. The first signs of infestation are distorted leaves with a slightly mottled appearance. In advanced stages of infestation, the spider mites will produce webbing over the top of the plants. Once the webbing is visible, plant damage is usually extensive and the problem is very difficult to control. In the initial stage of infestation spraying with water or insecticidal soap is recommended. In severe case neem oil or acaricides is beneficial.

10.2.4 Mealy Bugs

They are an occasional problem on impatiens. Mealy bugs are usually found in leaf axils, appearing as a white, cottony mass. These insects damage the plant by sucking plant juices and producing honeydew. Spray insecticidal soap directly onto visible mealy bugs for control.

Other pests like Broad mite, Cyclamen mite, Serpentine leafminer, Root knot nematode, Black vine weevil, Greenhouse whitefly also infest and damage the crop.

11.0 Harvesting and Postharvest Management

Impatiens is normally sold when in flower. The pots are usually watered before shipping to prevent drying out in transit.

In potted impatiens ethylene-induced petal and flower abscission is a significant problem during commercial transportation. Abscised flowers reduce visual impact of the potted plant and can increase the incidence of Botrytis and other saprophytic pathogens. Burana *et al.* (2013) studied the effect of 1-methylcyclopropene (1-MCP), modified atmosphere packaging (MAP) and their combination on the quality of potted impatiens after simulated transportation at 23°C for 48 h. The results revealed that carbon dioxide (CO_2) concentration increased while oxygen (O_2) decreased inside the MAP packages during treatments. Ethylene concentration was highest in those treated with MAP (0.06 µl/l) while lowest concentration was found in MAP combined with 1 µl/l of 1-MCP treatment (0.02 µl/l). It delayed the flower abscission and increased plant and floret longevity of potted impatiens cultivars 'Rouge' and 'Peach'. The results indicated that MAP combined with 1-MCP treatment during transportation can provide a feasible technique for extending the display life maintaining the quality and aesthetic value of potted impatiens.

Exogenous application of ABA before shipping had no effect on the postharvest performance of *Impatiens walleriana* as reported by Blanchard *et al.* (2007). Root dips of synthetic abscisic acid (s-ABA) have been shown to reduce the transpiration rate and subsequently prolong postharvest longevity in a selected group of herbaceous ornamental crops such as Impatiens. Drench the root substrate with synthesis ABA at the rate 250mg/l to prolong post production performance in low light conditions (Gibson and Crowley, 2006).

The frost and cool temperatures affect the marketing period of impatiens in temperate areas. The plants are usually marketed from a greenhouse-type structure. However, if the plants are kept outdoors, the retailer should monitor night temperatures to prevent frost or chilling damage. The plants should be moved to a warm, sheltered area during cool night temperature.

11.1.0 Seed Harvesting

The pods of *Impatiens walleriana* burst at the slightest touch. Therefore, the pods are collected individually when turned brown. Nylon pantyhose are used to harvest the pods. These are placed over the seed pods when it turns from green to brown. They are ideal as they trap the seeds and also maintain airflow around them. The pods should be properly dried before storage. The seeds in air-tight container can be stored upto 2 years if kept in a cool, dry and dark location.

12.0 Biotechnology

12.1.0 Molecular Markers

Impatiens is the leading bedding plant in the USA and several other countries. Carr and Korban (2004) evaluated the genetic variability present in this crop using amplified fragment length polymorphism technology to determine the level of polymorphism present across 20 commercial cultivars of seed impatiens and to investigate their genetic relatedness. A total of eight EcoRI and MseI primer combinations were used for polymerase chain reaction amplification. Fluorescence-labelled amplification products were subjected to electrophoresis and then analysed using an automated sequencer. High levels of polymorphism were detected among all 20 cultivars for all primer combinations tested. Gower's Genetic Dissimilarity estimates for the entire set of cultivars ranged from 1.000 to 0.316. The dendogram generated from these dissimilarity data revealed a number of groupings, including one major division which placed all of the cultivars into one of two groups.

Aquaporins comprise a large group of transmembrane proteins responsible for water transport, which are crucial for plant survival under stress conditions. During a study on molecular characterization of aquaporins in *I. walleriana* and their expression during drought stress and recovery four *I. walleriana* aquaporins were identified *viz.* IwPIP1;4, IwPIP2;2, IwPIP2;7 and IwTIP4;1. All of them had conserved NPA motifs (Asparagine-Proline-Alanine), transmembrane helices (TMh), pore characteristics, stereochemical properties and tetrameric structure of holoprotein. Drought stress and recovery treatment affected the aquaporins expression in *I. walleriana* leaves, which was up- or downregulated depending on stress intensity. Expression of *IwPIP2;7* was the most affected of all analyzed *I. walleriana* aquaporins. At 15 per cent and 5 per cent soil moisture and recovery from 15 per cent and 5 per cent soil moisture, *IwPIP2;7* expression significantly decreased and increased, respectively. Aquaporins *IwPIP1;4* and *IwTIP4;1* had lower expression in comparison to *IwPIP2;7*, with moderate expression changes in response to drought and recovery, while *IwPIP2;2* expression was of significance only in recovered plants. Thus, the molecular structure of *I. walleriana* aquaporins expanded knowledge about plant aquaporins, while its expression during drought and recovery contributed to *I. walleriana* drought tolerance mechanisms and re-acclimation (Marija *et al.*, 2021).

Impatiens downy mildew is a highly destructive disease of *Impatiens walleriana*. The disease is caused by a recently emerged pathogen *Plasmopara obducens*. Since both the host and pathogen are relatively less studied, there are only a few genomic resources available for both. Transcriptional changes in *I. walleriana* was analysed in response to *P. obducens* infection during different stages of disease development to identify candidate genes that may be involved in *I. walleriana* susceptibility to *P. obducens*. Since the genome of *I. walleriana* is not available publicly, a *de novo* transcriptome assembly consisting of 73,022 transcripts was constructed and optimized. Differential expression analysis based on this optimized *de novo* transcriptome assembly revealed 3,000 to 4,500 differentially expressed transcripts (DETs) at 0 hr, 12 hr, 48 hr, 120 hr, and 240 hr time points after infection. Functional annotation of these DETs revealed that numerous plant stress responsive genes are activated and deactivated throughout the infection cycle. Genes in the calcium signaling pathways, receptor-like kinases (RLKs) including 10 disease resistance associated RLK transcripts, powdery mildew resistance genes (*MLO*), and many other plant stress related genes were predominantly differentially expressed in *I. walleriana* in response to *P. obducens*. It has provided molecular insights into the disease susceptibility mechanism of the Impatiens downy mildew, and resistance breeding in *I. walleriana* (Suarez *et al.*, 2019).

12.2.0 Genetic Transformation

Dan *et al.* (2010) devised a simple and efficient *Agrobacterium*-mediated transformation system for *Impatiens*. They successfully conducted stable integration of T-DNA and of *Agrobacterium*-mediated transformation method for plants using *in vitro* maintained multiple bud cultures as explants. This transformation system has the advantages of 1) efficient, simple and rapid regeneration and transformation (with no need for sterilization or a greenhouse to grow stock plants), 2) flexibility (available all the time) for *in vitro* manipulation, 3) uniform and desirable green tissue explants for both nuclear and plastid transformation using *Agrobacterium*-mediated and biolistics methods, 4) no somaclonal variation and 5) resolution of necrosis of *Agrobacterium*-inoculated tissues.

Impatiens walleriana shoots were inoculated with *Agrobacterium rhizogenes* A4M70GUS and the effects of genetic transformation on the catalase (CAT), superoxide dismutase (SOD) and peroxidase (POX) activities in wounded region of stems and unwounded leaves were evaluated 10, 24, 240 and 720 hours after inoculation. Following *Agrobacterium* infection activities of plant antioxidative enzymes changed in a time dependent manner indicating that dynamic processes occurred during plant. *Agrobacterium* interaction, plant cell transformation and formation of hairy roots. Appearance of hairy roots on wound sites of shoots was observed ten days after inoculation with *A. rhizogenes* and the root induction frequency was 100 per cent. Among selected hairy root lines significant differences in growth rate and biomass production were observed and an average 3-fold increase in biomass production was observed for the best growing hairy root line compared with the untransformed roots. PCR analysis showed presence of uidA, rolB, rolC and rolD genes in all analyzed *I. walleriana* L. hairy root lines, while amplification fragment of rolA gene was detected in 83.3 per cent transformed lines. The efficient transformation protocol for *I. walleriana* offered possibilities to generate hairy root cultures for *in vitro* propagation of plant viruses (Miloševic et al., 2015).

13.0 References

Anonymous (1998) *American Nurseryman*, May, pp.18.

Arisumi, T. (1980) *J. Amer. Soc. Hort. Sci.*, **105**(5): 629–31.

Beck, A.R., Weigle, J.L. and Kruger, E.W. (1974) *Can. J. Bot.*, **52**: 923-925.

Blanchard, M.G., Newton, L.A., Runkle, E.S., Woolard, D. and Campbell, C.A. (2007) *Acta Hortic.*, **755**: 127-133.

Burana, C., Yamane, K. and Yamaki, Y. (2013) *Acta Horticulturae*, **970**: 223-228. DOI: 10.17660/ActaHortic.2013.970.25

Carpenter, W.J., Ostmark, E.R. and Cornell, J.A. (1994) *Hort Science*, **29**: 854–857.

Carr, J. and Korban, S.S. (2004) *Plant Breeding*, **123**(6): 577-581.

Christenhusz, M.J.M. (2009) *Balsaminaceae*. In: Flora Mesoamericana (ed. by Davidse G, Sousa Sánchez M, Knapp S, Chiang Cabrera F). St Louis, Missouri, USA: Missouri Botanical Garden.

Clevenger, S. (1971) *Evolution*, 25, **4**: 669-77. doi: 10.2307/2406947.

Dan, Y., Baxter, A., Zhang, S., Pantazis, C.J. and Veilleux, R.E. (2010) *BMC Plant Biol.*, **10**: 165.

Gager, F.C. (1987) *Graduate Thesis*, The University of Connecticut.

Gibson, J.L. and Crowley, S. (2006) *Hort Science*, **40**: 511–512.

Grey-Wilson C. (1980) *Impatiens of Africa*. A.A. Balkema., Rotterdam, Netherlands.

Howard, N.P., Stimar, D., Leon, N.De., Havey, M.J. and Martin, W. (2012) *J. Amer. Soc. Hort. Sci.*, **137**(1): 47-50.

http: / /www.flowerseedsuk.co.uk

http: / /www.gpnmag.com

http: / /www.plantsoftheworldonline.org

http: / /www.plantsrescue.com / impatiens-walleriana /

http: / /www.rhs.org.uk

http: / /www.shootgardening.co.uk

http: / /www.syngentaflowers-us.com

http: / /www.vitegreenhouses.com

https: / /www.cabi.org / isc / abstract

https: / /www.en.wikipedia.org

https: / /www.plants.ces.ncsu.edu / plants / impatiens-walleriana /

https: / /www.plants.ces.ncsu.edu

https: / /www.researchgate.net

Leue, E. (1999) *Bicolor Impatens*, US Patent No. 5986188.

Marija, J.Đ., Subotiæ, A.R., Prokiæ, L.T., Trifunoviæ-Momèilov, M.M., Cingel, A.D., Dragiæeviæ, M.B., Simonoviæ, A.D. and Miloševiæ, S.M. (2021) *Plants*, **10**(1): 154.

Mattson, N.S. and Erwin, J.E. (2003) *Acta Hortic.*, **624**: 191-197.

Milosevic, S., Cingel, A., Jevremovie, S., Stankovie, I., Bulajie, A., Krstie, B. and Subotic, A. (2012) *Pestie. Phytomed. (Belgrade)*, **27**(3): 203-211.

Miloševiæ, S., Lojiæ, M., Antoniæ, D., Cingel, A. and Subotiæ, A. (2015) *Genetika*, **47** (1): 71-84.

Mohammad, A.G., Jowkar, A., Salehi, H. and Zarei, M. (2019) *Plant Cell, Tissue and Organ Culture*, **138**: 299–310.

Parker, R. (1999) *Impatiens plant named '96-009-7'*. United States Plant Patent 10972.

Roberto, G.L. and Erik, S.R. (2008) *Hort Science*, **43**(7): 2052–2059.

Rojas-Sandoval, J. and Acevedo-Rodríguez, P. (2015) *Biological Invasions*, **17**(1): 149-163. http: / /rd.springer.com / article / 10.1007 / s10530-014-0712-3 / fulltext.html doi: 10.1007 / s10530-014-0712-3

Safari, M., Fard, S.M., Nejad, A.R., Sorkheh, K. and Sofo, A. (2021) *International Journal of Environmental Science and Technology*, DOI: 10.1007 / s13762-020-03092-2

Schroeter-Zakrzewska, A.; Janowska, B., Akademia, R.P. and Ozdobnych, K.R. (2007) *Roczniki Akademii Rolniczej w Poznaniu. Ogrodnictwo* (Poland) 41383: 195- 200.

Song, Y., Yuan, Y.M. and Küpfer, P. (2003) *Caryologia*, **56**(4): 463–481.

Suarez, S., Naveed, Z.A. and Ali, G.S. (2019) *bioRxiv*, 622480. doi: https: //doi.org/10.1101/622480

Subotiæ, A., Jevremoviæ, S., Cingel, A. and Miloševiæ, S. (2008) *Biotechnology and Biotechnological Equipment*, **22**(3): 817-819. DOI: 10.1080/13102818.2008.10817559

Uchneat, M.S. (2007) *Impatiens*. In: Anderson N.O. (eds) Flower Breeding and Genetics. Springer, Dordrecht. https: //doi.org/10.1007/978-1-4020-4428-1_10

Van-Went, J.L. (1981) *Acta Soc Bot Pol.*, **50**: 249-252.

Zinov'eva-Stahevitch, A.E. and Grant, W.F. (1984) *Canadian Journal of Botany*, **62**(12): 2630–2635. https: //doi.org/10.1139/b84-357

Impatiens balsamina

T. N. Saha, P. Naveen Kumar, Swati Saha,
Ganesh B. Kadam, D.V.S. Raju and K.V. Prasad

1.0 Introduction

Impatiens balsamina, commonly known as balsam is native to India and Myanmar. It is known by several names such as garden balsam, rose balsam, touch-me-not or spotted snap weed, jewel weed, *etc.* In India it is commonly cultivated as garden plant throughout tropical and sub-tropical parts. It is a short duration, semi hardy, erect, compact, annual herb which is one of the widely grown rainy season flowering plant in India (Singh and Singh, 2007). It is often grown as garden plant due to its attractive foliage and colourful flowers. The stem is glabrous or weakly pubescent when young, succulent and sparsely branched with swollen nodes. The flowers are used to prepare a red dye and used as a substitute for Henna (*Lawsonia inermis*) in cosmetics. The seeds of garden balsam are edible whereas the seed oil can be used for burning lamps and in the surface-coating industry (Chen *et al.*, 2010).

2.0 Importance and Uses

2.1.0 Aesthetic Uses

Balsam is ideal for both indoor and outdoor cultivation. For indoor cultivation it can be grown in pots as well as in hanging baskets. The seeds of different dwarf genotypes with varying colours can be planted in designer pots as well as in hanging baskets for beautification of balcony, dining table or study table. In landscape design it is widely used in beds, mixed borders, along walks, rock garden, *etc.* As mixed borders and in beds with different colours, it gives a beautiful look soothing to the eyes. In South India, balsam loose flowers (either loose or made in to garlands) are used for worship in temples.

2.2.0 Therapeutic and Other Uses

The medicinal uses of *I. balsamina* are very well mentioned in ancient books. Whole plant is having emetic, diuretic and laxative properties when taken internally, and externally it is used for relieving pains in joints (Jaykrishna, 1998). Traditionally it is used as diuretic, emetic, laxative, cathartic, demulcent, tonic and is reported to possess beneficial effects in lumbago, intercostal neuralgia and in burns. Studies have revealed presence of various phytoconstituents like flavanoids, triterpenoids, glycosides, fatty acids and alkaloids. Sanghi and Tiwle (2013) described balsam as an useful cosmetic herb. The juice extracted from leaf is used in treatment against warts. The crushed leaves are also used as a poultice on wounds and skin inflammations. The mucilaginous flowers are useful in inducing a cooling effect when applied to burns and scalds. Ethnomedicinally the flowers are utilized for the treatment of snake bites (Binorkar and Jani, 2012) or who got ingested with poisonous fish. The flowers, and their alcoholic extract, possess marked antibiotic activity against some pathogenic fungi and bacteria. The seed is expectorant and has been used in the treatment of cancer. The powdered seeds are given to women during labour in order to provide strength. A decoction of the roots is drunk to treat irregular menses. The roots are sometimes crushed and used to poultice wounds, inflammations of the skin and torn nails. Red flowers are most commonly used to obtain natural dye.

Uses of Balsam

In Pot | In Terrace | In Container Gardening

Mixed Colour Décor | Loose Flower for Rangoli | In Temple Offering

In Flower Bed | Along Garden Path | Along Roads as Borders

In As Landscaping | In Home Garden | In Home Garden

Source: www.alamy.com, www.scottarboretum.org, www.media.istockphoto.com, www.growsonyou.com, www.google.com, www.fittoniamania.com

3.0 Origin and Distribution

Impatiens balsamina is a native of Indian sub-continent spreading from Western Ghats in India to Myanmar in South-East Asia. It has been widely introduced to Europe, Asia, Americas, Oceania and some parts of tropical Africa as an ornamental plant. It is a genus with more than 1,000 species of flowering plants, widely distributed throughout the Northern Hemisphere and the tropics. They are generally found at elevation from close to sea level upto 3500 m. Two species namely *Impatiens turrialbana* and *Impatiens mexicana* occur in isolated areas in Central America (southern Mexico and Costa Rica). In Europe, only a single *Impatiens* species (*Impatiens noli-tangere*) occurs naturally. However, several neophytic species exist.

During 19th and 20th centuries, the voyagers introduced the North American orange jewelweed (*I. capensis*) to England, France, the Netherlands, Poland, Sweden, Finland and other areas of Northern and Central Europe. It was not recorded from Germany until 1996, but since then a population has naturalized in Hagen at the Ennepe River. The orange jewelweed is quite similar to the touch-me-not balsam (*I. noli-tangere*), the only *Impatiens* species native to Central and Northern Europe, and it utilizes similar habitats, but no evidence exists of natural hybrids between them. Small balsam (*I. parviflora*), originally native to southern Central Asia, is even more extensively naturalized in Europe. More problematic is the Himalayan balsam (*I. glandulifera*), a densely growing species which displaces smaller plants by devoiding them of sunlight. It is now an invasive weed in many places, and tends to dominate riparian vegetation along polluted rivers and nitrogen-rich spots. Thus, it exacerbates ecosystem degradation by forming stands where few other plants can grow and by rendering riverbanks more prone to erosion.

4.0 Taxonomy

Balsaminaceae is a family of dicotyledonous plants, comprising only two genera (*Impatiens* and *Hydrocera*) and 1001 species of fleshy herbs with more or less translucent stems, swollen nodes and usually spiral, toothed leaves. The growth habit is erect and branching to 75 cm in height with equal spread. The plants have broadly lanceolate leaves 3-15 cm long, the apex tapering to a point, deeply serrate to sometimes lobed; medium texture. The flowering plants may be annual or perennial. Flowers axillary, 2.5-5.0 cm across on a short pedicel, often overtopped by leafy shoots, fruit is an asymmetrical capsule, hairy and protandrous. Flower colours vary mostly from white, pink, rose to lilac. The seed capsules are sensitive to touch even before the seed is ripe, making seed collection difficult. They have seed capsules that spring open forcibly as the seed ripens to eject the seed at considerable distance.

Hydrocera is monospecific. Its calyx is pentamerous. *Hydrocera trifolia* is confined to the lowlands of Indo-Malayan region (Greywilson, 1980). While the genus *Impatiens* is very diverse and includes about 1000 species (Gupta, 2011). Although both the genera are seemingly rather similar in overall appearance, they differ in ecology, distribution range, habitat preference and morphology. Its calyx is trimerous. They are distributed in the highlands of old world tropics, sub-tropics besides Eurasia and North America (Janssens *et al.*, 2009; Stevens, 2012). *Impatiens balsamina* is variable in size, pubescence and available in varied flower colours and several varieties and cultivars have been developed (Anonymous, 2018).

5.0 Species and Cultivars

5.1.0 Species

The important species of *Impatiens* are described in Chapter 6 (*Impatiens walleriana*).

5.2.0 Cultivars

Tom Thumb Mix

It is a dwarf mix, growing to 30-40 cm tall. It prefers partial shade, however it also grows well in full to part sun. Plants bloom during summer and perform best in rich and well drained soil. The plants reseed readily.

Camellia Flowered Mix

It is a self-branching annual, growing upto 45-75 cm. The double flowers are found in rose, salmon, white, scarlet and violet. It grows in full to part sun, with moderate to moist conditions in rich and well-drained soils. It is suitable for both garden or containers.

Double Mix

These are showy annuals and are easily cultivated from seed. The plants are tall, bushy and produce double blooms in bright colours.

Dwarf Bush Flowered Mixed

It is a mix of compact, dwarf plants which produce flowers over a long season. Its dark green toothed leaves are a good contrast against the deep red, scarlet, pink and white flowers. It is bushy in habit, suitable for bedding plants, containers, short lived houseplants, hanging baskets and has good drought tolerance.

6.0 Genetics and Breeding

6.1.0 Cytology

The chromosome number of *I. balsamina* is 2n = 14 (Flora of China Editorial Committee, 2018). There is considerable cytological variation within the genus *Impatiens* with many species showing bimodal karyotypes (Song *et al.*, 2003). The flowers of *I. balsamina* are protandrous and as the anther walls break down and retract, the cellulose threads produced hold the exposed pollen in a lattice, from which the pollinating insects collect it (Vogel and Cocucci, 1988).

Male reproductive development in higher plants is an intricate biological process that includes the formation of anthers with differentiated tissues, in which microspores and pollen are formed. As the anther morphology becomes established, the anther typically consists of microsporocytes at the center of each anther locule surrounded by four different cell layers: the epidermis, the endothecium, the middle layer, and the tapetum from the outer to the inner. The cytokinesis in the

Cultivars of Balsam

Double-flowered | Tom Thumb | Tall Double Flowered Mix

Camellia Flowered Mix | Dwarf Bush Flowered Mix | Peppermint Sticks

Dwarf Multi-coloured Bush | Rose Double Mix | Blackberry Trifle

Red Garden Balsam | Rose Balsam | Rose Balsam Pink

Source: www.en.wikipedia.org, www.deargogo.com, www.greenmylife.in, www.cayman.desertcart.com, www.flickr.com, www.halifaxseed.ca, www.kingsseedsdirect.com, www.greenindia.garden, www.edenbrothers.com, www.rohrerseeds.com

microsporocyte meiosis in rose balsam is simultaneous type and microspore tetrads were mainly tetrahedral. The pollen exine began to form at the tetrad stage. The mature pollen grains were oval-shaped and bicellular. Accumulation of reserve substances begins at the late microsporogenesis stage, and results in an abundance of starch grains and lipids stored in pollen grains at anthesis. Polysaccharides and lipids change in different stages and play an important role in anther development. Moreover, the calcium oxalate crystals may protect the pollen and suggest that calcium distribution is related to anther development. The starch grains and lipids were abundant in the mature pollen grains of *I. balsamina*. Histochemical analysis of the anthers revealed smaller starch grain accumulation before the tetrad stage. After this stage, no starch grains were observed; however, starch grains reappeared at the late stage of bicellular pollen formation; notably, excessive amounts of starch grains and lipids were stored in mature pollen at anthesis. The anther biology of this plant provides crucial insights into the anther development processes, including the late microsporogenesis stage, late bicellular pollen stage, and anthesis. Importantly, Yang *et al.* (2018) confirmed that the accumulation of reserve substances began at the late microsporogenesis stage and abundant amounts of starch grains and lipids were stored in pollen grains at anthesis.

Stem morphology and characteristics of calcium oxalate crystals in seven species within the genus *Impatiens* have been studied. The results indicated that there were little anatomical differences between the seven species in the genus Impatiens, but the diversity of the characteristics of calcium oxalate crystals was large (Lersten and Horner, 2005). According to morphological characters of the raphides there are three types: focal raphides, clustered raphides, and scattered raphides. In addition, the shape, length, distribution and the degree of abundance are different among the seven species. The characteristic of calcium oxalate crystals can be of some significance in *Impatiens* taxa. Oxalate crystals are very common in angiosperms, but only few descriptions of them in pollen exist. Calcium oxalate crystals in *Impatiens* taxa have been described partly (Liu *et al.*, 2006), and most references have merely mentioned that calcium oxalate crystals occurred in the leaf, stem, root, or seeds. The calcium oxalate crystals in the pollen grains were observed through TEM. The shape of the calcium oxalate crystals, their arrangement and distribution were studied and found that the transversal section of the calcium oxalate crystal is square- or rectangle-shaped (Yang *et al.*, 2018).

6.2.0 Pollination

Impatiens psittacina, I. kerriae, I. racemosa and *I. daraneenae* were pollinated by bees, primarily *Bombus haemorrhoidalis*. *Impatiens chiangdaoensis* and *I. santisukii* had bimodal pollination systems which combined bee and lepidopteran pollination. Floral traits differed significantly among species with different pollination systems. Autogamous flowers were small and spurless and did not produce nectar; bee-pollinated flowers had short spurs and large floral chambers with a wide entrance; and bimodally bee- and lepidopteran-pollinated species had long spurs and a small floral chamber with a narrow entrance. Fragmentary evidence suggests that African *Impatiens* spp. with red flowers and trumpet-shaped lower sepals are specialized for sunbird pollination (Janecek *et al.*, 2015), whereas most studies performed outside Africa suggest pollination by bees (Mohandass, 2013). Some Asian species, such as *I. reptans* and *I. cuspidata* are characterized by relatively generalized pollination systems, including pollination by hawkmoths, bees and butterflies (Sreekala *et al.*, 2011).

Pollination and seed-setting in *Impatiens balsamina* (a long-day variety, originated from the Frankfurt Botanical Garden) was studied when flowers were isolated from insect pollinators or were openly pollinated and it was found that plants of both groups set seeds, but open pollinated had 10 per cent more seeds. Microscopic analysis showed that flowers could be self-pollinated before anthesis. The pollen tubes grew to the ovary and fertilized the ovules. It was found that selfing leads to development of normal endosperm and embryo. The viability of seeds was similar to that of cross pollination (Kulawiec *et al.*, 1996).

6.3.0 Polyploidy

According to morphological evaluation, seeds treated with 0.1 per cent aqueous colchicine solution for 10 min proved to be effective in induction of polyploidy in *Impatiens balsamina* (Manawadu and Dahanayake, 2014). It was, however, found that increasing doses of colchicine resulted in decreasing of the average seedling length, leaf length and increasing stem perimeter of *in vitro* seedlings.

6.4.0 Mutation Breeding

Chedeo and Wamaedeesa, (2021) investigated the effect of acute gamma irradiation on seed germination, seedling survival and morphology in garden balsam (*Impatiens balsamina*). The seeds were acute irradiated with gamma rays at the dose of 0, 50, 100, 200 and 300 gray. All the four gamma ray doses were found to decrease the seed germination rate and survival rate of seedlings of *Impatiens balsamina*. Gamma radiation dose of 50 gray resulted in darker fuchsia petal colour than control. Some flowers produced blemish petal colour after irradiation with 100 and 200 gray.

An attempt was made to create genetic variation in garden balsam using oryzalin and Gamma radiation (Defiani *et al.*, 2017). Seeds were treated with oryzalin (0 per cent, 0.01 per cent, 0.02 per cent) and then the seedlings were planted in the field four times to gain seeds generation M4. Seedlings of M4 (7 days of age) were radiated by gamma ray ^{60}Co (0, 5, 10, 15 Gy) and planted in the field to collect seeds for *in vitro* study to induce a new type of mutant plant. Growth percentage of seedlings reached 90 per cent by gibberellins (0.01 ppm). Leaf section was used for explant and cultured on MS media enriched by 0.5 ppm NAA and 0.5 ppm BAP aseptically. Shoots were regenerated 6 weeks after culture, however some shoots became vitrified. Number of chromosomes was found to be varied (mixoploid) in treated plants (M5). Form of secondary metabolites (alkaloids and terpenoids) in the roots extract was not changed by oryzalin and gamma radiation. Mixoploid explants showed variation in morphology and some treatments only had little shoots and a treatment has vigorous roots.

6.5.0 Interspecific Hybridization

Hybrids have been developed in some species of Impatiens *viz., I. sultani* and *I. holsti* which are useful as indoor plants only. But in balsam (*Impatiens balsamina L.*) which is self-pollinated crop having no male sterile line, there is no convenient method of producing commercial F_1 hybrids. However, it may be possible to isolate pure lines equalling or surpassing the yield in F_1 hybrids. A diallel experiment of 5 parents of balsam, *Impatiens balsamina* (double) was conducted to select most promising lines having more number of double flowers of large size. The characters included for investigation were plant height, number of branches per plant, number of flowers per plant, number of petals per flower and size of the flower. Heterosis was observed for all characters and the performance of F_1 hybrids was better than the top parent. The mean sum of square for G.C.A. was higher than that of S.C.A. Gene action was found additive in nature without any epistasis for all characters except flowers per plant where epistasis was of complementary nature. The association of excess of recessive genes with high values was observed for all characters. Though appreciable heterosis was observed for all characters, in the absence of convenient and economical method of F_1 seed production, it is better to fix the economic factors (especially because of the additive nature of the genetic variance observed for the characters) by selection in F_2 and further generations. After isolating the pure lines with superior economic characters, these lines may be further crossed to exploit the heterosis for commercial production of F_1 hybrids (Swarup *et al.*, 1975).

Pal *et al.* (2018) assessed the variability, heritability and genetic advance with 34 genotypes of Balsam (*Impatiens balsamina*) to identify superior genotypes for further crop improvement programme. The phenotypic coefficient of variation (PCV) and genotypic coefficient of variation (GCV) was found maximum for seed yield/plant, *i.e.* 29.02 and 26.34 g while, minimum in case of length of leaf, *i.e.* 8.35 and 6.48 cm, respectively. High heritability was observed for all the characters except plant height, width of leaves and duration of flowering. Maximum heritability was recorded for duration of flowering (90.68 per cent) and minimum was observed in width of leaf (30.04). High heritability (h^2= 90.64) with high genetic advance (GA= 71.36) as percentage of mean was observed for number of flowers/plant, which indicated that additive gene effects were more important for that trait. High genetic advance as per cent of mean was observed for number of flowers/plant (71.36), however, lowest for floral bud diameter (0.11) which indicates the preponderance of additive genes and selection will be effective for improvement of these traits.

The floral colours of *Impatiens balsamina* ranged from magenta to cream and white; and include red, purple, and various shades and tints of lavender and of pink. Several workers have reported on the inheritance of some or all of these colours. Kanna (1934) described a single factor difference between red and pink and another single factor difference between purple and pink. He also revealed that the combination of factors for red and purple gives magenta in the F_1; and, in the F_2, magenta, purple, red and pink in the ratio of 9:3: 3: 1 as expected.

A recessive gene for scorched (sc) has been shown to affect the colour of reds and magentas. This gene can be carried in the homozygous state without expressing itself phenotypically. It has been shown to be so carried by hermosa.

The genic composition of complex colours has been established as follows: Green-stemmed plants:-LH, hermosa-lavender; HPg, HPg′, or HPg″, rose; LPg, LPg′, or LPg″, lavender; LHPg, or LHPg′, rose-lavender. Red-stemmed plants:-HPr, HPr′, or HPr″, red; LPr, LPr′, or LPr″, purple; LHPr, magenta. All the flower colours of the garden balsam occur in duplicate series. Each colour in the non-cream series bears a dominant gene W, and has its counterpart in the cream series which bears the recessive genes *ww*. Of the non-cream series the lowest is pure white, and cream is the lowest member of the cream series. The occurrence of the following genes has been established: L, *l* for nearly white lavender versus white, with no dominance. H, *h* for hermosa versus white, with H partially dominant. Pg, Pg′ and Pg″ versus *p*, a series of gems/genes each producing its characteristic intensity of nearly-white pink on petals and of pink on spurs in contrast with the white of *p*. Pg, Pg′, Pg″ plants have green lower stems. Pr, Pr′ and Pr″ as additional alleles of *p*, each producing, not only its own intensity of pink colour on petals and spurs, but also, its own characteristic red on the stems in contrast with *p* and the Pg series (Davis *et al.*, 1958).

I. balsamina was in cultivation in Asia for long with several ethnic uses. It was introduced in the mid-fifteenth century by Portuguese explorers to the new world. It was about this time that the double, *I. balsamina* was being circulated which is not a true hybrid but really a cultivar of the species. It was only during 20th century the hybridization of Impatiens became popular. Claude Hope is regarded as the Farther of modern impatiens breeding (http://www.mrimpatiens.com/hybrid_impatiens.htm).

7.0 Growth and Flowering

Balsam is an annual and herbaceous flowering plant. It needs moist, but well-drained soil and performs best in partial shady locations. The plants are erect with good branching habit and reach upto a height of 75 cm. The plants have good spreading habit which makes them suitable for multi-utility ornamentals.

7.1.0 Flower Bud Initiation

Pouteou *et al.* (1998) studied the inflorescence of balsam during flower development and revealed that the response of axillary meristem primordial to the quantity of inductive signal, a response that is conditioned by the age or position of the primordia and allows un-differentiation of axillary primordial initiated before evocation to adopt different fates. Thus the vegetative and reproductive phases are not separate and antagonistic but interpenetrate each other to varying extent depending on the quantity of inductive signal

Tooke and Battey (2000) reported the role of leaves in flower development in *I. balsamina*. The leaf derived signal was found to have a role in floral maintenance. The increased petal number was neither a response to a stress effects associated with leaves removal nor result of alternation of petals for stamens. Rather, the petal initiation phase was prolonged when the amounts of a leaf derived signal were limiting. Therefore, leaf derived signal has a continuous and quantitative role in flower development. This may provide an explanation for the wide variety and instabilities of floral form seen among certain species in nature (https://core.ac.uk).

7.2.0 Temperature

The balsam plant grows best in full sun to partial shade. It thrives well in warm temperatures (28-32°C), but they do not tolerate frost and freezing temperature. In summer the leaves get scorched due to high temperature and plant may wilt in the heat wave. The water requirement of the plant is high and therefore they survive well in moderate humid areas. The plant can tolerate dry air, if there is adequate soil moisture around the plant.

7.3.0 Photoperiod

In general, balsams are qualitative short day plants (Nanda and Kumar, 1966) and floral bud reversion was observed in the plants when they were transferred to long days after exposure to 4 short-day (SD) cycles (Krishnamurthy and Nanda, 1968). Teske (1986) however found variation in photoperiodic response such as qualitative short-day plants, quantitative short-day plants and day-neutral plants in balsam varieties/hybrids of various locations. Nanda and Krishnamoorthy (1967) in a study on growth and development of *I. balsamina* found that while floral bud of this species are initiated with three short day cycles, at least eight such cycles are required for flowering. The numbers of floral buds and open flowers bear a linear relationship with the number of SD cycles. The induced floral buds revert to vegetative growth unless the plants receive the minimum number of SD cycles needed for flowering, this reversion occurring in a basipetal direction. The rate of extension growth of the stem increases with increasing numbers of SD cycles. The high rate is maintained longer in plants receiving 32 or more SD cycles, but the subsequent fall is also steeper in these plants than in plants receiving less inductive cycles. Senescence also occurs in these plants and appears to be related to the magnitude of reproductive development and the high rate of extension growth.

Seedlings of *Impatiens balsamina* raised under Normal Day (ND) and Long Day (LD) conditions were divided into two sub-groups when they had reached 5-leaf stage. While one sub-group was left under the same condition (NDND or LDLD), the other was transferred to the other photoperiod (NDLD or LDND). NDND plants were subdivided into 2 lots. One of these was transferred to Short Day (SD) in May. The dates of emergence of individual branches and floral buds were recorded and the vegetative period was calculated in each case. It was found that in NDND plants floral buds were produced from all the nodes except the lowermost which produced a single vegetative branch. In LDND plants the vegetative branches were produced from the lower 9 nodes but floral buds from those above these. Small leafy structures which ultimately dried up were produced from a few top nodes in both these cases. In contrast to this in LDLD plants only vegetative branches were produced from all the nodes. In NDLD plants floral buds were produced from the lower 3–5 nodes prior to transfer to LD condition, but vegetative branches were produced from the upper nodes after this transfer. Even some of the lower floral buds reverted to vegetative state under this condition. The production of floral buds or the vegetative branches as the case may be, occurred in acropetal succession under all the photoperiodic conditions and never in basipetal manner. LDLD and NDLD plants, which did not flower at all, continued to produce lateral branches without showing any sign of senescence, while LDND and NDND ones showed yellowing of the apical growing point which spread downwards and lead ultimately to the death of the plant. The senescence was hastened when these plants were transferred to SD condition towards the end of May. The senescence therefore, appears to be related with reproductive development (Nanda & Kumar, 1966).

7.4.0 Growth Regulators

The plant responds well to different plant growth regulators. Kanwar and Nanda (1986) in a study on *Impatiens balsamina* using GAs and Resorcinol found that application of former resulted in more number of floral buds under both inductive as well as non-inductive photoperiods. Nanda *et al.* (2006) found the effectiveness of GA_3 in inducing floral-bud initiation in *Impatiens balsamina* under non-inductive photoperiods. Bud initiation varies considerably at different parts of the year and appears to be determined by the prevailing temperature conditions. The period of floral-bud initiation is shortest in October and February and increases both with decreasing temperatures in November and with increasing temperatures in August and September. Floral buds do not initiate in March and April, during which time me temperature increases considerably.

8.0 Propagation

8.1.0 Seed Propagation

Balsam is easily propagated through seeds. The seeds sprout in 4-10 days in moist soil at 20-25°C. Seeds should be partially covered with soil as light hastens germination. Seed germinates within 10-15 days after sowing and seedlings become ready in 3-4 weeks.

8.2.0 Micropropagation

A reliable and efficient tissue culture system for *Impatiens* species was standardized in order to genetically engineer this plant for useful traits. Parts of cotyledons and hypocotyl derived from different age of seedlings, from surface sterilized seeds, were subjected to shoot induction on MS media containing BAP, TDZ or combination of BAP and NAA in different ratios. Roots were developed from the above shoots on full-or half-strength MS media containing IAA, IBA or NAA of different concentrations. Fully-developed plantlets were produced on hormone free MS media or full- or half-strength MS media containing specific hormones. It was observed that the most efficient shoot induction (93 per cent) was obtained from proximal part of cotyledon derived from 7 days old seedlings on MS media supplemented with 1 mg/l BAP. Roots were most efficiently produced (92 per cent), in term of percentage and morphology, on half-strength MS media supplemented with 0.1 mg/l IAA. Further regeneration of plantlets, approximately 8-10 cm in height, was achieved after 3 weeks on hormone-free MS media. A full regeneration system can be achieved within 8 weeks in culture (Taha *et al.*, 2009).

Mohamed *et al.* (2018) developed an efficient protocol for *in vitro* flowering in *Impatiens balsamina* cv Dwarf Bush through tissue culture techniques. Shoot, stem and petiole explants obtained from 4 week-old aseptic seedlings cultured on MS medium

Propagation and Growing of Balsam

In vitro **Flowering of** ***Impatiens balsamina cv.*** **Dwarf Bush**

(A) Four-month-old *in vivo* flowering of *Impatiens balsamina* (B) Aseptic seedlings (C) *In vitro* flowering of *Impatiens balsamina* cultured on MS supplemented with 1.0 mg/l GA_3 after 8 weeks. (D) *In vitro* flowering after 3 months (Mohamed *et al.,* 2018).

***Source*: www.image.shutterstock.com, www.colourbox.com, www.google.com, www.plantswithapurpose.net, www.itslife.in, www.agicagro.com**

supplemented with different concentrations of plant growth regulator (PGR) were used for in vitro flower induction. Gibberellic acid (GA_3), benzylaminopurine (BAP) and kinetin (Kin) treatment singly applied in MS media (pH 5.8), could all stimulate flowering at 23-26°C with photoperiod of 16 hours light and 8 hours dark. It was observed that shoot explants were more responsive than stem explants in floral formation. Regeneration was achieved *via* direct organogenesis. For shoot explants, the treatment that induced the highest rate of *in vitro* flowering (7.30 ± 0.16 flowers per plantlet) was 1.0 mg/l GA_3.

Mohamed *et al.* (2019) established the micropropagation of *Impatiens balsamina* from stem and shoot explants. In order to induce *in vitro* flowering, the effects of GA_3 and glutathione on the morphogenesis of this species was also investigated. It was found that the optimum *in vitro* plant regeneration was achieved on MS medium supplemented with 1.0 mg/l GA_3 and *in vitro* flowering was also obtained from the same medium after 4 weeks of culture. To understand cellular behavior during *in vitro* flowering, Mitotic Index (MI), chromosome counts, measurement of mean cell and nuclear areas, DNA measurements and ploidy levels were analyzed from *in vivo* plants, *in vitro* grown plants and plantlets that flowered *in vitro*. The chromosome count was the same for all, 2x=2n=14 or n=7. However, it was observed that *in vitro* flowering plants of *Impatiens balsamina* had the highest percentage of polyploid cells (30.7 per cent), based on a histogram plotted by the AxioVision 4.7 software. It was found that plant growth regulators, especially GA_3, increased the polyploidy level of the meristematic root cells.

9.0 Soil and Climate

9.1.0 Soil

Balsam can grow on a wide range of soil texture from heavy to light; soil types such as clay, sandy and loamy soils with pH in the range 5.6-7.5. However, it prefers organically rich, well-drained but moist soil. The plant is sensitive to salt. Suitable soil amendment with compost and breaking the soil lump before transplanting is recommended.

9.2.0 Climate

9.2.1 Temperature

The Balsam is widely naturalized through the tropics, subtropics and warmer areas of the temperate zone. Seed germination takes 8-14 days. Plants can flower all year round in warm climates. The optimum temperature for germination is 25-28°C and for the seedlings a night temperature of about 20°C is needed (van Valkenburg and Bunyapraphatsara, 2001). The plant does not tolerate temperatures below 0°C and are susceptible to frost (Gilman and Howe, 1999). They thrive well in warm temperatures but can wilt if summer temperature spike is too high.

9.2.2 Light

I. balsamina grows best in full sun to shade. Under full sun, 6 or more hours of direct sunlight in a day while under partial shade, direct sunlight for about 2-6 hours is required. Direct exposure to sunlight during afternoon may cause browning of the foliage. It survives even in full shade, except it will be sparse with few blooms.

10.0 Cultivation

The plant is widely naturalized through the tropics, subtropics and warmer areas of the temperate zone. It is grown as summer and rainy season annual for shaded beds as well as in containers and 'hanging baskets. The plants of this species are commonly cultivated as an ornamental in gardens across tropical and subtropical regions of the world. It is also found naturalized along roads, in rubbish dumps, fields, forest edges, grasslands, streambeds and other open places. It is also a common weed in wastelands near villages, dwellings and cemeteries (van Valkenburg and Bunyapraphatsara, 2001).

10.1.0 Sowing of Seeds

The seeds are sown in January-February for summer flowering and May-July for flowering in the rainy season. Seeds are sown thinly in nursery beds/seed pans/ seedling trays or can be sown directly into the garden. Fine sandy soil is required for seedbed (Outdoors), or coco peat is used as sowing medium for seedling trays (Indoors). Before sowing the seeds, soil/coco peat should be made slightly moist. Immediately after sowing the seeds, watering should be done carefully with a watering can. Germination takes about 8-14 days. Seedlings are allowed to grow on seedbeds/trays kept in semi-shade. Transplanting of seedlings is done in pots or flower beds at 6-leaf stage. Seedlings should be transplanted in rich soil or potting mixture. Pinching of apical meristems encourage canopy density and is generally practised when the plant reaches more than 10 cm.

10.2.0 Spacing

Overcrowded balsam plants do not produce good blooms. Therefore they are planted wide (30-45 cm) to grow them bushy. For pot culture, a single plant should be planted in a pot. Balsam can be grown in containers on balconies or terraces.

10.3.0 Manuring and Fertilization

Addition of well decomposed Farm Yard Manure not only improves the soil texture but also beneficial for plant growth and flowering. The plant requires moderate fertilizer application. Incorporation of slow release fertilizer prior to planting takes care of its nutritional requirement.

10.4.0 Irrigation

Plants in flower beds should be watered once a week while in pot, be judiciously watered. The crop may bloom for a longer time if sufficient watering is made during growth stage. Due to its moderate water requirements, balsam tolerates warm, humid weather well. Excessive water application may lead to rotting of the whole plant. The plant generally stop flowering if they do not get enough water.

11.0 Diseases, Pests and Disorders

11.1.0 Diseases and Pests

The details of diseases and pests affecting the crop are same as described in Chapter 6 (*Impatiens walleriana*).

11.2.0 Physiological Disorder

One of the most common problems with impatiens flowers is wilting. This is usually due to moisture stress. These plants need to be kept consistently moist, but not soggy. Water stress can also cause leaf and flower/bud drop. In addition to watering, wilting can be a result of heat stress, especially if the plants are under direct sun. It is therefore advisable to grow them in shady location.

12.0 Harvesting and Postharvest Management

Balsam is generally used as a garden and pot plants. However it is used as loose flower in some parts of India. Harvest the flowers when they are fully open during late in the afternoon or very early in the morning.

12.1.0 Seed Harvesting

The pods burst at the slightest touch. Therefore they are collected individually when turned brown. The pods are kept in a box for the matured seeds to dry. Keeping the seeds at low temperature enhances its viability. Seed storage under ambient temperature and relative humidity without deterioration in quality for a longer period is of immense importance for farmers. The seed yield per acre is around 300-320 kg.

13.0 Biotechnology

13.1.0 Molecular Markers

Nurul *et al.* (2010) during the RAPD analysis revealed that there is reasonable genetic diversity between and among red, pink, purple and white forms of *I. balsamina*. The quantitative selective chemo-profiling analysis has showed that different accessions (genomes) of *I. balsamina* have an influence on production of total phenolic and Gallic acid contents.

13.2.0 Genetic Transformation

Taha (2008) studied Biolistic- mediated transformation approach of gene transfer using particle bombardment method. Shoot regeneration from cotyledon explant of 7 day-old seedlings of *I. balsamina* was optimized by tissue culture method. The maximum average number of shoots (6.8 ± 1.05 shoots) per explants was obtained on full strength MS-based medium containing 1 mg/l BAP after three weeks of culture. For rooting induction, there was no significant difference on percentage of root regeneration on full strength (84 per cent -94 per cent) and half strength MS–based medium (78 per cent -96 per cent) supplemented with IAA (0.1 - 1mg/l), IBA (0.5 - 2 mg/l) and NAA (0.1 - 1 mg/l) after two weeks of culture.

Diseases and Pests of Balsam

Powdery Mildew | Downy Mildew | CMV

Necrotic Spots | Phytoplasma | Asian Balsam Aphid

Winged Balsam Twig Aphid | Mite | Aphids

White Flies | Thrips | Bubbles and Mushroom Mosquito

***Source*: www.img.farmforage.com, www.content.ces.ncsu.edu, www.pubmed.ncbi.nlm.nih.gov, www.influentialpoints.com, www.invasive.org, Plant Disease Journal (2019), Diningsih *et al.* (2020).**

Plantlets were successfully produced on full strength MS plates supplemented with 1 mg/l BAP (93 per cent) for shooting and half strength MS media supplemented with 0.1 mg/l IAA (92 per cent) for rooting in eight-week culture. Biolistic technique was used for the transformation of uidA and hph genes into *I. balsamina*. The 7 days old cotyledons explants were bombarded with pRQ6 containing uidA gene encoding β-glucuronidase (GUS) and hph gene conferring resistance to hygromycin and co-transformation was carried out using pRQ6 with pAHG11 (containbar gene conferring resistance to herbicide Basta). Transformation of *I. balsamina* with pRQ6 and co-transformed with pRQ6 and pAHG11 at 28 mm Hg vacuum using optimized bombardment conditions (9 cm target distance, 1100 psi helium pressure, one time bombardment, 1.0 µg DNA, 16 h pre-culture time with osmotic treatment of 0.4 M mannitol and sorbitol and 24 h post-bombardment incubation time) showed the highest average number of GUS spots of 149.3 and 128.1, respectively. Delay selection method was used to delay the timing of selection on shooted explants after bombardment to obtain transformed plants. Out of 160 bombarded explants with pRQ6, only 84 plants were successfully regenerated from selected 35-day old shooted explants after five weeks grown on MS –based medium containing 75 mg/l hygromycin. All regenerated plants (84 plants) were GUS positive. However, there was no plant regenerated on MS based medium containing 1 mg/l phosphinothricin (PPT) after co-transformed with pRQ6 and pAHG11. Furthermore, PCR results showed only 14 of 40 GUS positive plants survived at 75 mg/l hygromycin were *hph* gene positive. The PCR results also showed the possibility integration of *hph* gene into the genomic DNA of transformed plants.

13.3.0 Transcriptome

Foong *et al.* (2019) elaborated the first transcriptome of *I. balsamina* leaf using paired-end Illumina HiSeq sequencing which generated 10.79 GB of raw data. The leaf transcriptome significantly increased the amount of sequence information available for this plant and also useful as reference to other *Impatiens* species. The annotated transcripts against KEGG pathways could be useful for study on detailed protein-coding genes of this plant related to transcripts, genes and pathways involved in biosynthesis of secondary metabolites. It will serve as a useful transcriptomic resource for future studies including gene expression, RNAi induction analysis, genomics and functional genomics in *I. balsamina* and other *Impatiens* species.

De novo transcriptome sequencing and analyses of the leaf, flower and early-, mature- and postbreaker stage capsules allowed identification of all the annotated genes involved in the shikimate and DHNA pathways responsible for the production of MNQ in *I. balsamina*. Correlation between expression of shikimate- and DHNA pathway genes with MNQ pools, combined with knowledge of previous labeling experiments suggests that MNQ biosynthesis branches off the phylloquinone pathway. Significant upregulation of most genes of the DHNA pathway in early stage capsule compared to flower and leaf suggests that MNQ is synthesized de novo in a tissue-specific manner in the capsule of *I. balsamina*. A total of 11 candidate unigenes corresponding to the enzyme families of S- adenosylmethionine O-methyltransferases, oxidoreductases, and UDP glycosyltransferases postulated to catalyse the final reaction of MNQ production as well as lawsone stability were identified based on their expression levels being significantly and positively correlated with MNQ- and lawsone content in different tissues of *I. balsamina*. Knowledge and better understanding of the genes involved in these biosynthesis pathways (and their expression patterns) provided the required genomics resource for targeted manipulation of these pathways either via genetic engineering or synthetic biology (Foong *et al.*, 2020).

13.4.0 Suspension Culture

Kasem (2018) used the tissue culture technique for callus production and enhancing total phenolics (TPC) and flavonoids (TFC) content in suspension culture of garden balsam (*Impatiens balsamina*) which has many important secondary metabolites in its different organs. Callus was initiated on leaf discs cultured on MS medium supplemented with either 2, 4-D or NAA at different concentrations (0.5, 1.0 and 1.5 mg/l), besides the control (MS-free hormone medium) with a fixed BAP concentration (0.5 mg/l). Also, suspensions were supplemented with either yeast extract or chitosan (0, 50, 100 and 200 mg/l) for enhancing accumulation of phenolics and flavonoids content for two separated elicitation periods (24 and 48hrs). The highest callus formation percentage (100 per cent) was obtained from leaf explants cultured on MS medium supplemented with either 1.0 mg/l 2,4-D+0.5 mg/l BAP or 1.5 mg/l NAA+0.5 mg/l BAP. This superior treatment produced the biggest callus volume (9.38 and 9.19 mm^3, respectively). The heaviest callus fresh and dry weights (8.25 and 1.15 g/jar) were recorded for MS medium fortified with 1.0 mg/l 2,4-D+0.5 mg/l BAP. In addition, after 24hrs from the suspensions elicitation, the highest TPC value of 8.298 mg/gdw with 4.18 fold higher than the control was recorded for suspensions which received 50 mg/l yeast extract. At the same time, 50 mg/l chitosan increased TFC (3.929 mg/gdw). The matter was partially similar after 48hrs, with a slight increase in all the elicitation treatments, since the superior treatments after 24hrs resulted in higher TPC and TFC.

14.0 References

Anonymous (2018) *Flora of Pakistan/Pakistan Plant Database (PPD)*. Tropicos website. St. Louis, Missouri and Cambridge, Massachusetts, USA: Missouri Botanical Garden and Harvard University Herbaria. http: //www.tropicos.org/Project/ Pakistan

Binorkar, S.V. and Jani, D.K. (2012) *Journal of Pharmaceutical and Scientific Innovation*, **1**(5): 13-20.

Chedeo and Wamaedeesa, R. (2021) *Princess of Naradhiwas University Journal*, **13**(1): 363-376.

Chen, X.M., Qian, S.H. and Feng, F. (2010) *Chinese Chemical Letters*, **21**: 440-442.

Davis, D.W., Taylor, L.A. and Ash, R.P. (1958) *Impatiens balsamina* L.-The inheritance of floral colours. pp. 16-34.

Defiani, M.R., Astarini, I.A. and Kriswiyanti, E. (2017) *Curr. Agri. Res.*, 5(1).

Diningsih, E., Yanda, R.P., Soehendi, R. and Marwoto, B. (2020) *Earth and Environmental Science,* 468:012035.

Foong, L.C., Chai, J.Y., Ho, A.S.H., Yeo, B.P.H., Lim, Y.M. and Tam, S.M. (2020) *Scientific Reports*, **10**: 16123.

Foong, L.C., Ho,A.S.H., Yeo, B.P.H., Lim, Y.M. and Tam, S.M. (2019) *Data in Brief*, **23**: 103603.

Gilman, E.F. and Howe, T. (1999) *Impatiens balsamina*. Fact Sheet FPS-279. Florida, USA: Environmental Horticulture Department, University of Florida.

Grey-Wilson, C. (1980) *Impatiens of Africa*. Rotterdam, The Netherlands.

Gupta, A.K. (2011) *Hydrocera triflora*. The IUCN Red List of Threatened Species 2011: e.T168915A6547780. http: //dx.doi.org/10.2305/IUCN.UK.2011-2.RLTS. T168915A6547780.en

https: //www.applewoodseed.com

http: //www.cabi.org

http: //www.core.ac.uk

https: //www.gardenbeast.com

https: //www.genetics.org/content/genetics/43/1/16.full.pdf

http: //www.eprints.utm.my/id/eprint/10011/1/AishahMohdTahaMFSA.pdf

https: //www.groseeds.co.uk

https: //www.halifaxseed.ca

http: //www.mrimpatiens.com/hybrid_impatiens.htm

https: //www.thespruce.com

Janecek, S., Bartos, M. and Njabo, K.Y. (2015) *Biological Journal of the Linnean Society*, **115**: 127– 133.

Janssens, S.B., Knox, E.B., Huysmans, S., Smets, E.F. and Merckx, V.S.F.T. (2009) *Molecular Phylogenetics and Evolution*, **52**(3): 806-824. doi: 10.1016/j. ympev.2009.04.013

Jaykrishna, I.T. (1998) *Vanaspati Shashtra*. Pravin Prakashan Pvt.Ltd: Rajkot. 123-124.

Kanna, B. (1934) *J. Coll. Agric. Tokyo Imp. Univ.*, **12**(3): 421-477.

Kanwar, K. and Nanda, K.K. (1986) *Biologia Plantarum*, **28**: 202. doi:10.1007/ BF02894597

Kasem, M.M. (2018) *J. Plant Production*, Mansoura Univ., **9** (3): 241 – 248.

Krishnamoorthy, H. N. and Nanda, K. K. (1968) *Planta*, **80**(1): 43-51.

Kulawiec, M., Teske, E. and Sniezko, R. (1996) *Acta Societatis Botanicorum Polonidae*, **65**(1-2): 107-109.

Lersten, N.R. and Horner, H.T. (2005) *Amer. J. Bot.*, **92**: 1935- 1941.

Liu, C.W., Cong, K.M., Cai, Y.Y. and Zhen, X. (2006) *Life Sci. Res.*, **10**: 328 332.

Manawadu, I. P. and Dahanayake, N. (2014) *Proceedings of the International Symposium on Agriculture and Environment*, Ruhuna, Sri Lanka, 27 November 2014, pp.59-62.

Mohamed, N., Taha, R.M. and Razak, U.N.A.A. (2019) *Planta Daninha*, 37:e019170509.

Mohamed, N.; Taha, R.M.; Razak, U.N.A.A. and Elias, H. (2018) *Planta daninha* [online]., **36**: e018165357. http: //dx.doi.org/10.1590/s0100-83582018360100013.

Mohandass, D. (2013) *International Journal of Ecology and Environmental Science*, **39**: 59– 63.

Nanda, K.K. and Krishnamoorthy, H.N. (1967) *Planta*, **72** : 338-343

Nanda, K.K. and Kumar, P. (1966) *Planta*, **69**: 249–257

Nanda, K.K., Tory, K.L. and Sawhney, S. (2006) *Physiologica Plantarum*, **23**(6): 1085-1088.

Nurul, A.H., Nur, A.H., Bhore, S.J. and Shah, F.H. (2010) *Middle-East Journal of Scientific Research*, **5** (6): 454-463.

Ong, H.C., Zuki, R.M. and Milow, P. (2011) *Ethno Med.*, **5**(3): 175-185.

Pouteau, S., Tooke, F. and Battey, N. (1998) *Plant Physiology*, **118**: 1191–1201.

Sanghi, D.K. and Tiwle, R. (2013) *Int. J. Med. Pharm. Res.*, **1**(3): 308-317.

Singh, A.K. and Singh, N. (2007) *Journal of Ornamental Horticulture*, **10**(2): 128-130.

Song, Y., Ming,Y.Y. and Küpfer, P. (2003) *Caryologia*, **56**(4): 463-481.

Sreekala, A.K., Pandurangan, A.G., Ramasubbu, R. and Kulloli, S.K. (2011) *Journal of Threatened Taxa*, **3**: 1818– 1825.

Stevens, P.F. (2012) *Angiosperm Phylogeny* Website. http: //www.mobot.org/ MOBOT/research/APweb/

Sumit, P., Singh,A.K., Pal, A.K., Sisodia, A. and Tiwari, A. (2018) *Journal of Applied and Natural Science*, **10**(2): 810 -812.

Swarup, V., Raghava, S.P.S. and Balakrishnan, K. A.(1975) *Indian Society of Genetics and Plant Breeding*, **35**(1): 69-75. Print ISSN: 0019-5200. Online ISSN: 0975-6906.

Taha, A.M. (2008) *M.Sc. Thesis*: Universiti Teknologi Malaysia.

Taha, A.M., Wagiran, A., Ghazali, H., Huyop, F. and Parveez, G.K.A. (2009) *Biotechnology*, **8**: 1-12.

Teske, E. (1986) *Acta Societatis botanicorum poloniae*, **55**(1): 107-113

Tooke, F. and Battey, N.H. (2000) *The Plant Cell*, **12**: 1837–1847.

Valkenburg, J.L.C.H. van and Bunyapraphatsara, N. (2001) *Plant resources of South-East Asia No. 12*(2): medicinal and poisonous plants 2, [ed. by Valkenburg, J. L. C. H. van, Bunyapraphatsara, N.]. Leiden, Netherlands: Backhuys Publishers. pp. 782.

Vogel, S. and Cocucci, A. (1988) *Beiträge zur Biologie der Pflanzen*, **63**: 271-287.

Yang, S., Peng, L., Bao, H. and Tian, H. (2018) *J. Amer. Soc. Hort. Sci.* **143**(2): 95–100. https: //doi.org/10.21273/JASHS04303.

Catharanthus roseus

D.V.S. Raju and K. Ravindra Kumar

1.0 Introduction

Catharanthus roseus commonly known as Madagascar Periwinkle or Sadabahar is an evergreen ever blooming perennial sub-shrub with diploid chromosome number 2n = 16. It is grown as an annual in many parts of the world. The plant has ornamental value due to its year-round flowering. It blooms in a variety of colours like white, red, purple, magenta, orange, pink *etc.*

2.0 Importance and Uses

2.1.0 Aesthetic Uses

Catharanthus is suitable for growing as pot plant or for outdoor landscaping. In gardens it is planted as an edge and bedding plant. In case of home landscaping it can be planted in patios, balconies, hanging baskets and roof tops. In some countries the flowers along with stem are also used as cut flower.

2.2.0 Therapeutic and Other Uses

Different parts of this plant have been used in various forms of traditional medicine for the treatment of a wide range of ailments such as diabetes, fevers, malaria, menorrhagia, hypertension, cancer, stomach ailments, heart disease, leishmaniasis, amenorrhea, dysmenorrhoea, rheumatism, liver disease, *etc.* (Ross, 1999). Investigations have led to discovery of two alkaloids, vincaleukoblastine [vinblastine (VLB)] by Noble *et al.* (1958) and leurocristine [vincristine (VCR)] by Eli Lilly Company, USA, possessing anti-cancer property (Svoboda, 1975; Tyler, 1988). Apart from VLB and VCR, periwinkle produces more than 130 diverse groups of alkaloids (Zhu *et al.*, 2014) and has been termed as an alkaloid engine (Dugo de Bernonville *et al.*, 2015). The discovery of ajmalicine, used in the treatment of circulatory diseases, especially hypertension, in roots of periwinkle followed their detection as an adulterant in *Rauvolfia* roots exported from India. It was readily accepted as an alternate source of this alkaloid (Krishnan, 1995; Singh, 1996).

3.0 Origin and History

Periwinkle is considered to be a native of the West Indies but was originally described from Madagascar (Ross, 1999). Linnaeus named the Madagascar Periwinkle as *Vinca rosea* in 10th Edition of *Systema Naturae* of 1757. G. Don moved it to separate genus Catharanthus because he observed many botanical differences between *Vinca* and the Madagascar Periwinkle. Initially the seeds from Madagascar were sent to the Royal gardens in Versailles near Paris. Mr. Richard the gardener of King Louis XV sent his own plants to the Chelsea Botanic Garden in London in 1757. The seeds were grown by the famous botanist Philip Miller.

4.0 Taxonomy and Botany

The genus *Catharanthus* consists of eight species, seven of them *viz.*, *C. roseus*, *C. ovalis*, *C. trichophyllus*, *C. longifolius*, *C. coriaceus*, *C. lanceus*, *C. scitulus*, indigenous to Madagascar and one *viz.*, *C. pusillus*, indigenous to India (Stearn, 1975).

Uses of Catharanthus

In Flower Bed | In Mixed Flowers Basket | Under Tree

In Landscape Garden | In Mixed Border | As Border Plant

In Pot Plant | Display in Flower Show | Display in Pots

In Hanging Basket | In Wall Decoration | In Floral Display

Source: www.harrisseeds.com, www.plantsrescue.com, www.dhgate.com, www.in.pinterest.com, www.calloways.com, www.dhgate.com, www.lsuagcenter.com, www.plants.ces.ncsu.edu, www.panamseed.com, www.rotarybotanicalgardens.org, www.coolgarden.me

The plant grows up to 40-100 cm height. Because of its branched tap-root system, it thrives well in drought conditions. The sub-woody stem is solid, cylindrical (terete), longitudinally ridged or narrowly winged, green or dark red, pubescent at least when young. The leaves are opposite, borne on short petioles, 2.5-9.0 cm long, 1-3.5 cm in breadth, usually elliptical to obovate, dark green. The leaf tip is rounded to acute with a tiny point extending from the midrib. Stems and leaves usually with hairs (pubescent), sometimes hairless.

The inflorescence is a solitary axillary or dichasial cyme. Paired, hermaphrodite, pedicellate, actinomorphic, bracteate, hypogynous, pentamerous, and complete flowers are borne in axils with a 2.5–3 cm long cylindrical tube at the base on very short stalks (pedicels). The flowers are borne in leaf axils, either singly or paired. The calyx is polysepalous, composed of five velvet sepals, free to the base. Corolla is gamopetalous, made up of five petals, light to dark pink with a dark reddish-pink center, or white in colour with a diameter of 2.5–5 cm. Attached to the corolla tube, stamens are five in number with short filament. Anthers 5, attached to the inside of the corolla tube in the upper portion and concealed within it. Two distinct carpels, each with about 10–30 ovules in series of two, have long and slender style and capitate stigma. The fruit is a follicle, 2.0-4.7 cm long, with numerous small black seeds. Out of the two common cultivars of *C. roseus*, one blooms pink flowers named as 'Rosea' and another is 'Alba' with white flowers. Flowers of *Catharanthus roseus* are pollinated by butterflies and moths. This species is self-compatible, though self-pollination under normal conditions. Seeds are dispersed by ants, wind and water.

5.0 Species and Cultivars

There are eight species in *Catharanthus*. Out of which seven are endemic to Madagascar and one to India and Sri Lanka. Within the family Apocynaceae, the genus *Catharanthus* belongs to the sub-tribe Catharanthinae in the tribe Plumerieae, sub family Plumeroideae.

5.1.0 Species

There are eight species in *Catharanthus* which are described below.

Catharanthus coriaceus

Perennial herb, grows up to 40 cm high, erect. Stems very narrowly winged, glabrous. Leaves are mostly about twice as long as internodes, rounded at base. Leaf blade is fleshy and coriaceous. Flowers paired. The species is endemic to Madagascar.

C. lanceus

Perennial herb upto 1 m high, decumbent, glabrous with carrot like central tap root, 5-20 mm diameter. Stems and branches reddish, sub quadrangular. Internodes are much longer than leaves. Leaves are alternate at base. Flowers are axillary, solitary and fragrant. The species is endemic to Madagascar.

C. longifolius

Under shrub, 40-150 cm high, woody at the base and erect. Stems and branches straw coloured, ridged, very narrowly winged, glabrous and pilose. Leaves are

Species of *Catharanthus*

Catharanthus pusillus **(Murray)**

Catharanthus ovalis

Catharanthus roseus **var.** ***roseus***

Catharanthus lanceus **(Bojer ex A.DC.) Pichon**

Catharanthus longifolius **(Pichon)**

Catharanthus ovalis **(Markgr.)**

Catharanthus scitulus **(Pichon)**

Catharanthus trichophyllus

Catharanthus coriaceus

Source**: www.en.wikipedia.org, www.data:image.com, www.jungledragon.com, www.tropical.theferns.info, www.worldofloweringplants.com, www.powo.science.kew.org, www.inaturalist.org**

linear to oblong, not ciliate. Corolla tube is 11-22 mm long. Flowers are axillary and paired. The species is endemic to Madagascar.

C. ovalis

Perennial herb upto 1 m high, decumbent with many spreading stems from a tap root. Stems are reddish in colour and sub quadrangular. Leaves are sessile or short petiolate, blade ovate and cordate at base. Flowers are axillary and paired. The species is endemic to Madagascar.

C. pusillus

Annual herb of 5-45 cm high and erect. Stems and branches are winged and glabrous. Leaves are petiolate. Corolla is white in colour. Widest portion of corolla tube is distinctly above middle. Flowers are axillary, solitary or paired. The species is endemic to India and Sri Lanka.

C. roseus

Under shrub, 30-100 cm high, erect or decumbent. Stem terete, narrowly winged and pale green in colour. Leaves are petiolate, petiole 3-11 mm long. Flowers are axillary, solitary or paired. This species is endemic to Madagascar.

C. scitulus

Annual herb, 4-13 cm high, tender. Roots finely ramified, thin. Stems and branches are decumbent, winged and glabrous. Leaves are petiolate. Corolla lobes are pink or violet in colour. Corolla tube is upto 11 mm long and widest portion of the tube is at the middle. Flowers are solitary, axillary or terminal. The species is endemic to Madagascar.

C. trichophyllus

Under shrub upto 100 cm high, erect. Stems quadrangular, winged reddish or purple. Leaves sub sessile. Leaves are hairy on both sides. Flowers are axillary, solitary or paired.

5.2.0 Cultivars

Snoeijer (1996) proposed the following group cultivar classification of *Catharanthus roseus* cultivars.

Roseus Group (The corolla lobes are coloured)

'Apricot Delight', 'Blush Cooler', 'Dawn Carpet', 'Grape Cooler', 'Little Linda', 'Little Pinkie', 'Little Rosie', 'Madagascar Pink', 'Madagascar Rose', 'Pacifica Punch', 'Pacifica Red', 'Pink Carpet', 'Pink Panther', 'Pink Turkey', 'Pretty in Rose', 'Rose Carpet', 'Rose Cooler', 'Roseus', 'Sahara Pink', 'Sweet Elizabeth', 'Tall Crimson', 'Tropicana Blush', 'Tropicana Bright Eye'.

Ocellatus Group (The corolla lobes are white coloured)

'Birdie Yancy', 'Little Bright Eyes', 'Little Delicata', 'Morning Mist', 'Ocellatus', 'Parasol', 'Peppermint Cooler', 'Polka Dot', 'Sahara Bright Eye'.

Cultivars of Catharanthus

Viper Red

Pacifica Burgundy Halo

Sun Storm

Black Cherry Vinca

Cora Apricot

Cora Burgundy

Cora Red

Cora Cascade Shell Pink

Cora Cascade Polka Dot

Cultivars of Catharanthus

Cora Deep Lavender   Tattoo Raspberry

Cascade Pink Splash   Cora Strawberry

First Kiss Blueberry   Titan Icy Pink   Jaio Scarlet

Source: www.syngentaflowers.com, www.pinterest.ph, www.panamseed.com, www.rotarybotanicalgardens.org

Albus Group (The flowers have a green or white eye)

There are three subgroups within albus group

- **Albus sub-group** (corolla lobes are white coloured)
  'Albus', 'Little Blanche', 'Snowflake', 'Tropicana White', 'White Turkey'.
- **Unicolour sub-group** (Corolla lobes are in one colour)
  'Orchid Cooler', 'Passion', 'Pink Cooler', 'Tropicana Pink', 'Tropicana Rose'.
- **Bicolor sub-group** (Corolla lobes are in 2 colours)
  'Icy Pink Cooler', 'Pretty in Pink', 'Pretty in Soft Pink', 'Pretty in White'.

In addition to this classification, private seed companies released a number of F_1 hybrids and mixtures throughout the world according to the market requirement.

6.0 Genetics and Breeding

The chromosome number of periwinkle is 2n=16. It is self-compatible and self pollination is common. Out crossing may occur due to butterflies.

Flowering begins when plants are about 10-15 cm tall and is continuous. Flowers appear in pairs in the alternate leaf axils. The flowers are well adapted to self pollination according to various workers. Anthesis generally starts from 15.00-16.00 hrs. and continues until the next day. Anther dehiscence occurs just before anthesis and the pollen is shed as sticky mass. The stigma is found receptive between 06.00 and 10.00 hrs. and between 15.00 and 17.00 hrs. (Sreevalli, 2002). Automatic self-pollination does not occur in periwinkle and pollinators are necessary to bring about pollination (Sreevalli *et al.*, 2000). Simmonds (1960) considered that periwinkle is a self-pollinating species. Krishnan *et al.* (1979) used pink flower colour as the marker trait for determining out-crossing and found that natural out-crossing in periwinkle ranged from 1.7-13.8 per cent in white flowered plants. They opined that periwinkle is essentially a self-pollinating plant with not infrequent out-crossing, as flowers remain open up to 3 days for possible insect pollination.

6.1.0 Selection

A die-back resistant variety named 'Nirmal' was developed as a pure line selection from a single plant that survived in a severe die-back epidemic (Kulkarni *et al.*, 1999). Inheritance of resistance to die-back was studied using a die-back resistant dwarf mutant of variety, Nirmal, with a green stem, and a susceptible accession, 'OR', with a purple stem. From both quantitative and qualitative analyses of data, resistance to die-back appeared to be governed by a single gene and was inherited independently of genes governing dwarfness and stem pigmentation.

6.2.0 Hybridization

A flower bud which is about to open the next day is selected for emasculation. The adjacent flowers are removed. The pistil is situated just below the anthers. In case of seed parent anthers are removed. Due care should be taken so that latex does not stick to pistil during this process. The pistil is ready for pollination.

A flower which has started to wilt can be selected as pollen source. The corolla tube is to be cut off on top. The sap should be allowed to dry. The corolla tube should be cut lengthwise between the anthers. When the anthers are pushed they will release the pollen. The pollen can be collected and put on the stigmatic surface. The pollinated flower is labelled indicating seed parent and male parent. The remaining tube will fall off after 2-3 days just leaving the sepals with a small fruit. The seed can be collected when the fruit wall turns green to brown. Late harvesting is not possible as the fruit will burst open and the seed will scatter.

6.3.0 Polyploidy

Induced auto-polyploids generally have larger and thicker leaves, stems, roots, flowers and fruits. Seeds or apical buds of young seedlings were treated with different concentrations of colchicine solutions ranging from 0.01-1.0 per cent to induce tetraploidy. Colchicine treatment of apical buds was more effective in inducing tetraploidy than seed treatment. Increase in the length of stomata and pollen grain diameter have been observed as the two characteristic effects of tetraploidy. Auto-tetraploids were found to be more vigorous in growth with broader leaves, larger stomata, flowers, pollen grains and embryos but had low pollen fertility, poor fruit set and low seed production as compared with diploids (Janaki Ammal and Bezbaruah, 1963; Mohan Kumar, 1980). In auto-tetraploids, pollen fertility was low (32–43 per cent) leading to poor seed set (17.5–22.5 per cent). Tetraploids had higher nitrogen utilization efficiency than diploids and performed well under low fertility conditions. Induced auto-tetraploids were also found to be highly resistant to die-back, and collar and root rot.

6.4.0 Mutation Breeding

Three distinct reduced plant height mutants, 'dwarf', 'semi-dwarf' and 'bushy', respectively, about 60, 40 and 30 per cent shorter than their parental variety, Nirmal have been reported in periwinkle (Kulkarni *et al.*, 1999, 2009). Two other types of male sterile mutants *viz.*, indehiscent anthers (functional male sterility) and pollen-less anthers governed by a single recessive allele and duplicate recessive alleles, respectively, have also been reported (Sreevalli *et al.*, 2003; Kulkarni and Baskaran, 2008). The mutant with indehiscent anthers had relatively smaller anthers and about 30 per cent lesser number of pollen grains but was otherwise phenotypically normal, had high pollen fertility and exhibited high seed set on artificial selfing. A mutant described as 'leafless inflorescence', in which flowers are borne on nodes without leaves, has been reported and the locus 'lli' has been mapped (Chaudhary *et al.*, 2011). The mutant produced more number of flowers per plant than its parent and further enhancing its horticultural value. New improved horticultural genotypes were developed by crossing this mutant with other genotypes with different flower colours and plant habit (Kumar *et al.*, 2012).

7.0 Growth and Flowering

7.1.0 Temperature and Light

Catharanthus is adapted to wide range to climatic conditions. The plant thrives well in hot, humid as well as dry weather. But, it cannot tolerate frost and freezing conditions. The plant prefers day temperature of 27-30°C and night 20-22°C. Temperatures below 17°C causes leaf yellowing and stunted plant growth. If the plants are exposed to <15°C, the flowering is delayed.

The growth and development of vinca is optimal in full sunshine. The plant prefers to grow in partial shade. The lights conditions in which plant grow, however, affect the performance of the plant. In partial shade, the leaves tend to be bright green than direct exposure to sunlight. Ample sunlight however produces more flowers.

7.2.0 Growth Regulators

Catharanthus respond well to application of different plant growth regulators. Application of 25 ppm NAA (foliar spray) significantly increased the number of flowers per plant as well as flower size (Pahare and Das, 2020). Gibberellic acid @ 300 ppm was found to be superior in increasing the growth, flowering and yield of dry leaf (2784.16 kg/ha) compared to others (Sendhilnathan *et al.*, 2016). Uniconazole at 1 ppm effectively controlled plant height, but paclobutrazol causes leaf spotting and severe stunting (vtechworks.lib.vt.edu).

As the catharanthus crop is commercially cultivated for valuable alkaloids present in leaves and roots. Therefore most of the work is oriented towards increasing the content of secondary metabolites with PGR application. Pan *et al.* (2010) revealed that short-term application of ethephon and SA (Salicylic acid) significantly enhanced the accumulation of vinblastine, vindoline and catharanthine, while the application of ABA and GA3 have the opposite effects in catharanthus. CCC (Chlormequat chloride) significantly increases the vinblastine accumulation but greatly decreases the accumulation of its two precursors, vindoline and catharanthine. Jaleel *et al.* (2009) evaluated traditional and non-traditional plant growth regulators altering phytochemical constituents in *Catharanthus roseus* viz. paclobutrazol (PBZ), gibberellic acid (GA3) and *Pseudomonas fluorescens* elicitors (PF Elicitors). The effectiveness of PGRs in increasing serpentine contents is PBZ > PF Elicitors > GA3. They also reported differential action of PGRs in altering the alkaloids content.

8.0 Propagation

8.1.0 Seed Propagation

C. roseus is a fast-growing perennial plant that spreads principally by seed. Seed may remain dormant for several weeks after maturity. The optimum temperature for germination is 20–25°C, and the germination rate is in general over 95 per cent. The seeds remain viable for 3–5 years (Prota, 2015). The seeds will germinate in 2-3 days after sowing. The seedlings can be transplanted in 20-30 days depending on cultivar and season.

7.2.0 Vegetative Propagation

Catharanthus can also be successfully propagated by stem cuttings. Tip cuttings with two pairs of leaves from a non-flowering stem are collected. The second pair

Propagation and Growing of *Catharanthus*

Seeds

Seedlings in Protrays

Vinca Seed Pods

Open Field Cultivation

Mass Production in Pots

Seed Production in Polyhouse

Cryopreservation of Embryogenic Cells and Plant Regeneration in *Catharanthus roseus*

(a) Cryopreserved embryogenic suspended cells produced colonies on plated solid medium. (b) Induced callus (arrow, top left) and embryos (arrow, top right and bottom left) from cryopreserved culture. (c) Mature somatic embryo with shoot and root axis (arrow). (d) Plant regenerated from cryopreserved embryogenic cell. [*Source*: Fatima *et al.*, 2009).

Source: www.garden.org, www.commons.wikimedia.org, www.extension.uga.edu, www.123rf.com, www.depositphotos.com, www.link.springer.com

Micropropagation in *Catharanthus*

(a)Rooting : NN medium + 0.1 mg/l kinetin + 0.5 mg/l IBA after 1 week of culture. (b) Blooming : NN medium supplemented with 0.1 mg/l kinetin and 0.5 mg/l IBA after 6 passages. (c) Flowering in green house condition (d) Callus culture White line: B5 medium + 0.1 mg/l kinetin + 1.0 mg/l IAA (96 passage). (e) Callus culture Green line: B5 medium + 0.1 mg/l kinetin and 1.0 mg/l IAA (96 passage).

Source: Pietrosiuk *et al.,* 2007.

of leaves is removed and the first pair are cut into half. The rooting end is treated with rooting powder and placed in medium. Rooting occurs well in mist chamber. The rooting occurs in two to three weeks.

8.3.0 Micropropagation

Several micropropagation protocols were standardized and reported for the mass propagation of *Catharanthus* by using various explants *viz.*, nodal segments (Raouf Fard *et al.*, 2008; Bakrudeen *et al.*, 2011; Faheem *et al.*, 2011; Rahmatzadeh *et al.*, 2014), shoot tips (Taha *et al.*, 2008; Singh *et al.*, 2011; Al-Oubaidi and Amin 2014; Moghe *et al.*, 2016), internodes (Swanberg and Dai, 2008), petioles (Zulkepli and Samad, 2011), hypocotyles (Yuan *et al.*, 2011; Makhzoum *et al.*, 2015; Alam *et al.*, 2017), leaves (Verma and Mathur, 2011; Rashmi and Trivedi, 2015) and mature zygotic embryos (Dhandapani *et al.*, 2008). In most of the studies, MS medium was used as basal medium for both direct and indirect *in vitro* culture establishment, multiplication and rooting. Higher percentage of callus induction and shoot regeneration was reported in MS basal medium supplemented with BAP (0.5 to 7.0 mg/l) + NAA (0.2 to 4.0 mg/l). The concentration and combination of growth regulators varied based on the genotype and type of explant. Dhandapani *et al.* (2008) cultured the mature zygotic embryos on MS medium supplemented with TDZ (7.5 µm) and incubated in dark for somatic embryo induction. Organogenesis reported with TDZ (2.5 µM) or BAP (2.2 µM) + NAA (5.3 µM) when cultures were incubated in 10 h dark and 14 h light condition in *Catharanthus* cv. 'Little Bright Eye'. For higher *in vitro* shoot multiplication, 0.5 to 4.0 mg/l BAP was used in combination of 1.0 to 4.0 mg/l NAA in different genotypes of *Catharanthus*. In most of the studies ½ MS media supplemented with 3 to 5 mg/l IBA was reported as best rooting treatment. Some studies are also reported the synergistic effect of lower doses of NAA along with IBA. Supplementing the optimal media with 250-350mg/l tryptophan improved the shooting and rooting in this species (Rahmatzadeh *et al.*, 2014).

9.0 Cultivation

9.1.0 Soil and Climate

It prefers warm and sunny climate. It grows well throughout tropics and sub tropics. The optimum temperatures are 20-35°C. Night temperatures below 12°C causes leaf rolling, chlorosis, leaf drop or purpling of leaf tips.

It is susceptible to wet soil and water logging conditions. It grows well in any well drained soil. The EC of the medium should not be more than 1mmhos/cm. The soil and medium should have high porosity. They are susceptible to micro nutrient deficiencies when pH is above 6.3.

9.2.0 Manuring and Fertilization

The crop is susceptible to iron deficiency and low nitrogen levels (Campbell *et al.*, 2001). Fertigation with a solution of 50-100 ppm (NPK: 20-20-20) once in a week and calcium nitrate 100 ppm once in a month improved plant growth and flowering. Micronutrients can be applied as foliar spray and media application to counter nutrient deficiencies.

Normal Foliar Analysis Ranges for *Catharanthus roseus**

Element	*Range*	*Element*	*Range*
N	2.72 - 6.28	Fe	72 - 277
P	0.28 - 0.64	Mn	135 - 302
K	1.88 - 3.48	Zn	30 - 51
Ca	0.93 - 1.13	Cu	6 - 16
Mg	0.32 - 0.78	B	21 - 49
S	0.22 - 0.50	Mo	0.14 - 0.46
Na	0.01 - 0.156	Al	34 - 136

* Source: Plant Analysis Handbook II. M. Mills (ed.), Micro-Macro Publishing.

10.0 Diseases and Pests

10.1.0 Diseases

The most common diseases of vincas are black root rot (caused by *Thielaviopsis*) and damping-off (caused by *Rhizoctonia, Pythium* and *Phytophthora*).

10.1.1 Black Root Rot (*Thielaviopsis* sp.)

It appears by chlorosis followed by wilting and death. The roots look black from the black spores that reside within them. Roots are more susceptible to black root rot if they are exposed to high EC, high pH (>5.5) and low temperatures. The entire roots become black.

The disease can be controlled by proper sanitation and keeping the pH of the medium below 6.0 and reducing the nitrogen fertilization. Application of *Trichoderma* and *Mycorrhiza* formulations in the medium will also help in prevention of the disease.

10.1.2 Damping Off (*Pythium* sp.)

This disease affects the seedlings in nursery. Before emergence brown coloured spots appear on radicle and plumule causing pre-emergence death of the seedlings. After emergence, brownish necrotic spots appear on the collar portion of the seedlings. These necrotic spots in severe cases form a girdled portion leading to the collapse of seedlings. The disease can be controlled by prophylactic drenching of nursery beds with *Trichoderma* formulations or spraying chemicals like metalaxyl+mancozeb @ 2 g/l.

10.1.3 Phytophthora Rot

The infected plants have yellow and stunted growth and later die. Soil and media sanitation should be maintained. Over watering and higher nitrogen and application should be avoided. The fungus survives in infested soil and plant debris. The diseases is favoured by warm and wet conditions. Sanitation should be maintained and potting media should not be reused to prevent the disease. Fungicides like Fosetyl Al are recommended for control of Phytopthora.

10.1.4 Rhizoctonia Rot

Infected plants break or fall at soil line. Wilting and plant death is caused by brown cankers. Warm and humid conditions favour the disease. Sanitation should be maintained and potting media should not be used as the fungus survives in soil and infested debris. Application of *Trichoderma* formulations in the media or as a drench will help in prevention of disease. Fungicides like Pencycuron are recommended for control.

10.1.5 Leaf Spot and Blight (*Alternaria alternata*)

The symptoms appear as small brownish spots on leaves. In case of alternaria infection, the spots show concentric circles. These leaf spot diseases can be controlled by seed treatment and foliar spray with formulations like chlorothalonil or mancozeb.

10.1.6 Botrytis Blight (*Botrytis cinerea*)

The infected flowers and leaves give a gray appearance with water-soaked spots. It is favoured by cold and humid conditions. The fungus survives in infested plant debris. Sanitation is very important in prevention of this disease. The fungicide chlorothalonil was reported to be very effective against this disease.

Virus

10.1.7 Aster Yellows (Phytoplasma)

The disease is evident by yellowing, little leaf and phyllody of flowers. The disease is transmitted by leaf hoppers. The infected plants need to be immediately uprooted and burnt to further prevent the spread. Spraying proprietary insecticides will control the vector population and further prevent spread of the disease.

10.2.0 Pests

10.2.1 Aphids

Greenish or peach coloured sucking insects that suck sap and leave a sticky residue that attracts ants. Leaves become flecked and silvery due to damage. These can easily be controlled by spraying Acephate @ 0.5 g/l or Acetamiprid @ 0.2 g/l.

10.2.2 White Flies

They are white flying and sucking pests that suck sap and cause enormous damage to the ornamental value of the plants. In initial stages they can be effectively controlled by spray of neem oil or Azadirachtin 10000 ppm @ 1 ml/l. Application of Dimethoate @ 2ml/l or Imidacloprid @ 0.5 ml/l or Acetamiprid @ 0.2 g/l.

10.2.3 Spider Mites

These are minute that are invisible and colonize the underside of leaves. Their infestation is severe under hot and dry conditions. The foliage turns yellow and stippled. Webbing is visible on infested parts. They can be effectively controlled

Diseases and Pests of *Catharanthus*

Rust | Botrytis Blight | Stem Cankers

Root Rot | Tomato Spotted Wilt Virus | Stem Rot

Phytophthora Stem Blight | Big Bud | Black Root Rot

Vinca Leaf–folder Larva | Leaf Hopper | Aphids

Source: www.byf.unl.edu, www.hgic.clemson.edu, www.researchgate.net, www.pthorticulture.com, www.uaex.edu, www.pnwhandbooks.org

by spraying proprietary miticides like Abamectin @ 0.025 per cent or Propargite @1.5 ml/l.

11.0 Biotechnology

Catharanthus is one of the most important medicinal plants, which is having more than 130 alkaloids, including the powerful antitumor drugs vinblastine and vincristine which are used in the treatment of cancer. Many *in vitro* studies were conducted to produce the secondary metabolites in large scale through plant tissue, organ culture, cryopreservation and transformation. One of the most effective strategies for enhancing the biotechnological production of alkaloid compounds is elicitation. The biosynthetic pathway of terpenoidindole alkaloids (TIAs) in *C. Roseus* and characterization of the related genes encoding the enzymes involved in this pathway has been studied extensively (Rady, 2019). More than 50 biosynthetic events are composed of the involved genes, enzymes, regulatory genes, and intra/intercellular compartments (Zhao *et al.*, 2013).

Both *Agrobacterium tumefaciens* and *Agrobacterium rhizogenes* mediated Genetic transformation studies were extensively reported in *C. roseus* for enhancing the alkaloid/TIAs production (Das *et al.*, 2020). RAPD and ISSR molecular markers were used for testing the genetic fidelity of *in vitro* raised plants of Catharanthus. Kumar *et al.* (2013) used 20 RAPD markers for testing the clonal fidelity and found monomorphic banding pattern of regenerants with their mother plant. Verma *et al.* (2014) employed 17 ISSR primers for checking the genetic stability of transgenic *C. roseus* and found that 72.1 per cent were monomorphic and 27.9 per cent were polymorphic.

12.0 References

Alam, P., Khan, Z.A., Abdin, M.Z., Khan, J.A., Ahmad, P., Elkholy, S.F. andSharaf-Eldin, M.A. (2017) *3 Biotech.*, **7**: 26-36.

Al-Oubaidi, H.K.M. and Amin, A.S.M. (2014) *World J. Pharm. Pharm Sci.*, **3**(6): 2101–2107.

Bakrudeen, A.A.A., Shanthi, G.S., Gouthaman, T., Kavitha, M.S. and Rao, M.V. (2011) *Acta Bot.Hungar.*,**53**(1–2): 197–209.

Campbell, C.R., Whipker, B.E., Pitchy, D.S., Nelson, P.V. and Gibson, J.L. 2001. *Nutrient deficiencies of Vinca*. GM Pro Magazine, March 2001 pp. 36-42.

Chaudhary, S., Sharma, V., Prasad, M., Bhatia, S., Tripathi, B.N., Yadava, G. and Kumar, S. (2011) *Sci.Hortic.*, **129**: 142–153.

Das, A., Sarkar, S., Bhattacharyya, S. And Gantait, S. (2020) *Appl. Microbiol. Biotechnol.*, **104**(11): 4811-4835.

Dhandapani, M., Kim, D.H. and Hong, S.B. (2008) *In Vitro Cell Dev. Biol. Plant.*, **44**: 18–25.

Duge de Bernonville, T., Clastre, M., Besseau, S., Oudin, A., Burlat, V., Glevarec, G., Lanoue, A., Papon, N., Giglioli-Guivarc'h, N., Benoit St-Pierre, B. and Courdavault, V. (2015) *Phytochemistry*, **113**: 9–23.

Faheem, M., Singh, S., Tanwer, B.S., Khan, M. and Shahzad, A. (2011) *Adv. Appl. Sci. Res.*, **2**(1): 208–213.

Fatima, S., Mujib, A., Nasim, S.A. and Siddiqui, Z.H. (2009) *Plant Cell Tissue Organ Cult.*, **98**: 1–9.

http://www.vtechworks.lib.vt.edu

Jaleel, C.A., Gopi, R., Gomathinayagam, M. and Panneerselvam, R. (2009) *Process Biochmemistry*, **44**(2): 205-209.

JanakiAmmal, E.K. and Bezbaruah, H. P. (1963) *Proceedings of the National Academy of Sciences, India, Section B.*, **57**: 339–342.

Krishnan, R., Naragund, V. R. and Vasantha Kumar, T. (1979) *Current Sci.*, **48**: 80–82.

Krishnan, R. (1995) *Periwinkle*. In: Chadda, K.L. and Gupta, R. (eds) Advances in Horticulture, vol. 11. New Delhi: Malhotra Publishing House, pp. 409–428.

Kulkarni, R.N. (1999) *Plant Breed.*, **118**: 561–564.

Kulkarni, R.N., Baskaran, K. and Sreevalli, Y. (2008) *J. Applied Hort.*, **10**: 20–23.

Kulkarni, R.N., Baskaran, K., Shyamaprasad, D.V. and Kulkarni, S.S. (2009) *Euphytica.*, **170**: 309–316.

Kumar, A., Prakash, K., Sinha, R.K. and Kumar, N. (2013) *Applied Biochem. Biotechnol.*, **169**: 894–900.

Kumar, S., Chaudhary, S., Kumari, R., Sharma, V. and Kumar, A. A. (2012) *Proceedings of the National Academy of Sciences, India, Section B, Biological Science.*, **82**: 399–404.

Makhzoum, A., Bjelica, A., Petit-Paly, G., Mark, A. and Bernards, M.A. (2015) *All Res. J. Biol.*, **6**: 1–9.

Mills, H.A. and Jones, J.B. (1996) Bedding Plants. *Plant Analysis Handbook II.* Micro-Macro Publishing, Athens, GA. pp. 192.

Moghe, S., Laud, D., Dehankar, B., Moghe, R., Sadhu, P., Ade, G., Ishani, B. and Hadke, A. (2016) *Int. J. Life Sci.* Special Issue **A6**: 1–4.

MohanKumar, G.N. (1980) *M. Sc. (Hort.) Thesis*, University of Agricultural Sciences, Bangalore, India.

Noble, R.L., Beer, C.T. and Cutts, J.H. (1958) *Annals of the New York Academy of Sciences*, **76**: 882–894.

Pahare, P. and Das, J.N. (2020) *Int. J. Curr. Microbiol. App. Sci.*, **9**(6): 1961-1967.

Pan, Q., Chen, Y., Wang, Q., Yuan, F., Xing, S., Tian, Y., Zhao, J., Sun, X. and Tang, K. (2010) *Plant Growth Regul.*, **60**: 133-141.

Pietrosiuk, A., Furmanowa, M. and Lata, B. (2007) *Phytochem Rev.*, **6**: 459–473.

Prota (2015) PROTA4U web database. Grubben, G.J.H. and Denton, O.A. eds. Wageningen, Netherlands: Plant Resources of Tropical Africa. http: //www.prota4u.info

Rady, M.R. (2019) Plant Biotechnology and Periwinkle. In: *Plant Biotechnology and Medicinal Plants*. Springer, Cham.https: / / doi.org / 10.1007 / 978-3-030-22929-0_1

Rahmatzadeh, S., Khara, J. and Kazemitabar, S.K. (2014) *J. Sci. Kharazmi Univ.*, **14**(3): 249–260.

RaoufFard, F., Moieni, A. and Omidbaigi, R. (2008) *J. Agric. Sci. Technol.*, **10**: 337–344.

Rashmi, R. and Trivedi, M.P. (2015) *Biochem. Anal Biochem.*, **4**(4): 1–6.

Ross, I.A. (1999) *Medicinal Plants of the World: Chemical Constituents, Traditional and Modern Medicinal Uses*. New Jersey: Humana Press.

Sendhilnathan, R., Karuppaiah, P., Rajkumar, M. and Sureshkumar, R. (2016) *Int. J. Curr. Res.*, **8**(7): 35505-35506.

Simmonds, N.W. (1960) *Heredity.*, **14**: 253–261.

Singh, J. (1996) In: Handa SS and Kaul MK (eds) *Supplement to Cultivation and Utilization of Medicinal Plants*. Jammu-Tawi, India: Regional Research Laboratory, pp. 199–206.

Singh, R., Kharb, P. and Rani, K. (2011) *World J. Agric. Sci.*, **7**(6): 699–704.

Snoeijer, W. (1996) *Catharanthus roseus*, the Madagascar Periwinkle, a review of its cultivars. Wageningen Agricultural University Papers 96-3.

Sreevalli, Y., Kulkarni, R.N. and Baskaran, K. (2002) *J. Heredity.*, **93**: 55–58.

Sreevalli, Y., Baskaran, K. and Kulkarni, R.N. (2003) *Ind. J. Gen. and Plant Breed.*, **63**: 365– 366.

Sreevalli, Y., Baskaran, K., Kulkarni, R.N. and Kumar, S. (2000) *Current Sci.*, **79**: 1648–1649.

Svoboda, G.H. and Blake, D.A. (1975) In: Taylor RW and Farnsworth NR (eds) *The Catharanthus Alkaloids. Botany, Chemistry, Pharmacology, and Clinical Use*. New York: Marcel Dekkar, pp. 45–83.

Swanberg, A. and Dai, W. (2008) *Hort Science*, **43**(3): 832–836.

Taha, H.S., El-Bahr, M.K. and Seif-El-Nasr, M.M. (2008) *J. Appl. Sci. Res.*, **4**(8): 1017–1022.

Thomas, P.A. and Latimer, J.G. (1996) *J. Plant. Nutrition.*, **18**: 2127-2134.

Tyler, V.E. (1988) *Planta Medica.*, **54**: 95–100.

Van Bergen, M.A. (1996) Revision of *Catharanthus* G. Don Series of Revisions of ApocynaceaeXLI,Wageningen Agricultural University Papers., 96-3.

Verma, P., Khan, S.A., Mathur, A.K., Shanker, K. and Kalra, A. (2014) *Plant Cell Tissue Organ Cult.*, **118**: 257–268.

Verma, P. and Mathur, A.K. (2011) *Plant Cell Tissue Organ Cult.*, **106**: 401–408.

Yuan, F., Wang, Q., Pan, Q., Wang, G., Zhao, J., Tian, Y. and Tang, K. (2011) *Afr. J. Biotechnol.*, **10**(66): 14786–14795.

Zhao, L., Sander, G.W. and Shanks, J.V. (2013) *Adv. Biochem. Eng. Biotechnol.*, **134**: 23–54.

Zulkepli, A.Z. and Samad, A.A. (2011) *Int. J. Biosci.*, **1**(1): 31–35.

Zhu, X., Zeng, X., Sun, C. and Chen, S. (2014) *Frontiers of Medicine.*, **8**: 285–293.

Gazania

Ganesh B. Kadam, Tarak Nath Saha, P. Naveen Kumar, D.V.S. Raju, P.R. Jadhav and K.V. Prasad

1.0 Introduction

Gazania is a popular ornamental flowering plant known for its highly attractive an bright colour flowers. It is commonly known as treasure flower or botter bloom. This popular African daisy is similar in appearance to the genus Arctotis.

In India it is very popular plant among the gardeners, nurserymen and home growers because of its array of colours and extended duration of flowering. It is highly suitable for beds and pots. Easy to grow and require less maintenance for cultivation. In north Indian states it is popularly grow as winter season flowering ornamental plant which flowers till summer. It is easily grown by seeds and plants are compact mounding habit and reach 30 cm and spreads along the ground. The mix gazanias are popular for rock garden, containers, beds and pots. In Punjab (India) it is commercially cultivated for seed production.

2.0 Importance and Uses

2.1.0 Aesthetic Uses

Gazania is herbaceous ornamental plant produces large, daisy-like composite flowers in brilliant shades of yellow and orange. The trailing types of cultivars are particularly useful in hanging baskets, containers, beds and borders. They are often planted as drought-tolerant groundcover, and are commonly used in other areas as well. It helps control soil erosion and to add colour to the rock gardens, xeriscapes, commercial and home gardens.

It is a major annual component of plant exhibition during spring time mass-flowering in south-western Cape and Namaqualand regions of South Africa. It is locally seen after rainfall in the arid biomes of the Succulent Karoo and Nama-Karoo (Mucinaand Rutherford, 2006).

2.2.0 Therapeutic and Other Uses

Gazania has been reported in folk medicine to prevent miscarriage and tooth ache. It was also incorporated in purgative preparations especially with aloes. Tshabalala *et al.* (2016) reported the traditional use of the *Gazania krebsiana* plant among the Basotho tribe for treating diabetes mellitus and cardiac-related ailments. Balogun and Ashafa (2016) confirmed the antioxidant, antimicrobial, anthelmintic and hepato-protective properties of the aqueous leaf extract from *Gazania krebsiana*. Desoukey *et al.* (2016) reported that *Gazania longiscapa* is a richer source of phenolic acids as well as valuable sources of rutin beside having hepato-protective and antioxidant activity. The rhizome extract (cytotoxic n-hexane fraction) of *Gazania linearis* were found effective against MCF-7 and COLO-205 cancer cell lines.

The plant also has a range of medicinal uses and possible food source with health benefits (Reddy *et al.*, 2014). The flowers of *Gazania krebsiana* sub Sp. Serrulata's (Asteraceae) is reported to be useful in treating earache in babies and are also used for treating sterility as per Flora Database of Swazilands.

Uses of Gazania

In Path Divider

In Pot

In Rock Garden

As Ground Cover

In Garden Bed

In Hanging Pots

In Window Box

Along Walking Path

Near Beach

In Sand Garden

Step Planting

In Decorative Wall

Source: **www.images.app.goo.gl, www.in.pinterest.com, www.marketplace.secondlife.com, www.garden.org**

3.0 Origin and History

Gazania is a genus of flowering plants in the family Asteraceae, native to Southern Africa. The genus is named 'Gazania' to honour the 15th century Greek scholar 'Theodore de Gaza' (1398-1478). He is known for his invaluable contribution in translation of some important botanical works of Theophrastus from Greek into Latin. Possibly the genus name might have been derived from the Latin word *gaza* meaning 'treasure' in reference to the attractive flowers. Hence, also got the common name 'treasure flower' for this plant.

The genus is a mix of widely distributed species that are found throughout southern Africa and narrow endemics (such as *G. caespitosa* and *G. schenckii*) that have been rarely seen. Some species of Gazania are also found in the Fynbos Biome and Succulent Karoo Biomes belonging to the "Greater Capensis" region (Born *et al.*, 2007).

The species which were cultivated in Egypt as ornamental plants were *G. linearis, G. rigens* and *G. Nivea* (Kady *et al.*, 2015; Ahmed *et al.*, 1981 and Taha *et al.*, 2008). Other existing species includes *G. hypridus, G. krebsiana, G. lichenstenii* and *G. splendens.*

4.0 Taxonomy

Gazania is a common horticultural crop widespread across southern Africa. It is a small genus of 16 species, endemic to southern Africa. It includes seven other genera namely *Berkheya, Gorteria, Cuspidia, Didelta, Heterorhachis, Cullumia* and *Hirpicium*. However, very little is known about the biology and evolutionary history of the genus. The genus *Gazania* is placed within the tribe *Arctotideae*, subtribe *Gorteriinae*, based on the connate, acute involucral bracts, the four-lobed, neuter ray florets, the sclerified margins of the disc floret lobes and molecular data (Funk *et al.*, 2004; Karis, 2007).

Sixteen species with ten subspecies were identified by Roessler in 1959 based on morphological characters observed from herbarium specimens and geographical distribution of these samples. The plants were not studied in wild. Within the genus, two main growth forms were described *viz., Gazania* and *Hirpicium* (Roessler, 1959). Species with a shortened stem, dense rosettes at the base, from which a leafless un-branched stem arises bearing solitary inflorescences was described to be characteristics of Gazania group while the species with a well-developed stem, with evenly arised leaves along the stem was observed to be characteristic of *Hirpicium* species. The exact relationships between *Gazania* and *Hirpicium* are somewhat uncertain and requires further study (Funk *et al.*, 2004; Karis, 2006). Also, there are few studies on taxa from the Succulent Karoo Biome component of the Greater Cape taxa. Preliminary data suggest that it is more recent and have lineated from the semi-arid west coast flora as a result of increase in aridity in the Middle Miocene (Linder, 2003). More studies are required on lineages from this region.

Two new distinct species of Gazania were identified *viz., Gazania splendidissima* Mucina, Magee and Boatwr. from the arid Namaqualand coast (Northern Cape) and *Gazania lanata* Magee and Boatwr. from the surrounds of Robertson (Western Cape). These are provided with an updated key for recognition (Magee *et al.*, 2011).

Goldblatt and Manning, (2000) described species of *gazania* into annuals and perennials types. Annuals species are *G. lichtensteinii* Less, *G. pectinata* (Thunb.) Spreng and *G. tenuifolia* Less. Whereas, in perennials type gazania again these are two types, one is mat-forming perennials with stems rooting at nodes, or shrublets which includes *G. maritime* (Levyns), *G. Othonnites* (Th unb.) Less, *G. Rigens* (L.) Gaertn and another type is tufted, acaulescent perennials which includes *G.ciliaris* DC, *G. krebsiana* Less., *G. lanata* Magee and Boatwr., *G. linearis* (Th unb.) Druce, *G. rigida* (Burm.f.) and *G. serrata* DC.

5.0 Morphology

Gazania is a tender perennial that features decumbent stem, that spread along the ground. Narrow but variably-shaped, dandelion-like leaves (narrow spoon-shaped to lobed) are silvery green. Flowering stems typically rise 15-25 cm tall topped by solitary, daisy-like ray flowers (to 7.5-10 cm wide) with contrasting center disks. They produce large, daisy-like composite flower heads in brilliant shades of yellow and orange, over a long period in summer. Ray flowers of this species are orange with black eyes at the bases and orange-brown disks. *G. rigens* is a parent in many of the gazanias found today.

6.0 Species and Cultivars

6.1.0 Species

Several species and hybrids of gazanias are widely cultivated in gardens across the world due to its attractive flowers with strip of contrast colour on petals. *Gazania* was last revised by Roessler (1959) who recognised 16 species, all of which (with the exception of the wide spread *G. krebsiana* Less., extending into tropical East Africa) are endemic to southern Africa and mostly restricted to the GreaterCape Floristic Region. Later two distinct new species of Gazania were described one by Howis *et al.* (2009) from the Namaqualand coast in the Northern Cape (=Gazaniasp. nov.) and the other by Goldblatt and Manning, (2000) from the surrounds ofRobertson in the Western Cape (=Gazania sp.1).Magee *et al.* (2011) amended and updated list of species with distinctive characteristics to facilitate identification of the species is given in Table 1. It is based on the species concepts proposed by Roessler (1959) and on a study from the herbarium material.

6.2.0 Cultivars

Gazanias are easy to grow producing bright yellow, orange or red, daisy-like flowerswhich makes it a popular choice in gardens. The varieties with larger flowers, shorter peduncles, compact flowers, well suited for various climatic conditions are popular among professional gardeners and consumers. The popular cultivated varieties in clumping type gazania (*Gazania rigens*) are 'Aztec', 'Burgundy', 'Copper King', 'Fiesta Red', 'Goldrush' and 'Moonglow'. The popular variety of *G. rigens* are 'Xingbai', 'Hongwen', 'ZhongguoXunzhangju' and 'Richu' (Zeng *et al.*, 2016). The popular varieties of *G. spiendens* are 'Sundance Red', 'Sundance Yellow', 'Sundance Striped', 'Sundance Mixture', 'Minister Yellow', 'Minister Mixture', 'Golden

Margarita', 'Yellow Margarita', 'Grandiflora Mixture', 'Chansonette Mixture', 'Sunshine Giants' and 'Special Mixture'.

Table 1: Characteristics of different *Gazania* Species

Sl.No.	*Species*	*Characteristics*
1.	*G. lichtensteinii*	Plants annual, involucral base with a prominent collar.
2.	*G. tenuifolia*	Plants annual, involucre with several rows of linear-lanceolate parietal bracts extending from the base; base of involucre with afleshy truncate disc.
3.	*G. pectinata*	Plants annual, involucre distinctly connate below for at least half its length, parietal bracts irregularly placed, oblong; base of involucre obtuse, without a fleshy truncate disc.
4.	*G. rigens*	Plants perennial, leaves simple, alternate and spaced along the trailing stems; petioles not ciliate.
5.	*G. maritima*	Plants perennial, leaves pinnate, alternate and spaced along the trailing stems; petioles ciliate.
6.	*G. othonnites*	Leaves rosulate, basal or crowded at the tips of the branches, leaf surface glabrous beneath.
7.	*G. linearis*	Petioles not ciliate; involucre base somewhat intrusive;plants often with a basal tuft of persistent leaf bases.
8.	*G. ciliaris*	Petiole conspicuously ciliate; involucre base not intrusive, somewhat fleshy; leaf bases not persisting as a basal tuft.
9.	*G. jurineifolia*	Ray florets white; leaves pinnatifid, segments subtriangular; leaf segments and involucral bracts prominently spine tipped.
10.	*G. schenckii*	Woody shrublet; uppermost leaf surface densely whitefelted-araneose.
11.	*G. leiopoda*	Rhizomatous herb, uppermost leaf surface araneose. Involucre broadly campanulate, 10–18 mm wide; all leaves regularly pinnatisect.
12.	*G. heterochaeta*	Rhizomatous herb, uppermost leaf surface araneose. Involucre campanulate, 7–10 mm wide; leaves simple toweakly pinnatilobed
13.	*G. caespitosa*	Uppermost surface of the leaves glabrous to sparsely Echinate. Densely branched suffrutex; leaves simple, linear, <3 mm wide, spine-tipped.
14.	*G. krebsiana*	Rhizomatous herb; leaves simple, linear-lanceolate to pinnatisect, usually >3 mm wide, segments not prominently spine-tipped.
15.	*G. lanata*	Involucre densely pilose to densely echinate, also densely lanate.
16.	*G. splendidissima*	Uppermost leaf surface, in addition to being sparselyechinate, also prominently mealy; leaves semi-succulent; suffrutices with woody prostrate branches; involucre usually appearing distinctly maculate due to the contrast between the numerous dark hair bases and the prominent mealy involucre surface.
17.	*G. leiopoda*	Involucre broadly campanulate, 10–18 mm wide; upper surface of leaves densely echinate.
18.	*G. serrata*	Leaf lobes linear to oblanceolate (rarely obovate) withirregularly serrate margins (rarely entire); innermost terminal involucral bracts acuminate; plants sticky.
19.	*G. rigida*	Leaf lobes elliptic or sometimes linear to oblanceolatewith invariably entire margins; innermost terminal involucral bracts acute; plants not sticky.

Species of Gazania

Gazania rigens | *Gazania jurineifolia* | *Gazania krebsiana*

Gazania maritima | *Gazania pectinata* | *Gazania lichtensteinii*

Gazania longiscapa | *Gazania rigida* | *Gazania heterocyst*

Gazania linearis | *Gazania tomentosa* | *Gazania splendens*

Source: www.en.wikipedia.org, www.commons.wikimedia.org, www.gardensonline.com.au

Varieties of Gazania

Big Kiss Orange — Big Kiss Flame — Big Kiss Red

Big Kiss White — Big Kiss White Flame — Big Kiss Yellow

New Day Yellow — New Day Red Stripes — New Day Rose Stripes

New Day Red Shades — New Day Bronze Shades — New Day Pink Stripes

Source: www.syngentaflowers-us.com, www.panamseed.com

There are some types/series available in gazania which are as follows (www.thespruce.com/gazania-flowers-1315701).

Chansonette series: The varieties flower very early and are suitable for areas having short growing seasons.

Creamsicle series: They bear white flowers. The petals are look like ivory sport with a simple bronze center disk.

Daybreak series: It is early and plant bears large flower. It is easily propagated from seeds and flowers in sunset or stripes colours.

Sunbather's Sunset: Its flower remain open in the evening rather than closing by evening or in low light.

Sundrop gazanias: It bears attractive golden yellow flowers.

Tiger Stripe Mix: The flowers are very bright with red stripes on white or gold petals.

Kiss Bronze Star: The varieties are found in a riot of colour with orange and gold petals.

Talent Mix: It has silver foliage with bright colour flowers.

Some of the other popular cultivars which found in different colours and used for various purposes in garden are 'Aztec Queen', 'Burgundy', 'Christopher Lloyd', 'Colourama', 'Copper King', 'Cookei', 'Daybreak Bright Orange', 'Daybreak Garden Sun', 'Daybreak Red Stripe', 'Dorothy', 'Fiesta Red', 'Fire Emerald', 'Gold Rush', 'Michael', 'Mini Star White', 'Mitzuwa Yellow', 'Moon Glow', 'Northbourne', 'Sundance (Sundance Mixed Hybrids)', 'Sunburst', 'Sunglow', 'Sun Gold', 'Sunrise Yellow' and 'Talent Series'. The giant series which is heat resistant were developed includes 'Giant Bright Orange', 'Giant Deep Yellow', 'Giant Yellow Striped', 'Giant Red Bronze', 'Giant Gold Bronze', 'Giant Violet Bicolor and Giant White Tricolour'.

7.0 Genetics and Breeding

7.1.0 Cytology

The cytology of *G. rigens* is poorly known. Chen *et al.* (2003) reported that Gazania was diploid (2n = 20) and its chromosome cardinal number was x = 10, but Barkley (2006) suggested that the chromosome cardinal number was x = 9 (2n = 18). Further research and chromosome karyotyping of four Gazania varieties found that the chromosome number in three Hongwen, Xingbai and Richu is varieties was 2n = 10 and for the variety Zhongguo Xunzhangju, the number was 2n = 20. The chromosome cardinal number in Gazania was reported to be x = 5, and that Zhongguo Xunzhangju is a tetraploid (Jiashi *et al.*, 2016).

7.4.0 Hybridization

Ono (1986) developed the hybrid *Gazania rigens* "B" by controlled breeding using *Gazania rigens* "A"x *G. rigens* 'Mitsuwa White'. Egger and Beimel (1990) identified a natural outcross and named as Mitsuwa Better White. Its parents are unknown. However, it is a natural outcross within the *Gazania rigens* and the

Gazania rigens: var. leucoleana groups. Egger (1993) developed a hybrid and named as Moorpark Yellow. It was developed through controlled breeding of *Gazania rigens* plant #51-2 as the seed parent and plant #46B-4 as the pollen parent. Another natural hybrid was isolated from the same field and named as Moorpark Orange.

7.3.0 Polyploidy

Gazania 'Intrepid Hybrid Mix' is a modern hybrid polyploid variety of Gazania, which is an extremely large flowering variety (flowers >10cm diameter) and is perform well in beds. It has a healthy plant with dusted silver/dark green foliage. Polyploid gazanias are extremely large versatile flowering types, quick growing, floriferous, colourful, highly drought tolerant. The plants are suitable for both pots and garden. It is easy to grow and require low maintenance.

7.4.0 Mutation Breeding

In gazania the incidence of natural sports lead to the development of several varieties. Ono (1986) isolated spontaneous sport mutant and named as Mitsuwa Orange. Beimel (1986) identified and isolated the spontaneous sport mutant and named as Mitsuwa White which is a "sister" sport of Mitsuwa Orange. Ono (1986) identified spontaneous sport mutant from parent plant *Gazania splendens* aurantiacum and named as Mitsuwa Super Clump. This mutant has a much greater resistance to the many plant fungi and diseases which have decimated *Gazania splendens* aurantiacum and has proved commercially acceptable for its vigour and hardiness.

8.0 Growth and Flowering

The commonly cultivated gazanias are mostly annuals with large, bright colour flowers, and slightly larger foliage with silvery white undersides. They produce more flowers for a longer period of time. However, on the other hand perennials are simple in colour (solid) with minimal markings and blooms only in the summer. They have dense foliages with slightly smaller blooms.

8.1.0 Temperature

In temperate areas it is grown as annuals, whereas in tropics as perennials. The plant can tolerate light frost but not freezing. The optimum temperature for its cultivation is 15-20°C with 65-75 per cent Relative Humidity (RH). Plant tolerates low temperature, but it delays flowering.

8.2.0 Photoperiod

Gazania splendens is facultative long-day plant and long enhances flowering. Gazania flowers require full sunlight to grow. In shady areas or when light is less in morning or evening, flowers remain closed and may cause the plants to grow lanky. For good germination, gazania seed requires 16-18 hours of light every day.

8.3.0 Growth Regulators

The plant does not require application of any PGRs. However, spraying of B-Nine (2500 ppm) or cycocel (1500 ppm) delayed flowering. Treatment with

Giberellic acid (GA_3), Humic acid (HA) and Ascorbic acid (AA) positively influenced plant growth in gazania. The GA_3 at 100 ppm was the most active in stimulating the plant development followed by AA at 200 ppm and the HA at 600 ppm (Zulfiqar *et al.*, 2019).

Foliar application of citric acid and malic acid was investigated on flower stem height, plant height, flower performance and yield indices of Gazania. They had no significant effect on fresh weight and the flower count. The plant dry weight and the root fresh weight was significantly increased by Citric acid at 100 and 300 mg/l and 300 mg/l malic acid. It also has positive effect on the plant height, peduncle length and the flower senescence was delayed. The root to shoot ratio was increased significantly in 300 mg/l citric acid compared to all other treatments. Citric acid and malic acid were found to be effective on various aspects of growth and development of crop (Talebi *et al.*, 2014).

9.0 Propagation

The plants can be easily propagated by seeds as well as suckers.

9.1.0 Seed Propagation

Seeds are sown in nursery bed in the second week of October and transplanting is done from second to third week of November under north Indian conditions. The gazanias are transplanted at four leaf stage. Transplanting is generally carried out from October to November for bedding purpose, however for seed production appropriate time of planting is crucial as it affects the seed yield. Soaked seeds give the better germination. Seeds can be sown on raised bed in nursery and then transplanted to beds or pots. The seeds can also be directly sown into pot. It takes 15 to 20 days to germinate. During the germination period, soil should be kept moist. Gazania seeds can be successfully germinated at 21°C.

9.2.0 Vegetative Propagation

Gazania is easily propagated by dividing the clumps and sowing them in soil. The clumps can be dug from the soil and cut into smaller sections having three to five shoots with healthy intact roots. It should then be planted in suitable potting mix. It is necessary to keep the media moist until new growth appears.

9.3.0 Micropropagation

Although, gazanias are very popular bedding plant few micropropagation protocols were standardised. Yalcin-Mendi *et al.* (2009) reported first *in vitro* regeneration of *Gazania rigens* with different BA (0.0, 0.2, 0.5, 1.0 mg/l) and IAA (0.0, 0.5, 1.0 mg/l) concentrations and combinations. The best callus formation ratios were found on the media containing 0.2mg/l BA and 1.0 mg/l IAA and regeneration was obtained from themedia containing 0.2 mg/l BA and 1.0 mg/l IAA. The best rooting percentage was observed on the media containing 1.0 mg/l IAA.

Propagation and Growing of Gazania

Seeds

Germination of Seed

Seedling Plug

Selection of Cutting

Separation of Clumps

Planting

Planting in Pots

Mass Production

In Polyhouse

Seed Production

Matured Seed

Seed Collection

Source: www.uky.edu, www.images.app.goo.gl, www.growerfacts.com

10.0 Soil and Climate

10.1.0 Soil

Gazanias perform best in sandy, well-drained and slightly acidic soil (pH 5.5-7.0). However it can sustain different soil types and growing conditions. Although gazania is hardy plant, it may develop rot and powdery mildew in wet soil or humid conditions. The plants can thrive well in seaside conditions. The plant responds well to soils amended with organic matter. Compost is a good form of organic matter with a good balance of nutrients and pH level. The compost can be added in all the stages of growth.

10.2.0 Climate

In temperate regions, they are usually grown as half-hardy annuals and perennial in sub-tropical areas. The plant grows best in cool climatic conditions and can tolerate heat and drought. They require open area with full sunlight as blooming is severely affected in shade. Under mild climatic condition it can flower throughout the year. The plants can be successfully grown in different agro-climatic zones in India.

11.0 Cultivation

Gazania is a warm-weather perennial, usually grown as an annual. It grows well in full sun. It blooms in late spring and early summer. The trailing gazania (*Gazania rigens* var. leucolaena) is a commonly grown variety. They are preferred due to its fast growth rate and it can cover large areas or embankments.

11.1.0 Planting and Spacing

The planting of gazania should be done in the cool hours of the day. The seedlings are transplanted in beds at a spacing of 30 × 30 cm or 40 × 30 cm. In pot single seeding/sucker is planted as the plant spread within a short span of time and fill the pot. The seeds can also be sown directly in the field, if it is used for ground cover. However, the quantity of seed requirement shall be more and therefore in such cases close planting (20 × 20 cm) with seedlings or suckers is recommended. Ample watering should be given after transplanting or planting. Under Indian condition transplanting is done during the month of October-November.

11.2.0 Manuring and Fertilization

Gazanias are adopted to grow in soils with low fertility. Therefore compost and supplemental fertilizer is not necessary. However, application of slow release fertilizers is useful for plant growth in sandy soil. If needed, the plants can be given fertilizers after two weeks of transplanting. Application of complete balanced liquid fertilizers @ 200-250 ppm is beneficial. For the plants grown in pots granular application of NPK (10:10:10) is useful.

Hassan *et al.* (2015) studied the effect of different media on the vegetative growth and flowering characteristics of *Gazania rigens* grown in 25 cm size clay pots. The study revealed that calcareous soil alone was found to give the lowest values of parameters under study. While adding compost at 25 per cent or 50 per cent to the calcareous soil led to significant increase in almost all the measured parameters of Gazania plant. For pot culture, adding 25 per cent compost to the calcareous soil is recommended for growing *Gazania rigens* with high quality flowers.

The effect of different microbial inoculants on growth and floral responses of *Gazania rigens* (L.) Gaertn in pots was evaluated using arbuscular mycorrhizal fungi (AMF) – *Funelliformis mosseae* and *Acaulospora laevis*, along with *Trichoderma viride* and *Pseudomonas fluorescens*. Independent roles of these four microbes and their different combinations were used in sixteen treatments. Early flowering was noted in the combination of *F. mosseae, A. laevis* and *P. fluorescens*, which also showed the best results for flower head size, flower fresh and dry weight, total chlorophyll, carotene and phosphorus content. The number of leaves, number of buds and root length were maximum in the combined treatment (*F. mosseae* + *A. laevis* + *T. viride* + *P. fluorescens*). The study revealed the usefulness of microbial inoculants for the better yield and flower quality of *G. rigens* (Saini *et al.*, 2019).

11.3.0 Irrigation

The plants require moderate watering during its life cycle. They can very well tolerate drought condition. Excessive irrigation may lead to rotting of plant. Mulching with suitable materials not only conserve water but also controls the weed growth. Excessive dryness delay flowering and can causes foliage burn under high light intensity.

12.0 Diseases and Pests

12.1.0 Diseases

12.1.1 Downy Mildew (*Plasmopara*)

Dense white to greyish fungal growth develops on the undersides of leaves. The infected plants should be discarded and suitable fungicide may be applied for its control.

12.1.2 Pythium Root Rot

The infected plants turn yellow. In severe case it wilts and die. The growing media should be sterilized before planting. Clean cultivation should be adopted and infected plants should be discarded.

Virus

12.1.3 Impatiens Necrotic Spot Orthotospovirus (INSV)

The infected plants exhibit small necrotic spots, leaf yellowing and ringspots. In severe case the plant become dwarf and necrotic vein-clearing, wilting symptoms appears. The virus is transmitted by thrips.

12.1.4 Lettuce Mosaic Virus (LMV)

The virus-infected gazania plants exhibit symptoms of yellow mottling or mosaic, chlorotic spots and leaf distortion. The virus is transmitted by aphid.

Diseases and Pests of Gazanaia

Crown Rot

Blight

Phyllody

Monkey Beetle

Caterpillar

Diseases and Pests of Gazanaia

Aphids on Flower Bud

Slug

Hoverfly

Snail

***Source*: www.researchgate.net, www.picfair.com, www.gettyimages.in, www.alamy.com, www.sanbi.org, www.2.palomar.edu, www.canr.msu.edu, www.bibliomed.org**

12.2.0 Pests

The insects like aphids, spider mites, thrips and whiteflies attack gazania and can cause some damage if left unchecked. They can be suitably controlled by spraying insecticides.

13.0 Harvesting and Postharvest Management

The gazania takes 90-110 days to flower. It is not used as cut flower due to its habit of opening during day and closes at night. However, the pot grown flowering plants can be suitably placed both indoor and outdoor.

13.1.0 Yield

The inflorescence is single flower head with 5-10 cm diameter. A single plant clump can produce 30-40 flowers in a season.

13.2.0 Seed Harvesting

The plants produce flowers from end February to mid May. The fruit is a hairy cypsela with pappus. The seeds should be collected during March-April when the flower matures and petals have fallen from flower heads. Seed are harvested manually in 3-4 pickings. The seed is collected with broom and is cleaned by winnowing and picking of inert matter (flower stalks). The seeds should be dried in a warm and dry place in direct sunlight for 2-3 days. The completely died seed should be packed in an air-tight pack and store in a cool and dry place. The yield of clean seed is around 250 kg/acre.

14.0 Biotechnology

14.1.0 Molecular Markers

Kady *et al.* (2015) investigated genetic and botanical features of root, rhizome, leaves and flowers of the two Gazania species *viz., Gazania longiscapa* and *Gazania rigens* using RAPD and ISSR markers. RAPD primers OPB-09 and OPB-11 gave maximum polymorphic fragments. The protein profiling gave the total bands, among these fourteen are common to both species while the fragments with molecular weight 16, 8, and 7 kDa were absent in *G. longiscapa* and 115 kDa was absent in *G. rigens*. The similarity coefficient between both samples was 77.7 per cent.

A phylogenetic analysis of 15 species of Gazania was done using the DNA sequence data from four non-coding chloroplast regions and two nuclear regions. The nuclear regions used were the Internal Transcribed Spacer (ITS) and partial External Transcribed Spacer (ETS) andthe four plastid regions used weretrnL-intron,trnL-trnF spacer, the psbA-trnH spacer and the rps16 intron. As a range of mutation rates of ITS are available for other taxa in the Asteraceae, it allowed to estimate agefor key diversification events within Gazania. The dates were then related to a biogeographic scenario in a narrative manner.This studypartially resolved taxonomic issues within Gazania, and recognized seven species in the genus: *G. caespitosa, G. ciliaris, G. lichtensteinii, G. schenckii, G. tenuifolia, G. heterochaeta, G. jurineifolia* (which may comprise twosubspecies) and the large morphologically variable K-Rspecies complex. However, the accuracy and generality of this phylogenetic analysis needs more testing in correspondence with other similarly distributed plant groups (Howis *et al.*, 2009).

15.0 References

Ahmed, S.I. (1981) *M.Sc. thesis.* Faculty of Pharmacy, Assiut University.

Anonymous (2017) *AGM Plants - Ornamental.* Royal Horticultural Society. July 2017. pp. 39. Retrieved 27 February 2018.

Balogun, F.O. and Ashafa, A.O.T. (2016) *Transac Royal Soc. S. Afr.*, **71**(2): 145-156.

Balogun, F.O. and Ashafa, A.O.T. (2018) *Asian Pac. J. Tro.p Biomed.*, **8**: 52-58

Barkley, T.M. (2006) *Flora of North America.* Oxford University Press, New York, **19**: 196.

Beimel, S.H. (1986) Patents assigned to mitsuwa nursery, inc., https: //patents.justia.com/assignee/mitsuwa-nursery-inc

Born, J., Linder, H.P. and Desmet, P. (2007) *J. Biogeogr.,* **34**: 147-162.

Chen, R.Y., Song, W.Q., Li, X.L., Li, M.X., Liang, G.L. and Chen, C.B. (2003) *Chromosome atlas of major economic plants genome in China, vol. III*. Science Press, Beijing, pp. 207–208.

Desoukey, S.Y., Kady, W.M.E., Salama, A.A.A., Hagag, E.G., El-Shenaw, S.M. and El-Shanawany, M.A. (2016) *International Journal of Pharmacognosy and Phytochemical Research*, **8**(7): 1121-1131.

Egger and Beimel (1990) Patents assigned to mitsuwa nursery, inc., https: //patents.justia.com/assignee/mitsuwa-nursery-inc

Egger, J.N. (1993) Patents assigned to mitsuwa nursery, inc., https: //patents.justia.com/assignee/mitsuwa-nursery-inc

Elkhayat, E. S. (2016) *Bulletin of Faculty of Pharmacy*, Cairo University, **54**: 257-261.

Funk, V.A., Chan, R. and Keeley, S.C. (2004) *Taxon.*, **53**: 637–655.

Goldblatt, P. and Manning, J.C. (2000) *Cape Plants: a conspectus of the Cape Flora of South Africa*. Strelitzia, 9. National Botanical Institute, Cape Town and Missouri Botanical Garden, St. Louis.

Hassan, B.A., Hussein, M.K., El-Kiey, T.M. and Weheda, B.M. (2015) *Alexandria Science Exchange Journal*, **36**(1): 15-24.

Howis, S., Barker, N.P. and Mucina, L. (2009) *Taxon.*, **58** (3): 871–882

Jiashi, Z., Dongxu, W., Yangqing, W.U., Xiaoyu, G., Yazhou, Z. and Xiaoping, L. (2016) *Horticultural Plant Journal*, **2** (5): 279–283.

Justiss, E.S. (1985) *M.Sc. Thesis & Specialist Project*: The Performance of GazaniaSplendens Cultivars Under Kentucky Conditions,1735.

Kady, El.W.M., Salama, A.A.A, Desoukey, S.Y., Hagag, E.G., El-Shenawy, S.M. and El-Shanawany, M.A. (2015) *Bull Fac Pharm Cairo Univ.*, **53**: 129–45.

Karis, P.O. (2007) *Tribe Arctotideae*. In: Kadereit, J.W., Jeffrey, C. (Eds.), The Families and Genera of Vascular Plants VIII: Asterales. *Springer-Verlag, Berlin*, pp. 200–207.

Linder, H.P. (2003) *Southern Africa. Biol. Rev.*, **78**: 597-638.

Magee, A.R., Boatwright, J.S. and Mucina, L. (2011) *South African Journal of Botany*. **77** (1): 86-93.

Mucina, L., Jürgens, N., Le Roux, A., Rutherford, M.C., Schmiedel, U., Esler, K.J., Powrie, L.W., Desmet, P.G., Milton, S.J., Boucher, C., Ellis, F., Lambrechts, J.J.N., Ward, R.A., Manning, J.C. and Midgley, G.F. (2006) *Succulent Karoo Biome*. In: Mucina, L., Rutherford, M.C. (Eds.), The Vegetation of South Africa, Lesotho and Swaziland. Strelitzia, 19. SANBI, *Pretoria*, pp. 220–299.

Ono, T. (1986) Patents assigned to mitsuwa nursery, inc., https: //patents.justia.com/assignee/mitsuwa-nursery-inc

Reddy, K.D., Reddy, K.H., Brenda, M., Koorbanally, N.A. and Patrick, G. (2014) *Journal of Pure and Applied Microbiology*, **8**(6).

Roessler, H. (1959) *Mitteilungen der BotanischenStaatssammlung München*, **3**: 71–500.

Saini, I., Aggarwal, A. and Kaushik, P. (2019) *Gaertn Agriculture.*, **9**: 51. doi: 10.3390/ agriculture9030051

Taha, S.E., Mona, H.H., Shahira, M.E. and Amany, A.S. (2008) *Bull Fac Pharm Cairo Univ.*, **46**: 253–65.

Talebi, M., Hadavi, E. and Jaafari, N. (2014) *Advances in Agriculture*. http: //dx.doi. org/10.1155/2014/147278

Tshabalala, B.D., Alayande, K.A., Sabiu, S. and Ashafa, A.O.T. 92016) *Eur. J. Integr. Med.*, **8**(4): 533-539.

Yalcin-Mendi, Y., Unek, C., Eldogan, S., Akakacar, Y., Serce, S., Curuk, P. and Kocaman, E. (2009) *Romanian Biotechnological Letters.*, **14** (5): 4728-4732.

Zeng, J., Wang, D., Wu, Y., Guo, X., Zhang, Y. and Lu, X. (2016) *Horticultural Plant Journal*, **2** (5): 279–283.

Zulfiqar, F., Younis, A., Abideen, Z., Francini, A. and Ferrante, A. (2019) *Agronomy*, **9**: 773. doi: 10.3390/agronomy9110773.

Pelargonium

Madhu Bala and Reeta Bhatia Dey

1.0 Introduction

Pelargonium comprises the group of spring and summer flowering plants that includes geraniums or zonal pelargoniums, regal pelargoniums, ivy-leaf pelargoniums, scented leaf pelargoniums and miniature pelargoniums. The name *Pelargonium* has been derived from the Greek word *'pelargos'* meaning 'stork', due to its long-beaked fruits (seed case) resembling a beak of a stork (Miller, 2002). Geraniums (now classified as pelargoniums) are native to South Africa. Most of the cultivars are grown as a pot plant, in containers and for bedding purpose, in glasshouses or in open conditions during the summers, and these containers are shifted to protected structures during winters to protect from frost. A few species of temperate regions have the ability to withstand the outdoor low temperature conditions. Among the different species, the most widely cultivated species is *Pelargonium × hortorum* (Bailey), commonly known as 'zonal' Geraniums having peculiar dark pattern on leaves. Plants are compact with height up to 60 cm, succulent green to pale bronze stems, undulating leaves with distinct darker markings or 'zones'. The flowers are bright, single or double, mostly in clusters on long stems.

2.0 Importance and Uses

2.1.0 Aesthetic Uses

Pelargonium is an ornamental flowering plant grown as a pot plant for interior decoration, beautification of balconies, window sills and verandas. The longer blooming period and decorative leaves make this flower as a perfect bedding plant. In Southern and central California, it is grown as a herbaceous, perennial flowering shrub.

2.2.0 Therapeutic and Other Uses

Pelargoniums are also well known for their medicinal use and more than 120 different chemical compounds have been identified in its oil (Williams and Jeffrey, 2002). The plant species, indigenous to areas of South Africa are widely used by traditional healers of several tribes to treat dysentery, diarrhea, hepatic complaints, wounds, colds, fatigue, fevers, generalized weakness, and infections of the respiratory tract including tuberculosis.

3.0 Origin and History

It is reported that the name *Pelargonium* was introduced by Johannes Burman in 1738. More than 90 percent of pelargonium species are natives of South Africa. These species are found widespread throughout warm temperate areas, central South Africa, North Africa, adjacent Atlantic Islands and eastwards to Arabia and Southern India, and Australasia. The genus *Pelargonium* belongs to the family *Geraniaceae* and comprises of nearly 280 species (Kellen, 1983) with few species native to Australia, Eastern Africa, New Zealand, the Middle East and the islands of Madagascar, St. Helena, and Tristan de Cuhna (www.herbsociety.org). The first

Uses of Pelargonium

In Border

Bedding/Garden Plant

On Windows Sills

Indoor Decoration

Essential Oils

In Pot

Hanging Basket

Planted as Hedge

Balcony Decoration

pelargonium to reach Europe was probably *Pelargonium triste* which was brought from Cape by John Tradescant in 1632. The first evidence for pelargonium cultivation dates back to 1633.

The first ancestor of the zonal group, now known as *Pelargonium zonale*, was collected in the province of Western Cape in 1689 and was sent to Europe and was described by the Dutch botanist Jan Commelijn (1629–1692). Many species of pelargonium such as *P. inquinans*, *P. odoratissimum*, *P. peltatum*, *P. vitifolium*, and *P. zonale* had been introduced to Europe. It was only during the second half of the 18th century that the foundations for hybridization and floral biology of pelargoniums began through the work of two remarkable German botanists Kurt Polycarp and Joachim Sprengel. In 1793, Sprengel established that most hermaphroditic flowers could not be fertilized by their own pollen because their reproductive organs did not reach maturity at the same time (as is the case with *Geraniaceae*). A large number of pelargonium species were imported into Europe between the end of the 18th century and the beginning of the 19th century.

4.0 Taxonomy

Taxonomically *Pelargonium* belongs to the family *Geraniaceae*. It comprises of scented and unscented species, the odour of the leaves of various scented species range from pleasantly fruity or floral to rather oppressively balsamic. Initially, there was lots of ambiguity in the nomenclature of pelargoniums, they have been often misled by geraniums. The flower of the genus *Geranium* has five petals that are of same size and shape. The genus *Pelargonium* also contains five petals, however, unlike geraniums the upper two petals differ in shape and size from the lower three petals. The taxonomic revision of *Pelargonium* was published by Knuth in 1912 and according to him, there are 15 sections based on the morphological characters. The current subdivision of Pelargonium comprises 16 sections. The genus *Pelargonium* includes annuals and herbaceous perennials, shrubs and subshrubs, both evergreen and deciduous plants. The stems are erect, woody at base, soft-wooded, sometimes succulent, often aromatic, hairy, sometimes spine-like stipules are present. In many species stems are herbaceous initially but become woody with age. Some species grow erect while others have a trailing habit, and some have tuberous roots. The leaves of pelargonium are of different shapes and sizes and may resemble the fern leaves, grapevine or oak and may be deeply divided. The leaves may be small or large, sometimes covered with fine hairs and sometimes rough, sticky or velvety in texture.

Flowers are arranged in a pseudo-umbel consisting of 1-50 individual flowers, but most have 5-10 florets. The typical pelargonium flower is pentamerous and zygomorphic. The nectar tube opens at the base of the posterior sepal and the tube is formed by passive excavation through the conspicuous elongation of the receptacle (Japp 1909). Nectar is secreted at the base of the tube. The posterior (upper) petals are usually marked with nectar guides. Flower colour may vary from white, pink, mauve, lavender, pale yellow or burgundy, and upper petals may have purple, pink or red markings. Flowers of most species are not scented, except *P. triste*, which emits scent during night. The ovary consists of 5 mericarps - carpels *i.e.* it is 5-locular, which elongate on fertilisation. The fruit is a schizocarp which forms between dehiscent and indehiscent types, in which each carpel matures into a single-seeded unit. The fruit is composed of five mericarps, each containing one seed. On maturity, the mericarps split apart (dehisce), the plumed seed is ejected, and may be spread by wind or animals.

5.0 Species and Cultivars

5.1.0 Species

Important species of *Pelargonium* have been summarised below:

Pelargonium abrotanifolium

Plants are bushy type, erect, evergreen, woody stemmed and native of South Africa. This species grows up to 1-1.5 m in height, bearing white or pink flowers in clusters, 5 cm across, during spring to summer season.

P. acetosum

This is an evergreen, erect, perennial species and grows upward with strong stems up to 50-60 cm height with a spread of 20-25 cm. The stems are succulent bearing obovate, toothed, fleshy greyish-green leaves, often margined red. The flowers are star shaped, salmon-pink single type, 2-3.5 cm across with long narrow petals. Flowering time is from spring to summer or from May to July.

P. betulinum

Plants grow erect up to height of 90 cm with slender stems, and glaucous-tinted leaves. Flowers are white or pink in colour having carmine vein patterning at the base of two upper petals. Flowers appear in 2-3 clustered umbels during summer.

P. bowkeri

This is a succulent and tuberous type plant species with large underground tubers, named after Henry Bowker, soldier and naturalist of 19th Century who collected the plant. Leaves are pinnate, feathery with light yellow green flowers.

P. capitatum

It is a shrubby species native of South Africa, bearing white-hairy stems growing to 30-60 cm tall. Pink flowers with red-purple veins appear during summer months.

P. carnosum (P. crithmifolium, P. dasycaule)

A native from Namibia to NW Cape (South Africa), it is a shrubby succulent plant with white flowers comprises of 3-7-flowers in cluster, 2.5 cm wide with pink markings on the base of the upper pair of petals. *P. crithmifolium* and *P. daisycaule* are exactly similar to *P. carnosum* except that flowers in former case are lilac-pink while in later, these are white.

P. cucullatum

This freely branched species native of South Africa and grows up to 2 m tall, bearing purplish-red flowers with darker veins, 4-5 cm across, appear during summer.

Species of Pelargonium

Pelargonium zonale

Pelargonium. hortorum

Pelargonium capitatum

Pelargonium inquinans

Pelargonium grossularoides

Pelargonium quercifolium

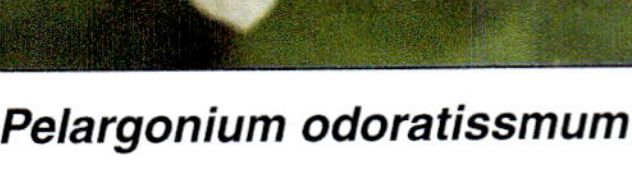

Pelargonium odoratissmum

Pelargonium denticulatum

Species of Pelargonium

Pelargonium longicaule

Pelargonium cucullatum

Pelargonium grandiflorum

Pelargonium austral

Pelargonium radens

Pelargonium tricolour

Pelargonium alchemilloides

Pelargonium flabelliforma

P. denticulatum (P. filicifolium)

This species is a native to South Africa and grows up to 1 m in height, bearing balsam-scented leaves. Lilac, rose-pink to pale-purple flowers having darker markings and 1.5-2.0 cm width are produced during summer.

Pelargonium × domesticum (P. grandiflorum, P. × macranthum, P. regale)

A group of hybrid pelargonium evolved by involving *P. angulosum, P. cucullatum, P. fulgidum, P. grandiflorum* and a few others. These are woody, evergreen and erect perennials, mostly compact, grow up to 40 cm to 1.5 m tall, flowering from May to September. The individual flowers are 4-5 cm wide, butterfly- or azalea-shaped of one colour to bicolored in white, pink, red and purple, *vis-a-vis* blotched with darker shades and/or picotee edges.

Pelargonium × fragrans

It is a hybrid of *P. extipulatum* and *P. odoratissimum*, morphologically very similar to the latter parent but leaves are almost 3-lobed with aroma of nutmeg, lemon or pine. Its flowers are small white where upper petal has red veins.

P. fulgidum

A native of South Africa, this is a sub-shrubby species growing to more than 60 cm tall, bearing bright scarlet flowers with darker veining, appear during summer, about 4 cm across.

P. gibbosum

A native to the coastal W. Cape (South Africa), it is a sprawling sub-shrub with thickened stems producing a mass of annual growth with glaucous semi-succulent pinnate leaves. The inflorescence is a panicle of night scented yellow flowers.

P. grandicalcaratum

A native to Namibia and the NW Cape (South Africa), woody shrub bearing small flowers white to pale-yellow with fine red markings on the upper petals.

Pelargonium × hortorum

It is a complex hybrid group involving *P. inquinans, P. zonale* and certain other species, and is popular with the name of zonal pelargoniums. They grow more than 1.5 m tall, bearing mid-green leaves with a darker brown or bronze horseshoe-shaped mark in the centre, rounded to bean-shaped, 6-13 cm across and with waved margins. White, orange, pink, red, purple or bicolored single to double flowers, some may have quilled petals or even picottee edges, 2-5 cm across, produced from May to October. There are cultivars where foliage is variegated with various colours, *viz.* green, dark green, white, yellow, crimson, purplish and brown.

P. paniculatum

It is an infrequently branching succulent shrub, native to S. Namibia, with a thick 90 cm trunk. The inflorescence which appears during summer is a panicle of small white flowers with pink markings on the base of the upper pair of petals.

P. papilionaceum

Native of South Africa, a shrubby species with well-branched stems growing up to 1 m in height, bearing multi-branched flowering stems and carries several umbels of 5-10 flowers, 2 cm long with a carmine blotch and veined pattern, though three lower petals are almost rudimentary.

P. peltatum (Geranium peltatum)

A native to the coastal tip of the Cape (South Africa), evergreen species with slender stems, growing up to 0.9-1.8 m. This is one of the most decorative popular geraniums, flowering duration is very long, *i.e.* from March to November in frost-free zones. It is suitable for hanging baskets, wall brackets, window baskets, terraces, balconies, against trellis or nettings, and in large receptacles. Flowers are carmine to pinkish with darker markings on the upper petals and appear at the end of a long stalk in the form of a loose umbel up to 10 cm in diameter. Flower buds are quite hairy.

Popular varieties are 'Abel Carrière' (semi-double, soft rose-purple), 'Gardenia' (large double, white), 'Gloire d'Orléans', (double, pink) 'La France' (double, mauve with upper surface of petals flecked maroon) *etc.*

P. radens (P. radula)

This shrubby species grows to about 1 m tall, native of South Africa, bearing pink to rose pink flowers with red-purple markings and 3 cm wide during summer.

P. saxifragoides

This was discovered in 1888, growing at Royal Horticultural Society's garden, Chiswick, having any record, hence is thought to be of garden origin, probably a hybrid of *P. peltatum*. The plant being dwarf, resembles a saxifrage, hence, its specific name *saxifragoides*. It is a succulent plant, bearing small dark green fleshy leaves, light pink flowers during summer.

P. scandens

A native of South Africa, sparingly branched shrubby species, grows to more than 1.5 m high, bearing pink to light red flowers, some 2.5-3.0 cm wide, appear during summer in umbels of 5-12. This species is suitable for hanging baskets, on walls and around a window embrasure.

P. tetragonum

A native to the Cape region of South Africa, scrambling-shrubby species grows to 1 m tall, produces cluster of cream to pink flowers, 2.5-3.5 cm across with four pink petals, upper two larger and deep-coloured than the lower and feathered with purple veining, during summer.

P. tomentosum

A native of South Africa, grows to a length of 1.5 m, if trained as a low climber. Flowers white with red markings are insignificant, narrow-petalled, 1.2-2.0 cm across and appear in umbels from the tip of the shoots from June to September.

Classification of Pelargonium

Zonal Pelargonium 'Fantasia Dark Red'

Zonal Pelargonium 'Appleblossom Rosebud'

Ivy-Leaved Pelargonium 'Pac Tommy'

Ivy-Leaved Pelargonium 'Pac Kate'

Regal Pelargonium 'Lord Bute'

Regal Pelargonium 'Dark Secret'

Scented-Leaf Pelargonium 'Bitter Lemon'

Scented-Leaf Pelargonium 'Shottesham Pet'

Angel *Pelargonium* 'Veronica Contreras'

Angel Pelargonium 'Carlos'

Unique Pelargonium 'Bolero'

Unique Pelargonium 'Voodoo'

Classification of Pelargonium

Pelargonium Tulip flowered 'Red Pandora'

Pelargonium Carnation Flowered 'Diana Palmer'

Pelargonium Phlox Plowered

Pelargonium 'Jay's Golden Bird's Egg'

Source: www.gardenersworld.com, https://plant-growing.com/

P. triste

A native to the West and South-West Cape (South Africa), it is a tuberous-rooted evergreen perennial growing up to 40 cm in height with hairy stems. Originally, it was described by Linnaeus as *Geranium triste*. It is one of the first South African pelargoniums to be brought to England by John Tradescant in 1632. Its pleasantly night-scented, dark maroon and spotted flowers appear during May-June.

P. zonale (Geranium zonale)

It was introduced to England in 1710. A native of South Africa, this woody-based shrubby species has succulent and somewhat hispid branches at younger stage. Inflorescence has many but almost sessile flowers in shades of white, orange, pink, red, scarlet or crimson and these have spathulate segments.

The cultivated pelargonium are classified into four broad groups. The most widely cultivated species is the 'Zonal' or 'fish' geraniums (*Pelargonium*× *hortorum* Bailey) that have dark band or 'Zone' within each leaf. The other species are 'regal' or 'show' geraniums (*Pelargonium*× *domesticum* Bailey) are also gaining popularity because of its beautiful, showy and unusual flowers. The ivy-leaved species (*Pelargonium peltatum*) have excellent cascading or trailing type plants, suitable for hanging baskets, windowsills and as wall climbers under sheltered conditions. The fourth group consisted of a series of scented species that have pungent or fragrant leaves, good for borders, edging or alongside a path and can also be grown in pots.

Brickell (1996) classified the cultivated pelargoniums into following horticultural groups:

I. Zonal Pelargoniums

This represents the largest and diverse group of cultivars of a hybrid classified under Zonal group, commonly known as *Pelargonium* × *hortorum* (Zonale hybrid with chromosome number of 2n=18) derived from *P. zonale*. The true ancestry of zonal pelargonium is though not well known but the species such as *P. inquinans*, *P. zonale*, *P. hybridum* and *P. frutetorum* have been considered to be their pedigree during the course of evolution. These are succulent stemmed and branched perennials with upright and bushy appearance, and bear single or double flowers with attractive foliage and commonly used for bedding purpose. The leaves are round, light to deep green with a conspicuous zone of bronze or maroon. The flowers are 1.25-2.5 cm across, in shades of scarlet, purple, pink, white, orange, red and rarely yellow appearing from the upper leaf axils during summer and autumn. Plants grow up to 60 cm, dwarf varieties growing up to 12-20 cm and miniature up to 12 cm tall. These are propagated by cuttings, many hybrids are propagated through hybrid seeds. The single flowered F_1 hybrids, grown as bedding plants, are mainly propagated through seeds. Whereas, large flowered types are raised through cuttings and generally used for growing in containers. The most prominent hybrids are'Carefree' which includes 'White', 'Bright Pink', 'Deep Salmon' and 'Scarlet' and their mixture; 'King of Denmark' (salmon pink, semi-double), 'Maxim Kovaleski' (orange), *etc.* F_1 hybrid includes, 'Deep Salmon', 'Light Pink' and 'Scarlet'. Other varieties are 'Always' (double, creamy-white, centrally flushed pink), 'A.M. Mayne' (double, magenta, centre flushed scarlet) and many hybrids have been developed with rich hues of different colours and attractive characteristics.

The zonal pelargoniums are further classified as:

Single flowered pelargoniums: They bear single flowers and mostly grown for bedding purpose.

Double and semi-double flowered pelargoniums: The flowers often comprises of six or more open petals.

Fancy leaf pelargoniums: Fancy leaf pelargoniums have attractive, variegated or coloured leaves ranging from yellow, gold and dark green. The leaves may be black, bronze, butterfly type, golden, golden tricolour, silver and silver tricolour. They may have single flowers, however to create mass effect, they are often maintained deflowered. The important cultivars are Cherry Cocktail, Cherry Sundae, Golden Chalice, Ice Queen, Frank Headley, Tricolour, *etc.*

Decorative pelargoniums: It is a diverse group in which flowers exhibit wide diversity in range of forms and colours. They are particularly good for conservatories or as single specimen plants on the patio.

Stellar pelargoniums: They were first bred in the 1970s in Australia and hence the one of the newest introductions in the Pelargonium category. They have bushy growth habit, flowers are fluffy in appearance due to the irregular pointed flower petals and the leaves have slightly pointed lobes. Flowers are star shaped, either single or double, delicate but

long lasting and able to tolerate scorching sun. Popular cultivars are Fresh Water, Pink Needles, Raydon,Snow Crystal, Vancouver Centennial,Shirley Gillam, *etc.*

Cactus flowered pelargonium: They are named due to their resemblance to the spines of the cactus. Petals are long, narrow, pointed resembles with cactus spines. The petals are curled longitudinally which give them a quill or tubelike appearance. The flowers may be single or double, the colour ranges from white, pink, red and scarlet. The popular cultivars are Fire Dragon, Red Spider, Noel, Pink Poinsettia, Pink Poinsettia, Star of Persia, *etc.*

Rosebud pelargoniums: A single flower has around 30-40 petals which give them rosebud appearance. The cultivars under this category are Apple blossom Rosebud, Plum Rosebud, Red Rosebud, Rosebud Black and Rosebud Supreme.

II. Regal Pelargoniums

They are botanically named as *P.* × *domesticum*, and are derived from *P. capitatum*, *P. cucullatum* (*P. angulosum*), *P. fulgidum* and *P. grandiflorum* species. These are the most beautiful flowering geraniums propagated through cuttings. These pelargoniums comprise a large group that are known as Show, Grandiflora, Large flowered, Decorative and Fancy and also named as Martha or Lady Washington pelargoniums in the United States. Plants are erect, evergreen, perennial and branched shrubs with bushy appearance with plant height of 38-60 cm. The leaves are mid-green, roundish to palmate, sometimes lobed and/or partially toothed or

Varieties and Sub-classes of Zonal Pelargonium

New Dynamo Dark Red Improved

Dynamo White

Dynamo Pink Flare

Dynamo Orange

Dynamo Hot Pink

Fantasia Neon Rose

New Dynamo Salmon Improved

Fantasia Fuchsia Berry

New Presto Rose Flare

Presto Deep Scarlet

Presto Light Salmon

New Presto Brilliant Red Improved

Varieties and Sub-classes of Zonal Pelargonium

Zonal-Rose Bud
'Apple Blossom Rose Bud'

Zonal-Rose Bud
'Red Rose Bud'

Zonal-Stellar
'Snow Crystal'

Zonal-Stellar 'Vectis Glitter'

Zonal-Fancy Leaf 'Jungle Leaf'

Zonal-Fancy Leaf 'Frank Headley'

Zonal-Cactus 'Red Spider'

Zonal-Cactus 'Spitfire'

***Source*: https://www.ballfloraplant.com**

deeply serrated, slightly hairy, odourless or with pleasant smell. Flowers are single type (rarely double) with shades of mauve, pink, purple or white, usually veined or blotched with darker shades. Most of the ancestral pelargoniums flower only in spring, however, in modern types flowering continue from May to October or and in some varieties even during winter.

Some of the promising hybrids are 'Alberts Choice' (orange-salmon, white throat), 'All My Love' (mauve, base creamy-white), 'Applause' (pink, centre white, edged frilled and pearly-white), 'Ashley Stephenson' (orange salmon, throat white), *etc.*

III. Ivy-leaved

The name ivy-leaved pelargoniums (*P. peltatum*) is derived from ivy (*Hedera helix*) due of close resemblance of leaves. Some cultivars such as 'Lilac Gem', also bear strong smell like ivy. The group includes evergreen perennials plants with trailing habit, ideal for hanging baskets, propagated through stem cuttings and only a few are seed-propagated. Leaves are stiff, round and lobed, fleshy, often shining and some cultivars having dark zone in the center. Flowers are single or double similar to zonal pelargoniums. The varieties usually grown as pot plant are 'Galilee' (large rose-pink), 'Madame Crousse' (pale-pink veined maroon), 'Scarlet Crouse' (scarlet sport of Madame Crouse), 'Souvenir de Charles Turner' (deep pink veined maroon), *etc.*

IV. Scented-leaved Pelargoniums

The characteristic feature of this group is the presence of scented foliage in varieties and hybrids. The flowers are small, and often irregularly star-shaped. These are evergreen perennial shrubs, propagated through cuttings, while some cultivars are seed propagated. Leaves are distinctly lobed, toothed or variegated. Cultivars may have aromas of rose, peppermint, lemon, lime, orange, strawberry, camphor, nutmeg, spice, apricot, apple, filbert, ginger, and coconut.

The common scented species incudes: *Pelargonium odoratissimum* (apple fragrance, white flowers), *P. crispum* (lemon scented, pale-lavender flowers, var. 'Variegatum'), *P. abrotanifolium* (artemesia scented), *P. limoneum* (strongly lemon scented, flowers pink), *P. radula* (rose fragrance, flowers pale-purple), *P. graveolens* (strong rose scent, flowers mauve pink), *P. capitatum* (rose-scented, flowers small pink), *P. tomentosum* (peppermint-scented, flowers small pink), *etc.*

V. Angel Pelargoniums

The angel pelargoniums are almost similar to regals and are classified as miniature regal pelargoniums with a difference that they are more compact and bushy in appearance. Angel name is considered to be derived from a dwarf pelargonium 'Angeline' which is not tall in cultivation and the group is mostly derived from *P. crispum*.

VI. Unique Pelargoniums

This group includes large evergreen perennials, single bright coloured flowers through the season. The leaves are often deeply lobed having pleasant smell. It produces sparse flowers in shades of white, pink, mauve and red. The varieties under this group are 'Apache' (darkest red), 'Better Times' (short-jointed plant, flowers soft salmon pink), 'Cal' (salmon), 'Corsair' (light red), 'Dark Red Irene' (very large dark red bloom), 'Electra' (rose tinged purple), 'Genie' (red), 'Irene' (crimson, parent of all Irenes) *etc.* These are insensitive to high temperature as plants do not lose vigour.

VII. Novelty Pelargoniums

Tulip Flowered Pelargoniums

These types have only 6-9 incurved petals that resemble tight buds of tulips. The first tulip flowered pelargoniums was discovered in the 1966 by Frank Andrea in his greenhouse in a Boston, Massachusetts, and was named after his wife, Patricia Andrea. Some of the popular varieties of this group are 'Emma Fran Bengtsbo', Happy Birthday 'Lilian Andrea', Linea Andrea 'Pandora Rouge', 'Patricia Andrea' and 'Marbacka tulpan' *etc.*

Carnation Flowered Pelargoniums

The carnation-flowered pelargoniums are characterized by flowers with fringed petals, serrated or toothed margin like carnations. Most of the cultivars with fringed or carnation-type flowers are likely to be 'sport' and produce ordinary flowers without any toothing of the petals. It is quite common in this type to have a 'sport' or 'reversion ' to a different colour and standard flower. Some cultivars includes: 'Diana Palmer', 'Jeanne', 'Lancastrian', 'Madame Thibault' and 'Skelly's Pride' *etc.*

Bird's Egg-flowered Pelargoniums

Flower petals are speckled with contrasting colours and resembling a bird's egg. Flowers are single or double. Important cultivars are 'Jay's Golden Bird's Egg' (Single, Soft pink with freckles) and 'Baby Birds Egg' (Single, pale pink flowers with purple spots).

Phlox Flowered Pelargoniums

Flowers are single type, having broad, flat petals giving the flower a round appearance. A darker colour band runs across each petal about midway and looks like a halo encircling the flower center like garden phlox.

Dwarf and Miniature Pelargoniums

These are existing for over a 100 years and very popular among the growers. The height should not exceeds 13 cm, whereas, a dwarf pelargonium may grow up to 20 cm. They may have small or even tiny leaves, but flowers are large and striking. Mostly suitably near a window due to their sensitivity to cold outdoors. Flowers are single as well as double type with range of colours. The miniature cultivars include: 'Red Black Vesuvius' (black leaves, bright red single flowers), 'Gay Baby' (ivy-leaved type; pale mauve flowers); 'Granny Hewitt' (double red flowers).

The dwarf pelargonium include 'Friesdorf' (single; deep rose pink with very narrow petals); 'Snowflake' (single; white flowers) and 'Bird Dancer' (pales pink flowers with narrow petals).

6.0 Genetics and Breeding

6.1.0 Genetics

The documented breeding work of Angel pelargoniums started in the 1920s by Langley-Smith in England (Brawner, 2003). Probably, the interspecific hybridization with *P. crispum* began much earlier, soon after the import of different *Pelargonium* species to Europe during the seventeenth century. The chromosomal studies of several species and cultivars indicated that most of the aromatic pelargoniums, derived from species, *P. crispum* and *P. graveolens* have x = 11 as their basic chromosome number (Daker, 1969). The basic chromosome number in zonal and the ivy-leaved pelargoniums are reported as x = 9, while in *P.* × *fragrans* x = 8. The karyotype analysis of 37 *Pelargonium* species revealed that the diploid (2n) chromosome number varies from 16, 18, 20, 22, 36, 45, 60, 72, 81, 88 and 90 with x = 8, 9, 10 and 11 (Federov, 1974). Further, Demame (1989) reported 2n = 8x = 88 in *P. radens*, *P. graveolens*, *P. vitifolium* and 2n = 6 x = 66 for *P. capitatum*. The chromosome numbers in other species are reported as: *P. ovale* (2n=20) *P. denticulatum* (2n=44), *P. exhibens* (2n= 22), *P. capitatum* (2n=66), *P. acetosum* (2n=18), *P. radens* (2n=88), *P. frutetorum* (2n=18), *P. gibbosum* (2n=22), *P. inquinans* (2n=18), *P. exstipulatum* (2n=16) *P. peltatum* (2n=18), *P. odoratissimum* (2n=16), *P. zonale* (2n=18), *P. endlicherianum* (2n=34) *P. echinatum* (2n=22), *P. cucullatum* (2n=22), *P. grandiflorum* (2n=22), *P. carnosum* (2n=22), *P. oblongatum* (2n=22), *P. fulgidum* (2n=22), *P. trifidum* (2n=22).

6.2.0 Breeding

Systematic breeding work has been undertaken by American and European breeders that resulted in evolution of new flower colours and shapes, early and continuous flowering, good postharvest and market quality and resistance to insect pest and disease. Recently, American and British plant breeders have developed the F_1 hybrids that can be grown as annuals and propagated through seeds thus overcome the need of vegetative propagation. McDonald & Van der Walt (1992) based on their observations on the pollination of *P. tricolour* Curt, concluded that there are embossed areas or false nectarines on their posterior petals, which are effective in attracting dipteran pollinators. Further, the pollination studies in the self-compatible, dichogamous flower of *Pelargonium dolomiticum* Knuth revealed that the male phase precedes the female phase, indicating protandry, dichogamy is complete and asynchronous. The major breeding objectives in pelargonium are earliness and profuse flowering, plant height and growth habit, leaf colour, size, shape and placement, novel flower colour and shapes, tolerance to biotic and abiotic stresses and ability to propagate through seeds. The breeding method varies among the sexually and asexually propagated cultivars.

6.2.1 Selection

Selection of promising suitable lines possessing desirable traits is the pre-requisite in any breeding programme. Selection criteria in pelargoniums includes, uniformity in different stages of growth, earliness and continuous flowering, plant vigour and growth habit, adaptive to abiotic stresses, morphological characteristics such as leaf shape, leaf size, colour and placement, flower shape, size, type and colour, plant productivity, pollen production, seed production, insect- pest and diseases resistance.

6.2.2 Hybridization

For production of F_1 hybrids in pelargoniums, the commercially cultivars or breeding lines are cross pollinated to produce hybrid seedlings. Most of the commercial cultivars are extremely heterozygous and usually tetraploid, hence development of inbred lines through conventional technique of inbreeding is a complex and tedious process. Crossing the parents with inherent heterozygosity often results in great variation among the hybrid seedlings. The breeder should grow and evaluate seedling and select type that are promising. Normally, the selected seedlings are vegetatively propagated and evaluated as cutting produced plants (clones). Initially, flower colour, size and various foliage characteristics are observed, however, rooting response must be evaluated early. After initial evaluation, more cuttings need to be propagated and tested for their growth habit, flower production, insect-pest and disease resistance and response to various environmental conditions including temperature, carbon dioxide and fertilisation. The cultivars having overall quality, including flower keeping quality and tolerance to stress and nutritional disorder must be evaluated. Later, promising lines tested at wider scale in greenhouse and in the field under garden conditions for confirmation and superior cultivar produced is introduced at commercialised level.

6.2.3 Polyploidy

Tetraploid cultivars of zonal pelargonium have been evolved with flowers exhibiting double types. However, the tetraploid cultivars exhibit a lower flower fertility due to disorders in segregation and dislocation of chromosomes in meiosis (Acquaah 2007). As a consequence, the flowers remain blooming for much longer period, whereas, after fertilization they would lose their petals and start forming seeds, which is an unfavourable attribute for ornamental flowering plants. Therefore, it is profitable to have breeding programs for tetraploid pelargoniums because of their market importance (Hofmann 1992).

6.2.4 Mutation Breeding

Mutation breeding is a valuable tool for inducing variations and creating new cultivars in vegetatively propagated cultivars of pelargonium. Several spontaneous mutants have been evolved in pelargoniums. However, irradiations can be exploited to induce the mutations in pelargoniums. Rooted cuttings are suitable for inducing mutations since new vegetative tissues from different cell layers give rise to bud sports. Bud sport type mutant selection is considered advantageous in the improvement of pelargoniums as it allows the breeders to recognise a new phenotype as distinct from the parent. The selected sports can be maintained and multiplied vegetatively. Recently, the carbon ion beam irradiations have been found effective in inducing genomic variations in pelargonium. Yu *et al.* (2016) successfully induced novel flower mutant in pelargonium within a shorter period of time by irradiating the samples with 30 Gy carbon ions. Hence, the ion beam irradiation combined with the vegetative propagation technique can be used as an efficient approach for improvement of pelargoniums. Some of cultivars such as 'Orchid Princess'; 'Royal Purple' and 'King Midas' have been developed through irradiations. Several families of *Pelargonium* × *hortorum* have been derived through sports, such as 'Better Times', 'Ricards', 'Fiats' and 'Irenes'.

7.0 Growth and Flowering

7.1.0 Temperature

Pelargoniums grow best at night/day temperatures of 10.0-15.5/18.3-23.8°C (Becker and Brawner, 1996). The gray-green-leaved pelargoniums are tolerant to high heat levels. The growth and development of most pelargoniums is hampered when temperature exceeds beyond 32.0°C. Trellinger (1997) stated that an average daily temperature of 20°C is best for ivy-geraniums as it allows for compact growth, while day/night temperatures of 24/16°C cause greater stem elongation. Zonal pelargoniums are tolerant to a wide range of temperatures with respect to floral initiation and development and if growth has occurred the flowering will also occur, however, 21°C night temperature proves better for promotion of flowering over 15°C or 10°C in cutting–grown plants (Carpenter and Carlson, 1970). In case of seed-geraniums the increase of temperatures from 10°C to 24°C progressively increased the rate of growth and flowering, optimum being 20/17°C day/night temperatures (Carpenter and Rodriguez, 1971; Erwin and Heins, 1992).

7.2.0 Photoperiod

Flower initiation in *P.* × *hortorum* is dependent on total cumulative light energy. During low light periods of winter, supplemental lighting hastens flowering if provided during early stages of seedling growth (Carpenter and Rodriguez, 1971; Armitage and Tsujita, 1979). Bethke and Carlson (1985) proved that supplemental lighting of at least 350 fc using high pressure sodium lamps accelerated floral initiation when applied 4-6 weeks after germination. Quatchak *et al.* (1986) reported hastened flowering when supplemental lighting was given after transplanting, though Kaczperskiand Heins (1995) stated that response of supplemental lighting was quite pronounced in case of 28-35 day old seedlings than those of 7-21 day old seedlings. Aimone (1985) stated that light intensity of 3,500–5,000 fc and 2,500-3000 fc are optimum for Zonal geraniums and ivy-leaved geraniums, respectively, and he further mentioned that for every 1 per cent increase in irradiance, 1 per cent growth increase should be expected if optimum temperature is available. Growing plants under 60 per cent shade causes development of 22-24 nodes prior to flower initiation and so requiring 110 days for flowering but in case of full ambient light only 16-18 nodes were found developed and the plants flowered only in 37 days, the difference being only of 8 days from floral initiation to visible floral buds between shaded and full sun treatments which strongly suggests that floral initiation is a light-driven process though floral development is a temperature-driven process (Armitage and Wetzstein, 1984), *i.e.* from visible bud stage to opening of the first floret of the inflorescence is less dependent on light but more on temperature (Armitage *et al.*, 1981).

7.3.0 Growth Regulators

The use of rooting hormones, such as IBA (indole butyric acid) have been shown to improve the uniformity of rooting and to speed up the root initiation, especially on slow rooted cultivars like Pink Camelia. IBA can be applied by using commercial dust or in solution form (not more than 0.10 percent). The growth regulators like A-Rest (ancymidol), Cycocel (chlormequat) and B-nine (daminozide) are efficient in retarding the growth and vigour. However, gibberellic acid can be used for internodal length elongation. Cycocel is the growth retardant commonly used on pelargoniums to reduce final height. The doses vary from 750 to 1,500 ppm, depending on the application method, cultivar vigor, and environmental conditions. Alar, at 500 ppm spray is applied 14 days after planting, with a second application 14 days later, only if needed on vigorous cultivars. Some growers prefer to use 750 ppm at beginning *i.e.* 14 days after planting and additionally 3 to 4 times at weekly intervals as needed.

8.0 Propagation

Pelargoniums are propagated by seed, division and stem cuttings

8.1.0 Propagation through Cuttings

The stem cuttings are are taken in late summer, from vegetative shoots, preferably with three nodes and planted under light conditions in sand, perlite or vermiculite media where new roots will develop within a fortnight and then these rooted cuttings are transferred to pots. The ideal size of terminal cuttings should be 5 to 7.5 cm with a good active growing tip and without any physical damage. Pelargonium stocks are often raised through tip-cuttings during active vegetative growth, *i.e.* during spring or in autumn. The cuttings should be obtained from health plants after following proper culture indexing and virus indexing protocols. Temperature plays an important role in rooting of cuttings. If air temperature is 10°C, the cuttings can be taken any time of the year, preferably from

February to October. However, it is suggested that temperature of rooting media should be 21-24°C for rapid callusing and rooting in all *Pelargonium* types (Dole and Wilkins, 1999). The most ideal time for induction of rooting is the active growth period, *i.e.* in early spring and late summer (Key, 1985). The use of rooting hormones by dusting the cut-bases is recommended for quick initiation of rooting, especially during winter months when temperature is quite low. In temperate regions, cuttings are taken in August or September as during this period bottom heating is not required. The cuttings should have three nodes (5-8 cm long) and the growing tips, and their leaves, *vis-a-vis* stipules should be trimmed leaving only the immature ones at the tips. Equal portions of damp peat + silver sand are the best for rooting, though others like perlite or vermiculite can also be used either singly or as a mixture. The cuttings will root in 12-21 days (Dole and Wilkins, 1999), Zonal pelargoniums root in about 10-14 days at a temperature of 18°C. whereas, Regal pelargoniums take 20-40 days for rooting. The cuttings should be protected from excessive sunshine so as to prevent excessive wilting.

Propagation and Growing of Pelargonium

Preparation of Terminal Cuttings

Prepared Cuttings of Pelargonium

Pelargonium Cuttings Planted in Plug Trays for Rooting

Pelargonium Seeds

Newly Emerged Seedling

Seedling Ready for Transplanting

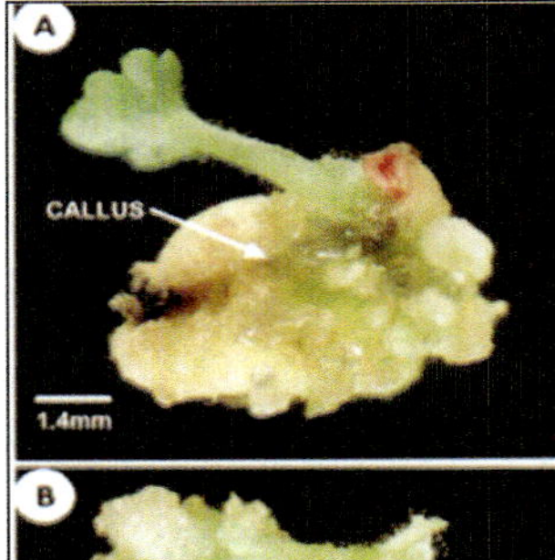

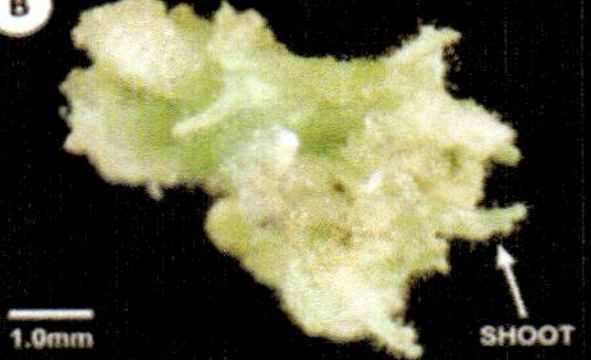

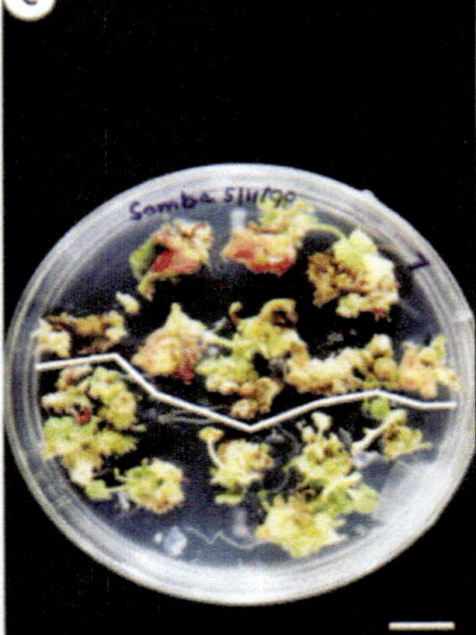

Micropropagation of Pelargonium

Stages of regeneration of Pelargonium hortorum cv. Samba from leaf and petiole: **(A) 3-week-old leaf explants; (B) 3- week -old petiole explant; (C) 4-week-old culture (D) Well-developed shoot with healthy roots. (*Source*: Agarwal and Ranu, 2000).**

8.2.0 Seed Propagation

Seed propagation is becoming increasingly popular in geraniums and pelargoniums. The seed germination in pelargoniums is influenced by temperature, light, and moisture. The seeds are sown in month of November under Northern plains. However, in hilly areas, April-May is ideal time for sowing of seeds. The seeds are sown in trays or pots at a depth of 0.8 to 10 mm. After sowing, the seeds are covered lightly with sieved soil mix. The seed beds are kept moist and not allowed to dry out excessively. The seeds normally germinate in one to three weeks when temperatures is maintained between 21.1-23.8°C. The commercial seeds are mostly F_1 hybrids and these are quite expensive. Approximately, the seed number per gram is 28-30. The temperature should be maintained at 16-18°C to ensure proper and uniform germination. Light brown coloured pelargonium seeds require a germination temperature of 24°C to 25°C in the initial stages and 21°C to 22°C in the later stages. Temperature above 26°C for longer duration may cause seed dormancy, resulting in poor germination. Key (1985) stated that seeds require a minimum temperature of 13°C to germinate and the ideal as 19°C. Vermiculite is used as a seed starter medium. Due to hard seed coat scarification either chemically or manually is done for improving the seed germination rates. Germination is enhanced when chamber is quite illuminated. Seeds germinate in 1-2 weeks time. Seeds can be sown in plug trays in environmentally controlled rooms or chambers. A mist or fog system in the germination room should provide optimum relative humidity of 70-75 percent for achieving best germination results.

Seedling plugs can be stored up to four weeks at 2°C (Heins *et al.*, 1994) under fluorescent light condition rather than dark-storage. However, plugs tolerate better storage or shipping when been previously irrigated with 150 ppm nitrogen for two weeks (Kaczperski *et al.*, 1996a). Areca *et al.* (1996) reported poor postharvest life of pelargonium cuttings when transported at warm temperatures due to build up of ethylene.

8.3.0 Micropropagation

The high value pelargonium plant has the advantage from large-scale propagation of pathogen indexed plants through tissue culture (Laughner, 1993). The advantages of *in vitro* mass propagation over conventional propagation techniques include: rapid multiplication, lower production costs, convenience in handling, storage and shipping, as well as the limited quarantine restrictions. The first tissue cultured pelargonium was developed from callus cultures (Mayer, 1956; Stephaniak and Zenkteler, 1982). *De novo* shoot organogenesis has been induced on pelargonium stem segments cultured in sterile and non-sterile conditions (Chen and Galston, 1967).

Meristem tip culture is extremely useful for the production of pathogen-indexed planting material. The process of *in vitro* culture of meristems involves the aseptic culture of shoot tips (0.1–0.5 mm) with attached leaf primordia (Horst and Klopmeyer, 1993). Culture indexing is a vigorous testing system combined with intense selection and sanitation procedures (Reuther, 1983; Horst and Klopmeyer, 1993). Vegetative cuttings are tested three times before entering the propagation system as pathogen indexed stock (Oglevee, 1993). The whole process takes a year with increased production costs. Since it is economically not viable to culture-index each cutting, multiplication for commercial propagation is done from selected plants and culture-indexing ensures clean stock for further propagation (Oglevee, 1993).

Organogenesis is the process by which cells and tissues undergo changes which lead to the production of a unipolar structure with either shoot or root primordium with a distinct vascular connection to the explant vasculature (Thorpe, 1993). Zonal geraniums (*Pelargonium* × *hortorum*) watered with low concentrations of the plant growth regulator thidiazuron (TDZ) formed outgrowths on the root tissues and prolific shoots from the crown (Sanago *et al.*, 1995). De novo shoot organogenesis was induced on intact seedlings of hybrid *P.* × *hortorum* in response to supplementation of the culture medium with BAP (Qureshi and Saxena, 1992). Adventitious shoots were induced from thin layer hypocotyl explants in *P.* × *hortorum.* Also, shoot regeneration was observed in fertilized ovules of a cross between *P.* × *hortorum and P. peltatum*, cultured *in vitro* (Fukuda *et al.*, 1997).

9.0 Soil and Climate

9.1.0 Soil

Pelargoniums can be grown in variety of soils, but prefer well aerated sandy loam soils for optimum growth. They do not like high moisture retentive soils as waterlogged conditions are not conducive for good vegetative growth and flowering. The optimum soil pH should be 6.0 to 6.5. Plants do not perform well in heavy manured soils, it is therefore advisable to add a minimum level of well decomposed farm yard manure in the soil before planting. Five kg of farm yard manure and 30 g of N, 29 g of P_2O_5 and 25 g of K_2O should be added per square meter area at the time of bed preparation. Well established plants can flower well even in dry conditions. However, the species have wide variations in their requirements as *P. graveolens* prefers moist soil and *P. scabrum* is well suited to dry and sandy areas (Amidon and Brobst, 2005). The pH of soil should be 5.6-6.0 though for *P. peltatum* the pH should be slightly lower, *i.e.* 5.3 (Tellinger, 1997).

9.2.0 Climate

Pelargoniums are sun-loving plants as most of the species flower profusely in full sunlight, however, some species like *P. odoratissimum*, *P. graveolens*, *P. grossularioides* and *P. tomentosum* prefer shady conditions (Amidon, and Brobst, 2005). At least three to four hours of direct sunlight will be sufficient for getting good bedding crop in hybrid pelargoniums. Eastern exposure are better that the western aspects. They thrive well if they avail two-thirds of the daylight (at least six hours) where double and semi-double types may also be grown. Such flowers last longer, are larger, have proper standing, and shattering is minimum even during inclement weather conditions. Some of the species perform well in partially shaded conditions but the flowering is not prolific. However, singles prefer sheltered or filtered situations. Shading against high temperature (>32 °C) and high light levels during summer months will be required to prevent scorching of plants. Various groups of pelargoniums differ slightly in their light requirements, such as ivy-leaf pelargoniums require high quantum of light whereas Regal pelargoniums require shading from strong sunlight, especially during flowering. The flowering of many Zonal pelargoniums also last longer, if lightly shaded.

10.0 Cultivation

10.1.0 Planting

Pelargoniums are highly valued as potted plants. Under controlled environment, rooted cuttings or seedlings are transplanted during January-February, so that they are ready for sale in April-May. However, under natural field conditions, transplanting is done in the month of March after the cold spells are over. After transplanting, they generally take six to eight weeks to become ready for sale. While planting in pots, proper ventilation should be maintained as these plants dislike a stuffy, stagnant atmosphere. Both, low and high levels of humidity adversely affect plant growth and development. Excessive dry air during summer is highly detrimental to both, vegetative and reproductive phases, so misting of greenhouse is recommended to increase humidity. High humidity in winters cold days causes rotting of flowers and flower buds, so it is necessary to lower atmospheric humidity by heating the air to prevent the occurrence of fungal diseases. Ivy-leaved pelargonium is more prone to a physiological disorder called oedema under high humidity conditions. CO_2 concentration in the greenhouse should be 800-1,500 ppm for stock plants, propagation through cuttings and seed geranium production and so the vents should be kept closed (Shaw and Rogers, 1964).

10.2.0 Pinching

Pinching is done to break apical dominance to promote lateral branching to have for compact and dwarf plant stature. Pinching is done when the plants have 2 to 3 pairs of leaves and attained a height of 5-8 cm. Morval and Deacon series pelargoniums naturally produce neat and beautifully dome-shaped plants without being pinched. Pinching is practiced when the plant is in active growth, *i.e.* in spring or early summer. When the plants developed 2-3 shoots with 2 nodes, growing tips should be removed. This practice should be repeated further as frequent pinching will make the plants producing more shoots, and compact and bushy plant growth, full of flowers. Any flower buds appearing should be removed until a three fortnights before scheduled date of flowering and then stopping is not done so that flowers appears in time. Normally, pinching is carried out till May end, and in the end of June the plants require another potting in a 15-cm pot where it remains till the end of the season.

Spent flowers are continuously removed to conserve the energy being spent towards seed formation so that more flowers are formed. In October beginning, these plants are headed-back to almost 10 cm stump with all the leaves removed if still any, and the pots are kept closely on the benches of the greenhouse where these are given occasional watering for their growth. Alternative to heading-back, pruning is done to produce healthy, attractive, sturdy, much-branched, bushy (by preventing upright leggy growth) and well-shaped plants. The geraniums are

Growing of Pelargonium

Well Established Plants of Pelragonium

Pots under Protected Structure

Under Protected Structure

more commonly grown as pot plants, hence, to control the plant height dwarfing chemicals are also used.

10.3.0 Manuring and Fertilization

Pelargonium groups, like Zonal and Ivy-leaved have high requirement of nitrogen whereas, Regal and Unique require more of potassium. Their requirement of Ca and Mg is also high, therefore their nutrient regimes are mostly based on KNO_3 and $Ca(NO_3)_2$ combinations, the phosphorus is added to the medium as superphosphate or in the water as phosphoric acid so that proper pH is maintained. The stock plants should regularly be monitored for their nutrient requirements as they are frequently in production. The container-grown plants require regular fertilization for better presentation but plug trays, especially of *P.* × *hortorum* are sensitive to high soluble salts due to restricted drainage in the plug-flats. Cutting-propagated Zonal pelargoniums have requirement similar to *P.* × *hortorum*. Seedlings should be fertilized with 100-150 ppm N but after proper rooting of the transplants the dose should be increased to 250 ppm N, while at inflorescence stage the dose should be decreased to 150 ppm N, and it is further reduced to 50 ppm prior to transportation provided the plants are grown at 13 °C (Dole and Wilkins, 1999). Craig (1986) reported that *P. peltatum* and *P.* × *domesticum* should not be fertilized with more than 200 ppm of nitrogen. Iron and manganese toxicity is reported in seed geraniums due to low pH and then plants become stunted, show necrotic spots on leaves and leaf edges, and lower leaf necrosis. Hence it is recommended to raise pH of the medium above 6 so that they are less available to the plants. Foliar tissue levels of Fe and Mn should not exceed 500 ppm (Erwin, 1997). A fertilizer dose containing 1:0:2 or 1:1:2 NPK is found most suitable during winter and summers for Regal pelargoniums. In case of low potassium levels, a dose of 1:0:3 NPK can occasionally be used as frequent use will lead to magnesium deficiency, though it can be overcome by spraying $MgSO_4$ @25g/25 l water (Clark, 1988). Zonal and Ivy-leaved pelargoniums require more nitrogen and less potassium than Regal pelargoniums. A fertilizer dose of 1:1:1 NPK is most suitable with an occasional NPK dose of 2:1:1. For having brightly coloured pelargoniums, the cultivars with gold or tricoloured leaves should be fertilized with high potassium fertilizers similar to that of Regal pelargonium (Clark, 1988).

11.0 Diseases and Pests

11.1.0 Diseases

11.1.1 Botrytis Blight (*Botrytis cinerea*)

It appears during cool and humid conditions and it causes brown to tan lesions on leaves, stems and petioles in the form of fuzzy fungal growth. To control this disease, proper plant density should be adopted to permit proper ventilation and maintenance of optimum relative humidity and hygienic conditions. Use of copper hydrochloride, iprodione or carbendazim is suggested.

11.1.2 Downy Mildew (*Peronospora* spp.)

It causes pale-green or light brown discolouration of leaves throughout the year, mostly in the form of angular patches, and off-white slightly fuzzy fungal patches develop beneath. In severe infection, the leaves wither and die completely. This disease is caused due to high moist or humid conditions and poor air circulation.

11.1.3 Powdery Mildew (*Oidium* spp.)

The infection mostly occurs on the leaves, leaf petioles, stems and buds in the form of white powdery coating, especially during spring and summer. Usually, upper leaf surfaces are affected first most severely and then spreading to lower surfaces and other parts. Its infection is encouraged by humid atmosphere. To prevent this disease, proper ventilation should be maintained and plants should be sprayed with carbendazim, copper sulphate pentahydrate, mancozeb.

11.1.4 Alternaria Leaf Spots (*Alternaria tenuis*)

It appear in the form of water soaked spots on the old leaves below but later on new ones are also engulfed. These spots turn brown to dark brown forming concentric rings. Diseased leaves should be collected and destroyed, proper sanitation should be maintained, cuttings should be collected only from the healthy plants and 0.2 per cent Dithane M-45 should be sprayed.

11.1.5 Bacterial Leaf Spots (*Xanthomonas campestris, X. pelargonii, Pseudomonas cichorii, P. syringae* and *Acidovorax* spp.)

Among the causal agents, *Xanthomonas campestris* is most serious as it spreads through propagules, irrigation water and for many years its infection remains mute and with chemicals it is difficult to control (Roberts, 1997). To prevent this problem, use of healthy planting material, proper sanitation, avoiding overhead irrigation, pruning of the diseased part and use of copper hydroxide, copper sulphate pentahydrate may be beneficial.

11.1.6 Anthracnose (*Amerosporium trichellum* or *Colletotrichum trichellum*)

It is characterized by brown and sunken lesions on foliage and sometimes with black specks in the centre of the lesions. Use of healthy planting material, no-overhead irrigation, pruning of the infected portion, proper sanitation of the field, maintaining of proper plant density and use of copper hydroxide, manzate or iprodine will keep this disease at bay.

11.1.7 *Rhizoctonia* Root Rot and Aerial Blight (*Rhizoctonia solani*)

The infection causes brown irregular lesions, blighting and wilting of plants. To prevent the occurrence of this pathogen, potting media should be sterilized, planting material should be healthy, overhead watering and over-fertilization should be avoided, affected portion should be removed and burnt, field should be clean and the plants should be sprayed with etridiazole + thiophanate methyl, oriprodione or PCNB (terrachlor). *Glomerella cingulata* and *Sphaceloma hederae* also cause leaf spot diseases in ivies (Horst, 1990).

Diseases and Pests of Pelargonium

Diseases

Pelargonium Rust

Fungal Leaf Spots

Bacterial Infection Caused by *Xanthomonas*

Bacterial Infection Caused by *Ralstonia*

Botrytis Blight

Brown Root Rot Caused by *Pythium* & *Phytophthora*

Diseases and Pests of Pelargonium

Diseases

Sunburn

Alternaria Leaf Spot

Pests

Aphids

Spider Mites

Caterpillars

Whitefly

11.1.8 Pelargonium Rust (*Puccinia pelargonii-zonalis*)

It forms concentric rings of dark brown fungal pustules on lower leaf surface with corresponding yellow blotches on the upper surface. The spores spread through air and water and are favoured by humid atmosphere and poor ventilation. Infected leaves should be removed, air circulation should be improved and spraying with fungicides such as mancozeb, penconazole or triforine with bupirimate should be done.

Vascular bacterial blights (*Pseudomonas solanacearum*), verticillium wilt (*Verticillium alboatrum*), root and stem rots (*Pythium, Rhizoctonia, Fusarium* and *Thielavapsis*) and bacterial stem fasciation (*Clavibacter fascians*) also occur in geraniums (Horst, 1990; Daughtrey and Chase, 1992).

Pelargonium viruses cause yellow markings in the form of flecks, ring spots, streaks and distortion of leaves and flowers may also have pale streaks on the petals. Some of the viruses are transmitted by aphids, soil nematodes or mechanically. Such plants should be uprooted and destroyed.

11.2.0 Pests

11.2.1 Aphids (Blackfly, Greenfly; *Myzus persicae*, Greenflies, Plant Lice and Certain others)

They suck cell sap and transmit the virus diseases. They most commonly feed on young, tender parts such as newly growing stems, leaves, flowers, *etc.* They also excrete the excess sugar obtained from plant sap and this sticky material provides a food source for fungi which lead to development of sooty mould. Infested plants should be sprayed with a diluted soap solution, growing marigold side by side in the greenhouse will attract hoverflies whose larvae feed on aphids, and malathion spray kills pests. Insecticidal soaps, pyrethrum and derris are the effective organic insecticides.

11.2.2 Caterpillars

The larvae of various types of moths and butterflies, infesting badly on outdoor Zonal pelargoniums and others. Budworms feed on the floral buds as well as young shoots. An introduction of *Bacillus thuringiensis* will kill the caterpillars. Permethrin insecticide spraying is also very effective.

11.2.3 Whitefly (*Trialeurodes vaporariorum*)

They infests Zonal and Ivy-leaved pelargoniums in general and Regal pelargoniums in particular and excrete a sugary substance which entices sooty mould. Flies breed continuously throughout the year (except winter) on indoor plants and their adults as well as immobile, oval and scale-like nymphs suck cell sap on underside of leaves whereas upper surface becomes sticky with honeydew excretion enticing sooty mould. Their infestation causes small yellow spots on the leaves, followed by yellowing and sooty mould. Hanging of yellow sticky trap is quite effective.

11.2.4 Mealy Bugs (*Planococcus* spp. and *Pseudococcus* spp.)

They are sucking insects which infest young soft plant tissues. Nymphs and female adults may form cluster on the shoots in the leaf axils, at the tips, peduncles and pedicels, and on tender leaves. The infested plants become sticky due to excretion of honeydew, new growth is distorted and flowers aborted. These can be killed by wiping the affected part with a cotton swab dipped in alcohol.

11.2.5 Spider Mite (*Tetranychus urticae*)

It is found mostly infesting on the undersides of the leaves, spinning a fine web there. Spider mites cause speckling damage on the leaves and sometimes cause the ivy leaves to form brown corky areas. Red spider mite is now known to be resistant to most organophosphorus insecticides such as diazinon, dimethoate and malathion. Introduction of the predatory mite, *Phytoseiulus persimils* proves very effective.

11.2.6 Scales (*Lichtensia viburni*)

Scales are sap sucking insects, feeding on the upper leaf surface and making the surface sticky with the honeydew excreted by them but live on the lower leaf surface next to a leaf vein. In mid-summer, the underside of leaves should be sprayed with malathion to control this pest.

11.2.7 Ladybird Predator (*Cryptolaemus montrouzieri*)

Ladybird predator over plants during summers but at other times thorough spray of insecticidal soap or malathion should be done.

12.0 Harvesting and Postharvest Management

All the pelargoniums are highly sensitive to ethylene. The apical meristem can totally abort, leaves may become thickened and cupped, veins exaggerated, stems thickened and swollen and flower buds abnormally developed or aborted due to ethylene buildup (Dole and Wilkins, 1999). Even the seed geranium plants shatter the petals during transportation due to build up of ethylene. Silver Thio-sulphate (STS) sprays at 175 ppm 14-21 days prior to transportation is quite effective in case of Zonal and *P.* × *domesticum*, and then these should be transported at 5°C (Deneke *et al.*, 1990). Unrooted cuttings can be stored for 4-6 weeks at -0.5°C (Eisenberg *et al.*, 1978) while rooted ones for up to 8 weeks at 5°C and 250 fc illumination for 9 h per day (Kaczperski *et al.*, 1996b).

13.0 Biotechnology

13.1.0 Genetic Transformation

Novel technologies have been developed for pathogen detection, such as the Enzyme Linked Immuno Sorbent Assay (ELISA) and colourimetric assays are available commercially (Oglevee, 1993). There have been several recent reports of genetic transformation of Pelargonium cultivars for the development of value-added crops. Robichon *et al.* (1995) were the first to report the production of transgenic *Pelargonium* × *hortorum via Agrobacterium tumefaciens*. These two reports provided the foundation for the development of a system for somatic embryogenesis and

Agrobacterium-mediated transformation for scented geraniums (*Pelargonium* sp. 'Frensham') (Krishna Raj *et al.,* 1997). The protocol developed in this study was unique in that it coupled high-frequency somatic embryogenesis with efficient gene insertion and expression for the production of relatively large numbers of transgenic plants (Krishna Raj *et al.,* 1997). In subsequent studies, this protocol was used to produce transgenic plants containing a gene encoding for the antimicrobial protein Ace-AMP1. The incorporation of this gene into *Pelargonium* sp. 'Frensham' conferred an increased resistance to infection by *Botrytis* (Bi *et al.,* 1999). Pelargonium cultivars with relative levels of resistance to *Botrytis* have recently been identified and it has been reported that diploid genotypes possess greater resistance than tetraploid cultivars (Uchneat, 1999). These genotypes may provide additional genetic information for incorporation into resistance breeding programs for Pelargonium. An improved protocol for genetic transformation of *P.* × *domesticum* was reported using acetosyringone to induce vir genes on the *Agrobacterium tumefaciens* Ti plasmid (Boase *et al.,* 1998).

14.0 References

Acquaah, G. (2007) Principles of Plant Genetics and Breeding. Oxford, Blackwell Publishing: 218–220.

Agarwal, P.K. and Ranu, R.S. (2000) *In vitro Cellular and Dev. Biol.* **36**: 392-397.

Aimone, T. (1985) *Grower talks,* **48**(11): 128-134.

Amidon, C and Brobst, J. (2005) *The Herbarist,* **71**: 4-10.

Areca, R.N., Arteca, J.M., Wang, T.W. and Schlagnhaufer, C.D. (1996) *J. Amer. Soc. hort. Sci.,* **121**: 1063-1068.

Armitage, A.M. and Wetzstein, H.Y. (1984) *Hort Science,* **19**: 114-116.

Armitage, A.M. and Tsujita, M.J. (1979) *J. hort. Sci.,* **54**: 194-198.

Becker, J. and Brawner, F. (1996) *Scented Geraniums: Knowing, Growing and Enjoying Scented Pelargoniums.* Loveland, CO: Interweave Press.

Bethke, C.L. and Carlson, W.H. (1985) *GrowerTalks,* **49**(6): 58, 60, 62, 64, 66.

Boase, M.R., Bradley J.M. and Borst, N.K. (1998) *Plant Sci. Shannon.* 139: 59-69.

Carpenter, W.J. and Rodriguez, R.C. (1971) *Hort Science,* **6**: 206-207.

Carpenter, W.J. and Carlson, W.H. (1970) *Hort Science,* **5**: 183-184.

Chen, H.R. and Galston, A.W. (1967) *Physiol. Plant.* **20**: 533–539

Clark, D. (1988) *Kew Gardening Guides: Pelargoniums,* p. 124. Published by The Royal Botanic Gardens, Kew, in association with Collingridge.

Craig, R. (1986) *Bedding Plant News,* 17(**6**): 6-10.

Daker, M. G. (1969) *J. Roy. Hort. Soc.,* **94**: 346.

Daughtrey, M. and Chase, A.R. (1992) *Geranium.* In: *Ball Field Guide to Diseases of Greenhouse Ornamentals,* pp. 97-113.

Deneke, C.F., Enensen, K.B. and Craig, R.(1990) *Hort Science*, **25**: 937-940.

Dole, J. M. and Wilkins, H. F. (1999) *Floriculture Principles and Practices*, pp. 451-460, 613. Published by Prentice Hall, Inc., USA.

Eisenberg, B.A., Staby, G.L. and Eretz, T.A. (1978) *J. Amer. Soc. hort. Sci.*, **103**: 732-737.

Erwin, J. (1997) *Minnesota Com. Fl. Grs Assoc. Bull.*, **45**(6) & **46**(1): 1-10.

Erwin, J.E. and Heins, R.D. (1992) *Minnesota Com. Fl. Gr. Assoc. Bull.*, **41**(1): 1-9.

Fukuda T, Kakihara F & Kato M (1997) *J. Soc.High Technol*. Agric. **9**: 255–262.

Heins, R., N. Lange, T.F. Wallace, Jr. and Carlson, W. (1994) *Plug Storage, Cold Storage of Plug Seedlings*. Meister Publishing, Willoughby, Ohio, USA.

Hofmann, C.H. (1992) *Schöne Pelargonien* (Geranien). Stuttgart, Eugen Ulmer: 57.

Horst RK & Klopmeyer MJ (1993) Controlling virus diseases. In: White JW (ed) Geraniums IV (pp 286–292). Ball Publishing Co., IL

Horst, R.K. (1990) *Geranium*. In: *Westcott's Plant Disease Handbook* (5th ed.), pp. 653-654. Nostrand Reinhold, New York, USA.

Kaczperski, M.P., Armitage, A.M. and Lewis, P.M. (1996a) *Hort Science*, **31**: 361-363.

Kaczperski, M,.P., A.M. Armitage and P.M. Lewis, (1996b) *Hort Science*, **31**: 656. Kruger Brentt Publisher UK

Kaczperski, M.P. and Heins, R.D. (1995) *Hort Science*, **30**: 760.

Kellen, V. (1983) *Florists' Review*, **173**: 19–24.

Key, H. (1985) *Pelargoniums*, 63 pp. Published by Cassell Educational Limited, London for Royal Horticultural Society, London.

Krishna Raj, S., Bi, Y. and Saxena, P.K. 1997. *Planta*, **201**: 434–440

Laughner LJ (1993) History. *In*: White JW (ed) Geraniums IV (pp 363–371). Ball Publishing Co., IL.

Li-Xia, Y.,Wen-Jian, Li., Yan, Du., Gang, Chen., Shan-Wei, Luo., Rui-Yuan, Liu.,Hui, Feng., Li-Bin, Zhou. (2016) *Nucl. Sci. Tech.*, **27** : 112-119.

Mayer L (1956) *Planta*, **47**: 401–446

Miller, D. M. (2002) *The taxonomy of Pelargonium species and cultivars, their origins and growth in the wild*. In: *The Genera Geranium and Pelargonium* (ed. Lis-Balchin, M).

Oglevee WA (1993) *Culture-indexing for vascular wilts and viruses*. In: White JW (ed) Geraniums IV (pp 277-285). Ball Publishing Co., IL

Quatchak, D.J., White, J.W. and Holcomb, E.J. (1986) *J. Amer. Soc. hort. Sci.*, **111**: 376-379.

Qureshi, J.A. and Saxena, P.K. 1992. *Plant Cell Rep*. **11**: 443–448.

Reuther, G. (1983) *Acta Horticulturae*, **212**: 311–319

Roberts, D.L. (1997) *Prof.Plant Growers Asso. News*, 28(**1**): 21-22.

Robichon, M.P., Renou, J.P. and Jalouzot, R. (1995) *Plant Cell Rep.* **15**: 63–67.

Sanago, M.H.M., Murch, S.J., Slimmon, T.Y., KrishnaRaj, S. and Saxena, P.K. (1995) *Plant Cell Rep.*, **15**: 205–211.

Stephaniak, G. and Zenkteler, M. (1982) *Acta. Soc. Pol.*, **51**: 167–172

Thorpe TA (1993) *In vitro organogenesis and somatic embryogenesis: Physiological and biochemical aspects.* In: RoubelakisAngelakis KA & Tran Thanh Van K (eds) Morphogenesis in Plants (pp 19–38). Plenum Press, New York

Trellinger, K. (1997) *GrowerTalks*, **61**(8): 73, 75.

Uchneat, M.S. (1999) *Hort Science*, **34**: 718–729.

Williams, C. A. and Jeffrey, B. H. (2002) *Phytochemistry of the genus Pelargonium.* In: *Geranium and Pelargonium: The genera Geranium and Pelargonium* (ed. Lis-Balchin, Maria), pp. 99-115. Taylor and Francis, New York.

Iris

M.R. Dhiman and Raj Kumar

1.0 Introduction

Iris L. is one of the most attractive flower crops known worldwide for centuries for their beautiful flowers, which hold many mythological, artistic, and historical traces (Kandeler and Ullrich, 2009). Consequential of the broad spectrum of flower colours that exist in nature, the name iris is derived from the Greek word for 'rainbow'. Iris species have been valued as an ornamental and medicinal plants since ancient times, but more recently it has mostly been used as an ornamental plant. Irises are great garden plants growing from creeping rhizomes (rhizomatous irises) or, in drier climates, from bulbs (bulbous irises). They have long, erect flowering stems which may be simple or branched, solid or hollow, and flattened or have a circular cross-section. The rhizomatous species usually have 3–10 basal sword-shaped leaves growing in dense clumps. The bulbous species have cylindrical, basal leaves. The inflorescences are in the shape of a fan and contain one or more symmetrical six-lobed flowers. The three sepals are usually spreading or droop downwards, are known as 'falls'. They expand from their narrow base (the 'claw' or 'haft'), into a broader expanded portion ('limb' or 'blade') and can be decorated with veining, lines or dots. In the centre of the blade, some of the rhizomatous irises have a 'beard' (a tuft of short upright extensions growing in its midline), referred as the plant's filaments. The three, sometimes reduced, petals stand upright, partly behind the sepal bases called 'standards'. Some smaller iris species have all six lobes pointing straight outwards, but generally limb and standards differ markedly in appearance.

2.0 Importance and Uses

2.1.0 Aesthetic Uses

The Iris flower is a popular gifting bloom and is highly valued for its appealing colours and unique structure. This can be used as cut flower as well as for bedding purposes & pot culture. They can be found growing in arid and semi-desert regions, on rocky slopes, grassy banks, in meadows, bogs, and along riverbanks. They are also an excellent choice for steep or difficult slopes, to help prevent erosion. Although iris has very short vase life, their importance as cut flower continues since the last several centuries.

2.2.0 Therapeutic and Other Uses

Irises were valued by the early Indian and Egyptian people for their medicinal uses and for manufacturing of perfumes. These perfumes were used in the offerings to the god. Beside this Iris has medicinal value, as rubbing the extract of flowers will help to cure skin and hair problems like acne signs, split ends, irritated scalp. Iris flowers are rich in substances that enhance body's potential and most significant active compounds of this flower are: vitamins, acids, volatile oils, tannin, minerals (zinc, sodium, selenium, copper, iron, magnesium, potassium and manganese), nutrients, natural enzymes and antioxidants. The peeled and dried rhizomes of *Iris germanica* as well as its variety *florentina* and *Iris pallida* are known as orris root (Troy

Uses of *Iris*

In Pot

As Companion Plant

Along the Pathway

Iris laevigata **Planting Near Water Body**

As a Bedding Plant

In Flower Vase

In Garden for Mass Display

Field View

As a Aquatic Plant

and Beringer, 2006) or as *rhizoma iridis* and were used for centuries in medicinal treatments while today they are most important particularly for production of orris butter. *Iris* plants (the flowers) were used for obtaining 'iris green' and 'iris blue' pigments used by medieval painters and manuscript illuminators (Eastaugh *et al.*, 2004).

3.0 Origin and History

The history of *Iris* is rich, dating back to Ancient Greek times when the Greek Goddess Iris, the messenger of the gods and the personification of the rainbow, acted as the link between heaven and earth. Purple irises were planted over the graves of women to summon the Goddess to guide the dead in their journey. Ancient Egyptian kings marvelled in the iris's exotic nature and drawings have been found of the flower in a number of Egyptian palaces. During the middle ages, the meaning of irises became linked to the French monarchy, and the Fleur-de-lis eventually became the recognized national symbol of France. Several reports note that iris cultivation originated around 1749 BC in Egypt when King Thutmose III had conquered Syria. Back then, Syria was home to many irises and King Thutmose III, an avid gardener, was fascinated with the beauty of these flowers. He immortalized irises through sculptures at the Temple of Amon at Karnak and in the gardens of Egypt.

The iris flower was included on royal banners as the fleur-de-lis, a well-known symbol of French royalty. King Clovis I, the first Frank King, adopted the iris flower as an emblem in the early 500s when he was converted to Christianity. In 1147, King Louis VII used the purple iris as well for his image. Fleurs-de-lis were destroyed by the revolutionaries as a symbol of their wrath toward the monarchy during the French Revolution. In England, Edward III used three fleurs-de-lis to his coat of arms in 1376. Also, a famous painting of Elizabeth I of England presents the queen in a gown embroidered with iris flowers. The bearded iris flower (*Iris germanica*) is the official state cultivated flower of Tennessee. The well-known symbol fleur-de-lis, which took inspiration from the iris flower, is New Orleans' emblem.

With striking uniqueness and beauty, irises have rich meanings and when given as gifts, they can convey deep sentiments. With over 200 varieties in a wide spectrum of colours, the iris, which fittingly takes its name from the Greek word for 'rainbow' can be found in virtually every part of the world, growing both naturally and in farms. While garden irises can come in any of these many varieties, the flower's cut versions are mostly blue (the most popular type), white and yellow.

Through its intricate history, the meanings of the iris have come to include faith, hope and wisdom. Depending on factors such as colour and region, irises may bear additional meanings as well. In some parts of the world, the dark blue or purple iris can denote royalty, whereas the yellow iris can be a symbol of passion. Irises may also express courage and admiration. The many meanings of the iris make the flower a great choice for an array of gift giving occasions: corporate, sympathy, get well, just because and birthdays are just some of the occasions for which irises might be the perfect choice. Today, the iris is the state flower of Tennessee and the Fleur-de-lis is the emblem for the city of New Orleans. Irises are cultivated all over the world and they can be found naturally in Europe, the Middle East, Northern Africa, Asia and North America.

4.0 Taxonomy and Morphology

Iris is the largest genus of the family Iridaceae with up to 300 species – many of them natural hybrids. Modern classifications, starting with Dykes (1913), have subdivided them. Dykes referred to the major subgroupings as sections. Subsequent authors such as Lawrence (1953) and Rodionenko (1987) have generally called them subgenera, while essentially retaining Dykes' groupings, using six subgenera further divided into twelve sections. Of these, section Limneris (subgenus Limneris) was further divided into sixteen series. Like some older sources, Rodionenko moved some of the bulbous subgenera (*Xiphium*, *Scorpiris* and *Hermodactyloides*) into separate genera (*Xiphion*, *Juno* and *Iridodictyum*, respectively), but this has not been accepted by later writers such as Mathew (1989), although the latter kept *Hermodactylus* as a distinct genus, to include *Hermodactylus tuberosus*, now returned to Hermodactyloides as *Iris tuberosa*.

Rodionenko also reduced the number of sections in subgenus Iris, from six to two, depending on the presence (Hexapogon) or absence (Iris) of arils on the seeds, referred to as arilate or nonarilate. Taylor (1976) provides arguments for not including all arilate species in Hexapogon. In general, modern classifications usually recognise six subgenera, of which five are restricted to the old world; the sixth (subgenus *Limniris*) has a Holarctic distribution.

4.1.0 Morphology

The rhizome has cord- or string-like roots. The leaves are flat, ensiform, thin, two-rowed and sometimes linear with a wax coating. Flowers of irises are single, sometimes in small inflorescences, large, often fragrant; they are characterized by an elegant shape and come in a variety of colours. The flower has six petals (in fact, they are the lobes of the perianth). The three outer lobes are slightly turned downwards and they usually differ in colour from the upper lobes that form a tube by merging at the base. Irises bloom from May to July, two or three flowers bloom simultaneously from one to five days. The fruit of iris is a capsule that opens up in three parts.

5.0 Species and Cultivars

5.1.0 Species

The genus *Iris* consists of about 300 species with rhizomes or bulbs (Wang *et al.*, 2010; Cantor, 2016), which may be perennial or deciduous. The diversity of the genus is centred in the North Temperate Zone, though some of its most attractive species are native to the Mediterranean and central Asian areas. *Iris* species can be divided into groups: (1) rhizomatous types and (2) bulbous types. Some important species are described below.

5.1.1 Bulbous Iris

Bulbous species are fewer in number and are commercially important. Bulbous *Irises* are sometimes the 'forgotten child' when it comes to the world of iris. They

Species of Bulbous Iris

Iris latifolia | ***Iris xiphium* (Spanish Iris)** | ***Iris juncea***

Iris filifolia | ***Iris magnifica*** | ***Iris reticulata***

Iris winogradowii | ***Iris domestica*** | ***Iris × hollandica***

can be very effective when used in annual or perennial flower beds, lending an extra splash of colour and then fading away until the next year. Once planted, the bulbs generally do not need to be lifted or divided for a number of years. Bulbous irises are subdivided into two sections: *Xiphium* and *Reticulata*. The first section comprises 8 species with long-stemmed flowers, up to 60 cm; the second section includes 10 species with short-stemmed flowers.

Iris latifolia

It is also known as the *Iris xiphiodes* and *Iris anglica,* native to the Pyrenees of South-western France and North-western Spain. It is generally cultivated in temperate regions for its blue flowers which appear in early summer. It grows to a height of 50 cm and produces 2-3 deep blue flowers with yellow marks in the centre of the lower petals.

I. filifolia

It is a species in the subgenus *Xiphinum*. It is a perennial type and native of North Africa and Europe. It has thin leaves, summer flowers in shades of red-purple with orange spots or yellow stripe on the falls. It flowers between March and June.

I. juncea

This species commonly known as rush iris, is a smooth-bulbed type. The name is derived from '*juncea*' from the Greek word meaning 'rush-like' (William, 1972). Its flowers are light yellow and fragrant, normally 2 per stem. It flowers between June and July. It grows upto a height of 1 to 2 feet. The bulbs are reddish-brown in colour. It can be found in North Africa, Southern Spain and Sicily.

I. xiphium

Commonly known as the Spanish iris and native to Spain and Portugal. This species is also known as the small bulbous-rooted iris. It bears blue, violet, white or yellow flowers of around 6 cm wide during May-June. The plant height reaches up to 80 cm. The Spanish iris is favoured by florists for its striking colour combinations. This species has many varieties popular in horticulture, among them var. *lusitanica* whose flowers are yellow all over. This is hardier among the bulbous iris and can be grown in northern Europe. It is sensitive to frost.

I. colletti

It was found in North Burma, Thailand, Tibet and the province of Yunnan and Sichuan. It was named after Sir Henry Collett (1836-1901), who collected plants in most of those regions. It has been found growing in a range of habitats, including wood edges, clearings, shrubby areas, and sunny grasslands. This species has 3–7 lilac-blue flowers on a 5–15 cm tall stem, flowers in May – June. The flower has an orange caterpillar-like beard on the midrib. Two hybrids have been found: *Iris collettii* var. *collettii* and *Iris collettii* var. *acaulis*.

I. magnifica

It is native to the mountains of Central Asia, including the Zarafshan Range in Uzbekistan. It is cultivated as an ornamental plant in temperate regions – growing to 60 centimetres, and producing pale lilac and white flowers during spring.

I. reticulata

This bulbous iris species commonly known as the netted iris or golden netted iris. It is native to Russia, the Caucasus, and northern Iran, but cultivated widely in temperate regions. The reticulata group of irises is characterised by a fibrous net surrounding the bulb. They are small plants to 15 cm, with tubular, sharply-pointed, ribbed leaves, and flowers of yellow, blue or purple with an orange blaze on the falls, appearing in early spring. They are hardy, but prefer a well-drained sunny position in soil which dries out in summer; they are therefore suitable for a rock or gravel garden.

I. winogradowii

This is a fragrant species of bulbous iris native of Caucasus Mountains of Azerbaijan and Georgia. It has pale primrose yellow flowers with green spots on the falls and emerges between April and May. The plant reaches at a height of 10–15 cm. The leaves emerge after flowering and grow up to 30–40 cm.

I. domestica

Commonly known as leopard lily, blackberry lily, and leopard flower. The plant grows 60–90 cm tall in full sun and is often found blanketing hillsides; the flowers can range from red to orange to yellow, or mixed, and bloom in summer to early autumn (fall). The flowers are typically orange spotted with red, although yellow-flowered varieties are in cultivation.

Iris × hollandica

Dutch irises are the hybrids of *Iris xiphium* and *Iris tingitana* and *Iris latifolia*, also called as *Iris × hollandica*. They are taller in habit and flowered early. They are available in blue, yellow and white colours. The bulb generally is smooth in texture and somewhat teardrop shaped.

5.1.2 Rhizomatous Iris

Iris × germanica

It is commonly known as the bearded iris or the German bearded iris. It is one of a group of hybrid origin. *Iris × germanica* grows up to 120 cm high and 30 cm wide.

I. kashmiriana

It is a perennial species from Kashmir, India. It has straight, sword-shaped, glaucous leaves, tall, thick stem with up 2 short branches, which hold 2–3 flowers of white, cream or pale blue, lilac, lavender or blue-purple. It is cultivated as an ornamental plant in temperate regions.

I. lutescens

It is native to North East Spain, Southern France and Italy. It is found on rocky or sandy hillsides or in woodlands. It grows up to 30 cm tall, with broad leaves (about 2.5 cm wide), producing yellow or violet flowers in spring (March–April). It

Species of Rhizomatous Iris

Iris × germanica | *Iris kashmiriana* | *Iris pallida*

Iris auranitica | *Iris hookeriana* | *Iris kemaonensis*

Iris afghanica | *Iris stolonifera* | *Iris hartwegii*

Iris sibirica | *Iris japonica* | *Iris tectorum*

prefers full sun, a slightly acid soil, and dry conditions during its dormant period in the summer. It is very hardy and tolerates a temeprature down to –20 °C or less.

I. pallida

This species of rhizomatous bearded iris also called sweet iris. It is native to the Dalmatian coast. This iris prefers rocky places. It grows to a stem height of 50–80 cm. The leaves are bluish-green in colour, and sword-shaped, 40–50 cm in length, and 2.5–3 cm in width. The inflorescences appear in May or June is fan-shaped and contain two or three flowers which are usually pale purplish to whitish.

I. variegata

It is a perennial iris species native of Eastern Europe. It has dark green, ribbed leaves. The branched flowering stems can be as tall as the leaves; they can hold 2–3 flowers in summer. They are yellowish-white, with brown-purple veins on the drooping falls. It is very hardy and it is commonly cultivated as an ornamental plant in temperate regions.

I. auranitica

This rhizomatous perennial iris originated in the Jaba-al-Druze region in Syria. It has thin and long, greyish-green, semi-evergreen leaves. In May, it has fragranced flowers (between light yellow to mustard), with a dark signal patch and yellow beard with purple tips. It is rarely cultivated as an ornamental plant in temperate regions, as it needs very dry conditions during the summer.

I. hookeriana

It is a rhizomatous perennial species from the Himalayan Mountains of India including the states of Himachal Pradesh, Jammu and Uttarakhand (Chowdhery and Wadhwa, 1984) and Pakistan. It has long pale green or yellow green leaves, long slender stem and fragrant blue, purple or lilac flowers, that are mottled with a darker colour. It flowers in late spring between April and May. It is cultivated as an ornamental plant in temperate regions.

I. sikkimensis

It is a perennial rhizomatous iris species from Sikkim. It has pale green or light green thin leaves, slender stem, 2 - 3 lilac or purple flowers, with a white beard with orange tips. It is a hybrid of *Iris hookeriana* and *Iris kumaonensis.*

I. kemaonensis

The Kumaon iris is from Tibetan China, Bhutan, India, Kashmir and Nepal. It has light green or yellowish green leaves that extend after flowering time. It has a short stem, 1–2 fragrant flowers that are purple, lilac, lilac-purple or pale purple. They also have darker coloured blotches or spots. It is cultivated as an ornamental plant in temperate regions.

I. afghanica

This species is from Afghanistan, with thin bluish-green leaves and creamy yellow or white flowers, which are veined with purple-brown. It has yellow-green or purple beards. Although, in the wild, it can vary in colour and size. It is cultivated as an ornamental plant in temperate climatic regions.

I. heweri

Iris heweri species is from Afghanistan. It has tall, green curved leaves, tall slender stems and purple blue or violet-blue flowers, with white and purple or lilac beard. It also is cultivated as an ornamental plant in temperate regions.

I. hoogiana

This species has long green leaves which are slightly purple at the base and a long slender flowering stem. The flowers are blue, ranging from sky-blue to lavender blue and blue purple. It has orange or yellow beards. This species is mostly found in the grassy mountainsides of Turkestan and cultivated as a temperate ornamental plant.

I. stolonifera

It from the mountains of Turkestan between Tajikistan, Uzbekistan and Afghanistan. It has red-skinned stolon roots and rhizomes, glaucous, long, blue-grey leaves, and bi-coloured flowers, in various shades from milky white, to blue, purple, pale lilac, lavender and brown. It normally has blue to yellow beards on all the petals (Khassanov and Rakhimova, 2012).

I. hartwegii

This iris species is endemic to California, where it can be originate on low-elevation mountain slopes in the central counties. It is also called as foothill iris, rainbow iris Sierra iris or Hartwegs iris. It bears one to three flowers on a slender stem in a shade of purple or yellow to almost white. It has lavender veining.

I. graminea

This iris species belongs to subgenus *Limniris*. It has purple or violet blue flowers almost hidden by narrow, grass-like leaves, and a plum scented fragrance. It is cultivated as an ornamental plant in temperate regions. This species naturally occurs in the southern half of Europe, from Spain and France in the West to Russia and the Caucasus in the East.

I. sibirica

Commonly known as Siberian iris or Siberian flag, from Europe (including France, Italy, Switzerland, Austria, Czech Republic, Slovakia, Germany, Hungary, Poland, Romania, Bulgaria, Former Yugoslavia, Belarus, Estonia, Latvia, Lithuania, Moldova, Ukraine and northern Turkey) and Central Asia (including Armenia, Azerbaijan and Siberia). It has long green grass-like leaves, tall stem, 2–5 violet-blue, to blue, and occasionally white flowers. It is generally cultivated as an ornamental plant in temperate regions.

I. laevigata

It is known as Japanese iris, rabbit-ear iris, or shallow-flowered and native to Japan. It is found growing in shallow waters and seems to prefer marshy and still ponds, although it can also be grown in damp soil if conditions are right. Flowers are generally blue, purple or violet and have unique colour patterns including some types with predominantly white flowers with blue spots and dark purples bordered with white.

I. tenuifolia

It is a beardless iris species from a wide region over central Asia, including Afghanistan, Pakistan, and in China. It has long greyish-green leaves, short stem and pale violet, lilac, pale blue or purple flowers.

I. cristata

This species is also known as dwarf crested iris and crested iris. It is endemic to the eastern United States. It has pale lavender flowers with a white patch and orange or yellow crest. It is a close relative to *Iris lacustris* (Dwarf lake iris), the only other crested iris native to North America. It is cultivated as an ornamental plant in temperate regions.

I. japonica

Also known as 'fringed iris', 'shaga' or 'butterfly flower' is a native of China and Japan. It has pale blue, lavender or white flowers with an orange or yellow crest. It is commonly cultivated as an ornamental plant in temperate regions.

I. tectorum

This iris species is also known as roof iris, Japanese roof iris and wall iris. It is native of China, Korea and Burma. It bears lavender-blue, bluish-violet, purple-blue, blue-lilac or sky-blue flowers. It is a compact plant and is cultivated as an ornamental plant in many of the temperate climatic regions of the world.

5.2.0 Cultivars

There have been about 60,000 iris cultivars in the world. However, few of them bloom in the hot summer from the middle of July to the end of August (Luo *et al.*, 2016). Most of the commercial cultivars are the Dutch irises (*Iris* × *hollandica*) that were derived from crosses between the various bulbous species. Some important cultivars are:

Sweet Princess: This cultivar is selected from the hybrids of *I. dichotoma* × *I. domestica*. It exhibited purple coloured attractive, big flowers and bloom from the middle of July to the end of August. It a promising summer flowered iris cultivar for landscaping as well as a parent for further breeding.

Butterfly Dreams: It produces dark purple flowers, the central crest is yellow-orange around the crest on the outer perianth. It has purple stems, with 30 to 44 flowers per stem. The plant height is 76.5 cm.

Butterfly Dawn: This cultivar has a violet-blue flower with a striking yellow patch and a purple circle on the outer perianth. It has green stems with 42 to 56 flowers per stem. The plant height is up to 71.0 cm.

Cultivars of Iris

Day Light — Miss Saigon — Mount Everest

Nova Blue — Saturuns — Casablanca

Sweet Princess — Butterfly Dreams — Butterfly Dawn

Alizes — Kaboom — Kachina Spirit

Cultivars of Iris

Hail Mary

Sarah Taylor

Sable

Kah-Nee-Ta

Dakota Smoke

T-Rex

Butterflies in Bloom: The cultivar has a purple stem, and its flower has a yellow patch and a purple circle that present a striking contrast to its white perianth, and it has a pale purple arm. Its plant height is more than 60 cm, with 58 to 68 flowers per stem.

Butterfly Veil: The cultivar Butterfly Veil produces a graceful blue and white porcelain flower. The crest on the outer perianth is surrounded by violet and gray-orange spots. It has green stems, with 24 to 30 flowers per stem. It has the shortest plant height up to 50 cm.

Bearded Iris cultivars

Miniature: Bantam, Scribe, Zipper, Bay Ruffles, Software, Starlight, Violet Lulu, Watercolour, Just Jennifer, Pink Bubbles, Predictions, Tulare.

Intermediate: Baby Blue Marine, Brighten Up, Piece of Cake, Red Zinger, Chickee, Disco Jewel, Loreley, New Wave, Rosemary's Dream.

Tall Bearded: Dusky Challenger, Silverado, Jesse's Song, Beverly Sills, Edith Wolford, Before the Storm, Laced Cotton, Superstition, Bountiful Harvest, Champagne Elegance, Eleanor Roosevelt.

Beardless Iris: Butter and Sugar, Halcyon Seas, Pink Haze, Sky Wings, Flight of Butterflies, Caesar's Brother, Eric the Red, Orville Fay, Snow Queen, Super Ego.

Japanese Iris: Light in Opal, Rikki Pikki, Variegata, Acclaim, Azure Perfection, Ruffled Dimity, Sapphire Star, Summer Moon, Cascade Crest, Geisha Obi.

Dutch Iris/Bulbous Iris cultivars: Blue Magic, Telstar, Discovery, Blue Diamond, Casablanca, Superstar, Danfordiae, Professor Blaauw, Purple Sensation, Sapphire Beauty, White Excelsior, Nove Blue, Saturuns, Mount Everest, Day Light, Miss Saigon *etc.*

6.0 Genetics and Breeding

6.1.0 Genetics

There are variations in chromosome numbers of iris species found growing in all over the world. *I. tingitana* (2n = 28), *I. latifolia* (2n = 42), *I. juncea* (2n = 32), *I. serotina*, *I. xiphium* (2n = 34) and *I. boissieri* (2n = 36). *I. lactea* (2n = 40), *I. lactea* var. *chinensis* (2n = 40), *I. chrysographes* (2n = 40), *I. japonica* (2n = 36), *I. tectorum* (2n = 28), *I. confusa* (2n = 32), *I. germanica* (2n = 40), *I. sichuanensis* (2n = 26), *I. leptophylla* (2n = 26), *I. goniocarpa* (2n = 26). Development of new colour combination is the major challenge for the breeders. Blue coloured cultivars are in good demand. The major disadvantages of Dutch irises are low vase life and high light requirement during flower forcing. Breeding of irises majorly focused on early flowering, longer vase life and chromosome doubling. The major selection criterions are: Low light requirement for forcing, rapid growth capability for bulb production and forcing, maximum flower bud opening, longer vase life, pot type cultivars and longer bulb storability.

6.2.0 Breeding

Jean-Nicholas Lemon began hybridizing irises in France in 1840. English gardener John Salter introduced 'Queen of May' around 1859. Sir Michael Foster collected 'Amas' in Turkey around 1885. It was a tetraploid iris, having twice as many chromosomes. James Gibson of the USA developed many important irises in the plicata pattern. Red irises developed from the purple pigment found in the old species iris. Most red irises have a plum or bronze cast. A true scarlet red is the goal of many modern hybridizers. In the 1950s, iris hybridizers began breeding with changing the form of an iris, as well as the colour. The smooth falls and standards became wider and wavier as *'ruffles'* were introduced. Also beginning in the 1950s, hybridizers like Lloyd Austin began developing irises with long beards that had *'horns'* at the tips. These evolved into *'spoon'* and *'flounce'* shapes and became known as *Space Age* irises. These are classified as *Novelty Irises*.

The cultivation history of the Dutch irises started at the end of the nineteenth century when the firm C. G. van Tubergen of Haarlem, The Netherlands crossed two varieties of *Iris xiphium* (var. *praecox* from Spain and var. *lusitanica* from France) with *I. tingitana* from North Africa (Hekstra and Broertjes, 1968). 'Wedgwood' and its mutants 'Ideal' and 'White Wedgwood' are the hybrids of these species and varieties. 'Blue Magic' and 'Professor Blaauw' inherited the characteristics of *I. fontanesii*. A fertile spontaneous tetraploid of 'Wedgwood' was found in about 1952 (Eikelboom and Van Eijk 1990). It was crossed with cultivars of *I. xiphium* in order to produce several triploid cultivars, for example, 'Telstar'.

It is known that *Iris aphylla* had a great contribution in obtaining dark shade cultivars of today and *Iris reichenbachii* had a contribution in obtaining first cultivars of Amoena type flowers admired in many current hybrids. *Iris pumila* genotypes from Romania were documented contributors to the obtaining of fertile dwarf garden irises. Also, recently in Romanian villages were identified nameless tetraploid old garden irises similar with 'Amas' that can present importance for breeding disease resistant ornamental irises (Crinan *et al.*, 2017).

Hekstra and Broertjes (1968) irradiated bulbs of sterile cultivar of Iris 'Wedgwood' with 1000 rad X-rays after lifting during August when the growing points are in the youngest possible ontogenetic stage. They obtained about 160 flower mutants, ranging from paler to darker blue with or without purple colour.

Fan *et al.* (2017) hybridized seven male parents and two female parents to estimate the combining ability, genetic advance components and heritability. They found that height of individual flower and length of fall had the largest number of F_1 populations gaining heterosis, thus more likely to generate taller progenies with larger blossoms. The estimates of combing ability revealed that *I.* 'White and Blue' and *I.* 'Sugar Blues' appeared to be ideal pollen donators. The combination of *I.* 'India Chief' × *I.* 'Champagne Elegance' exhibited positive special combining ability (SCA) on all the tested characters, becoming the most potential populations of all. Width of fall and number of flower per scape had the highest broad-sense heritability (91.47 per cent) and strict-sense heritability (53.96 per cent), respectively. Correlation analysis showed that plant height was closely associated with number of bud per scape and number of flower per scape, while length of leaf had significant correlations with height of individual flower and length of fall, making it possible to select taller individuals in the vegetative stage to obtain those with abundant flowers.

Multivariate analysis of morphological characteristics in *Iris germanica* hybrids were studied by Azimi *et al.* (2018). They evaluated the genetic difference between 43 F_1 hybrids and 8 parents, seven of *Iris germanica* and one of *I. spuria* and to estimate genetic parameters and the morphological variation. The mean comparisons showed that the Iris hybrids 10 and 11 had higher flower size than other Iris hybrids and the parental. The Iris hybrid 8 showed the most significantly difference in crown diameter, flower branches diameter and flower peduncle diameter. Relatively high heritability values, in broad sense, were found 97 per cent for plant height, fall width and length, flower size and leaf width.

Liu *et al.* (2018) studied the variation of flower opening and closing times in hybrids of evening flowering species *Iris dichotoma* and daytime flowering species *Iris domestica*. They reported that the flower opening and closing times are controlled by different polygenes. For flower opening time, opening in the early morning as *I. domestica* is partial dominant to opening in the afternoon as *I. dichotoma*. For flower closing time, flower closure at dusk is dominant to closure in the evening. Mitotic doubling to restore F_1-fertility has been used in crosses *I. laevigata* × *I. ensata* and *I. hollandica* × *I. tingitana* (Yabuya, 1985).

7.0 Growth and Flowering

7.1.0 Bulb Size

Flowering of Dutch iris depends on its bulb weight, so an iris bulb must reach a critical size to produce a flower (De Munk and Schipper, 1993). On the other hand, its flowering can be manipulated using storage temperature treatments (forcing) to obtain cut flowers out of their natural flowering season (Elphinstone *et al.*, 1987). Fernández *et al.* (2009) studied the influence of four bulb sizes (>2 g, 2-4 g, 4-6 g, >6 g) and two bulb temperature treatments 20°C (control) and 9°C for 7 weeks on sprouting, earliness, yield and quality parameters of *I. xiphium*. They found that the quality and vegetative parameters improved with increased bulb size but not by increased temperature treatments, but flowering features and yield were improved by low temperature treatment. For 'Wedgwood' and 'Sports' the critical bulb size is 7-8 cm in circumference (Kamerbeek, 1965).

7.2.0 Temperature

Commercial flower production, year-round, involves temperature storage treatments which are applied to unplanted or 'dry' bulbs. These treatments have been developed to give a rapid flower initiation and subsequent development for early markets or to retard growth and development to produce flowers later than the natural season. Following retarding treatments, initiation and growth must be rapid so that the glasshouse periods remain short. Hartsema (1961) established that an initial high temperature treatment of 1 week at 31°C, given before a treatment in which flower initiation could occur, delayed flowering but increased the number of bulbs in which a flower was initiated. The heat curing treatment also appears to be responsible for the breaking of dormancy, or earlier sprouting of the bulb, (Tsukamoto and Ando, 1973; Sano, 1973; Le Nard, 1980) once they are placed in lower temperatures. The earlier breaking of dormancy could be responsible for the shortened chronological time to flower harvesting. High temperature was found to be most successful for retarding growth and increasing subsequent flowering performance, presumably because of the prevention of the formation of a daughter bulb at the apex. The optimum temperature for the shortest chronological time to initiation is 9-13°C. Halevy (1984) classifies iris as a plant with autonomous induction. Dutch irises could have a quantitative low temperature/vernalization response, where a low temperature promotes but is not essential for flowering.

7.2.1 Standard Temperature Treatments

The requirement of a high temperature treatment to obtain bulbs of adequate maturity, followed by a cool treatment for flower induction and initiation, coupled with the discovery of high temperature retarding treatments have lead to the development of standard treatments for year round forcing. It was revealed that an intermediate treatment at 17°C between the heat curing and cooling improved the changes of flower production, mainly due to a reduction in the occurrence of blasting (Kamerbeek, 1965). By reversing the treatment, 17°C was applied after the cool treatment, the time to harvest was reduced and the occurrence of blasting further diminished (Kamerbeek and Beijer, 1964). Several variations of standard treatments now exist, but basically they can be represented by the following examples:

Very early flowering	Dec-Jan	35°C (2 weeks); 40°C (3 days); 17°C (2 weeks) and 9°C (6 weeks) or 35°C (2 weeks); 40°C (3 days); 9°C (6 weeks) and 17°C (2 weeks)
Early flowering	Feb-April	30°C (4 weeks); 40°C (3 days); 17°C (2 weeks) and 9°C (6 weeks) or with the reversal of the 17 and 9°C treatments
Normal, late andvery late flowering	May-Dec	30°C followed by 17°C (2 weeks) and 9°C (6 weeks)

7.3.0 Photoperiod

Irises are sensitive to insufficient light intensity. Too little light can result in culls due to flower blast. During the winter, plant irises in greenhouses with good light levels. Make sure that the glass or plastic film is clean and replace old plastic sheets. The relative humidity in the greenhouse can be reduced by adding heat combined with improving ventilation. This increases transpiration rate and decreases the risk of flower blast. Flower blast can occur during any stage of growth. It does not become evident until the flower bud fails to swell and a hollow leaf bud can be felt. Although the exact cause of flower blast is not known, periods of too little light, high temperatures and low relative humidity will increase the risk. Insufficient irrigation during production also leads to an increased risk of flower blast and pale flowers. Long day conditions can induce floral abortion in irises (Elphinstone *et al.*, 1986). Besides photoperiod, light intensity can also play a role in flower differentiation and anthesis. In Dutch iris, higher flowering percentages were observed when supplementary lighting was provided 40–50 days prior to the predicted date of flowering. In Dutch iris, Hartsema and Luyten (1955) found higher flowering percentages when supplementary lighting began 40–50 days prior to the predicted date of flowering.

7.4.0 Growth Regulators

The chemical control of plant height for potted irises has been reported. The immersion of bulbs in 20 mg/l of paclobutrazol reduced the plant heights of 'Casablanca' and 'Professor Blaauw' by 41–44 per cent against the controls. Flowering, however, was delayed by 3–4 days (Francescangeli, 2009). In order to promote reproductive development, bulbs are exposed to ethylene. The concentrations of ethylene and duration of treatments vary according to the cultivar. For example, ethylene application in a concentration of 10 µl/l for 3 h prior to cold treatment is effective for 'Blue Magic' (Yue and Imanishi, 1990). It has been demonstrated that cytokinins stimulate the transportation and distribution of assimilates toward the floral buds, and that the deficiency of cytokinins, in conditions of cultivation with low luminosity, can lead to 'blasting' or flower abortion (Vonk *et al.*, 1986). Lee *et al.* (2005) found a direct relationship between the combination of gibberellins (GA_{4+7}), and cytokinins [6-benzylaminopurine (BAP)] and increased rates of flowering of the iris cultivar 'Blue Magic'.

8.0 Propagation

8.1.0 Bulbs and Rhizome

For commercial cultivation, irises are propagated through bulbs of sizes 6-8 cm and rhizomatous irises are propagated by division as well as from rhizome cuttings. Since a bulb possesses a limited number of axillary buds, only two to four new bulbs can be formed annually (Shibli and Ajlouni, 2000). At this rate it takes up to 10 years to build up commercial quantities of new cultivars (Jehan *et al.*, 1994; Radojevic and Subotic, 1992). As viruses pose serious problems in their multiplication, it became essential to go for tissue culture to produce virus free plants. Rhizomatous irises are generally multiplied by rhizome divisions in late summer (Kohlein, 1987). For rhizomatous irises, adventitious shoot regeneration occurred on flower stem sections (Meyer, 1984) and from axillary shoot regeneration from shoot apices (Laneri and Lucretti, 1983; Kromer, 1985).

8.2.0 Seed

Propagation through seed is not difficult but requires three to four seasons for flowering. The iris seeds exhibit physiological dormancy. It requires different physical and chemical methods to break the physiological dormancy. Wang *et al.* (2018) developed a regeneration system for propagation of *I. sanguinea* using seeds.

8.3.0 Micropropagation

Conventional propagation of the bulbous and rhizomatous irises is by asexual means. Daughter bulbs produced by the original mother bulb are removed and

Propagation of Iris

Rhizome Division

Separation of Rhizomes

Rhizome

Iris Bulbs

Iris Seedpod and Seeds

Germinating Iris Seed

Source: Peter (2012).

Propagation of Iris

(a) Shoot initiations from callus after 1 month of culture; (b) Multiple shoots formation *I. sari* and (c) *I. schachtii* after 3 months of culture; (d) root formation of *I. sari* and (e) *I. schachtii.*

Source: Uzun *et al.* (2014)

Bulbs for planting

Planting in Greenhouse

Field Cultivation of Iris

Iris in Greenhouse

Source: www.edepot.wur.nl, www.pinterest.ru, www.vws_flowerbulbs, www.fragrantica.com, www.shutterstock.com

grown on. However, this is a slow process, as most species produce less than 5 daughter bulbs annually and takes 10 years to build up commercial quantities of new varieties. Therefore, *in vitro* propagation methods have been pursued to increase multiplication rates. The majority of *in vitro* propagation work on this genus has concentrated on Dutch Irises (De Munk and Schipper, 1993), although rhizomatous irises have also been propagated *in vitro* (Meyer, 1984; Jehan *et al.*, 1994). The primary explants of choice have included: shoot apices (Anderson *et al.*, 1990), scale pieces and basal plate sections, axillary buds, inflorescence stem sections (Yabuya *et al.*, 1991), embryos (Laublin *et al.*, 1991), small bulblets from the mother bulb basal plate, rhizomes (Kromer, 1985), shoots developing naturally on rhizomes, leaves (Gozu *et al.*, 1993) and leaf bases, flower pieces and rhizome apices (Jehan *et al.*, 1994). *In vitro* studies of the *Iris* genus have mainly focussed on increasing the propagation rate, virus eradication, embryo culture and the physiology of flower initiation. For shoot regeneration MS medium at full or half strength, with a sucrose concentration between 2-4 per cent has been successful (De Munk and Schipper, 1993). For rhizomatous irises, a modified LS medium has also been used (Meyer, 1984). For optimal regeneration of shoots, NAA (0.03-2.0 mg/l) was a necessary component of the media, or equal amounts of NAA and BAP or kinetin (0.5-1.0 mg/l). Activated charcoal also enhanced shoot regeneration. Hussey (1976) found that a propagation rate of 100 plantlets per original bulb was possible, with a potential rate of up to 1000 or more.

Apical meristems have been used as the primary explants type in studies of floral initiation (Doss and Christian, 1979) and virus eradication (Anderson *et al.*, 1990). The relationship between bulb size, apex size and flowering were studied in the bulbous iris cultivar 'Ideal'. A complete micropropagation system beginning with shoot apices and including virus indexing by TEM and ELISA was also developed to produce virus-free Dutch Iris (Anderson *et al.*, 1990). In an effort to further increase the multiplication rate of irises *in vitro*, somatic embryogenesis has been studied using commercially important rhizomatous species. Using embryos of *I. setosa*, *I. pseudacorus* and *I. versicolour* as the primary explant type, following shoot and root growth from the embryo, root and leaf pieces were excised and cultured (Laublin *et al.*, 1991). A method of rapid multiplication by somatic embryogenesis associated with multibudding has also been developed for *I. pallida* and *I. germanica* (Jehan *et al.*, 1994). The rare black iris (*Iris nigricans*) propagated via somatic embryogenesis using callus, cell suspension and protoplast culture systems (Shibli and Ajlouni, 2000).

9.0. Soil and Climate

9.1.0 Soil

Irises can be produced in different types of soils, but sandy and fertile soils are preferable. The soil must be well drained, with pH ranges from 6.0-7.0 and poorly drained soil may lead to the low performance due to diseases. Cultural rotation of fields with a minimum of 4-years intervals is recommended. Iris bulb should be planted at a depth of at least 10 cm and 3-5 cm layer of compost should be incorporated in the beds of poorly drained soils to improve the drainage. Soil should be prepared from 2-3 weeks prior to the planting of bulbs. The iris is a salt-sensitive crop. Excessively high salt concentrations in the soil often occur after growing a crop requiring large quantities of fertilizer. An excessively high salt concentration in the soil can be reduced by flushing the soil with water before planting.

9.2.0 Climate

Once planting is completed, keeping the soil at the right temperature is important. Aerial growth ceases at about <6°C, while high soil temperatures cause short stems. It should not be lower than 5 to 8°C but not higher than 20°C. The optimum soil temperature is 15°C and low soil temperatures will increase production time. High soil temperatures will reduce the forcing time but will also produce shorter plants. In the field, the optimum air temperature is 15 to 17°C; in the greenhouse is 12 to 17°C. The minimum temperature that irises can tolerate is 0°C and the maximum average daily temperature is 20°C. In the Dutch iris, bulb storage at 30°C induced a higher number of flower buds than lower temperatures, and no flowers formed under 10°C (Yue and Imanishi, 1990).

10.0 Cultivation

10.1.0 Planting and Planting Distance

Planting is generally done during mid-July to early-September. The planting density selected varies according to bulb type, cultivar, planting date and production region. The soil in which the iris bulbs will be planted must be thoroughly cultivated. Plant the bulbs so that they are covered by 5 to 7 cm of soil. The higher the temperature, the deeper the bulbs should be planted, but never deeper than 7 cm. Preventing excessively high soil temperatures can be achieved by mulching with a layer of straw, wood chip or sawdust after planting. Rhizomes of bearded iris should be planted in mid to late summer, preferably in July and August. Rhizomatous irises are set 25-30 cm apart while Dutch irises are set 10-15 cm apart. It requires high amount of irrigation during the vegetative growth.

10.2.0 Manuring and Fertilization

Due to the salt sensitivity of iris, applying a base dressing before planting is not advisable. Depending on the analysis of a soil sample, a small application of calcium nitrate may be applied. Under conditions of high light intensity, the crop remains shorter. This is why a base dressing of 750 kg/ha of 12-10-18 is often applied to the soil prior to field production. Four weeks after planting, 250 kg/ha of calcium nitrate is applied to the irises. Nutrient requirements per hectare are 150 kg nitrogen, 70–100 kg calcium, 20–30 kg magnesium, 25–40 kg phosphorus, and 100–200 kg potassium (De Munk and Schipper, 1993). Verma *et al.* (2019) reported that plants supplied with 30 g N/m^2 spaced at 20 × 30 cm apart recorded maximum values in terms of most of the growth and flowering parameters in *Iris orientalis* cv. 'Frigia'.

10.3.0 Weed Management

For good yields, it is important to control weed growth during production. Weeds can be hosts for many diseases and increase the risk of damage to the crop.

Ideally, bulbs should be planted in soil free from pathogens and weeds. Weed control is achieved by applying a pre-emergence application of a herbicide. They should be used in autumn and early spring (Schipper *et al.*, 1988). Weeds can also germinate after the emergence of the irises and hand weeding is recommended.

11.0 Diseases and Pests

11.1.0 Diseases

Iris can be affected by different pathogens like *Fusarium oxysporium, Pencillium* sp., *Rhizoctonia solani, Botrytis cinerea* and *Phytophthora* sp. Field infections by *Rhizoctonia solani* can survive during storage (Doornik, 1982). She however recorded a sharp decrease in the viability of the disease after harvest, but the surviving mycelium grew better on bulbs stored at 30°C in contrast to those stored at 20°C.

11.1.1 Blue Mould or Bulb Rot

On the bulbous irises, blue or green mould, or bulb rot caused by *Penicillium* is mainly a storage problem. It is prevalent if bulbs are stored in damp and poorly ventilated conditions and can destroy a stock if not controlled. Thiabendazole and benomyl effectively controlled *Fusarium oxysporum* rot of iris. Optimum results required soaking bulbs within 48 hours after digging using at least 1000 ppm at 20°C for 30 minutes. Adjuvants improved control. Dusting bulbs was often effective. Both fungicides controlled *Fusarium* during the hot water treatment for nematodes in iris. Proper storage of treated bulbs was essential to control *Penicillium corymbiferum* rot. Best storage of retarded iris bulbs was 30°C and > 95 per cent RH for 1–2 days before cleaning and treating, followed by storage at lower relative humidity (Gould and Miller, 1971).

11.1.2 Bacterial Soft Rot

On rhizomatous irises, bacterial soft rot, caused by *Erwinia carotovora* can be a problem. The symptoms are small water-soaked lesions on the leaf blades developing into large irregular spots. With disease progress, leaf sheaths and rhizomes rotted completely so that the shoot collapsed at soil level (Choi *et al.*, 1990). Eventually the rhizome is reduced to a wet, evil smelling pulp. Rhizomes planted too deeply are mostly attacked. Wet locations result in poor drainage, which provides favourable conditions for the disease. Damaged portions should be cut out with a knife and the uses of chemicals are recommended.

11.1.3 Viruses

The occurrence in Dutch bulbous irises (*Iris hollandica*) of two viruses – iris mild mosaic virus (IMMV) and iris severe mosaic virus (ISMV) – in association with two diseases – mosaic and grey was reported so far. In the Netherlands, three virus diseases have been distinguished: mild mosaic, mild yellow mosaic, and severe mosaic. These diseases were associated with IMMV, IMMV plus iris mild yellow mosaic virus (IMYMV, a newly recognized virus), and IMMV plus ISMV, respectively. The viruses are antigenically distinct and their presence could be established serologically. Tobacco mosaic virus (TMV), tobacco rattle virus (TRV),

Diseases and Pest

Iris Leaf Spot

Iris Basal Bulb Rot

Iris Rust

Iris Borer

Mealy Bugs

Bacterial Rot Due to Borer

Harvesting and Postharvest

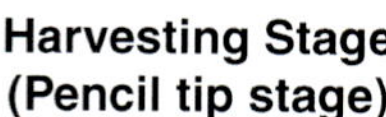

Harvesting Stage
(Pencil tip stage)

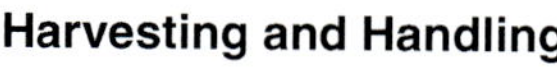

Harvesting and Handling

Packaging of Iris
for Sale

Source: www.gingerandspiceflorist.co.uk, www.underatinroof.com, www.lovethatimage.com

and tobacco ringspot virus (TRSV) were also detected in irises, but not in association with particular symptoms (Asjes, 1979). A new tospovirus was identified in Iris cultivation and was named as Iris Yellow Spot tospovirus (IYSV) (Cortes *et al.*, 1998). Symptoms are yellow spots on the leaves. Aphids found on the bulbous Dutch Iris spread different viral diseases.

11.2.0 Pests

Iris borer is the major pest. The pest over winters as eggs in the previous year leaves and eggs will hatch during April. The young larvae will feed on the leaves and later they make small tunnel to the bulbs/rhizome. This feeding will result into clear water streaks on the leaves and they completely destroy the whole bulb. Proper sanitation in the field and removal of leaves from the plant early can reduce the incidence of borer. Single application of systematic insecticide imidacloprid @2-3 per cent will completely controls this pest.

11.2.1 Aphid and Mealy Bug

Aphids also the pest of irises and also the vector for transmission of viruses. Mealy bug infestation was also found during storage at temperature 30 °C. Aphids and bugs are controlled by the spray of systemic fungicide (imidacloprid @2-3 per cent).

12.0 Harvesting and Postharvest Management

12.1.0 Harvesting

Harvesting of irises at the right time is crucial for flower opening and keeping quality. The correct time to harvest is as soon as the iris buds shows colour. During the winter when light levels are low, irises have to show more colour before harvesting than during the summer. Under warm conditions during the summer harvesting twice a day may be necessary to obtain a more uniform product. The flowers are harvested by pulling the plants up bulb and all. Irises sometimes produce leaves that are too long or withered, yellowish-brown leaf tips. These should be removed because they reduce ornamental value. Leaf tipping is easiest to do after bunching. It is usually done with a small knife. After bunching (and possibly binding) the flowers, it may still be necessary to remove dirt and mud from the lower parts of the stems by rinsing in water. The flowers are then put into buckets of water and placed in cold storage. The optimum storage temperature is about 2°C. Having a cold store is absolutely essential. Keep the cold storage period to a minimum as it will reduce quality.

12.2.0 Postharvest Management

Dutch iris (*Iris×hollandica*) is a commercially important cut flower species (Jerardo, 2007), although their marketability is limited by a short vase life of 2–5 d. The failure of iris flowers to open fully after dry transport and storage is also a major postharvest problem. Incomplete flower opening results from insufficient growth of the pedicel and ovary tissues that otherwise force the bud free of the constraining sheath leaves. Iris flower senescence is first visualized as discolouration, wilting and inward rolling of the distal edges of tepals (van Doorn *et al.*, 1995). A postharvest pulse with 0.2–1.0 mM TDZ for 6–24 h at 0 or 20°C extended the vase life of flowers by up to 1.5 d relative to control (0 mM TDZ) stems in water (Macnish, 2010).

Pulsing of cut irises for 12 hours in solution comprising of sucrose (10 per cent) + 8-HQC (400 ppm) improved all the postharvest qualities. Cut flowers can be stored in darkness for as much as about 11 days if placed in water at about 3 °C. Iris cut flowers are very sensitive to ethylene and bud sleepiness/blasting will occur when they expose to high level of ethylene. The vase life and bud development of dry shipped flowers improve after conditioning in solution containing 2g/l citric acid, 2g/l sucrose for 12 hours at 20°C. After shipment the flowers should be placed in warm (40-50°C) water for 3 hours. Thakur *et al.* (2019) reported that pulsing of cut spikes (*Iris orientalis* Mill) with solution comprising of sucrose (10 per cent) + 8-HQC (400 ppm) for 16 hours resulted in best treatment for most of the postharvest parameters studied. Ahmad *et al.* (2013) suggested that wet and dry storage of premature scapes of Irises (*Iris germanica, Iris reticulata* and *Iris kashmiriana*) for 72 hours at 5°C, and placing them in sucrose improves the cut flower performance.

13.0 Biotechnology

13.1.0 Genetic Transformation

Flower colour in tall bearded irises (*Iris germanica*) is determined by two distinct biochemical pathways; the carotenoid pathway, which imparts yellow, orange and pink hues and the anthocyanin pathway, which produces blue, violet and maroon flowers. Red-flowered *I. germanica* do not exist in nature and conventional breeding methods have thus far failed to produce them. Jeknic *et al.* (2014) transformed a pink iris *I. germanica*, 'Fire Bride', with a bacterial phytoene synthase gene (crtB) from *Pantoea agglomerans* under the control of the promoter region of a gene for capsanthin-capsorubin synthase from *Lilium lancifolium* (Llccs) developing iris cultivars with red flowers. This increases the flux of metabolites into the carotenoid biosynthetic pathway and lead to elevated levels of lycopene and darker pink or red flowers. Iris callus tissue ectopically expressing the crtB gene exhibited a colour change from yellow to pink-orange and red, due to accumulation of lycopene. Transgenic iris plants, regenerated from the crtB-transgenic calli, showed prominent colour changes in the ovaries (green to orange), flower stalk (green to orange), and anthers (white to pink), while the standards and falls showed no significant differences in colour when compared to control plants. Fan *et al.* (2020) studied the reblooming mechanisms of *Iris germanica* revealed by transcriptome profiling. They found that the four genes, PHYTOCHROME A(PHYA), GIGANTEA (GI), SHORT VEGETATIVE PERIOD (SVP) and AUXIN RESPONSE FACTOR (ARF), to be involved in the second floral initiation of the rebloomers.

13.2.0 Molecular Markers

Despite the large variation of flower colour in iris, little is known about the phylogenetic and ecological contexts of floral colour. Roguz *et al.* (2020) studied the diversification of flower colour and intraspecific colour variations in genus Iris. They created a phylogenetic tree for the genera first, based on a database of sequences. The database using six sequences, five plastid genomes (matK, trnL, trnK, NADPH, and rbcL) and one nuclear internal transcribed spacer (ITS), to reconstruct the ancestral state of flower colour and other pollination-related traits (*e.g.*, the presence of nectar and mating system), and also tracked the evolution of colour variation. In iris, 6530 ESTs were produced from leaf and root cDNA to develop EST-SSR molecular markers. Tang *et al.* (2009) investigated genetic diversity among ecotypes and cultivars. The dihydroflavonol 4-reductase (*DFR*) gene from iris was used ectopically in rose to alter flower colour in that crop (Katsumoto *et al.*, 2007).

14.0 References

Ahmad, S.S., Tahir, I. and Shahri, W. (2013) *J.Agr.Sci. Tech.*, **15**: 323-331.

Anderson, W.C., Miller, P.N., Mielke, K.A. and Allen, T. (1990) *Acta Horticulturae*, **266**: 77-81.

Asjes, C.J. (1979) *Netherlands J.Pl.Path.*, **85**: 269-279.

Azimi, M.H., Jozghasemi, S. and Barba-Gonzalez, R. (2018) *Euphytica*, **214**: 161.

Cantor, M. (2016) *Special floriculture*, Ed. Academic Pres Cluj-Napoca.

Cassidy, G.E. and Linnegar, S. (1987) *Growing Irises* (Revised ed.). Bromley: Christopher Helm. pp. 144–145.

Choi, J.E., Han, K.S. and Kang, J.C.(1990) *Kor. J.Pl.Path.*, **6(2)**: 289-291.

Chowdhery, H. J. and Wadhwa, B. M. (1984) *Flora of Himachal Pradesh*.

Cortes, L., Livierato, I.C., Derks, A., Peters, D. and Kormelink, R. (1998) *Phytopathology*, **88**(12): 1276-1282.

Cri'an I., Vidican R., Stoian V. and Stoie, A. (2017) *Romanian Journal of Grassland and Forage Crops*, **16**: 21-32.

De Munk, W.J. and Schipper, J. (1993) In: *The physiology of flower bulbs*. De Hertogh, A. and Le Nard, M. (Eds). Elsevier Scientia Publishers. Amsterdam, pp. 349-379.

Doornik, A.W. (1982) *Netherlands J.Pl.Path.*, **88**: 185-190.

Doss, R.P. and Christian, J.K. (1979) *Physiologia Plantarum*, **45**: 215-218.

Dykes, W. R. (1913) *The Genus Iris*. Cambridge: The University Press. Photo-offset reprint by Dover Publications, New York, 1974.

Eastaugh N., Walsh V., Chaplin T. and Siddall R. (2004) *Pigment Compendium: A Dictionary of Historical Pigments,* Elsevier Butterworth-Heinemann, Oxford, pp. 198.

Elphinstone, E.D., Rees, A.R. and Atherton, J.G. (1986) *Acta Horticulturae*, **177**: 613–618.

Elphinstone, E.D., Rees, A.R. and Atherton, J.G. (1987) In: *Manipulation of flowering*. Atherton, J.G. (Editor). Butterworths, Sevenoaks, pp. 393-405.

Fan, Z., Gao, Y., Guo, Y., Liu, R. and Zhang, Q. (2017) *Euphytica*, **213**: 1-11.

Fan, Z., Gao,Y., Ren, Y., Guan, C., Liu, R. and Zhang, Q. (2020) *BMC Genomics*, **21**: 2-18.

Fernández, J.A., Peñapareja, D., López, J., González, A. and Bañón, S. (2009) *Acta Horticulturae*, **813**: 605-608.

Francescangeli, N. (2009) *Chilean J. Agric. Res.* **69**: 509–515.

Gould, Ch.J. and Miller, V.L. (1971) *Acta Horticulturae*, **23**: 178-186.

Gozu, Y., Yokoyama, M., Nakamura, M., Namba, R., Yomogida, K., Yanagi, M. and Nakamura, S. (1993) *Plant Cell Reports*, **13**: 12-16.

Hartsema, A. M. and Luyten, I. (1955) *Acta Bot. Neerl.* **4**: 370–375.

Hartsema, A.M. (1961) In: *Encyclopedia of plant physiology. External factors affecting growth and development.* Ed. W. Ruhland. **16**: 123-61. Springer-verlag, Berlin.

Hekstra, G. and Broertjes, C. (1968) *Euphytica*, **17**: 345-351.

Hussey, G. (1976) *Scientia Horticulturae*, **4**: 163-165.

Jehan, H., Cortois, D., Ehret, C., Lerch, K. and Petiard, V. (1994) *Plant Cell Reports*, **13**(12): 671-675.

Jeknic, Z., Jeknic, S., Jevremonic, S., Sabotic, A. and Chen, T.H.H. (2014) *Plant Cell Rep.*, **33**(8): 1307-1321.

Jerardo, A. (2007) *Economic Research Service, United States Department of Agriculture*, Washington, DC, USA.

Kamerbeek, G.A. (1965) Rep. *1st Int. Symp. Iris. Florence*, 1963, 459-475.

Kamerbeek, G.A. and Beijer, J.J. (1964) *Meded. Dir. Tuinb.* **27**: 298-604.

Katsumoto, Y., Fukuchi-Mizutani, M., Fukui, Y., Brugliera, F., Holton, T. A., Karan, M., Nakamura, N., Yonekura-Sakakibara, K., Togami, J., Pigeaire, A., Tao, G.Q., Nehra, N. S., Lu, C.Y., Dyson, B. K., Tsuda, S., Ashikari, T., Kusumi, T., Mason, J. G. and Tanaka, Y. (2007) *Plant Cell Physiol.* **48**: 1589–1600.

Khassanov, F. O. and Rakhimova, N. (2012) *Stapfia*, **97**: 177.

Kim, K. W. and De Hertogh, A. A. (1997) *Hort. Rev.*, **18**: 87–169.

Kromer, K.D. (1985) *Acta Agrobotanica*, **38**: 65-87.

Laneri, U. and Lucretti, S. (1983) *Genetica Agraria*, **37**: 178-179.

Laublin, G., Saini, H.S. and Cappadocia, M. (1991) *Plant Cell, Tissue and Organ Culture*, **27**: 15-21.

Lawrence, G. H. M. (1953) *Gentes Herb.*, **8**: 346–371.

Le Nard, M. (1973) *Ann. Amelior. Plantes.* **23**: 265-278.

Lee, A.K., Rhee, S.R., Suh, J.K. and Cha, H.C. (2005) *Acta Horticulturae*, **673**: 315-321.

Liu, R., Gao, Y., Ruan, L., Fan, Z. and Li, C. (2018) *Plant Breeding*, **137**: 920-927.

Luo, G.J., Xiao, Y.E., Xu, W.J., Bi, X.Y., Li, H., Lian, X.Y., Yu, F.Y., Zheng, Y. and Lei, J.J. (2016) *J. Plant Genet. Resources*, **17**(2): 266–272.

Macnish, A.J., Jiang, Cai-Zhang. and Reid, M.S. (2010) *Postharvest Biol. Technol.,* **56**: 77-84.

Mathew, B. (1989) *The Iris*. London: Batsford.

Meyer, M.M. (1989) *Hort Science,* **19**: 575.

Peter J.M. Knippels (2012) Advanced *in Vivo* Propagation Techniques for Specialty Bulbs, *Floriculture and Ornamental Biotechnology, Global Science Books.*

Radojevic, L. and Subotic, A. (1992) *Journal of Plant Physiology,* **139**: 690-696.

Rodionenko, G. I. (1987) *The Genus Iris L.*: The British Iris Society. (English translation).

Roguz, K., Gallaghar, M.K., Senden, E., Bar-Lev, Y., Lebel, M., Heliczer, R. and Sapir, Y. (2020) *Front. Plant Sci.,* **11**: 569811: doi: 10.3389/fpls.2020.569811.

Schipper, J.A., Van der Weijden, J.A., Bruin, P.N.A. and Van der Valk, G.G.M. (1989) *Vakwerk,* **63**(47): 7-9.

Shibli, R.A. and Ajlouni, M.M. (2000) *Plant Cell, Tissue and Organ Culture,* **61**: 15-21.

Stearn, W. (1972) *A Gardenerer's Dictionary of Plant Names*. London: Cassell. pp. 184.

Tang, S.X., Okashah, R.A., Cordonnier-Pratt, M.M., Pratt, L.H., Johnson, V.E., Taylor, C.A., Arnold, M.L. and Knapp, S. J. (2009) *BMC Plant Biol.,* **9**: 72(1–11).

Taylor, J. J. (1976) *Proc. Biol. Soc. Washington,* **89**: 411–420.

Thakur, M., Sharma, B.P. and Verma, T. 2019. *Int.J.Curr.Microbiol.App.Sci.,* **8(5)**: 2116-2124.

Tsukamoto, Y. and Ando, T. 1973. *Environ. Control. in Biol.,* **11**: 69-78.

Uzun, S., Ilbas, A.I., Ipek, A., Arslan, N. and Barpete, S. (2014) *Turkish Journal of Agriculture and Forestry*. DOI:10.3906/TAR-1306-47.

van Doorn, W.G., Harkema, H. and Song, J.S., 1995. *Postharvest Biol. Technol.* **5**: 345–351.

Verma, T., Sharma, B.P. and Thakur, M. (2019) *Int.J.Curr.Microbiol.App.Sci.,* **8**(5): 2101-2109.

Vonk, C.R., Davelaar, E. and Ribot, S.A. (1986) *Plant Growth Regul.* **4**: 65-74.

Wang H., Cui Y., Zhao C. (2010) *Mini-Reviews in Medicinal Chemistry,* **10**: 643-661.

Yabuya, T. (1985) *Japan. J. Breed.,* **35**: 136–144.

Wang, L., Du, Y., Rahman, Md. M., Tang, B., Fan, Li-Juan and Kilaru, A. (2018) *Scientific Reports,* **8**: 17100 | Doi: 10.1038/s41598-018-35281-y

Yabuya, I., Ikeda, Y. and Adachi, T. (1991) *Euphytica,* **57**: 77-81.

Yue, D. and Imanishi, H. (1988) *J. Japan. Soc. Hor. Sci.* **57**: 289–294 (In Japanese with English summary).

Yue, D. and Imanishi, H. (1990) *Scientia Hortic.,* **43**: 331–337.

12

Hemerocallis

C. Aswath, Rajiv Kumar, A.S. Kakote and
M. Mitra and V.A. Parthasarathy

1.0 Introduction

Hemerocallis, commonly known as Daylily, is a fine garden flower. Daylilies are a lucrative crop used for its floral beauty, medicinal proprieties, landscaping, fire prevention, nutritional value, and research (Matand *et al.*, 2020). Day lilies are one of the most important perennial ornamental plants used in gardens and parks (Mosonyi *et al.*, 2019). It is well known for its hardiness and showy blooms produced during spring and summer. The glory of individual flower is, however, short-lived and lasts for a day or so only. But the flower buds on the spike continue to open in the beds and vases as well for several days.

2.0 Importance and Uses

2.1.0 Aesthetic Uses

Hemerocallis is an indispensable choice for woodland and landscape gardens. When grown in the background of a herbaceous border, in front of a shrubbery or along the side of ponds or streams, a highly delightful effect is produced. It can be grown in pots for outdoor or indoor decoration and also for cut flower production. There is variety of colours and great diversity of forms in the flowers which look very impressive. Both early and late cultivars are available and thus, by careful selection, a long spell of bloom can be obtained.

2.2.0 Therapeutic and Other Uses

Daylily is also used for some other purposes. For hundreds of years it has been an important food in China (Yue and Shen, 1991) and Japan. In many of the Chinese markets the dried buds and flowers are sold in large quantity. They are used in soups and stews to provide gelatinous quality and a delicious flavour. Before the flowering scapes appear, the central tender leaves are sliced for salads and eaten like asparagus which is quite palatable. Flower bud and flowers are also eaten in various ways. In Europe, tube-like roots are eaten raw which are sweet in taste and has nut-like flavour (Flory, 1965).

Hemerocallis is a potential source of alkaloid (Antsupova, 1983). Some varieties of hemerocallis contains toxic hemerocallin, which is neurotoxic to man and animal (Barlow and Middleton, 1989; Liu and Zou, 1990), hemerocallin is also reported to cause blindness in sheep (Barlow *et al.*, 1989). It contains some metals like Cu, Pb and Zn (Lee *et al.*, 1983). *Hemerocallis longituba* possess antifebrile and diuretic property (Yoshikawa *et al.*, 1994). Klumpp *et al.* (1995) reported that the crop acts as a bioindicator of air-borne fluoride pollution.

The flower buds are most widely consumed with young shoots and roots also being edible but with some toxicity (Zhang *et al.*, 2004). Hemerocallis "Yellow Flowers" (*Hemerocallis citrine*, *H. lilio-asphodelus*, *H. minor*) as vegetables have over 2,000 years of horticultural history in China (Geng *et al.*, 2012). The dried flower buds are quite commonly used as a vegetable in the Far-East which is incorporated into soups and stews, but can also be eaten uncooked. Flower buds and flowers of

Uses of Hemerocallis

In Chinese Cuisine, In Flower Arrangement, In Flower Bouquet, In Vase Decoration, In Pot, In Garden, Around Trees, Rock Garden, Flower Centerpiece, As Borders, Along Walkways, In Road Dividers

Source: www.wikipedia.org, www.pinterest.at, www.waltersgardens.com, www.in.pinterest.com, www.alamy.com, www.gardendesign.com, www.perennialresouce.com, www.garden.org, www.gradengatemagazine, www.wordpress.com, www.flickr.com

Hemerocallis enriched with antioxidants compared with many other vegetables and has considerable potential as 'nutraceutical' foods (Rodriguez-Enriquez and Grant-Downton, 2012; Rop *et al.*, 2012).

Traditionally, hemerocallis has been used to treat sleep disorders and to alter mood; in Japan and China it is known as 'forget one's sorrow plant' (Rodriguez-Enriquez and Grant-Downton, 2012). Edible lily (*Lilium lancifolium* Thunb) is an important economic crop in Huoshan, Anhui Province, China (Zhao *et al.*, 2016), while *Hemerocallis citrina* Baroni flowers are commonly used as a vegetable and in medicine in China (Li *et al.*, 2018).

3.0 Origin and History

Hemerocallis has a remarkable track record of domestic usage by humans. There is evidence that points to the genus being cultivated in China for several thousands of years (Valder, 1999). It was suggested that ancestors of *H. middendorfii* originated at latest 25 million years ago when the Japanese Archipelago and the Continent had joined; that is, before the formation of the Sea of Japan (Noguchi *et al.*, 2004). The first reference of of *H. fulva* (daylily) was found in a writing in the Chou Dynasty of China (ca. 112–255 BCE) (Li *et al.*, 2020).

Hemerocallis is a native to eastern Asia, chiefly China and Japan. However, the triploid clone, *H. fulva* has been known to the West since 1576 which was believed to have spread from China. It has been in cultivation in France and Belgium as early as 1570 (Fosler and Kamp, 1954). Both *H. thunbergii* and *H. middendorffii* are reported to be widely distributed in eastern Asia (Kang *et al.*, 1998; Park *et al.*, 1998), while *H. exaltata* is restricted to Western Japan (Kang *et al.*, 1998). Liu and Shen (1990) identified *Hemerocallis fulva* var. 'distichia' (a wild species) from Hunan and Hubei. *H. hongdoensis*, a new species of section *Fulvae*, is reported from the islands of southwestern coast of Korea (Chung and Kang, 1994). Large populations of *H. lilioasphodelus* was found in western Hungary (Kovacs and Takacs, 1995). Although daylilies have been known and grown since before the time of Christ, it has been only within the last 80 years that plant breeders in England and lately in America have given this plant the attention it deserves (Flory, 1965).

4.0 Morphology

The monocotyledonous genus *Hemerocallis* is restricted to herbaceous perennial species. It belongs to the family Liliaceae and is related to *Hosta* and *Phormium*. Plants tall (60-90 cm or more) and erect having tuberous and fleshy roots; leaves numerous, both radical and 2-ranked at the base of the scape and yellowish green; flowers like lily, mostly horizontal or oblique, borne in tall scapes and the inflorescence may branch above; common flower colours are orange, yellow or reddish; tepal tubes short enclosing the ovary; segments 6 in number, much exceeding the tube, oblong or spatulate; stamens 6, inserted in the throat, and the filaments slender; style simple, ovary oblong and 3-celled seeds black and spherical.

Xiong *et al.* (1998) reported that in three Chinese endemic species *Hemerocallis plicata*, *H. forrestii* and *H. nana* pollens were of verrucate sexine pattern and that of *H. nana* were boat-shaped-elliptic. Palynotaxonomic studies of pollen grains of 8

Hemerocallis sp. revealed to be monad, boat-shaped and monosulcate. The species investigated were divided into two types by surface pattern : rugulo-reticulate in *H. middendorffii* and reticulate in *H. minor*, *H. flava*, *H. fulva* var. *Kwanso*, *H. thunbergii*, *H. littorea*, *H. taeanensis* (Ahn *et al.*, 1999).

5.0 Taxonomy

In their review, Rodriguez-Enriquez and Grant-Downton (2012), highlight the taxonomic problems in the genus *Hemerocallis* which was first classified as a member of the lily family in its broadest sense–the Liliaceae sensu lato. According to them, 20th century treatments of the monocotyledonous plants recognized that this classification was actually flawed and Hemerocallis was placed in a separate family, *Hemerocallidaceae* (Dahlgren *et al.*, 1985). A large number of molecular studies have shown evidence supporting this placement of the genus away from the Liliaceae sensu stricto. Despite the great resolution that such studies provide, the exact position has remained rather problematic to determine. While the genus itself is easily delimited the taxonomy of the genus remains rather confused and there has been no recent formal monographic treatment of the genus (Rodriguez-Enriquez and Grant-Downton, 2012). The primary monographic study of the genus was published by Stout (1934), although several new species have been described since then. Recently, Murakami *et al.* (2020) proposed a transfer of *H. hakuunensis* and *H. major* from *H.* sect. *Fulvae* to *H.* sect. *Hemerocallis*. *Hemerocallis minor* appears to be conspecific with *H. lilioasphodelus* and is better regarded as *H. lilioasphodelus* var. *minor*. *Hemerocallis lilioasphodelus* var. *thunbergii* is placed in synonym under *H. lilioasphodelus* var. *ilioasphodelus*. The studies on molecular taxonomy is being dealt under Biotechnology.

6.0 Species and Cultivars

6.1.0 Species

Daylily is one of the extensively bred plant genera and the cultivars are highly diverse in flower colour and form. Genus *Hemerocallis* has 13-15 species (Brickell, 1996). Recently, Cao *et al.* (2013) reported that there are 18 species in *Hemerocallis*, of which 11 are native to China. Some important species are described below.

Hemerocallis aurantiaca

This is very popular orange flowered species native to Japan and commonly known as 'golden summer daylily' or 'orange daylily'. Leaves green, about 60-90 cm long; inflorescence many flowered, scape erect, about 90 cm long; flowers tubular at base, lobes wide, fragrant, bright orange and not opening widely.

H. citrina

This is a tall growing fine species native to China. Flowers fragrant, lemon-yellow or pale sulphur-yellow.

Species of Hemerocallis

Hemerocallis dumortierii

Hemerocallis aurantiaca

Hemerocallis citrina

Hemerocallis flava

Hemerocallis fulva

Hemerocallis hongdoensis

Hemerocallis minor

Hemerocallis washingtonia

Hemerocallis dumortieri

Hemerocallis yezoensis

Hemerocallis lilioasphodelus

Hemerocallis thunbergii

***Source*: www.commons.wikimedia.org, www.en.wikipedia.org, www.wallpapers13.com, www.enacademy.com, www.plantwordseeds.com, www.shutterstock.com, www.nativebotany.com, www.mygarden.org**

H. dumortierii

This dwarf lily is native to Siberia and Japan. Leaves long (up to 45 cm); flowers fragrant, deep-yellow on the inside, bronzy-yellow outside. This is an early flowering species.

H. flava

Commonly known as 'yellow daylily' or 'lemon daylily', distributed from E. Siberia to Japan, China. Leaves large; flowers many, fragrant, in different shades of yellow, orange and red, darken with age; scape 60-90 cm. Early flowering. May-June; var. *Major* : leaves darker; petal tip reflexed, wavy.

H. fulva

This species grows widely in China. Leaves large, flowers several, fragrant, orange-red to brick-red to coppery, becoming crimson with age; var. *Kwanso flore pleno* : both variegated, double flowered; var. *Cypriana* : glossy green foliage, flowers more in number, than in type, about 10 cm across, brown outside, July to September.

H. hongdoensis

Native to south-western coast of Korea. Root tuberous; plants deciduous, dichotomous or trichotomous inflorescence 30-98 cm long; perianth stout tube, 2.4-3.2 cm long, 5.5-6.0 mm wide, orange-yellow, lacking a red tinge; ovate lanceolate; bracts 1.2-3.5 cm long, July to August (Chung and Kang, 1994).

H. minor

Leaves 35-45 cm long, 5-8 cm wide, dark green; flowers golden yellow, scapes branched shortly, 25 cm long; corymb 3-6, wavy at the margin.

H. × washingtonia

A colchicine-induced tetraploid synthetic species. Leaves numerous, narrowly lanceolate, about 81 cm long; scapes are 102 cm tall, branching at 67 cm; flowers fragrant, upto 35 in number, wide open, brilliant bronzy-rose, peach or apricot; tepal tube 2.1-2.2 cm long; sepal 7.7-8.3 cm long, broadly lanceolate and obtuse; petal 7.9-8.5 cm long, broadly oblanceolate; stamens 6.7 cm long, style 2.5 cm longer than stamens (Traub, 1951).

6.2.0 Cultivars

Daylily (*Hemerocallis* spp.) is widely cultivated due to its showy flowers and its adaptation to a wide range of soil and climates. There are large numbers of registered cultivars. The publications of the American Hemerocallis Society are an ideal source of information on the modern hemerocallis. Some attractive and popular cultivars of daylily are as follows.

Early	'Aquamarine', 'American Revolution', 'Apricot', 'Bertie Ferris', 'Charles-Johnston', 'Elizabeth', 'Flavinia', 'Florepleno', 'Green Glitter', 'Gold Dust', 'Happy Return', 'Little Deeke', 'Martha Adams', 'Penny's Worth', 'Purple Rain', 'Sebastian', 'Yellow Lollipop' and 'Wind Song'.

Intermediate	'Carrice', 'Duncan' and 'Queen of Gozales;.
Summer blooming	'Amadcus', 'Ann Kelley', 'Betty Slick', 'Betty Woods', 'Black Hills', 'Brocade', 'Camden Gold-Dollar', 'Chestnut Lane', 'Comet,Constance', 'Crimson Star', 'Janice Brown', 'Joan Senior', 'Journey's End', 'Majestic', 'Mellon Balls', 'Moonlight Mist', 'Naranja', 'Nova', 'Painted Lady', 'Patricia', 'Pygmalion', 'Pandora's Box', 'Pardon Me', 'Red Joy', 'Red Sun', 'Scarlet Sunset', 'Show Amber', 'Silver Stars', 'Superlative' and 'Super Purple'.
Late	'Apple Court Champagne', 'Autumn Blaze', 'Autumn Prince', 'Beauty To Behold', 'Bright Eye', 'Little Business', 'Lullaby Baby', 'Martie Everest' and 'Nile Plum'.

Some of the registered clones of *H. washingtonia* are Allegiance (red), Bengaleer (deep yellow), Caption (lavender), Golden Prize (gold), Major (gold), Pink Darling (pink), Prolific (golden yellow), Tetracarmine (carmine) and Yellow Beauty (yellow). Few promising cultivars grown commercially throughout the world are Betty Bennet, Edna Spalding, Ella Pettigrew, Globetrotter and Nancy Hicks. The cultivars Gold Cargo, Golden West, Little Cherub, Mildred Orpet, Old Paint Samrock and Sundown showed good cold resistance (Box *et al.*, 1962). Registered dwarf clone yellow 'Tinkerbell' have commercial potential as a perennial border or pot plant (Gries-bach, 1989).

Tombolato *et al.* (2009) reported seven new daylily cultivars from Instituto Agronômico (IAC), Campinas, Brazil. They are IAC Bárbara (IAC 06A) with red greenish with yellow throat, IAC Canário (IAC L) with light yellow, IAC Estrela (=IAC Castanho) (IAC A civ) with maroon with golden throat, IAC Guaratiba (IAC E. 16182) with salmon with yellow throat, IAC Jundiaí (IAC 08) with red with yellow throat, IAC Neusa (=IAC Campinas) (IAC 13) with golden yellow with red eye, IAC Olga Ulmann (IAC Olga) with pink with yellow throat. They are well adopted and the flowering season at least six months (October to March) in the Southeast part of Brazil. These are suitable for landscape use and also tolerant to rust (*Puccinia hemerocallidis*).

According to Li *et al.* (2020) that in the US, over 83,000 daylily varieties have been registered in the American Hemerocallis Society online database. Another way to classify daylilies is by flower colour. Missouri Botanical garden recognize six types of colour patterns in daylilies (https://www.missouribotanicalgarden.org/Portals/0/Gardening/Gardening per cent 20Help/Factsheets/Daylilies14.pdf).

- Self: the petals and sepals are the same colour.
- Blend: the petals and sepals blend two colours into each other.
- Polychrome: the flowers are a blend of many colours.
- Bitone: the petals and sepals are the same basic colour but differ in the colour.
- Shade Bicolor: the petals and sepals are completely different in colour.

Cultivars of Hemerocallis

Bright Sunset — Happy Returns — Happy Ever Appster™ 'Stephanie Returns'

Red Hot Returns — Frans Hals — Little Grapette

Children's Festival — Double Pardon Me Daylily — Ruby Spider

American Revolution' Maslenica — Bonanza — Rosy Scenario

Source: www.hgtv.com, www.bloemenhuis.be, www.trajnice-strgar.si, www.provenwinners.com, www.boulevard-flower-gardens.shoplightspeed.com, www.seedpantry.co.uk, www.garden.org, www.growjoy.com, www.wagon-wheel.shoplightspeed.com, www.gardencrossings.com

Cultivars of Hemerocallis

Hold Your Horses Startle Yellow Bouquet

Lady Lucille Wayside King Royale Yasmin

Peacock Alley Red Volunteer Lilting Belle

Source: www.hgtv.com

Cultivars of Hemerocallis

Oakes Love

Super Coral

Out of the Blue

Myth and Magic

My Special Angel

Siloam Double Classic

Source: www.hgtv.com

☆ Banded or Eyezone: these are flowers that have a distinctively different colour or colour shade at the throat compared to the ends of the petals and sepals.

Josue (2020) recommends the following popular daylily varieties.

'Stella de Oro': Perhaps the most popular and widely grown daylilies in the world. This hybrid is known for its adaptability to a variety of planting conditions and the ability to rebloom for long periods. It requires full sun but tolerates part shade.

'Chicago Silver': It is a semi-evergreen daylily that grows about 60-90 cm in height. The flowers are purple with darker eye zone, silver edges, and green throat. Individual bloom reaches up to 15 cm in diameter.

'Crimson Pirate': It is one of the many favorite red selections. The blooms appear in bright crimson red with lighter midrib and darker eye. The throat comes in yellow. An individual flower grows up to 19 cm in width and lasts until 16 hours, but no more than 24 hours. 'Crimson Pirate' daylilies are mid-season bloomers and re-bloomers.

'Barbara Mitchell': This cultivar bears soft pink blooms with full petals and measures up to 16 cm, semi-evergreen, early mid-season bloomer and a rebloomer. The flower form is single, grows up to 50 cm.

'Catherine Woodbury': It produces fragrant lavender-pink trumpet-formed blooms with a yellowish throat and slightly ruffled petals. Flowering occurs late mid-season with an extended bloom period, plants reach about 60-90 cm tall, with flowers growing up to 15 cm in full size.

'Ice Carnival': This multi-awarded daylily features huge and ruffled pale yellow to white blooms with yellow-green throat. Each flower reaches up to 6 inches in full size and blooms mid-summer with a slight rebloom. Mature plants grow up to 45 cv in height and 75 cm in width.

Moonlit Masquerade: It has large, cream blooms with unique purple eye zone and green throat. It blooms early mid-season, highly reblooming. This tetraploid and semi-evergreen daylily has won multiple awards over the years, including Stout Silver Medal in 2004 and Award of Merit in 2002, among others.

7.0 Genetics and Breeding

7.1.0 Cytology

Ploidy level was estimated for nine species and 94 cultivars in the genus *Hemerocallis* by flow cytometry (FCM) analysis (Saito *et al.*, 2003). *Hemerocallis* genotypes were clearly classified into three groups: 1.84–2.13 (group 1), 2.94–3.10 (group 2), and 3.74–4.26 (group 3). Since the diploid *H. fulva* var. *littorea* ($2n$=22) and the triploid *H. fulva* var. *kwanso* ($2n$=33) belonged to groups 1 and 2, respectively, the other *Hemerocallis* genotypes belonging to groups 1, 2 and 3 were considered to be di-, tri- and tetraploid, respectively. Chromosome counting in root tip cells revealed that *Hemerocallis* genotypes belonging to groups 1, 2 and 3 had 22, 33 and 44 chromosomes, respectively. No cytochimeras and polysomaty were observed. Thus, FCM analysis has proven useful for simple and rapid estimation of the ploidy level of *Hemerocallis* species and cultivars.

Cytological investigations, made by Stout (1932), revealed that a somatic number of 22 chromosomes is the rule for all species of the genus. Triploidy exists in several cultivated clones of fulvous daylilies and diploids, triploids, aneuploids and para-triploids are obtained among the seedlings of diploids crossed with triploids. It was concluded that as long as the seedlings obtained have one parent which is diploid there is little chance of advancing to higher levels of polyploidy. Stout (1934) during a study on karyotypes of the daylily indicated 11 as basic chromosome number. Chang *et al.* (1984) reported that the chromosome numbers of *H. fulva* and *H. fulva* var. *longituda*, *H. fulva* cv. Kwans were 2n=33, 2n=22 and 2n=33, respectively. The karyotype of the species and its cultivars were similar. Studies on somatic chromosome number and karyomorphology of 52 cultivars also revealed the occurrence of diploid, triploid, tetraploid and aneuploid races (Zadoo *et al.*, 1976; Vijayavalli and Mathew, 1990). Microsporogenesis was studied by Karihaloo (1994) in a selfed clone of *Hemerocallis* 'Sleeping Beauty', it showed regular meiosis under normal cultural conditions and desynapsis when subjected to low-water/ high temperature stress. The progeny segregated in the ratio 3 non-desynaptic : 1 desynaptic plants, indicating that the parental clone was heterozygous and that desynapsis is controlled by a single recessive gene.

Several preliminary studies have been made concerning the inheritance in Hemerocallis. Traub *et al.* (1968) concluded that only the full recessive of 'melon' flower colour exhibits the 'melon' phenotype. Flory and Phillips (1968) reported that in less than 1 per cent of the slow-growing, delayed flowering seedlings that appear when tetraploids are crossed, called static dwarfs, the loss or gain of chromosomes and perhaps particular chromosomes, would seem to be ultimately responsible for the abnormal growth habits. In this connection, it should be noted that new aneuploid *Hemerocallis* species, with lesser or greater than the euploid number of chromosomes, 2n=40, 42, 46, could be established. These might not inter-cross with the diploids, 2n=22 or the tetraploids, 2n=44.

Ren *et al.* (2019) studied inheritance by observing flowering processes among species, cultivars and F_1 hybrids. Daily flowering rhythm was a relatively stable character that could be influenced through a breeding program. In F_1 generations of crosses between daylilies and night lilies, there was an obvious tendency toward nocturnal opening. The result suggests that in hemerocallis, opening is characterized by mixed major loci and polygenic inheritance, with dominant nocturnal opening. Closure was controlled by multiple genetic loci. Inheritance of opening and closure is controlled by different nuclear circadian clock genes and perhaps photoreceptor-binding motifs. By manipulating cultivar diversity, single flower duration could be prolonged effectively through hybridization. In addition, it was assumed that daily flowering rhythm was not a completely conserved character, and could be partly affected by environment.

7.2.0 Breeding

7.2.1 Historical Perspectives

According to Rodriguez-Enriquez and Grant-Downton (2012), outside of the Far-East, Hemerocallis is primarily cultivated as a garden ornamental, with currently very little interest in culinary and medicinal usage. This is despite horticultural material–including presumed Far-East cultigens, the virtually sterile triploid forms of *H. fulva* having been cultivated for centuries (Valder, 1999). Although, one of the earliest plants of the Far-East to reach the gardens of Western Europe, serious horticultural attention only began after George *et al.* (1988) made the first artificial interspecific hybrid, which precipitated a new era of deliberate breeding. Subsequently, selection of the fertile hybrid breeding lines resulting from interspecific crosses, and the discovery and introduction of new species and variants to add to the gene pool, generated a complex hybrid gene pool. Stout of New York Botanic Garden was not only responsible for describing new species and maintaining them in cultivation, but also for a large amount of experimental hybridization work (Stout and Chandler, 1933). By repeating artificial crosses many hundreds of times, Stout was able to incorporate the genetics of almost sterile triploid clones of *H. fulva* into hybrid lines. No tetraploid individuals have ever been found in nature (Xiaobai, 1996) and anyway there are significant barriers to interploidy crosses in hemerocallis, with triploids only reliably produced from diploid–tetraploid combinations via *in vitro* embryo rescue (Arisumi, 1973; Li *et al.*, 2009). Presumably, they were formed by extremely rare events involving unreduced gametes; By the 1950s, this pioneering work meant that the gene pools of hybrid lines in cultivation contained a number of species such as *Hemerocallis altissima*, *H. citrina*, *H. fulva* and *H. multiflora*. Intensive selection of the fertile hybrid lines–primarily by horticulturists in the USA–resulted in great changes to plant characteristics, in particular to the floral morphology, with astonishing rapidity. Transgressive characteristics resulted, such as alterations to the placement and dimensions of the tepals–plant breeders selected away from the original lily-like, funnel-shaped to trumpet form of the speciesto generate variants with long, linear floral segments (the so-called 'spider' forms) and, at the other extreme, more flat-opening flowers with rounded, broad and blunt tepals (Petit and Peat, 2000, 2004). From the 1950s onwards, diploid hybrid breeding lines were joined by man-made tetraploids generated by treating plant material with anti-mitotic agents such as colchicine.

The first important advance in the development of the genus *Hemerocallis* as an ornamental plant began in the early 1890s with the production of hybrids by Yeld in England. More extensive breeding experiments have been continued since then in Europe and the United States (Norton *et al.*, 1949). The progress was greatly accelerated with the induction of polyploidy as suggested by Traub (1949). *Hemerocallis washingtonia* is the first named and published colchicine-induced tetraploid plant species and is thus unique (Traub, 1951). To preserve a record of its development it was planned to publish reports of the species under cultivation at ten-year interval and the same has already been detailed for the first (1949-58), second (1959-68) and third (1969-78) decades. The interest in the species greatly increased and it is now grown in large number, particularly in many of the gardens of United States (Traub *et al.*, 1973). The establishment of the American Hemerocallis Society and its publication related to this plant are other achievements in the history of the improvement of daylily.

Cui *et al.* (2019) conducted numerical classification of flower colour and provide valuable baseline data and foundational theory for normalization and precision of Hemerocallis. The throat was predominantly green-yellow, light yellow and yellow; green-yellow accounted for the largest proportion of sepals (67.76 per cent) and petals (69.40 per cent). The limb was clustered into five colour groups (orange, yellow, pink, red and purple); the yellow group had the most varieties for both sepals and petals, containing 57.38 per cent and 55.74 per cent, respectively. Both sepals and petals had significant differences in colour, redness and colour angle for the throat, eye and limb. According to "Dual Classification", the colour classification standard was proposed as a 3-level standard. The colour of sepal and petal consistency served as the first standard, and the colour of limb was the second standard. The colour pattern types of pure, gradual change, watermark and eye spot, served as the third standard. It has been proposed that all the 183 experimental varieties were divided into two categories, five groups and finally four types. This study provides a classification basis and reference for numeric and standardized colour phenotype description for Hemerocallis.

7.2.2 Plant Genetic Resources

The natural distribution of daylilies covers from East Asia to Russia's Siberia and is mainly in China. There are problems of classification and evolutionary relationships among populations. These include naming and classification confusion, unclear distinction between species and cultivars, variation between different geographical groups, natural hybrid populations, and the difference between day and night flowering in hemerocallis. Of all 383 collections of wild daylilies of herbarium specimens in the Herbarium, Institute of Botany, CAS (PE), 5 species and 1 variant of wild daylilies were collected in Shandong, Shaanxi, Hebei Province, and Beijing. In addition, a field investigation was conducted in Saihanba National Forest Park in Hebei Province. The maximum average coefficient of variation within populations is the length of the bracts (86.28 per cent) in *Hemerocallis lilio-asphodelus* and number of buds (21.39 per cent) in *Hemerocallis citrina*. The main factors among populations are the height of plants in *Hemerocallis minor* (Ren *et al.*, 2017).

Hemerocallis × *hybrida* hort. collection of Nikita Botanical Gardens (NBG) 86 cultivars of foreign breeding and four own hybrid forms were investigated. All cultivars were perennial herbaceous polycarpic with assimilative non-succulent shoots. The cultivars were classified into groups according to type of leaf development (3 groups), height of vegetative (3 groups) and generative shoots (4 groups), height of generative shoots concerning the position of leaves (3 groups), size (5 groups), shape (5 groups) and flower colour (7 groups), blooming term (4 groups) and remontant ability (2 groups). Perspective cultivars and hybrid forms possess valuable ornamental characters and economically biological properties (Plugatar and Ulanovskaya, 2018).

7.2.3 Breeding Objectives

McGarty (2008) suggested the following objectives for breeding of Hemerocallis.

- ✰ Expanding bicolor flowers.
- ✰ Increasing branching.
- ✰ Maximizing bud count.
- ✰ Extending flowering time.

Some of the other traits the present day breeders have in mind for ideal plant type are as follows (McGarty, 2008).

- ✰ The plants should have winter hardiness.
- ✰ The plant should bloom for a long season.
- ✰ Flower colour should not bleach out and petals and sepals should not curl or wilt prematurely.
- ✰ Flowers must drop quickly after bloom on their own.
- ✰ Flowers should stay open in the evenings.
- ✰ Flowers must sit high enough above the foliage so as to be seen.
- ✰ Scapes should be neither too heavy to overwhelm the plant or too thin to allow drooping.
- ✰ Foliage must be full, lush and green.

Wooten (1975) suggested breeding for warmer conditions should be another area to be considered by breeders. According to him hybridization of named and unnamed daylilies has been producing plants with a marked decrease in chill requirement and with a notable improvement in heat and sun tolerance. These characteristics, added to excellent form and colour, indicate the possibility of superior new daylilies for tropical regions, comparable in every way to the best of those grown in temperate areas. Those daylilies which made it in Miami were those first hybridized. These included (a) named *cvs*. which were performing well and were fertile; (b) a number of unnamed hybrids selected in the field for their evident merits but with little information as to their parentage; (c) some daylilies bought under hybridizers' numbers, many of which were dormant; and (d) very fine, recent Florida evergreen introductions. For clarity Wooten (1975) considered them in three groups, tetraploids, diploid miniatures and other diploids.

7.2.4 Pollination

Flowers of members in sections *Fulvae* Nakai and *Capitatae* Nakai in the genus *Hemerocallis* are orange-yellow and usually visited by bees and butterflies. On the other hand, fragrant lemon-yellow flowers in section *Hemerocallis* are pollinated by moths (Kang and Chung, 2000). However, Hirota *et al.* (2013) do not support a hypothesis that yellow flower colour and strong scent intensity, the distinctive floral characteristics of *H. citrina*, having evolved in adaptations to hawk moths. It was suggested that the key trait that triggers the evolution of nocturnal flowers is flowering time rather than flower colour and scent. To trace the fate of individual pollen grains through pollination processes, Hirota *et al.* (2013) determined genotypes of single pollen grains deposited in *Hemerocallis* stigmas in an experimental mixed-species array. *Hemerocallis fulva*, pollinated by butterflies, has diurnal, reddish and unscented flowers, and *H. citrina*, pollinated by hawk moths, has nocturnal, yellowish and sweet scent flowers. It was observed pollinator visits to an experimental array of 24 *H. fulva* and 12 F_2 hybrids between the two species (*H. fulva* and *H. citrina*) and collected stigmas after every trip bout of swallowtail butterflies or hawk moths. Hirota *et al.* (2013) measured selection by swallowtail butterflies or hawk moths through male and female components of pollination success as determined by single pollen genotyping. As expected, swallowtail butterflies-imposed selection on reddish colour and weak scent: the number of outcross pollen grains acquired is a quadratic function of flower colour with the maximum at reddish colour, and the combined pollination success was maximal at weak scent (almost unrecognizable for human). This explains why *H. fulva*, with reddish flowers and no recognizable scent, is mainly pollinated by swallowtail butterflies. However, no evidence of hawk moths-mediated selection on flower colour or scent was found.

7.2.5 Hybridization

The breeding of daylilies generally involves the crossing of heterogeneous genotypes followed by the selection of desirable floral and plant growth phenotypes. Selected plants are vegetatively propagated and the new cultivar is registered with the American Hemerocallis Society and introduced into commerce (Tomkins *et al.*, 2001). While extensive breeding has been undertaken by numerous breeders, very little genetic study has been accomplished in the daylily. Since the American Hemerocallis Society began acting as the official registry of daylily cultivars in 1947, more than 40,000 registrations have been processed (Tomkins *et al.*, 2001). McGarty (2008) wrote detailed account on breeding of hemerocllis. According to him, the hybridizing of the genus *Hemerocallis* has been performed for just over a hundred years and within the past thirty there has been an explosive growth in the effort. The hybrids being introduced have a significant amount of variation in colour and form.

Ahn *et al.* (2003) carried out intergeneric crossing with *Lilium* for development of hemerocallis as a new cultivar with ornamental value. The capsule formation by the cut style pollination was 40-60 per cent while that of stigmatic pollination was lower than 3 per cent. After pollination, the enlarged capsule died after ten or twelve days in natural environment. Therefore, *in vitro* culture should be performed within twelve days. In *in vitro* pollination, the process of cut style pollination showed higher capsule formation than that of the stigmatic pollination. Medium containing IBA 1.0 + 2,4-D 1.0 + zeatin 0.05 mg/l was found the most effective materials for seed formation.

Hasegawa *et al.* (2006) reported that time of flower anthesis in a day is thought to evolve in response to the time of pollinator activities. They studied blooming and withering time in natural populations of daylily (*Hemerocallis fulva*), nightlily (*Hemerocallis citrina*) and their hybrids, and also in an artificially obtained array of the F_1 hybrids. Blooming time of *H. fulva* varied from 4:30-7:30 and *H. citrina* varied from 16:30-20:30. In a natural hybrid population, blooming time and withering time showed discontinuous bimodal distribution in spite that morphological traits of flowers showed continuous unimodal variation. Most F_1 hybrids showed diurnal flowering. These findings indicated that only a few genes have strong phenotypic effect on the determination of flowering time in hemerocallis, and suggest that the evolution from a *H. fulva*-like ancestor to *H. citrina* was not a continuous process by accumulation of minute mutations (Hasegawa *et al.*, 2006).

Yasumoto and Yahara (2006) reported that post-pollination isolation does exist between *H. fulva* and *H. citrina*, although it is partial, asymmetric, and weakened in sympatry suggesting that floral and post-pollination isolation may develop independently, and reinforcement may not be a general phenomenon in plants.

Gao and Qi (2012) provide a reference for the hybridization of hemerocallis. Artificial hand-pollinations were made, where-after the seed-setting rate and seed number per fruit were investigated, using 21 modern hemerocallis varieties, including diploids, triploids and tetraploids as well as two original species from China as parent material. The results showed that the level of compatibility was dissimilar in the various crossing combinations. The average seed-setting rate between diploids was 48.75 per cent. There were significant segregations between diploids and tetraploid crossings and the average seed-setting rate was 11.34 per cent. It was difficult to obtain seed from crosses between triploids and diploids and the seed-setting rate was as low as 1.56 per cent.

Ren *et al.* (2017) reported that crossing between diploid and tetraploid may produce triploid hemerocallis. Moreover, the resistance and adaptability of triploid hemerocallis are much stronger than those of diploid and tetraploid plants. However, the crossing between diploids and tetraploids is not easy and the abortive rate is high. To overcome this problem, the immature embryos were collected from the ovaries at 5, 7 and 9 d after pollination. A small number of embryos were collected from the first day of embryo rescue, and paraffin sections were made and compared with untreated embryos. The results showed that the highest survival rate (31.68 per cent) was obtained with ovaries 7 days after pollination, with the medium that contained MS + 0.01 mg/l NAA + 0.032 mg/l BA. Out of 21 cross combinations, crosses between triploids and tetraploids displayed the highest survival rate (38.57 per cent).

7.2.6 Polyploid Breeding

Interploidy crosses between diploid daylily (Hemerocallis) cultivars (2n = 22) and tetraploid cultivars (2n = 44) were performed via *in vivo* hybridization aiming to produce triploid hybrid germplasm (3n = 33) (Li *et al.*, 2009). Plant growth regulator-free MS based medium containing 5 levels of sucrose was used to optimize the embryo rescue medium. It was determined that the medium supplemented with 3 per cent sucrose gave highest rate of immature hybrid embryo rescue. Thirty-seven hypothetically triploid genotypes were subjected to verification of ploidy status by root tip. The newly developed triploid genotypes could open new horizons for further polyploidy breeding (Li *et al.*, 2009).

8.0 Growth and Flowering

8.1.0 Environmental Factors

Tuber size and its stage of development at the time of planting and also environmental conditions markedly influence the growth and flowering of hemerocallis. Large sized tubers hasten flowering and give higher yield of flowers. Insufficient cold during the winter to break dormancy was reported to be one of the factors causing irregular and often sparse blooming of certain early cultivars like Apricot. Moist warm weather in the latter part of June and early July was very favourable for daylily bloom. During this period when many of the mid-season cultivars as San Juan, Mayor Starzynski, Lidice, Dr. Stout, Patricia, Corliss, Caballero and many others produced good quality flowers. A few warm bright days in the latter part of June following moist and cool weather resulted in severe burning of many cultivars of daylilies. It was interesting to note that during a period of unfavourable weather for the development of good red colour in daylilies some cultivars were much less affected by the adverse conditions than others. Berwyn is a good example of a cultivar that has good quality red colour even during a spell of cold weather (Cooley, 1951).

Six cultivars exposed to 4-hour supplementary light in addition to natural day light bloomed earlier than those receiving no extra light. The earliest flowering occurred with 4-h supplementary light in the middle of night increasing the length of time the plants were held at 4.4°C (40°F) before forcing at 23.9°C (75°F) delayed flowering (Anon., 1963).

Treder *et al.* (2013) attempted to accelerated growth and improve quality of plantlets of two cultivars of daylilies 'Fire Agate' and 'Polish Truffle'. Enrichment with CO_2 of and higher irradiance level significantly improved growth of both cultivars. Plantlets of 'Polish Truffle' grown at higher irradiance level had more leaves with higher length, irrespectively of CO_2 enrichment. Longer leaves had plants grown in medium consisted of peat and perlite. Fresh and dry weight of aerial part and root system were significantly higher when plants were grown with CO_2 enrichment and higher irradiation level. Better root system with longer roots of both cultivars had plantlets grown in medium consisting of peat and perlite.

Mosonyi *et al.* (2019) found that flower forcing of daylily hybrids in early spring or autumn could broaden their application and marketing possibilities. They stated that the differentiation of inflorescence is not depending on the day length rather on the leaf surface area. Growing degree-day is the main factor which influences the speed of inflorescence development once it has already differentiated, so temperature is the key factor in flower bud opening and it also defines the number of the developing flower buds. The bare-rooted daylily plants are suitable for long-term cold storage and offers late season flower forcing.

8.2.0 Growth Regulators

Foliar spray with GA_3, IAA, NAA, ethrel (ethephon), maleic hydrazide and CCC (chlormequat) at 100 or 200 ppm done 30 and 50 days after planting. GA_3, NAA and IAA treatments increased plant height, while ethephon, MH and CCC retarded it.

Maximum flowers spike length (110.4 cm) and highest number of flowers/spike was obtained with NAA 200 ppm (Das *et al.*, 1992). Exogenous benzyl adenine and zeatin significantly prolonged the life of flower of *H. minor* (Zhang *et al.*, 1991).

Leclerc *et al.* (2006) reported that application of the seaweed/PGR mixture at 3000 mg/l, Cycocel at 3000 mg/l or BA at 2500 mg/l applied at flowering, increased the number of plants producing two divisions in Hemerocallis 'Happy Returns'. However, Amling *et al.* (2007) found that foliar application of benzyladenine (BA) 2500 or 5000 ppm for 1, 2, 3, 4 or 5 consecutive weeks increased ramet production of all cultivars. Increasing number of BA applications resulted in 'Lavinia Love' and 'Beguine' forming 1.6-3.2 and 0.7-1.1 more ramets, respectively, than control plants at 9 weeks after initial treatment, regardless of concentration. The var. 'Sarah Sikes hybrid' formed 1.1 and 1.3 more ramets when treated with 2500 ppm and 5000 ppm BA, respectively.

9.0 Propagation

Hemerocallis can easily be propagated by seeds, division of clumps, or by cutting. In recent years, tissue culture techniques are also being employed. Many of the species are self-sterile and seeds can be obtained easily by crossing.

9.1.0 Seed Propagation

Seed germination of *H. dumortiesii* was promoted by chilling for 14 or 28 days with germination percentage of 97 per cent and 93 per cent after 30 days from sowing, respectively (Sim *et al.*, 1996).

9.2.0 Vegetative Propagation

9.2.1 Cutting and Division

The roots of daylily are fleshy tubers which are often catalogued as bulbs. They are divided after the flowering is over. In this way, one can have manifold increase in the planting material. In another method, the crowns are cut horizontally and treated with kinetin or SD 8339 (synthetic cytokinin) for shoot production. Four weeks after treatments, 10 cm long shoots, separated from mother crowns are treated with hormodin No. 2 and rooted under mist. This process is repeated as new shoots on the mother crowns reached 10 cm in length (Apps and Heuser, 1975).

9.2.2 Micropropagation

Propagation of daylily through tissue culture is a recent method. Two methods of tissue culture of *Hemerocallis citrina* were employed by Zhou *et al.* (1983). In the first, stamens were cultured on MS medium with 2,4-D at 2 ppm. The resulting adventitious buds formed at the base of the filaments were then transferred to the same medium without 2,4-D. The second method involved culturing nearly mature seeds on MS medium plus 2,4-D at 2 ppm, kinetin at 1 ppm and 500 ppm yeast powder. The resulting plants had 22 chromosomes. Fitter and Krikorian (1981) were able to get plantlets from suspension cultures in about 240 days in modified MS medium. The optimum temperature was found to be 22°C with 50 per cent relative humidity. Maintaining culture in dark, stimulates production of plantlets in

Propagation of Hemerocallis

Seed Pod and Seeds

Germination in Pots

Raising Seedling in Plug Trays

Raising Nursery in Protected Condition

Transplanting in Pots and Beds

Open Cultivation

Proliferation with Roots Attached to the Scape

Proliferations in Soil

Division of the Crown

Callus

Shoot Regeneration

Rooting and Hardening

Micropropagation of Day Lily

Source: www.howmytexasgardensgrow.com, www.daylilygarden.com, www.smokeysgardens.com, www.midaylilysociety.com, www.howmytexasgardensgrow.com, www.taglilien-hemerocallis.de, www.americandaylily.com, www.bmcplantbiol.biomedcentral.com, www.garden.org, Matand *et al.* (2020)

suspension culture. Cell suspension of daylily *cv.* Autumn Blaze was obtained from larger tissue masses by culturing on a MS medium along with 10 per cent coconut water and 2 mg/l 2,4-D. Cells were transferred for 3 weeks to the same medium without 2,4-D and sub-cultured in a modified White's medium + 200 mg/l casein hydrolysate. An additional 3 weeks in a Schenk and Hildebrandt medium (PSH) followed by 6 weeks in the same medium + 100 mg/l inositol stimulated root and shoot growth (Krikorian and Kann, 1981).

A reliable protocol for *in vitro* propagation of *Hemerocallis aurantiaca,* was developed by Saker *et al.* (1998). Shoot proliferation was greater from shoot tips than from bulb scales when cultured on MS medium supplemented with 4 mg NAA/l and 8 mg BA/l. Adding adenine sulphate at 60 mg/l to a medium containing 4 mg NAA + 8 mg BA/l increased shoot development. MS medium with 2.0 mg IBA/l was the most effective for rooting.

It involves removing immature flower buds (5-7 mm) and placing their petals and sepals on the culture medium to induce callus formation. The plantlets are produced from these calli by transferring to an agar medium containing kinetin (1.0 mg/l) and IAA (0.5 mg/l). Shoots and roots developed when the callus was sub-cultured on a medium lacking 2,4-D. This method produces thousands of plants in a short period (Heuser and Apps, 1976). Flower buds of *H. fulva* with 5-10 mm size provided the best source of filament explants when cultured in MS medium supplemented with cytokinins and grown at 25±2°C under 16 h days (Carter and Dhir, 1999).

The regeneration of plantlets from flower stalk callus of *H. fulva* was also obtained on half strength MS medium supplemented with 0.3 mg/l BA. After 20 days the rooted plantlets were transplanted in soil (Nixin and Ma Yu, 1984).

The challenges of daylilies micropropagation are exacerbated because a few good protocols that have been developed are generally difficult to reproduce across genotypes (Matand *et al.,* 2020). This difficulty might be partly because the most valuable varieties are interspecific hybrids. Further, the difficulty of *de novo* plant regeneration *in vitro* might also explain the broad complaints by daylilies farmers and enthusiasts about the inconsistencies of true-to-type inflorescences and overall profiles of several plants purchased as *in vitro* tissue culture-produced.

There are quite a few reports on daylily micropropagation using agar cultures have only been useful (Krikorian and Kann, 1979a, b; Meyer, 1976, 1979). Recently Matand *et al.* (2020) developed an efficient protocol, which can be applied for mass multiple shoots formation using the daylilies stem tissue as the main explants. In less than three months, using thidiazuron, the use of the stem tissue as the *in vitro* experimental explant was successful in inducing multiple shoots several folds greater than current daylilies shoot organogenesis protocols. The study showed that tissue culture can be conducted successfully under unrestricted room environmental conditions as well as under the controlled environment of a growth chamber. It also showed that splitting lengthwise stem explants formed multiple shoots several folds greater than cross-sectioned and inverted explants. Shoot conversion rate was mostly independent of the number of shoots formed per explants.

9.2.3 Bioreactors

Scaled-up liquid cultures in flasks or bioreactors, together with mechanization and automation of the micropropagation process, contribute to overcoming the limitations imposed by existing conventional labour-intensive methods (Sluis and Walker, 1985; Aitken-Christie, 1991; Chen *et al.*, 2005). Chen *et al.* (2005) were the first to report the use of liquid cultures for scaled-up micropropagation of both ornamentally and medicinally useful daylilies by *in vitro* culture, followed by *in vivo* acclimatization. They induced meristematic clusters from daylily scape explants (pedicel–scape junction) in the presence of the growth retardant Paclobutrazol on semisolid agar medium. Liquid shake culture was used to proliferate meristematic clusters. Highly efficient regeneration of adventitious shoots occurred on clusters after subculture on a 0.8 per cent agar strength semisolid medium with the addition of activated charcoal. The use of liquid shake cultures for mass proliferation of meristematic clusters followed by regeneration of adventitious shoots on semisolid agar culture could be an efficient system for large-scale micropropagation of daylily.

10.0 Soil and Climate

10.1.0 Soil and other Growing Media

Hemerocallis grows in any garden soil having pH 6.5 to 7.0 (Jana and Mukherjee, 1979). Plants can be grown in fertile, moist but well-drained soil. Hemerocallis can also be grown in other growing media other than soil. Plants of hemerocallis 'Stella de Oro' were grown in containers, containing peat moss : sandy loam : perlite (2:2:3) or peat moss : pine bark : perlite (1:1:1) for 2 consecutive years, recorded significant differences in dry weight (DW), fresh weight (FW) root, DW and FW gain over 90 days in the medium containing sandy loam (Perry and Adam, 1990). Turkey litter compost added to container grown *Hemerocallis* cv. Red Magic grown in pine bark substrate, on a gravel pad increased container capacity and available water by 12-16 per cent and 17-30 per cent respectively, though there was no sufficient water in the container to maintain optimal growth under reduced irrigation frequencies. It also enhanced photosynthesis and stomatal conductance and proved to be a better alternative to commercial substrate (Tyler *et al.*, 1993a).

10.2.0 Climate

It flourishes very well in a wide range of climate from temperate to tropical climate of Japan, India and China. Plants are most luxuriant along the borders of lily ponds or moist places and partial shade. However, sunlight should be available for couple of hours for proper flowering. Red and purple flowered cultivars are intolerant to heavy rainfall and very hot sun. Daylilies are frost hardy and may die by alternating frost and thaw.

11.0 Cultivation

11.1.0 Planting and Spacing

The soil will have to be prepared well before planting. It is important to dig the ground to a depth of 50 cm and mix well rotted compost in the soil. Planting is done in February to March and plant density varies with the type of planting materials used. Normally, the seedlings are planted closer (15 cm apart) compared with the ones derived from division of clumps (30 cm apart). Plants take long time to establish properly and, therefore, should not be disturbed frequently. Locke and Altland (2012) evaluated wheat straw and horse manure compost, locally available resources, as components of container media for establishing and overwintering daylily, *Hemerocallis fulva* 'Stella D'Oro'.

11.2.0 Irrigation

Sufficient moisture in the soil is beneficial for proper growth and flowering in hemerocallis. Daylily needs liberal watering for proper growth and flowering. However, it should not be watered after planting as the fleshy underground stem contains sufficient food materials to develop initial growth of root and shoot. Before root formation, watering proves injurious to the tubers and may cause rotting.

Li *et al.* (2016) reported that drip-irrigation was effective in salt leaching. The average values of soil EC for five treatments were 1–4 and 2–6 dS/m in 0–35 cm soil profiles of sandy loam and silt soils, respectively, after 18 months with drip irrigation. The survival rate and dry mass decreased by 0.63 per cent and 17.14 per cent for each unit of EC increase in the irrigation water in sandy loam saline soil, and the corresponding values were 0.70 per cent and 25.63 per cent in silt saline soil. This study implied that Daylily is a suitable plant in landscape construction of coastal saline soils using drip-irrigation at ECiw < 7.8 dS/m.

11.3.0 Manuring and Fertilization

Proper nutrition improves the plant growth and flowering. Higher levels of nitrogen and phosphorus were found to be very effective in increasing the growth of the plants and flower production in *H. fulva* cv. Double Orange. Deficiencies of these nutrients, on the other hand, resulted in marked reduction of plant height and number and size of flowers. The effect of potassium was less pronounced (Jana, 1979).

Chaturvedi *et al.* (1988, 1991) reported that application of 20g N and 40 g P/ bed (90 x 90 cm) improved the plant growth, flowering and clump weight (Table 1).

Perry and Adam (1990) added NH_4NO_3 (0 to 600 ppm) and KCl (125 ppm) in container grown *Hemerocallis* x 'Stella de Oro' at weekly intervals, showed linear increase in plant size, grade, FW and DW with increase in N rate. Application of nitrogen at 185.10 kg/ha and phosphorus at 164.60 kg/ha increased flower yield (75.79 million flowers/ha). The entire dose of P and half of N were applied at the time of planting with the rest of N, top dressed before flowering (Pandey *et al.*, 1993).

Matheus da Silva *et al.* (2015) found that highest tillering gained with the use of 150 kg/ha N combined with 400 kg/ha of P in *Hemerocallis fulva* var. Flore Full. Intermediate doses of N (75-170 kg/ha of N) showed higher production of plant fresh and dry mass and root dry mass. Plants fertilized with 300 kg/ha N and 273 kg/ha P led to the highest levels of N in the leaves and doses of 80 kg/h N and 68 kg/ha P, the highest content of P in the leaves. In the roots, the highest N content was obtained with 300 kg/ha N in the presence of P and doses of 225 kg N/ha combined with 300 kg/ha P provided the highest P content in the roots.

Table 1: Response of Daylily (*Hemerocallis fulva* L.) to Nitrogen and Phosphorus Fertilization*

Treatment	*Plant Height Before Initiation of Flower Stalk (cm)*	*Size of the Leaf*		*Length of Flower Stalk (cm)*	*Days to Flowering from Stalk Initiation (Heading)*	*Number of Flowers*	*Weight of Clump (g)*
		Length (cm)	*Width (cm)*				
N0	22.16	33.47	1.38	43.05	13.07	5.57	7.73
N1	23.11	35.54	1.53	48.57	15.83	5.84	9.44
N2	25.58	44.14	1.73	53.81	17.81	6.40	11.10
N3	23.98	43.63	1.57	53.66	17.44	5.74	10.50
SEm ±	0.25	0.63	0.03	0.26	0.94	0.26	0.270
C.D. at 5 per cent	0.52	1.31	NS	0.54	1.95	0.55	0.560
P0	23.35	38.73	1.53	48.97	15.53	5.75	9.21
P1	23.66	39.04	1.57	49.23	15.70	5.93	9.88
P2	24.12	39.81	1.56	51.12	16.90	5.97	9.98
SEm ±	0.22	0.54	0.03	0.05	0.82	0.23	0.24
C.D. at 5 per cent	0.45	NS	0.06	0.11	1.69	NS	0.49
P0	21.67	32.30	1.32	42.78	12.00	5.22	7.05
N0--P1	22.30	33.46	1.33	42.99	12.22	5.67	7.97
P2	22.50	34.46	1.50	43.39	15.00	5.82	8.16
P0	22.67	34.76	1.52	46.44	15.44	5.82	8.18
N1--P1	23.30	35.00	1.53	49.45	15.55	5.83	10.00
P2	23.35	36.85	1.54	49.83	16.50	5.87	10.13
P0	24.11	43.89	1.59	50.67	16.89	6.09	10.96
N2--P1	25.30	44.17	1.72	51.78	17.67	6.51	11.11
P2	27.33	44.35	1.87	59.00	18.89	6.61	11.22
P0	24.94	43.97	1.70	56.00	17.77	5.89	10.66
N3--P1	23.72	43.54	1.68	52.72	17.33	5.72	10.44
P2	23.28	43.39	1.32	52.27	17.22	5.60	10.38
SE ±	0.43	1.09	0.06	0.45	1.63	0.46	0.47
C.D. at 5 per cent	0.90	2.27	0.12	0.93	NS	NS	0.98

* Chaturvedi *et al.* (1991).

Tyler *et al.* (1993b) planted hemerocallis *cv.* Red Magic in pine bark substrate, on a gravel pad, amended with composted turkey litter (0-16 per cent by vol.) resulted in an increase in both foliar N, P, Mg, Mn, Cu, Fe and B concentration and content with increase in turkey litter rate.

Sun *et al.* (2006) found significant effects of magnesium applied to soils in increasing Daylily (*Hemerocallis citrina* Baroni) yield, improving its growth, and strengthening its antivirus property as well as increasing the amount of exchangeable Mg, N, P, and K in red soil. In particular, the effects of magnesium compound fertilizer II (MCF2) with higher Mg content were better in increasing yield by 57.4, 32.8, and 14.5 per cent compared to that of control treatment. It increased soil Alkali N, available P, exchangeable K, and exchangeable Mg by 94.9, 46.5, 31.1, and 35.3 per cent, respectively, compared with that of CK treatment.

11.4.0 Weed Control

Pre-emergence herbicide ronstar (oxadiazon) and goal (oxyfluorfen) caused severe injury (28-55 per cent) to *Hemerocallis* cv. Sammy Russell (Whitwell and Kelly, 1989). Norcini and Aldrich (1992a) suggested application of granular pre- emergence stakeout (dithiopyr) at 1.1-2.2 kg/ha and snapshot TG (isoxaben + trifluralin) at 5.6 kg/ha for effective control of *Chamaesyce maculata* in container grown *Hemerocallis* x Aztec Gold. Pre- and post-emergence application of chlorimuron at 0.03 and 0.06 kg/ha suppressed the growth of *Cyperus esculentus* and *C. rotundus* in yellow daylily fields, but showed phytotoxicity on crop (Derr and Wilcut, 1993).

No detrimental effect was observed on growth, development or flowering of *Hemerocallis* x cv. Pink Charm by pre-emergence herbicide products dacthal (DCPA, chlorthal dimethyl), gallery (isoxaben), pennant (metolachlor), PRE-M (pendimethalin), stakeout (dithiopyr), derby (metolachor + simazine) and snapshot (isoxaben + oxyzalin). Most of the products were effective in controlling *Digitaria sanguinalis, Conyza canadensis* and *Euphorbia humistrata* (Skroch *et al.*, 1994).

Bing (1983) evaluated 7 herbicides for weed control in hemerocallis field, and observed that with light weed infestation all chemicals gave satisfactory control. Oxyfluorfen granules at 2 and 4 lb/acre caused burn from which plants did not recover. Combinations of Rout (oryzalin + oxyfluorfen) or Ornamental Herbicide 2 (pendimethalin) at 1.12 kg/ha with Goal (oxyfluorfen) 2.24 kg/ha caused acute phytotoxicity to containers grown daylily (*Hemerocallis*) hybrid Aztec Gold. Surflan (oryzalin) at 4.48 kg/ha and ornamental herbicide 2 at 1.12 kg/ha inhibited flowering in daylily by 98 per cent and 65 per cent respectively. After 3 applications, surflan, Pre-M and pendimethalin controlled weed effectively (Stamps and Neal, 1990). Bentazone at 1.1-2.2 kg/ha along with crop oil at 2.3 l/ha was found to be phytoxic to *Hemerocallis* cv. Aztec Gold and therefore unacceptable (Norcini and Aldrich, 1992b). Herbicide residues in irrigation water containing 1 mg/l and 10 mg/l of oryzalin, oxyfluorfen and isoxapen resulted in reduced root and shoot growth in daylily (*Hemerocallis* hybrida) when grown in fine pine bark medium in greenhouse. Oryzalin was most phytotoxic to daylily, followed by isoxaben and oxyfluorfen (Bhandary *et al.*, 1997).

12.0 Diseases and Pests

12.1.0 Diseases

12.1.1 Leaf Spot

The plants are usually free from diseases and pests. Leaf spots may be caused by *Cercospora hemerocallis* or *Heterosporium iridis*. Such leaves should immediately be collected and burnt. Greenish yellow spots on the leaves which gradually enlarge and turn orange-brown are common on *H. fulva rosea* and cultivars derived from it. It was reported that plants grown in partial shade were free from this disease (Pirone *et al.*, 1960).

A leaf streak disease caused by *Collecephalus hemerocalli* was reported in the United States (Spencer, 1968). The disease caused necrotic spotting and streaking of leaves, and the lesions which varied in colour from dark brown to whitish tan, were usually surrounded by chlorotic tissue.

12.1.2 Rust

Emmitt *et al.* (2016) detected daylily rust, caused by *Puccinia hemerocallidis*, for the first time in the United States. Spray of azoxystrobin at 14-day intervals had significantly lower disease progress curve (AUDPC) than when applied at 21- or 28-day intervals.

Buck and Williams-Woodward (2003) reported that protectant fungicides chlorothalonil and mancozeb significantly reduced pustule developed by daylily rust caused by *Puccinia hemerocallidis*. Of three sterol-biosynthesis inhibiting compounds, triadimefon provided greatest reduction in pustule development. Mueller *et al.* (2003) evaluated daylily cultivars for resistance to daylily rust. From a greenhouse evaluation of 84 inoculated cultivars, 14 (17 per cent) were classified as resistant; 13 (15 per cent) were moderately resistant; 22 (26 per cent) were moderately susceptible; and 35 (42 per cent) were fully susceptible.

12.1.3 Flower Rot

Li *et al.* (2018) reported daylily flower rot caused by *Fusarium proliferatum* first time in China. Brown and rotted flowers, some of which were covered with white mycelia, were observed in commercial daylily field in Yanqing District, Beijing, China. Disease incidence was 40 per cent.

Inoculating *Bacillus subtilis* into the microplants of *Hemerocallis* cv. Stella d' Oro, grown under high humid conditions in fogging glasshouses, could prevent phytopthora and pythium damping off (Berger *et al.*, 1996).

12.2.0 Pests

12.2.1 Thrips

Hemerocallis multiflora, *H. citrina* and *H. thunbergii* may severely be injured by thrips (*Franklinielia hemerocallis*). Thrips feed on leaves and flower buds, causing brown scarifications. The entire flower cluster may die without developing. Late

Diseases and Pests of Hemerocallis

Root Rot | Rust | Leaf Streak Fungus

Bacterial Soft Rot | Root Knot Nematode | Snails

Damage by Leafminer | Tarnished Plant Bug | Aphids

Gall Midge | Thrips Damage on Petals

Source: www.daylilies.org, www.viettes.wordpress.com, www.daylilies.org, www.influentialpoints.com, www.citybugs.tamu.edu, www.hgic.clemson.edu, www.extension.usu.edu, www.wieseacres.blogspot.com, www.canr.msu.edu, www.apsnet.org, www.upload.wikimedia.org

flowering cultivars are most seriously affected. Spraying weekly with malathion at 4 per cent was found beneficial to control the pest (Smith, 1967).

12.2.2 Other Pests

Halstead and Harris (1990) first recorded midge pest (*Contarinia quinquenotata*) of daylily (*H. fulva*) in U.K. *H. yezoensis* is attacked by *Liriomyza hemerocallis*, a seed feeding pest (Iwasaki, 1993; Halstead, 1995). The plants are also infested by aphid, *Myzus hemerocallis*, reported from Europe (Remaudiere and Munos, 1992).

12.2.3 Nematodes

Daylilies are occasionally affected by root-knot nematode (*Meloidogyne incognita*). Lehman (1996) reported the nematode infection and identified as *Pratylenchus holdemani* and *P. breviculus*. Nematodes also affected the plants in nursery (Inserra *et al.*, 1995). On the basis of root gall index, Montasser (1995) reported hemerocallis to be slightly resistant to *M. incognita*.

13.0 Harvesting and Postharvest Management

13.1.0 Harvesting

Daylily flowering scape should be cut, when the first flower bud is just showing colour, with a sharp knife and placed immediately in water. Harvesting should be done on the previous evening and marketing in the next morning.

13.2.0 Yield

Flower production varies with cultivars, density of planting and cultural practices adopted. Pandey *et al.* (1993) recorded 75.79 million flowers/ha, by using 185.10 kg N/ha.

13.3.0 Postharvest Management

Pulsing of *H. fulva* flower buds in cycloheximide solution for one hour during flower opening, delayed senescence by preventing the degradation of petal proteins by ubiquitin proteins during floral development and senescence (Courtney *et al.*, 1994). Delayed senescence by cycloheximide treatment (10-50 µM) was also reported by Bieleski and Reid (1992) in *Hemerocallis* x Cradle Song.

13.3.1 Senescence of Flower and Vase Life

The petals of daylily (*Hemerocallis* hybrid) have a genetically based programme that leads to senescence and cell death, 24 h after the flower opens. Six cDNAs, whose levels increase during petal senescence, were isolated and sequenced and designated DSA 3, 4, 5, 6, 12 and 15. All six DSAs are members of gene families. With the exception of DSA 4 accumulation of the DSA mRNAs is increased 3.2 to 43 times by a concentration of abscisic acid that cause premature senescence of the petals (Panavas *et al.*, 1999).

The rapid senescence of daylily flowers was accompanied by a transient respiratory climateric but not by any appreciable evolution of ethylene. The application of silver thio-sulphate or amino-oxy-acetic acid did not delay senescence of daylilies, it was thus concluded that ethylene is not involved in this process (Lukaszewski and Reid, 1989). Flowers of *Hemerocallis* hybrid cv. Stella d' Oro are insensitive to ethylene, but ABA is a constituent of the signal transduction chain leading to cell death of daylily petals. Endogenous ABA increases before flower opening and continues to increase during petal senescence (Valpuesta *et al.*, 1995; Panavas *et al.*, 1998).

Bieleski (1993) studied the flower senescence of *Hemerocallis* hybrid *cv.* Cradle Song. Timing of events was related to the time when flower expansion was 60 per cent complete (hour 0), tight bud (h 62), fully developed flowers (h 10) and senescence advanced (23 h). Soluble carbohydrate constituted 50 per cent of petal dry weight up to h 10, then decreased during senescence to reach 4 per cent by h 34. Up until h 14, fructan accounted for 80 per cent of the soluble carbohydrate, whereas hexose only 2 per cent. Fructan hydrolysis began just before bud crack at h 14, reaching completion by h 10 when no detectable fructan remained. Osmolarity of peal cell sap increased significantly during fructan hydrolysis, from 0.300 to 0.340 osmolal. Cycloheximide applied to excised buds between h 38 and h 14 halted both fructan hydrolysis and flower expansion. The results suggest that the onset of fructan hydrolysis with the concomitant large increase in osmoticum, is an important event driving flower expansion in daylily. Flower senescence in cv. Cradle Song appears to be independent of ethylene (Lay *et al.*, 1992). Ethylene production was low throughout flower development, and there was a transient increase, up to 2.5 nl/flower/h, as flowers senesced. Flowers detached from the plant at bud development stage, placed in deionised (DI) water for 7 h and then transferred to fresh DI water or actinomycin D (25 mg/l), showed an increase in fresh weight (FW) up to a maximum at about 12 h after transfer, followed by a rapid decline. By 48 h the flowers had completely senesced. In contrast, senescence of flowers treated during opening with cyclo-hexamide (1 mM) was greatly delayed, with little loss of FW or reduction in flower diameter, even up to 181 h after transfer. Inhibition of flower senescence by cycloheximide suggests that senescence requires *de novo* synthesis of specific proteases or other proteins facilitating protein break-down.

Islam *et al.* (2011) found that pre-treatment of scapes with cycloheximide (CHI) (0.01 and 0.05 mM) as well as spraying isolated flowers with CHI (0.5 mM) before transfer to holding solutions resulted in an enhancement of vase life/ flower longevity, maintenance of membrane integrity and improving postharvest performance of cut scapes and isolated flowers of *Hemerocallis fulva*.

14.0 Biotechnology

14.1.0 Tissue Culture

14.1.1 Protoplast Culture

Zhou (1989) isolated pollen protoplasts from cold-pre-treated (5-6°C for 4 or 7 days) anthers and cultured on K3 medium containing 1 mg 2,4-D and 2 mg NAA/l. The formation of multicellular clusters, and equal and unequal first cell divisions were observed.

Mesophyll protoplasts were isolated from leaf tissues of a diploid daylily (*Hemerocallus* × Red Magic) by enzymatic digestion with a solution containing 0.5 per cent pectolyase Y-23, 0.1 per cent cellulase R-10, 0.1 per cent driselase, 0.6 sorbitol and half-strength MS inorganic salts. When cultured on MS medium supplemented with 0.5 mg NAA and 0.5 mg benzyladenine, the protoplasts underwent sustained division to produce multicellular colonies. The optimal planting density for cell division was 0.5 x 105 protoplasts/ml. The highest planting efficiency was obtained in cultures grown in media solidified with 0.2 per cent gelrite. Under these conditions, formation of colonies occurred from 14 per cent of cultured protoplasts. Calluses were recovered from 9 colonies only after the cultures were treated with a conditioned medium. Intact plants were regenerated from protoplast-derived calluses through organogenesis (Ling and Sauve, 1995).

14.1.2 Somatic Embryogenesis

Callus of an embryogenic cell of *Hemerocallis cv.* Autumn Blaze was cultured on a medium containing growth regulator for at least 12 weeks. Cultures was transferred to a low pH (4-4.5) growth regulator-free medium and maintained and multiplied for over a year. When transferred to a growth regulator-free medium at pH 5.8 they developed through a range of somatic embryo stages. Up to 20 per cent of the somatic embryo produced did not develop into well-formed plantlets. Placing filter paper impregnated with activated charcoal on top of the agar medium enhanced the germination of somatic embryos into plantlets (Smith and Krikorian, 1991).

Stamens were excised from sterilized young flower buds (5-20 mm) of *H. fulva* plants, between July and September, to provide filament explants culture were grown on MS medium supplemented with various cytokinins and auxins at 25 + 2°C under 16-h days. After 6 weeks of culture, 63, 53, 17 and 12 per cent of explants excised from 5-, 10-, 15 and 20-mm buds, respectively, had formed somatic embryos. The number of embryos per explant was 24.00, 30.83, 6.33 and 3.67 for 5-, 10-, 15- and 20-mm buds, respectively. In 10-12 weeks, upon further subculture, the somatic embryos developed into heart and torpedo shaped structures and were converted into complete plantlets. It was concluded that buds 5-10 mm in size provide the best source of explants (Carter *et al.*, 1999).

Matand *et al.* (2013) investigated leaf, flower, flower bud, stem, and root tissues for shoot organogenesis after different growth regulator treatments on MS medium. The variety 'Summer Echoes' was the best with greatest average shoots per explant (118) and shooting explant rate 87 per cent when 6-benzyl-aminopurine (BA) and kinetin were used singly. This protocol is the most improved for shoot organogenesis in daylily.

14.1.3 Somaclonal Variation

Griesbach (1989) developed a procedure for inducing somaclonal variation in *Hemerocallis* tissue culture. The frequency of genetic variation, however, was very low.

14.2.0 Molecular Markers

14.2.1 Diversity and Phylogeny

Han *et al.* (1998) used RAPD markers to detect genetic polymorphism among 3 species (*H. fulva, H. dumortieri* and *H. thunbergii*) and 1 botanical variety (*H. fulva* var. Kwanso). It was possible to identify a high level of intraspecific polymorphism within species and interspecific polymorphism among different species. The results showed that *H. dumortieri* was phylogenetically different from other *Hemerocallis* lines.

Tomkins *et al.* (2001) used AFLP primers to detect the genetic polymorphism among nineteen primary genotypes (species and early cultivars) and 100 modern cultivars. When comparing cultivar groups from different time periods (1940–1998), genetic similarity was initially increased, compared to the primary diploid genotypes, remained constant from 1940 to 1980, and then steadily increased as breeding efforts intensified and hybridizers began focusing on a limited tetraploid germplasm pool derived by colchicine conversion. Among modern (1991–1998) daylily cultivars, genetic similarity has increased by approximately 10 per cent compared to the primary genotypes.

Cao *et al.* (2013) used six ISSR primers to analyse the diversity among twenty-one accessions of *Hemerocallis* spp. Seven accessions were triploid, rest were diploid. Genetic diversity was high among the collected genotypes with 95 per cent polymorphic loci (PPL), but was relatively low within commercial cultivars (PPL=89 per cent). The genetic diversity among commercial cultivars within each country were low (PPL=72 per cent for USA and 51 per cent for New Zealand). These results demonstrated that ISSR profiling is a powerful method for the identification of genetic variation of daylily cultivars and wild collections.

14.2.2 Evolution

Allozymes were used to study the evolution of Daylilies. Kang and Chung (2000) analyzed thirty populations from five species of *Hemerocallis* in Korea by starch gel electrophoresis to measure genetic diversity and to determine genetic population structure and the amount of genetic divergence within and between species at 12 isozyme loci. Populations of five Korean species maintain high levels of genetic variation and little differentiation among populations and species.

The origin and evolution of the Japanese nocturnal *Hemerocallis citrine* var. *vespertina* (Asparagales: Hemerocallidaceae) were studied (Noguchi *et al.*, 1995; 2004; Noguchi and DeYuan, 2004).

Flower longevity and 13 external morphological characters were measured and compared with a molecular phylogenetic tree obtained from the nucleotide sequences of three noncoding chloroplast regions, the *rbc*L *atp*B intergenic region, the *trn*L (UAA) intron, and the spacer region between *trn*L (UAA) 32 exon and *trn*F (GAA). Japanese *H. citrina* var. *vespertina* had multiple origins and originated from at least three different lineages that invaded the Japanese Archipelagoes separately through different routes. Based on morphological, phenological, and molecular analyses, it was suggested that pollinator mediated evolutionary changes in floral characters were accompanied by adaptive evolution for flower longevity (Noguchi and DeYuan, 2004).

14.2.3 Genomics

Ou *et al.* (2020) reported the first chloroplast genome sequence of *H. citrina*. The chloroplast genome size is 156,088 bp with GC content of 37.3 per cent, including a large single-copy (LSC) of 84,843 bp, a small single copy (SSC) of 18,507 bp, and a pair of 26,369 bp IR (inverted repeat) regions. The phylogenetic analysis revealed that *H. citrina* belongs to the Hemerocallis genus in Asphodelaceae family.

Podwyszyñska *et al.* (2015) studied the phenotypic and genome size changes (variation) in synthetic tetraploids of daylily (*Hemerocallis*) in relation to their diploid counterparts. Compared to diploids, in tetraploids cultivars, percentage of flowering plants and bud number/scape was lower. Tetraploid leaves and flowers of two cultivars were significantly larger, chlorophyll concentration index was higher by approximately 40 per cent, and stomata were longer by 35 per cent.

14.2.4 Tagging Traits

A cDNA clone encoding a daylily (*Hemerocallis* spp.) thiolprotease (*SEN11*), whose expression is strongly upregulated in flower tepal senescence, has been isolated (Guerrero *et al.*, 1998). SEN102 and SEN11 transcripts were not detectable in flower buds at the opening stage, but two peaks of transcripts were seen after 9 h and 19 h, in both petals and sepals, when wilting symptoms were apparent. The expression of SEN*102* was lower in the senescent leaf than in the green leaf. The pattern of expression of these genes suggests their involvement in the protein hydrolysis occurring in tepals at the late senescence stage, whereas in leaves they could be involved in the constitutive protein turnover machinery. Exogenous gibberellic acid application to cut flowers increased transcripts of both genes (Guerrero *et al.*, 1998).

Zhu *et al.* (2015) reported that daylily is once blooming (OF) when the genotype is dominant, whereas a reblooming daylily (RF) has homozygous recessive genotype. A single gene-Hemerocallis Terminal Flower 1 (HTFL1) is found to govern the reblooming trait. The low HTFL1 expression is associated with the reblooming trait in daylily. Low HTFL1 expression is homozygous recessive (tt), high HTFL1 expression is homozygous dominant (TT), and medium HTFL1 expression is heterozygous (Tt). Therefore, reblooming daylily can be identified by HTFL1-marker assisted selection. The application of HTFL1-marker in parent selection can aid in creating new reblooming daylilies.

15.0 References

Ahn, M.S., Lee, K.J., Jeong, J.S., Choi, D.C., Lim, H.C., Choi, S.R., Choi, J.S. and Choi, Y.G. (2003) *Acta Hortic.*, **620** : 305-310.

Ahn-Y.H., Kim, S.H., Lee, C.H. and Lee, S.T. (1999) *J. Korean Soc. Hort. Sci.*, **40** : 505-510.

Aitken-Christie, J (1991) *Automation*. In: (Eds.) Debergh, P. C. and R.H. Zimmerman. *Micropropagation*. Dordrecht: Kluwer Academic Publishers; pp.363–388.

Amling, J.W., Keever, G.J., Kessler Jr, J.R. and Eakes, D.J. (2007) *J. Environ. Hortic.*, **25**(1) : 9-12.

Anonymous (1963) *Agric. Res. Wash.*, **11** : 15.

Antsupova, T.P. (1983) *Ulan-Uda, USSR*, pp. 3-16.

Apps, D.A. and Heuser, C.W. (1975) *Comb. Proc. Int. Pl. Prop. Soc.*, **25** : 362-367.

Barlow, R.M. and Middleton, D.J. (1989) *Neuropathology and Applied Neurobiology*, **15** : 602.

Barlow, R.M., Bedford, P.G.C. and Wang, J.H. (1989) *Neuropathology and Applied Neurobiology*, **15** : 592.

Berger, F., White, D., Frazer, R., Leifert, C. and Li, H. (1996) *Phytopath.*, **86** : 428-433.

Bhandary, R.M., Whitwell, T. and Briggs, J. (1997) *Weed Technology*, **11** : 793-797.

Bieleski, R.L. (1993) *Plant Physiology*, **103** : 213-219.

Bieleski, R.L. and Reid, M.S. (1992) *Plant Physiology*, **98** : 1042-1049.

Bing, A. (1983) *Proc. 37th Annu. Meet. Northeast Weed Sci. Soc.*, pp. 352-356.

Box, C.O., Gordon, W.C. and Hance, K.P. (1962) *Miss. Fm. Res.*, **25** : 7.

Brickell, C. (1996) *A-Z Encyclopedia of Garden Plants*, pp. 511-516.

Buck, J.W. and Williams-Woodward, J. L. (2003) *Crop Prot.*, **22**(1) : 135-140.

Cao, D.M., Zhang, C., Zhang, X.C., Kang, L.F., Duan, J.J., Ma, X.L., Yan, G.J. and Wang, Y.S. (2013) *Acta Hortic.*, **977** : 299-306.

Carter, J. and Dhir, S. (1999) *Camellia J.*, **54** : 24-25.

Chang, Q., Dong, X. and Yang, W. (1984) *Acta Hortic. Sinica*, **11** : 265-270.

Chaturvedi, O.P., Singh, R.P., Shukla, H.S. and Pandey, K.S. (1988) *Prog. Hort.*, **20** : 337-340.

Chaturvedi, O.P., Singh, R.P., Shukla, H.S. and Pandey, K.S. (1991) *Haryana J. Hor. Sci.*, **20** : 77-80.

Chen, J., Hall, D.E. and De Luca, V. (2005) *In Vitro Cell. Dev. Biol. Plant*, **41** : 58–62.

Chung, M.G. and Kang, S.S. (1994) *Novon*, **4** : 94-97.

Cooley, J.S. (1951) *Plant Life*, **11** : 265-270.

Courtney, S.E., Rider, C.C. and Stead, A.D. (1994) *Physiol. Plantarum*, **91** : 196-204.

Cui, H., Zhang, Y., Shi, X., Gong, F., Xiong, X., Kang, X., Xing, G. and Li, S. (2019) *PLoS One.*, doi: 10.1371/journal.pone.0216460.

Dahlgren, R.M.T., Clifford, H.T. and Yeo, P.F. (1985) The families of the monocotyledons. Berlin Springer.

Das, S.N., Jana, B.K. and Das, B.C. (1992) *South Indian Hort.*, **40** : 336-339.

Deer, J.F. and Wilcut, J.W. (1993) *Weed Technol.*, **7** : 112-117.

Emmitt R.S., Stevenson K.L, Brenneman, T.B. and Buck, J.W. (2016) *Plant Dis.*, **100**(1) : 188-191.

Fitter, M.S. and Krikorian, A.D. (1981) *Ann. Bot.*, **48** : 591-597.

Flory, W.B. (1965) *Hortic.*, Vol. XLIII, **7** : 16-17, 28.

Flory, W.B. and Phillips, R.P. (1968) *Plant Life*, **24** : 84-86.

Fosler, G.M. and Kamp, J.R. (1954) *Day-lilies for Every Garden*, Circular 716, Univ. Illinors.

Gao, Y. and Qi, S. (2012) *Acta Hortic.*, **937** : 547-554.

Geng, J., Gao, S., He, Q., Gao, Y. and Zhang, Q. (2012) *Acta Hortic.*, **953** : 255-260.

George E.F., Puttock D.J.M. and George, H.J. (1988) *Plant culture media*, vol. 2. Exegetics Ltd Westbury, England.

Griesbach, R.J. (1989) *Hort Science*, **24** : 1027-1028.

Guerrero, C., Calle, M. de la., Reid, M.S. and Valpuesta, V. (1998) *Plant Mol. Biol.*, **36** : 565–571.

Halstead, A.J. (1995) *Cecidology*, **10** : 16-17.

Halstead, A.J. and Harris, K.M. (1990) *British J. Entomol. Natural History*, **3** : 1-2.

Han, H.N., Cho, U.H. and Kim-Lee, H.Y. (1998) *J. Korean Soc. Hort. Sci.*, **39** : 218-221.

Hasegawa M., Yahara, T., Yasumoto, A. and Hotta, M. (2006) *J. Plant Res.*, **119** : 63–68. doi 10.1007/s1026500502413

Heuser, C.W. and Apps, D.A. (1976) *Canadian J. Bot.*, **54** : 616-618.

Hirota S. K., Nitta, K., Suyama, Y., Kawakubo, N., Yasumoto, A.A. and Yahara, T. (2013) *PLoS One* **8**(12) : e85601. https: //doi.org/10.1371/journal.pone.0085601

Inserra, R.N., Robinson, W.L. and Smith, W.W. (1995) *Nematol. Circular Gainesville*, **211** : 2.

Islam, S. T., Tahir, I., Shahri, W. and Bhat, M. A. (2011) *J. Plant Sci.*, **6**(1) : 14-25.

Iwasaki, A. (1993) *Japanese J. Ento.*, **61** : 303-306.

Jana, B.K. (1979) *Orissa J. Hort.*, **7** : 42-46.

Jana, B.K. and Mukherjee, A. (1979) *Indian Hort.*, **24** : 17-27.

Josue, M. (2020) https: //florgeous.com/daylilies/

Kang, S.S. and Chung, M.G. (2000) *American J. Bot.*, **87**(11) : 1634–1646

Kang, S.S., Park, K.B. and Chung, M.G. (1998) *Nordic J. Bot.*, **18** : 581-587.

Karihaloo, J.L. (1994) *Cytologia*, **59** : 483-491.

Klumpp, G., Klumpp, A., Domingos, M. and Guderian, R. (1995) *Environ. Monitoring Assement*, **35** : 27-42.

Kovacs, J.A. and Takacs, B. (1994) *Kanitzia Res. Rep. Series, Bio-Tar.*, **2** : 43-46.

Krikorian, A. D. and Kann, R.P. (1979a) *Hem. J.*, **33**(1) : 44–61.

Krikorian, A. D. and Kann, R.P. (1979b) *Clonal micropropagation of daylilies*. In: (Eds) Sharp, W. R., P.O. Larsen, E.F. Paddock and V. Raghaven. *Plant cell and tissue culture: principles and applications.* Columbus: Ohio State University Press, pp. 835–836.

Krikorian, A.D. and Kann, R.P. (1981) *Ann. Bot.*, **47** : 679-686.

Lay, Y.M., Stead, A.D. and Reid, M.S. (1992) *Physiologia Plantarum*, **86** : 308-314.

Leclerc, M., Caldwell, C.D., Lada, R.R. and Norrie, J. (2006) *Hort Science*, **41**(3) : 651-653.

Lee, M.H., Cho, J.K., Kim, K.S., Kim, S.K., Kim, B.Y. and Park, C.S. (1983) *Res. Rep. Rural Development, South Korea, Soil Fert. Crop Protect. Myco. and Farm Prod. Utilization*, **25** : 69-74.

Lehman, P.S. (1996) *Tri Ology Technical Report*, **35** : 10-11.

Li, B.J., Yan, W.X., Shi, Y.X., Chai, A.L. and Xie, X.W. (2018) *Plant Dis.*, **102**(4). doi: 10.1094/PDIS-07-17-0997-PDN.

Li, S., Ji, F.F., Hou, F., Cui, H., Shi, Q., Xing, G., Weng, Y. and Kang, X. (2020) *Front. Plant Sci.* **11** : 686.doi: 10.3389/fpls. 2020.00686

Li, X., Kang, Y., Wan, S., Xu, J. and Li, N. (2016) *Sci. Hortic.*, **205** : 39-44.

Li, Z., Pinkham, L., Campbell, N.F., Espinosa, A.C. and Conev, R. (2009) *Euphytica*, **169** : 313–318.

Ling, J. and Sauve, R.J. (1995) *Plant Cell Rep.*, **15** : 293-296.

Liu, Y. and Zou, K.N. (1990) *Acta Veterinaria et Zootechnica Sinica*, **21** : 54-58.

Liu, Y.Q. and Shen, M.J. (1990) *Acta Hortic. Sinica*, **17** : 45-50.

Locke, J. C. and Altland, J. E. (2012) *J. Environ. Hortic.*, **30**(4) : 207–210.

Lukaszewski, T.A. and Reid, M.S. (1989) *Acta Hortic.*, **261** : 59-62.

Matand, K., Shoemake, M. and Li, C. (2020) *BMC Plant Biol.*, **20** : 31. https: //doi. org/ 10.1186/s12870-020-2243-7.

Matand, K., Wu, N., Conley, S. and Acquaah, G. (2013) *Afr. J. Biotechnol.*, **12**(8) : 820-825.

Matheus da Silva, F. P., Giacon, G.M., Soares, J.S., Rosa, Y.B.C.J., Rosa, D.B.C.J., Lemes, C.S.R., Pereira, S.T.S. and Gutierrez, R.S. (2015) *Ornam. Hortic.*, **21**(3) : 351-362.

McGarty, T.P. (2008) *Classic Genetics, Hybridizing of Hemerocallis and Current Techniques.* In : *Hemerocallis Species, Hybrids, and Genetics* pp. 319-392. http: //citeseerx.ist. psu.edu/viewdoc/download?doi=10.1.1.486.5979&rep=rep1&type=pdf.

Meyer, M. M. Jr. (1976) *Hort Science*, **11** : 485–487.

Meyer, M.M. Jr. (1979) *Hem. J.*, **33**(3) : 20–23.

Montasser, S.A. (1995) *Pakistan J. Nematol.*, **13** : 99-102.

Mosonyi, I.D., Tilly-Mándy, A., Kohut, I. and Honfi, P. (2019) *Acta Hortic.*, **1237** : 177-184.

Mueller, D.S., Williams-Woodward, J.L. and Buck, J.W. (2003) *Hort Science*, **38**(6) : 1137-1140.

Murakami, S., Takayama, K., Fuse, S., Hirota, S.K., Koi, S., Ideno, T., Yamamoto T. and Tamura, M. N. (2020) *Acta Phytotax. Geobot.*, **71**(1) : 1–11.

Nixin and Ma, Y. (1984) *Acta Hortic. Sinica*, **11** : 202-206.

Noguchi, J. and De Yuan, H. (2004) *Internatl. J. Plant Sci.*, **165** : https: //doi. org/10.1086/380983

Noguchi, J., Hong, D. and Grant, W. (2004) *Plant Syst. Evol.* **247** : 1–22.

Noguchi, J., Tasaka, M. and Iwabuchi, M. (1995) *J. Plant Res.*, **108** : 41–45. https: // doi.org/10.1007/BF02344303.

Norcini, J.G. and Aldrich, J.H. (1992a) *J. Environ. Hortic.*, **10** : 14-17.

Norcini, J.G. and Aldrich, J.H. (1992b) *Hort Science*, **27** : 1196-1198.

Norton, J.B.S., Stuntz, M.F. and Ballard, W.R. (1949) *Descriptic Catalog of Hemerocallis Clones 1893-1948, The Amer. Plant Life Soc.*, Stanford, California.

Ou, X., Liu, G. and Li, H.W. (2020) *Mitochondrial DNA part B.*, **5**(1) : 1109–1110. https: //doi.org/10.1080/23802359.2020.1726227

Panavas, T., Pikula, A., Reid, P.D., Rubinstein, B. and Walker, E.L. (1999) *Plant Molecular Biol.*, **40** : 237-248.

Panavas, T., Walker, E.L. and Rubinstein, B. (1998) *J. Experimental Bot.*, **49** : 329.

Pandey, K.S., Chaturvedi, O.P., Singh, R.P. and Shukla, H.S. (1993) *Prog. Hort.*, **23** : 73-75.

Park, K.B., Chung, J.M. and Chung, M.G. (1998) *J. Japanese Bot.*, **73** : 137-142.

Perry, L.P. and Adam, S.A. Jr. (1990) *J. Environ. Hort.*, **8** : 19-21.

Pirone, P.P., Dodge, H.C. and Rickett, H.W. (1960) *Diseases and Pests of Ornamental Plants*, Constable and Co. Ltd., London.

Plugatar, Y.V. and Ulanovskaya, I.V. (2018) *Acta Hortic.*, **1201**. doi: 10.17660/ ActaHortic.2018.1201.72.

Podwyszyñska, M., Gabryszewska, E., Dyki, B., Steposwka, A.A., Kowalski, A. And Jasinski, A. (2015) *Euphytica*, **203** : 1-16.

Remaudiere, G. and Munos-Viveros, A.L. (1992) *Revue Francaise d' Entomologie*, **14** : 172.

Ren, Y., Gao, Y., Gao, S., Yuan, L., Wan, X. and Zhang, Q. (2019) *Sci. Hortic.*, **250** : 19-26.

Ren, Y., Gao, Y.K. and Yang, J. (2017) *Acta Hortic.*, **1171** : 245-252.

Ren, Y., Gao, Y.K., Liu, J.J., Fu, M. and Zhang, Q.X. (2017) *Acta Hortic.*, **1185** : 65-72.

Rodriguez-Enriquez, M.J. and Grant-Downton, R.T. (2012) *AoB Plants*, **5**: pls055; doi: 10.1093/aobpla/pls055.

Rop, O., Mlcek, J., Jurikova, T., Neugebauerova, J. and Vabkova, J. (2012) *Molecules,* **17** : 6672–6683.

Saito, H., Mizunashi, K., Tanaka, S. and Nakano, Y.A.M. (2003) *Sci. Hortic.,* **97** : 185-192.

Saker, M., Rady, M. and El-Bahr, M. (1998) *Egyptian J. Hort.,* **25** : 113-128.

Sim, Y.G., Han, Y.Y., Song, I.K., Kwon, T.Y., Jung, J.S., Yoon, J.T. and Choi, B.S. (1996) *RDA J. Agri. Sci. Hort.,* **38** : 700-704.

Skroch, W.A., Catanzaro, C.J., Hertogh, A.A.de, Gallitano, L.B. and De-Hertogh, A.A. (1994) *J. Environ. Hort.,* **12** : 80-82.

Sluis, C. J. and Walker, K. A. (1985) *IAPTC Newsl.* **47**: 2–12.

Smith, D.L. and Krikorian, A.D. (1991) *Annals Bot.,* **67** : 443-449.

Smith, F.F. (1967) *Agri. Inf. Bull.* No. 237, U.S. Deptt. Agri., pp. 39.

Spencer, J.A. (1968) *Plant Dis. Reptr.,* **52** : 751-753.

Stamps, R.H. and Neal, C.A. (1990) *J. Environ. Hort.,* **8** : 52-57.

Stout AB. 1934. *Daylilies: the wild species and garden clones, both old and new, of the genus Hemerocallis.* New York: Macmillan Press.

Stout, A.B. (1932) *Cytologia,* **3** : 250-259.

Sun, N., Zeng, X., Li, J., Gao, J. and Wang, B. (2006) *J. Integr. Agric.,* **5**(2) : 123-129.

Tombolato, A. F. C., Matthes, L. A. F., Uzzo, R. P., Castro, C. E. F., Costa, A. M. M., Fochi, C. G. and da Silva, F. (2009) *Acta Hortic.,* **813**. doi : 10.17660/ ActaHortic.2009.813.48.

Tomkins, J.P., Wood, T.C., Barnes, L.S., Westman, A. and Wing, R.A. (2001) *Theor. Appl. Genet.,* **102** : 489–496.

Traub, H.P. (1949) *Sci.,* **110** : 686-687.

Traub, H.P. (1951) *Plant Life,* **7** : 83-116.

Traub, H.P. (1968) *Plant Life,* **24** : 24-48.

Traub, H.P., Bohn, G.H. and Whitaker, T.W. (1968) *Plant Life,* **24** : 163-164.

Traub, H.P., Buck, W.R. and Lloyd, H.C. (1973) *Plant Life,* **27** : 125-140.

Treder, J., Borkowska, A. and Gabryszewska, E. (2013) *Acta Hortic.,* **1002**. doi: 10.17660/ActaHortic.2013.1002.24.

Tyler, H.H., Warren, S.L., Bilderback, T.E. and Fonteno, W.C. (1993a) *J. Environ. Hort.,* **11** : 131-136.

Tyler, H.H., Warren, S.L., Bilderback, T.E. and Perry, K.B. (1993b) *J. Environ. Hort.,* **11** : 137-141.

Valder, P. (1999) *The garden plants of China.* Portland, OR: Timber Press.

Valpuesta, V., Lange, N.E., Guerrero, C. and Reid, M.S. (1995) *Plant Mol. Biol.,* **28** : 575-582.

Vijayavalli, B. and Mathew, P.M. (1990) *Cytologia*, **55** : 335-338.

Whitwell, T. and Kelly, J. (1989) *J. Environ. Hort.*, **7** : 29-31.

Wooten, J. (1975) *Proc. Fla. State Hort. Soc.*, **88**: 435- 439.

Xiong, Z.T., Chen, S.C., Hong, D.Y. and Luo, Y.B. (1998) *Nordic J. Bot.*, **18** : 183-189.

Yasumoto, A.A., Yahara, (2006) *J. Plant Res.*, **119** : 617–623.

Yoshikawa, K., Nagai, M. and Arihara, S. (1994) *Phytochemistry*, **35** : 1057-1058.

Yue, Q. and Shen, J.S. (1991) *Chinese Vegetables* **5** : 31-33.

Zadoo, S.N., Roy, R.P. and Khoshoo, T.N. (1976) *La Cellule*, **71** : 71.

Zhang Y., Cichewicz R.H. and Nair, M.G. (2004) *Life Sci.*, **75** : 753–763.

Zhang, W., Zhang, H., Gu, Z.P. and Zhang, J.J. (1991) *Acta Hortic.*, **33** : 429-436.

Zhao, W., Wang, T., Chen, Q.Q., Chi, Y.K., Swe, T.M. and Qi, R.D. (2016) *Plant Dis.*, **100**(11) : 2328. 10.1094/PDIS-04-16-0517-PDN.

Zhou, C. (1989) *Plant Sci. Lim.*, **62** : 229-235.

Zhou, P., Hu, J. and Hongzhi, F. (1983) *Acta Hortic. Sinica*, **10** : 273-276.

Zhu, L., Jia, H.Y., Gao, Yike and Zhang, Q. (2015) *Acta Hortic.*, **1104** : 367-374.

13

Spathiphyllum

Simrat Singh

1.0 Introduction

Spathiphyllum is an important tropical foliage ornamental highly valued for its large attractive foliage and long lasting inflorescence consisting of a showy white spathe and spadix. Commonly called as Peace lilies, these plants are popular houseplants and indoor landscaping plants. It produces beautiful white spathes at the end of erect stalks, resembling white flags being waved, that is why they are named as the Peace lily. The word 'Spathiphyllum' originated from two Greek words– 'spathe' referring to the leaf-like bract and 'phyllon' meaning leaf. These are herbaceous perennials, typically grown as a houseplant and have now become a cherished pot plants in the home gardens and conservatories for its creamish white spathes. The plant possesses large leaves, 12–65 cm long and 3–25 cm broad. The flowers are produced in a spadix, surrounded by a 10–30 cm long, white, yellowish, or greenish spathe. *Spathiphyllum* is one of the major import/export ornamental plants (Rout *et al.*, 2006).

2.0 Importance and Uses

2.1.0 Aesthetic Uses

Spathiphyllum is an adaptable and low-maintenance houseplant. These are widely used as pot plants for indoor landscaping. They are symbolism of peace as the white 'spathe' represents a white flag which is recognized worldwide as a 'truce' signal. Due to its low maintenance requirement, the plants are very suitable for vertical landscape and vertical wall under shade. Its ability to remove indoor air pollutants and its high transpiration rate make it one of the most valuable indoor plants. Spathiphyllum is beneficial in promoting restful sleep hence, it is considered as an ideal plant for bedroom. The NASA has listed this plant among the top ten 'Household Air Cleaning Plants' and found that spathiphyllum cleans certain gaseous environmental contaminants, including benzene, formaldehyde, xylene, carbon monoxide, alcohols, acetone and trichloroethylene.

2.2.0 Therapeutic and Other Uses

Spathiphyllum is only mildly toxic to humans and other animals when ingested. It contains calcium oxalate crystals, which can cause skin irritation, a burning sensation in the mouth, difficulty swallowing, and nausea, but it does not contain the toxins found in true lilies, which could cause acute kidney failure in cats and some other animals (Ann King, 1997). Spathiphyllum helps to remove the mildew present in high moisture rooms and eliminates the mold spores in the air. It is beneficial, especially to those people who are sensitive to the mold spores.

3.0 Origin and History

Spathiphyllum, commonly known as 'peace lily' is a member of the Araceae (Arum) family. It is a tropical plant originated in Central and South America with some of its species native to South East Asia. Its natural ecosystems are the shady areas of rainforests usually found in wet habitats. The spathiphyllums were believed

Uses of Spathiphyllum

Indoor Pot Plant | In Flower Bed | In Pot

Border Plant | Private Garden | Public Garden

In Terrarium | Vertical Wall | Around Waterpool

Source: www.greenmylife.in, www.madplantladies.com, www.in.pinterest.com, www.mydomaine.com, www.joyusgarden.com, www.bishanhomegardener.wixsite.com, www.mybageecha.com

to be domesticated by Victorians during late 1800s. Later, it spread to Colombia, Venezuela, South and Central America, extending to Southeast Asia.

4.0 Taxonomy

The genus *Spathiphyllum* belongs to the division-Magnoliophyta; class-Liliopsida (Monocotyledons); from the order- Arales and family-Araceae (Arum family). The most prominent species is *'wallisii'* Regel, commonly known as peace lily. The genus *Spathiphyllum* comprises of about 47 species of monocotyledonous flowering plants from Araceae family. The plant is generally 30-50 cm in height and spread exhibiting slow growth rate. The species are classified based on vegetative characteristics such as arrangement of leaves which is usually simple, with undulate type of leaf margins and pinnate venation often appearing rippled or puckered over the leaf lamina. The length of leaf blade may vary from 30-40 cm with no distinction of leaf colour change noticed during autumn. Variations in floral characteristics in different species also arise in form of flower and fruit colour (usually in shades of white), shape and length of spadix and spathe. Species also exhibit variation in tolerance to light requirement, drought and salts in the substrates.

Botanically, the inflorescence of spathiphyllum consist of a 'spathe' (bract-modified leaf) enclosing a cluster of flowers or 'spadix' (fleshy spike). Some species produce small inflorescences while, others exhibit a quite large inflorescence. The inflorescence is solitary and erect, peduncle shorter than the petiole, spathe is herbaceous and not fully enclosing the spadix, ascending, white, with distinct mid-vein and pinnate laterals. The spadix is sessile or short-stipitate, cylindrical and flowering from base to apex. Spathiphyllum is a terrestrial herb with a short, erect or creeping rhizomatous stem. The plant forms clumps with a cluster of many stems emerging from the ground level. Leaves arise in a loose rosette form and are simple, consisting of a long-petiole which is slightly geniculate at apex enclosed with a long sheath. The nodes often produce roots, leaves and the inflorescence arise from the leaf axils known as 'spathe' which is supported on a stalk known as peduncle. The stem sometimes may spread through rhizomes, creeping along the ground but is often found just beneath the surface. Stems may either grow above ground, underground or partially above the soil. The white spathes produced by spathiphyllum plants are referred to as their flowers.

5.0 Species and Cultivars

Spathiphyllum is a genus of about 41 species (Bunting, 1960) which are most widespread species in Colombia and certain other parts of South America. Different species and varieties may vary in size of the leaves and overall size of the plant. Some species and hybrids have often small and narrow lanceolate leaves while others have much larger and broader leaf blades. Some of the prominent species have been described as follows.

5.1.0 Species

Spathiphyllum wallisii

It is a herbaceous perennial which produces flowers in the typical aroid structure: a densely crowded inflorescence called a spadix is subtended by one large bract called a spathe (occasionally two spathes are produced, with the upper spathe smaller). The spadix is generally cream or ivory when young, and turns green with age; the spathe is generally white or white with green nerves distally from the margin, turning green with age. Leaves are basal, glossy and somewhat deeply veined, ovate and acuminate. The petioles are long and the leaves arch gracefully. The plant produces offsets at the base and in time becomes a dense clump. A number of cultivars, many of hybrid origin, are commercially available, such as the larger hybrids 'Mauna Loa', named after the Hawaiian volcano, and the even larger 'Sensation', both popular as indoor plants. 'Domino' is a finely variegated variety of intermediate size.

S. floribundum

It is a herbaceous perennial plant. The leaves are oval to lanceolate, 12–20 cm long and 5–9 cm broad. The flowers are produced in a spadix, surrounded by a 10–20 cm long, greenish- or yellowish-white spathe.

S. friedrichsthalii

This is a sturdy plant, without an obvious stem, grows over 1 m tall. Its leaf stalks are 30 to 60 cm long. Leaves are 28 to 70 cm long and 7 to 22 cm wide. Its inflorescences are almost as long as the leaves. The inflorescence is formed of a white spathe protecting the spadix, which carries the flowers. The fruit is green, 1 cm long and contains 2 to 11 seeds. This is a highly fragrant plant with a distinctive perfume.

S. cannifolium

This species is an evergreen perennial species also commonly known as a Peace Lily. It produces lance-shaped, shiny leaves and white flowering spathes. Leaves can reach up to 50 cm in length and around 20 cm in width. They are found growing at 200 to 1,000 meters elevation in tropical wet forest zones and produces an inflorescence almost any time of the year.

Spathiphyllum lancaefolium is found in tropical wet rainforests at 200 to 300 meters, while *S. silvicola* is found in both rain forests and tropical wet rain forests.

5.2.0 Cultivars

Various popular varieties of spathiphyllum are mentioned with their prominent characters as follows:

Emerald Star: Thick and dark green foliage and pure white 10 cm long flowers

High Five: Bear dark green foliage with 10-13 cm long pure white flowers

Milkyway: Broad and variegated (white and light green over green base) foliage

Power Petite: Bearing tough, thick, dark and shiny green leaves with usual white flowers

Sensation: Leaves dark green, tough and attractively ribbed with white flowers.

Supreme: Disease resistant and tolerant to adverse conditions, bear deep green and wide leaves.

Species of Sapthiphyllum

Spathiphyllum wallisii

Spathiphyllum floribundum

Spathiphyllum friedrichsthalii

Spathiphyllum cannifolium

Spathiphyllum wendlandii

Spathiphyllum silvicola

Spathiphyllum ortgiesii

Spathiphyllum wendlendii

Spathiphyllum montanum

Source: **www.species.wikimedia.org, www.en.wikipedia.org, www.commons.wikimedia.org**

Sweet Chico: A highly preferred cultivar for small pots, exhibit profuse branching and blooming.

Sweet Dario: Possesses dark green symmetrical foliage even during hottest days of summer.

Sweet Lauretta: A large structured plant with glossy foliage snow-white spathes.

Sweet Pablo: Quick growing cultivar having white flowers and with consistent flowering habit.

The other promising cultivars are Alpha, Chris, Claudia, Cervin, Codys Colour, Daniel, Double Take, Londonii, Lyskamm, Mascha, Petite, Queen Amazonica, Rica, S1007, S1008, S18, S4002, Showpiece, Sierra, Sonya, Tasson, Textura, Vanessa, Vickylynn, *etc.*

Some of the other commercially cultivated varieties of spathiphyllum have been presented in Table 1.

Table 1: Commercial Cultivated Varieties of Spathiphyllum

Cultivars	*Morphological Characteristics*
Annette	Petite with shiny, dark green foliage.
Calypso	Curly rounded leaves.
Ceres	Dense, standard green foliage.
Cody's Colour	Smaller leaves and shorter bloom stalks.
Cupido	Fountain-like growth habit with hooded spathes.
Debbie	Mid-sized with broad leaves.
Del Rio	Very wide, low-growing spathes.
Deneve	Low but wide habit with the occasional green spathe.
Domino	The only variegated *Spathiphyllum* on the market.
Figaro	Dark, glossy foliage with wide spathes.
Flower Power	Well-rounded habit with tall, white and green spathes.
Hi Ho Silver	Light-coloured veins, this is a variation of 'Ceres'.
Jetty	Fast growing with many blooms. More cold and heat tolerant than others.
Knockout	Ribbed, wide-spreading foliage with tall blooms.
Lynise	Standard green foliage with narrow spathes.
Mauna Loa Supreme	Grow 3 feet tall with 7-inch spathes.
Pablo	Olive coloured leaves with ovate spathes.
Patrice	Sparse habit with varying lengths in bloom stalks.
Petite	Stout and compact habit.
Piccolino	Dark green foliage and early blooming time.
Power Petite	Dense growth for small containers.
Sensation-Mini	Just like 'Sensation', only bred for smaller containers.
Sensation	Enormous, dark green, ribbed leaves. Plant can reach 4 feet tall.
Sonya	Extremely compact.

Cultivars	Morphological Characteristics
Starlight	Dark lanceolate leaves and stout bloom stalks.
Sunlight	Compact and robust with green-tipped spathes.
Supreme	In high demand with dark foliage and cupped spathes.
Sweet Pablo	Fast growing with consistent flowering.
Taylor's Green	Deep green and rounded foliage.
Ty's Pride	Large, oblique spathes in the midst of dark foliage.
Vicki-Lynn	More columnar than most *Spathiphyllum*.
Viscount	A fast-growing, year-round bloomer that has deeply ribbed foliage.
Viscount Prima	Large white spathes tower over glossy foliage.
WallisiiWhite stripe	Papery lanceolate foliage with similar spathes.Leaves have matte finish and a white stripe running up the center.

* UF Extension ENH 958, (Chen *et al.*, 2003).

6.0 Genetics and Breeding

6.1.0 Genetics

The chromosome number in spathiphyllum is 2n=30, 60 or more. Chen *et al.* (2006) proposed the strategies for increasing genetic diversity of cultivated peace lily for future breeding programmes. A study was conducted to access the genetic relationships of 63 spathiphyllum cultivars and breeding lines by AFLP Markers. All cultivars were clearly differentiated by their AFLP fingerprints, and the relationships were analyzed using the unweighted pair-group method of arithmetic average cluster analysis (UPGMA). This study provides genetic evidence as to why cultivars from cluster I and III are not readily crossable because the Jaccard's similarity coefficient between the two clusters was only 0.35. Results also indicate that commercial cultivars were genetically close.

6.2.0 Breeding

Breeding for ornamental tropical foliage plants is a time consuming process due to the limited number of viable seed set and prolonged germination time. Several years are required to raise sufficient number of hybrids with desirable traits for evaluation. Once a hybrid with desirable character(s) is selected and tested for its adaptability to varying climatic conditions, it can be maintained or perpetuated vegetatively for obtaining true to type progeny. Micropropagation methods (*in vitro*) may be utilized to mass propagate 'aroid hybrids' under aseptic conditions following standardized protocols to reach commercial production within 2-3 years. The novel variants can be 'introduced' and assessed for their suitability in different agro-climate conditions. Such 'introductions' keep the potted foliage market fresh and profitable. The production of new hybrids with novel traits in foliage and flowering crops like spathiphyllum enrich variation in existing germplasm.

6.2.1 Hybridization

Breeding new varieties is a continuous process, to cater the demand for novelty

Cultivars of Spathiphyllum

White Stripe

Sensation

Calypso

Piccolono

Starlight

Viscount

Sweet Pablo

Sonya

Supreme

Petite

Cupido

Knockout

Cultivars of Spathiphyllum

Taylor's Green

Mauna loa Supreme

Domino

Debbie

Allison

Jetty

Clevelandii

Power Petite

Wallisii

Source: www.plantexchange.com, www.in.pinterest.com, www.balconygardenweb.com, www.davesgarden.com, www.plants.ces.ncsu.edu, www.agristarts.com, www.flowerpower.com, www.candidegardening.com, www.costafarms.com, www.benaranurseries.com, www.joshsfrogs.com, www.chooseyourplant.com

in flower colour, leaf morphology, inflorescence form and other traits in ornamental crops. However, hybridization within the aroid family can be difficult to achieve because of sporadic flower production and specific requirements for successful pollination. Research by the 'Foliage Breeding Program' at the University of Florida [Mid-Florida Research and Education Center (MREC), Apopka] has resulted in standardization of new techniques for plant breeders to manipulate flowering and pollination in ornamental aroids for hybridization purposes. These techniques have made breeding easier and practically possible for spathiphyllum and other aroid genera those were previously difficult to hybridize.

Henny (1981) reported that potential barrier to breeding caused by flower unavailability in spathiphyllum can be overcome through single foliar spray of 250 ppm GA_3 to stimulate flowering throughout the year, and the plants start flowering within 3-5 months of spraying, though with seasonal variation. Differences in the natural flowering cycles, even within the same genus makes breeding difficult.

The spathe unfurls and exposes the spadix several days before flowers are receptive for pollination. The unfurling of the spathe reveals numerous uniformly sized bisexual flowers located along the length of entire spadix. All flowers on a spadix mature simultaneously. The stigmatic surfaces become dry and brown before pollen is dehisced and hence floral emasculation is not required. A spathiphyllum inflorescence will produce pollen over a period of 3-4 days. The pollen of spathiphyllum is light and may be dispersed by air, hence a soft brush should be used for transferring the pollen. It is always best to use fresh pollen. However, if pollen is in short supply, it can be stored in a container in a high humidity environment in a refrigerator. Spathiphyllum pollen can be stored for several days or weeks, if necessary (Henny, 2000). To ensure seed set for spathiphyllum, there is no need for manipulation of environmental parameters such as temperature and relative humidity after pollination and subsequent fertilization. Pollinated flowers develop into mature fruits within 4-6 months. The spadix begins to turn yellowish and becomes soft with maturation of seeds.

6.2.2 Mutation

An alternative tool to conventional system is mutation breeding and polyploidization to create variation. The cultivar 'Domino' was selected as a result of spontaneous mutation from cultivar 'Daniel' (disclosed in U.S. Plant Pat. No. 8,655). It has non-variegated leaves, compared to the population of plants of the cultivar 'Daniel' that consistently had variegated leaves. Asexual propagation by division, of the new cultivar has shown that the unique features of this new spathiphyllum plant were stabilized and were able to get reproduced true to type in successive generations of asexual propagation.

6.2.3 Polyploidy

Eeckhaut *et al.* (2004) obtained polyploid *Spathiphyllum wallisii* var. 'Speedy' plants from somatic embryos induced on anther filaments exposed to mitosis inhibitors. It was observed that the primary embryos yielded less polyploidy plantlets than secondary embryos. The results concluded that colchicine could be efficiently replaced by oryzalin or trifluralin (10 μM), to increase in proportion of polyploids. These mitosis inhibitors were directly added to the induction medium, with an objective to evolve tetraploids with an altered morphology, or triploids with a reduced fertility. A study was conducted to analyze morphological, anatomical and physiological effects of polyploidization in *S. wallisii* (Laere *et al.*, 2011). It was observed that the stomatal density was inversely proportional to increased ploidy level. However, the stomatal size increased at higher polyploidy levels. Tetraploid spathiphyllum plants had more ovate and thicker leaves with more ovate and thicker spathes and a more cylindrical spadix. It was also observed that the biomass production of the tetraploids was reduced compared to their diploid progenitors. It was interesting to find that the tetraploid spathiphyllum plants were induced with drought and stress resistance compared to diploid plants. The findings were confirmed by the analysis of measurements of stomatal resistance, leaf water potential, relative water content and proline content. Similarly, Vanstechelman *et al.* (2009) induce tetraploids in *S. wallisii* (2n = 2x = 30) through *in vitro* application of mitosis inhibitors.

7.0 Growth and Flowering

Flowering in spathiphyllum is influenced by plant size, maturity, environmental conditions and cultural conditions such as light, irrigation and nutrition. Different *Spathiphyllum* species and cultivars may not flower simultaneously. Natural flowering of spathiphyllum involves three parameters: (i) photoperiod, (ii) temperature, and(iii) plant genetic factors (including age/maturity and cultivar genetics).

An experiment was carried by Pavlovie *et al.* (2019) to understand the physiological mechanism of greening of white spathe during a post-anthesis period and fruit development, possibly elucidating source–sink relationship. The greening of spathe was examined in fertilized vs unfertilized and decapitated (with spadix removed) plants. Results revealed that spathe of fertilized plants is viable for a longer period of time and greening process is more intensive in comparison with decapitated or unfertilized plants. Some of the physiological mechanisms involved in greening were chloroplast differentiation and chlorophyll accumulation which resulted in development of photosynthetically active organs. Photosynthesis performance index of spathe in fertilized flowers was significantly higher than the unfertilized and decapitated plants, indicating its better overall photosynthetic performance. Spathe removal during fruit development revealed approximately 30 per cent smaller fruits incomparison with intact ones, implicating that green spathe contribute significantly as assimilate source for developing fruits. The cytokinin analysis showed that t-Z-type cytokinins were the most abundant in the spadix (t-Z and t-ZR) and the greening spathe (t-ZR), while overall cytokinin pool was significantly lower in the spathe of the decapitated plants. Exogenous treatment of spathe in decapitated plants with t-Z enhanced the intensity of greening. Therefore, it was concluded that the spathe greening in peace lily is genetically programmed process that may be mediated by fructification and, in part, by cytokinins originated in the developing fruit.

7.1.0 Temperature

Spathiphyllum plants tend to bloom continuously once they mature, similar in their growth habit to their close relative Anthurium. McConnell *et al.* (2003) reported that growing temperature above 30°C results into narrow leaves and delayed flowering in two cultivars. 'Petite' and 'Tasson' which were exposed to 29°C, 35°C and 41°C for 12 h daily during day time and 21°C in the night up to 12 weeks. The leaves were found narrow when exposed to above 29°C temperature and the growth rate was found considerably decreased, hence there need to evolve cultivars that are heat tolerant, that can exhibit glossy and healthy even at exposure to relatively higher temperature.

7.2.0 Photoperiod

Heemers and Oyaert (2001) in Belgium, planted the spathiphylum *cvs*. 'Alfa', 'Lyskamm' and 'Cervin' in pots on February 14, 2000 at 20/21°C (day/night) and placed the light covers over them on April 27 and May 29. The plants were left as such at the same place till July 3 which was the expected initiation date. They observed that cv. 'Lyskamm' initiated flowering much earlier after light reduction, *i.e.* 47 days in case light covers placed on April 27 while, 71 days in case when light covers were placed on May 29, though in case of cv. 'Cervin' the initiation was found more pronounced by biomass than external environmental factors.

7.3.0 Growth Regulators

Two growth regulators, *viz.* benzyl adenine (BA) and gibberellic acid (GA_3) are commonly used to regulate the growth in spathiphyllum. BA has been quite effective in enhancing the branching and fullness of the plant and is generally used when the plant is young or at linear stage. Sometimes, the young plants are subjected to BA treatments soon after these have been transplanted into larger pots. For spathiphyllum pot plant production, enhanced branching and leafiness are important and desirable traits, since shorter production periods limit the impact of natural branching. BA can be applied as a spray or drench at 250 to 1000 ppm. However, it has been observed that BA treatments can inhibit root development if it is applied before roots are well established. The overall effect of BA application is dependent on its concentration, the cultivar response, stage of growth, application method and season.

GA_3 is used extensively to force early or scheduled flowering in spathiphyllum. It flowers consistently in the spring and sporadically during the rest of the year. Since the market demand for peace lilies with attractive shiny foliage accompanying blooms is always higher, therefore, the growers use GA_3 as a foliar spray to schedule flowering at special occasions or weekly orders. With GA_3, growers can also force early flowering to allow the production of smaller potted plants. A standard treatment consists of a single foliar spray of 150-250 ppm, 8 to 15 weeks prior to sale. The dose and frequency of application depend upon the cultivar, scheduled flowering period and season.

Elbohy (2018) studied the effect of foliar spraying of growth regulators gibberellic acid (GA_3) at 100 and 200 ppm, kinetin (Kin) at 100 and 200 ppm and a microelements mixture at (0, 1, 2 and 3 g/l) on vegetative growth, root system and quality of peace lily plants. It was revealed that spraying peace lily plants with Kin at 100 ppm in addition to the microelements mixture at 2 g/l gave the highest values regarding number of leaves/plant, stem diameter and leaf area, number of roots/ plant in both seasons, fresh weight of roots in the 2 season and dry weight of root, fresh and dry weight of shoots, number of both shoots and flowers/plant in both seasons. The interaction treatment between spraying plants with GA_3 at 200 ppm and 2 g/l microelements mixture gave the tallest plants in both seasons (Elbohy, 2018).

Some cultivars produce good quality flowers after treatment with growth regulators while, others exhibit unusual morphological symptoms such as narrowing of new leaves, stretching of the petioles and distorted flowers. Therefore, it is always advisable to test the efficacy of the growth regulator for phytotoxicity and response, prior to any large scale treatment. Growth regulators should be applied carefully and uniformly over the entire crop to ensure consistent results. It is strictly advised to never apply growth regulators when plants are under stressful conditions.

8.0 Propagation

8.1.0 Division

Spathiphyllum is easily propagated by division of the baby plants that are formed at the base of the plant with their root system intact. The plants can be divided at any time of the year, preferably in February with the onset of spring. Mother plants are gently removed by tapping the pot upside down. Baby plants are gently separated from the parent plant taking care that each of the baby plants has rhizomes and has at least two or three leaves attached. The crown of baby plants should be properly rooted in pots containing mixture of equal parts peat moss/ cocopeat and perlite or coarse sand.

8.2.0 Micropropagation

Several studies have documented successful micropropagation of spathiphyllum through axillary bud culture (Fonnesbech and Fonnesbech, 1979; Wand *et al.*, 1995; Werbrouck *et al.*, 1995; Watad *et al.*, 1997; Malagon *et al.*, 2001; Dewir *et al.*, 2006). Parkinson *et al.* (1996) and Malagon *et al.* (2001) observed that decontamination of the culture media and establishment of explants are major problems in spathiphyllum *in-vitro* propagation. To overcome the above problems, an efficient *in vitro* propagation protocol for *Spathiphyllum floribundum* cv. 'Petite' was evolved after conducting a series experiments with varying concentrations of growth regulators (Bandyopadhyay *et al.*, 2011). The controlled use of cytokinin concentration at the multiplication stage ensured mass multiplication with a potential of one million plantlets within one year.

In order to enhance the formation of quality plantlets, *Spathiphyllum* was cultured *in vitro* on sugar-free liquid medium under higher concentrations of CO_2. The growing environment enriched with CO_2 (3000 µmol/mol 24h/d) at a low photon flux density (PPFD of 45 µmol/m^2/s). This novel technique involving integration of light (photoautotrophic) and CO_2 for the micropropagation of

Propagation and Growing of Spathiphyllum

Division

Separated Plantlets

Planting in Pot

Pot-bound Plant

Separation of Clump

Replanting

A B C D E

Micropropagation

(A) Alternate application of 10 & 2.5 μM BA+MS media producing shoot clusters, elongated shoots (B) Rooted plants. (C) Elongation and rooting in MS medium + BA (0.5 μM) + NAA (1.0 μM) (D) Hardening in Cocopeat+ Perlite(1:1). (E) Acclimatization in Soil:Sand:FYM (1: 1: 1). [*Source*: Bandyopadhyay *et al.* (2011)].

Spathiphyllum in Greenhouse

Spathiphyllum in Shadenet

Source: www.plantindex.com, www.joyusgarden.com, www.dengarden.com, www.in.pinterest.com, www.houseplantsexpert.com, Fun gardening, Chen *et al.* (2003).

Spathiphyllum has been coined as '*Vitron*' culture system (Teixeira da silva *et al.*, 2006). Four types of culture media and 12 combinations of plant growth regulators, *viz.* 6-BA and NAA at different concentrations were tested for *in vitro* culture of shoot tips of spathiphyllum 'Xiangshui'. MS medium containing 2 ppm 6-BA and 0.1 ppm NAA was concluded as the best medium for culturing and sub culturing of spathiphyllum. The in vitro multiplied shoots were rooted on ½ strength MS media supplemented with 0.5 mg/l IBA + 0.5 g/l active carbon was found to be the best medium. A growing medium of rice husk ash and bran of coconut shell (3:1) was adjudged best for establishment of transplanted micro-propagated plantlets (Hua, 2000).

Kaçar *et al.* (2005) sub-cultured the shoots of spathiphyllum cv. 'Sweet Pablo' on MS medium supplemented with two formulations of cytokinins (BA and PBA). The multiplication rate began to increase in first sub-culturing, and reached its maximum at the 4th sub-culturing phase, with a mean count of 8.49 shoots per plant. The cytokinin formulation PBA was found best compared to BA. Dewir *et al.* (2005) recorded greater fresh and dry weights of roots and shoots, number of leaves and roots per plantlet, and shoot length of *S. cannifolium* in mixtures consisting of peat moss: carbonized rice hull: vermiculite (1:1:1) cultured in microponic system in case of micropropagated shoots. The advancements made in micropropagation protocols has led to rapid multiplication in number of true to type plants and has considerably reduced the time period to release the new cultivar in flower market.

9.0 Soil and Climate

9.1.0 Soil

Spathiphyllum is adapted to grow in a variety of soils with varying proportions of sand, silt and clay. The soil should not be too clayey and must be well drained besides retaining adequate moisture for the roots to maintain turgidity in the leaves. They thrive well in acidic to slightly acidic soils with pH in the range 5.8-7.0. The ideal pH for spathiphyllum cultivation is 5.8- 6.5. Potted plants require well drained potting mix with good drainage and water holding capacity. An ideal growing media mixture comprises peat, perlite and bark (1:1:1). Pogroszewska and Hetman (2004) concluded that hotbed soil mixed with highmoor peat and pine bark as the best substrates to grow spathiphyllums in Poland.

9.2.0 Climate

Spathiphyllum are tropical, evergreen plants that thrive on the forest floor, where they receive dappled sunlight through the trees and consistent moisture The optimum range of temperature for successful production of spathiphyllum is 20°C (night) and 32°C during day. However, the cultivars can tolerate 7-35°C lower and higher temperature extremes. It has also been observed that occurrence of occasional frost or even short duration freezing temperatures lead to foliar damage and under adverse climate could led to crop loss. Plants grown at temperatures above 32°C for extended periods can exhibit narrow leaves (strap leaf), loss of chlorophyll from leaf, inhibited root development and reduction in flower quality and flower number produced per plant. The optimal temperature for flowering is 22°C. It has been observed that the growth, development, plant height and leaf size were found enhanced by higher ambient air temperatures up to 23°C. Elevated root-zone temperatures increased the shoot/root ratio and decreased the number of side shoots, however, the dry weight per shoot was found increased despite reduction in the number of side shoots. Therefore, it can be inferred that the plant shape can largely be influenced by the root-zone and air temperatures (Vogelezang, 2003). However, exposure to a lower temperature (12°C) can induce early flowering by 3-4 weeks.

Spathiphyllums can be successfully grown under varying light intensities. Although, the optimum light intensity is a cultivar dependent, the plants are well adapted to grow in range of 9,000–27,000 lux. It has been found that the plants cultivated under lower light intensities develop longer petioles with reduced branching. Leaves exhibit soft and succulent appearance with darker green colour. It has been observed that the plants exposed to higher light intensities, become more compact, exhibit profuse branching and appear lighter in colour. The leaves also exhibit curled appearance with pale or chlorotic spots over the leaf lamina. Fewer flowers are produced in such plants compared to those grown under optimum light intensities. The spathiphyllum cultivars exhibit three types of flowering responses *viz.* (i) flowering after attaining a specific age or maturity, (ii) flowering after attaining a specific age but flower only during spring and summer seasons (iii) flowering more than once in a year.

10.0 Cultivation

10.1.0 Growing Media

The growth of *Spathiphyllum wallisii* plants was evaluated in growing media formulations mixed in volumetric proportions by Kakoei and Salehi, (2013). The growth characteristics such as leaf area, leaf number, shoot and root fresh and dry weights and root length were observed significantly better in media containing only perlite. Mean length of shoots was found higher in growing medium containing perlite: sand (1:3) mixture; and root number was found maximum in the medium comprising perlite: sand (3:1) mixture and only perlite. The highest number of suckers were obtained in leaf-mold: sand (1:1) mixture.

10.2.0 Planting Density

Heemers (1999) in Belgium observed that spathiphyllum cultivated at higher planting density led to deterioration in its vegetative development. The three cultivars *viz.* Cervin, Alfa, Lyskamm were planted in the greenhouse at the plant density of 16, 20 and 24 plants/m^2.

10.3.0 Manuring and Fertilization

It has been observed that the application of slow release dry fertilizers, water soluble fertilizers (WSF) or combinations of both have been found equally effective methods of applying nutrients in potted spathiphyllums. Newly potted plants of spathiphyllum are not fertilized for three months. N-P-K- based WSFs in the ratio 3:1:2, applied as a slow release or liquid feed produces high quality plants in spathiphyllum. Some of the micronutrient deficiencies have also been observed in spathiphyllums. The symptoms of magnesium (Mg) deficiency, appears as golden-yellow margins on lower leaves, which can be prevented with the application of $MgSO_4$. Iron (Fe) and Manganese (Mn) deficiencies exhibit reduced growth rates and chlorotic leaves can occur during winter months when the soil temperature is below 18°C. Sulphur (S) deficiency, exhibits as overall chlorosis of foliage. The symptoms become more prominent with the application of highly refined sulphur fertilizers. Boron (B) deficiency is commonly seen causing longitudinal ribbing of the leaves, often seen on new growth. Yeh *et al.* (2000) attempted to raise spathiphyllums *cv.* 'Sensation' under hydroponic conditions to record the growth response under varying macro and micronutrient levels. The results of deficiency symptoms of various nutrient have been summarized in Table 2 below.

Table 2: A Diagnostic Key to N, P, K, Ca, Mg, B and Fe Deficiencies in Spathiphyllum 'Sensation'

Deficiency Symptoms	*Nutrient Deficiency*
A. Chlorosis or necrosis not expressed:	
A1. Plants grow slowly. No visible deficiency symptoms	P
A2. Yellow specks (at 2 mm) on the interveinal of lower leaves	K
B. Chlorosis is expressed:	
B1. Entire leaf blades are chlorotic. Small leaves prematurely defoliated. Plant growth is strongly affected	N
B2. Only recently fully developed or lower leaves are slightly chlorotic. Some old leaves are interveinal chlorotic	Mg
B3. Interveinal chlorosis on young leaves. Plants has increased root hairs on young lateral roots	Fe
C. The dominant symptom is necrotic leaf	
C1. Marginal necrosis on middle and basal blade of young leaves. Roots are short, brown, thickened and fail to elongate	Ca
C2. Marginal necrosis on distal blade of young leaves, sometimes leaving a blunt end. Crinkled petioles. Main roots elongate, with many short lateral roots	B

Yeh *et al.*, 2000.

10.4.0 Irrigation

Irrigation system should be designed to keep the soil medium evenly moist during all phases of the crop cycle. Spathiphyllums easily tolerate overhead irrigation and do exceptionally with drip irrigation and fertigation. They are susceptible to saturated soil conditions for extended period of time. Various diseases can easily infect in over-watered spathiphyllums causing wilted or collapsed leaves, necrosis along leaf margins and extensive root damage.

11.0 Diseases, Pests and Disorders

Henny *et al.* (2006) gave a detailed account of insect-pests and physiological disorders of spathiphyllum. Simon *et al.* (1994) gave a detailed description of symptoms of diseases infestation in spathiphyllums. The arthropod pests (caterpillars, mealy bugs, scales, and thrips) of spathiphyllum are of relatively minor importance. Mealybug and scale insect infestations often begin by carrying infested plant material into the greenhouse. Moths and thrips usually inhabit the weeds inside the greenhouse or are introduced by carrying other infested plants from outside.

11.1.0 Diseases

Fungal Diseases

11.1.1 Aerial Blight (*Phytophthora* sp.)

The symptoms of aerial blight appear as large (up to 2.5 cm wide) black or brown necrotic spots are formed on leaf margins. Spots appear wet and soggy under moist conditions but may dry if the foliage is kept dry. Keep plant foliage dry and grow plants in a sterilized potting medium on raised benches.

11.1.2 Cylindrocladium Root Rot (*Cylindrocladium spathiphylli*)

The symptoms appear as yellowing of lower leaves accompanied by slight wilting. Elliptical dark brown spots may be seen on leaves and petioles followed by rotting of petioles. Severely infected plants show rotting of roots and are easily removed from the pot without any adhering roots. *C. spathiphylli* is most prevalent during warm summer months, at a temperature range of 20-26 °C, pH 5-9 and at a higher relative humidity (90 percent), The spore of this fungi germinate above 34 °C and below 15 °C temperature is quite unfavourable for this fungus (Wu *et al.*, 2004). Since spores tend to disperse via water, therefore care must be taken that water from the roots of infected plants do not come in contact with healthy plants. Affected plants should be discarded and trifumizole should be drenched in the beds of spathiphyllum plants (Horiuchi *et al.*, 2000). Select pathogen-free plants and always use sterilized potting medium and keep the pots on raised benches to keep the disease under control. Most of the spathiphyllum cultivars are highly susceptible to *Cylindrocladium* root rot with the exception of *S. floribundum*.

11.1.3 Myrothecium Leaf Spot (*Myrothecium roridum*)

Brown to black circular lesions are formed on the leaf margins and center of the leaf. The lesions appear as water-soaked in concentric rings of light and dark tissue or they may be uniformly black in colour. The fungal fruiting bodies can be seen easily in the lower surface which is irregularly shaped, black having a white fringe around the borders. Tissue-cultured plantlets are highly susceptible to this disease and may be lost if timely preventive measures are not taken. Disease severity is highest when the air temperatures are between 15 and 29.4°C.

11.1.4 Pythium Root Rot (*Pythium* sp.)

Roots appear blackened and soggy and the outer cortex is easily removed from the inner core. Foliage of plants affected with *Pythium* root rot appear chlorotic and wilted.

Diseases and Pests

Aerial Blight

***Cylindrocladium* Root Rot**

***Myrothecium* Leaf Spot**

***Pythium* root rot**

Dasheen Mosaic Virus

Improper Watering

Catterpillar

Fungus Gnats

Thrips Infestation

11.1.5 *Phytophthora parasitica*

It causes wilting, root die-back, leaf chlorosis and discolouration. *P. parasitica* colonizes soil and its zoospores swim across the wet leaf surfaces causing black lesions on leaves and finally die back. Plants in the nursery should be kept on raised benches to avoid contact of water and ground surface. Leaf surfaces should be kept dry and healthy plants should be treated with either fosetyl-aluminium or metalaxyl.

Viral Diseases

11.1.6 Dasheen Mosaic Virus (DMV)

A typical mosaic pattern of light green to yellow appear on the new leaves of infected plants. These symptoms can be seen at different times during the year. There are no signs of leaf distortion and the disease does not appear to cause economical losses in spathiphyllum production. Other genera belonging to aroids (dieffenbachia, anthurium and philodendron) are also susceptible to DMV and can be sources of infection for spathiphyllum. Plants exhibiting these symptoms should be discarded and destroyed and population of aphids should be checked since it acts as the vectors of the virus.

11.2.0 Pests

11.2.1 Aphids

Aphids in spathiphyllum appear as pear-shaped, light green to dark brown soft-bodied insects can cause distortion of new growth or in acute infestation causes stunting in the infested plants. Sometimes, the infested plants may go undetected until honeydew or sooty mould is observed. The aphids infesting the roots can be controlled by soil drenching with tobacco decoction. Aphids infecting the aerial parts can be controlled through forced water spraying or Dimethoate (Rogor) at 0.1-0.2 per cent dilution.

11.2.2 Caterpillars

Caterpillars feed on succulent fleshy leaves and usually make holes in the leaves and buds which are yet to unfurl. The caterpillars also leave excreta on plants while crawling from one part to the other.

11.2.3 Fungus Gnats

These are small black flies measuring 3 mm long. The larvae of fungus gnats spin webs similar to spider webs, on the soil surface and feed on small fibrous root hairs, stem tissues and on the lower leaves touching the soil surface. Soil drenching with 0.2 per cent Metasystox has been found effective to control fungus gnats. Also, care should be taken to water the plants adequately ensuring proper drainage and pore space in the potting media.

11.2.4 Mealybugs

Mealybugs appear as white wooly growth in the terminal leaf axils and on the lower surface of leaves. Infested plants become stunted, with appearance of honeydew and sooty moulds. The soil should be sterilized properly before planting to kill eggs of mealybugs. Spray of Metasystox (0.3 per cent) has been found effective to control this pest.

11.2.5 Scales

The scale insects feed on leaves, petioles or stems causing the stunting of plants and loss in vigor. Scale insects are effectively controlled by application of Rogor at 0.2 percent.

11.2.6 Thrips

Thrips suck the juice of the succulent portion of plant making the leaves curl and distort. These also act as a vector of 'tomato spotted wilt virus'. The thrips can be controlled by spraying 0.2 per cent Metasystox.

11.3.0 Physiological Disorders

Some of the physiological problems associated with spathiphyllum production are listed in Table 3.

Table 3: Physiological Problems Associated with Spathiphyllum Production

Symptoms	*Probable Cause and Severity*	*Treatment*
Leaves wilt or collapse and sometimes develop necrosis along leaf margins. Roots develop black tips.	Saturated growing medium. Symptoms generally appear during cold weather when the plant's watering requirement is less	Reduce irrigation frequency to improve soil aerationFormulate a potting medium with a higher porosity
Reduced growth and chlorotic leaves. This disorder frequently occurs	Nutrient deficiency.Symptoms appear during winter months when soil is cold	Foliar sprays of micro-nutrients (Fe, S, B or Mn) are beneficial. Mg salt can be added if chlorosis is only on the lower leaf margins
Leaves appear curled, pale, and chlorotic to necrotic.Leaves have scorched margins and tips, and flowering is sparse.	Excess light or temperature	Reduce the light level to 2500 foot-candles or below and/or the temperature to 90°F (32.2°C) or lower. Remove the plant from direct sunlight.
Lack of flowering in young plants	Plants are immature	Plants grown for 9 to 15 months will usually bloom in spring season (March-April).The younger plants may be induced to flower using a foliar spray of 250 ppm of GA_3
Burnt tips and roots.	Application of excessive fertilizer	Plants kept interior require less fertilizer than greenhouse plants. Therefore, adjust fertilizer regime to ¼ strength as suggested for commercial growers.
Stunted growth, necrotic areas on leaves and death of some roots. Mature leaves are more sensitive than new leaves.	Exposure of plants to cold or extended periods below 7.2°C Amount of damage and tolerance to cold temperature vary with cultivars	Keep daytime temperatures between 20 and 29.4°C with 5°C drop at night temperature

Symptoms	Probable Cause and Severity	Treatment
Severe wilting of plant to soil level.	Lack of adequate water	Thorough watering the plants will ensure turgidity in leaves. Sometimes, the plants may drop in the noon in bright light and temperatures at are above 26.6-29.4°C even though growing media is moist. Frequent drooping will result in brown tips and edges.

12.0 Harvesting and Postharvest Management

The flowers of spathiphyllum can be cut through the peduncle with or without the leaf sheath. The flowers can be cut off the spathe, to avoid it dropping pollen and the spathe will continue to look good for weeks after. The Peace Lily is one of the few plants that will reliably bloom indoors. Spathiphyllum plants are severely injured when exposed to very low temperatures. The ideal temperatures for shipping of spathiphyllum foliage and whole plants are 10-16°C.

13.0 Biotechnology

Eeckhaut *et al.* (2001) tried to produce doubled haploids in different cultivars of *Spathiphyllum wallisii* with different combinations of growth regulators. Thidiazuron (TDZ) at the rate 0.25–1 μM was induced in ovary cultures of spathiphyllum. It was observed that high TDZ concentration induced diploid parthenogenesis in ovules of 'Alfa'. The use of the imidazole fungicides Imazalil (IMA), Prochloraz (PRO) or Triflumizole (TRI) during ovary culture was not critical, though they enhanced the swelling of ovules, making the isolation of the ovules easier. Flow cytometry and AFLP-patterns showed that two doubled haploid genotypes could be obtained from 'Stefanie' that showed a normal phenotype. Therefore, it was concluded that the induction of homozygous spathiphyllum can be done through gynogenesis.

14.0 References

Ann King, F. (1997) http: //www.plantsciences.ucdavis.edu/ce/king/PoisPlant/ Tox-COM.htm

Bandyopadhyay, T.K, Bandyopadhyay, M. Teixeira da Silva, J.A., Paul, S., Dam, A., Ghosh, P.D. (2011) *Floriculture and Ornamental Biotechnology*, **5 (1)**: 57-63.

Bunting, G.S. (1960) *Mem.New York bot. Gard*, **10**: 1-53.

Chen, J., McConnell, D.B., Henny, R. J. and Everitt, K.C. (2003) *UF Extension Newsletter*, ENH958, 1-4.

Chen, J., Henny, R.J. and Devanand, P.S. (2006) *Hort Science*, **41**(4): 968.

Dewir, Y. H., Chakrabarty, D., Hahn, E. J. and Paek, K. Y. (2005) *J. Hort. Sci. Biotechnol.* **80**: 581–587.

Dewir, H.Y., Chakraborty, D., Hahn, E.J., Paek and K.Y. (2006) *In vitro Cell. Dev. Biol. Plant*, **42**: 291-297.

Eeckhaut T, Werbrouck, S., Dendauw, J., Van Bockstaele, E. and Debergh, P. (2001) *Plant Cell, Tissue and Organ Culture* **67**: 181–189, 2001. © 2001 *Kluwer Academic Publishers. Printed in the Netherlands.*

Eeckhaut, T.G.R., Werbrouck, S.P.O., Leus, L.W.H., Van Bockstaele, E.J. and Debergh, P.C. (2004) *Plant Cell, Tissue & Organ Cult.*, **78** (3): 241-246.

Elbohy, N. (2018) *Scientific Journal of Flowers and Ornamental Plants,* **5**: 275-29.

Fonnesbech, M. and Fonnesbech, A. (1979) *Scientia Hort.*, **10**: 21-25.

Heemers, L. (1999) *Verbonsnieuws*, **43** (14): 30-32.

Heemers, L. and Oyaert, E. (2001) *Verbondsniwuws*, **45** (15): 28-29.

Henny, R.J. (1981) *Hort Science*, **16** (4): 554-555.

Henny, R.J., Chase, A.R. and Osborne, L.S. (2006) *CFREC-RH*, pp. 91-32.

Henny, R.J. (2000) *In*: *Breeding Ornamental Plants* (eds Callaway, D.J. and M.B. Callaway, pp. 121-132. Timber Press, Portland, Oregon (USA).

Horiuchi, H., Hagiwara,H., Izutsu, S. and Taguchi, Y. (2000) *Proc. Kansai Pl. Protect. Soc.* **42**, pp. 15-22.

Hua, C.B. (2000) *J. Fujian Coll. Forestry*, **20**(3): 273-275.

Kaçar, Y.A., Mazmanoðlu, M., Mendi, Y.Y., Serçe, S. and Çentiner, S. (2005) *Asian J. Pl. Sci.*, **4** (4): 401-404.

Kakoei F and Salehi, H. (2013) *Journal of Central European Agriculture,* DOI: 10.5513/JCEA01/14.2.1242.

Laere KV, Franc¸S.C.,Vansteenkiste, H., Van Huylenbroeck, J., Steppe, K., Van Labeke, M.C. (2011) *Acta Physiol Plant,* **33**: 1149–1156.

Malagon, R.R., Borodanenko, A., Barrera-Guerra, J.L., Ochoa-Alejo, N. (2001) *Scientia Hort*, **89**: 227-236.

McConnell, D.B., Chen,J.,Henny,R.J., Pennisi, S.V. and Kane, M.C. (2003) *Proc. XXVI intern. hort. congr.*, held at Toronto (Canada) on August 11-17, 2002. *Acta Horticulturae*, **620**: 273-279.

Parkinson, M., Prendergast, M., Sayegh, A.J. and Blakesley, D. (1996) *Plant Growth Regulation*, **20**: 61-66.

Pavlovie, I., Tarkowski, P., Prebeg, T., Lepeduš, H. and Salopek-Sondi, B. (2019) *South African Journal of Botany*, 124: 54–62.

Pogroszewska, E. and Hetman, J. (2004) *Folia Universitatis, AgriculturaeStetinensis, Agricultura*, No. 94, pp. 153-157.

Rout, G.R., Mohapatra, A. and Jain, S.N. (2006) *Biotech Adv.*, **24**: 531-560.

Simon G, Elliott, M. and Mullin, R. (1994) *Florida Plant Disease Control Guide*, Vol. 1. Plant Pathology Dept., University of Florida, Gainesville FL. pp. 362.

Teixeira Da Silva, J.A., Giang, D.D.T. and Tanaka, M. (2006) *Photosynthetica*, **44** (1): 53-61.

Vanstechelman, I., Vansteenkiste, H., Eeckhaut, T., Huylenbroeck, J.V. and Van Labeke, M.C. (2009) *Acta Horticulturae,* **836**: 79-84. DOI: 10.17660/ ActaHortic.2009.836.10

Vogelezang, J.V.M. (2003) *Scientia Horticulturae,* **49**(3-4): 311-322.

Wand, J.F., Cai, L., Jia, C.L. and Dewi, Z. (1995) *Acta Horticulturae,* **404**: 157-160.

Watad, A.A., Raghotama, K.G., Kochba, M., Nissim, A. and Gaba, V. (1997) *Hort Science,* **32**: 307-308

Werbrouck, S.P.O., Van der Jeut, B., Dewitte, W., Prinsen, E., Van Onckelen, H.A., Debergh, P.C. (1995) *Plant Cell Reports,* **14**: 662-665.

Wu, H.X., Wang, J., Wang, J., Zhan, B., QianZhu, C. and Yao, J.D. (2004) *J. South China agric.Univ.,* **25**(4): 30-34.

Yeh, D.M., Lin, L. and Wright, C.J. (2000) *Sci. Hort.,* **86**(3): 223-233.

Eucharis

M. Ganga and M. Visalakshi

1.0 Introduction

Eucharis, popularly known as Amazon lily, Eucharis Lily or Star of Bethlehem is a monocotyledonous tropical bulbous plants. It forms tight, compact rosettes of deep green, glossy leaves and highly valued for its long-lasting, waxy, snow white flowers. It is a graceful bulbous plant that produces fragrant white flowers that closely resembles daffodils. The word 'eucharis' derived from the Greek words '*Eu*'-, meaning 'good, true or original,' and '*Charis*', meaning "loveliness, grace, and favor'. *Eucharis amazonica* Linden ex Planchon is widely known in horticulture, often erroneously as *E. grandiflora* Planchon & Linden (Meerow and Dehgan, 1988). It is a beautiful shade loving plant inhabiting understorey rainforest niche. *Eucharis grandiflora*, is a greenhouse florist crop in temperate regions and occasionally flowers in autumn to spring under ordinary growing conditions. Eucharis flowers are considered as a high-grade material for making wedding bouquets, and hence there is a great demand for this flower in spring and autumn.

2.0 Importance and Uses

2.1.0 Aesthetic Uses

Amazon lily forms tight, compact rosettes of deep green, wide glossy leaves and have tall fragrant spikes with long-lasting, waxy elegant white flowers. The appeal of their leaves makes it a favourite ornamental in gardens in semi-shade (Gilman, 1999). It is a good container plants for a shaded deck or patio. It can be planted among other small plants in a rock garden. They have also been used as a ground cover. It is widely grown in country like Japan for cut flower purpose. The sweet aromatic and beautiful flowers are widely used for decoration and preparation of wedding bouquets.

2.2.0 Therapeutic and Other Uses

Apart from the aesthetic value, bulbs of the plant are known to contain medicinal properties. Lowland native people are known to use bulbs as cataplasm. The bulbs are mashed, heated and applied to sores and tumours. Lewis (1986) has documented the application of mucilage derived from eucharis bulbs by the Jivaro Indians of Peru for treating facial blemishes and acne.

3.0 Origin and History

The genus *Eucharis* is native to Central America and South America, from Guatemala south to Bolivia. *Eucharis* × *grandiflora* belongs to Colombia and Peru. Some of the species have become naturalized in Mexico, the West Indies and scattered in tropical islands. The major center of distribution for *Eucharis* is in the western Amazon basin (inclusive of major tributary systems, *e.g.*, the Napo, Pastaza, and Huallaga) and the adjoining lower slopes of the eastern Andean cordillera. They are mainly found growing in forest area at 50-1800 (rarely to 2,000) m elevation on soils of high fertility. Large-scale deforestation has proven catastrophic to these plants, which are unable to adapt to the higher light intensity of the clearings and

Uses of Eucharis

Indoor Plant

Potted Plant

Border Plant

As Cut Flowers

As Display

Bedding Plant

Floral Arrangements

Source:https://alchetron.com/Eucharis-(plant); https://www.gardenexpress.com.au/product/eucharis-lily/;https://wiki.nurserylive.com/t/the-indoor-specialist-the-beautiful-eucharis-lily/374

soon perish. At least several species are probably extinct and this number will only increase with time as rain forest destruction continues unabated.

4.0 Taxonomy

The genus *Eucharis* belongs to the family Amaryllidaceae and contains 15-20 species mostly inhabiting rainforest understory with a range from Guatemala to Bolivia. The genera under the sub order of 'Amaryllideae' were unified by the presence of the single character of staminal connection. The tiny, white, funnel shaped flowers were marked by the presence of a bristle or slender tooth at either side of the filament. Planchon and Linden (1852) was first to introduce the genus *Eucharis*. *E. candida* Planchon and Linden was the first species to be described and characterized on the basis of its crateriform flowers with a conspicuous staminal cup and a broadly spreading perianth limb. Traub (1971) later combined *Eucharis* with *Urceolina*, a small Andean genus with petiolated leaves and markedly urceolate, brightly coloured flowers. He designated five subgenera: *Urceolina*, *Eucharis*, *Caliphruria*, *Mathieua*, and *Plagiolirion*. Meerow (1989) has divided genus *Eucharis* into two subgenera *viz.*, *Eucharis* and *Heterocharis*. Seventeen species and two natural hybrids with two subgenera are recognized in *Eucharis*. While, subgenus *Heterocharis* represents three relict species and two natural hybrids with many ancestral characters of the genus. Subgenus *Heterocharis* is widely spread from Colombia to Peru, but each of its three species are narrowly distributed. *E. moorei* is limited to Ecuador. It is the only species of *Eucharis* found on both sides of the Andes. *Eucharis amazonica* occurs endemically within a narrow region of the Río Huallaga valley in Peru. *Eucharis sanderi* is found in the Chocó region of Colombia.

The Amazon lilies are marked by their evergreen, shiny dark green highly attractive foliage. Leaves are petiolate with distinctive cuticular striations. Eucharis is marked by its crateriform or campanulate flowers, curved perianth tube, pigmented androecium usually developed into a staminal cup, and unicellular stigmatic papillae. They have a turgid seed covered with a glistening black testa.

5.0 Species and Cultivars

Eucharis amazonica

It is a species of flowering plant native to Peru. It is cultivated as an ornamental flower in many countries and naturalized in Venezuela, Mexico, West Indies, Ascension Island, Sri Lanka, Fiji. An evergreen bulbous perennial, *Eucharis amazonica* grows up to 75 cm tall and 50 cm wide, with long narrow dark leaves and umbels of fragrant white flowers. The stamens are fused into a single cup. Staminal cup measures 11.2-13.8 mm long and 28-30 mm wide, margins are slightly recurved, interior is stained green particularly along filamental traces and shallowly cleft between stamens (ref: Eucharis Copyright Bulb Society). It is not hardy and requires a sheltered spot with a protective winter mulch in colder areas. This species has a narrow distribution from Huallaga valley near Moyobamba and Tarapoto, north eastern Peru. This species is partly sterile (Meerow, 1989) and the plants flower at least twice a year, July-August and December-March, fragrance is sweet and spicy and seems stronger during the day.

Species of Eucharis

Eucharis amazonica

Eucharis×grandiflora

Eucharis formosa

Eucharis astrophiala

Eucharis plicata

Eucharis formosa

Eucharis astrophiala

Eucharis plicata

Source: https://www.pacificbulbsociety.org/pbswiki/index.php/Eucharis

E. grandiflora

It is a natural hybrid between *E. moorei* and *E. sanderi,* a perennial bulbous plant with 2–6 cm bulb diameter and grows to a height of 60 cm. The broad-stalked leaves are 20–55 cm long and 10–20 cm broad. The large showy white flowers are borne in an umbel of three to ten on an erect scape, 40–80 cm long. The flowers resemble narcissus flowers by having a prominent central cup or corona, which is sometimes tinged with green. (Planchon, 1853). The flowers are often confused with *E. amazonica,* usually differing in its size and pigment colouration in the centre of the flower. The seeds contain black head, which is not viable for propagation. The plant has a bulb that can measure up to 5 cm and has two types of roots *viz.,* branched contractile and monofilament (Estrada *et al.,* 2010)

E. astrophiala

It is native to Ecuador. Owing its name "*astrophiala*" due to the 'star-shapped cup' appearance of the flowers. The flowers are small and distinctive from other species (ref: Eucharis.Pacific Bulb Society).

E. formosa

It is a variable species, native mostly to Ecuador, Peru and Colombia, usually found in rich, moist soil in pre-montane and lower montane rainforest 100-1800 m (330-5900 ft) elevation. It flowers from January-March. Floral fragrance is slightly stinking. Flowers are pendent like, staminal cup is less than 15 mm long to apex of teeth or lobes, with a cleft of 3-5 mm between each stamen, style extends 1 cm beyond the anthers (Meerow, 1989). It is an excellent house plant with very little light requirement, however it requires more than average indoor light for blooming. Potted plant requires a well-drained, high organic mix soil. It is advisable to water frequently with occasional fertilizers. The strain in cultivation is self-compatible.

E. plicata

It is a native from Peru to Bolivia, found on the understorey of wet primary forests. The species is named because of the plicate (pleats) folding of the staminal cup. This species of the genus easy to grow and can bloom several times a year (ref: Eucharis.Copyright Pacific Bulb Society).

6.0 Genetics and Breeding

6.1.0 Genetics

The somatic chromosome number of the genus Eucharis has been reported as 2n=2x=46. Only two tetraploid (2n = 92) species of *Eucharis* are so far known, *E. bouchei* and *E. bonplandii,* the northernmost species of *E.* subg. *Eucharis. E. amazonica* is interpreted as triploid-derived (3x) and have somatic chromosome number as 2n = 3x= 68. Of the four species of subg. *Eucharis,* that occur above 2°N latitude, three are polyploids. However, chromosome number is unknown for *E. lehmannii. Eucharis bonplandii* (2n = 92) is rare and limited to the Cordillera Oriental and Cordillera Central of Colombia. The *E. bouchei* complex (2n = 92) is a series of geographically isolated and morphologically distinct population clusters (Meerow, 1987) and is native to Central America, majorly Panama, while *Eucharis caucana* is hexaploid (2n = 138).

All species of *Eucharis* exhibit some degree of self-incompatibility and thus most species are predominantly outcrossing. Moreover, protandry has been observed in most of the Eucharis species (excluding *E. castelnaeana*, in which stigma receptivity coincides with anthesis). The presence of putative interspecific hybrids in nature (*E. candida* x *E. formosa* (Meerow and Dehgan, 1988), *E. plicata* subsp. *plicata* x *E. ulei* (Meerow, 1989) further suggests outcrossing in Eucharis. Of all species in cultivation, only *E. castelnaeana* sets capsules with self-pollen. Autogamy in *E. castelnaeana* is accompanied with a number of other varied characters for subg. *Eucharis*. This species displays mass-flowering phenology; bears the smallest flowers in the genus; a green, often tardily dehiscent capsule; a less turgid seed with a dull, rugose testa; and telocentric chromosomes.

Reproductive biology of *Eucharis* and *Caliphruria* is thus still largely conjectural. Annual flowering patterns, short-term pollen viability and susceptibility to systemic diseases in cultivation impede greenhouse exploration of these questions and the rarity of the plants is an obstacle to field studies. The disappearance of populations of these genera and their pollinating agents, concurrent with rain forest destruction, may be the final impediment to investigation of this aspect of their biology.

7.0 Growth and Flowering

Adams and Urdahl (1973) showed that flowering of eucharis can be controlled by manipulation of bulb temperature and that 20.6°C for 21 days or 29.4°C for 12 days is required to induce flowering. Van Bragt and Sprenkels (1983) and Van suggested the importance of a warm temperature (27°C) followed by a moderate temperature (21°C) for the induction of floral initiation. They also found that morphological floral initiation occurs only after removing plants to 21°C (Van Bragt *et al.*, 1986). For harvesting four cut flowers per plant per year, they proposed repeated treatment with 27°C for 4 weeks followed by 21°C for 8 weeks, although they only succeeded in obtaining a second flowering from the same plant. Kawamura *et al.* (1998) revealed that a temperature drop from 30 to 15°C is more effective to induce flowering than that from 30 to 20°C. Doi *et al.* (2000) clearly indicated that a short chilling (15°C) exposure is a requisite for the initiation of flower buds, *i.e.* apical meristem enlargement, but prolonged moderate (20°C) period has the effect to convert the apical meristem from reproductive to vegetative growth. They developed an energy saving temperature schedule for regulating soil temperature and recommended that soil should be chilled at 15-18°C for 10 days with or without subsequent heating at 25°C or above, once in every 2 months. Plants in soil without temperature regulation produced 1.1 flowering stems per bulb per year. While, the plants grown in the soil chilled without subsequent heating and in the soil chilled followed by heating produced 1.6 and 2.9 flower stems respectively.

8.0 Propagation

Eucharis grandiflora is mainly propagated vegetatively through bulbs and offset bulbs. The daughter bulbs arise in the axils of bulb scales. The production of daughter bulbs is very poor in eucharis and is highly influenced by size of the mother bulb. Van Bragt and Sprenkels (1983) revealed that the bulbs with a diameter of 20 mm produce on an average of 1.5 side bulbs, whereas a bulb of 40 mm can produce around 3.0 side bulbs after one year of planting. The plant growth and development is directly correlated to the size of the bulb, nutrient availability and environmental factors. Bulb size of minimum 35 mm diameter is optimum for the propagation (Table 1, Van Bragt and Sprenkels, 1983).

Table 1: Diameter of Dry Eucharis Bulbs at Time of Planting and Growth after 1 Year

Diameter of planted bulbs (mm)	19	22	26	29	32	35
Diameter after 1 year growth (mm)	34	35	36	40	40	44
Number of side bulbs	1.4	1.5	1.9	2	2.3	2.8
Flowering per cent after 27°C treatment	0	0	0	29	76	90

8.1.0 *In vitro* Propagation

Pierik *et al.* (1983) developed *in vitro* protocol for mass multiplication of *E. grandiflora* with an aim to increase the propagation rate and to produce homogeneous planting material for commercial flower production. The excised bulb scale when cultured *in vitro* successfully induced adventitious bulblets formation on the basal sides. Nine weeks after excision of the scales, the newly formed bulblets were sub-cultured for rooting, bulb growth and leaf formation. The complete plants were successfully transferred to soil for development.

9.0 Soil and Climate

9.1.0 Soil

Eucharis grows well in humus rich soil at a pH range of 7.6 to 7.8. Soil amended with coarse sand is highly suitable for its cultivation. Ideal soil composition: 1 part of the sod land, 1 part of the compost land, 2 parts of the leaf land, 1 part of coarse sand or perlite. It grows well in other types of soils also. However, good drainage is required.

9.2.0 Climate

The Amazon lily grows best in warm tropical climates upto 1800 m above sea level. It can also be grown in cooler climates but may become dormant during winter, whereby it is often best to lift the bulbs and store over the winter period for planting again in Spring. *Eucharis × grandiflora* requires an optimum temperature of 18–23°C and relative humidity of 40-50 per cent. It is suggested to place pot in a tray with wet gravel to maintain medium humidity. Eucharis loves bright sunlight but direct sun light should be avoided. Well illuminated space with bright diffused light for 3-4 hours a day is desirable. It is known to grow well even under artificial light. To mimic tropical humid climate, the greenhouse is kept very humid and warm. The plants are put in flower boxes with clay granules and are cooled with a water system.

Propagation and Growing of Eucharis

Stages of Growth in Eucharis: Bulb Planting to Flowering

Source: **https://typicalgardener.wordpress.com/2010/11/04/eucharist-lily-eucharis-amazonica/**

10.0 Cultivation

10.1.0 Planting

Amazon lily can be planted during February to May. The bulbs are planted at 8 - 10 cm depth. The tip of the bulb neck should be above the soil. The bulbs can also be planted in containers at three to five bulbs in containers of 15 cm width. In the field the bulbs can be planted directly or can be grown in protrays rooted and transplanted at a spacing of 45 – 60 cm between plants.

10.2.0 Irrigation

Amazon lilies need copious irrigation throughout the growing season and can spare irrigation when plant is at resting stage. Water abundantly with lukewarm water every 4 days, avoiding dry out between watering, but the plant should not also stand in water. During the autumn, watering is reduced, but complete drying of the soil should be avoided. In winters, water once a day in 10-12 days, to avoid complete drying out of soil. In early spring, desiccation stress could be beneficial to stimulate flowering.

10.3.0 Manuring and Fertilization

Fertilize the plant at bimonthly intervals with a balanced liquid fertilizer. High content of potassium in the feed is good for plant growth. Evangelista-Lozano *et al.* (2015) evaluated the vegetative development of the *E. grandiflora*, in response to the treatment with of the Steiner solution at different concentrations. It was revealed that treatment of plants with Steiner solutions of –0.054 and –0.072 MPa induced a greater vegetative growth, 29.5 per cent increased stomata developed and synthesized 29 per cent more chlorophyll-b than the control.

10.4.0 Flower Forcing

Flowering largely depends on the quantity of water received and temperature prevailing during bud development stage. Flower formation is induced after drying out to wilting and then re-watering the plants. As a result, 25 per cent of the plants flower, 95 days after withholding water. Stress is thought to be the common cause for the induction of flower formation after a change in temperature or after wilting. If the plants do not flower, withhold irrigation during early autumn or spring to induce drought stress. To stimulate blooming, the growing plants must receive an optimum of 27-29°C for 4 weeks followed by a minimum temperature drop of 8°C. Constant temperatures are imperative for the bloom. These temperatures must be very constant or bulbs may not flower. After 12 weeks at the lower temperatures, the plants should flower.

Amazon lily blooms at least once in a year, in the winter. To get second bloom in a year, dry the soil for a month, then irrigate the plants and new growth developed and put forth flower. Flower formation is induced when plants are exposed to 4 weeks at 27°C followed by 21°C. Maximum of 80 to 100 per cent of the plants flower after 90 to 95 days at 21°C. The bulb has one terminal growing tip. At 21°C it changes from vegetative to generative and a vegetative lateral growing tip is formed. All

Commercial Cultivation of Eucharis under Greenhouses

E. grandiflora inside Poly house

Flowering in *E. grandiflora*

Harvesting of *E. grandiflora* Flowers

Grading of *E. grandiflora* Flowers

E. grandiflora in Full Bloom

Source: https://www.florapodium.com/index.php/en/eucharis-grandiflora-en;https://wiki.nurserylive.com/t/the-indoor-specialist-the-beautiful-eucharis-lily/374

Disease of Eucharis

Symptoms of *Tomato Spotted Wilt Virus* (TSWV) Infection on *Eucharis grandiflora*

Symptoms of *Hippeastrum Mosaic Virus* (HMV) Infection on *Eucharis grandiflora*

Source: https://www.ndrs.org.uk/article.php?id=030013#; Alexandre *et al.* (2011).

flowers of the inflorescence (two helicoid cymes) are completed after 28 days at 21°C. The vegetative lateral growing tip is morphologically complete after 19 days at 21°C. After 56 days at 21°C, it can be induced to flower, if plants are exposed again to 4 weeks at 27°C, followed by 21°C. After 90 to 95 days at 21°C the plants will flower again. Generally, a temperature drop of about 6°C is a prerequisite for floral induction. The low temperature should, however, not be too low, to avoid growth arrest. Temperature-controlled flower formation allows the production of 4 cut flowers per plant per year (Van Bragt *et al.*, 1986).

11.0 Diseases and Pests

11.1.0 Diseases

Amazon lily is occasionally attacked by viruses (although not lethal) and bulb rot. Common viral pathogens include, *Amazon lily mosaic virus, Hippeastrum mosaic virus, Nerine latent virus, Cucumber mosaic virus, Impatiens necrotic spot virus* (INSV) and a member of the *Anulavirus* genus, collectively called as Amazon lily mild mottle virus (Alexandre *et al.*, 2011; Fuji *et al.*, 2013). Interestingly, INSV was previously recognised as the impatiens strain of TSWV (TSWV-I). Later, it was recognised by ICTV (International Committee on Taxonomy) as a distinct species and thus correctly reclassified as INSV (Murphy *et al.*, 1995). Chlorotic rings and spots were observed on *E. grandiflora* plants from a commercial nursery in Atibaia county, São Paulo State, Brazil. Disease symptom incidence was estimated at 60 per cent. The causal agent was diagnosed following transmission tests, electron microscopy (EM), RT-PCR and sequencing (Alexandre *et al.*, 2014).

11.2.0 Pests

The common pests that attack are scale bug insects, caterpillars, bulb mites, thrips and spider mites. These common pests love to feed on the soft luscious leaves of eucharis lily during winter. At low humidity, spider mite problem occurs.

12.0 Harvesting and Postharvest Management

The cut spikes are harvested at 90 days after planting. The flowers are graded according to the flower diameter. The cut flowers have shelf life of about 8 days provided clean fresh water is supplied on a regular basis.

12.1.0 Bulb Storage

The bulbs are stored in peat soil at 20°C. It exhibit dormancy if the temperature reachs freezing.

13.0 References

Adams, D.G., Urdahl, W.A. (1973) *J. Amer. Soc. Hort. Sci.*, **98**: 29-30.

Alexandre, M.A.V., Duarte, L.M.L., Harakava, R., *et al.* (2014) *New Disease Reports*, **30**: 13.

Alexandre, M.A.V., Duarte, L.M.L., Rivas, E.B., *et al.* (2011) *J. Plant Pathol.*, **93**: 643-649.

Doi, M., Kawamura, N., Sugimoto, T., and Imanishi, H. (2000) *Sci. hortic.*, **86**(2): 151-160.

Estrada, G.M., Bermejo, C.M., Leszczynska, B.H. *et al.* (2010) Distribution of root potted *Eucharis grandiflora*. In: *Selected Topics in Tropical Agronomy*, eds. M. Estrada-Botello, R. Osorio-Osorio, pp. 121–125. Villahermosa, Mexico: Colection Jose N Rovirosa.

Eucharis: Copyright Pacific Bulb Society, www.pacificbulbsociety.org.

Evangelista-Lozano, S., Estrada-García, M., Escobar-Arellano *et al.* (2015) *J. Plant Nutr.*, **38**(4): 551-567.

Fuji, S., Kikuchi, M., Ueda, S., *et al.* (2013) *Archives of Virology*, **158**: 201-206.

Gilman, E. F. (1999) Eucharis amazonica ´. Gainesville, FL: Institute of Food and Agricultural Sciences, University of Florida.

Kawamura, N., Doi, M., Imanishi, H. (1998) *J. Jpn. Soc. Hort. Sci.* **67** (Suppl. 1), 260.

Lewis, W.H. (1986) *Discovery*, **17**: 2-6.

Meerow, A. (1987d) *Ann. Missouri Bot. Gard*, **73**(1): 278-296.

Meerow, A. (1989) *Ann. Missouri Bot. Gard.*, **76**(1): 136-220.

Meerow, A. and Dehgan, B. (1988) *Amer. J. Bot.*, **72**: 1857-1870.

Murphy, F.A., Fauquet, C.M., Bishop, D.H.L., *et al.* (1995) Virus taxonomy: classification and nomenclature of viruses. *Sixth Report of International Committee on Taxonomy of Viruses*. (*Archives of Virology* Supplement). NY, USA: Springer Verlag.

Pierik, R.L.M., Sprenkels, P.A. and van Bragt J. (1983) *Acta Horticulturae*, **147**: 179-186.

Planchon, J. É. (1853) *Catalogue des Plantes Exotiques*, **8**: 3.

Planchon, J. É. and Linden, J.J. (1852) *Ann. Cat. Hort.*, **8**: 3.

Traub, H.P. (1971) *Pl. Life*, **27**: 57-59.

van Bragt, J. and Sprenkels, P.A. (1983) *Acta Horticulturae*, **147**: 173-178.

van Bragt, J., Luiten, W., Sprenkels, P.A. and Keijzer, C.J. (1986) *Acta Horticulturae*, **177**: 157-164.

Pancratium

Dhwani A. Patel, Shreya and Reeta Bhatia

1.0 Introduction

Pancratium spp., commonly known as 'Sea Daffodil', is a perennial geophyte belonging to the family Amaryllidaceae. Other vernacular names for this includes 'Sand Daffodil', 'Sand Lily' and 'the Lily of St. Nicholas'. The word pancratium is derived from two Greek words '*Pan*' means 'all' and '*Cratos*' means 'potent', together which means 'omnipotent' in English (Gledhil, 2008). The Latin name of the species '*maritimum*' means 'of the seashore'. It is mostly found growing on the sandy shores of the Mediterranean region, Canary Islands, tropical Africa and tropical Asia. *P. maritimum* L. is widely used as a decorative plant in gardens and also has medicinal and cosmetic utility. The population of this species is rapidly decreasing due to urbanization, tourism development, tourist trampling, alteration, destruction of dune systems and overcollection. Presently it is categorized under endangered species.

2.0 Importance and Uses

2.1.0 Aesthetic Uses

P. maritimum is famous because of its aesthetically beautiful and fragrant white flowers. The flowers are eye catching elements on the sand dunes at beaches. It plays an important role in the preservation of coastal landscape keeping back the sand through the roots and preventing the dune erosion. So the plant is gaining importance as an ornamental decorative plant. It is used as pot plant, bedding plant in garden or even as focal point. The plant can also be used in cactus and succulent garden. Besides, its use as a commercial ornamental, Sea daffodils are receiving much attention from the international scientific community owing to its potential industrial value of its chemical compounds and as a bioindicator plant (Giovino, 2015).

2.2.0 Therapeutic and Other Uses

The bulbs of *Pancratium trianthum* are used for culinary purpose in Western Sahara and its decoction is sometimes taken for cough. The bulbs should be used judiciously in cooking as they contain some poisonous traces. Seeds of *P. maritimum* are used as condiment. It is used in the medicinal industry because of its purgative properties. Till date, around 53 different alkaloids have been isolated from the *Pancratium* genus. Crinine, haemanthamine and lycorine are predominant alkaloids isolated from this genus, are used for treating asthma and lowering blood pressure. Extracts from *Pancratium maritimum* are found to have properties of antioxidant and works as inhibitor of melagenosis. Its extracts can be used in cosmetic and pharmaceutical industries for preparing the drugs which can cure signs of ageing, helps in skin whitening and clarifying the skin. Alkaloids like lycorine, haemanthamine and galanthamine are extracted from *P. maritimum* have antiviral properties and also used against Alzheimer's disease.

Uses of Pancratium

Pancratium in Landscaping

P. maritimum **on Kalathas Beach, Akrotiri peninsula, Greece**

P. maritimum **on Black Sea coast, Georgia**

P. zeylanicum **as Ground Cover**

P. zeylanicum **Grown under Trees**

P. maritimum **as Pot Plant**

3.0 Origin and History

It is believed to be originated from Southern Europe. *P. maritimum* is a native to the Mediterranean region. It is found growing on sandy foreshores on both sides of the Mediterranean region and Black Sea from Portugal, Morocco and the Canary Islands to Turkey, Syria, Israel and Caucasus. It can also be seen on the South Bulgarian, North Turkish and Georgian coasts of Black Sea, where it is threatened with extinction. It is also naturalized in Southern California, Bermuda and the Azores. *P. maritimum* has been known for at least 2,500 years in Greece and cultural aspects indicated that the plant ornamentation has played an important role in the early civilization. The beautiful flowers of the sea daffodil were represented through frescos in Crete and Santorini; they were depicted on archaeological murals, known from excavations in Greece, which displayed detailed naturalistic portraying animals and plants.

4.0 Taxonomy

The genus *Pancratium* was established by Linnaeus (1753) and is represented by 21 species, 7 of them in Mediterranean region, 4 in Africa and 11 in tropical Asia (De Castro *et al.*, 2012, The Plant list 2012). The taxonomical classification of the genus was initially given by Engler and Prantl Pax, (1887), and Pax and Hoffman, (1930) and was later revised by Hutchinson in 1934 and 1959. The genus *Pancratium* belongs to the family Amaryllidaceae and sub-family Amaryllidoideae and the genus *Pancrtium* falls under *Pancratieae* tribe and sub tribe *Eucharidinae.* It is related to *Narcissus* and *Galanthus* flowers.

5.0 Morphology

Pancratium is an evergreen bulbous plant with globose or oval shape bulbs with long neck which are borne solitary or gregarious with length of 5- 10 cm and 5- 9 cm diameter. Leaves are 15-50 cm long, 5- 20 mm wide, tongue shaped, often twisted spirally, narrow and having sub-acute apex, having deep glaucous appearance. Inflorescence is umbel having 10-30 cm long, naked and glaucous scape. Plants generally produce single inflorescence per season having 3- 15 white delicate flowers which are surrounded by a 2- valved spathe, which are 5- 8 cm long, 1- 2 cm wide and acute. Flowers are funnel shaped from above and cylindrical below, tepals lanceolate and spreading. Pedicels are 0.5-1 cm stout, awl shaped slender filaments have length of 6mm. Ovary is inferior having ellipsoid shape. There are numerous ovules per locule. It is a cross pollinated plant. Anthesis of flower takes place during the dusk hours and dehiscence is completed within few minutes of flower opening. Fruit is three sided capsule having round to broad ellipsoidal shape having 2- 3 cm length and 1.5- 2.5 cm diameter, located on the fructiferous pedicels, leaned towards the floor. Seeds have black colour and cone-angular shape, 10- 13 mm length and 6-10 mm diameter. Around 1000 pancratium seeds weight around 50 gm.

6.0 Species and Cultivars

Around 21 species have been identified in this genus which are widely distributed around Canary Islands, North Africa, Mediterranean Sea and Indic area. *P. maritimum* is the most important species in the genus *Pancratium* and has been extensively studied. The other economically important species include *P. canariense*, *P. foetidum* Pomel, *P. illyricum* L., *P. landesii* Traub., *P. longiflorum* Buch.-Ham. ex Roxb., *P. malabathricum*, *P. maximum* Forssk., *P. parvicoronatum* Geerinck, *P. sickenbergeri*, *P. tenuifolium* Hochst., *P. tortuosum* Herb., *P. trianthum* Herb., *P. triflorum* Roxb. and *P. zeylanicum* L.

Pancratium canariense

This is commonly known as 'Canary Sea Daffodil', native to the Canary Islands and mostly seen growing on the edges of rock. In autumn, it produces a tuft of bluish green, straplike leaves and a tall inflorescence with highly fragrant, white flowers from a large underground bulb. In the drier summer months, the bulbs become dormant. It makes a carefree ornamental for the Mediterranean and warm temperate garden.

P. foetidum

This species is also known as 'Stinking Sea Daffodil', found growing in the surrounding areas of Morocco to Tunisia. It has broader leaves and bigger flowers with musky scent and bold seeds as compared to *P. canariense*. It blooms during September in Northern Hemisphere and remain dormant in summer season.

P. illyricum

It is endemic of Sardinia, Corsica and the Tuscan archipelago (Italy) and is considered to be the hardiest pancratium, mostly found growing on the rocky slopes and sparse woodland areas, upto 1300 m above mean sea level. Leaves are large, glaucous which can be seen during winter season. Flowers are fragrant, blooming during months of April to June after which the plant becomes dormant. The scape is up to 45 cm long, and the flowers are clustered in umbels of about 12, white with a short corona and very fragrant.

P. linosae syn. *P. angustifolium*

It is native to the small island of Linosa in the Sicilian Channel. Its habitat is black volcanic cinder. Bulbs are small, round and without the long neck. This species has a rather short growing cycle and is slow to reach flowering size. Leaves last for three months only and the flower stalk quickly elongates together with the emerging leaves.

P. maritimum

P. maritimum is mostly seen along the coast of Mediterranean Sea. It is an evergreen bulbous plant, growing 30- 70 cm tall with white, magnificent and scented flower. Flowers have mild scent like that of lily which is more prominent in the summer season during night time. The specific epithet *'maritimum'* comes from Latin language which means 'from seaside'. It will grow well in ordinary garden soil and needs full sun. Seeds are thick due to spongy outer layer which helps them float on water.

Species of Pancratium

P. martimum

P. foetidum

P. parvum

P. canariense

P. maximum

P. zeylanicum

P. sickenbergeri

P. illyricum

***Source*: https://www.pacificbulbsociety.org/pbswiki/index.php/Pancratium**

P. maximum

It has been used as a medicinal plant in traditional medical systems in southern Yemen. The bulb is large, globose; outer tunics dull brown. Leaves 4–5 to a bulb, contemporary with the flowers, linear, glabrous, not crisped, 20 cm long, 1-2 cm broad. The flowers are borne solitary, resembling a white daffodil, consisting of 6 narrowly lanceolate lobes containing similar to cup, the corona of the slender tube which is 10-15 cm long, the whole flower is borne on a short peduncle and subtended by a large sheathing scarious spathe.

P. sickenbergeri

This species is originated from Egypt and Middle East. This species found on sand dunes in Israel is subjected to high levels of grazing. When there is no above ground biomass, bulbs of *P. sickenbergeri* are consumed for eating (Ward and Saltz, 1994).

P. zeylanicum

P. zeylanicum is tropical species from Asia, India and the Islands of the Indian Ocean. Rain induces flowering and development of the glossy foliage. The flowers are particularly graceful. It can be grown indoors or in warm greenhouse during autumn- winter- spring and kept outside during summer. The small bulbs may be planted in pots with good drainage. The plant undergoes dormancy if water is withheld.

P. bhramarambae

This is native to India, found in the Nallamala hills of Telangana State and Seshachalam hills of Andhra Pradesh. The species is named after the goddess Bhramaramba Devi. Requires moist deciduous habitat for growth. White flowers are seen blooming during April- June.

7.0 Genetics and Breeding

The somatic chromosome number of *P. maritimum* has been reported as 2n= 2x= 22 (Constantinidis and Kamari, 1995). Medrano *et al.* (1999) studied the reproductive behaviour of *P. maritimum* population and reported the self- compatible nature of this species. The plants were capable of setting seeds on spontaneous selfing and recorded the value of the Schoen and Lloyd self-compatibility index as 0.91, indicating that fertilization capacity of self- pollen is almost as high as that of nonself pollen. However, their results were contrasting with the results of a study of populations of *P. maritimum* in Israel, where plants were found to be self-incompatible. Thus the breeding system of this species varies among populations.

8.0 Growth and Flowering

Flowering takes place between June and September and in some species its blooming is extended upto October which is followed by leaf desiccation. New leaves emerge at the end of flowering which lasts until next flowering. It requires warm climate for flower initiation. Flower opening takes place during the night time giving a fruity fragrance. Duration of flowers in bloom is generally 3-15 days

Different Growth Stages of *P. maritimum*

Seeds of *P. maritimum*

Germination of *P. maritimum*

Flowering in *P. maritimum*

Flower Bud of *P. maritimum*

P. maritimum Plant in Bloom

Fruits of *P. maritimum*

Ripe Seeds of *P. maritimum*

Bulbs of *P. maritimum*

Source: El- Hadidy *et al.* (2011) https://www.pacificbulbsociety.org/pbswiki/index.php/Pancratium

depending on the number of flowers in single inflorescence as it is noticed that one flower blooms per day. Medrano *et al.* (1999) studied the flowering phenology, breeding system and temporal variation in reproductive success in the population of *P. maritimum*. They observed that the flowering period in the population of *P. maritimum* was late June to early September, with the flowering peak in mid- July. The plants which flowered in July and August had higher fruit set than plants which flowered in June. However, no differences were observed in mean number of seeds per fruit or mean seed- to- ovule ratio.

9.0 Propagation

Pancratium is propagated by both sexual and asexual methods. Seed propagated plants generally take 3- 4 years for flowering. Use of bulbs, bulbils and division of offsets is done for propagation by asexual means; plants grow faster and flower earlier than seed propagated ones.

9.1.0 Seed Propagation

Seeds can be sown in spring in a warm greenhouse in a light sandy soil. Pre-soaking of seeds in warm water for one hour can facilitate the germination process of sea daffodils. Freshly collected seeds should be sown 8-10 cm deep for good germination. The viability of seeds reduced after a storage period of 8-9 months. It takes 1-3 months for seed germination. Place the seed in a bright position at 15-22°C for germination. Germination per cent is more in dark as compared to altering light condition providing 12 hour light/dark. Seeds require the temperature of 10-30°C for germination. Frost and extreme heat affect germination. GA_3 increased the seed germination and reduced the lag time (Balestri and Cinelli, 2004). The seedlings should be left undisturbed in the pot for their first two years of growth. The seedlings should be occasionally fertilised with liquid feed during the growing season so that they do not become nutrient deficient. When the plants become dormant in late summer, the small bulbs can be transplanted in new pots (placing 2 bulbs in each pot). They are allowed to grow for another one or two years in the greenhouse before planting them out when they are dormant in late summer.

9.2.0 Vegetative Propagation

9.2.1 Bulbs

Pancratium can also be propagated through bulbs. The propagation ratio of bulbs is low, since each bulb produces only 1-3 bulblets in a 2 year period. Bulbs are planted deep in the soil. It can be stored in pots by placing two bulbs in single pot when it is in dormant stage followed by germination in next season. Bulbs should be given enough space for their growth and multiplication.

9.2.3 Division

Another method for multiplication of pancratium is by division of the off shoots that are seen from the vegetative growth. The plants are separated and planted at desired place or in the field.

Propagation of Sea Daffodils (*P. maritimum*)

Propagation through Seeds

Vegetative Propagation through Bulbs

***Source*: https://www.pacificbulbsociety.org/pbswiki/index.php/Pancratium**

9.3.0 Micropropagation

Pancratium have become endangered because of the sea erosion and human activities. To prevent it from extinction and to use it as an ornamental plant, *in-vitro* multiplication is practiced. Freshly collected seeds are sterilized with 70 per cent ethanol and 0.5 per cent hypochlorite and allowed to germinate on 0.7 per cent agar medium without salts under fluorescent lights for 16-hour photoperiod at 25°C temperature. Then the seedlings are sub-cultured on MS medium containing BAP (2 mg/l) and NAA (0.15 mg/l) for direct shooting. Proliferation of bulblets takes after three months in liquid medium containing 3 per cent sucrose solution. *In vitro* rooting also takes place in hormone free MS medium containing 3 per cent sucrose and 0.7 per cent agar. Growing of bulblets directly in liquid culture followed by sub-culture on solid medium can be helpful for rapid multiplication. Multiplication can be easy through indirect organogenesis.

10.0 Soil and Climate

10.1.0 Soil

In their natural habitats, sea daffodil grows on the shorelines and coastal sand dunes. It prefers light and well drained sandy soil for its optimum growth.

Pancratium is considered resistant to drought and spin drifts, its roots help in preventing dunes erosion. It can be grown both in acidic and alkaline sand.

10.2.0 Climate

Pancratium is found on the coastal belt in its natural habitat. It prefers sunny locations and hot and dry weather. It can be cultivated from sea level to 1300 m above the mean sea level and requires an annual rainfall between 300-900 mm. It can withstand the soaring temperatures of Mediterranean summer as well low temperature upto -5°C, adaptive to thrive in the harsh, salt-laden conditions of the sandy beaches. It requires full sunlight for blooming and during the winter season only foliage can be seen. Pancratium requires hot summer in order to induce flowering and is often considered as a shy bloomer in cooler climates. It can also be grown under a deciduous tree so that plants will receive winter sun or in a decorative pot.

11.0 Cultivation

It is not cultivated on commercial scale for flower production but because of its fragrant white blooms it is in demand as decorative pot plant and for the alkaloids present in the plant. Pancratium can also be grown in green house, in pots containing well drained media. It is a treat for eyes during scorching heat of summer season. The plant is easy to grow and does not require extra care for its cultivation. The bulbs are hardy to tolerate high and low temperature. This species has the same growing cycle like *Amaryllis* or *Brunsvigia*.

11.1 Manuring and Fertilization

Organic manure like farm yard manure (FYM) should be added before planting the bulbs of pancratium for getting good vegetative and floral growth. Approximately 3 kg FYM/plant is added along with chemical fertilizers, N:P:K, @ 10:10:8 gm/plant which is given in three splits. The first dose is given at the time of planting, second dose is given after thirty days of planting and last dose is given after cutting the flower spikes. Better production can be obtained by spraying fulvic acid at the rate of 3-5ml/litre of water which can also reduce the chemical fertilization upto 25-50 per cent (Sharaf El-Din *et al.*, 2011).

12.0 Pests

It is usually affected by aphid, mealy bug and snail.

13.0 Biotechnology

13.1.0 Protoplast Culture

Balestri *et al.* (2001) developed the protocol for the isolation of leaf protoplasts from *P. maritimum.* The influence of the donor plant source (field collected versus cultivated plants) and season on protoplast yield was assessed. A combination of cellulase R-10 Onozuka, hemicellulase and pectinase enzymes were found useful for isolation of protoplasts from younger leaves. They observed that field-grown plants harvested in autumn or winter yielded approximately six times more protoplasts per g of fresh tissue than those plants collected in spring. Irrespective of seasons, 3.6×10^6 yield of protoplasts per gram fresh tissue was successfully obtained in the cultivated plants. The viability of these protoplasts was 89.2 per cent and within 2-3 days in liquid culture approximately 88.6 per cent of the protoplasts were able to regenerate cell-walls. First divisions were detected after 5-7 days of culture.

13.2.0 Molecular Markers

El- Hadidy (2012) assessed the taxonomic status of the *Pancratium* species of Egypt, and examined morphological and genetic variation within and between different species. Species identification was confirmed by two phenetic dendrograms generated with 26 quantitative morphological characters and RAPD analysis, while species delimitation was verified by principal component analysis. The retrieved RAPD polymorphic bands showed better resolution of the morphologically and ecologically closely allied Pancratium species (*P. arabicum* and *P. maritimum*), and also confirm the morphological and ecological divergence of *P. tortuosurn* from the other studied species. Genetic diversity has significant value to enable a population's existence in the future while environmental conditions change. Libol and Bilgen (2017) assessed the genetic structure and genetic diversity of four natural *P. maritimum* populations using eight RAPD and four nrSSR primers. All RAPD and nrSSR loci, except SSR-20, were found to be polymorphic. They observed a high proportion of the genetic diversity (81 per cent for nrSSR, 72 per cent for RAPD) mainly due to within population variation and the remaining part was due to variation between populations.

The comparative transcriptional analysis conducted by de Felice *et al.* (2013) revealed differential gene expression between Sand Daffodil tissues. Sequencing of 2,000 differentially screened clones from the SSH libraries resulted in 136 unigenes. Functional annotation and gene ontology analysis of up-regulated EST libraries showed several known biosynthetic genes and novel transcripts that may be involved in signalling, cellular transport or metabolism.

14.0 References

Balestri and Cinelli (2004) *J. Coast. Res.*, **20**(3): 761-770.

Balestri, E., Luccarini, G., and Cinelli, F. (2001) *J. Coast. Res.*, **17**(1): 188-194.

Bogdanova Y.; Pandova B.; Yanev S; Stanilova M. (2009) *Biotechnol. & Biotechnol. Equip.*, **23**: 919-922.

Constantinidis, Th. and Kamari, G. 1995: *In* Kamari, G., Felber, F. and Garbari, F.(eds), *Mediterranean chromosome number reports* – 5, Fl. Medit. **5**: 265-278.

D'Amato, G. F. and De Dominicis, R.I. (1996) *Cytobios*, **85**: 185-194.

De Felice, B., Manfellotto, F., D'Alessandro, R., *et al.* (2013) *Genetica*, **141**(10-12): 443-452.

El- Hadidy, A., El-Ghani, M. A., Amer, W. and Hassan, R. (2011) *Not. Sci. Biol.*, **3**(2): 24-38.

El-Hadidy, A., El-Ghani, M. A., Amer, W., and Hassan, R. (2012) *Acta Biologica Cracoviensia. Series Botanica*, **54**(1).

Elibol, C., and Bilgen, B.B. (2017) *Turk. J. Bot.*, **41**(6): 569-578.

Georgiev V., Ivanov, I. and Pavlon, A. (2020) *Molecules*, **25**.

Giovino, A., Domina, G., Bazan, G., Campisi, P., and Scibetta, S. (2015) *Acta Botanica Gallica*, **162**(4): 289-299.

Gledhill, D. (2008) *The Names of Plants*, Cambridge University Press: Cambridge, U.K.; p. 289

http: //www.llifle.com/Encyclopedia/BULBS/Family/Amaryllidaceae/33079/ Pancratium_maritimum.

https: //www.pacificbulbsociety.org/pbswiki/index.php/Pancratium.

Hutchinson, J. (1934) *Families of Flowering Plants*. Vol. **2**. Monocotyledons, 1st Ed. MacMillan, London.

Hutchinson, J. (1959) *Families of Flowering Plants*, Vol. **2**. Monocotyledons, 2nd Ed. Clarendon Press, Oxford.

Medrano, M., Guitian, P. and Guitian, J. (1999) *Flora*, **194**(1): 13-19.

Paradiso, R., Buonomo, R., De Pascale S. and Cardarelli, M. (2010) *Acta Horticulturae*, **881**: 563-566.

Pax, F. (1887) Amaryllidaceae, Engler A and Prantl K (Eds.). Die NatürlichenPflanzenfamilien, 1st Ed. vol. **2**(5): 97-124. Verlag von Wilhelm Engelmann, Leipzig.

Pax, F. and Hoffmann, K. (1930) Amaryllidaceae, In: Engler A and Prantl K (Eds.). Ibid., 2nd Ed., 15a 391-430. Verlag von Wilhelm Engelmann, Leipzig and Beriln.

Pouris J., Chinmona, C. and Rhizophoulou, S. (2020) *Agronomy*, **10**: 539.

Pouris, J. and Rhizopoulou, S. (2018) *Horticult. Int. J.*, **2**(3): 116-118.

Sadasivaiah, B. and Karuppusamy, S. (2018) *Species* Discovery Publication, **19**: 132-139.

SevimDemir, S. and Çelikel, F.G. (2017) *Food Sci. Technol.*, **5** (13): 1726-1731.

Sharaf, EL-Din M.N., Abd Al-Kafie, O.M., El-Bably, S.Z. and Aboukamar, A.N. (2011) *J. Plant Production*, **2**(8): 1037-1045.

Haemanthus

K. Dutta

1.0 Introduction

Haemanthus is a beautiful flowering bulbous plant in the family *Amaryllidaceae*. It is known as 'blood lily' or 'paintbrush lily'. Most of the species have brush- like flower heads enclosed in four or more bracts which protect the flower heads from damage and desiccation. The two species of the genus *Haemanthus, viz., H. coccineus* and *H. sanguineus* possess brilliant reddish flowers, give rise to the generic name as well as 'blood flower'. It is found wild in coastal dunes, mountain tops, rocky ledges, gravel plains and bogs of tropical and subtropical of the world.

2.0 Importance and Uses

2.1.0 Aesthetic Uses

Haemanthus are best for growing in a rockery or as container plants, in full sun or partial shade. It is used in front of a mixed border, beds or naturalised in lightly-shaded areas. In general, *Haemanthus* are not grown for their displaying in vases because of its toxic exudates. Instead, species such as *H. coccineus, H. unifoliatus, H. nortieri, H. crispus* and *H. deformis* are probably more appreciated for their elegant or unusual leaves (Manning *et al.*, 2002).

2.2.0 Therapeutic and Other Uses

Early Cape records mention that haemanthus leaves were traditionally used in the treatment of ulcers and the extracts of the bulb were taken as a diuretic and also in treatment of asthma (Watt *et al.*, 1962). *Haemanthus coccineus* extracts (HCE) have traditionally been used to treat a variety of diseases, like febrile colds or asthma (Fuchs *et al.*, 2015).

3.0 Morphology

The flowers of haemanthus are very attractive, produce abundant nectar and pollen. Fruits are mostly globose and usually aromatic, found in wide range of colours *viz.*, bright red, pink, orange, white *etc.* Haemanthus generally have one to six leaves, ranging from broad, leathery, prostrate to narrow, succulent or erect, with a variety of surface textures from smooth to extremely hairy or even sticky. A few species such as *H. unifoliatus* and *H. nortieri*, usually produce only a single erect broad leaf. The genus produces relatively large bulbs that act as food and water storage organs, and consist of fleshy leaf bases or tunics.

4.0 Origin, Species and Cultivars

The genus *Haemanthus* was created by Linnaeus in 1753. The name is derived from Greek words *haima* and *anthos*, meaning 'blood flower'. There are some 22 known species, native to South Africa, Botswana, Namibia, Lesotho and Swaziland. About 15 species occur in the winter rainfall region of Namaqualand and the Western Cape, rest is found in the summer rainfall region. Three of the species, *H. albiflos, H. deformis* and *H. pauculifolius* are evergreen. Description of some species is given below:

Uses of Haemanthus

In Garden | In Pot | In Shade under Tree

Along Border | In Open Field | In Rock

Morphology of Haemanthus

Bulbs | Leaves | Flower

Haemanthus albiflos

It is native to the coast and mountains of South Africa. It is an evergreen bulbous perennial species, found in winter and summer rainfall regions, mainly coastal areas, prefers cool and shady places.

H. multiflorus

This is commonly called as blood lily, ball lily, poison root and powderpuff lily. This species is naturalized in Mexico and in Chagos Archipelago. It is also found in Indian peninsula. It is grown as an ornamental plant for its brilliantly coloured flowers, either in containers or in the ground in where the climate is suitable.

H. coccineus

It is a bulbous perennial with short brown stems surmounted by red flowers. It is native to Southern Africa, growing to 35 cm tall. The flowers appear in spring and summer.

H. pubescens

It is found at elevations up to 300 m from Cape Town along the coast to southern Namibia. It has 4- 7 large fleshy spathes. Leaves are broad, flat on the ground and usually pubescent. Bulbs are deep, with fleshy distichous tunics.

H. sanguineus

It is a bulbous geophyte of South African origin. It remains dormant during the summer. In the autumn, a striking crimson inflorescence emerges from the ground, consists of tiny red, crimson and yellow individual flowers. The plant then produces two circular leathery leaves in the autumn and winter.

H. crispus

It is a small plant, almost 15-20 cm in height; bulbs more or less globose with a long slender neck. Leaves are usually two, dry at flowering. Flower cluster are compact with red or pink flowers. Flowering time is March to April. (Snijman,1984).

Some other species are *Haemanthus amarylloides, Haemanthus pauculifolius, Haemanthus humilis* and *Haemanthus tristis.*

5.0 Genetics and Breeding

Using conventional stains, quinacrine fluorescence and C- banding the chromosome analysis of nine species of *Haemanthus* (2n = 16) and four species of *Scadoxus* (2n = 18) were analysed by Vosa and Marchi (1980) and found that the two genera do not possess significant amounts of constitutive heterochromatin. The two genera are closely related and differ in respect of a translocation.

6.0 Growth and Flowering

Environmental factors affecting growth and flowering of haemanthus are light, temperature, water, humidity, nutrients and plant growth regulators. Any environmental stress in growth and flowering in haemanthus makes it more susceptible to diseases or insect pests.

Species of Haemanthus

Haemanthus albiflos

Haemanthus coccineus

Haemanthus pubescens

Haemanthus sanguineus

Haemanthus pauculifolius

Haemanthus multiflorus

Haemanthus crispus

Haemanthus barkerae

Haemanthus canaliculatus

7.0 Propagation

Haemanthus is propagated by seeds, bulbs and separation of offsets depending upon the species.

7.1.0 Seed Propagation

Seeds are sown once the ripe berries have become soft and turn bright orange or red colour from the end of winter to early summer. The hard oval-shaped seeds are removed from the berries, sticky pulpy layer surrounding the seeds washed and sown after drying.

7.2.0 Vegetative Propagation

It is also propagated by offsets. Offsets (adventitious bulblets) with adequate roots are separated from thick clumps and planted in early spring. They should not be forcibly broken away from the mother bulbs but should be separated gently (Duncan, 2000).

7.3.0 Micropropagation

In vitro adventitious shoot formation was initiated from bulb and leaf explants of *Haemanthus deformis, H. humilis,* and *H. albiflos.* The MS medium containing 2.0 mg/l kinetin and 1 mg/l IAA stimulate shoot and root development. *H. humilis* produces shoots from bulb explants. *H. deformis* had 20 per cent shoot and root formation from bulb explants and 30 per cent shoot formation from leaf explants. *H albiflos* produced 50 per cent shoot formation from leaf explants and the callus developed numerous shoots (Rabe and Staden, 1999).

8.0 Soil and Climate

Haemanthus is easy to grow in tropical and subtropical climate. It is not suited to extremely cold places. The ideal temperature for its growth and development is 30°C from early autumn to mid- spring. During the summer dormancy, the bulbs are survived in the soil. It grows well in a pot with proper drainage facilities. A soil mixture for pot cultivation is half loamy soil and half perlite or coarse sand. Keep the soil regularly watered (but not soggy) during the growing season. It grows well both in full sun as well as in filtered sun. The leaves grow well in shady conditions.

9.0 Cultivation

9.1.0 Planting

Haemanthus prefers loamy soil. The seeds are sown directly in soil or seed trays in a medium of equal parts of vermin- compost sand and soil. The seed germinates within 3-4 months. Seedlings are allowed to remain in the trays, then transplanted in the garden or pots in spring during which the first flowers can be expected under ideal conditions.

9.2.0 Irrigation

After sowing of seeds, the trays are placed in a shady position protected from heavy rain and other adverse climatic conditions. Soil should be kept moist by

Propagation of Haemanthus

Micropropagation of Haemanthus

A. Whole plant of *Haemanthus albiflos;* B. Adventitious shoot and root formation on a leaf expl ant of *H. albiflos;* C. Bud and root formation from the basal plate of a single scale bulb explant of *H. deformis;* D. Shoot formation in-between a twin scale bulb explant of *H. humilis;* E. Adventitious shoot from the adaxial surface of a single bulb scale of *H. humilis;* F. *H. albiflos* shoots giving rise to rooted plants; G. Hardened off plantlet of *H. humilis;* H. Acclimatized plantlets of *H. deformis.* (*Source*: Rabe and Staden, 1999).

Propagation of Haemanthus

Seed Germination **Propagation by Offshoot Separation**

watering at least two times a week for germination of seeds. The soil is kept moist when the plants grow and flower, watering is withheld during dormancy.

10.0 Diseases and Pests

These plants have a toxic principle which prevents them from being eaten by moles and mole rats. They suffer very few other problems (Jane, 2003).

11.0 Harvesting

Generally fully opened flowers are harvested. Haemanthus seeds are harvested when they are red, soft and ripe.

12.0 Biotechnology

12.1.0 Phylogenetic Study

DNA sequences from the nuclear ribosomal internal transcribed spacer (ITS) and the plastid trnL- F regions were used for phylogenetic study (Bay- Smidt *et al.*, 2011). Maximum parsimony analyses provide increased support for the sister relationship of *Haemanthus* and *Scadoxus*. Within haemanthus, a well supported clade (89 per cent BS) corresponds to a summer rainfall group (mainly Eastern Cape) with white-pale pink flowers. A second strongly supported clade (100 per cent BS) corresponds to a winter rainfall group (mainly Western Cape) with red-pale pink flowers. *Haemanthus montanus*, which is from the summer rainfall region, is sister to the winter rainfall group.

13.0 References

Bay-Smidt M.G.K., Jäger, A.K., Krydsfeldt, K., Meerow, A.W., Stafford, G.I.,Van Staden, J. and Ronsted N. (2011) *South African Journal of Botany*, **77**: 175–183.

Dressler, S., Schmidt, M. and Zizka, G. (2014) *"Haemanthus". African plants – a Photo Guide.* Frankfurt/Main: Forschungsinstitut Senckenberg.

Du Plessis, N. and Duncan, G. (1989) *Bulbous plants of Southern Africa.* Tafelberg, Cape Town.

Duncan, G.D. (2000) *Grow Bulbs. Kirstenbosch Gardening Series*. National Botanical Institute, Cape Town.

Jane, F. (2003) *Haemanthus coccineus* L., Harold Porter National Botanical Garden.

Manning, J., Goldblatt, P. and Snijman, D. (2002) *The colour encyclopedia of Cape bulbs.* Timber Press, Portland, Cambridge.

Rabe, T. and Staden, J.V. (1999) *Afr. J. Bot.* **65**: 438-440.

RHS Plantfinder (2018) *Haemanthus coccineus".*

Simone, F., Louise, T.H., Werner, S., Clemens, A.J.E., Thomas, A.W., Liliana, S., Egon, K and Robert, F. (2015) *J. Cell Mol. Med.*, **19**: 1021–1032.

Smith, C.A. (1966) *Memoirs of the Botanical Survey of South Africa* No. 35.

Snijman, D.A. (1984) *Journal of South African Botany*, Suppl. Vol. 12.

Watt, J.M. and Breyer-Brandwijk, M.G. (1962) *The medicinal and poisonous plants of southern Africa and eastern Africa.* Livingstone, Edinburgh and London.

WCSP (2011) *World Checklist of Selected Plant Families*, The Board of Trustees of the Royal Botanic Gardens, Kew.

Wikipedia (2015) *The Plant List: A Working List of All Plant Species.*

Habranthus

Babita Singh, Ritu Jain, Varun M.H and A.K. Tiwari

1.0 Introduction

Habranthus, is a genus of summer blooming bulbs, belonging to the family Amaryllidaceae. It is commonly known as rain lily or copper lilies, since they bloom several times during a season, usually following rainfall. The name 'Habranthus' comes from the Greek word *'habros'*, meaning graceful and *'anthos'* meaning 'flower'. They typically appear in late spring and bloom through autumn in response to rain. Bloom is erratic because it is dependent on rain, but they may bloom several times during the summer. These can be used like any other bulb or small flowering perennial in the gardens. It is found in scattered areas, or it occurs in lawns, in cemeteries, and on roadsides. The plants will continue to flower all summer after periods of drought followed by heavy rains. Rain lilies can produce flushes of star-shaped, crocus-like flowers that are bronze, copper, white, pink or yellow depending on the species. Flowers of some new hybrids are in shades of peach, orange and red and some have multicoloured flowers in striped or picotee patterns.

2.0 Importance and Uses

2.1.0 Aesthetic Uses

Habranthus robustus is widely grown as an ornamental. It is one of the most prolific summer flowering rain lilies. It's easy care, broad adaptability and beautiful, starry flowers make it ideal for gardens. The narrow, grass like foliage blends well in rock gardens and can be used as borders in gardens and containers. These small perennials are among the easiest bulbs to grow and if left undisturbed will develop over several seasons into truly impressive clumps. A new trend emerging is to inter-plant them with mondo grass to create feature flowering 'garden lawns' or plant directly into grass areas. In their natural environment they are found growing in pastures amongst grasses. Often they are used to create borders in planter boxes or small 'hedges' to define garden beds. For maximum impact they are grown in masses in broad ribbons or blocks rather than scattering throughout a garden bed. They look good throughout the year, forming a lush green grassy cover (Gary, 2006). Rain lily is very effective when planted as part of perennial beds, flower gardens, rock gardens, ground cover beds and containers. It mixes well with other bulbs as the evergreen foliage of most of the rain lilies is an effective foil for other flowering bulbs. Since, rain lily foliage resembles liriope and mondo grass, so they can substitute for these plants as a border, as an edging for beds, or to line pathways and sidewalks.

2.2.0 Therapeutic and Other Uses

It has no therapeutic uses. However, some species like *H. robustus* contains toxic alkaloids. Though the poison severity is low but if ingested may cause vomiting, diarrhea, convulsions and sometimes death in humans and other animals. All its part (root,leaves, flowers and fruits) are poisonous. The plant contains galanthamine-type alkaloids which can be lethal for humans.

Uses of Habranthus

In Flower Beds | Flowering in Garden Lawn | Bedding Plant

In Pot | Edge Plant | In Garden

Side of Water Pool | In Container | Border Plant along Path

Source: www.pacificbulbsociety.org; www.picclick.com; www.henrikhakansson.com; www.alpinegardensociety.net

3.0 Origin and History

The genus *Habranthus* is a small genus of bulbous, herbaceous perennials with showy flowers that is native to North America and South America. In North America, it is found in Alabama, Florida, Louisiana, and Texas. It is also found in Argentina, Brazil, Chile and Uruguay in South America. There are a small handful of other species with this unusual bicentric distribution between North and South America (Holmes and Wells, 1980). Harold Hume, a botanist and horticulturalist at the University of Florida, suggested that Spanish missionaries introduced this species to North America. Many of the known sites for this plant are around old Spanish missions or Spanish colonial cities. Walter S. Flory, professor of botany at Wake Forest University and an expert on the Amaryllis family, pointed out that the Spanish missions in Texas were founded by Catholics from Mexico (Flagg *et al.*, 1982).

4.0 Taxonomy

Genus *Habranthus* belongs to the tribe Hippeastreae, under the subfamily Amaryllidoideae of family Amaryllidaceae. *Habranthus robustus* is a perennial bulb covered with a dark tunica, contractile roots, deciduous leaves with sheathing basis, linear blades and isolated flowers (Theron and De Hertogh, 2001). The flowers of *Habranthus robustus* are very similar to those of *Zephyranthes* and because of this resemblance it has been earlier included in the genus *Zephyranthes*. Both are commonly called 'rain lilies' because of their tendency to bloom after rain periods (Fellers, 1996). *Habranthus* have zygomorphic flowers which point upward at an angle. The flowers of habranthus are clearly distinguished by filaments of 4 different lengths and always longer than the perianth tube (Fernandez-Alonso and Groenendijk, 2004). Habranthus seeds are more openly obliquely winged.

5.0 Species and Cultivars

5.1.0 Species

The genus *Habranthus* contains more than 40 species, and among them, species of horticultural importance are *H. tubispathus* and *H. robustus*, which are native to Brazil (Read, 2004). Some of the major species are described hereunder.

Habranthus robustus

It is commonly known as the 'pink fairy lily', 'pink rain lily', 'pink zephyr lily', 'pink magic lily', 'pink fawn lily' and 'Colombiaanse mini-amaryllis'. It is among the most common species of *Habranthus* found grown in gardens. *Habranthus robustus* is a small, semi-evergreen herbaceous perennial with flattened leaves. The plant grows to a height of about 30 cm, and the leaves are linear in outline, entire, erect, and glabrous. They grow from bulbs having a brown, papery outer covering, ovate to obovate around 3.5 to 5 cm in diameter. Its leaves emerge in the autumn and remain green throughout the winter months. In early spring, the foliage dies down and the bulb remains dormant until heavy rains occur in the summer months. They bear solitary lavender to pale pink, funnel-shaped flowers with 6 tepals, 5–6 cm long, on 15–30 cm scapes, with a leaf-like bract 1.8–4 cm long at the base. The

Species of Habranthus

Habranthus robustus

Habranthus brachyandrus

***Habranthus tubispathus* (syn. *H. texana*)**

Habranthus gracilifolius

Habranthus ruber

Habranthus sanavirone

***Source*: www.plantdelights.com; www.pacificbulbsociety.org**

fruit is a three-angled capsule containing many thin, wafer-like, black seeds. It is propagated easily by seed or division.

H. brachyandrus

It is favored by those who admire its large pink-to-burgundy flowers, which may reach nearly 10 cm in diameter. This is a small, semi-evergreen herbaceous perennial, growing upto twelve inches tall or more. The flowers are light pink with a deep pink center. It grows well in a sunny, well-drained site. It is propagated by seeds and division. They flower from mid-June through August, usually after a rain and it spreads very slowly compared to *H. robustus*.

H. tubispathus (syn. *H. texana*)

It is native to Texas and parts of Louisiana. It has lovely blossoms of bronze-orange on the outside and yellow on the inside of the petals. The Texas-Louisiana form (*var. texensis*) is referred as copper lilies (Diamond *et al*., 1998). This is a small, deciduous herbaceous perennial with slender, dark green, grass-like leaves. Flowers are zygomorphic, funnel shaped with six tepals, pink, yellow to reddish-brown or nearly white in colour. Fruits are capsular, 3 loculated, sub-globose or oblate.

H. gracilifolius

It is a bulbous herb with leaves that shed months before flowering time and are not produced until the flowers decay. It has wire-like leaves (2-3 mm wide), scape grows upto 20 cm tall and usually has pale purplish pink coloured with one or two scentless flowers that close at night. The filaments are arranged in four heights and are shorter than the style (Herbert 1824). This species occurs in Uruguay and Argentina (Arroyo, 1990), and in Brazil has only been found in Rio Grande do Sul state (BFG, 2015).

H. sanavirone

It is a newly described and illustrated species, found in sandy soils of central Argentina. It has glaucous leaves and recurved light yellow tepals. It resembles *H. robustus* in size, but it does not have bulblets.

H. ruber

It is also known as 'Red Brazilian Rain Lily', is a very rare rain lily species found in open fields and woodland edges in the southern Brazilian province. It bears 1-2 large scarlet orange flowers that open outfacing before drooping slightly the second day. It is the only species producing reddish colour flower among the genus *Habranthus*, which is mainly known for white and pink flowers.

H. cardenasianus

It is a very attractive species of *Habranthus* with white flowers that turns pinkish colour and is very common in dry areas.

5.2.0 Cultivars

Habranthus* × *floryi: It is a hybrid of *H. robustus* × *H. brachyandrus*. The plants are robust and flower beautifully outdoors. The flowers are large, with overlapping

Cultivars of Habranthus

***Habranthus × floryi* 'Amazing Jumbo'** | ***H. robustus* 'Dees Lavender'** | ***Habranthus* 'Candy Cane'**

***H. × floryi* 'Purple Base'** | ***H.× floryi* 'Green Base'** | ***Habranthus* 'Pink Flamingos'**

***H. robustus* 'Albatross'** | ***H. robustus* 'Russell Manning'** | ***H. robustus* 'Pink Splendor'**

Cherry Pink | **Texanus** | **Roseus**

***Source*: www.plantdelights.com ; www.seedhunter.net; www.pacificbulbsociety.org**

pointed tepals, bright pink over white and with prominent longitudinal veining.

***Habranthus × floryi* 'Purple Base'**: It has flat blue-green foliage with 5-8 cm long, giant rosy pink flowers produced on the top of 35-40 cm stalks. The flowers are horizontal-facing and reveal a gorgeous crimson purple deep throat on opening. The entire clumps come into bloom at once, then rest until the next rain, when they flower again.

***Habranthus × floryi* 'Green Base':** This hybrid of *Habranthus robustus* × *Habranthus brachyandrus* grows up to 35-40 cm tall. It has very large outward facing flowers resembling *H. robustus*, each with a nice green throat. As with other *H. × floryi* clones, this is a fast multiplier and flowers from mid-summer until frost.

***Habranthus* 'Candy Cane'** : It is a sterile *H. robustus* hybrid seedling. The flowers are large, dark pink streaked petals toward the tip at the time of emergence, which lighten to pink with maturity. The flowers are dark pink with dark raspberry central eye.

***Habranthus robustus* 'Russell Manning':** It is an old selection of rain lily that has the largest flowers among the Habranthus. The unopened flower buds are over 7 cm long, and as the horizontally-held, cone-shaped flowers open, they reveal a green base in the throat of the flower. The leaves are boat-shaped up to 25 cm long, arches gently upward, eventually forming large clumps.

Habranthus 'Pink Flamingos': It is a floriferous hybrid of *H. robustus* producing pink coloured trumpet flowers. This fast-multiplying clump starts flowering in midsummer and continues even after each rain shower.

6.0 Genetics and Breeding

6.1.0 Genetics

The basic chromosome of *Habaranthus gracifollius, H., tubipathus* and *H. brachyandrus* is 2n=4x=24. The reproductive biology of *Habranthus* has revealed that this is a highly versatile genus (Raina and Khoshoo, 1972), includes both self-compatible (Fernandez *et al.*, 2013) and self-incompatible, as well as apomictic species (Brown, 1951).

6.2.0 Breeding

Studies on breeding system, seed production, rate and type of pollination are crucial for the comprehension of the gene flow mechanisms within and between populations. Fernandez *et al.* (2013) made self- and cross-pollinations on plants collected from a natural population. Fruits set were recorded in both self- and cross-pollinated plants, as well as in open-pollinated plants from the natural population. Based on the fructification ratio, the reproductive success (RS) and the compatibility index, it was found that both self- and cross-pollinated plants set fruits, and fruit setting was lower in self-pollinated plants (59.4 per cent) as compared to cross-pollinated ones (91 per cent), which showed similar setting as open-pollinated plants (93 per cent). The growth of the pollen tubes (more than 100 tubes reaching the ovary) and the quantity of seeds per fruit (53) did not differ among self and cross-pollinated plants. Another evidence has shown the presence of herkogamy in *H. tubispathus*, indicating outbreeding behaviour, despite that this species exhibits partially self-compatibility with the compatibility index value of 0.65 (Paula, 2006).

7.0 Growth and Flowering

Most Habranthus species are native to dry regions. Rain lily bulbs are also well adapted to growing in containers, provided plants are watered regularly to maintain adequate moisture. They flower in summer within 3-4 days after rain. Among the different environmental factors, temperature, light and moisture significantly affect bulb growth and development. These three factors are often manipulated to force bulbs, since they act directly on rooting, flower development, shoot elongation, and bulbing (bulb elongation).

7.1.0 Temperature

In bulbous plants, temperature is the major external factor controlling growth, development and flowering (promoting or delaying it). Temperature also affects bulb dormancy in some genera and the physiological maturity of the bulbs (Bryan, 1989). Unlike other rain lilies, they are not tolerant to cold conditions, and do not survive below 10°C. *Habranthus* spp. become dormant during the cold season and require very little water during this period. In colder regions, they may be grown in pots if kept frost-free in winter. Artificial heating of seed trays or plug trays can prevent seed germination.

7.2.0 Photoperiod

Paula (2006) assessed the responses of *Habranthus robustus* to different light levels including full sun, 30 per cent and 60 per cent shade. It was revealed that Habranthus flowered for a longer period under full sun and 30 per cent shade than 60 per cent shade, but plants had more flowers under 30 per cent shade. With the increase in light level, growth of bulbs increased in both the species. The number of leaves, number of offsets, bulb size, total fresh bulb weight and number of flower buds showed linear responses and increased as light levels increased. However, the plants grown under full sun had shorter roots with no ramifications, plants grown under 30 per cent shade had medium roots with few ramifications, while plants grown under 60 per cent shade had very long roots with several ramifications.

8.0 Propagation

Habranthus can be easily propagated by seeds, bulbs and division of bulbs (including offsets).

8.1.0 Seed Propagation

The shiny black seeds should be collected from the capsules when they turn yellow and start shattering. The seed viability decreases rapidly after harvest, hence it is recommended to sow fresh seeds on a moist potting soil. Seeds can be sown directly and lightly covered after sowing. Fresh seeds germinate in 1-2 weeks' time.

Propagation and Growing

Seed Pod — Seeds — Seedlings

Bulb — Bulb Propagation — Clump

Mass Production in Pot — In Shadenet — Flowering in Natural Habitat

Source: www.pacificbulbsociety.org; www.redbuttegarden.org; www.tylerpaper.com; www.semanticscholar.org, www.plantsaya.blogspot.com; www.pinemountainnursery.com.au

Bulbs can be transplanted after 2 years of growth. Allow plants to go into natural dormancy cycles. Give the plants a light fertilizer during growth. Under ideal conditions, seedlings will flower in eight to twelve months.

8.2.0 Vegetative Propagation

Small clumps resulting from dividing large clumps or plants can be planted in the garden anytime, and often flower and multiply faster than plants established from dormant bulbs. Depending on the species, bulbs range in size from 1.5 cm to more than 5 cm in diameter. The easiest way to propagate rain lily is to divide clumps. Large clumps of bulbs need to be divided every few years. Divisions of rain lily clumps establish and flower more quickly if each division consists of a group of three or more bulbs. The best time to divide clumps of bulbs is when they are dormant (winter for maximum types). Most of the species rapidly produce offsets or bulblets, small bulbs at the base of the parent bulb. Each bulblets can be detached and planted separately. However, individually planted bulbs or bulblets often take longer to develop into flowering clumps. The number of growing seasons required for a bulblet to reach flowering size depends on the kind of bulb and size of the bulblet.

9.0 Soil and Climate

9.1.0 Soil

Rain lily grows best in rich, moist, well-drained soil with a slightly acid to neutral pH. They are sensitive to saline soils.

9.2.0 Climate

They grow well under tropical and semi-tropical regions. Rain lily bulbs adapt to seasonal dry weather by losing leaves until rainfall resumes. It flowers throughout the year, if kept alternatively wet and dry. Rain lily adapts to drought and other harsh conditions by going dormant. Leaves reappear when conditions are again suitable. Most *Habaranthus* species flourish under full sun to partial shade. Their bulbs are well adapted to containers provided plants are watered regularly to maintain adequate moisture. They prefer full sun (6 or more hours of direct sunlight a day) for their growth. If grown under partial shade, direct sunlight of 2-6 hours is required.

10.0 Cultivation

It is one of the easiest bulbs to grow when planted at an appropriate site, needs very little care after planting. It tolerates most soil conditions and does not require fertilizer, irrigation or replanting. Rain lily seems to flower best when clumps are left undisturbed.

10.1.0 Planting

Rain lily bulbs should be planted, when dormant, about 8 cm deep and at least 15 cm apart with the tops of the bulbs at the soil level.

10.2.0 Manuring and Fertilization

Habranthus does not need much fertilizer in open cultivation as it grows naturally in meadows and forests. However, fertilizer application promotes growth and flowering of container grown plants. Paula (2006) studied the influence of different fertigation treatments on flower and bulb development in potted habranthus. The potted rain lily plants were fertilized with 250 ml of water with N at 150 mg/l using a water soluble fertilizer with 20N-4.7P-16.6K (twice a week, once a week, and every other week). Fertigation once or twice a week seemed to be effective for flowering in *H. robustus*, however, when fertigated every other week they had a shorter flowering period. Plants fertigated twice a week had the earliest flowering peak, and 40 per cent more flowers compared to the other two groups. The regression analysis of the factors influencing bulb development such as number of leaves, number of offsets, bulb size, total fresh bulb weight, and number of flower buds showed linear responses, and all factors increased as fertigation frequencies increased. These results demonstrated that optimum fertilizer rates for habranthus bulb propagation by separation was 300 ppm of N using Peter's Excel 20-10-20, since bulbs produced more offsets under this treatment. However, the optimum fertilizer rates for growing bulbs to the market were 75 or 150ppm of 20-10-20, which produced largest bulbs. Additionally, optimal fertilizer rates for flowering would be 150ppm since this treatment produced the greatest amount of flower buds (Paula, 2006).

11.0 Diseases and Pests

The bulbs and other plant parts of *Habranthus* species contain the poisonous alkaloid, lycorine, which makes it resistant to damage from deer, rodents, squirrel and insects. Snails and slugs are the only pests reported to feed on rain lily leaves.

11.1.0 Diseases

Red blotch (*Stagonospora curtisii*): It occurs occasionally in rain lily during cool, damp weather. As the name of this disease implies, it appears as a streak or blotch of dark red on leaves. This disease does not significantly harm rain lily, but primarily affects the plant's appearance, and leaves forming at times other than cool, damp weather will not show red-blotch symptoms.

12.0 Storage of Bulbs

True bulbs are very well protected from drying out by their papery coverings. Keep bulbs in planting medium and store at a temperature range of 5-10°C.

13.0 References

Arroyo, S.C. (1990) *Parodia,* **6**: 11–30

BFG (2015) *Rodriguesia,* **66**: 1085–1113

Brown, W.V. (1951) *Am. J. Bot.,* **38**: 697–701

Bryan, J.E. (1989) *Bulbs,* Volume I, A-H. Timber Press. Portland, Oregon.

Diamond, A.R., C.P. Chapman. and Brummett, J. (1998) *Sida,* **18**: 353-355.

Fellers, J.D. (1996) *Herbetia*, 51: 78-112.

Fernandez, A., Pablo, M., Gabriela, F. and Curvetto, N. (2013) *Acta Hortic.*, 1000: 183-188.

Fernandez-Alonso, J.L. and Groenendijk, J.P. (2004) *Rev. Acad. Colomb. Cienc.*, **28** (107): 177-186.

Flagg, Raymond O., Smith, Gerald L. and Flory, Walter S. (1982) *Flora of North America*, Oxford University Press, pp. 55.

Gary, W.K. (2006) Environmental Horticulture Department, UF/IFAS North Florida Research and Education Center, Quincy, FL.

Herbert, W. (1824) In: *Curtis* W (ed) *Curti's Botanical Magazine*, **51**: 2464.

Holmes, W.C. and Wells, C.J. (1980) *Sida*, **8**: 328-333.

Paula, C.B. (2006) University of Florida.

Raina, S.N. and Khoshoo, T.N. (1972) *Euphytica*, **21**: 317–323

Read, V.M. (2004) *Hippeastrum: the gardener's amaryllis*. Timber Press. Portland, Oregon. In association with the Royal Horticultural Society. Cambridge, England.

Theron, K.I. and De Hertogh, A.A. (2001) *Hort. Rev.*, **25**: 1-70.

Billbergia

Prativa Anand, Vanlalruati and Reeta Bhatia

1.0 Introduction

Billbergia is a genus of flowering plants in the family Bromeliaceae, subfamily Bromelioideae. The genus contains around 60 species and is named after the Swedish botanist, zoologist, and anatomist Gustaf Johan Billberg. They are rosette-forming, evergreen perennials, usually epiphytic in habit, often with brilliantly coloured flowers (Brickell *et al.*, 2008). They grow comfortably attached to trees and shrubs and a few grow among rocks. However, if they fall from their host they will nest in the ground.

2.0 Importance and Uses

2.1.0 Aesthetic Uses

While Billbergia grows naturally attached to a substrate, they can also grow terrestrially (*B. nutans, B. pyramidalis*). Billbergia tolerates a wider range of temperature than many other bromeliads. Its unique adaptations make it especially appropriate for outdoor landscapes in climates that do not regularly experience hard frosts. Several species are grown indoors as decorative plants for their colourful foliage, flowers or bracts (leaflike clusters beneath the flowers). Flaming Torch bromeliads (*Billbergia pyramidalis*) can be grown in containers or in ground beds. They are good competitors with overhead trees and large shrubs. Being epiphytic, these bromeliads can also be grown in the forks of trees or on tree ferns. Mass plantings form a good groundcover in dry, shady positions. Billbergia 'Windii' with pendulous flower spike, is ideal for tall pots, pedestal display or hanging baskets.

2.2.0 Therapeutic and Other Uses

The ground fresh leaves of *B. pyramidalis* contain colloid, tannin, flavonoid compounds, amino acids and are used to disperse swelling and discharging pus (School of Chinese Medicine).

3.0 Origin and History

They are native to forest and scrub, up to an altitude of 1,700 m, in southern Mexico, the West Indies, Central America and South America, with many species endemic to Brazil (Kew World Checklist of Selected Plant Families). *B. nutans* is a native of South America, and is one of several bromeliads found in 19th century gardens. *B iridifolia*, with similar flowers arranged down a hanging stem, was listed at Elizabeth Bay House's famous gardens. Presently, Billbergias are being grown in Brazil, Mexico, Central America, along the Atlantic coast, through Uruguay and as far south as Argentina, and along the Pacific coast through Ecuador and Peru.

Billbergia was first described in 1821 by Carl Peter Thunberg, who honoured the late, Gustaf Johan Billberg, a Swedish botanist and lawyer in the early 19th century. Billbergia were first introduced in the United States in 1897. The species, *B. nutans*, was first described by Hermann Wendland in 1869, using the Latin word for 'drooping' with reference to the hanging nature of its flowers. However,

Uses of Billbergia

Billbergia pyramidalis
Used as Bedding Plant

***B. pyramidalis* Used in**
Garden Display

Billbergia pyramidalis
Used as Ground Cover

***Billbergia nutans* on**
the Front Porch

Billbergia Used for Indoor Landscaping

***B. pyramidalis* as Epiphyte**

***Billbergia nutans* Grown as House Plant**

Billbergia as Ground Cover

they were not popularized until 1940 when Mulford Foster began collecting and hybridizing the genus. Since then they can be found in home collections and landscapes throughout the United States.

4.0 Taxonomy

Billbergia belongs to the family Bromeliaceae and subfamily Bromelioidae. The genus is divided into two subgenera *Helicodea* and *Billbergia*. The subgenera *Billbergia* is most commonly found in cultivation. The difference between the two subgenera can be seen in their flowers. *Helicodea* petals coil back like a spring and while, *Billbergia* petals bend back slightly at the tips. Billbergia, with its two subgenera (*Billbergia* and *Helicodea*), is a predominantly epiphyte bromeliad genus with 65 species distributed from Mexico to Argentina and Uruguay (Gaiotto *et al.*, 2010). Billbergias grow mostly in Brazil, but some species are found in Mexico, Central America, along the Atlantic coast, through Uruguay, and as far south as Argentina, and along the Pacific coast through Ecuador and Peru.

Billbergia come in many shapes, sizes and colours. Their leaves either stand up to create a sort of tube shape or spread outward to make a central cup with their rosette. Billbergia often has fewer leaves than most other genera of bromeliad. Most Billbergia leaves also have spikes around the margins. The foliage are of different colours and even have variegations. The plants grow to a height of 20cm (*Billbergia nutans*) to 90cm (*Billbergia amoena*). Billbergia flowers are magnificent and showy, though not as long lived as many other genera of bromeliad. They are usually found on a flower stalk that emerges from the central tank. The stalk or scape either stands upright or droops in a pendant form. The inflorescence and flowers come in many different colours from reds and pinks to purples and blues. *Billbergia nutans* have fruits classified as berries, presenting exocarp and hypodermis as mechanical layers and many vascular bundles in each carpel (Fagundes and Mariath, 2010).

5.0 Species and Cultivars

5.1.0 Species

The genus *Billbergia* contains around sixty species. Characteristics of *Billbergia* species can often be determined by their scientific name. Penrith Goff of the S.E. Michigan at Bromeliad Society has given a list of scientific names and descriptions:

Leaves- *vittata* means "striped" or "banded", *zebrina* "like a zebra", *maculata* "spotted", *chlorosticta* "dotted with green", *marmorata* "marble-like", *horrida* "rough, bristly". *Rosea* and *decora* describe the colourful bracts.

Scape- *nutans* "nodding," *pyramidalis* "pyramidal" (as opposed to the usual scattered array of flowers), *leptopoda* "thin-stemmed," *filicaulis* "thread-like stems"

Flowers- *viridiflora* "green-flowered"

Some of the *Billbergia* species that would be an excellent addition to any bromeliad collection are:

Species of Billbergia

B. nutans | *B. pyramidalis* | *B. saundersii*

B. nutans | *B. pyramidalis* | *B. saundersii*

B. porteana | *B. zebrina* | *B. amoena*

Source: www.en.wikipedia.org, www.orchidboard.com,www.facebook.com, www.logees.com, www.wikiwand.com, www.glasshouseworks.com

Billbergia nutans

It has green to silvery green leaves that stand upright to make a sort of vase before dropping over. The plant can grow up to 2 feet tall and a foot wide. It has beautiful flowers that hang in pendant form. The inflorescence is typically varying shades of pink. The flowers often drip a nectar when touched, giving this plant the common name "queen's tears." Two varieties are recognized- *Billbergia nutans* var. *nutans* and *Billbergia nutans* var. *schimperiana.*

B. pyramidalis

Commonly known as the flaming torch and foolproof plant, is an excellent landscape plant. It is cold tolerant and requires very little care. The plant clumps readily and grows in shade or full sun. The amount of sunlight affects the colour of the leaves. It produces full, rounded pink and purple flowers during late summer which lasts roughly for a month. Varieties- *Billbergia pyramidalis* var. *bicolor, Billbergia pyramidalis* var. *concolour, Billbergia pyramidalis* var. *croyana, Billbergia pyramidalis* var. *farinosa, Billbergia pyramidalis* var. *lutea, Billbergia pyramidalis* var. *pyramidalis, Billbergia pyramidalis* var. *striata, Billbergia pyramidalis* var. *vernicosa, Billbergia pyramidalis* var. *zonata.*

B. saundersii (Synonym- *Billbergia chlorosticta*)

It blushes reddish pink in full sun areas. In partial sun the leaves are deep green with white spots. The plant grows about a foot tall. It produces spectacular pendant inflorescence with pink bracts and blue and white flowers. It is commonly known as Rainbow plant.

B. zebrina

It is a unique plant. Its name comes from the silvery horizontal banding on the foliage. The plant produces a large pendant flower with pink bracts and has tightly recurved bright green petals.

5.2.0 Cultivars

A total of 1104 cultivars have been registered with the Bromeliads International Society (BSI). Some of the popular Billbergia hybrids includes 'Caramba', 'Catherine Wilson', 'De Nada', 'Ellen', 'Fantasia', 'Fascinator', 'Hallelujah', 'Luna Blanca', 'Muriel Waterman', 'Pink Champagne', 'Poquito Mas', 'Sangrea', 'Tequila Sunset'.

Some of the common Billbergia cultivars are Billbergia 'Windii' (*B.* × *windii*), 'Fantasia' (*B. pyramidalis* × *B. saudersii*), Billbergia 'Casa Blanca'.

Billbergia 'Casa Blanca': Foliage tubular, upto 2 inch broad and about 17 inches tall. The colour takes time to develop, as with many billbergias. When mature, the ivory base has green marbling throughout, with an overall pink wash.

Billbergia 'Hallelujah': It is one of several striking cultivars of Billbergia belonging to *B. vittata* species. It has an upright growth habit and red-green foliage marked with white spots to pollock-like effect. The flowers are an unusual colour combination of aubergine-purple, tea green and fuchsia.

Cultivars of Billbergia

Casablanca

Fantasia

Hallelujah

Albomarginata

Pink Champagne

Sangre

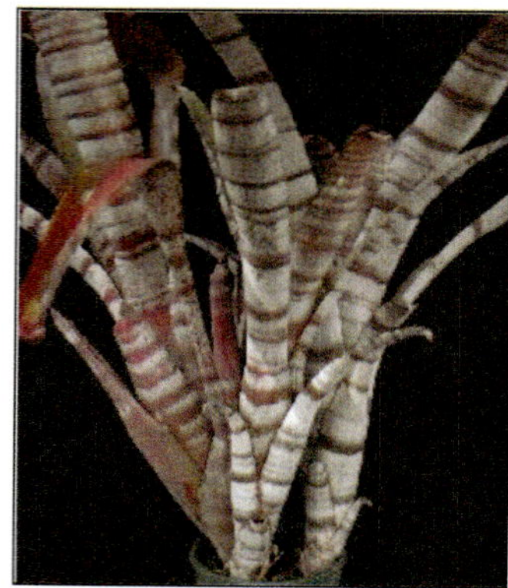

Muriel Waterman

Caramba

Island Fire

Source: www.bromeliad.org.au, www.registry.bsi.org

Billbergia 'Fantasia': It is a cross between *Billbergia pyramidalis* and *Billbergia saundersii*. It has foot-long, celadon-green leaves with cream-coloured splotches, bright red bracts and purple flowers.

Billbergia pyramidalis 'Albo Marginata': It is a white-margined variety with two-foot leaves, reddish bracts, and red and blue flowers. The eight- to ten-inch Billbergia nutans (queen's tears, friendship plant) is especially easy to grow under lights. Its rosy bracts, yellow flowers and easy blooming make it a real favorite.

Billbergia 'Pink Champagne': It has Richly patterned hot pink to red flaring curly leafed 12- 15" tube with rose scape bracts and green to rose sepals and strong blue violet petals - intense colour best developed in bright light at low fertilizer levels.

Billbergia 'Sangre': It is a cross of *B. horrida* v. 'tigrina' x 'Baton Rouge') × *B. stenopetala*. Base colour is crimson with semi-glossy red with white-silver barring and patterning of white spots throughout the plant.

Billbergia 'Muriel Waterman': It is a cross between *B. horrida* cv. tigrina× *B. euphemiae* cv. purpurea. It has rich plum-wine maroon heavily knuckled open rosette with vivid silvery scurfing and banding. The large nodding inflorescence has many dull rose-pink bracts and steel blue flowers.

Billbergia 'Poquito Blanco': It has small 4" white flared few leafed tube with tightly wrapped base and flaring upper section and green to black. Scape, bracts, ovary and sepals red having clear green petals with purple margins.

Billbergia 'Poquito Mas': It is a cross between Fantasia × Poquito Blanco. The plants are compact growing to about 11 inches tall in a tubular form with many stages of colour. The plant develops heavy white blotches with a red veneer. The inflorescence is on a spike of bright red bracts forming yellow flowers with purple petals.

Billbergia 'Caramba': It is a hybrid of 'Mandas Othello' x 'Poquito Blanco', with white flaring graceful small tube with hot pink tones. Plant always appears striated and variegated. The spike are erect with rosy red scape bracts and green flowers tipped purple.

Billbergia 'Grand Pink': Maroon leaves heavily marbled/spotted bright pink with narrow silver cross-banding. Erect spike of scarlet scape bracts, pale green ovaries and sepals tipped blue, plus shiny.

Billbergia 'Island Fire': Dark maroon leaves with hot pink spotting/marbling and silvery grey cross-banding on the reverse. Erect spike of rosy red scape bracts, pale green ridged ovaries, mid pink sepals tipped purple and pale green flowers tipped purple.

6.0 Genetics and Breeding

6.1.0 Genetics

The two *Billbergia* species analyzed by Cotias-de-Oliveira *et al.* (2000) showed 2n=50. *B. morelii* had small chromosomes, 0.23-0.30 µm, while *B. chlorosticta* had large ones, 0.86-1.07 µm. The count of 2n = 50 for *B. chlorosticta* differs from the first report of 2n = 54 by Brown and Gilmartin (1986). Polyploidy seems to play an important role in the chromosome evolution of the family, from the original basic number x = 25 (Brown and Gilmartin, 1989) to the diversification of some genera (Oliveira *et al.*, 2000).

6.2.0 Breeding

Pollination biology and reproduction are essential to understand the evolutionary history of pollination systems. In Billbergia, variation in the pollinator spectrum has been recorded for different habitats among Brazilian forests. Selfing has been found in *Billbergia*, (Martinelli, 1997; Siqueira Filho and Machado, 2001), yet incompatibility systems are also found (Martinelli, 1997). *B. distachia* is completely self-incompatible, and a biotic vector (*i.e.* hermit hummingbirds) is necessary for pollen transference. *Billbergia* species are pollinated mainly by hummingbirds (Pansarin and Pedro, 2016). Matallana *et al.* (2016), after studying 13 species of different genera (*Aechmea, Billbergia, Quesnelia, Tillandsia* and *Vriesea*), found 28 per cent success in crosses, including the ones using parents from different sub-families (*Aechmea* and *Quesnelia* – Bromelioideae) and (*Tillandsia* and *Vriesea* – Tillandsioideae).

Billbergia pyramidalis was first introduced into cultivation in 1815, but it took over 80 more years before the first hybrid, 'Herbaultii,' a cross between *B. amoena* and *B. leopoldii* (now named *B. brasiliensis*), was developed in France in 1897. About the time the first hybrid was made, *Billbergia nutans* was introduced into southern California. Bromeliads in general were not very popular in the United States until the early 1940s when Mulford Foster started collecting and hybridizing the family. His Billbergia hybrids include Gerda, Henry Teuscher, Horena, Muriel Waterman, and Olive Baldwin.

7.0 Propagation

Billbergia are prolific producers of pups or basal offshoots.

7.1.0 Division

Propagation is by dividing off the young 'pups' of the plants in spring or autumn and potting these in a shady spot until they develop a root system. After the plant flowers, it will produce "pups" or young plants then die. The young pups will take over the next generation. The pups should not be removed until visible root structures can be seen at their base or they are at least 1/3 to 1/2 the size of the mother plant. The basal offsets can be separated once they have a sufficient root system and attained almost half the height of the mother plant'. Bromeliad Society/ Houston suggests using a mixture of vermiculite, perlite and Canadian peat for growing potted billbergia. Alternatively, the plants can be mounted on a substrate such as driftwood. Smaller species grow well in a small 4 inch pot. Larger species grow in a pot up to 6 inches.

7.2.0 Micropropagation

Simao *et al.* (2016) determined the best conditions for shoot production from leaf explants of *B. euphemiae*, being the first report on micropropagation of this species.

Propagation and Growing of Billbergia

Root-bound Plant Ready for Division

Cutting into Half

Plants after Dividing

Replanting in Pots

Source: **of-familithebelmontrooster.com/families-ar-plants/bromeliaceae-family/billbergia-nutans**

The best *in vitro* responses were observed in the presence of 1 μM NAA after 60 days of culture, which induced the best production of shoots per explant (16.39), as well as the highest rates of shoot length (1.08 cm), number of leaves per shoot (5.00) and the longest leaf length (0.56 cm). The use of silicon in *Billbergia zebrina* cultivation *in vitro* is an alternative for optimizing micropropagation of this important ornamental plant species (Martins *et al.*, 2018). Calcium silicate promotes better performance of *B. zebrina* plants *in vitro* and mainly during the acclimatization, especially at 1.0 mg/l, therefore is recommended for micropropagation of this species. *In vitro regenerative* system based on the induction, proliferation and development of nodule cultures of *B. zebrina* was described by Vesco Dal *et al.* (2011).

8.0 Soil and Climate

8.1.0 Soil

Billbergia requires humus rich, well drained and slightly acidic soil. If the soil drains poorly, organic material such as compost or peat moss should be mixed in 2 to 3 inches layer of the soil. The potted billbergia prefer a well-drained medium such as a mixture of ½ small bark chips and ½ peat based soil less mix. A mix of vermiculite, perlite, and Canadian peat in equal parts also works well. In any event the mix should be light, well drained, and slightly acidic.

8.2.0 Climate

They usually grow in tropical and sub-tropical regions and can withstand periods of drought. Billbergia can be grown in a wider range of temperature than many other bromeliads. Hence, they are the ideal choice for outdoor landscapes in areas free from hard frosts. Billbergia can thrive in a wide range of temperatures *i.e.* 7-18°C (night) and 21-35°C during the daytime. The night time temperature should be 10-15 degrees lower than the daytime temperature for best growth and colour. The plants of *B. nutans* require warm temperatures of 18-27°C during the summer months and slightly cooler temperatures of 16-24°C throughout the remainder of the year. Extremely low temperatures will cause some damage, but the plant will usually survive. When grown under indoor conditions, they prefer a temperature of 18-24°C. High temperatures, up to 48°C degrees, will cause the plants to bleach out, but the plants will regain their colour when the temperatures moderate.

Billbergia prefers to grow in a spot receiving filtered sunlight. Hence, the ideal location for growing billbergia is any spot where sunlight does not directly fall on the plant. The plants cannot stand sun rays falling on its leaves directly. So the plant rejoices in moderate to adequate light but not direct sun (50-70 per cent shade which is about 3000 to 5000 foot candles). In apartments, heat sources such as radiators should be placed far from the plants because moisture is what this tropical plant needs most. East or West-facing windows are thus usually the best for billbergia. Billbergia also thrive well in areas with plenty of air circulation around them. Spraying/Misting the plant regularly and placing the pot atop a bed of wet gravel helps to recreate the plant's natural habitat. Having fresh air helps the plants avoid fungal infections and pests.

9.0 Cultivation

Billbergia are easy to grow houseplants. They can be grown in single pots or large hanging baskets with three plants average per basket grouped by commonality. Alternatively, they can also be attached to trees to resemble their natural habitat. Billbergia usually flowers during end of winter or summer. It is not a long-lived inflorescence but provides a welcome splash of colour in a shaded spot.

9.1.0 Planting

When grown in containers, they can be planted in 4 inch pots; while the larger varieties are best potted in 5 inch pots. Plants that are allowed to clump can be potted in 6 inch hanging baskets or 6 inch pots. They can be even grown mounted on wood or cork bark. It is advisable to start with a pup and not a mature plant when billbergia are mounting a plant. Well drained light potting mixture such as ½ part of small bark chips and ½ part of peat is preferred. Alternatively, a mix of vermiculite, perlite, and peat in equal parts can also be used. The potting mixture contains ingredients such as tree and plant fiber, peat moss and bark chips. While planting in the ground, always plant offsets in well-drained soil.

9.2.0 Manuring and Fertilization

A small amount of slow release fertilizer can be added to the potting mix when potting pups. Over fertilization can lead to oversized, lanky, green plants that are more susceptible to insect and disease damage. Well established plants growing in bright light with good air circulation can be fertilized monthly with dilute (¼ strength) general purpose fertilizer during warm weather.

9.3.0 Irrigation

Billbergia do not require much water, but they prefer high humidity. When grown outdoors, billbergia should be watered only when the top of the soil feels dry. Because of the shallow nature of its roots, Billbergia gathers most of its moisture from the air and directly from the leaves and flowers. During the summer months, the leaves, flowers, and roots should be watered daily to keep them moist. During the fall, winter, and cooler months, the plant should be misted every few days and watered enough to prevent the soil to completely dry. Over watering can be detrimental to the plants and may result in yellowing or browning of inner leaves. Distilled water or rain water should be preferred to avoid mineral build up on the leaves. The tanks should also be regularly flushed and rinsed to avoid stagnate water. Stagnate water can result in pests and diseases.

9.4.0 Pruning and Maintenance

The yellow or dying leaves and any other plant debris should be removed to encourage better growth of the plants. While pruning, make clean incisions and never cut through yellowed tissue as this may cause further damage in the likes of diseases or bacterial infections. The spent bracts (flower stalks) should be removed once they begin to wilt.

9.5.0 Repotting

It is done to split new shoots from the mother plant. As Billbergia does not tolerate water stagnation, the pots chosen for repotting should have hole at the bottom. Drainage can be increased with a layer of gravel or clay pebbles along the bottom of the pot, to make water flow through more easily. When planted indoors the plant needs soil mix because it is the only source for the nutrients it needs. Low but constant moisture levels must be maintained, which can be ensured by frequent spraying of leaves. The pot should be rested on a bed of gravel, rocks or clay marbles doused in water.

10.0 Diseases and Pests

If proper air circulation is maintained and plants are not allowed to become crowded, Billbergia are seldom infected by any pests or disease. The incidence of powdery mildew is reported sometimes.

Insect problems are usually minimal, but if the plants are crowded, and there is little air circulation scale infestation can develop. The common pests are red spider mite, scale insects and aphids. The two most common types of scale found in Billbergia are the brown soft bodied scale and the black tea scale. A treatment or two of insecticidal soap will usually bring the brown scale under control. However tea scale is much difficult to control. The infected plants can be destroyed or appropriate insecticide should be used.

11.0 Biotechnology

Plants in the Bromeliaceae family provide excellent models for the study of genetic diversity, population genetics as well as the divergence between populations and species (Sheu *et al.*, 2017). Polymorphism and usefulness of amplified fragment length polymorphism (AFLP) markers in assessing the genetic diversity in populations of *Billbergia zebrina* were investigated in nodular cultures and adult plants of the species (Vesco *et al.*, 2011). AFLP markers revealed moderate-to-high genetic diversity based on the estimations of Nei's gene diversity (mean $H_e = 0.28$), Shannon index of diversity (mean $H_S = 0.48$) and the number of polymorphic fragments (mean of 56.17 polymorphic fragments over six primer pairs). The results obtained reveal that AFLP markers comprise a powerful tool in order to assess the levels of genetic diversity in natural populations of this endangered species. Integrating AFLP markers with *in vitro* propagation techniques is understood as an adequate strategy for conservation of this species.

12.0 References

"Billbergia" Bromeliad Society/Houston.

<http: //bromeliadsocietyhouston.org/genera-intro/billbergia> Brickell, C., Cathey and Marc, H. (2008) *RHS A-Z encyclopedia of garden plants*. United Kingdom: Dorling Kindersley, pp.1136.

Brown, G.K. and Gilmartin, A.J. (1986) *Selbyana,* **9**: 88-93.

Brown, G.K. and Gilmartin, A.J. (1989) *Am. J. Bot.,* **76**: 657-65.

Cotias-de-Oliveira, A.L.P., Assis J.G.A., Bellintani, M.C., Andrade, J.C.S. and Guedes, M.L.S. (2008) *Genet. Mol. Biol.,* **23**(1): 173-77.

Fagundes, N.F. and Mariath J.E.A. (2010) *Acta Bot. Bras.,* **24**(3): 765-79.

Gaiotto D.F., Tardivo R.C. and Cervi A.C. (2010) *Fontqueria,* **56**: 81-100.

Martinelli, G. (1997) In: Lima, H.C. and Guedes-Bruni, R.R. (Eds). *Biologia Reprodutiva de Bromeliaceae na Reserva Ecológica de Macaé de Cima. pp.* 213-250.

Martins, A.D., Martins, J.P.R., Batista, L.A. and Dias, G.M.G. *et al.* (2018) *An. Acad. Bras. Ciênc.,* **90**(4): 3449-62.

Matallana, G., Oliveira, P.E., Silva, P.R.R. and Wendt, T. (2016) *Bot. J. Linn. Soc.,* **181**: 521–531.

Pansarin, E.R. and Pedro, S.R.M. (2016) *Plant Biol.,* **18**(6): 920-27.

Simao, M.J., Faria, D.V., Werner, E.T., Soares, T.C.B. and Gontijo, A.B.P.L (2016) *Acta Scien.,* **38**(2): 207-13.

Siqueira Filho, J.A. and Machado, I.C.S. (2001) *Acta Bot. Bras.*, **15**(3): 427-43.

Sheu Y., Cunha-Machado A.S., Gontijo A.B.P.L., Favoreto, F.C., Soares, T.B.C. and Miranda, F.D. (2017) *Genet. Mol. Res.*, **16**(2): 1-8.

Vesco Dal L.L., Stefenon, V.M., Welter L.J., Steiner, N. and Guerra, M.P. (2011) *Plant Genetic Resources,* **10**(1): 20-23.

Vesco Dal L.L., Stefenon, V.M., Welter L.J., Scherer, R.F. and Guerra, M.P. (2011) *Sci. Hortic.*, **128**: 515-522.

Shrubs: Beautiful and Perennial Ornamental Plants

K. Dutta

1.0 Importance and Uses

Shrubs are usually woody or semi-woody perennial plants with branches arising from the base of the plants and grow up to a height of 4 m. They constitute a large group among the perennial ornamentals grown in gardens, private or public.

Availability of a large number of species and cultivars, a wide range of variation in shape and size of the plants, diversified types of leaves and growth habits, remarkable ranges in size, shape and colour of flowers and the variety of fragrance have made the shrubs as most important garden plants. Several species of shrubs have foliage with unusually attractive colour and those having colourful fruits are also not uncommon. Semi-woody dwarf or prostrate shrubs are grown in edges, as hedge or ground cover.

Shrubs are used in shrub border (shrubbery), in groups with one species of plants of different cultivars or using different species of shrubs, and also as hedge and topiary. Many shrubs are very popular pot plants throughout the world. Well shaped tall shrubs flowering for several months is also planted as specimen plants *viz., Codiaeum variegatum, Erythrina cristagalli, Hibiscus rosa-sinensis, Ixora singaporensis, Mussaenda erythrophylla, Portlandia grandiflora, Tecoma gaudichaudi, etc.*, in tropical gardens. Species and cultivars of *Camellia, Fuchsia, Hydrangea, Rhododendron* (azalea), *etc.*, are extensively grown as specimen plants and in shrub border in subtropical and temperate climates. Cultivars of *Brugmansia* spp., *Euphorbia pulcherrima* (poinsettia), *Gardenia* spp., *Hibiscus rosa-sinensis* and many others are grown both in tropical and subtropical gardens in most parts of the world. *Jasminum auriculatum, Jasminum grandiflorum* and *Jasminum sambac* apart from beautiful fragrant flowers, essential oils of commercial importance are extracted from these flowers.

2.0 Introduction of Ornamental Shrubs in India

Large number of shrubs grown in tropical and sub-tropical region of India have been introducing in the 19th and 20th century by the Agricultural and Horticultural Society of India, established in 1820 and Indian Botanic Garden. Some of these species and cultivars include *Abelia* spp., *Acalypha wilkesiana* and its cultivars, *Allamanda* sp., *Aphelandra squarosa, Adenium obesum, Caesalpinia* spp. and cultivars, *Calliandra* spp., *Callistemon* sp. and many cultivars of *Codiaeum variegatum, Camellia* spp. and cultivars with variegated foliage, *Euphorbia pulcherrima* (Poinsettia) and many cultivars, *Gardenia* spp., *Fuchsia, Pelargonium, Begonia* spp.; and cultivars, *Cestrum elegans*, species and cultivars of *Hibiscus* including *Hibiscus rosa sinensis, Hibiscus syriacus, Hibiscus mutabilis, Hydrangea* spp. and many cultivars of Ixora *chinensis, Ixora coccinia, Ixora singaporensis, Ixora duffii, Kopsia fruiticosa, Jasminum* spp., *Magnolia* spp., *Nerium odorum* and cultivars of *Mussaenda phillipica, Mussaenda erythophylla, Pentas* spp., *Polyscias* spp., *Petrea arborea, Randia* spp., *Rhododendron* spp. and cultivars (Azalea) and many others. They were introduced from tropical America, Africa, Australia, south East Asia, Europe *etc. Bougainvillea* variety 'Million Dollar' and *Mussaenda* variety 'Queen Sirikit' were introduced in India by Dr. T.K. Bose at the Agri-Horticultural Society of India, Kolkata in 1967.

Importance and Uses of Shrubs

Shrubs in Garden

Shrubs in Border

As Topiary in Pots

Hedge of *Ixora*

Topiary of Shrubs

Shrubs in Rock Garden

Mixed Border of Shrubs

Coleus in Border

Hydrangea in Pot

Adenium in Pot

Clerodendrum inerme Topiary

Importance and Uses of Shrubs

Jasminum multiflorum in Garden

Lantana in Garden

Caesalpinia pulcherrima in Garden

Euphorbia milli in Pot

Fruits of *Ochna serrulata*

Codiaeum variegatum in Border

Importance and Uses of Shrubs

Hibiscus in Pots

Poinsettia in Greenhouse

Azalea in Pot

Fuchsia in Pot

Bougainvillea as Border

Gardenia in Pot

Codiaeum variegatum **in Greenhouse**

Hydranzea **in Greenhouse**

3.0 Growth and Flowering

Most of the species and cultivars of ornamental shrubs are hardy plants, and grow in various types of soil with little attention, but well drained soil rich in organic matter is most desirable. Usually, they cannot tolerate soggy soil and waterlogged condition. Although some plants flower almost throughout the year, in most cases they are seasonal, and a selection can be made in such a way that some plants in the shrubbery will be in bloom in different seasons of the year. Shrubs flower regularly every year in almost all attractive colours like white, light yellow, golden yellow, pink, scarlet, salmon, crimson, rose, violet, light blue, deep blue, and some are even sweet-scented. Most of these plants are fairly quick growing and when vegetatively propagated, may start flowering from the first year after planting. Plants like *Bauhinia acuminata, Cassia alata, Catharanthus rosea, Thryallis glauca,* which are usually grown from seeds also produce flowers in the first year. Shrubs can be classified into three groups according to the requirement of sunlight for growth and flowering: (i) plants that show normal growth and flowering in full sun, (ii) grow only in partial shade, and (iii) can grow well in both sun and semi-shady location.

Most of the tropical flowering shrubs are sun-loving; growth and flowering in these plants are adversely affected in shade. It is also observed that absence of morning sun shows much more adverse effect than the afternoon sun. Plants like *Beloperone amherstae, Calliandra brevipes, Calliandra houstonii, Magnolia fuscata, Magnolia mutabilis, Magnolia pumila,* and *Mussaenda erythrophylla* and cultivars of *Nandina domestica, Pentas, Olea fragrans* and *Wormia* (*Dillenia*) *burbidgii* grow better in partial shade. In full sun, these plants show signs of scorching during hot weather and suffer in dry heat. *Brunfelsia americana, Brunfelsia undulata,* species of *Crossandra, Cestrum* and *Daedalacanthus, Eranthemum laxiflorum, Hamelia patens, Jatropha panduraefolia, Jatropha rosea, Kopsia fruticosa,* speices of *Mussaenda, Ravenia spectabilis, Turnera erecta* and *Thunbergia ulmifolia* grow in full sun and also in semi-shade. Foliage shrubs like *Codiaeum variegatum* (croton), species and cultivars of *Eranthemum, Pentas, Polyscias* grow well in semi-shade where most of the flowering plants fail to grow properly; bright filtered light, however, promotes growth and improves colour of the foliage.

In mild subtropical climate, apart from those grown in tropical climate, species and cultivars of *Acalypha, Bougainvillea, Brugmansia, Camellia, Cestrum, Duranta, Euphorbia pulcherrima* (poinsettia), *Fucshia, Gardenia, Hibiscus, Hydrangea, Rhododendron* (azalea), *etc.*, are popular flowering shrubs.

4.0 Propagation of Shrubs

Hundreds of semi-woody and woody shrubs grown in tropical and subtropical gardens can be easily propagated from cuttings in a favourable environment, using suitable rooting media and with proper management. Application of root promoting chemicals and planting in mist or fog chamber enhance root formation (Bose, 1967).

Cuttings are taken from the terminal part of healthy shoots, 8-15 cm in length below a node with 2-6 leaves depending on the size of cuttings; and planted in the rooting media in rooting tray or bed. Media should be porous, consisting of

Propagation of Shrubs

Cuttings and Rootings of Ixora

Cuttings of Azalea in Plugtray

Propagation by Cuttings

Growth of Seedlings in Nursery

Grafting on Hibiscus

Micropropagation of Rhododendron

peat, vermiculite, sand, perlite, cocopeat, coirdust, vermicompost, *etc.*, singly or in combination in a warm humid atmosphere and semi-shade in a rooting chamber covered with shade-net allowing more than 50 per cent light on the cuttings and 80-90 per cent relative humidity; use of root promoting substances at the basal part of the cuttings would improve root formation in a shorter period. Propagation of difficult-to-root cuttings is done by air- or ground-layering, grafting or budding. After root formation in woody and semi-woody cuttings in a nursery, the rooted cuttings are transferred from the rooting medium and planted in porous compost rich in organic matter in poly-packets or small pots and arranged with adequate space between the rooted plants for sale or for future use. Grafting and budding are also practiced in the propagation of shrub s like *Bougainvillea, Hibiscus, Euphorbia pulcherrima, Jasminum, etc.* Micropropagation *in vitro* is a successful method of propagation in many species and cultivars of shrubs *viz., Euphorbia pulcherrima, Hibiscus, Bougainvillea, Hydrangea, Camellia* and many others.

Several species of shrubs like *Bauhinia acuminata, Thryalis glauca etc.*, which produce seeds, and seedlings often flower in one year, may be propagated by seeds (Bose *et al.*, 1991).

Bose (1965) reported marked improvement in the propagation of more than 200 species and cultivars of garden shrubs from cuttings in mist chamber, using IBA or NAA in talc. It was also possible to multiply them throughout the year. Retention of leaves in cuttings promoted rooting.

5.0 Selection of Shrubs

5.1.0 Shrub Border

Shrubs show a marked variation in height. Tall shrubs may reach a height of 3 to 4 m and dwarf shrubs grow only up to 1 m or less. Thus plants can be chosen from a large selection for growing in various places according to their suitability and utility. For planning a shrub border and planting, plants are selected according to their height, flowering season and flower colour, so that the shrubbery looks attractive throughout the year. Shrubs with variegated leaves should also be planted. Depending on space available; a shrub border may consist of several rows of plants, taller ones at the back, shorter ones towards the front rows. Shrubs can also be planted in one or two rows in front of a wall if space is limited. Planting in a group with different cultivars of one species like *Camellia, Ixora, Hibiscus rosasinensis*, poinsettia, *etc.*, or with different species of shrubs makes a good display. In subtropical gardens, shrubs suitable for the soil and climate are selected for planting.

5.2.0 Shrubs for Hedge and Topiary

In addition to their utility as flowering or foliage plants, shrubs are also used as hedge, in topiary and for screening. *Acalypha* forms a colourful hedge; *Barleria* sp., *Cesturm* spp., *Clerodendrum inerme, Dodonaea viscosa*, variegated forms of *Duranta macrophylla, Duranta repens, Ixora chinensis, Jasminum pubescens, Lantana camara* 'Dcpressa', *Lantana camara* 'Variegata', *Malpighia coccigera, etc.*, can withstand frequent pruning and often produce showy flowers, are usually grown as hedge of low to medium height. *Caesalpinia pulcherrimma*, hardy cultivars of *Hibiscus rosa-*

sinensis, Murraya exotica, and cultivars of *Duranta* spp. can be trained as tall hedge often serving the purpose of screening. Bougainvellia cultivars are also grown as tall hedge; maintenance of shape, however, is often difficult. The practice of clipping shrubs into ornamental shapes, known as topiary, is of great antiquity. Shrubs suitable for this purpose are species of *Clerodendrum inerme, Duranta, Ixora, Murraya, Phyllanthus, etc. Cestrum* sp., *Duranta, Rhododendron* are used as hedge plants in subtropical gardens.

5.3.0 Shrubs in Rock Garden

In addition to large number of hardy cacti and other succulents, dwarf shrubs like *Acalypha, Catharanthus rosea, Gardenia jasminoides* 'Fortuniana', *Hydrangea, Lantana camara, Lantana sellowiana, Malpighia coccigera, Plumbago capensis, Rhododendron, Ruselia juncea* and *Turnera ulmifolia* are also grown in rock gardens to add diversity of plants and colour. Cultivars of *Adenium obesum* and *Euphorbia milli,* though classified as succulents, flower freely in many colours in a rockery.

5.4.0 Shrubs with Showy Fruits

Shrubs are usually grown for their showy flowers or beautiful foliage. While *Ochna wightiana* bears black fruits on red thalamus, many yellow berries are found on *Duranta repens* and *Duranta macrophylla* in the rainy season; blue fruits of *Cestrum diurnum* and numerous red fruits of *Rauwolfla canescens* give a showy effect to the shrubbery.

5.5.0 Shrubs with Colourful Leaves

Variegated forms of *Acalypha, Excoecaria, Phyllanthus, Strobilanthes, Sanchezia,* cultivars of *Codiaeum varigatum,* species and cultivars of *Eranthemum, Polyscias, Ravenia spectabilis, Manihot esculenta, Duranta macrophyla, Duranta repens, Hibiscus rosa-sinensis, Malvaviscus, Tabernaemontana* and *Nerium oleander* having colourful or variegated leaves are important garden shrubs. A yellow leaved cultivar of *Duranta* without spine is now widely used as an attractive hedge.

5.6.0 Shrubs as Pot Plants

Flowering and foliage shrubs are extensively grown in tropical and subtropical climate for decoration of gardens, residential areas including balcony, terrace, patio, *etc.* Species and cultivars of *Bougainvillea,* cultivars of *Codiaeum variegatum* (croton), *Euphorbia pulcherrima* (poinsettia), *Gardenia, Hibiscus rosasinensis, Ixora, Jasminum, Ruttya* and *Tabernaemontana* are popular and attractive pot plants in tropical climate. Apart from a number of shrubs mentioned above, cultivars of *Camellia, Fuchsia, Rhododendron, Hydrangea, etc.* are extensively grown in pots in mild climate (Bose *et al.,* 2015).

6.0 Cultivation of Shrubs

6.1.0 Selection of Site

Selection of shrubs in the garden depends on the climate, availability and duration of sunlight at the site of planting, maximum height and spread of the plant, utility of plants, *i.e.*, for foliage or flower, deciduous or evergreen, colour of flowers and season of flowering. Except hydrophytes, garden plants are susceptible to waterlogging, and a well-drained site should always be selected. Waterlogging for a few hours or a very wet soil adversely affects growth of most of the shrubs and a well-grown plant cannot be obtained in such a location. Plants like *Cestrum, Hibiscus, Lantana, Plumbago, Tecoma, etc.*, do not survive in poorly drained soil.

6.2.0 Planting and Care

The growth and flowering of garden plants depend on the initial preparation of pit before planting and for those which will grow above 1.5 m, it is recommended to dig pits of 60 cm^3 and mix about 5 kg well-rotted farmyard manure (FYM) or 2 kg vermicompost, and raise the soil of the pit to about 15 cm from the ground level. Use of cocopeat and neem cake is also suggested. It should be done before or at the beginning of the monsoon so that with few showers the manure decomposes further, mixes with soil, and the loose soil settles down. When farmyard manure (FYM) is not available, use vermicompost, adequate leaf mould and about 100 g of bonemeal per pit. The distance of planting in a shrubbery depends on the size of the plants. Tall and bushy shrubs are generally planted at a distance of 2.5-3.0 m. Shorter plants with less spreading habit can be grown at a distance of 1.0- 1.5 m. It is always safe to plant shrubs in early monsoon. In the next 2 or 3 months during the rainy season, they will establish well, start new growth and thrive much better during hot months in the following year.

Shrubs should be manured annually during the rainy season. Well decomposed FYM or vermicompost can be applied in the soil around the root region; the use of bone meal once or twice a year also gives excellent result. Mixed fertilizer (NPK) spread around the root zone of the shrubs, after pruning in the rainy season or in spring depending on the growth habit, followed by light hoeing will improve growth and flowering. Shrubs need watering in winter and more frequently in summer. Temporary wilting due to dry soil may cause defoliation, and affects the growth of the plant. It is always advised to give a thorough soaking to all plants rather than to sprinkle every day. Soaking of surface soil leads to development of a shallow root system which not only prevents utilization of nutrients from the lower level of the soil, but also makes the plant more susceptible to drought and strong wind.

6.3.0 Shrubs in Greenhouse

Apart from rose, cultivars of *Bougainvillea, Codiaeum variegatum* (croton), *Euphorbia pulcherrima* (poinsettia), *Hibiscus rosa-sinensis, Hydrangea, Rhododendron, Solenostemon* (coleus), *etc.*, are grown in uncooled and environment controlled greenhouses for regulating growth and flowering. Pot plants of flowering and foliage shrubs have enormous demand in all parts of the world.

6.4.0 Growing in Pots

Perennial ornamental plants, particularly shrubs, are widely grown in pots due to limited ground space available in cities and towns. Most of the flowering shrubs require full sun and do well on terraces or a sunny location facing east or south. Though all shrubs can be grown in pots but medium-sized and dwarf plants of bushy growth are usually selected for this purpose. It should be remembered that normally plants grow better and flower more profusely in the ground. Small plants obtained from a nursery should be first planted in small or medium-sized pots (15-20 cm) and when the roots become almost pot bound they are shifted to larger pots. Practice of changing the plants to bigger pots as they grow helps to produce more well-shaped bushy plant than to grow a small plant in a large pot right from the beginning. For most of the shrubs, a soil mixture consisting of 2 parts of loamy soil, l part each of leaf mould and decomposed farmyard manure or vermicompost produces healthy plants in pots; 100 g of bone meal should be applied in a 20 cm pot. The potting compost should be fertile and well-drain. Planting should not be done in dry compost with undecomposed organic matter. Before potting, odd branches at the base of the stem and diseased shoots be removed. After establishment, careful pruning of the shoot tips will produce a well-shaped bush.

7.0 Pruning and Training

7.1.0 Pruning

All woody ornamentals, particularly shrubs and climbers and to some extent trees need pruning. These perennials grow for many years, produce flowers and often fruits, and it becomes essential to regulate the growth of the plants in limited space in a garden and maintain them in a condition favourable for growth and flowering. Pruning is one of the important practices to achieve these objectives. Fertilization and irrigation are also necessary to promote plant growth but the necessity of pruning and training cannot be replaced by any other treatment.

7.2.0 Time, Extent and Method of Pruning

The rate of growth, time of flowering, age of flower-bearing shoots and the response of plants to pruning are the major factors determining the time and extent of pruning. The formative pruning in case of trees and shrubs aims at developing well-shaped plants of attractive appearance.

Perennials can broadly be classified into two groups with regard to flowering. One group of plants like *Acalypha, Barleria, Cestrum, Daedalacanthus, Jasminum, Lagerstroemia indica, Lantana, Pentas, Malvaviscus, Thunbergia* and many others flower in 1 to 6 months after pruning, on shoots emerging from pruned branches. These shoots grow fast. In *Bougainvillea,* new axillary shoots arising from old wood and also the terminal growth are the organs for flowering. Several species of plants including *Brunfelsia hopeana, Ixora singaporensis, Ixora duffii, Oncoba spinosa, Ochna wightiana, Portlandia grandiflora, etc.,* flower on mature shoots which are 6 months to 1 year- old. Pruning in such plants should be done carefully and the new shoots are allowed to mature for flowering.

Apart from the variation in growth and flowering habit, plants also vary in their response to regeneration of shoots after pruning and they can be classified into three groups: (i) very responsive, when emergence of shoots starts within 5-10 days after pruning under favourable climatic conditions, as in plants like *Acalypha, Brugmansia, Cestrum, Daedalacanthus, Eranthemum,* variegated *Manihot, Polyscias,*

Pruning of Shrubs

Pruning of *Cestrum* sp.

Rejuvenation Pruning of Bougainvillea

Manual Pruning

Pruning by Machine

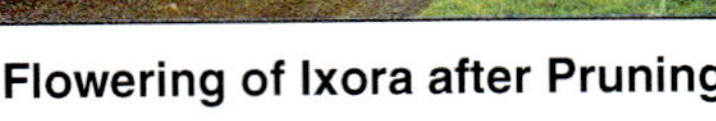

Flowering of Ixora after Pruning

Flowering in Bougainvillea after Pruning

poinsettia, *etc.*; shoot growth of these plants is also faster; (ii) moderately responsive as in *Gardenia, Ixora, Jasminum, Tabernaemontana, Buddleja, Jatropha* and others, (iii) while the next group consisting of *Ochna wightiana* and hybrid cultivars of *Hibiscus rosa-sinensis, Calliandra emarginata, Petrea arborea, Ravenia spectabilis* and *Rondeletia speciosa, etc.*, not only produce axillary shoots a few weeks after pruning, number of shoots and their growth are also less in comparison to those of the earlier groups. These shoots also require much longer period to bear flowers. Species of plants in a genus also show variation in their response to pruning, *Ixora coccinea* and *Ixora lutea* are more responsive to pruning than *Ixora singaporensis* and *Ixora macrothyrsa*. Similarly, *Calliandra hybrida* and *Calliandra haematocephala* respond moderately to pruning, while *Calliandra brevipes* and *Calliandra emarginata* show very slow response to regeneration and shoot growth after pruning (Bose and Bhattacharyee, 1975).

The diseased or dead branches should be removed first. The branches close to the ground and those growing irregularly and affecting the shape of the plant should be pruned. Old branches usually produce weak shoots, so it is necessary to prune them from the base. Thin, weak and interlaced shoots are not capable of bearing flowers and should thoroughly be removed. To induce compact growth, the centre of the bush should be opened by pruning a few shoots from the base. Healthy shoots pruned above an axillary bud tend to produce branch in an outward direction. The height and time of pruning of shrubs depend on the height of the plant expected during next flowering, its growth habit and time of flowering. General cleaning can be done during monsoon. Spring- and summer-flowering plants are pruned at the onset of monsoon. Deciduous and semi-deciduous plants are pruned after the leaves have fallen. It is safe to prune the plants which continue to flower up to August-September, after flowering is over. These principles of general nature cover a large number of perennial garden plants. Some fast-growing plants like *Acalypha, Cestrum, Eranthemum, Beloperone, Pentas, Polyscias, Sanchezia, Thunbergia, etc.*, respond to pruning and they can be pruned even twice a year, low pruning at the onset of rainy season and again at the end of monsoon to maintain the shape and also to encourage flowering provided the growth is adequate. Climatic conditions, therefore, partly determine the time and extent of pruning, since it markedly influences the growth and flowering of plants.

8.0 Diseases, Pests and Disorders of Shrubs

A number of diseases and pests are reported in shrubs during their growth and development when planted in tropical and subtropical regions. Several fungal and viral diseases commonly infecting hibiscus are *Phyllosticta* leaf spot, sooty mould, root rots and collar rots, hibiscus chlorotic ringspot virus *etc.*, bud drops and yellow leaves are two main physiological disorders of hibiscus. The pests of hibiscus are mainly aphids, jassids, leaf hoppers, mealy bugs, scale insects, bugs and mites *etc.* Bougainvillea is also susceptible to various fungal diseases like leaf spot, powdery mildew, root rot and insect pests which infest this shrub are loopers, aphids, leaf miners, thrips, spider mites *etc.* Chlorosis and irregular flowering are very common physiological disorders of bougainvillea. The diseases affecting ixora are *Alternaria* blight, bacterial leaf spot, cucumber mosaic virus and aphids, scale insects, and mealy bugs. Powdery mildew, sooty mould, canker and leaf spots are common diseases

Diseases, Pests and Disorders of Shrubs

Leaf Spots of Bougainvillea

Bougainvillea Loopers

Bougainvillea Chlorosis

Hibiscus Bacterial Leaf Spots

Mealy Bug Infestation in Hibiscus

Hibiscus Mites

Ixora Leaf

Ixora Aphids

Sooty Mould of Gardenia

Stem Canker of Gardenia

Mealy Bugs of Gardenia

Camellia Leaf Gall

Diseases, Pests and Disorders of Shrubs

Camellia Flower Blight

Scale Insects of Camellia

Anthracnose of Hydrangea

Leaf Scorch of Azalea

Bark Scale of Azalea

Powdery Mildew of Poinsettia

affecting gardenia. Spider mites, thrips, aphids and mealy bugs infest gardenia in different seasons. Camellia is affected by flower blight, root rot, dieback, canker, leaf gall, *Camellia yellow mottle virus etc.* and aphids and scales among the pests. Bud drops, sunscald and oedema are important disorders in camellia. Some common problems of hydrangea are botrytis blight, leaf spots, anthracnose, powdery mildew, rust, phoma leaf spot, *Alternaria* leaf spot, bacterial leaf spot, hydrangea ringspot virus *etc.* Among the pests infesting hydrangea are black vine weevil, four lined plant bug, Japanese beetles, two spotted spider mite *etc.* Calliandra is sometimes damaged by mites and dieback diseases. Among the diseases in azalea, *Phytophthora* root rot, leaf spots, petal blight, powdery mildew and dieback are common. Other minor diseases of azalea are leaf gall, leaf folding, *botryosphaeria* canker and dieback *etc.* Insect pests affecting azalea are lace bug, weevil, bark scale, borer, leaf miner, white fly *etc.* Lace bug, *Stephanitis* spp., adults and nymphs with piercing and sucking mouthparts; azalea bark scale (*Eriococcus azalea*) also damages by sucking plant fluids; borer (*Synanthedon rhododendri*) is a clearwing moth borer affecting rhododendron and deciduous azalea (Mills, 2017). Poinsettia is susceptible to some foliar diseases, such as *Botrytis* blight, scab, *Alternaria* leaf spot, *Xanthomonas* leaf spot, and anthracnose *etc.*, as well as many common root and stem diseases like *Pythium* root rot, *Phytophthora* and *Thielaviopsis* root rot. A wide variety of insect and spider mites have been reported in association with poinsettia cultivation both in net houses and in greenhouses. Poinsettia is also affected by whiteflies, thrips, aphids and leaf miners (Huelsman, 2000).

Systemic fungicides are applied to manage most of the fungal pathogens of shrubs. Bacterial diseases are managed by applying streptomycin or tetracycline. The various pests of shrubs can be controlled by applying systemic as well as contact insecticides. Physiological disorders are often corrected by proper water and nutrient management, applying improve package of practices in cultivation of shrubs.

9.0 References

Bose, T.K. (1965) *Bull. Agri. Hort. Society*, **4**: 65-75.

Bose, T.K. (1967) *Bull. Agri. Hort. Society*, **6**: 52-55.

Bose T.K. and Bhattacharyee, S.K. (1975) In: *Garden plants*, Oxford IBH Publisher, New Delhi.

Bose, T.K., Sadhu, M.K., Mitra, S.K. and Das, P. (1991) In: *Propagation of Tropical and Subtropical Horticultural crops*, **2**: 23-35.

Bose, T.K., Choudhury, B. and Sharma, S.P. (1991) In: *Tropical Garden Plants in Colours*, Hort. and allied publishers.

Mills, J. (2017) In: *Plant Pathology & Plant-Microbe Biology*, Cornell Univ., **24**: 1-2.

Huelsman, M.F. (2000) In: *Crop Profile for Poinsettia in Ohio.*, pp. 1-21.

Bose, T.K., Singh, L.J., Sadhu, M.K., Maity, T.K. (2015) In: *Ornamental Plants and Garden design in Tropics and Subtropics*, **2**: 215-222.

Hydrangea

Arindam Bhunia and Sk. Masudul Islam

1.0 Introduction

Hydrangea is a genus of erect or climbing woody shrubs, in the family Hydrangeaceae, native to the Western Hemisphere and eastern Asia. The name hydrangea comes from two Greek word 'hydor' and 'angos' which means 'water vessel', due to the shape of its seed capsules (Gledhill, 2008). This is commonly names as hydrangea or hortensia. The earlier name, Hortensia, is a Latinised version of the French given name Hortense, referring to French astronomer and mathematician Nicole-Reine Hortense Lepaute (Oxford Dictionaries). Hydrangea grows naturally in temperate areas of eastern Asia, North America and in subtropical Middle and South America. Hydrangea was introduced in Europe by explorers since 1736 (McClintock, 1957). The genus is divided into the sections Hydrangea, which contains mainly temperate climate species, and Cornidia, which consists of climbing species from tropical and subtropical area. The section Hydrangea has a number of ornamentally attractive characteristics and commonly cultivated (Kudo *et al.*, 2008). McClintock (1957) evaluated that the cultivated species are all members of the section Hydrangea and have been placed into subsections Americanae (*H. arborescens* and *H. quercifolia*), Asperae (*H. involucrate* and *H. aspera*), Calytranthae (*H. anomale*), Heteromallae (*H. paniculata*) and Macrophyllae (*H. macrophylla*). The mountain hydrangea (*H. serrata*) is considered by some separated species and closely related to *H. macrophylla*. Hydrangea was the second most popular deciduous shrub across all horticultural markets in 2014 and was produced in more than 1,500 U.S. nurseries, with more than 10 million plants sold, accounting for 13.5 percent ($91.2 million) of total annual U.S. shrub sales ($676.6 million; Owen *et al.*, 2015). It is grown as both a garden and greenhouse plant. It is valued for its large corymbs that, depending on soil pH and cultivar, range in colour from pink to blue and in intensity from pale to deeply coloured. Recently it is becoming a very popular ornamental plant as garden decoration, but also as interior decoration; for this purpose, it is commercialised as cut flower branches (Pecchioli *et al.*, 2011). The hydrangea represents gratitude, grace and beauty. It also radiates abundance because of the lavish number of flowers and the generous round shape. Its colours symbolize love, harmony and peace. The flower lends itself naturally to floral arrangement for weddings and birthdays, but is also equally suitable for memorial bouquets.

2.0 Importance and Uses

2.1 Aesthetic Uses

Hydrangea is a popular decorative plant because of its big flower heads which attract attention wherever it is planted or used. Because of showiest flower it used as a focal point in the garden, in a vase, and in wedding bouquets. Hydrangeas are appropriate for various special occasions and events like engagements, weddings, asking for forgiveness and reconciliation, rejecting a suitor's proposal *etc.* Large hydrangea flowers are also loved by the cottage gardeners due to their lovely old-fashioned charm. Sometimes it also planted along the edges of a drive for a relaxing entry to a property. The large blossoms of hydrangea plants are very easy to dry

Uses of Hydrangea

As a Herbaceous Border

As a Hedge in Landscaping

As a Specimen Plant in Garden

In Pot

In Wedding Bouquet

In Flower Vase

indoors. Drying flowers will bring the outdoors into the home for use in dried flower arrangements. Hydrangeas can symbolize many different things, depending on where around the world and their culture. According to a Japanese legend, the emperor gave hydrangeas to the family of a girl he loved to show how much he cared for her. In Victorian era, white hydrangeas represent boastfulness, bragging and vanity. In United States hydrangeas are used for 4th wedding anniversaries to symbolize appreciation and heartfelt emotion.

French hydrangea has long been used as a small flowering pot plant that is discarded after flowering. Many cultivars have been developed exclusively for this use. Container grown French hydrangeas are very popular at holidays such as Mother's Day and Easter. Some hydrangea species like *Hydrangea paniculata* and *Hydrangea quercifolia* are also suitable for used in a mixed border or as a mass planting or as a deciduous hedge. *Hydrangea macrophylla* is included in the Tasmanian Fire Service's list of low flammability plants, indicating that it is suitable for growing within a building protection zone (www.theflora.in; www.plantaddicts.com; www.edis.ifas.ufl.eduu; www.fire.tas.gov.au; The United States National Arboretum).

2.2.0 Therapeutic and Other Uses

The young leaves, when dried and rubbed between the hands, become very sweet and are used to make a sweet tea called 'tea of heaven', it is used in Buddhist ceremonies. The leaves contain phyllodulcin, a very sweet substance that can be used as a sugar substitute. One small leaf is sufficient to sweeten a cup of tea. The older leaves can be dried, powdered and used as flavouring on foods. The young leaves and shoots are also eaten cooked. Amacha is a Japanese beverage made from fermented leaves of *Hydrangea macrophylla* var. *thunbergia* (Facciola, 1990; Duke *et al.*, 1985; Tanaka, 1976; Kunkel, 1984).

Hydrangea paniculata, *Hydrangea macrophylla*, and *Hydrangea arborescens* are the most popular when it comes to medicinal properties. Hydrangea root is rich in a compound called coumarin. Both coumarin and its derivative skimmin may offer anti-inflammatory properties. It is used in case of urethritis, cystitis, prostatitis, urinary calculi, cystitis, gravel and enlarged prostate gland. In Nepal, its leaves are used for treating cold, cough and bronchitis and root is used as a cure for indigestion and fevers. The decoction made from stem bark is useful for fevers, while the decoction of leaves useful for treating malarial fever. It helps to soothe irritated mucus membranes and provides relief from backache caused due to kidney problems. Hydrangea provides relief from backaches, rheumatism, scurvy, paralysis, dropsy and use externally as a poultice for burns, wounds, sprains, sore muscles and tumors. In Western herbal medicine to treat inflammation of bladder, prostate, urethra, enlarged prostate and kidney stones. Hydrangeae Dulcis Folium is a drug made from the fermented and dried leaves of *H. macrophylla* var. *thunbergii* with possible antiallergic and antimicrobial properties. It also has a hepatoprotective activity by suppression of D-galactosamine-induced liver injury in vitro and in vivo. Hydrangeic acid from the leaves is being investigated as a possible anti-diabetic drug as it significantly lowered blood glucose, triglyceride, and free fatty acid levels in laboratory animals (www.healthbenefitstimes.com; Nakagiri *et al.*, 2003; Zhang *et al.*, 2009).

3.0 History

The relationship between humans and hydrangeas has been filled with beauty, mystery, and misnomers since the very first description of Hydrangea was published in Gronovius' 1739 publication of North American plants, 'Flora Virginica'. The genus name was soon to be adopted by Linnaeus in 1753 (McClintock, 1957). Nearly 30 years later in 1784 Carl Peter Thunberg had discovered the genus in China and gave the first binominal names to these Hydrangeas but he referred them wrongly to the genus *Viburnum* (Wilson, 1923). This discovery established a wide range of nativity for hydrangeas ranging from North America to China, Japan, and Taiwan. In the years following these discoveries, even more Asiatic species would emerge and earn their names: *Hydrangea anomala*, *H. aspera*, and *H. heteromalla* were discovered by Francis Buchanan-Hamilton and Nathaniel Wallich in the Himalayas (McClintock, 1957). In Europe, hydrangeas didn't appear until 1736, while around 1788 introduced as a cultivated ornamental from Asia by Sir Joseph Banks and caught the attention of gardeners and exploded in popularity across Europe (Wilson, 1923). Hydrangeas would continue to have an uncertain place in botanical nomenclature – first filed in the family Saxifragaceae, then in 1830 designated Hydrangeaceae (Dirr, 2004). 1830 was an auspicious year for the genus Hydrangea, as Seringe completed the first taxonomic treatment of the genus that incorporated all North/South American and Asian species known at the time (McClintock, 1957).

4.0 Origin and Distribution

Many believe that hydrangea was first cultivated in Japan and is often referred to as Ajisai. The flowers bloom during the rainy season from June to July and turn places like the Meigetsuin Temple into a mystical, fragrant garden (www.proflowers.com). Ancient fossils dating back to 40-65 million years ago show that the plant has also been growing in North America for quite some time (www.proflowers.com). Although the plants are native to Asia and the Americas, but the greatest species diversity is in eastern Asia, notably China, Korea, and Japan (The United States National Arboretum). Different types of hydrangeas are scattered throughout eastern Asia, especially China and Japan, but found in many other countries in that region including Korea, Taiwan, Tibet, Philippines, Vietnam and as far south as Indonesia. They also inhabit large parts of the Americas, from the U.S. Northeast via Central America in countries such as Mexico and Costa Rica down to the Andes in Ecuador and Peru. In India, places having cool climates like Kashmir, Himachal Pradesh, Nainital, Kalimpong, Darjeeling, southern and northern hills, Shillong, *etc.* has the potential for growing good quality hydrangeas as cut flowers or container plants (Bhattacharjee, 2006).

Hydrangea macrophylla mainly found in Japan and Korea but also distributed in eastern Himalayas, north-east Burma and western China. *Hydrangea paniculata* found to be cultivated throughout the Southeastern China, islands of Kyushu, Honshu and Hokkaido in Japan. H. paniculata is a fairly uniform species, showing no particular variation in any of its parts throughout its geographical range. *Hydrangea quercifolia* distributed in the Southeastern United States States (Georgia, Florida, Alabama, Missouri). This is also found chiefly in the area of the Piedmont and coastal plain of western Georgia, Alabama, and Mississippi. *Hydrangea anomala* distributed in the woods and mixed forests of Japan from sea level to 2,000 meters, on Honshu, Kyushu and Hokkaido islands; in Korea on Dagelet and Quelpaert Islands; in the eastern Himalaya Mountains of northern India, from 1,200 to 2,500 meters, and in west central to east central China, from 600 to 3,800 meters. *Hydrangea aspera* grown in the woods or thickets, eastern Himalaya Mountains from 1,500 to 2,300 meters; western China from 700 to 4,000 meters; central and eastern China from 250 to 2,100 meters; Formosa, Sumatra and Java from 1,000 to 2,500 meters. *Hydrangea heteromalla* found to be grown in Nepal, northeastern China, and the Himalayas (www.heritagehydrangeas.com; www.rhs.org.uk; Dirr, 2004; van Gelderen, *et al.*, 2004).

5.0 Morphology

Hydrangeas are either deciduous or evergreen, though the widely cultivated temperate species are all deciduous (The United States National Arboretum). Most of these species are shrubs 1 to 3 meters tall, but some are small trees, while others are of the climbing variety (Bailey, 1998). The stems are sturdy but little woody and cylindrical. The leaves are in opposite pairs or ternate, deciduous or evergreen. Leaves are exstipulate, simple, entire, serrate or pinnately lobed, ovate or obovate, acuminate, cuneate. The upper leaf surface is glossy, smooth, matt, hairy, according to the species; the lower dull with a conspicuous network of pinnate veins. The inflorescence of hydrangeas is terminal, a corymbose cyme, umbellate cyme, panicle or thyrse; occasionally axillary, bracts deciduous with peduncle. Fertile flowers are usually very numerous, bisexual, small; pedicel short. Calyx tube is adnate to ovary, 4- or 5-dentate, persistent. Corolla lobes 4 or 5, free, rarely connate and forming a calyptra, ovate or spatulate, valvate. Stamens usually 8 or 10 (but up to 25), inserted on disc; filaments linear; anthers oblong to sub-globose, apex sub-rounded to obtuse. Ovary is inferior to 2/3 superior, imperfectly or perfectly 2-4(or 5)- loculed; placentation parietal or axile; ovules numerous. Styles 2- 4(or 5), free or basally connate, persistent; stigmas terminal or decurrent along style adaxially. Sterile flowers few (more numerous in cultivated varieties), sometimes absent, borne at margin of inflorescence, with no stamens or pistils but with sepals 2 to 5, entire, serrated, elliptical, oval, rounded, petaloid, and enlarged. Under cultivation, and sometimes in a wild state, the corymbs may consist entirely of sterile flowers and are then very showy. Whilst the majority has white sterile flowers, some cultivars of *H. macrophylla* and *H. macrophylla* ssp. *serrata* are pink and several have a tendency to blue. Fruits are brown colour small 2- to 5-celled poricidal, hemispheric to turbinate capsules. Seeds are presented in capsule, dehiscing apically among styles, apex projected or truncate. Seeds numerous per locule, small, fusiform or ellipsoid, winged or wingless; seed coat thin, with reticulate or striate veins (www.almanac.com; www.Plantdatabase.kpu.ca; www. heritagehydrangeas.com).

6.0 Species and Cultivars

6.1.0 Species

There are more than seventy species of hydrangea reported (www.rollmagazine.

Species of Hydrangea

Mophead Hydrangea

Lacecap Hydrangea

Mountain Hydrangea

Hydrangea arborescens

Hydrangea paniculata

Hydrangea quercifolia

Species of Hydrangea

Hydrangea petiolaris

Hydrangea anomala

Hydrangea aspera

Hydrangea heteromalla

Hydrangea integrifolia

Hydrangea involucrata

com). Among them several species are cultivated world-wide, but the most popular types are *Hydrangea macrophylla, H. arborescens, H. paniculata, H. quercifolia* and *H. anomala* subsp. *petiolaris*.

Hydrangea macrophylla

It the most widely grown with over 500 named cultivars commonly known as Bigleaf, Florist, Garden, House, French or Japanese hydrangeas. Plants are grown as both garden and pot plants. Commercial greenhouse growers force them into bloom for Easter and Mother's Day in Tennessee (Halcomb *et al.*, 2010). This Japanese native is rated as hardy to cold-hardiness. It produces large inflorescences of white, pink or blue flowers in early summer (The United States National Arboretum). There are 3 different categories of *H. macrophylla*; mophead, lacecap and mountain hydrangeas.

I. Mophead Hydrangeas (Hydrangea macrophylla var. macrophylla)

These are the most popular Bigleaf hydrangea. It has showy flowers are arranged on the outside of the rounded inflorescence. On the interior of the inflorescence, a few small flowers are present; these are the flowers that produce seed (The United States National Arboretum). In general leaves are relatively thick and crisp, somewhat shiny, and often heart-shaped. Their edges are coarsely toothed. It is approximately 4"-6" long and 3"-5" wide, but in some cases, it may grow even larger. Leaf stems (petiole) are short, causing the leaves to hug close to the main stems in most cases. Stems often have tiny black or red streaks or speckles (www.plantaddicts.com).

II. Lacecap Hydrangeas (Hydrangea macrophylla var. normalis)

Lacecaps are identical to mopheads in every way except the shape of their blooms. The little buds in the center of the lacecap are the fertile flowers, and the large showy blossoms around the outer edge are the sterile flowers (www.plantaddicts.com). Both mophead and lacecap hydrangeas can be used as specimen plants, in mixed borders or in mass plantings (The United States National Arboretum).

III. Mountain Hydrangeas (Hydrangea macrophylla ssp. serrata)

These are the least common bigleaf hydrangea, have smaller leaves and flowers than other *H. macrophylla* cultivars, and their foliage may develop some red colouration in fall. Their inflorescences are generally of the lacecap type (Halcomb *et al.*, 2010). The plants are extremely hardy and built to survive harsh winters and climates (www.plantaddicts.com). Mountain hydrangeas are delicate in appearance and particularly well suited for placement at the edge of a wooded area (The United States National Arboretum).

Hydrangea arborescens

These are also called as smooth hydrangea or wild hydrangeas. These shrubs are native to the Eastern United States (Halcomb *et al.*, 2010). These bushes can tolerate hotter climates and rated as hardy from USDA Cold Hardiness Zones 4 to 9. These are sometimes planted as hedges because of their size. The leaves of *H. arborescens* are generally heart shaped, thin, and floppier than the mopheads (macrophylla). They have a matte surface and a courser texture than the smooth leaf of the mophead. The leaf stems (petiole) are long and hold the leaf away from the main stem (www.plantaddicts.com). Plants typically have a lacecap type inflorescence consisting of a combination of a few large sterile flowers and many small fertile flowers. Flowering occurs during early to mid-summer. At the peak of flowering, smooth hydrangea flowers are a pure white. At the time of ages, it's become develop a pale green colour. Smooth hydrangea is extremely striking in mass plantings (The United States National Arboretum).

Hydrangea paniculata

Panicle hydrangeas are known for their cone shaped flower heads (www.plantaddicts.com). These are the coldest hardy member of the genus. It can be reliably grown in USDA cold-hardiness zones 4 to 7 (The United States National Arboretum). It is native to Japan, eastern and southern China. It grows 3- 6 m tall and spread (Halcomb *et al.*, 2010). Large creamy- white flowers, which are borne in 15- 45 cm long panicles, contain both fertile and sterile flowers. Flowering occurs during mid-summer. As flowers mature, they may turn pink (The United States National Arboretum).

The leaves of panicle hydrangeas are relatively easy to identify when compared with other hydrangeas. They are smaller, thinner, and rougher than leaves of the mophead hydrangea. The edges are finely toothed in some varieties and more coarsely toothed in others. These are medium green with a matt finish. In *H. paniculata* three leaves grow from a stem-node and are distributed around the node in a whorl, which is an important feature for their identification. These are the only hydrangeas that will form trees. Their central stems can be developed into very attractive trunks (www.plantaddicts.com). They are also suitable for use in a mixed border or as a deciduous hedge (The United States National Arboretum).

Hydrangea quercifolia

H. quercifolia is popularly known as oakleaf hydrangea as their leaves are shaped much like those of a red oak. The foliage on these hydrangeas also changes colour in the fall and are the only type of hydrangeas that do that. The colour will change from orange to red to mahogany (www.plantaddicts.com). It can be reliably grown in USDA cold-hardiness zones 5 (Halcomb *et al.*, 2010). These hydrangeas are native to United States, found growing primarily in moist woodlands in the southeastern United States. Plants generally grow 1.8-2.4 m in height, although a few cultivars with smaller and larger plant habit are available. Flowers are creamy white in colour, borne in 10- 30 cm long panicles during early summer. As flowers age, they often turn a medium- to deep-rose colour. Oakleaf hydrangea is best suited used in a mixed border or as a mass planting (The United States National Arboretum).

Hydrangea petiolaris

This is commonly known as climbing hydrangea, but not as well-known as the previous four species. These are easy to spot because they are actually true clinging vines (www.plantaddicts.com). Climbing hydrangeas are native to the woodlands of Japan, the Korean peninsula, and on Sakhalin island of easternmost Siberia in the Russian Far-East (www.missouribotanicalgarden.org). These are becoming more popular due to the uniqueness of growing up structures and having large blooms. While initially slow growing, the plant can eventually cover tall (up to 24 m) structures. White, corymbs type inflorescences up to 25 cm diameter are produced in early- to mid-summer. The stems of climbing hydrangea leave a residue that is very difficult to remove. Climbing hydrangea grows well in shade, but can also tolerate a sunny location (The United States National Arboretum).

Hydrangea petiolaris is sometimes treated as *Hydrangea anomala* subsp. *petiolaris*, a subspecies of the closely related *Hydrangea anomala* (also known as Japanese climbing-hydrangea) from the woodlands of the Himalaya, southern and central China and northern Myanmar. The *Hydrangea anomala* species differs in being smaller (4- 9 m) and having flower corymbs up to 15 cm diameter (Botanical Society of Britain and Ireland).

Hydrangea aspera

This species is shrub or small tree in stature, 1- 4m tall; branch lets and inflorescence pubescent with stiff and upwardly appressed or curling and erect hairs. Inflorescences are flat-topped, many-branched cluster, 10-30 cm broad; bracts lanceolate, densely hairy on outer surface, glabrous on inner, caducous or remaining; fertile flowers white or blue, numerous, hypanthium 1-1.5 mm long; sepals 5, 0.4-0.5 mm long; petals 5, 2-2.5 mm long, truncate at base, deciduous before stamens; stamens 10, 2-7 mm long, usually alternately long and short; ovary completely inferior; styles 2, 3 or 4, 0.5-1.5 mm long, sterile flowers always present, white, 4-lobed, lobes 1-3.5 cm long, usually slightly longer than wide, with entire or fimbriate margins (www.heritagehydrangeas.com; www.rhs.org.uk).

Hydrangea heteromalla

It is also known as *H. vestita*, *H. sungpanensis*, *H. mandarinorum*, *H. khasiana* and *H. dumicola*. This is a much- branched shrub or small tree, 0.5 to 7 m tall; branch lets and inflorescence pubescent with upwardly appressed hairs 0.5-1 mm long. Inflorescence rounded many-branched cluster, 8- 30 cm broad; bracts linear to lanceolate, chartaceous, 1-3 cm long, 2-5 mm wide, few pili on back and along margins; fertile flowers white (www.heritagehydrangeas.com; www.rhs.org.uk).

Hydrangea integrifolia

This species is also known as *H. cuneatifolia*, *H. glandulosa*, *H. integra*. Distributed in the Philippine Islands and Taiwan. Evergreen woody climbers or shrubs, pubescence on inflorescence of coarse stellate hairs (0.15-0.25 mm long) borne on short (0.05 mm long) epidermal papillae, pubescence on stem dense, of fine stellate hairs with occasionally coarser hairs interspersed (www.heritagehydrangeas.com; www.rhs.org.uk).

Hydrangea involucrata

This species is mainly grown in Japan, Taiwan. These are Shrub, 1-2 m tall; branchlets, inflorescence and pedicels pubescent with upwardly appressed hairs 0.5 mm long. Inflorescences are rounded, compound, many-branched; involucres bracts subtending the inflorescence, 1.5-2.5 cm long, broadly obovate, pubescent with upwardly appressed white hairs on outside, glabrous on inside. Sterile flowers always present, lavender, 4- lobed, lobes rounded, 1-3cm in diameter (www.heritagehydrangeas.com; www.rhs.org.uk).

6.2.0 Cultivars

Several species of hydrangea of ornamental importance have large number of cultivars which differ in colour, size, shape and flowering season. Important cultivars according to the species are the following:

Cultivar	*Species*	*Characteristics*
All Summer Beauty	*Hydrangea macrophylla*	One of the most reliable mophead cultivars. Continues to produces a few flowers throughout the summer. Good blue colour in acidic soil.
Alpengluhen	*Hydrangea macrophylla*	Also sold as 'Glowing Embers'. Attractive dark green foliage. Thick sturdy stems. Not suitable for low pH soils. Susceptible to damage from late spring freezes.
Altona	*Hydrangea macrophylla*	Gained the Royal Horticultural Society's Award of Garden Merit. Compact plant with large rose-red sterile florets, turning to green then deep red in autumn. Blues well on acid soils.
Ami Pasquier	*Hydrangea macrophylla*	Floriferous, wine pink to blue mophead type cultivar. Have gained the Royal Horticultural Society's Award of Garden Merit.
Ayesha	*Hydrangea macrophylla*	Mophead type cultivar bears domed clusters of small, cupped, pink, lilac-like flowers in summer.
Blaumeise	*Hydrangea macrophylla*	One of the 'Teller Series' lacecaps. Sterile flowers are much larger and showier than in other lacecaps. Beautiful pure blue colour in acidic soil.
Blue Billow	*Hydrangea macrophylla*	Small plant with coarse foliage. Leaves burn in sun. Blooms moderately well.
Blue Bird	*Hydrangea macrophylla*	Commonly available lacecap. Looks great planted at edge of woodland. Fairly reliable flowering
Blue Wave	*Hydrangea macrophylla*	Commonly available lacecap cultivar. Performs very poorly following late spring freezes.
Blushing Bride	*Hydrangea macrophylla*	Cold-hardy, ever-blooming white mophead
Coerulea	*Hydrangea macrophylla*	More tolerant to late spring freezes than most cultivars. Attractive lacecap flowers. May be difficult to find.
Cocktail	*Hydrangea macrophylla*	Bushy shrub with ovate, serrated sepals.
Dooley	*Hydrangea macrophylla*	Reported to flower well even following late spring freezes. Large blue mophead flowers.
Eldorado	*Hydrangea macrophylla*	This is medium-sized mop head hydrangea bears red colour flowers.
Endless Summer	*Hydrangea macrophylla*	Non-stop flowering, even after spring freezes. Small plant with mophead flowers.
Europa	*Hydrangea macrophylla*	Compact, deep pink mophead type cultivar. Have gained the Royal Horticultural Society's Award of Garden Merit.
Harlequin	*Hydrangea macrophylla*	This is a picoted pink to purple mophead type cultivar.

Cultivar	Species	Characteristics
Jogasaki	*Hydrangea macrophylla*	Bushy, deciduous, rounded shrub, to 1.5m tall and wide. Flattened clusters of pink or lavender-blue flowers form flowerheads 7-12cm across; the double ray florets are held in a ring around the fertile flowers on long, stiff stalks.
Lady in Red	*Hydrangea macrophylla*	This is a rounded, deciduous shrub with toothed dark green leaves. In mid and late summer, it bears large, flattened flower heads of rosy-red sterile flowers surrounding pinkish-green fertile flowers.
Lemon Wave	*Hydrangea macrophylla*	Green, yellow and white variegated foliage. Attractive even without flowering.
Lemon Zest	*Hydrangea macrophylla*	Lime green foliage and pink mophead flowers. Unusual contrast. Attractive even without flowering.
Lilacina	*Hydrangea macrophylla*	Cold-hardy, disease-resistant pink to blue lacecap type cultivar.
Lime Rickey	*Hydrangea macrophylla*	Lime green mophead blooms sport showy pink pollen and take on a marbleized jade effect as they age. Super strong stems make this fantastic in the landscape and as a fresh or dried cut flower.
Love You Kiss	*Hydrangea macrophylla*	A robust deciduous shrub with glossy dark green red-flushed leaves with good autumn tints and bearing large, distinctive flattened lacecap flower-heads of strongly red-margined white ray florets in mid and late summer that gradually turns pink. Also gained the Royal Horticultural Society's Award of Garden Merit.
Madame Emile Mouillere	*Hydrangea macrophylla*	White-flowered mophead cultivar. Large plant with slender stems.
Mariesii Variegata	*Hydrangea macrophylla*	Green and white variegated foliage. Attractive even if lacecap flowers destroyed by cold weather.
Masja	*Hydrangea macrophylla*	Mophead type bushy and compact cultivar with dark pink to pink colour flowers.
Nikko Blue	*Hydrangea macrophylla*	Most commonly available cultivar. Large blue mophead flowers. One of the most reliably flowering cultivars.
Nigra	*Hydrangea macrophylla*	Black stems. Light green foliage and large mophead flowers.
Penny Mac	*Hydrangea macrophylla*	Reported to flower well even following spring freezes. Attractive blue to pink flowered mophead.
Pia	*Hydrangea macrophylla*	Dwarf cultivar that rarely exceeds 0.6 m in height. Produces small bright pink mophead inflorescences throughout the summer. Very susceptible to Cercospora leaf spot. Leaves scorch in sun. Should not be planted in low pH soil.
Preziosa	*Hydrangea macrophylla*	Thought to be a hybrid between var. macrophylla and subsp. serrata. Small mophead inflorescences open white then gradually age to bright fluorescent pink. Not very cold hardy, but very attractive and unusual
Sister Therese	*Hydrangea macrophylla*	White-flowered cultivar. Not considered to be very cold-hardy.
Sumida-no-hanabi	*Hydrangea macrophylla*	Rounded, bushy, deciduous shrub with flattened flowerheads of double, star-shaped, white florets surrounding smaller, fertile flowers.

Cultivar	*Species*	*Characteristics*
Tokyo Delight	*Hydrangea macrophylla*	This is a compact lacecap hydrangea with mauve-pink inner fertile florets surrounded by showy white sterile florets which turn purplish with age. Have gained the Royal Horticultural Society's Award of Garden Merit.
Twist-N-Shout	*Hydrangea macrophylla*	Ever-blooming, hardy pink to blue lacecap type cultivar.
Veitchii	*Hydrangea macrophylla*	Lacecap type cultivar with small blue fertile flowers surrounded by large white sterile florets which turn pink with age. Have gained the Royal Horticultural Society's Award of Garden Merit.
Annabelle	*Hydrangea arborescens*	Commonly available cultivar with large ball shaped inflorescences that are comprised primarily of sterile flowers. Flowers initially a pale green, but at maturity become a pure white colour. Stems often pulled to ground by weight of flowers.
Incrediball	*Hydrangea arborescens*	This is an improved 'Annabelle' hydrangea with substantially sturdier stems and massive blooms. Each mophead inflorescence has roughly four times as many florets as 'Annabelle'. Lightly fragrant.
Incrediball Blush	*Hydrangea arborescens*	Incrediball Blush hydrangea adds a new colour to the series, with big, bodacious, blush-pink blooms that change to an attractive green colour as they age.
Invincibelle Ruby	*Hydrangea arborescens*	Dark, near-red flower buds open to a two-toned combination of bright ruby red and silvery pink. Deep green foliage and strong stems ensure outstanding performance in the container as well as in the garden. It is a strong rebloomer.
Invincibelle Garnetta.	*Hydrangea arborescens*	Blooming several weeks later than Invincibelle Ruby hydrangea, the flowers emerge dark burgundy then transform to a deep garnet-pink at maturity. Popular due to its unique, glossy, dark green foliage and dwarf-compact habit.
Invincibelle Limetta	*Hydrangea arborescens*	Beautiful rounded dwarf variety with dark green leaves and showy round flower heads that emerge a lush lime green, lighten to soft greenish-white, then age to green again.
Invincibelle Mini Mauvette	*Hydrangea arborescens*	A whole new colour for smooth hydrangea. This dwarf rounded plant has dark green leaves and showy, mauve-purple mophead flowers on rigid stems.
Invincibelle Wee White	*Hydrangea arborescens*	The smallest *H. arborescens* cultivar. This tiny, tidy reblooming hydrangea bears abundant flower heads that flush soft pink, age to white, then finish green. Strong, supportive stems prevent flopping.
Big Ben	*Hydrangea paniculata*	Panicles have a distinctive half-open outline, conical and tapered. Sterile florets open pale green, paling to greenish-white, maturing to deep pink. Very floriferous and strongly scented.
Brussels Lace	*Hydrangea paniculata*	Very lacy-looking flowers. Flowers from mid-summer till frost. Relatively small plant (to 5 feet) with dark green foliage.
Bobo	*Hydrangea paniculata*	Flowers are large white colour, held upright on strong stems, and take on nice pink tones as they age. This early bloomer is one of the smallest *H. paniculata* on the market.
Chantilly Lace	*Hydrangea paniculata*	Compact, dense plant with dark green leaves.
Fire Light	*Hydrangea paniculata*	Upright panicles are packed with florets that transform from pure white to pink and finish red by late summer. Its thick, sturdy stems present well in containers as well as the garden.

Cultivar	Species	Characteristics
Kyushu	*Hydrangea paniculata*	Flowers heavily from a young age. Pure white conical flower heads on upright stems.
Limelight	*Hydrangea paniculata*	Lime green flowers are held upright on strong stems, maturing to an incredible combination of green, pink and burgundy in autumn.
Little Lime	*Hydrangea paniculata*	Same as ever-popular 'Limelight' except in size. It is just a third the size of the 'Limelight'. It's a great choice for container gardens and smaller landscapes.
Pee Gee	*Hydrangea paniculata*	Rounded panicles of primarily sterile flowers. Large plant, often grown tree-form. Most commonly available cultivar.
Phantom	*Hydrangea paniculata*	Strong, stiff stems supporting large, very dense panicles, tapering above, rounded below. Leaves are yellow-green, stems green, maturing to mid-brown. Fertile florets are pale pink and white. Sterile florets white with some lime green flushing, deep pink at maturity.
Pink Diamond	*Hydrangea paniculata*	Very large inflorescences that age rapidly to pink. Both fertile and sterile flowers visible.
Pillow Talk	*Hydrangea paniculata*	Bright lime green flowers mature to pure white, then blush with pink as summer progresses. Large, full panicles are held upright on sturdy stems.
Quick Fire	*Hydrangea paniculata*	The earliest blooming *paniculata* on the market. White blooms appear early in the season in Michigan and turn red-pink before other paniculatas even begin to bloom. Noticeably high-quality foliage takes on excellent fall colour.
Silver Dollar	*Hydrangea paniculata*	Very dense, rounded, broadly conical panicles. Fertile florets are few, so the panicle is coloured by the crowded sterile florets, which are white becoming pink flushed. Slight fragrance. Suitable for small gardens.
Tardiva	*Hydrangea paniculata*	Later flowering than 'Pee Gee'. Panicles held upright on plant. Commonly available. Both sterile and fertile flowers visible, creating a lacy effect.
Unique	*Hydrangea paniculata*	Early flowering type. Flowers are age to pink where large sterile flowers hide most of the fertile flowers.
Alice	*Hydrangea quercifolia*	Large plant, 12 feet by 12 feet. Panicles are14" in length. Flowers age to rosy-pink. Selected by Dr. Mike Dirr at the University of Georgia.
Alison	*Hydrangea quercifolia*	Another Mike Dirr selection. Slightly smaller than 'Alice' and with more upright panicles. Fall foliage turns a fluorescent burgundy red.
Gatsby Gal	*Hydrangea quercifolia*	An elegant, compact oakleaf with lots of impressively large flowers. Full panicles of pure white blooms are held upright above the foliage, and are quite large relative to the plant's size, making it an ideal choice for smaller gardens.
Gatsby Moon	*Hydrangea quercifolia*	Densely packed, doubled flowers, with stronger stems than 'Harmony'. The pure white florets create full, dense mophead panicles with a unique 'quilted' effect. As the bloom ages, it turns a nice shade of green that lasts through most of the summer.
Gatsby Star	*Hydrangea quercifolia*	Doubled blooms are similar to those of 'Snowflake', but the individual petals are pointed instead of rounded. The result is a beautiful, lacy panicle and a very showy plant.
Harmony	*Hydrangea quercifolia*	Long panicles composed of primarily sterile flowers. Weight of flowers often pulls stems to the ground.

Cultivar	Species	Characteristics
Little Honey	*Hydrangea quercifolia*	The foliage of this branch sport of the dwarf selection 'Pee Wee' emerges gold and then changes to a pleasing chartreuse as the season progresses. Brilliant red autumn foliage.
Munchkin	*Hydrangea quercifolia*	Very compact with medium pink flowers. Developed by Dr. Sandra Reed. A commercial nursery producer.
Pee Wee	*Hydrangea quercifolia*	Compact plant, only growing to 3 feet by 3 feet. Good plant for small garden. Should be relatively easy to find.
Ruby Slippers	*Hydrangea quercifolia*	Compact with flowers that age to deep rose.
Snowflake	*Hydrangea quercifolia*	Double-flowered form. Long panicles often droop to the ground, but more graceful than 'Harmony'. Very attractive plant.
Snow Queen	*Hydrangea quercifolia*	Probably the most popular cultivar. Panicles are held upright. The sterile flowers are larger and more numerous, giving the inflorescence a denser solid appearance. Flowers age to a medium pink.

Source: Halcomb *et al.*, 2010; www.thespruce.com; www.rhs.org.uk; www.springmea downursery.com.

7.0 Genetics and Breeding

7.1.0 Genetics

Chromosome number within the genus ranges from 2n = 2x = 30 in *H. involucrata* Sieb. to 2n = 4x = 72 or 2n = 6x= 108 in *H. paniculata* Sieb. (Funamoto, T. 1988). Asiatic species have significantly larger genomes than North and South American species (Cerbah *et al.*, 2001). The Asiatic species *H. involucrata* (2n = 2x = 30) has approximately twice the nuclear DNA content of the American species *H. arborescens* (2n = 2x = 38) although both are diploid species (van Gelderen *et al.*, 2004). A considerable difference in chromosome size between these two species was strikingly apparent in their hybrid (Jones *et al.*, 2006). A somatic chromosome number of 2n = 2x = 36 was first reported for *H. macrophylla* in 1957 (Haworth-Booth, 1984). Kudo *et al.* (1999) reported 52 chromosomes in *H. macrophylla* ssp. *macrophylla* 'Blaumeise'. DNA content of *H. macrophylla* has been investigated using flow cytometry. Working with 42 ,6-diamidino-2-phenylidole (DAPI)-stained nuclei, Demilly *et al.* (1998) found 100 diploid and 21 triploid *H. macrophylla* ssp. *macrophylla* cultivars. All 23 *H. macrophylla* ssp. *serrata* taxa examined were diploids. Unfortunately, these authors did not list names of any cultivar evaluated in this study. They reported that flow cytometric results were supported by chromosome counts with diploids having 36 and triploids having 54 chromosomes, but no details as to cultivars examined or numbers of cells counted were provided. Flow cytometric measurements of nuclear DNA content of five *H. macrophylla* ssp. *macrophylla* and five *H. macrophylla* ssp. *serrata* plants were made using DNA intercalating, GC-specific and AT-specific fluorochromes (Cerbah *et al.*, 2001). The *H. macrophylla* ssp. *macrophylla* plants had 11.7 per cent more nuclear DNA than the *H. macrophylla* ssp. *serrata* plants. Both subspecies had similar GC percentages of 41.0 and 40.6 per cent, respectively.

Information on ploidy level of specific *H. macrophylla* cultivars will help in selecting parents for both intra- and interspecific hybridizations. Jones *et al.* (2006)

Cultivars of Hydrangea

'All Summer Beauty' 'Alpengluhen' 'Altona'

'Ami Pasquier' 'Ayesha' 'Blaumeise'

'Blue Bird' 'Blushing Bride' 'Dooley'

'Endless Summer' 'Eldorado' 'Europa'

Cultivars of Hydrangea

'Annabelle' 'Invincibelle Ruby' 'Invincibelle Limetta'

'Invincibelle Wee White' 'Brussels Lace' 'Bobo'

'Fire Light' 'Silver Dollar' 'Phantom'

'Gatsby Gal' 'Little Honey' 'Snow Queen'

evaluated the ploidy level in a diverse group of *H. macrophylla* cultivars. Ploidy was first estimated using flow cytometry and then confirmed in selected diploid and triploid plants using chromosome counts. The potential for using stomatal guard cell length and pollen grain size to identify ploidy level of *H. macrophylla* plants was also examined. Finally, fertility of triploid cultivars was estimated through pollen staining and controlled pollinations.

7.2.0 Breeding

Historically, hydrangea breeding has involved intraspecific crosses within *H. macrophylla*, *H. paniculata*, or *H. serrata*. Intra-specific crosses involve two plants from the same species, while interspecific crosses involve plants from two different species in the same genus. Of the three above-mentioned species, the primary focus has been improvement of *H. macrophylla*; which was first bred by the French in the early 1900s. Only limited germplasm was available during that time period to European breeders and, according to Haworth-Booth (1984), for 40 years *H. macrophylla* cultivars 'have been bred one to another, and in that time almost every possible combination of the available genes has been given birth.' Almost hundreds of *H. macrophylla* cultivars were bred in Europe in the early 20th century. Most of the early breeding focused on the production of cultivars with improved flower size and colour for the pot plant or florist market (Lawson-Hall *et al.*, 2005). Many of these cultivars are used today in the landscape. During its several centuries of cultivation in Japan, numerous cultivars of *H. macrophylla* ssp. *serrata* (including double-flower, variegated, and coloured leaf forms) were selected (Dirr, 2004).

Breeding programs in hydrangea usually have specific goals depending on how the plants will be used. Breeding landscape hydrangeas focuses on whether the plant forms a rounded or mounded shrub composed of erect, unbranched stems, whereas florist hydrangea breeding programs target stem strength and durability. However, flowering traits are always considered as a priority for any hydrangea breeding program, as flowers present the key horticultural attraction for hydrangea as for many other ornamental crops. Breeders should strive to develop plants that are adapted to garden culture and provide multiple seasons of interest. Traits of interest for *H. macrophylla* breeding include: the ability to rebloom (remontancy), inflorescence type, improved flower colours, fragrance, showy fruits, improved foliage, fall colour, pigmented stems, strong stems, compact habit, cold hardiness, drought tolerance, disease resistance, and pest resistance (Kardos, 2006).

Wild hydrangea accessions with superior characteristics have been used as breeding parents to create attractive cultivars, and it has a long history of use as an ornamental garden plant in temperate regions. In hydrangea, there are both decorative and non-decorative flowers within groups of flowers on the stems of the Hydrangea inflorescence. Decorative flowers have large ornamental petaloid sepals that attract pollinators, whereas non-decorative flowers have inconspicuous perianths with normal sepals that instead of play a major role in seed production (Uemachi *et al.*, 2004; Uemachi *et al.*, 2012). In hydrangea, there are two decorative flower phenotypes: single flower and double flower. Double flowers do not have stamens or petals in the decorative flower. Therefore, petals and stamens could be converted to petaloid sepals since the number of petaloid sepals increases, and stamens and petals are lost. Because of their high ornamental value, the production of double flowers is an important breeding target in hydrangea cultivation. The double flower cultivars 'Sumidanohanabi' and 'Jogasaki' have been used as breeding parents in Japan to obtain double flower progenies (Suyama *et al.*, 2015).

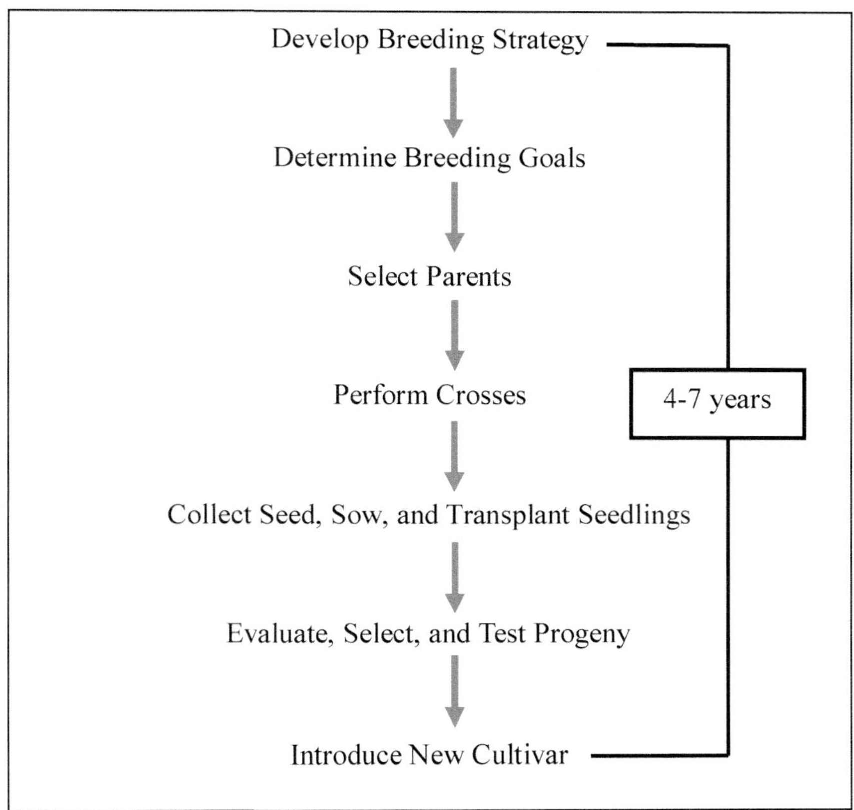

***Source*: Kardos (2006).**

Traditional breeding of hydrangeas for trait improvement takes at least two years before promising plants can be selected, cloned, and evaluated in large numbers. The time-to-market for new cultivars can be reduced greatly by identifying molecular markers associated with specific traits. In the present study, a SNP associated with inflorescence type was identified and converted to both CAPS and KASP-SNP markers for small and large population selection in *H. macrophylla*. Inflorescence type was detected with 100 per cent accuracy using the CAPS marker in a validation panel. KASP-SNP genotyping revealed that 76 per cent of progenies in an F_2 breeding population produced lacecap inflorescences. These plants could be discarded as seedlings before further container or field evaluation in a breeding program targeting mophead inflorescence type. Furthermore, the SNP markers developed here will be useful to understand the biological mechanisms behind inflorescence architecture as improved genome resources for hydrangea become available (Wu *et al.*, 2020).

Interspecific crosses may prove important sources of genetic diversity and unique characteristics. Recently, Rinehart *et al.* (2006) developed simple sequence repeat or microsatellite (SSR) markers for hydrangeas. These markers may reveal genetic relationships among species, assess genetic diversity among and within species, verify hybridity of progeny from intraspecific and interspecific crosses, and analyze parentage. Research is ongoing to identify markers linked to traits of interest, such as remontancy and disease resistance, for use in marker-assisted selection (MAS). Several researchers have determined chromosome number, genome size, and ploidy level for many different species and cultivars of Hydrangea (Demilly *et al.*, 1998).

8.0 Propagation

Hydrangeas can be propagated through seed, cutting and grafting. As hydrangeas produce spectacular flowers, they rarely, if ever, produce seeds. This can create a problem for a gardener who wants to propagate hydrangea shrubs. Because of this, propagating hydrangeas is typically done from cuttings.

8.1.0 Seed Propagation

Seed propagation is desirable when breeding or when looking for novel characteristics such as flower colour or compact branching structure. Collect seeds in late fall or early winter, crush capsules and clean excess trash from the seeds. The seeds are small, so take care to sow evenly. Generally, moist pre-chilling (stratification) is not required for germination; however, 30-90 days of stratification may increase the speed of germination and uniformity. Sow seeds in flats on surface or medium and keep moist, preferably in a greenhouse or other high-humidity enclosure where the temperature can be controlled. Germination occurs within 10-30 days. The seeds of *H. arborescens* and *H. macrophylla* sown on milled sphagnum will germinate in two to three weeks. *Hydrangea paniculata* seeds require no pretreatment; plants will flower the same year seeds are sown. Similarly, fresh *H. quercifolia* seeds, sown directly after collection, germinate at high percentages two weeks later (Dirr, 2004).

8.2.0 Vegetative Propagation

8.2.1 Cuttings

Cuttings are commonly used to mass propagate hydrangeas for nursery production. In general, stem cuttings are collected earlier in the season (May) will yield more robust, faster-growing plants than those collected later in the summer (July), but cuttings collected over that time frame all root readily. Hydrangea cuttings can be taken from the tip of the vegetative shoots or from a mid- stem section of the shoot. Tip cuttings and mid-stem cuttings are both more successful if they are taken from softwood stems. Softwood stems are neither new and green nor old and woody. If new growth is used as a cutting it will rot before it has time to root, while older stems will be slow to root. Hydrangea stems that are ready for cutting

Propagation of Hydrangea

Seed Pod **Germinating Seeds** **Seedlings Arise from Seeds in Pots**

Propagation through Cutting

Propagation of Hydrangea

Propagation through Ground Layering

Propagation through Air Layering

Propagation of Hydrangea

Propagating through Division

In vitro Propagation

a) The leaf on the right shows the way the explant was cut before cultivation on the media. b) Adventitious shoots regenerated from the cut edge of leaf explants. c) Induction of adventitious shoots via direct organogenesis on B5 medium supplemented with 2.25 mg/l BA and 0.1 mg/l IBA 50 days after culture. d) Comparison of rooted plantlets (i) cultured on perlite: liquid half strength B5 basal medium = 2:1 (v/v) plus 0.5 mg/l NAA and (ii) cultured on half strength B5 basal medium containing 0.5 mg/l NAA, 30 g/l sucrose and solidified with 7 g/l agar; e) Acclimatized plantlets growing ex vitro conditions Collapse (Liu *et al.,* 2011).

should snap easily when broken. When choosing stems from which to make cuttings, it is also important to choose shoots that are healthy and vigorous. Hydrangeas, especially *H. quercifolia*, have large leaves that can be cut by half or more to limit leaves from transpiring as well as from blocking mist during rooting. Next for better rooting cuttings are treated with IBA through quick dips method or through foliar applications and then placing them in a moist rooting media that should consist of 1 part peat/potting soil and 1 part medium to fine grade perlite, vermiculite, or sand. Since the hormone used is the potassium (K) salt of indole 3-butyric acid (IBA), or K-IBA, which is water-based, there is no phytotoxicity associated with other solvents like alcohol. When IBA was applied via foliar spray to stem cuttings of *H. paniculata* at 50 ppm K-IBA or 50 ppm K-IBA + 25 ppm naphthalene acetic acid (NAA), rooting percentage, root length and shoot length were similar to stem cuttings treated traditionally with a basal quick dip of either 1,000 ppm IBA or 1,000 ppm IBA + 500 ppm NAA.

Cuttings should be watered by misting. The cuttings cannot take over watering, but the soil also cannot be dry. Keeping a moist soil is essential, so cuttings should be checked every day. Most people don't have greenhouses so cover the cuttings with plastic. This can be done by adding stakes to keep the plastic propped up above the cuttings. Then cuttings should be kept in an area out of full sunlight and wind, where the temperature is between 15.5 and 26.6 °C. If everything is done correctly cuttings should produce new growth within 2 weeks and by the end of week 4 the new roots should be formed and ready for transplant (Ed. Jim Nau, 2011; Grillo, S. 2012; Howe, K, E. 2005; Blythe *et al.*, 2003; www. extension.tennessee.edu; www. ndsu.edu).

8.2.2 Layering

This method of propagation is probably the easiest one of all. However, the only drawback is that one can only propagate a few plants at a time. This method is mainly adopted in *Hydrangea paniculata* during winter month. First a branch selected that is close to the ground. A small amount of bark on the underside of the branch should be scraped off also to encourage rooting. Probably the most important thing while propagating this way is to make sure there will be at least one leaf node under the ground when the branch is finally placed. This is important because this node or nodes is where roots will usually form. Before placing the branch under the ground, a trench should be dug to the length of the leafless branch. This trench provides the spot for the branch to be placed. Once the branch is in the trench it should be covered thoroughly and watered. A brick or stone should be placed on the newly watered soil to help hold moisture in and keep the branch under the soil. It should be watered occasionally and eventually roots should form. When it does, it can be cut from the mother plant, potted up, and treated like a mature cutting. To Propagation by ground layering decrease the chance of rot or disease all leaves located on the length of the branch that will be placed under the ground must be removed (www.ndsu.edu; www. heritagehydrangeas.com).

8.3.0 Micropropagation

Tissue culture method is part of biotechnology that is used for massive propagation especially for horticulture crops and ornamental plants. So many factors such as growth regulator, plant and explants type, environmental condition (temperature, light) influence organogenesis and in vitro multiplication. It seems that among these factors, growth regulators have the most effect on plant in vitro micropropagation (Jain, S. M. 2002). Sebastian *et al.* (1987) reported the first micropropagation protocol for *Hydrangea quercifolia* Bart. The influence of thidiazuron (TDZ) on in vitro shoot proliferation was demonstrated by Preece *et al.* (2002) and Ledbetter *et al.* (2004). Low concentrations of this growth regulator induced a low number of long shoots but high concentrations of TDZ ensured many short adventitious shoots. Abou Dahab, T. A. M. (2007) set up a protocol for micropropagation of *Hydrangea macrophylla* for commercial production. To optimize in vitro adventitious shoot regeneration in *H. macrophylla* 'Nacthtigall', experiments on salts compositions (MS or B5 - full or half strength), on different cytokinins applied (BAP or TDZ or m-Top) and onto the propagations way (Solid Medium or Temporary Immersion System or Permanent Immersion System) were performed by Doil *et al.* (2008). Sacco *et al.* (2012) reported that *H. quercifolia* 'Snow Queen' showed an efficient in vitro propagation aptitude combined with a good propagation rate when culturing the explants onto agarised MS medium with BA 0.25 mg/L.

Excise apical and axillary buds of young branches of hydrangea ornamental genotypes from mature greenhouse-grown plants, maintained at 20°C in 14-cm diameter pots, under natural light conditions; excising explants during April-May (northern hemisphere) produce better results. Firstly, rinse explants in a detergent-warm aqueous solution with a few drops of liquid dish soap. Then, sterilize with 70 per cent ethanol for 30 s, treat with NaOCl solution (5 per cent active chlorine) for 15 min, and rinse twice with sterilized distilled water. Transfer explants on the medium for in vitro culture establishment for 30 days (Ruffoni *et al.*, 2012). Vigorous shoots of hydrangea were multiplied in vitro onto Murashige and Skoog (MS), 3 per cent sucrose, supplemented with 0.25 mg/l of 6-benzyladenine (BA) and agarized with 8 g/l of technical Agar (Sacco *et al.*, 2012). Arafa *et al.* (2017) repoted that application MS medium on shoot tip of *H. macrophylla* containing 1.0 mg/l Kin and 1.0 mg/l BA produced higher number of shoots and leaf, while 2.0 mg/l Kin and 3.0 mg/l BA produced longest shoot. They also reported 1.0 mg/l BA and 0.5 mg/l Kin concentration on nodes explants developed highest shoot number, leaf number.

After shoot multiplication singularize the shoots, 2.5 cm long, by a transverse cut, discarding the explants base of the clusters coming from the multiplication media. Then transfer the shoots onto the rooting medium containing aqueous K-IBA solutions (5,000 or 10,000 ppm) for 30 s. Immediately transfer the dipped explants to the acclimatization glasshouse in plastic alveolar pots under mist system, 10 s vaporization every 30 min, with 70 per cent relative humidity. After 3 months transfer the acclimatized plants to 11-cm diameter pots filled with the same substrate (Ruffoni *et al.*, 2012). Shoot organogenesis was also successful from leaf explants of *H. macrophylla* 'Hyd1' and the genetic stability of regenerates using ISSR markers was recently reported (Liu *et al.*, 2011).

9.0 Soil and Climate

9.1.0 Soil

Well-drained, moist soil, preferably a fertile sandy or silt loam with greater than one percent organic matter, is recommended for field planting. Nevertheless, hydrangeas have been produced successfully in clay soils with proper water management. Hydrangeas generally prefer slightly acidic soils in the pH range of 4.5-6.5, and manipulation of the soil pH is one key factor in controlling bloom colour. A medium to high level of phosphorus and potassium is desirable. Soil analyses must be performed to determine the specific needs of each soil (Halcomb *et al.*, 2010; www.extension.tennessee.edu; www.bartlett.com).

9.1.1 Impacts of Soil pH on Blossom Colour

Hydrangeas are nature's litmus test which means it determine the pH of the soil where it is growing in. *Hydrangea macrophylla* mostly planted in the garden for this purpose. The flowers of this plant are sensitive to soil pH and change colour to reflect changes in acidity. Pink Hydrangeas indicate soil that is alkaline to neutral. Blue flowers indicate soil that is acidic (www.naturalnavigator.com).

The true cause of colour variation in hydrangea blooms is the amount of aluminum found in the flower tissue. This element binds with certain pigment complexes, causing them to appear blue in the presence of aluminum, and pink in its absence. Changing bloom colour is not as simple as adding aluminum to the soil, because the plants' ability to take up aluminum is largely dependent upon the soil pH, and to a lesser extent dependent upon the presence of phosphorous.

Blue flowers: Hydrangea blooms will be blue under most natural conditions due to the fact that most soils have some aluminum present, and have the appropriate pH for uptake of that element. The ideal pH for blue flowers is in the range of 4.5 to about 5.5 because this is the ideal range for aluminum uptake. The presence of phosphorous can bind aluminum making it unavailable for uptake, and therefore prevent blue colour even when ample aluminum is present in the soil. This should be a consideration when choosing fertilizer formulations. If a soil test shows that pH is in the correct range but aluminum is lacking, it can be added to the soil in the form of dissolved aluminum sulfate. This will add aluminum and acidify the soil, but care must be taken because aluminum can become toxic at high levels. Also, direct contact of this solution with foliage will cause burn.

Pink flowers: In order to produce pink blooms on hydrangea, it is necessary to inhibit the uptake of aluminum. This can be achieved by raising pH to levels where uptake is inhibited, normally in the range of 5.7-6.2. Exact pH levels will vary and intermediate colours will result from pH levels in the middle of these ranges. Soil pH can be raised using lime, but care must again be taken to avoid direct contact with foliage. Raising the pH too high will lead to iron deficiency because uptake of this element is reduced under basic conditions. Exact rates of aluminum sulfate or hydrated lime will vary depending on existing pH, aluminum levels, and the buffering capacity of the soil. The only way to determine the pH, aluminum level, and nutrient needs of a soil is to submit a sample for a laboratory analysis. Soil nutrient analyses must be conducted before soil amendment takes place (www.bartlett.com).

9.2.0 Climate

9.2.1 Temperature

In hydrangeas flower initiation occurs over a wide range of temperature but a temperature of 15-18 °C for at least 6 weeks is optimum for the flower bud's formation. Subsequently, a six-week period at 5 °C is required to start flower forcing. Generally, it requires a cool environment with high humidity and mild temperatures. It is necessary to protect the plant because it does not tolerate frost and temperatures below 4 °C.

9.2.1.1 Flower Forcing

It consists in maintaining the plants in dormant state. It is carried out at temperatures above 15 °C for 70-90 days. If flower forcing is too much, it may reduce the quality of flowers. At first it is important that the relative humidity is high, so it will be convenient to water and sprayed the surroundings. During this period nitrogen fertilization is necessary.

9.2.2 Light

The hydrangea cannot live under the sun all year round except in coastal areas with mild climate. Shades must be available especially in the middle of the day and if it is possible, it should be a partial shade. A location with an eastern or southern exposure that provides afternoon shade offers the optimal amount of sunlight for healthy hydrangeas.

9.2.3 Moisture

Soil moisture results in healthy hydrangea shrub growth and flowering. Regular irrigation helps provide the needed moisture, but it's also important to select a well-draining planting site. After rain or irrigation, water tends to collect in low-lying sites with dense or heavy soils and is absorbed too slowly back into the soil. Standing water near the surface can also cause rot problems in the roots or near the bottom portion of the trunk (Litlere *et al.*, 1975; www.agriculture.infoagro.com; www.homeguides.sfgate.com).

10.0 Cultivation

10.1.0 Manuring and Fertilization

Hydrangeas are heavy feeder of major nutrients *i.e.*, N, P, K particularly during vegetative growth phase. So, use of an optimum dose of N P K for obtaining good vegetative growth, profuse flowering, improved cut flower yield and quality attributes of hydrangeas for commercialization (Thaneshwari, 2014). Fertilizing once or twice yearly provides the soil nutrients needed for hydrangea growth. A balanced fertilizer, such as a 10-10-10 variety, is appropriate in spring, when new growth begins. A second fertilization after flowering replenishes the nutrients used during the bloom period (www.homeguides.sfgate.com). Each type of hydrangea

Cultivation of Hydrangea

In Open Condition

In Greenhouse Condition

In Containers

has different needs and will benefit from different application timing. Bigleaf hydrangeas can benefit from several light fertilizer applications in March, May and June. While, Oakleaf and panicle hydrangeas do best with two applications in April and June. Smooth hydrangea plants only need fertilization once, in late winter (www.almanac.com).

Nutrient management in nursery production of hydrangeas is largely depending on manipulation of N rate as it is often considered the limiting factor. Increasing N rate increased plant growth index, leaf area, and plant dry weight of *H. macrophylla* 'Merritt's Supreme' (Li *et al.,* 2019). High N fertigation rates of 210 and 280 mg/l were found to increase the number of flowers and flower size in *H. macrophylla* 'Merritt's Supreme' (Bi *et al.,* 2008a). High N rate increased N content in H. *macrophylla* 'Berlin', where a good N status was considered important in the fall (Bi *et al.,* 2008b). Stored N in the fall will be remobilized the following spring to facilitate new growth (Millard, 1993). Hydrangeas may develop chlorosis, a yellowing of the leaves, if the soil lacks sufficient iron, but a chelated iron spray applied in spring or summer remedies the deficiency (www.homeguides.sfgate.com).

In case of growing hydrangeas in containers, dolomite [$CaMg(CO_3)^2$] is used to supply calcium and magnesium to the plant and is amended to the substrate to have the desired pH. Micronutrients in addition to sulfur are also added. Gypsum ($CaSO_4$) can be used as an alternative to dolomite to supply calcium if no pH adjustment is desired (www.extension.tennessee.edu).

10.2.0 Watering

For the first or second year after planting and during any drought, hydrangeas required plenty of water. Water provided at a rate of 1 inch per week throughout the growing season. It's better to deeply water 3 times a week than sprinkle water in a shallow manner. This encourages root growth. Bigleaf and smooth hydrangeas require more water, but all varieties benefit from consistent moisture. Leaves will wilt if the soil is too dry, and flowering will be hampered by a lack of water. Use a soaker hose to water deeply and keep moisture off the flowers and leaves. It's best to water in the morning to prepare hydrangeas for the heat of the day and to avoid disease. Add organic mulch underneath your hydrangeas to help keep the soil moist and cool, add nutrients over time, and improve soil texture (www.almanac.com). Water use by hydrangea started to decrease at a higher VWC (0.28 m^3/m^3) and water uptake ceased at 0.16 m^3/m^3 in hydrangea (O'Meara *et al.,* 2014). For hydrangea, relatively higher water content in the growing substrate should be maintained to maximize plant growth and avoid water stress.

10.3.0 Pruning

Pruning is probably the most important step in taking care of the hydrangea. It is good to follow the four D's of pruning, *i.e.,* removing any dead, diseased, dying or damaged plant material. But also, it should be known that some hydrangea bloom on dead wood, which shouldn't be pruned (www.ugaoo.com). The first step of pruning hydrangeas is to identify the variety, as this determines how, when, and even if it needs pruning. There are 2 main groups of hydrangeas:

Pruning of Hydrangea

Pruning in Group 1 (in spring)

Pruning in Group 2 (in late winter)

10.3.1 Group 1

Those that bloom on last year's growth, or old wood, and should be pruned in late summer: Oakleaf hydrangeas (*H. quercifolia*), Bigleaf hydrangeas (*H. macrophylla*), Mountain hydrangeas (*H. serrata*) and Climbing hydrangeas (*H. petiolaris*).

Methods

Flower buds actually form in the late summer and flower afterwards the following season, so avoid pruning after August 1. Only cut away dead wood in the fall or very early spring. Cut one or two of the oldest stems down to the base to encourage branching and fullness. If the plant is old, neglected, or damaged, prune all the stems down to the base. As a result, plants lose the flowers for the upcoming season, but also rejuvenate for future years. It's best not to deadhead (remove faded blooms) on the big Mopheads; leave them over the winter and cut them back in early spring (to the first healthy pair of buds). It's fine to deadhead the lacecaps; cut down to the second pair of leaves below the flower head. When growing *H. macrophylla* and *H. serrata* varieties in Zones 4 and 5, do not prune unless absolutely necessary, and then do so immediately after blooming. Otherwise, remove only dead stem in the spring.

10.3.2 Group 2

Those that bloom on new growth, or new wood, and should be pruned in late winter to early spring: Smooth hydrangeas (*H. arborescens*) and Panicle hydrangeas (*H. paniculata*).

Methods

Prune in the late winter when the plant is dormant. So that if the buds are killed during the winter, the plant will produce new buds in the spring which will produce blooms. Cut branches back by one-half to one-third, cutting just above a node. Next, remove any weak or spindly branches. For *H. arborescens*, minimal pruning promotes large vigorous shrubs with numerous, smaller flower heads. Hard pruning 12 to 18 inches from the ground or even all the way to the ground, will produce fewer, but larger, flower heads that may flop without propping. For *H. paniculata*, in order to create a strong framework, prune out surrounding smaller wood, leaving the larger stems.

There are also new varieties called 'reblooming' or 'remontant' that bloom on both old and new wood. These types don't require pruning at all, just maintenance for plant health such as cutting out dead, diseased, damaged or crossing branches (www.almanac.com; www.gardendesign.com).

10.4.0 Re-blooming

Potted hydrangeas add a burst of spring colour to winter and early spring arrangements. The florist hydrangea varieties which are usually grown indoors can be transplanted outdoors after they stop blooming, it's also possible to force them into bloom a second time as a potted plant. Re-blooming hydrangeas requires proper pruning and care, along with a dormant period, to produce attractive flowers.

First cut back the hydrangea after the last blooms wilt and die. Trim each stem so only the bottom two sets of leaves or leaf buds remain on the stem, using clean pruning shears. Mix together equal parts potting soil and peat moss. Fill a pot one size larger than the old container with the soil mixture. Lift the hydrangea out of the old pot, and brush off as much of the old soil as possible. Replant it into the new container at the same depth it was growing at previously. Set the pot in a location that receives all-day sunlight. Move the plant outdoors after all frost danger has passed in spring, and place it so the hydrangea receives full morning sun and light afternoon shade. Water the hydrangea when the top 2.5 cm of soil begins to feel dry. Allow the excess water to drain from the pot. Then dilute 1/2 teaspoon of 24-8-16 fertilizer in 3.78 l of water, and irrigate with the fertilizer solution every two weeks. After that pinch back the stems to the topmost leaf at four-week intervals beginning when you repot and continuing until early July. Prune the hydrangea so it only has three or four main stems if you want larger flower clusters. Remove the interior stems back to their base and leave the exterior stems to grow on. Then bring the pot indoors in fall before the first frost. Continue to water the soil as needed until the beginning of November, and then remove any remaining foliage on the plant. Move the hydrangea to a dark, 1.66 to 4.44 °C location and stop all watering for six weeks. Move the hydrangea out of storage at the end of the dormancy period. Supply it with all-day sunlight, and resume regular watering and fertilization. Hydrangeas typically begin to bloom within four months after dormant storage (Harrington, 2017; www.gardenguides.com).

10.5.0 Indoor Cultivation

Hydrangea macrophylla, mainly the mophead hydrangea is best for growing indoors. An ideal location for indoor hydrangeas is somewhere that gives the plant morning shade and afternoon sun, is free from drafts and away from direct heat sources. It's the reverse of outdoor growing conditions because it's warmer indoors. Rooms with average to high humidity are favorable. Plants are selected with just a few blooms and lots of buds still developing. The plant should fill out the pot with healthy dark green leaves. The ideal temperature in flowering season is 21 °C, dropping to 15.5 °C during the night and for bud development, 6- weeks of temperatures at 18 °C.

For indoor plant soil, use a potting soil mix with some peat moss then apply a thin layer of shredded bark across the top of the soil to help with water retention. They require bright light, but not intense, direct light that could scorch their leaves. It is better, applying a weak fertilizer or a slow-release fertilizer. When adding a slow-release fertilizer at planting, add a bit of compost to the soil later during the growth and blooming season. With the weak solution, need to repeat applications every 7-14 days, depending on dosage recommendations. For watering indoor hydrangea, the rule to follow is not to let the soil become waterlogged. It's best to finger test the top inch of soil twice daily and water the plant before it dries. Hydrangeas need consistently moist soil but never too moist that the soil waterlog. Signs of watering problems include wilting, yellowing leaves and leaf drop. If any of those are present, it's likely the plant requires watering more frequently (www. guide-to-houseplants.com; www.plantindex.com; www. thepracticalplanter.com).

11.0 Diseases, Pests and Disorders

11.1.0 Diseases

11.1.1 Botrytis Blight (*Botrytis cinerea*)

Botrytis blight, also called gray mold, is a serious disease problem in the production and postharvest of floral crops. All species of hydrangea are susceptible to this disease, but bigleaf hydrangea is most susceptible. This fungus can severely affect the flower buds and even kill them before they open. In addition, infected flower parts can fall on the leaves and infect them. The first symptoms are water-soaked spots on the flowers. However, these grow into reddish brown lesions. Botrytis is more likely to be a problem under cool and damp conditions, such as several days of cloudy, humid, and rainy weather. To prevent this infection, keep the humidity low. Avoid watering late in the day and only water at the roots, so that the flowers and leaves are not wet. Keep good airflow around the plants. Space them properly, and prune branches that are closely spaced. Treat pruning shears with bleach to prevent accidentally spread disease. Also remove dead or damaged flowers and leaves to prevent the fungus from gaining egress into the plant. Clean up debris around the plant, so that Botrytis can't live on the dead tissue. Some effective fungicides include iprodione, mancozeb, and thiophanate-methyl also recommended for controlling this disease. Some biofungicides, such as Mycostop (*Streptomyces griseoviridis* Strain K61), actinovate (*Streptomyces lydicus*) found effective to control Botrytis blight (Li *et al.*, 2016; George, 2019).

11.1.2 Leaf Spots (*Cercospora* sp. and *Phyllosticta hydrangea*)

This disease generally affects bigleaf and smooth types of hydrangea as well as oakleaf hydrangea in both landscape and nursery settings. *Cercospora* manifests as circular purple or brown spots on the bottom of the plant. As the lesions get larger, the leaves can turn yellow and fall off the plant. Watering without getting the leaves wet will help to prevent these diseases. Fungicides containing chlorothalonil, myclobutanil, or thiophanate-methyl are highly effective if applied at the onset of the disease (Mmbaga *et al.*, 2012; Li *et al.*, 2016; George, 2019).

11.1.3 Anthracnose (*Colletotrichum gloeosporioides*)

Anthracnose of hydrangea found periodically in landscape and field plantings of bigleaf hydrangea. This disease affects both leaves and blossoms of hydrangea plants. Plants that have been heavily fertilized are more likely to contract this common fungal pathogen. Continued rainy weather or heavy fog produces the conditions that favor infection. The fungus produces large brown spots on the leaves or flowers that will become more lightly coloured in the centers. One distinctive symptom is that spots by the veins develop at an angle. The use of protective fungicide treatments is a viable option to protect susceptible hydrangea from anthracnose. The application should be done at 10-14-day intervals during the summer for good efficacy. Some of the recommended fungicides include chlorothalonil and thiophanate-methyl *etc.* (Li *et al.*, 2016; George, 2019).

Diseases of Hydrangea

Botrytis **Blight infection**

Cercospora **Leaf Spot infection**

Anthracnose infection

Powdery Mildew infection

Rust infection

Bacterial Leaf Spot infection

Hydrangea Mosaic Virus infection

Hydrangea Ringspot Virus infection

11.1.4 Powdery Mildew (*Golovinomyces orontii* DC)

The disease is a common problem in bigleaf hydrangea (*H. macrophylla*) in greenhouses, nurseries, and landscapes when environment conditions are warm and humid. The disease also affects *H.* serrata as well as smooth (*H. arborescens*), bigleaf (*H. macrophylla*), and panicle (*H. paniculata*) hydrangeas. Powdery mildew manifests as a white powdery substance on the surface of the leaves. Developed white, cottony growth on the bottom of the leaves. Left unchecked, the fungus can infect the newly developing buds and stunt their growth. Powdery mildew is most likely to be a problem on hydrangeas when the days are warm and the nights cool. This disease can be control by reducing humidity and increasing air circulation. One way to control this disease is to apply a fungicide include azoxystrobin, fenarimol, paraffinic oil, and thiophanate-methyl as soon as it reported. Another option is to use neem (Li *et al.*, 2016; George, 2019).

11.1.5 Rust (*Pucciniastrum hydrangea*)

Like other rusts, hydrangea rust needs two hosts to survive and does not kill either of them. This rust only infects the smooth hydrangea, *H. arborescens* and hemlock as its alternate host. The first symptoms are orange pustules on the bottoms of the hydrangea leaves and yellow spots on top. This disease is difficult to control, can be manage by cleaning up infected leaves and debris that has fallen to the ground around both hosts. Planting resistant cultivar 'Frosty' (George, 2019) can be followed.

11.1.6 Phoma Leaf Spot (*Phoma* spp.)

This disease can occasionally cause leaf spots in hydrangea. *Phoma exigua* has been reported to cause leaf spot disease in hydrangea in different parts of the United States and central Italy. Phoma leaf spot is characterized by small necrotic spots surrounded by chlorotic haloes on the upper side of infected leaves. As a result infected leaves become abscised leading to premature defoliation. Infected plants rarely die, but the presence of lesions on mature plants decreases aesthetic quality and subsequent market values. Sanitation and removal of diseased leaf debris help to eliminate source of infection (Garibaldi *et al.*, 2006; Li *et al.*, 2016).

11.1.7 Alternaria Leaf Spot (*Alternaria* spp.)

Alternaria leaf spot of hydrangea can be caused by several Alternaria species that has been reported in Italy and the United States. Infected plants rarely die, but the presence of lesions reduces the aesthetic quality and subsequently the commercial value. Initial symptoms of Alternaria leaf spot are small brown spots with yellow halos that appear on the upper surface of the infected leaves. And then, lesions become dark brown or black and often coalesced into large necrotic areas. Symptoms are commonly observed on the leaf margins and near the petioles. Severely affected plants are defoliated. To prevent this disease spacing plants to minimize dense canopy formation and improve air circulation. Avoiding overhead irrigation. Appropriately heating and ventilating greenhouses to reduce humidity. Removing infected leaves from infected plants and destroying them or putting them in trash (Garibaldi *et al.*, 2006, 2008).

11.1.8 Bacterial Wilt (*Ralstonia solanacearum*)

This important bacterial disease first manifests as blight in the leaves and flower clusters. However, more severe infestations can cause both wilting and root rot. Bacterial wilt occurs mainly in hot weather and heavy rains. There are no chemical options to control this disease (George, 2019).

11.1.9 Bacterial Leaf Spot (*Xanthomonas campestris*)

Bacterial leaf spot is a common problem on oakleaf hydrangea, smooth hydrangea, and bigleaf hydrangea. The bacteria that cause this disease can enter the plant through natural openings like stomata or through wounds. The first symptoms are water-soaked spots. The spots darken and become angular in shape. These spots become larger lesions and can kill the leaves. Normally, the disease starts on the lower leaves and spreads upward on the plant. To control this, Avoid overhead irrigation, space planting. Promptly remove and discard severely infected leaves during the growing season. At the end of a season, remove and destroy fallen leaves. Kaolin clay (Surround) and neem oil (Neem Gold) were reported to reduce disease severities of bacterial leaf spot in the field experiments. Copper-containing fungicides may provide some control when applied in late spring, but will not be effective in plants that are grown under conditions favoring the disease (Mmbaga *et al.*, 2007; Li *et al.*, 2016; George, 2019).

11.1.10 Hydrangea Ringspot Virus (HRSV)

The early symptoms of HRSV on hydrangea leaves are brown spots or rings, and then plants show leaf distortion, leaf roll, and stunted growth. The virus can be transmitted mechanically through saps, but aphids or seed does not transmit HRSV. Cultural control of HRSV includes buying clean stock and sanitizing pruning tools (Daughtrey *et al.*, 1995).

11.1.11 Hydrangea Mosaic Virus (HdMV)

Hydrangeas infected with this virus will have a pattern of yellow mosaics on their leaves. This is another virus that is not transmitted by aphids. However, it can be spread through tools. So, disinfect the pruning shears to avoid inadvertently spreading this virus (George, 2019).

11.1.12 Tomato Ringspot Virus (ToRSV)

This virus causes the leaves to turn yellow and become distorted, and the growth of the plant will be stunted. Nematodes, not pruning tools, spread this virus. If growing hydrangeas in containers, then avoid tomato ringspot virus by using a soil mix that is free of nematodes (George, 2019).

11.2.0 Pests

11.2.1 Aphids

Aphids can sometimes build up to high populations on the new growth of hydrangeas. At high numbers, aphid feeding may cause some leaf yellowing or distortion. In addition, aphids excrete a liquid waste high in sugars called honeydew.

Pests of Hydrangea

Two Spotted Spider Mite Infestation

Black Vine Weevil

The Four Lined Plant Bug Infestation

Japanese Beetles Infestation

Hydrangea Leaf Tiers Infestation

Sticky, clear honeydew can coat the leaves and other surfaces below the aphids. Sooty mold, a black coating of fungal growth, may form on the honeydew. While these fungi do not directly infect the plant, a lot of sooty mold can block sunlight from the leaves, resulting in decreased photosynthesis. Ants are also attracted to honeydew and some ant species will groom aphids by rubbing their antennae to stimulate honeydew production. They also defend aphids against parasites and predators. To control an aphid infestation, hose them off with a strong spray of water or apply insecticidal soap (www.ladybug.uconn.edu).

11.2.2 Black Vine Weevil (*Otiorhynchus sulcatus*)

The larvae of this weevil injure hydrangea by feeding on the roots. After devouring the small roots, the small grubs move on to chew the bark from larger roots, often girdling them. The tops of girdled plants turn yellow then brown, followed by the death of the plant. The adult is 1/2″ long, black, with a beaded appearance on the thorax and scattered yellow hairs on the wing covers. Adults are flightless and feed on the leaves at night, leaving notches along the margins. Adults and large larvae overwinter, emerging from May-July. After feeding for 3-4 weeks, egg-laying begins. Insect pathogenic nematodes can be used for control of larvae in the soil. If necessary, chemical control can be used when there is adult feeding and before egg laying starts. The usual timing is during May, June, and July (www.ladybug.uconn.edu).

11.2.3 The Four Lined Plant Bug (*Poecilocapsus lineatus*) and The Tarnished Plant Bug (*Lygus lineolaris*)

These pests cause round, brown sunken areas around the feeding puncture wounds. Eggs are laid in soft stem tissue. They hatch around mid-May and the young bugs suck sap from the leaves. They molt five times and mature around the middle of June. Adults are about 1/3″ long. The four-lined plant bug is yellow with four lengthwise black stripes on its wings. Leaf spots can become holes. The tarnished plant bug is brown, 1/5″ long, with a yellow 'Y' shape pattern on its back. There is one generation per year of the four-lined plant bug and two to five generations per year of the tarnished plant bug. Control is not usually necessary for these pests but if needed, use insecticidal soap or ultrafine horticultural oil (www.ladybug.uconn.edu).

11.2.4 Hydrangea Leaf Tiers (*Olethreutes ferriferana* syn. *Exartema ferriferanum*)

These are quite noticeable when present. While this pest does not have a big impact on the health of the plant, it can reduce the number of flowers. After overwintering in the pupal stage, small brown and white moths emerge in the spring, mate and lay eggs on the tips of hydrangea branches (all types). The newly hatched caterpillars web a few leaves together and feed on the inner (top) surfaces of the leaves and, if present, the flower bud. The webbed together leaves look like little tents and there can be quite a few on a single plant. During the summer, larval feeding is completed and the insect drops to the ground to pupate. There is only one generation per year. If control is desired or needed, the leaves can be pulled gently apart and the caterpillar can be removed and killed. Biological insecticides containing the active ingredient *Bacillus thuringiensis kurstaki* (Btk) can be used during the caterpillar stage (www.ladybug.uconn.edu).

11.2.5 Japanese Beetles

These beetles are problematic on oakleaf hydrangea but seldom bother the other types. Adults are metallic green and consume leaf tissue while leaving behind the veins, resulting in a lacy looking 'skeleton' of the leaf. A good way to get rid of these beetles is to knock them off the leaves into a container of soapy water to kill them. Insecticides are also effective but should never be applied during bloom to protect pollinators and other beneficial insects (www.ladybug.uconn.edu).

11.2.6 Rose Chafer (*Macrodactylus subspinosus*)

These pests sometimes feed on hydrangea and when abundant can injure plants by skeletonizing the leaves. They will also feed on the flower petals, especially of white flowers. They occur in large numbers where soils are sandy. Adults appear around mid-June and feed for about a month to 6 weeks. Females lay 24- 36 eggs singly in the ground a few inches beneath the surface. They soon hatch and the grubs feed on the roots of grasses and other plants. In late autumn they burrow several inches into the ground to overwinter. The next April or May they return to near the surface and pupate. Beetles emerge 2- 4 weeks later. Chemical control is not usually effective because more rose chafers can quickly move into treated areas (www.ladybug.uconn.edu).

11.2.7 Two Spotted Spider Mite (*Tetranychus urticae*)

They feed on leaves presents as a stippled or freckled appearance and/or webbing. Spider mites are most active during hot, dry weather. Like aphids, they have piercing and sucking mouthparts and each little 'freckle' is an injury associated with a feeding site. In most cases, spider mites on outdoor plants are controlled by naturally occurring predators. Spider mite outbreaks often occur after the application of a broad-spectrum insecticide that kills beneficial insects along with the pests. Plants that are well watered during dry periods may withstand spider mite injury (www.ladybug.uconn.edu).

Other insects and pests that can sometimes cause damage to hydrangeas include oyster shell scale (*Lepidosaphes ulmi*), and stem, leaf, and root-knot nematodes.

11.3.0 Disorders

11.3.1 Chlorosis

Leaves yellow between the veins or sometimes the entire leaf is yellowed. It is thought to be due to too much time in the soil. It may also be correlated with very high levels of nitrates and calcium. Soil analysis is indicated (www.portal.ct.gov).

11.3.2 Winter Injury

Winter hardiness problems are frequently encountered on hydrangeas (particularly *Hydrangea macrophylla*) growing away from the coast. This is often exhibited by lack of flowers. Since the flowers are set on previous season's growth, they are lost when shoots are killed by winter injury. This situation can be minimized by planting in sheltered locations (www.portal.ct.gov).

11.3.3 Blindness

It can be caused by several factors. Most of them occur during the fall and include poor light, too early storage, too early defoliation and too early onset of cool temperatures. Also at low temperatures bud kill can occur (www.plantgrower.org).

12.0 Harvesting and Postharvest Management

12.1.0 Harvesting

Harvest hydrangea flowers in late summer, from August to October. Choose flowers that have opened completely. Cut the stems on the diagonal with sharp pruning shears. Remove the leaves from the hydrangea stems and place the stems immediately into cool water. Allow them to rest in the water for 1 to 2 hours. Ready a vase with water for arranging the fresh flowers. Hold the hydrangea stems under water and cut again about 3 cm above the original cut. Arrange the flowers in the vase and keep them watered (www.gardenguides.com).

In Chile, hydrangeas are harvested at 3 different stages based on the floret development stage and sepal colour: 'fresh' (more than 50 per cent of florets are open and sepals show a bright colour), 'transition' (all florets are open and sepals show greenish shades) and 'antique' or 'classic' (past the 'transition' stage). The stems cut in this last stage are the most durable and have the longest postharvest life because they are less susceptible to dehydration (Schiappacasse *et al.*, 2014).

12.2.0 Postharvest Handling

12.2.1 Defoliation

During production of florists' hydrangea, defoliation before cold storage is required for prevention of diseases such as botrytis bud rot. Many hydrangea growers manually remove leaves before natural leaf abscission. This practice is time-consuming, expensive, and can result in damage to stems and buds. An alternative to this practice is chemical defoliation. A wide variety of chemicals have been tested for their effectiveness in defoliation of hydrangeas, but all have some limitations (Bailey, 1989). Some chemicals such as ethephon (2-chloroethylphosphonic acid) adversely influence performance during forcing by retarding growth or reducing inflorescence size (Shanks, 1969).

As an ethylene source, mature apples (3-liter apples/m^3 of storage space) are effective for defoliating hydrangeas held in a fairly airtight enclosure. According to Blom *et al.* (1994) gaseous ethylene at 5-10 µl/liter at 16 °C for 1 week in darkness was an effective alternative to BD at 10 g/liter or the use of apples for defoliating florists' hydrangea cultivars ('Kasteln', 'Mathilda Gutges', and 'Todi') at minimal cost.

Application procedures for chemicals such as Vapam (sodium N-methyldithiocarbamate) and ethylene require an air-tight storage unit. Some

Harvest and Postharvest Management of Hydrangea

Hydrangea in Full Bloom

Store in Vase Solution

A: Fresh B: Transition C: Antique or Classic

Different Stages of Harvesting

Drying and Storage of Hydrangea

Harvest and Postharvest Management of Hydrangea

Packaging of Hydrangea

Aalsmeer Royal Holland Flower Auction, 2016

Exibition of Double Petal Flower in US

chemicals such as 2-butyne-1,4-diol (BD) are effective defoliants at high concentrations but are poisonous to humans and require special safety regulations for application. The influence of only a few combinations of chemicals on defoliation of hydrangea has been evaluated. Gibberellic acid (GA_3) has been found to be able to enhance chemical [BD and TPTA (tributyl phosphorotrithioate)] defoliation of hydrangeas (Bailey, D.A. 1990).

12.2.2 Storage

After harvesting, hydrangea cut flowers are kept in a holding solution containing 3 per cent sucrose + 200 ppm 8-HQC + 10 ppm BA was better for improving the post-harvest life (Chhune, 2015). Boonyakiat *et al.* (2016) reported that when 'White 027' hydrangeas placed in a vase containing citric solution with a pH of 3 had 5.1 days vase life. While 'Old Blue' showed longer vase life in a vase containing liquid as compared to those flowers with the traditional postharvest handling. In case of long-term storage hydrangeas dried for a couple of weeks and store in a dry location out of direct sunlight (www.almanac.com).

12.2.3 Drying

After harvesting, hydrangea flowers cut the stem with sharp pruning shears, remove the leaves and arrange the flowers in a vase or display container without water. Leave the flowers to dry. Flowers with thin, weak stems can be hung upside down to dry. They will dry out and retain their colour for a long time, but will eventually turn brown. Spray the hydrangea blooms with a mixture of glycerin and water to help keep them supple. Spray with hairspray to help them retain their colour.

In another method hydrangea blossom branches that are 18 inches or less in length and have the desired shape and size, place into a container fill with the mixture of warm water and glycerin for better drying (Loucks, 2017; www.gardenguides.com).

13.0 Biotechnology

13.1.0 Genetic Transformation

The double flower phenotype is a recessive trait controlled by a single major gene. The crosses between the progeny of 'Sumidanohanabi' and the progeny of 'Jogasaki' produced only single flower descendants, which indicated that the genes controlling the double flower phenotype are different (Suyama *et al.*, 2015). Nashima *et al.* (2021) described the double flower locus of 'Sumidanohanabi' and 'Jogasaki' as d_{su} and d_{jo} respectively. Waki *et al.* (2018) identified d_{su} on the genetic linkage map. They also reported that the DNA marker STAB045 was the closest marker to d_{su} and that STAB045 could help distinguish flower phenotypes with a high degree of agreement. Genomic information is essential for DNA marker development and identification of genes controlling specific phenotypes. The genome assembly of hydrangea (1.6 Gb) is so fragmented that it comprises 1,519,429 contigs with an N_{50} size 2,447 bp (Tränkner *et al.*, 2019). So, improved, advanced long-read technologies and bioinformatics methods are required to determine the sequences of complex genomes. Nashima *et al.* (2021) developed DNA markers linked to d_{jo} using SNP information, which was obtained by double digest restriction site associated DNA sequence (ddRAD-Seq) analysis of the F_2 mapping population 12GM$_1$. It segregated double flower phenotypes of djo. They developed DNA markers J_{01} and S_{01} which distinguished recessive double flower alleles for d_{jo} and d_{su}, individually. These DNA markers are very useful in marker-assisted selection (MAS) of double flower progenies. Identification of flower phenotype at the seedling stage by MAS helps to disposing of single flower individuals and permits the development of double flower individuals. These DNA markers would speed up the breeding of double flower phenotypes.

The mechanism of colour change of hydrangea sepals was established by the analytical method of single- coloured cell analysis. Vacuolar pH measurements of single- coloured cells clarified the difference between blue and red cultivars. The pH of the blue cells in the blue cultivar was approximately 4.1, which was significantly higher than that in the red cells (pH 3.3) (Yoshida *et al.*, 2003). Ito *et al.* (2009) quantified the differences in the concentrations of delphinidin 3-O-glucoside, three copigments and aluminum in approximately 150 coloured cells. These results showed that the molar ratios of 5-O-acylquinic acid (neochlorogenic acid and 5-O-p-coumaroylquinic acid) and Al^{3+} to delphinidin 3-O-glucoside were much higher in the blue cells than that in the red cells. The concentration of aluminum in blue cells in matured sepals is higher than 10 mm (Toyama- Kato *et al.*, 2003). These information's indicated that an Al-transport and storage system exists in hydrangea. Similar to tea plants and buckwheat, hydrangea is an Al-hyper accumulator (Ma *et al.*, 2001) that exhibits tolerance to acidic soils. High concentrations of Al were detected in the sepals, leaves, and other tissues of hydrangea, suggesting that this plant could possess some mechanisms to translocate high amount of aluminum from the cytosol into the vacuoes for detoxification. Negishi *et al.* (2013) reported two Al- transporters, HmVALT and HmPALT1, from hydrangea through candidate gene searching, using sequence data obtained from cDNA libraries in silico and microarray analyses, followed by an Al- tolerance assay using yeast. Both transporters belong to the aquaporin family, and HmPALT1 was expressed only in the sepals. Later in 2013 they also identified a new plasma membrane- localized aluminum transporter, HmPALT2, in hydrangea sepals, which plays a role in transporting aluminum from outside the cell into the cytosol in both the shoot and root.

14.0 References

Abou Dahab, T.A.M. (2007) *Arab J. Biotechno.l*, **10**: 161-178.

Arafa, A.M., Nower, A.A., Helme, S.S. and Abd-Elaty, H.A. (2017) *Int. J. Curr. Microbiol. Appl. Sci.*, **6**(5) : 776-778.

Bailey, D.A. (1989) *Hydrangea production*. Timber Press, Inc.

Bailey, D.A. (1990) *Hort Science*, **5** : 25

Bailey, D.A. (1998) Commercial Hydrangea Forcing. *Department of Horticultural Science. University of Georgia. Published by North Carolina Cooperative Extension Service.*

Bhattacharjee, S.K. (2006) Flowering shrubs and seasonal ornamentals. *Advances in ornamental horticulture* (*Volume 1*). Pointer Publisher, Jaipur, India, pp.201- 216.

Bi, G. and Scagel, C.F. (2008a) *Hort Science*, **43**(7) : 2151-2154.

Bi, G., Scagel, C.F. and Harkess, R. (2008b) *Hort Science*, **43**(2) : 472-477.

Blom, T.J. and Smith, R.B. (1994) *Hort Science*, **29**(6) : 636-637.

Blythe, E.K., Sibley, J.L., Tilt, K.M. and Ruter, J.M. (2003) *J. Environ. Hortic.*, **21**(3) : 131-136.

Boonyakiat, D., Poonlarp, P. and Chuamuangphan, C. (2016) Research and development on postharvest management of the Royal Project flowers. In *VIII International Postharvest Symposium: Enhancing Supply Chain and Consumer Benefits-Ethical and Technological Issues 1194*. pp.587-594.

BSBI List (2007) Botanical Society of Britain and Ireland.

Cerbah, M., Mortreau, E., Brown, S., Siljak-Yakovlev, S., Bertrand, H. and Lambert, C. (2001) *Theor. Appl. Genet.*, **103**(1) : 45-51.

Chhune, T. (2015) *Studies on post harvest handling of cut hydrangea* (*Hydrangea macrophylla Thunb.*) *as influenced by pre-harvest fertilization* (Doctoral dissertation).

Daughtrey, M.L., Wick, R.L. and Peterson, J.L. (1995) *Compendium of flowering potted plant diseases*. American Phytopathological Society (APS Press).

Demilly, D., Lambert, C. and Bertrand, H. (1998) Diversity of nuclear DNA contents of Hydrangea. In *XIX International Symposium on Improvement of Ornamental Plants 508* (pp. 281-284).

Dirr, M. (2004) *Hydrangeas for American gardens*. Timber Press.

Doil, A., Zhang, R., Schum, A., Serek, M. and Winkelmann, T. (2008) *Propag. Ornam. Plants*, **8**(3) : 151-153.

Duke, J.A. and Ayensu, E.S. (1985) *Medicinal Plants of China*. Reference Publications.

Facciola, S. (1990) *Cornucopia - A Source Book of Edible Plants*. Kampong Publications.

Funamoto, T. (1988) *La Kromosomo*, **49** : 1583-1594.

Garibaldi, A., Bertetti, D. and Gullino, M.L. (2008) *Plant Dis.*, **92**(1) : 173-173.

Garibaldi, A., Gilardi, G., Minerdi, D. and Gullino, M.L. (2006) *Plant Dis.*, **90**(8) : 1113-1113.

George, H. (2019) *How to identify and treat hydrangea diseases*. https: / / gardenerspath. com.

Gledhill, David (2008) *The Names of Plants*. Cambridge University Press. ISBN 9780521866453 (hardback), ISBN 9780521685535 (paperback). pp 50, 206.

Grillo, S. (2012) *Propagate Your Shrubs from Softwood Cuttings*. *Fine Gardening*.

Halcomb, M. and Sandra, R. (2010) Hydrangea production. *United States: University of Tennessee*.

Harrington, J. (2017) *How to get potted hydrangeas to re-bloom*. https: //www.gardenguides.com

Haworth-Booth, M. (1975) *The hydrangeas*. Constable.

Hortensia (2017) *Definition of hortensia in English by Oxford Dictionaries*.

Howe, K, E. (2011) 'Propagation of Hydangea'. *Hydrangea and Country Gardens*.

https: //www.agriculture.infoagro.com

https: //www.almanac.com

https: //www.bartlett.com

https: //www.edis.ifas.ufl.edu

https: //www.extension.tennessee.edu

https: //www.fire.tas.gov.au

https: //www.gardendesign.com

https: //www.guide-to-houseplants.com

https: //www.healthbenefitstimes.com

https: //www.heritagehydrangeas.com

https: //www.homeguides.sfgate.com

https: //www.ladybug.uconn.edu

https: //www.missouribotanicalgarden.org

https: //www.naturalnavigator.com

https: //www.ndsu.edu

https: //www.plantaddicts.com

https: //www.plantdatabase.kpu.ca

https: //www.plantindex.com

https: //www.proflowers.com

https: //www.rhs.org.uk

https: //www.rollmagazine.com

https: //www.springmeadownursery.com

https: //www.theflora.in

https: //www.thepracticalplanter.com

https: //www.thespruce.com

https: //www.ugaoo.com

Hydrangea FAQ (2013) The United States National Arboretum.

Ito, D., Shinkai, Y., Kato, Y., Kondo, T. and Yoshida, K. (2009) *Biosci. Biotechnol. Biochem.*, **73**(5) : 1054-1059.

Jain, S.M. (2002) Feeding the world with induced mutations and biotechnology.

Jones, K.D. and Reed, S.M. (2006) *Hort Science*, **41**(3) : 564-566.

Kardos, J. (2006) The Principles and Practices of Breeding Hydrangeas. In *Combined Proceedings International Plant Propagators' Society* (Vol. 56, p. 534).

Kudo, N. and Niimi, Y. (1999) *J. Jpn. Soc. Hort Science*, **68**(2) : 428-439.

Kudo, N., Matsui, T. and Okada, T. (2008) *Plant Biotechnol. J.*, **25** : 529-533.

Kunkel, G. (1984) *Plants for Human Consumption*. Koeltz Scientific Books.

Lawson-Hall, T. and Rother, B. (1995) *Hydrangeas: A gardeners' guide*.

Ledbetter, D.I. and Preece, J.E. (2004) *Sci. Hortic.*, **101**(1-2) : 121-126.

Li, T., Bi, G., Harkess, R.L., Denny, G.C. and Scagel, C. (2019) *Hort Science*, **54**(1) : 167-174.

Li, Y., Mmbaga, M., Zhou, B., Joshua, J., Rotich, E. and Parikh, L. (2016) *Springer, Cham, Alemania*, 1-19.

Litlere, B. and Stromme, E. (1975) The influence of temperature, daylength and light intensity on flowering in *Hydrangea macrophylla* (thunb.) ser. In *Symposium on Protected Cultivation of Flowers and Vegetables 51* (pp. 285-298).

Liu, F., Huang, L.L., Li, Y.L., Reinhoud, P., Jongsma, M.A. and Wang, C.Y. (2011) *Plant Cell Tiss. Org.*, **104**(1) : 111-117.

Loucks, J. (2017) *How to dry hydrangeas with glycerin*. https: / /www.gardenguides.com.

Ma, J.F., Ryan, P.R. and Delhaize, E. (2001) *Trends Plant Sci.*, **6**(6) : 273-278.

McClintock, E. (1957) A monograph of the genus Hydrangea. *Proc. Calif. Acad. Sci.*, **29** : 147-256.

Millard, P. (1993) Internal cycling of nitrogen in trees. *Mineral Nutrition of Deciduous Fruit Plants*, **383** : 3-14.

Mmbaga, M.T. and Oliver, J.B. (2007) *Arboriculture and Urban Forestry*, **33**(3) : 210.

Mmbaga, M.T., Kim, M.S., Mackasmiel, L. and Li, Y. (2012) *J. Phytopathol.*, **160**(2) : 88-97.

Nakagiri, R., Hashizume, E., Kayahashi, S., Sakai, Y., and Kamiya, T. (2003) *Biosci. Biotechnol. Biochem.*, **67**(12) : 2641-2643.

Nashima, K., Shirasawa, K., Ghelfi, A., Hirakawa, H., Isobe, S., Suyama, T. and Namai, K. (2021) *DNA Res.*, **28**(1) : dsaa026.

Nau, J. (2011) *Ball Redbook: Crop Production*, vol.2. 18th ed. Illinois: Ball Publishing.

Negishi, T., Oshima, K., Hattori, M. and Yoshida, K. (2013) *Genes. Cells*, **18**(5) : 341-352.

O'Meara, L., Chappell, M.R. and Van Iersel, M.W. (2014) *Hort Science*, **49**(4) : 493-498.

Owen, J.S., Fulcher, A., LeBude, A. and Chappell, M. (1840) Hydrangea production: Cultivar selection and general practices to consider when propagating and growing hydrangea. UT Ext.

Pecchioli, S., Lazzereschi, S., Nesi, B., Antonetti, M., Grassotti, A. and Salazar-Orozco, G. (2011) Morphological characterization of a Hydrangea spp. Collection. In *VII International Symposium on New Floricultural Crops*, **1000** : 99-105.

Preece, J.E. and Ledbetter, D.I. (2002) The influence of thidiazuron on in vitro shoot proliferation of oakleaf hydrangea (Hydrangea quercifolia Bartr.). In *XXVI International Horticultural Congress: Biotechnology in Horticultural Crop Improvement: Achievements, Opportunities*, **625** : 233-236.

Rinehart, T.A., Scheffler, B.E. and Reed, S.M. (2006) *J. Am. Soc. Hort Science*, **131**(6) : 787-797.

Ruffoni, B., Sacco, E. and Savona, M. (2012) In vitro propagation of Hydrangea spp. In *Protocols for Micropropagation of Selected Economically-Important Horticultural Plants* (pp. 231-244). Humana Press, Totowa, NJ.

Sacco, E., Savona, M., Antonetti, M., Grassotti, A., Pasqualetto, P.L. and Ruffoni, B. (2010) In vitro propagation and regeneration of several hydrangea genotypes. In *XXVIII International Horticultural Congress on Science and Horticulture for People (IHC2010): International Symposium*, **937** : 565-572.

Schiappacasse, F., Moggia, C. and Contreras, R. (2014) *Idesia*, **32**(4) : 71-76.

Sebastian, T.K. and Heuser, C.W. (1987) *Sci. Hortic.*, **31**(3-4) : 303-309.

Shanks, J.B. (1969) *Hort Science*, pp. 56-58.

Suyama, T., Tanigawa, T., Yamada, A., Matsuno, T., Kunitake, T., Saeki, K. and Nakamura, C. (2015) *Hortic. J.*, **84** : 253–60.

Tanaka, T. (1976) *Tanaka's Cyclopaedia of Edible Plants of the World*. Keigaku Publishing.

Thaneshwari, N. (2014) *Effect of nitrogen and potassium on growth and flowering of hydrangea (Hydrangea macrophylla Thunb.)* (Doctoral dissertation, M. Sc. Thesis submitted to Dr. Yashwant Singh Parmar University of Horticulture and Forestry, Nauni, Solan (HP), India).

Tjia, B. (1976) *Hort Science*, **11**(5) : 487-488.

Toyama-Kato, Y., Yoshida, K., Fujimori, E., Haraguchi, H., Shimizu, Y. and Kondo, T. (2003) *Biochem. Eng. J.*, **14**(3) : 237-241.

Tränkner, C., Krüger, J., Wanke, S., Naumann, J., Wenke, T. and Engel, F. (2019) *BMC Genet.*, **20**(1) : 1-12.

Uemachi, T. and Okumura, A. (2012) *J. Jpn. Soc. Hort Science*, **81**(3) : 263-268.

Uemachi, T., Kato, Y. and Nishio, T. (2004) *Sci. Hortic.*, **102**(3) : 325-334.

van Gelderen, C.J. and van Gelderen, D.M. (2004) *Encyclopedia of hydrangeas*. Timber Press.

Waki, T., Kodama, M., Akutsu, M., Namai, K., Iigo, M., Kurokura, T. and Yagi, M. (2018) *Hortic. J.*, **87**(2) : 264-273.

Wilson, E. H. (1923) *J. Arnold Arbor.*, **4**(4) : 233-246.

Wu, X. and Alexander, L.W. (2020) *Hortic. Res.*, **7**(1) : 1-9.

Yoshida, K., Toyama-Kato, Y., Kameda, K. and Kondo, T. (2003) *Plant Cell Physiol.*, **44**(3) : 262-268.

Zhang, H., Matsuda, H., Yamashita, C., Nakamura, S., and Yoshikawa, M. (2009) *Eur. J. Pharmacol.*, **606**(1-3) : 255-261.

Hibiscus

Solanki Bal and Debmala Mukherjee

1.0 Introduction

Hibiscus is a genus of flowering plants in the mallow family, Malvaceae. The genus is quite large, comprising several hundred species that are native to warm temperate, subtropical and tropical regions throughout the world. Several species are renowned for their large, showy flowers and those species are commonly known simply as 'hibiscus', or less widely known as rose mallow. Other names include hardy hibiscus, rose of sharon, and tropical hibiscus. The genus includes both annual and perennial herbaceous plants, as well as woody shrubs and small trees. The species which are grown as ornamental plants because of their beautiful flowers include *H. rosa-sinensis* L., *H. syriacus* L., *H. mutabilis* L., *H. schizopetalus* Hook., *H. moscheutos* L., *H. tiliaceous*. The important ones with large number of cultivars varying in shape and size of plants, flower size and colour are *H. rosa-sinensis* and *H. syriacus*. These are popular flowering shrubs throughout the world and grown in areas with suitable environment. Though *H. sabderiffa* is not popular as an ornamental plant, but it has many important uses and cultivated in many locations for its beneficial products.

2.0 Importance and Uses

2.1.0 Aesthetic Uses

The hundreds of cultivars of *H. rosa-sinensis* are extensively grown in tropical and mild sub tropical gardens, in shrub border, in groups and as specimen plant in garden with the spectacular, large colourful flowers. This species is very popular as pot plants, as hedge bearing flowers and for screening. The varieties with red small to medium sized flowers are commonly used as loose flowers, garlands and floral ornaments for offering to Goddess. The plant is having enormous demand as a flowering shrub. Cultivars of *H. syriacus* are used as in *H. rosa-sinensis*, mentioned above and extensively grown in gardens in subtropical and mild temperate climate. *H mutabilis*, *H. schizopetalus*, *H. moscheutos* are also grown as ornamental plants in suitable climate.

2.2.0 Therapeutic and Other Uses

2.2.1 *H. rosa-sinensis*

Though, *H. rosa-sinensis* is commonly cultivated as ornamental plant because of its beautiful flowers, each part of *H. rosa-sinensis* contains a wide range of medicinal compounds. The root is used to treat cough. The leaves are used to treat constipation, skin disease, cardiac disease, nerve disease, and diarrhea. The flowers are used to treat bronchial catarrh, menorrhagia, and ischemic disease. It has well documented pharmacological effects such as anticancer, antidiabetic, antiinflammation, and antimicrobial activities. Phytochemical analysis showed that this plant contains flavonoids, flavonoid glycosides, hibiscetin, cyanidine, cyanidin glucosides, taraxeryl acetate, βsitosterol, campesterol, stigmasterol, ergosterol, citric acids, tartaric acids, oxalic acids, cyclopropenoids, and anthocyanin pigments

Uses of Hibiscus

Hibiscus rosa-sinensis **in Hedge**

Hibiscus syriscus **as Specimen Plant**

Hibiscus schizopetalus **on Wall**

In Pot

As Specimen Plant in Garden

Along Border of Garden

Along Path

As Garland in Worship

Indoor Decoration

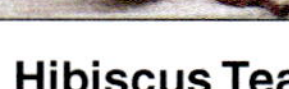

Hibiscus Tea

Hibiscus Oil

Window Box

(Khristi and Patel, 2016; Maganha *et al.*, 2010). It was reported that phlobatannins, glycosides, saponins, flavonoids, terpenoids including other compounds such as thiamine, riboflavin and niacin are present in leaves, flowers, stem and roots (Salem *et al.*, 2014). Drinking hibiscus tea might be as effective as the prescription drugs captopril and more effective than the drug hydrochlorothiazide for reducing blood pressure in people (Wee, 2003); taking hibiscus extract by mouth can lower levels of cholesterol and other blood fats in people with metabolic disorders such as diabetes; it reduces level of sugar and fats in the blood; decrease spasms in the stomach, intestines, and uterus; reduce swelling; and work like antibiotics to kill bacteria and worms. The flower extracts of *Hibiscus rosa-sinensis* has been shown to function as an anti-solar agent by absorbing ultraviolet radiation (Nevade *et al.*, 2011), hence used in skin disease management. Its leaves, barks, roots and flowers have been used in the Indian traditional system as medicine to treat various diseases (Missoum, 2018). In India, flowers and leaves are used for abortion, antifertility, contraceptive, diuretic, menorrhagia, bronchitis, emmengogue, demulcent, cough (Jadhav *et al.*, 2009).

2.2.2 Therapeutic Uses of *H. syriacus*

Epidemiological evidence suggests an inverse relationship between dietary intake of flavonoids and the risk of coronary heart disease. The leaves are diuretic, expectorant and stomachic. A decoction of the flowers is diuretic, ophthalmic and stomachic. It is also used in the treatment of itch and other skin diseases. Furthermore, it is also a medicinal plant used as antipyretic, antihelminthic, anti-inflammatory activity, cholesterol- lowering effects or serum cholesterol levels, prevent heart attacks, mild laxative, urinary tract problems, cleansing and detoxification, skin and vascular health and antifungal agent in the orient. There have been some reports on the active constituents of its stem and root bark. However, there are few investigations into the antioxidant activity of the pigment from its petals (Geng *et al.*, 2012).

3.0 Origin and History

The exact origin of *Hibiscus rosa-sinensis* is unknown, although it has been cultivated in China, Japan and the Pacific islands for a long time. Two white-flowered species, *Hibiscus arnottianus* and *Hibiscus waimeae,* are believed to be native to Hawaii (Wee, 2003). Hybrid hibiscus flowers are available today in many different colours, such as pink, yellow, orange, purple, lavender and even in multi-colours (Chin, 1986). *Hibiscus rosa-sinensis* is the national flower of Malaysia (Ewe, 2016). The announcement of the decision on the national flower was made by then Prime Minister Tengku Abdul Rahman on 28 July 1960. The *Hibiscus rosa-sinesis,* also known as 'Bunga Raya', was chosen because it was a popular flower that could be found throughout Malaya and was also known by the same name in the whole country. The national flower was the basis for the intricate design adorning the Yang di-Pertuan Agong's *baju maskat* (the Malaysian King's royal costume) worn during the installation ceremony of Sultan Azlan Shah of Perak in 1989. In addition, the thrones of the Yang di-Pertuan Agong and Raja Permaisuri Agong (the Queen) were decorated with carvings of the flowers, save for the legs and armrests. Interestingly, the first diesel train of the Malayan Railway and the first ship of Malaysia's national shipping line were also named 'Bunga Raya' in 1957 and 1970, respectively.

4.0 Taxonomy

Hibiscus rosa-sinensis was first described in 1753 by Carl Linnaeus in *Species Plantarum*. The specific epithet *rosa-sinensis* literally means 'rose of China', though it is not closely related to the true roses (Fouda, 2007).The genus *Hibiscus* is placed in the tribe Hibisceae, subfamily Malvoideae of the family Malvaceae (Lee, 2007). The genus is quite large, comprising several hundred species that are native to warm- temperate, subtropical and tropical regions throughout the world.

5.0 Species and Cultivars

5.1.0 Species

There are 300 species of *Hibiscus* widely distributed in tropical and sub- tropical parts of the world (Akpan, 2006). The species are characterized by pentamerous and regular flowers that exhibit a variety of colours. The leaf morphology is very diverse in the genus. They vary from alternate and petiolated to simple or entire, lobbed and parted (Lawton, 2004). *Hibiscus* species produce a wide range of growth habits ranging from small trees and shrubs to annual and perennial herbs (Akpan, 2006). They are generally classified as either tropical or hardy types. The most appreciated tropical cultivars belong to the *H. rosa-sinensis* group, however, they have limited use as outdoor plants in temperate regions due to their chilling sensitivity (Lawton, 2004, Vazquez- Thello *et al.*, 1996). The hardy *Hibiscus* plants include cultivars derived from a few species that are mainly native to North America such as *H. mutabilis* and *H. moscheutos* and a few Asiatic species, which comprise *H. syriacus* (Lawton, 2004, Van Laere *et al.*, 2007). Only five ornamental species, *H. rosa-sinensis*, *H. syriacus*, *H. mutabilis*, *H. schizopetalus* and *H. moscheutos* are grown almost all over the tropics and sub- tropics and mild temperate climate. Among these ornamental species, *H. rosa-sinensis* is the most important and beautiful among all of them. A large number of types are found freely growing in India and majority of these types are hardy- and tolerant to drought conditions compared to the Hawaiian varieties.

A brief characteristics of important ornamental species are described below:

Hibiscus rosa-sinensis (China rose, Chinese hibiscus)

Hibiscus rosa-sinensis is a perennial shrub having tap root system. Its leaves are 5 to 12 cm long and 5 to 8 cm wide. Leaves are simply ovate or lanceolate which are entire at the base and coarsely toothed at the tip/margins. Flowers are actinomorphic, pedicillate, complete, and pentamerous. Corolla contains five petals and is 8- 12 cm in diameter (Rao *et al.*, 2014). A number of varieties differing in colour and size of corolla are available. They produce vibrantly coloured showy flowers, throughout the year. Various cultivars having flowers (single or double) in shades of red, peach, white, pink, and orange are available (Gilman,1999). Although *H. rosa-sinensis* appears in variety of different corolla sizes, shapes and colours such as yellow, orange, pink, and white (Hammad, 2009; Kadve *et al.*, 2012; San Pascual *et al.*, 2017), the wild type for this plant remains as bright red flowers that grow in

Species of Hibiscus

Hibiscus rosa-sinensis

Hibiscus schizopetalus

Hibiscus moscheutus

Hibiscus syriacus

Hibiscus mutabilis

Hibiscus platinifolius

Hibiscus sabdariffa

Hibiscus tiliaceus

Hibiscus scotii

Hibiscus trionum

Hibiscus heterophyllus

Hibiscus clayi

single forms (Kumar and Singh, 2012). Fruit is a capsule having length of 3 cm and forms very rarely (Ross, 2003). *Hibiscus rosa sinensis* shows best growth in well-drained and slightly acidic soils. In sandy soils, it uses fully decomposed organic matter to maintain water holding capacity, aeration and drainage of the soil. Plants require full sunlight because inadequate light limits flowering. However, they can tolerate partial shade.

H. syriacus

H. syriacus is very important species in sub- tropical and mild temperate region bearing spectacular colourful flowers and with higher yield potential. It is native to south-central and southeast China, but widely introduced elsewhere, including much of Asia. It is commonly known as 'Rose- of- Sharon'. It is valued for large flowers produced in summer when few other shrubs bloom. It is useful as a garden accent due to its strict, upright habit. The open, loose branches and light green leaves make Rose- of- Sharon ideally suited to formal or informal plantings, and with a little pruning makes an attractive, small specimen tree. The plant grows in sun or partial shade and in any soil (Punasiya *et al.*, 2014). It is the national flower of South Korea and is mentioned in the South Korean national anthem.

Hibiscus syriacus is a hardy deciduous shrub. It is wild on mountain slopes, though the original habitat is obscure. It is upright and vase- shaped, reaching 2- 4 m in height, bearing large trumpet-shaped flowers with prominent yellow- tipped white stamens. The flowers are often pink in colour, but can also be dark pink (almost purple), light pink or white. Individual flowers are short-lived, lasting for a day. However, numerous buds produced on the shrub's new growth provide prolific flowering over a long summer blooming period. This species is highly tolerant of air pollution, heat, humidity, poor soil and drought. It has naturalized very well in many suburban areas. The species thrives well and flowers profusely in a wide range of climate from tropical to temperate; prefers subtropical and mild temperate climates; most of the cultivars flower all the year round in subtropical regions; like porous soil rich in humus, and adequate sun; propagated by cuttings.

Hibiscus mutabilis

This species is native to China, also known as 'Cotton Rose', is a large shrub or small multi- stemmed tree that grows to 4 m in height and spread. Dropping its leaves in winter, the shrub's slender stems seem to disappear among neighbouring plants until midsummer, when it leafs out into a big bushy mass. The large leaves are 12- 18 cm long, bright green, hairy on the undersides and deeply lobed. They impart a coarse texture that gives the plant a distinctive eye- catching appeal. Cotton rose is very impressive when in full bloom starting in autumn, continues still early summer in India. The large double or single- flowered blooms that are 12-16 cm across, open pure white and change colour over a three- day period until they are deep pink and then as they die assume a dark 'blue- pink' hue. The most notable characteristic of this flowering shrub is that flowers of three distinct colours appear on the bush simultaneously as the blooms colour cycle independent of one another. In fact, the species name *mutabilis* means, changeable. After flowering, a round, hairy capsule forms which dries and releases fuzzy seeds, a trait that inspired one of the plants common names, cotton rose as the buds resemble the boll of that famous member of the hibiscus family.

H. schizopetalus

This is popularly known as Coral Hibiscus, Fringed Hibiscus, Skeleton Hibiscus, Chinese Lantern, Japanese Lantern, and Pagoda Flower, is a native of tropical East Africa (Kenya, Tanzania, and Mozambique) with unusual, lacy flowers. Instead of upward or outward- facing flowers, this species has pendant flowers that dangle from slim, arching stems. In the tropics, this evergreen perennial is used as specimen plant, in ornamental borders and for hedges. In temperate areas it tends to be a curiosity in botanic gardens and specialty nurseries but can be grown as a houseplant. The delicate flowers are a bright red colour or streaked red and pink, with 3 cm long petals that are deeply pinnately- lobed, cut and fringed. The 5 petals are reflexed, curving back to form a sphere shape. The flowers are produced singly from leaf axils on upper stems. Plants will bloom intermittently throughout the year if they have sufficient sunlight. If pollinated, flowers may be followed by oblong seed capsules. This species is typically grown as a container plant in the Midwest, although it could be planted in the ground during the warm summer months. As a tropical plant, it does best under warm conditions and should be protected whenever temperatures approach freezing. Although the plant may survive a light freeze, it will lose all its leaves. *H. schizopetalus* prefers full sun and copious moisture.

Hibiscus moscheutos

This species, popularly called the rose mallow, swamp rose- mallow, crimson eyed rosemallow, or eastern rosemallow, is a species of flowering plant in the family Malvaceae. It is a cold-hardy perennial wetland plant that can grow in large colonies. The hirsute leaves are of variable morphology, but are commonly deltoidal in shape with up to three lobes. It is found in wetlands and along the riverine systems of the eastern United States from Texas to the Atlantic states, its territory extending northward to southern Ontario. There exist in nature numerous forms, and petal colours range from pure white to deep rose, and most have an eye of deep maroon. Taxonomic consensus is lacking for the nomenclature of the multiple subspecies. The flowers are borne apically, whereas the related *Hibiscus laevis* carries bud and bloom along the stem. This is very popular as garden plant. It can be propagated by seed, or by crown divisions during winter dormancy, and some success can be achieved by hard-wood stem cuttings. Numerous hybrids of the native North American *Hibiscus* species have been released by the commercial nursery trade. They are attractive and have wildlife value for nectar-feeders and birds.

Hibiscus sabdariffa

H. sabderiffa or commonly called 'Roselle' is an important species of *Hibiscus*. It is used for the production of bast fibre and as an infusion, in which it may be known as carcade. Though, it has no such ornamental value but it can be quite attractive and planted beside the farm. Cobley (1949) suggested roselle is a native plant of West Africa and from there it was carried to other parts of the world such as Asia and America, whereas in others opinion, Roselle was originated from India (Mat *et al.*, 1985) and Saudi Arabia (Abu-Tarboush Saif Aldin *et al.*, 1997). It is a medicinal plant with a worldwide fame. Roselle can adapt to a variety of soil in a warmer and more humid climate. The flowers of roselle are axillaries or in terminal racemes, the petals are white with reddish center at the base of the staminal column and this species is widely used as food (Abu- Tarboush Saif Aldin *et al.*, 1997).

Roselle is rich in organic acids including citric, malic, tartaric and allo-hydroxycitric acids. The plant is also known for its β- carotene, vitamin C, protein and total sugar. Roselle, having various medically important compounds called phytochemical, is well known for its nutritional and medicinal properties. Many parts of roselle including seeds, leaves, fruits and roots are used in various foods as well as in herbal medicine as a potential non- pharmacological treatment. Different extracts from roselle play crucial role in treating different medical problems including many cardiovascular disorders, helmenthic disease and cancer. Roselle is valued for its mild laxative effect, ability to increase urination, relief during hot weather and treatment of cracks in the feet, bilious, sores and wounds (Yadong *et al.*, 2005). Traditionally in Sudan, roselle has been used for relief of sour throat and healing wounds (Aziz *et al.*, 2007). In African folk medicine, roselle leaves are used for their, antimicrobial, emollient, antipyretic, diuretic, anti- helmentic, sedative properties and as a soothing cough remedy, whereas in India, leaves are poultice on abscesses (Neuwinger, 2000).

A number of post harvest products can be prepared from this species. In Sudan, the dry calyx is used to produce a flavorsome and healthy drink and dried calyces are used for tea, jelly, marmalade, ices, ice cream, sorbets, butter, pies, sauces, tarts, and other desserts (Duke and Ayensu, 1985). Cookies incorporated with roselle seed powder exhibited improved antioxidant properties. Cookie with 20 per cent roselle seed powder added showed higher content of total dietary fibre (5.6 g/100 g) as compared with control cookie (0.90 g/100 g) (Nyam *et al.*, 2012).

Hibiscus tiliaceus

This species is commonly known as 'coastal hibiscus' or 'sea hibiscus', native to old world. It reaches up to a height of 4- 10m; flowers are bright yellow with a deep red center upon opening but gradually it deepens to orange over the course of the day and finally red before they fall. *Hibiscus tiliaceus* can be found at elevations from sea level to 800 m in areas that receive 90- 250 cm of annual rainfall. It is commonly found growing on beaches, by rivers and in mangrove swamps. Sea Hibiscus is well adapted to grow in coastal environment in that it tolerates salt and waterlogging and can grow in quartz sand, coral sand, limestone (Elevitch and Lex, 2006) and crushed basalt (Allen, 2003). It grows best in slightly acidic to alkaline soils (pH of 5- 8.5) (Elevitch and Lex, 2006). *Hibiscus tiliaceus* is widely used in Asian countries as a subject for the art of bonsai, especially Taiwan. Besides, its ornamental importance, its wood is widely used to make furniture and leaves are also used as vegetables.

H. platanifolius

Hibiscus platanifolius, the maple- leaved mallow, is a species of mallow family, Malvaceae. In Sri Lankan texts, the plant is widely known by its synonym *H.*

eriocarpus. The tree is about 8m tall. Leaves are cordate at base, hairy, trilobed. Flowers are in axillary panicles where flowers show typical hibiscus flower colours, pink with dark center. Fruit is a capsule.

5.2.0 Cultivars

There are several hybrids of the *Hibiscus rosa-sinensis* type with single and double flowers in varying shades. Their brief characteristic features are as follows:

'Agni', flowers large, reddish purple, centre- red;

'Agnes', large flowers with cyclamen- pink and deep pink centre.

'Ali pore Beauty', a medium-sized, hardy bush; flowers double, reddish pink;

'Anamika', flowers large, reddish mauve, centre- red;

'Annanya', large light mauve flowers, white rays;

'Ananda', an attractive combination of colour, light yellow, pink and deep crimson centre;

'Anindita', large purple flowers, centre- red;

'Anurag', flowers orange- red, centre red;

'Apurba', flowers large, combination of light yellow, pink and red;

'Aurora': Very large flowers with flesh pink colour.

'Australian Single': Very large flowers and deep rose with maroon centre.

'Bahurupi', large flowers with unique combination of colours yellow, purple and crimson;

'Beauty Queen', a contrasting combination of red and white, flowers large;

'Bichitra', large glossy, reddish purple flower, centre dark;

'Biswarupa', flowers creamy- white, semi- double;

'Blue Ballerina', flowers light blue in colour;

'Blue Beauty', large flowers showy, blue- violet colour, centre dark;

'Blue Bird', flowers bluish white, centre red;

'Carneoplenus', a medium- sized shrub with large number of branches; leaves deep green, broadly ovate; flowers large, double, 12 cm across, bright crimson fading to rose; anthers yellow;

'Chandramukhi', an attractive cultivar, reddish orange, centre radiating pink;

'Chitra', flowers single, large, marigold- orange;

'Cooperi', a compact bush with colourful foliage, leaves ovate, variegated with white, pink, pruple and deep green, flowers small scarlet;

'Daffodil', a medium-sized shrub, flowers large double, up to 15 cm diameter, true daffodil- yellow, rather slow growing but fairly free flowering;

'Décor', a colourful large-flowered cultivar with rosy-red flowers, edge golden;

'Deep Rose Soundaraya', a popular and free- flowering cultivar bearing profuse large, single, deep rose flowers throughout the year;

'Delight', an outstanding large flowering hybrid cultivar, flowers up to 20 cm across, bright orange-red with phlox-pink centre, petals curved, margin slightly frilled;

'Durga', flowers large, red, double;

'Geeta', a showy cultivar, pink, deep red centre;

'Ganga', large mauve flowers, margin curled, centre deep red;

'Glowing Sunset', deep glowing salmon orange

'Golden Gleam', very huge attractive shade of buttercup yellow.

'Hawaii White', an excellent cultivar, medium in size, flowers snow white, petals narrow, spreading, very susceptible to excess moisture;

'Indrani', flowers large, pinkish white;

'Juno', dwarf bush, flowers double, compact, cherry- red;

'Kali Jaba', a tall, much- branched shrub, may reach a height of 3 m, flowers numerous, bright scarlet, drooping on long stalk, petals curving outward;

'Kalimpong', large pinkish white flowers, centre red;

'Kalyani', a hardy, tall- growing plant, producing flowers almost throughout the year; flowers large, reddish orange in centre and stamina) tube red, 15- 18 cm across;

'Kaveri', large reddish mauve flowers, petals curved downward;

'Lipstick': Bright red with dark centre.

'Magestic', large flowers, unusual combination of colour from centre to edge of petals;

'Mahatma', big double flowers, cadmium orange with red centre.

'Maywaykar', a cultivar with largeg lossy mauve flowers, centrered;

'Mini Gold', flowers vivid gold, centre lemon- yellow;

'Monalisa', flowers light orange, centre deep red;

'Nazrul', large pinkish- mauve flowers, centre red;

'Netaji', white flowers with crimson centre.

'Pallisree', large flowers pink-purple, centre whitish;

'Percy Lancaster Hybrid', dwarf bushy plant, flowers single, reddish orange, centre maroon, very floriferous;

'Pink Card', flowers large, glossy, reddish purple, centre-red;

'Prolific', a hardy cultivar, well-shaped bush, leaves roundish, cordate at base, undulate margin, thick, up to 15 cm long; flowers yellow, with deep centre and band of white, 12- 16 cm in diameter;

Cultivars of *Hibiscus rosa-sinensis*

'Andersonii' 'Cooperi' 'Dainty Pink'

'Dainty White' 'Double Psyche' 'Madeline Champion'

'Mango Dainty' 'Orange El Capitolio' 'President'

'Brilliant' 'Rainbow Sherbet' 'Sylvia Goodman'

'Ragini', ivory white flower, centre-red;

'Rashmani', attractive deep yellow flowers, centre red;

'Red Rose', large crimson flowers, dark centre, semi-double;

'Red Gold', large flowers, petal red with golden patches, centre dark;

'Red Glory', flower deep red, patches of pink- purple;

'Red Satin', a spectacular cultivar with large double blooms, 15- 18 cm across, pendulous, attractive red in colour;

'Rekha', flowers delicate lemon- yellow;

'Rupa', multi-coloured, orange, pink and deep red, very attractive;

'Snow Flake', an attractive cultivar of *Hibiscus rosa-sinensis,* medium-sized shrub, leaves variegated with irregular patches and stripes of greyish green and cream- white, flowers small, scarlet;

'Snigdha', large yellow flowers, centre radiating- white;

'Sunset', large yellow flowers radiating red, centre dark;

'Shri Krishna', a magnificent cultivar, flowers large, reddish orange, centre dark red;

'Sipra', large, light pink flowers, centre white;

'Sindhu', large, reddish orange flowers, petals curved downward, centre red;

'Srimayee', large flowers, purple- mauve, centre dark;

'Sristi', large flowers, petals reddish orange, blue- violet centre;

'Strange', large flowers, colour ranging from dark red, to crimson and light pink;

'Subarna', flowers golden- yellow, red centre, outer pinkish;

'Suvra', a pure white cultivar, flowers large, centre red. Flowers of red hibiscus commonly known as **'Kali Jaba'** are extensively used throughout the country as loose flower and garland by the Hindus for worshipping the Goddess of Power (Ma Kali).

'Viceroy', small deep rose-red flowers.

'Waimeae', snow white with slightly fragrant flowers.

Out of the nearly 1200 intra- varietal hybrids and open- pollinated seedlings of *H. rosa-sinensis,* the following 25 new varieties with their flower colour were selected and released between 1972 to 1979 by IIHR, Hessaraghatta.

'Aikta' (Post Office Red), 'Anuradha' (Golden Buff), 'Arunodaya' (Nasturtium Orange), 'Ashirwad' (Yellow), 'Basant' (SulphurYellow), 'Benazeer' (Bright Yellow), 'Bharat Sundari' (Deep Neyron Rose), 'Chitralekha' (China Rose with white variegated petal), 'Dilruba' (Mark Golden Buff), 'Geetanjali' (Turkey Red), 'Jogan' (Azalea Pink), 'Nartaki' (Marigold Orange), 'Nazneen' (Tangerine Orange), 'Neelofer' (Magenta Rose), 'Pakeezah' (Carmine Red), 'Phulkari' (Delft Rose with yellow border), 'Priya' (Rose Bengali), 'Queen of Hessaraghatta' (Orange), 'Ratna' (light yellow with orange stripe), 'Red Gold' (Dutch Vermilion), 'Red Saturn' (Signal

Cultivars of *H. syriacus*

'Oiseau Bleu'

'Red Heart'

'Peppermint Smoothie'

'Hamabo'

'Diana'

'Woodbridge'

Red), 'Shanti' (Primrose Yellow), 'Smt. Indira Gandhi' (Indian Yellow), 'Smt. Kamala Nehru' (Rose Bengal) and 'Tribal Queen' Cardinal Reds. Some of the very popular ones are briefly described below:

Cultivars of *Hibiscus syriacus* grown in different parts of the world include-

'Diana': single, white in colour; **'Hamabo'**: pale pink with red centre; **'Meehanii'**: pink with variegated leaves; **'Oiseau Bleu'** ('Blue Bird'): blue-violet with maroon centre; **'Red Heart'**: white with red centre; **'White Chiffon'**: white in colour, double; **'William R. Smith'**: white, single; '**Woodbridge**': deep pink in colour; **'Aphrodite':** rose flowers; **'Blue Bird'** :Blue vase-shaped flowers; **'Blushing Bride'**; **'Collie Mullens'**; **'Diana'**: A triploid with large, pure white flowers which remain open at night, foliage is dark green and waxy; **'Freedom';** '**Helene**': white/ purple flowers; **'Lavender Chiffon'**; **'Minerva'**: A heavily bloomer with lavender flowers overcast with some pink, red eye at base, foliage is higher quality than the species; **'Peppermint Smoothie'**; **'Pink Giant'**; **'Purpureus Variegatus'**; **'Red Heart'**; **'Strawberry Smoothie'**; **'Tri-colour'**: Double flowers are pink, red or purple and are mixed over the plant; **'White Chiffon'**.

6.0 Genetics and Breeding

6.1.0 Genetics

Hibiscus rosa-sinensis is one of many plant species with a genetic characteristic known as polyploidy, in which there are more than two complete sets of chromosomes, unlike most other species. A side effect of polyploidy is a condition where the phenotype of the offspring may be quite different from the parent, or indeed any ancestor, essentially allowing possibly random expression of all (or any) of the characteristics of all the generations that have gone before. Because of this characteristic, *H. rosa-sinensis* has become popular with hobbyists who cross and re-cross varieties, creating new named varieties and holding competitions to exhibit and judge the many resulting new seedlings and often strikingly unique flower.

A large number of forms are referable in *H. rosa-sinensis* (Bates, 1965). These range from single, semi-double to fully double and in colour from maize- yellow to scarlet red. Most of these forms are of hybrid origin, raised over a number of years and named differently by different workers and nurserymen. More often than not, there is no authentic record of their ancestory and they show a great deal of morphological and cytological polymorphism. A cytological survey of a collection of *H. rosa-sinensis* ($x = 9$) revealed the following chromosome numbers $2n = 46, 54, 63, 68, 72, 77, 84, 90, 96, 112, 132$ and 144. The 12 cytotypes could be divided into two distinct groups, a polyploid group ($2n = 54, 63, 72, 90, 144$) and a dysploid group ($2n = 46, 68, 77, 84, 96, 112, 132$). In the former, flower diameter was found to be directly proportional to chromosome number, while in the latter it was inconsistent. The probable mechanism involved in the origin of *H. rosa-sinensis* complex seems to be intervarietal and interspecific hybridization involving different basic numbers and levels of polyploidy. These together with segretational errors have been responsible for the origin of dysploids. Sexual sterility is very common. This may be genic, chromosomal and environmental but is of no consequence because of the vegetative multiplication (Singh and Khoshoo, 1970).

Pollination studies on *Hibiscus rosa-sinensis* reveal that 50 per cent pollination is possible (Abirami and Uma, 2018). Reasons for cross pollination in a *Hibiscus rosa-sinensis* flower, if both the sex organs mature at the same time, the pollination is prevented by the arrangement of the sex organs at different heights. (Kulkarni, 2002). The endangered *Papilio homerus* butterfly, the largest in the western hemisphere, is known to feed on the nectar of the Hibiscus (Lehnert *et al.*, 2017). Extensively grown in gardens as shrub border, as specimen plant and in pots, in well drained soil rich in organic matter and full sun; do not respond to annual pruning; propagated by cuttings, layering and grafting.

6.1.1 Inheritance of Qualitative Traits

Common rosemallow (*Hibiscus moscheutos*) is an attractive native plant known for its showy blooms and newer hybrids have incorporated stunning red-purple foliage. From crossing and selfng parents of contrasting foliage colour, as well investigation of red foliage colour in F_2 plants, Barrios and Ruter (2019) found the appearance of red foliage, in any amount, to be controlled by a single locus with a dominant allele for red foliage. It follows that parent with green foliage, a specimen of *Hibiscus moscheutos* subsp. *lasiocarpos* (Cavanilles) O. J. Blanchard, unless crossed with a red-foliage *Hibiscus*, should only yield all-green foliage progenes. As red foliaged hibiscus is highly desirable in the nursery trade, ornamental breeders would benefit from understanding the type of inheritance of red foliage in *Hibiscus moscheutos*.

The segregation of petal colour and eyespot in *Hibiscus syriacus* was studied by Lattier and Contreras (2020). The petal body colour is controlled by a single gene called *geisha*, named for the recessive allele that produces white to blush-pink petal bodies and dark red eyespot. This trait exhibits incomplete dominance and is under epistatic control by spotless. *Geisha*-type flowers lack pigment in the petal body, or exhibit a blush pink, likely produced by low levels of cyanidin, peonidin, and pelargonidin.

6.2.0 Breeding

Breeding work has been intensively exploited in the *Hibiscus* genus. Most of the improvement work through breeding in *Hibiscus rosa-sinensis* has been taken up in sub-tropical areas such as Mauritius, Hawaii, Fiji, India, California and Florida. In India, work on improvement of ornamental *Hibiscus* was mainly taken up at the Indian Institute of Horticultural Research, Hessaraghatta; Lalbagh Botanic Garden, Bangalore; Tamil Nadu Agricultural University, Coimbatore and Kerala Agricultural University, Trichur, besides some leading nurserymen. Many of the *Hibiscus rosa-sinensis* cultivars were proved to be completely sterile. Only a limited number combined desirable characters with a reasonable degree of fertility, could be used as parents for the breeding programme. The major breeding objectives include plant with good growth habit, floriferous nature, desirable flower colour, size, shape and good keeping quality.

6.2.1 Hybridization

Many hybrids were developed at the Agri- Horticultural Society of India, Calcutta and Indian Institute of Horticultural Research, Bangalore. Hawaii has earned the reputation of being the world centre for evolving magnificent cultivars of hibiscus through planned hybridization. In India also, a large number of cultivars with spectacular flowers have been evolved from the Hawaiian cultivars. Hybridization among different species and genera is one of the most important breeding approaches used for improvement of ornamental plants. The main aim of this strategy is to merge distant gene pools, hence broadening the genetic variability. However, hybridization barriers often occur in interspecific crosses hampering production of hybrids. Traditionally, two types of hybridization barriers have been defined (Morgan *et al.*, 2011, Van Tuyl and De Jeu, 1997). The first group, pre- fertilization barriers, includes lack of stigma receptivity or pollen viability at the time of pollination, low adhesion and germination of pollen grains, and abnormal growth of pollen tubes in the style and ovary (Rieseberg and Carney, 1998). Moreover, different abnormalities in the growth of pollen tubes can influence their ability to successfully deliver sperm cells to ovules (Winkelmann *et al.*, 2010). The second group comprises post-fertilization barriers that affect development of hybrid embryos and further development of hybrid plants. In some cases, failure in hybridization process is due to the breakdown of endosperm development. Moreover, post-fertilization barriers often involve deleterious hybrid characteristics, collectively called hybrid incompatibilities (HIs), such as unviability, lethality, sterility, albinism and abnormality in phenotypic traits of hybrids (Johnson, 2010, Kinoshita, 2007).

Interspecific hybridization was performed using several tropical *Hibiscus* species (Menzel and Wilson, 1969; Kuwada, 1970). In addition, crosses using cold hardy *Hibiscus* plants yielded progenies. Interspecific hybrids were obtained from crossing between *H. coccineus* and *H. moscheutos* (Kuwada, 1960). *H. syriacus* was successfully crossed with *H. paramutabilis* and *H. sinosyriacus* (Eeckhaut *et al.*, 2004; Van Laere *et al.*, 2007). *H. rosa-sinensis* has limited use as an outdoor plant due to chilling sensitivity. Cold hardiness is available in the genus but is limited to a few species. For a long time, improvement of *H. rosa-sinensis* with respect to resistance to low temperatures has been of great interest. However, a complicated genetic background and frequently occurring hybridization barriers make wide hybridization difficult in this species. The results of study (Kuligowska *et al.*, 2013) proved the occurrence of incompatibility reactions in interspecific crosses between cultivars of *H. rosa-sinensis* and cold hardy *Hibiscus* species. The main drawback in production of hybrid progeny was the presence of post-fertilization barriers associated with the occurrence of hybrid incomptibities that limit the usefulness of the obtained plants for further breeding purposes.

Ornamental Hibiscus is a highly polymorphic species (Gast, 1971). *H. rosa-sinensis* (show flower), *H. mutabilis* and *H. schizopetalus* hook exhibited cross-compatibility barriers. Crosses between *H. mutabilis* (HM) and *H. schizopetalus* (HS) as the maternal parent did not produce any ovary stimulation, while on selfing HM gave 82.6 per cent capsule set. While, *H. rosa-sinensis* (HR) was used as the maternal parent, there was better fruit set. In case of HR × HM, fruits that were initially set did not come to maturity, while HR × HS, in fruits which were carried to maturity, seeds did not germinate (Markrose, 1984).

6.2.2 Mutation Breeding

Mutation induction is a method to increase genetic divergence associated with selection, recombination, or a combination of these approaches in plant breeding. One of preliminary procedures for an excellent mutation breeding program is the radio-sensitivity study to determine the optimal doses for irradiation. One of preliminary procedures for an excellent mutation breeding program is the radio-sensitivity study to determine the optimal doses for irradiation in *Hibiscus* (Hossain *et al.*, 2011). A total of 10 different doses of acute gamma rays (0, 100, 200, 400, 800, 1000, 1200, 1500, 1700 and 2000 Gy) from ceasium-137 source were applied to the seeds of V 36 kenaf variety. The irradiated seeds including the control were planted in trough for 30 days. The gamma irradiation effects on several parameters such as seedling survival percentages, plant height, root length, shoot fresh weight, root fresh weight, shoot dry weight and root dry weight were analysed. From the radio-sensitivity curve, the LD_{50} and LD_{25} values were estimated at 810 and 310 Gy, respectively. Two doses (200 Gy and 300 Gy) based on LD_{25} were chosen for evaluating the effects of gamma irradiation on morphological traits in M1 generation (Basri *et al.*, 2014). Irradiated and non-irradiated seeds were planted in the field at Beseri, Perlis. About 10 phenotypic traits of irradiated plants were observed and evaluated against the controls. Cluster analysis on M_1 progenies showed that the mutation could be classified into eight genotypic groups. The first two components from principal component analysis explained about 77.99 per cent of variation (Faruq *et al.*, 2013). Number of seeds per pod, weight of seeds per plant and dry stem biomass play an important role in explaining the variation since they showed positive correlated values for the first component analysis. Morphological changes such as flower shape, flower colour, and leaf shape were also observed in M1 generation. The findings are important in determining the effectiveness of these doses in generating mutations on kenaf plant and subsequent breeding program to develop new kenaf varieties with enhanced quality traits (Wong, 2001).

The pink tea rose is an example of a somatic or bud mutant. It developed when a colour gene mutated in a stem cell of the white tea rose plant. All of the cells of the branch that developed had the mutant trait (Singh and Khoshoo, 1970).With Hibiscus many strains have resulted from bud mutation. Mutations can also be caused by external influences, such as toxic chemicals or ionizing radiation. Most mutations in the functional part of genes will produce mutant, altered versions of the proteins for which these genes are coding. These altered proteins lead to adaptations and changes in characteristics that may or may not enhance the fitness of an organism. 'Colour genes' that mutate seem to be an inherited trait in some instances *e.g.*, 'Dorothy Brady' (Bates, 1965).

Semi- acute gamma rays have developed flower colour and form mutations in hibiscus (*H. rosa sinensis*) cvs. Alipur Beauty and Cruenthus (Das *et al.*, 1974 and Das *et al.*, 1977). 'Shirasagi-no-Yume' a new induced mutant variety has been commercialized in Japan (Maluszynski *et al.*, 1995). 4 krad gamma ray treated five petalled single flower type mutant 'Anjali' has been developed from double flower type cv. 'Alipur Beauty' and commercialized (Banerji and Datta, 1986a,b).

Semi- hardwood cuttings of hibiscus were treated with three doses of gamma rays *viz.*, 25, 30 and 35 Gy and three doses of EMS *viz.*, 0.8 per cent (64.43 mM), 0.9 per cent (72.48 mM, 1.0 per cent (80.54 mM) separately. The highest mutation rate in terms of effectiveness and efficiency was observed in gamma rays than the ethyl methane sulphonate treatment in hibiscus (Naveena *et al.*, 2020).

7.0 Growth and Flowering

7.1.0 Day length

Hibiscus requires a minimum of 12 hours of day length to flower. Flowering is faster when day length is 14 hours or longer. Commercially grown *Hibiscus* are both day-neutral (*H. rosa sinensis*) (Dole and Wilkins, 1999) and long-day plants [*H. syriacus* (Thomas and Vince-Prue, 1997), *H. moscheutos* (Runkle *et al.*, 1996)]. *Hibiscus coccineus* flowers naturally in July (Bates, 1965), suggesting that it may also be a long-day plant. *Hibiscus hiernianus, H. mastersianus*, and *H. mechowii* are short-day plants (Menzel *et al.*, 1986). Hibiscus requires 4-8 months with night-time temperatures not below 20°C. In addition, it requires 12-13 hours of sunlight during the first 5 months of growth to prevent premature flowering.

7.1.1 Plant Growth Regulators

Plant growth regulators (PGRs), such as paclobutrazol, uniconazole and flurprimidol were widely used in inducing flower of ornamental species. These PGRs inhibit the formation of growth-active gibberellin (GA) in plants. GA is known to stimulate cell division and elongation, and the growth of stem, leaf, flower stalk, and fruit. It can also inhibit the development of lateral buds (Salisbury & Ross 1992). Plants treated with the PGRs usually showed reduction in growth (Fletcher *et al.*, 2000, Rademacher 2000). Paclobutrazol reduced the plant height and leaf area, but increased the chlorophyll content in the leaves of *H. rosa-sinensis*. This compound also increased the number of flower buds of this species at 75 days after application. Paclobutrazol at this rate, however, delayed the flowering as compared to uniconazole. Increased root length and number of blooms were also recorded with uniconazole treatment. Therefore, 2 mgL^{-1} uniconazole is recommended for flower induction of potted *H. rosa-sinensis*. This approach could serves as an alternative method in improving the unpredictable flowering habit of H. rosa-sinensis in the urban landscapes (Nazarudin, 2012).

8.0 Propagation

Hibiscus can be propagated in following ways:

8.1.0 Cuttings

8.1.1 Softwood Cuttings

Softwood cuttings are taken from the new soft succulent growth on the plant. They are taken in the spring to early summer and should be about 7.5 to 12.5 cm in length. Before planting they should be treated with IBA or NAA.

Propagation of Hibiscus

Seeds

Softwood Cutting

Air Layering

Grafting

Budding

In vitro Propagation of Hibiscus

8.1.2 Semi-Hardwood Cuttings

Semi-Hardwood cuttings are taken from the partially mature wood on the current season growth. They are taken in the late spring to late summer and should be about 7.5 to 15 cm in length. Before planting they should be treated with IBA or NAA.

8.1.3 Hardwood Cuttings

Hardwood cuttings are taken from the mature, dormant hardwood stems of the plant. They are taken during the late fall to early spring and should be about 10 to 76 cm in length. Before planting they should be treated with IBA or NAA.

All cuttings should be placed in a well draining medium, such as coco peat, vermiculite,. Cuttings will usually begin rooting in about six weeks, and flower in about nine months. Maintaining a high humidity level is best for the cuttings.

8.2.0 Grafting and Budding

This method is used to propagate hibiscus varieties that are susceptible to disease or difficult to root.

8.2.1 Grafting

The main thing to remember when grafting is that in order to have a successful graft one must make sure to line up cambium layers of the stock and scion. Some of the most popular types of grafts that can be used on hibiscus are the cleft graft, bark graft, whip-and-tongue graft and side grafts. The grafted area should be wrapped firmly and evenly with grafting tape after successfully lining up the cambium layers.

8.2.2 Budding

This method requires a little more patience and skill than grafting. Budding utilizes only one bud at a time and a small section of the rootstock. Budding may result in a stronger union than grafting and is used commercially in many nurseries. Some of the more popular types of budding are the t-bud and chip bud.

8.3.0 Air Layering

Air Layering is an excellent method of propagation for hibiscus varieties that are harder to root. By placing the scion directly onto the ringed stock helps in inducing root formation quickly and easily. After taping the union one should make sure to wrap it in moist sphagnum moss and seal it with a plastic wrapper. The wrapper should preferably be black to block out direct sunlight and prevent scalding of the new roots.

8.4.0 Seed

Hibiscus moscheutos is propagated through seed. But other species are commonly propagated through vegetative propagation. After pollen is applied to the female parts a pod develops and ripens within a few months. They then can be planted and grown. This method is the hardest and requires very specialized facilities and equipment.

8.5.0 Micropropagation

Propagation by vegetative means is the predominant technique in *H. rosa-sinensis*, but it does not ensure healthy and disease- free plants. Moreover, dependence on season and slow multiplication rates are some of the other major limiting factors in conventional propagation methods. Significant features of in vitro propagation are the enormous multiplicative capacity in a relatively short time span, the production of healthy and disease-free plants and the ability to generate propagules throughout the year independent of seasonal change (Kasseanis, 1957; Baker and Phillips, 1962; Lloyd and McCown, 1980; Martin, 1985; Chalupa, 1988). An in vitro propagation system for *H. rosa-sinensis* L. has the potential for efficient and rapid multiplication. Production of disease-free plants is a major concern due to viruses present in *H. rosa-sinensis* L. as reported by many authors (Adkins *et al.*, 2003; Huang *et al.*, 2004; Rajeshwari *et al.*, 2005). Methods are available to eradicate virus from plants (Kasseanis 1957; Baker and Phillips 1962), and if these methods are employed to obtain virus-free material to be used in the initial culture, certified virus-free plants of *H. rosa-sinensis* L. can be produced in large numbers using in vitro methods. In vitro methods have been reported for some related species of *H. rosa-sinensis* L. such as *H. cannabinus* L. (Zapata *et al.*, 1999; Srivatanakul *et al.*, 2000; Herath *et al.*, 2004), *H. moscheutos* L. (West and Preece 2004), *H. acetosella* Welw. exHiern (Reynolds and Blackmon 1983) and *H. syriacus* L. (Yang *et al.*, 1995; Jenderek and Olney, 2001). In these studies, somatic embryos, callus and explants from seedlings were used for various purposes such as regeneration and genetic transformation. Initial contamination resulting from mite infestation was also reported to be a problem (West and Preece 2006). Moreover, regeneration through somatic embryogenesis and callus appears complicated and it involves long processes. Although, multiple shoot regeneration was reported to be efficient using explants from seedlings of *H. cannabinus* L. and *H. syriacus* L. (Yang *et al.*, 1995; Srivatanakul *et al.*, 2000; Jenderek and Olney, 2001; Herath *et al.*, 2004), seedlings could not be used in the current study because of the hybrid nature of *H. rosa-sinensis* L.

9.0 Soil and Climate

Hibiscus plants prefer sandy loam but not heavy soils. Growers can get many flowers with soil rich in organic matter. If these plants are grown in pots or containers, regular quality potting soil with added garden compost is an ideal base. Composed cow manure, compost bark, and coarse peat should be mixed. The soil must be well- drained as these plants cannot tolerate water logging conditions. Hibiscus prefers neutral to slightly acidic, coarse soil rather than fine tilth. These plants thrive best in the soils with pH level between 5.5 and 6.5. If this crop is taken up as commercial on large scale, it is advised to go for soil testing. Any micro-nutrient deficiencies should be filled based on the test results.

Hibiscus plants require about 6 hours of full sunlight per day in order to bloom to its full potential. It can tolerate partial shade, but cannot grow or bloom properly. The more sunlight hibiscus gets, the better it will bloom, up to a point. Requirement of monthly rainfall of 127 cm during the first 4 months of growth is highly needed.

Hibiscus flowers best in temperatures between 16- 32°C and tropical species, *i.e. H. rosa-sinensis, H. schizopetalus etc.* cannot tolerate temperature below 0°C. Though, *H. syriacus* is cold- hardy plant and can be grown in sub- tropical and temperate region.

10.0 Cultivation

10.1.0 Planting Time

Hibiscus can be grown throughout the year provided there is a sufficient irrigation and sun light. Usually, hibiscus is propagated through both cuttings and seeds. For proper formation of roots, wet the bottom end of the cutting with liquid fertilizer. This soft wood can be planted in pots or containers having well-drained soil. After covering the container/pot having cutting with plastic bag and place in a partial shaded area as direct sunlight can dry out the cuttings. It takes 8 to 10 weeks for cuttings to get roots. In case of seed propagation, make small nick in seeds using utility knives and soak the seeds in water overnight and then plant into the prepared pot or container. When buying hibiscus in nurseries, select disease free plants that have lots of new buds and healthy leaves.

10.2.0 Manuring and Fertilization

Hibiscus plant needs lots of nutrients for good blooming. In the summer time, use a high potassium fertilizer. Supplementing the soil with garden compost or applying low phosphorous based fertilizer will result in high blooming. Hibiscus do not tolerate phosphorus well, and in high doses, it will slowly damage hibiscus plants over time (Hidden Valley Nature Arts., 2021). In winter, there is no need of fertilizing the plants. Adequate K nutrition can improve drought resistance and root longevity in *Hibiscus rosa-sinensis* (Egilla, 2001). Application of organic fertilizers at higher rate showed positive effects on the growth, yield and antioxidant content of roselle (*H. sabdariffa*). Therefore, organic fertilizer should be frequently applied in the farms to get the best growth and yields of the plant (Yusuf *et al.*, 2019).

10.3.0 Irrigation

As Hibiscus plants are tropical, they prefer constant moisture and sunlight for good blooming. Irrigating or watering depend on the soil type and season. In case of hot and dry seasons, it requires daily watering. When the top leaves turning to yellow and started dropping indicates that the plant is not getting enough water. In rainy season, there shouldn't be any water stagnating as it causes the root rot. *H. sabdariffa* exhibited low water demand and the plant kept growing even at irrigation treatment of 20 per cent ETo (Seghatoleslami *et al.*, 2013). Seghatoleslami *et al.* (2013) also found that water supply level had nosignificant effects on Roselle plant height, stem diameter, number of branches and on stomatal closure but irrigation treatment was only connected with chlorophyll contents in leaves. Drought stress or overwatering exhibited a decrease in flower number, fresh weight and dry weight (El-Boraie *et al.*, 2009). In addition, the same authors announced an increment in anthocyanin subsequent to the increase of water supply and added that stress condition is an elevator for carbohydrate accumulation that amassed into anthocyanin and other secondary metabolites. In another research, water schedule under light soil cultivations did not affect Roselle plant height and branches number but instead number of leaves and calyx yield significantly increased under longer intervals only (Babatunde and Mofoke, 2006). Longer water supply interval articulates a level of stress on plants even though equal amount of water was supplied with shorter intervals. However, water stress led to reductions in plant height, calyx yield and quality of Roselle (Khalil and Yousef, 2014). In heavy soil lands, water supply depends mainly on flooding irrigation. Increase water supply or reduction of water intervals were connected with a surge increase in the final yield of Roselle calyx (Mandour *et al.*, 1979). They also found that increasing intervals in between irrigation gave a significant increase in the total soluble solids in roselle beverage.

10.4.0 Pruning

Pruning of fruits is an important practice in promoting new growth and encouraging flowering in ornamental plant in general, and shrubs, climbers and herbaceous plants in particular. In case of *Hibiscus* of ornamental importance, fruit pruning is done depending on their growth and flowering, habit and variation in response to pruning. All perennial hibiscus ornamentals need pruning, time and extent of pruning, however vary.

11.0 Diseases and Pests

11.1.0 Diseases

A number of fungal and viral diseases commonly infect hibiscus plants; in addition, physiological disorders also occur.

Fungal Diseases

Phyllosticta Leaf Spot (*Pyllosticta hibiscina*)

Large circular or irregular shaped leaf spots with tan centers, yellow or white halo. The infected tissue becomes very thin, and breaks out reaching a diameter of 1 cm or larger. Small black fruiting bodies (pycnidia) can be seen scattered over the infected area, and usually on the upper side of the leaves.

Keeping leaf wet for a long time is a major cause of infection primarily through water-splashed spores. Manage overhead irrigation with minimal chances of continued long periods of leaf wetness. If the problem is a continuous one, use protective fungicides before any infection can happen. Fungicide management may be difficult once the symptoms develop. In that case, remove the infected branches that will help in reducing the number of spores available for further infection, and spread.

Sooty Mould (*Leptoxyphium kurandae*)

Sooty mould is a black fungus on the upper surfaces of leaves, growing in the secretion of aphids, mealybugs, many scales, and immature whiteflies. The mould spoils the appearance of foliage but is not particularly injurious to the plant. Oil emulsion or miscible oil sprays will loosen the mould and help clean up the plants. By spraying this material at about sundown, the dew helps to free the sooty mould from the leaf, and a forcible stream of water from a hose nozzle next morning before the sun dries the leaf will wash away 90 per cent of the sooty mould. A fungicide such as Mancozeb may also be used.

Root Rots and Collar Rots (*Phytophthora parasitica*)

The first noticeable symptom of this disease is a loss of leaf colour. The plant may stay in this unhealthy state for several days and then suddenly wilt and die. A characteristic water-soaked, slimy brown area will be found in the collar and root region when the plant is removed from the soil. The rotten bark around the collar has a pungent odour. If the water soaked area is allowed to dry out the outer layers of bark will peel off as a papery layer. The bark which is left has a characteristic stringy appearance.

In general, root rots are associated with over wet conditions brought about by overwatering or poor drainage. Study conductd by Hassan *et al.*, 2014 revealed that, microelements (copper and manganese), antioxidants (salicylic acid, ascorbic acid, and EDTA), a fungicide (Dithane M- 45) and biological control agents (*Trichoderma harzianum* and *Bacillus subtilis*) were significantly reduced the linear growth of the causal pathogens. Additionally, application of the previous microelements, antioxidants, a fungicide and biological control agents significantly reduced disease incidence of root rot and wilt diseases under field conditions. Copper, salicylic acid, and *Trichoderma harzianum* showed the best results in this respect. In conclusion, microelements, antioxidants, and biocontrol agents could be used as alternative strategies to fungicides for controlling root rot and wilt diseases in roselle.

Viral Diseases

Leaves infected with viral disease, Hibiscus chlorotic ringspot virus have chlorotic ring spots surrounded by a yellowish halo, was reported from Taiwan (Li and Chang, 2002). Two hibiscus-infecting tobamoviruses have been reported in Singapore and in Florida, USA. The viruses were originally designated as Hibiscus virus S (HVS) (21) and Florida hibiscus virus (FHV) (1) and have been officially renamed as *Hibiscus latent Singapore virus* (HLSV) and *Hibiscus latent Fort Pierce virus* (HLFPV), respectively (Adkins *et al.*, 2004; Kamenova and Adkins, 2004; Liu *et al.*, 2004; Wong and Srinivasan, 2004).

11.2.0 Pests

The pests of hibiscus can mainly be grouped as chewers or suckers according to the way they feed. In addition, there are a few miscellaneous pests that damage plants in other ways.

Chewing Pests

These include caterpillars and grasshoppers. Caterpillars are the larval stage of moths and butterflies. They devour foliage, sometimes stripping a plant of all its leaves. The *Heliothos* grub and the cabbage looper (commonly called inch worm) are the most prevalent worm pests of hibiscus buds, flowers and foliage. The *Heliothos* grub is nearly 5 cm long when full grown, yellowish, green, or brown or lengthwise light and dark stripes. It is usually found in the bud or flower. The cabbage looper is greenish, the body tapering to the head, with a thin white line along the body. It is more often found on the leaf.

Grasshoppers and katydids consume large quantities of foliage. Katydids are green and feed at night. They are neither found in great numbers. Control to eliminate immediately, before they become too numerous, they can frequently be removed by hand.

Sucking Pests

These include aphids, jassids or leaf hoppers, mealy bugs, scale insects, bugs and mites. Their mouth parts limit their feeding to piercing the surface tissue of the plant and sucking the sap.

Mealy bugs are important pests. They excrete honeydew which attracts ants and serves as a medium for the development of sooty mould. Mealy bugs are soft bodied and scale like insects usually covered with powdery or cottony, wax like material. They are around 6 mm long when mature and make cottony nests at twig joinings and under leaves.

Spider mites or red spiders are tiny, less than 0.5 mm long, and one of the most common pests. They are found by checking the undersides of leaves with a magnifying glass. They may be tan, red or purple. They suck plant juices causing tiny white spots on the leaves. Control early before infestation is great with Kelthane.

Aphids or plant lice are small, soft bodied insects about the size of a pinhead which attack new growth, causing leaves to curl and blossoms to be malformed. They are green to brownish, with one type being black. Ladybugs and aphid lions are natural enemies of aphids.

Whiteflies are about the size of a gnat, and their young are circular, flat, almost translucent, and hard to detect. Thrips are tiny, fast moving, yellow winged insects. They leave silvery patches on leaves and cause buds to drop. Stink bugs are 1 to 2.5 cm long, of a variety of colours and markings. They suck sap from buds, leaves and stems. They leave an odour on plants.

Miscellaneous Pests

Various other pests also cause damage to hibiscus. Snails and slugs are night feeders and injure plants, especially in damp, shady areas. Shiny trails may be seen, and snail baits should be applied when the pests are most active, *e.g.* dewey mornings or in wet weather. Beetles are chewing, hard shelled insects. Some feed on leaves, others on flowers. Some feed at night and hide during the day, others feed during the day. Their larvae feed on roots and bore through stems and branches. Carbaryl will protect against many of them. Leaf miners are small insects that feed between the upper and lower surfaces of the leaves. They leave distinctive trails on the leaves.

11.3.0 Physiological Disorders

Physiological disorders are the result of some unsuitable factor in the growing conditions such as over fertilizing, trace element deficiency, over or under watering, soil too acid or alkaline, position too hot or too shady *etc.*

Diseases and Pests of Hibiscus

Leaf Spot Infection

Crown Rot Infection

Mealy Bugs Infestation

Aphids Infestation

White Flies Infestation

Bud Drop

This is one of the most common physiological disorders of hybrid hibiscus. The problem is more severe in some seasons than in others and some varieties are more severely and more regularly affected than others. Sometimes an improvement can be effected by shifting the plant to another part of the garden. Sometimes if a variety is severely and continually affected by bud drop, it is better removed and replaced with another variety. The most common causes of bud drop appear to be lack of food or lack of water. Excessive amounts of nitrogen, particularly when associated with foliar fertilizers, have been known to cause bud drop. Double flowered forms seem more affected than singles. Regular watering and fertilizing is the best cure.

Yellow Leaves

Hibiscus bushes naturally discard their old leaves several times a year. The larger older leaves on the bottom of the bush will turn bright yellow a few days before they drop. This is perfectly natural and should cause no concern, unless of course the yellowing and dropping continues up the stems to the top of the bush. If this happens it is usually caused by excessive fertilizer or moisture around the root zone. The yellowing is usually triggered off by an excessively hot day, cold snap, or by spraying with unsuitable insecticides such as Malathion or Lannate. A number of the systemic insecticides will also cause yellow leaves, and could lead to complete defoliation if used constantly.

12.0 Harvesting

As the flowers of *Hibiscus rosa-sinensis, H. syriscus, H. mutabilis* and *H. schizopetalus*, commonly grown in garden do not last long as cut flowers, they are not harvested to use in vases and the faded flowers on plants are removed from the flower stalk. Red flowers of *H. rosa-sinensis*, used for worship are harvested and sold in flower market as floral ornaments and loose flowers. The bark, flowers, calices and fruits of *H. sabderiffa* are processed and used as fibre, food and herbal medicine.

13.0 Biotechnology

13.1.0 Molecular Markers

The Random Amplified Polymorphic DNA technique has been successfully used in a variety of taxonomic and genetic diversity studies (Jain *et al.*, 1993). Its popularity stems from the ease, cost effectiveness and applicability of the technique. Only a small quantity of DNA is required to generate a great amount of genetic markers, even without cloning, sequencing, or any knowledge of the genome being examined (Bassam *et al.*, 1991). Genetic variation among different Hibiscus species and varieties was determined using the RAPD method.

The most important characters which helped in characterizing the different species or varieties of *Hibiscus* is its flower colour, shape and form. The quantitative results obtained were not effective to differentiate between the species or the varieties. Despite the difficulties of characterization based on the morphology of the plants, morphological characterization remains a quick and easy method of identification (Zhou *et al.*, 2002).

Randomly Amplified Polymorphic DNA markers are simple, versatile, and relatively inexpensive (Crawford *et al.*, 1993) and are able to detect minute genetic differences in different Hibiscus species and varieties. The morphological analysis together with RAPD markers are useful for classification and indication of relationships among the different Hibiscus and can be useful for Hibiscus breeding program. Sequence Characterized Amplified Regions (SCAR) primers can be easily designed for many of these Hibiscus varieties and species. Close relationships could be found between the species using the RAPD primers.

13.2.0 Genetic Transformation

Christensen *et al.* (2009) observed differences in tissue susceptibility to *A. rhizogenes* infection, and differences in the virulence of *A. rhizogenes* strains. Their study showed that different types of explants of *H. rosa-sinensis* differed in their susceptibility. The difference in susceptibility to infection may be explained by the regenerative capacity of these plant organs and the genetic interaction between the bacterium and the plant. Differences in the virulence of *A. rhizogenes* strains have been documented in other infection studies (Balen *et al.*, 2004; Mihaljevic *et al.*, 1996; Porter, 1991). Spontaneous regeneration of shoots from hairy roots grown on hormone-free media has been reported in several species (Mugnier, 1988; Pellegrineschi and Davolio-Mariani, 1996). In *H. rosa-sinensis*, however, no spontaneous regeneration was observed. Besides, somatic embryogenesis in *H. rosa-sinensis* helped in regeneration of plants from hairy roots. In plants, which present difficulties for direct organogenesis, indirect somatic embryogenesis *via* a callus phase has been achieved (Feher *et al.*, 2003; Jimenez, 2005).

14.0 References

Abirami, S., Uma, G. (2018) *Int. J. Sci. Res.*, **7**(6): 1878-1886.

Abu-Tarboush, H.M., Ahmed, Saif Aldin B. and Kathani, H.A. Al. (1997) *Cereal Chem.*,**74**: 352-355.

Adkins, S., Kamenova, I., Achor, D. and Lewandowski, D.J. (2003) *Plant Dis.*, **87**: 1190-119.

Adkins, S., Kamenova, I., Chiemsombat, P., Baker, C. A. and Lewandowski, D. J. (2004) In: Abstract book of the 11th International Symposium on Virus Diseases of Ornamental Plants, Taichung, Taiwan.

Akpan, G.A. (2006) In: *Flower Breeding and Genetics: Issues, Challenges and Opportunities for the 21st Century* (Ed. N.O. Anderson), Springer, Dordrecht, pp. 479-489.

Allen, J.A. (2003) "Hibiscus tiliaceus L."(PDF). Tropical Tree Seed Manual. Reforestation, Nurseries & Genetics Resources.

Anonymous (2010) Guidelines of the use of herbal medicine in family health care. Edn 6, Ministry of Health Republic of Indonesia.

Aziz, E.E., Nadia, G. and Nadia, M.B. (2007) *Aust. J. Basic & Appl. Sci*, **1**(2): 73-78.

Babatunde, F.E. and Mofoke, A.L.E. (2006) *Pak J Nutr.*, **5**(4): 363-367.

Baker, R., Phillips, D.J. (1962) *Phytopathology*, **52**: 1242–1244

Balen, B., Leljak-Levanic, D., Miha;jevic, S., Jelenic, S. and Jelaska, S. (2004) *In Vitro Cellular & Developmental Biology - Plant*, **40**: 182-187.

Banerji, B.K. and Datta, S.K. (1986a) *J. Nucl. Agric. Biol.*, **15** : 237-240.

Banerji, B.K. and Datta, S.K. (1986b) *J. Nuclear Agric. Biol.*, **17**(2) : 113-114.

Barik, S., Senapati, S.K., Aparajita,S., Mohapatra, A. and Rout, G.R. (2006) *Z. Naturforsch* **61**: 123-128.

Barrios, K. and Ruter, J.M. (2019) *Bot. Stud.*, **60**: 3.

Basri, M.H.A., Abdu, A., Junejo, N., Hamid, H.A. and Ahmed, K. (2014) *Scientific Research and Essays*, **9**(11): 458-470.

Bassam, B.J., Caetano-Anollés, G. and Gresshoff, P.M. (1991) In: *DNA amplification fingerprinting and its potential application for genome analysis*, Plant Biotechnology and Development, pp. 1-9.

Bates, D.M. (1965) *Baileya*, **13**: 56–130.

Blanchard, O.J. Jr (1977) *M.Sc. Dissertation*, Cornell University, USA.

Blanchard O.J. Jr (2008) *Novon.*, **18**(1): 4–8.

Caballero-George, C. and Gupta, M.P. (2011) *Planta Med.*, **77**: 1189-202.

Chalupa, V. (1988) *Biol Plantarum*, **30**: 414.

Cobley, L.S. (1975) *Econ Bot.*, **3**: 89-103.

Crawford, D.J., Brauner, S., Cosner, M.B. and Stuessy, T.F. (1993) *American Journal of Botany*, **80**: 89- 92.

Chin, H. F. (1986) In: *The hibiscus: Queen of tropical flowers*. Kuala Lumpur: Tropical Press, p. 64.

Cribb, A.B. and Wild, J.W. (1976) In: *Food in Australia*, Fontana, Sydney, Australia.

Das, P.K., Ghosh, P., Dube, S. and Dhuha, S.P. (1974) *Technology* (Coimbatore, India)**1112**(3) : 185-186.

Das, P.K., Ghosh, P., Dube, S. and Dhuha, S.P. (1977) *Indian J. Hort.*, **34**(2): 169-174.

Dole, J.M. and Wilkins, H.F. (1999) Floriculture: Principles and species.Prentice-Hall, Inc., Upper Saddle River, N.J.

Duke, J.A. and Ayensu, E.S. (1985) In: *Medicinal Plants of China*. Reference Publ. Inc., Algonac, Michigan, 2, S705.

Eeckhaut, T.G.R., Van Huylenbroeck, J.M., De Riek, J. and Van Bockstaele, E. (2004) *Acta Hort.*, **630**: 85-90.

Egilla, J.N., Davies, F.T. and Drew, M.C. (2001) *Plant Soil.*, **229**: 213- 224.

El-Boraie, F.M., Gaber, A.M. and Abdel-Rahman, G. (2009) *World J. Agric. Res.*, **5**(4): 504-514.

Elevitch, C.R., Lex, A.J.T. (2006) "Hibiscus tiliaceus (beach hibiscus)" (PDF). *The Traditional Tree Initiative.*

Ewe, P.L. (2016) https: //www.nst.com.my/news/2016/08/168254/our-national-flower.

Faruq, G., Alamgir, M.A., Rahman, M.M., Motior, M.R., Zakaria, H.P., Marchalina, B., and Mohamed, N.A. (2013) *J. Animal Plant Sci.*, **23**(1): 60-67.

Feher, A., Pasternak, T. P. and Dudits, D. (2003) *Plant Cell, Tissue and Organ Culture*, **74**: 201–228.

Fletcher, R.A., Gilley, A., Sankhla, N. and Davis, T.D. (2000) *Horticultural Reviews*, **24**: 55-138.

Fouda, A.M., Daba, M.Y. and Dahab, G.M. (2007) *Can.J.Physiol.Pharmacol.*, **85**: 1020-1031.

Fukui, H, Koshimizu, K and Mitsui, T (1971) *Phytochemistry*, **10**(3): 671-673.

Fu, S., Luo, S., Zhou, L. and Zhang, F. (2004) *Guangxi Sci.*, **11**(**2**): 131–133.

Gast, R.H.(1971) *Lasca leaves*, **21**: 85-90.

Gautam, R., Saklani, A. and Jachak, S. (2007) *J. Ethnopharmacol.*, **110**: 200–234.

Geng, M., Ren, M., Liu, Z. and Shang, X. (2012) *Afr. J. Biotechnol.*, 11(2): 429-435.

Gilman, E.F. (1999) In: *Hibiscus rosa-sinensis*, Institute of Food and Agricultural Science, Univrsity of Florida, p. 254.

Gilman E.F. and Watson D.G. (1993) In: *Hibiscus syriacus Rose-of-Sharon*, Fact Sheet ST-295, a series of the Environmental Horticulture Department, Florida Cooperative Extension Service, Institute of Food and Agricultural Sciences, University of Florida.

Ghogare, P.B., Bhalke, R.D., Girme, A.S., Nirmal, S.A., Jadhav, R.S. and Tambe, V.D. (2007) *Dhaka University Journal of Pharmaceutical sciences*, **6**(1): 55-57.

Godfrey, R.K. and Wooten, J.W. (1981) In: *Aquatic and wetland plants of southeastern United States: dicotyledons*. Georgia: The University of Georgia Press, pp. 320–329.

Hammad, I. (2009) *Australian Journal of Basic and Applied Sciences*, **3**(1): 113–125.

Herath, S.P., Suzuki, T. and Hattori, K. (2004) *Plant Cell Tiss. Org. Cult.*, **77**: 49–53.

Hossain, M.D., Hanafi, M.M., Jol, H. and Hazandy, H.A. (2011) *African J. Biotech.*, **10**(50): 10087-10094.

Houghton, P.J. and Osibogun, I.M. (1993) *Journal of Ethnopharmacology*, **39**(1): 1-29.

Huang, J.G., Fan, Z.F., Li, H.F, Tian, G.Z., Hu, J.S. (2004) *Plant Dis.*, **88**: 683.

Huxley, A., Griffiths, M. and Levy, M. (1992) In: *The new royal horticultural society dictionary of gardening*. Vol. 2. Macmillan Press, Great Britain, UK.

Ishikura, N. (1982) *Agricultural and Biological Chemistry*, **46**: 1705-1706.

Ishiguro, K., Takahashi, Y., Oku, H., Mishima, A., Matzno, S., Juman, S. and Nakabayashi, T. (2006) In Fourth international workshop for anthocyanins, p.6.

Jadhav, V.M., Thorat, R.M., Kadam, V.J. and Sathe, N.S. (2009) *J. Pharm. Res.*, **2**(8): 1220-1222.

Jain, S.M., Das, S.C. and Barman, T.S. (1993) *Proc. Indian National Science Academy*, B 59: 623-628.

Jenderek, M.M. and Olney, A.J. (2001) Int. Plant Propag. Soc. Prod., **50**: 565–568.

Kamenova, I. and Adkins, S. (2004) *Plant Dis.*, **88**: 674-679.

Kasseanis, B (1957) Ann. App. Biol., **45**: 422–427.

Kadve, S., Yadav, M. and Tiwari, A. (2012) *International Journal of Biomedical and Advance Research*, **3**(6): 473–485.

Khalil, S. and Yousef, R.M.M. (2014) *Int. J. Adv. Res. Publ.*,**2**(5): 738-750.

Kim, J.H., Nonaka, G.I., Fujieda, K., and Uemoto, S. (1989) *Phytochemistry*, **28**(5): 1503.

Kim, S.H., Kim, W.J., Lee, G.J., Song, H.S., Kim, D.S., Kim, J.B., Kang, S.Y. (2009) *Hort. Environ. Biotechnol.*, **50**: 555-565.

Kemp, M.S., Burden, R.S., and Brown, C. (1979) *Phytochemistry*, **18** (10): 1765-1766.

Kinoshita, T. (2007) *Genes Genet. Syst.*, **82**: 177-186.

Kirtikar, K.R. and Basu, B.D. (1991) In: *Indian Medicinal Plants* (Eds. Bishen Singh and Mahendra Pal Singh), 2nd ed., Vol. I, Dehradun, India, pp. 325-341.

Khristi, V., Patel, V.H. (2016) *Int. J. Nutr. Dietetics.*, **4**(2): 105-123.

Kuligowska, K., Simonsen, M., Lütken, H.and Müller, R. and Christensen, B. (2013) *Acta Hort.*, 990: 235-241.

Kulkarni, G.N. (2002) *Principles of seed technology*.1- 390.

Kumar, A. and Singh, A. (2012) *International Journal of Research in Pharmaceutical and Biomedical Sciences*, **3**(2): 534–538.

Kuwada, H. 1960. *Kagawa Daigaku Nogakubu Gakujutsu Hokoku/Tech. Bull. Fac. Agric. Kagawa Univ.*, **12**: 1-8.

Lattier, J.D. and Contreras, R.N. (2020) *J. Amer. Soc. Hort. Sci.*, **145**(4): 215–227.

Lawton, B.P. (2004) Hibiscus: hardy and tropical plants for the garden. Timber Press; Portland, OR: p.160

Lee, D.W. (2007) Nature's Palette: the Science of Plant Colour. University of Chicago Press. p. 183.

Lehnert, M.S., Kramer, V.R., Rawlins, J.E., Verdecia, V. and Daniels, J.C. (2017) *Insects.*, **8** (3): 68.

Li, C., Liu, S., Wu, D., *et al.* (2009) Sci. Technol. Food Ind., **3**(11): 97–99.

Li, S.C. and Chang, Y.C. (2002) First report of *Hibiscus chlorotic ringspot virus* in Taiwan. Plant Pathol., **51**: 803.

Liu, F.L., Chen, T.C., Yeh, S.D., Hsu, H.T., Chen, C.C., Bau, H.J., and Chen, Y.K. (2004) *Plant Pathol. Bull.*, **13**: 283-290.

Lloyd, G. and McCown, B. (1980) *Int. Plant Propag. Soc. Prod.*, **30**: 421–427.

Lowry, J.B. (1971) *Phytochemistry* **10**(3): 673- 674.

Jimenez, V. M. (2005) *Plant Growth Regulation*, **47**: 91–110.

Jalan, P. (2002) In: *Compendium of Medicinal Plants in Malaysia*, Herbal Medicine Research Center Institute for Medical Research, Kuala Lampur, P.14.

Johnson, N.A. (2010) *Trends Genet.*, **26**: 317-325.

Jose, E.A. and Vijayan, K.K. (2006) *Indian J. Chem. Sect.* B;

Maganha, *e.g.*, Halmenschlager, R.C., Rosa, R.M., Henriques, J.A.P, Ramos, A.L.L.P. and Saffi, J. (2010) *Food Chem.*, **118**: 1-10.

Malauszynski, M., Van Zanten, L., Ashri, A., Brunner, H., Ahloowalia, B., Zapata, F.J. and Weck, E. (1995) In: *Induced Mutation and Molecular Techniques for Crop Improvement*. Proc. of a Symposium, 19-23 June, IAEA, Vienna.

Mandour, M.S., Abou-Zied, E.N. and Hassib, M. (1979) *Plant Soil*, **52**(4): 485-490.

Markrose, B.U. (1984) *M.Sc.(Hort.) thesis* submitted to the Kerala Agricultural University, Trichur, Kerala, India.

Martin, C. (1985) *Endeavour*, **9**: 81–86.

Mat, I.A, Isa, P.M. and Abd Aziz, A.R. (1985) *Mardi Res Bull.*,**13**: 68-74.

Menzel, M.Y., Richmond K.L., Contolini C.S. and Huang, P. (1986) *Amer. J. Bot.*, **73**: 304–309.

Mihaljevic, S., Stipkovic, S. and Jelaska, S. (1996) *Plant Cell Reports*, **15**: 610–614.

Maganha, *e.g.*, Halmenschlager, R.C., Rosa, R.M., Henriques, J.A.P., Ramos, A.P., Saffi, J. (2010) *Food Chem.*, **118**(1): 1–10.

Marais, E., Jacobs, G. and Holcroft, D.M. (2001) *Acta Horticulture*, **553**: 91-93.

Menzel, M.Y. and Wilson, F.D. (1969) *Brittonia*, **21**: 91-125.

Missoum, A. (2018) *Journal of Ayurvedic and Herbal Medicine*, **4**(3): 135-146.

Morgan, E.R., Timmerman-Vughan, G.M., Conner, A.J., Griffin, W.B. and Pickering, R. (2011) *Plant Breed. Rev.*, **34**: 161-220.

Mugnier, J. (1988)*Plant Cell Reports*,**7**: 9–12.

Nadkarni K.M. (1982) In: *Indian Materia Medica*. 3rd ed., Vol. I, Bombay Popular Prakashan, Bombay, pp. 115-116.

Naveena, N., Subramanian, S., Jawaharlal, M., Iyanar, K. and Chandrasekhar, C.N. (2020) *Electronic Journal of Plant Breeding*, **11**(4): 1187-1193.

Nazarudin, M.R. (2012) *J. Trop. Plant Physiol.*, **4**: 29-40.

Neuwinger, H.D. (2000) *African Traditional Medicine*. Stuttgart: Medpharm Scientific Publication, 156-169.

Nevade, S.A., Sachin G.L. and Kalyane, N.V. (2011) *Res J Pharm Technol.*, **4**(3): 472-473.

Nyam, K.L., Leao, S.Y., Tan, C.P. and Long, K. (2012) *J Food Sci Technol.*, DOI 10.1007/s13197- 012-0902-x.

Ong, H.C. and Nordiana, M. (1971) *Phytochemistry*, **10**(3): 661.

Pal, S., Sarkar, J., Bhattacharya, S., Biswas, M. (2011) *Pharmacology Online,* **2**: 1431-1436.

Patil, D.A. (2010) *Deccan Curr. Sci.,* **3**: 104-111.

Pellegrineschi, A. and Davolio-Mariani, O. (1996) *Plant Cell, Tissue and Organ Culture,* **47**: 79–86.

Perez, G.R., Zavala, S.M., Perez, G.C. (1998) *Phytomedicine,* **5**(1): 55–75.

Porter, J. R. (1991) *Critical Reviews in Plant Sciences,* **10**: 387–421.

Punasiya, R., Devre, K. and Pillai, S. (2014) *Int. J. of Pharm. Life Sci.,* **5**(6): 3617-3621.

Rahmatullah, M., Kabir, A.A.B.T., Rahman, M.M., Hossan, M.S., Khatun, Z., Khatun, M.A., Jahan, R. (2010) *Adv. Nat. Appl. Sci.,* **4**: 45-51.

Rao, K., Geetha K. and Banji, D.(2014) *Journal of Pharmacognosy and Phytochemistry.*3(4)

Rademacher, E. (2000) *Annual Review of Plant Physiology and Molecular Biology,* **51**: 501-531.

Rajeshwari, R., Reddy, R.V.C., Maruthi, M.N., Colvin, J., Seal, S.E. and Muniyappa, V. (2005) *Ann. Appl. Biol.,* **147**: 15-25.

Reynolds, B.D. and Blackmon, W.J. (1983) *J. Amer. Soc. Hort Science,* **108**: 307-310.

Rieselberg, L.H. and Carney, S.E. (1998) *New Phytol.,* **140**: 599-624.

Rahmatullah, M., Hassan, R., Hossan, S., Jahan, R., Chowdhury, M.H., Seraj, S. *et al.* (2010) *American-Eurasian Journal of Sustainable Agriculture,* **4**(3): 309-325.

Ross, I.A. (2003) In: *Medicinal Plants of the World*, Springer, pp. 253-266.

Runkle, E.S., Heins, R.D., Cameron, A.C. and Carlson, W.H. (1998) *Hort Science,* **33**: 672–677.

Salisbury, F.B. and Ross, C. (1992) In: *Plant Physiology*, 4th Edition. Freeman and Worth, New York, P. 682.

Sajib, N.H. and Uddin, S.B.(2015) *Pertanika J.Trop. Agric. Sci.,* **38**: 197-210.

Salem, M.Z.M., Olivares-Pérez, J., Salem, A.Z.M. (2014) *Life Sci.,* **11**(5): 1-8.

San Pascual, A.O., Magdalita, P.M., Medina, N.G. and Apacionado, B.V. (2017) *Australian Journal of Crop Science,* **11**(12): 1508-1519.

Seghatoleslami, M.J., Mousavi, S.G. and Barzgaran,T. (2013) *J Anim Plant Sci.,* **23**(1): 256-260.

Shvart, M., Borochov, A. and Weiss, D. (1997) *Physiologia Plantarum,* **99**(1): 67-72.

Singh, F. and Khoshoo, T. N. (1970) *Caryologia,* 23: 19-27.

Skovsted, A. (1935) *Indian J. Genet.,* **31**(2): 263-296.

Small, R.L. (2004) *Syst. Bot.,* **29**(2): 385-392.

Srivatanakul, M., Park, S.H., Sanders, J.R., Salas, M.G. and Smith, R.H. (2000) *Plant Cell Rep.,* **19**: 1165-1170.

Spira, T.P. (1989) In: *The evolutionary ecology of plants* (Eds. J.H.Bock and Y.B. Linhart), Westview Press, Boulder, CO. pp. 247–256.

Thomas, B. and Vince-Prue, D. (1997) In: *Photoperiodism in plants*. 2nd ed. Academic Press, San Diego, California, USA.

Runkle, E.S., Heins, R.D., Carlson, W. and Cameron, A. (1996) *Grower Talks*, **60**: 66–70.

Van Laere, K., Van Huylenbroeck, J.M. and Van Bockstaele, E. (2007) *Euphytica*, **155**: 271-283.

Van Tuyl, J.M. and De Jeu, M.J. (1997) In: *Pollen Biotechnology for Crop Production and Improvement* (Eds. K.R. Shivanna and V.K. Sawhney), Cambridge Univ. Press, London, UK, pp.273-292.

Van de Laar, H.J. (1997) *Dendroflora*, **34**: 43-60.

Vazquez-Thello, A., Yang, L.J., Hidaka, M. and Uozumi, T. (1996) *Plant cell Rep.*, **15**: 506-511.

Yadong, Qi., Kit, L.C., Fatemah, M., Mila, B., and Janet, G.Y. (2005) *Southern University and A & M College System.*, **225**: 771-2262.

Yang, L.J., Hidaka, M., Masaki, H. and Uozumi T. (1995) *Plant Tisue Culture Letters*, **12** (2): 173-177.

Yang, L.J., Tello, A.V., Hidaka, M., Masaki, H. and T. Uozumi. (1996) *Plant Tisue Culture Letters*, **13** (2): 161-167.

Yasin Nasir, J. (1979) In: *Flora of West Pakistan* - Biebersteiniaceae. National Herbarium, Pakistan Agricultural Research Council, Islamabad, P.12.

Yun, Cheung Kong, Jing-Xi and Paul Pui-Hay (1986) *Journal of Ethnopharmacology*, **15**(1): 1-44.

Yun, B.S., Lee, I.K., Ryoo, I.J. and Yoo, I.D. (2001) *J. Natl. Prod.*, **64**(9): 1238–1240.

Yusuf, N., Ng, W.H. and Adzemi, M.A. (2019) *Malays Appl Biol.*, **48**(1): 177-184.

Wee, Y. C. (2003) In: *Tropical trees and shrubs: A selection for urban planting*, Sun Tree Publishing, Singapore, P. 65.

West, T.P. and Preece, J.E. (2004) *Acta Hortic.*, **630**: 293-297.

Welsh, S. (1980) *Great Basin Naturalist*, **40**(1): 27-37.

Winkelmann, T., Doil, A., Reinhardt, S. and Ewald, A. (2010) In: Plant Cell Culture: Essential Methods, (Eds. M.R. Davey, and P. Anthony), John Wiley & Sons, London, UK, pp. 79-95.

Winters, H.F. (1970) *Econ Bot.*, **24**(2): 155–164.

Wise, D.A. and Menzel, M.Y. (1971) *Brittonia*, **23**(4): 425-437.

Wong, C.C., Nizam, A.M., Vijaysegaran, S., Mah, S.Y. and Ghawas, M. (2001) In: First Technical Review Meeting on the National Kenaf Research project. MARDI, pp. 13-24.

Wong, S.K., Lim, Y.Y. and Chan, E.W.C. (2009) *J. Trop. For. Sci.*, **21**: 307-315.

Wong S.K., Lim Y.Y. and Chan E.W.C. (2010) *Ethnobot. Leaflets,* **14**: 781-796.

Wong, S.M. and Srinivasan, K.G. (2004) In: Abstract book of the 11th International Symposium on Virus Diseases of Ornamental Plants, Taichung, Taiwan.

Zahid, H., Rizwani, G.H., Shareef, H., Ahmed, M., Hina, B. (2012) *Int. J. Pharm. Pharm. Sci.,* **4**: 218-221.

Zahid, H., Rizwani, Z.H., Shareef, F. (2014) *J. Med. Plant Res.,* **8**: 802-810.

Zapata, C., Srivatanakul, M., Park, S.H., Lee, B.M., Salas, M.G. and Smith, R.H. (1999) *Plant Cell Tiss. Org. Cult.,* **56**: 185–191.

Zhou, C., Bao-Rong, L., Brian, S.B., Kazuhiko, S., Jia-Kuan, C. (2002) *Hereditas,* **136**: 231-239.

Bougainvillea

Debmala Mukherjee

1.0 Introduction

Bougainvillea belongs to Nyctaginaceae family which is also popularly known as four-o'clock family and is native to South America. Bougainvillea cultivars are diploid with a somatic chromosome constitution 2n= 34, irrespective of the species or hybrid group to which they belong (Zadoo *et al.*, 1975). The name, 'Bougainvillea' comes after Louis Antoine de Bougainville, a French navigator and military commander who was the first European to take note of the plant, in Brazil in the year 1768. The genus *Bougainvillea* has about 18 species of shrubs, vines, or small trees. Many species are thorny. Only the woody vines have attained wide popularity; several species have produced very showy cultivars. This climbing shrub produces stems up to 7 metres or more long. Its long arching thorny branches bear heart-shaped leaves and masses of papery bracts in white, pink, orange, purple, and burgundy. Many cultivars, including double- flowered and variegated, can also be seen. The cascading bloom of bougainvillea is being considered as one of Florida's signature tropical images. When in a more tree form it has a very dense ellipsoid crown. The plant can support itself on other plants by means of thorns carried in the leaf axils, though in cultivation it might need tying in to pergolas. Bougainvillea provides hedges, barriers, slope coverings and its growth habit along with beautiful showy bracts make it a identical plant for landscapes.

2.0 Importance and Uses

2.1.0 Aesthetic Uses

Bougainvillea is a popular ornamental plant for its colourful attractive bracts. It is popular among parks, homes and institutional gardens grown mainly as bush, climber, hedge, topiary, pot plant, bonsai, on pergolas and trees (Roy, 1987; Sharma and Roy, 2001). Bougainvillea can also be used as a houseplant, in hanging basket or in protected structures in cooler climates. It has wide adaptability to various agro- climatic conditions and easy multiplication method has made it a popular ornamental plant. Moreover, as it is a drought and pollution tolerant plant, it is well suited for industrial places and on road dividers (Kumar and Prasad, 2002). The drought tolerance of bougainvillea makes it for year round cultivation in warm climate and its high salt tolerance makes it a natural choice for cultivation in coastal region. It can be pruned into a standard, but is also grown along fence lines, on walls, in containers and hanging baskets, and as a hedge or an accent plant. A study conducted by Muhammad (2019) in Jazan Area of Saudi Arabia stated that *B. glabra*, despite the polluted plant being under some biochemical stress; it shows a high tolerance to the vehicular pollution depicted by a high air pollution tolerance and anticipated performance index. Hence this plant species can be recommended for inclusion in the urban greenbelt development plan and city landscaping The plant is generally spreading horizontally or hanging downwards But it can be trimmed and grown as a formal hedge, or over the ground as a ground cover. The plant is traditionally grown in living fences in the north- western Himalayas, where it helps

Uses of Bougainvillea

In Pot | In Beds | In Hanging Basket

In Hedge | As Specimen Plant | Against Wall

In Trellis | On Pergola | On Fence

Arch | Topiary | In Road Divider

Source: www.in.pinterest.com, www. pixels.com, www. gardentia.net, www. lexiconworld. in, www.tourmet.com, www.insightsgreece.com, www.madisongardensnursery.com

to exclude livestock and other animals by marking out land boundaries; whilst also providing a range of medicinal and other uses.

2.2.0 Therapeutic and Other Uses

Bougainvillea is one of the traditional medicinal plants with potential antifertility activity (Ghogar, 2017). The aqueous extract and decoction of this plant have been used as fertility control treatment among the tribal people in many countries. Besides, it has exhibited to possess anticancer, anti-diabetic, antihepatotoxic, anti-inflammatory, antihyperlipidemic, antimicrobial, antioxidant, and antiulcer properties. The phytoconstituents such as alkaloids, essential oils, flavonoids, glycosides, oxalates, phenolics, phlobotannins, quinones, saponins, tannins, and terpenoids present in it were reported as the basis of its efficacious therapeutic properties. The other important constituents which contribute to the remedial properties are bougainvinones, pinitol, quercetagetin, quercetin, and terpinolene. *B. spectabilis* flowers extracts exert protective effect against the toxic effects of rotenone on brain, liver and kidney. *B. spectabilis* flower extracts decreased brain lipid peroxidation and prevented neuronal death due to rotenone and might thus prove the value in treatment of Parkinson's disease (Abdel- Salam, 2017). The infusion of *B. glabra* plant's tender leaves and bracts is used orally to treat gastrointestinal problems and respiratory diseases. By keeping the view of its multiple uses there is always demand for the new coloured and forms of bracts in bougainvillea. All the present day colourful bracted bougainvilleas are developed through bud sports, mutations and inter and intra specific hybridization (Salam, 2017). A study conducted by Ji *et al.* (2015) investigated and evaluated the biomethane potential (BMP) of *Bougainvillea spectabilis* waste in Yunnan, China, when subjected to mesophilic (AD). This species is an excellent feedstock for anaerobic digestion.

3.0 Origin and History

Bougainvillea is originated in South America and was first collected by Commerson, a French Botanist, at Rio-de-Janeiro, Brazil. The genus 'Bougainvillea' belongs to family Nyctaginaceae and was discovered by French Botanist Commerson in 1766- 69. The generic name Bougainvillea Commers. was first published by A. L. De Jussieu in his work *Genera Plantarum* in 1789 (Jussieu, 1789). It is not clear whether the cultivars outside South America were derived from plants already cultivated in Brazil or from truly wild plants (Holttum, 1938). Much of the evolution in bougainvillea took place outside its native home, primary factor being natural hybridisation and showy bracts accompanied by self incompatibility (Zadoo *et al.*, 1975). The *B. glabra* and *B. spectabilis* are widely used species and most of the present bougainvillea cultivars are thought to have originated from them. Third species, *B. peruviana* is also of horticultural importance.

The year 1910 was marked by a major discovery in cultivated bougainvillea when Mrs. R.V. Butt brought crimson coloured bougainvillea cuttings from Cratagena (Columbia) to Trinidad. Over the years the cultivar turned out to be ever mutating, being the source of various colour mutants that have added materially to the richness of the colour in the genus. The plant was named after Mrs. Butt. This plant reached Kew in 1915 and from there it got distributed to India, Africa,

Australia, Malaya and Singapore in 1923 (Holttum, 1955). The cv. 'Scarlet Queen' was introduced to India directly from West Indies by Mr. Tomlinson in 1920. Percy Lancaster gave the name 'Scarlet Queen' to this variety. According to him, the coloured bracts of this plant are slightly darker than 'Mrs. Butt' but the actual flower is malformed, so instead of there being three white or cream coloured flowers in the bracts, there are merely little bunch of anthers (Holttum, 1955; Zadoo *et al.*, 1976). In 1931, Mrs. McClean of Trinidad produced apricot orange bracts instead of the normal crimson bracts of 'Mrs. Butt'. This variant was propagated by R.O. Williams and named as cv. 'Mrs. McClean' (Holttum, 1955). The cv. 'Scarlet Queen' produced orange coloured variety in 1932 at Madras in the garden of Mrs. Louis Wathen. It was named as 'Louis Wathen' by Mr. B.S. Nirody. Subsequently majolica yellow sports appeared from 'Louis Wathen' and 'Mrs. McClean' by spontaneous bud variation and were named as 'Enid Lancaster' and 'Mary Baring' (Pal and Swarup, 1974). S. Percy Lancaster had separated a purple coloured sport from 'Scarlet Queen' in 1942 and named it as 'Alick Lancaster'. The variegated leaved bud sport from 'Scarlet Queen' has been released as cvs. 'Scarlet Queen Variegata' and 'Rao' (Pal and Swarup, 1974). An important floriferous and recurrent blooming seedling was raised by P.S. Swaminathan at Madras from 'Princess Margaret Rose' and named it 'Mrs. H.C. Buck' (Lancaster, 1951; Holttum, 1955). The famous bicolored variety 'Mary Palmer' was isolated from 'Mrs. H.C. Buck' in 1949 by S. Percy Lancaster (Holttum, 1957). 'Thimma' with variegated foliage and 'Shubra' with pure white bracts were evolved as a sport of 'Mary Palmer' (Pal and Swarup, 1974).

The first milestone in the history of bougainvillea cultivation through the evolution of a bicolored cultivar 'Mary Palmer' bearing white and magenta coloured bracts on the same plant in the year 1949. It appeared as a bud sport of 'Mrs. H.C. Buck' following its drastic pruning, resulting in a complex chimera. It was named after 'Mrs. Palmer', in whose garden at Alipore, Calcutta, it appeared first. This is one of the most outstanding and popular cultivars of bougainvillea even today (Salam, 2017). The second milestone also created an impact among the bougainvillea lovers was the evolution of bougainvillea cultivars having variegated foliage which remains attractive even when plants are not in bloom. These cultivars are very much preferred for pot culture. Some of the outstanding varieties which have variegated foliage are 'Archana', 'Arjuna', 'Louis Wathan Variegata', 'Marietta', 'Parthasarthy', 'Scarlet Queen Variegata', 'Surekha', 'L. N. Berta' and 'Thimma'. The third milestone was the evolution of multi- bracted cultivars commonly known as 'Million Dollar' group as a budsport of *B.* × *buttiana* in Phillipines namely 'Cherry Blossom', 'Los Banos Beauty', 'Mahara' and 'Roseville's Delight' having 18- 21 bracts as against to common occurrence of three per flower. This group has further brightened the usefulness of this plant for ground as well as pot culture. The fourth milestone was the evolution of bougainvillea cultivar 'Shubhra'. It has fulfilled the long felt desire of having parchment white bracts, with perpetual blooming habit. It is also hardy and easy to multiply. Another footprint of introduced bougainvillea cultivar by Prof. T. K. Bose was 'Mohan' from Thailand. But progress in bougainvillea breeding was hampered all over the world largely because of the extensive pollen and/or seed sterility. Thus the choice of female and male parents is limited only to a few relatively fertile types which, more often were not attractive cultivars. After detailed studies of the reasons for sterility, fertility was restored by colchiploidy. This has been the fifth milestone, which has enabled immediate broad basing of germplasm by inclusion of such cultivars in the breeding programme that were out for reach of Bougainvillea breeders all over the world (Sharma, 1986). The bougainvillea improvement work in India started with the introduction of a few varieties by the Agri Horticultural Societies at Calcutta and Madras. 'Scarlet Queen' named by Sir Percy Lancaster in 1920 is probably the first bougainvillea raised in India. In all, more than 150 cultivars have been developed in bougainvillea. Of these, a large number of cultivars have been developed at IARI New Delhi, IIHR Bangalore and NBRI, Lucknow.

4.0 Distribution

Some species of the *Bougainvillea* are distributed worldwide and without being specific to any single place, species, cultivars, or hybrids (cited in Abarca-Vargas and Petricevich, 2018).

(i) *B. buttiana*: this species was found in India, Mexico, and Thailand. (ii) *B. glabra*: it was found in Italy, Spain, France, Bangladesh, India, China, Egypt, Israel, Thailand, Philippines, Madagascar, Nigeria, Hawaii, Bolivia, Colombia, Costa Rica, Cuba, Ecuador, El Salvador, United States of America, Guatemala, Honduras, Virgin Islands, Mexico, Nicaragua, Puerto Rico, Dominican Republic, Venezuela, and Brazil. (iii) *B. spectabilis*: this species exist in Nigeria, Bahamas, Bolivia, Colombia, Costa Rica, Ecuador, Guatemala, Honduras, Jamaica, Mexico, Panama, Puerto Rico, Dominican Republic, Tanzania, Trinidad and Tobago, India, Montenegro, Pakistan, Australia, Brazil, and Vietnam. (iv) *B. spinosa*: this plant species is a very poorly documented plant species, but it was reported in Argentina. (v) *B. peruviana*: this plant is scarce and only noted in the scientific literature in China, India, and Peru.

5.0 Morphology

Bougainvillea is a tropical and subtropical dicotyledonous, woody, evergreen, shrubby vine. Bougainvillea is deciduous when grown in areas with a long dry season. It has a spreading, round plant habit with a height and spread of up to 6 meter. It climbs by slender arching canes armed with stiff curved thorns. As they age, the stems turn from mid-green to dull green-brown. Numerous cultivars are available, with a striking array of colours. Colourful flowers are botanically called as bracts, or modified leaves and are attached to the mid rib through 2-4 cm long structures. Inflorescence occurs on short stem axes borne laterally in the axils of leaves. From their initial emergence to full maturity, the bracts of many cultivars change colour. Bracts may retain their colour for several months after the flowers have finished, gradually fading to resemble the colour and texture of paper (Kobayashi, 2007). Leaves are simple and alternate, with an undulate leaf margin and are mid- to deep green in colour, although some cultivars have variegated foliage. The leaf blade is 5- 10 cm long, with much variation in shape: globular, elliptical, obviate, ovate, or cordate. The true, perfect flowers are small, tubular, and surrounded by showy, colourful petaloid bracts and hermaphrodite in nature. The flowers borne in clusters of three, each flower subtended by a brightly coloured bract, which helps to attract insects for cross pollination. The vibrant colours of bougainvillea are not from the inconspicuous, commonly white or yellow, flowers. Bract colour in bougainvillea is contributed by betacyanins and betaxanthins *i.e.* betalains (Mabry and Dreiding, 1968). The tip of the flower is conspicuous with a star. There is a solitary carpel at the base surrounded by ring shaped nectar. Fruit is an elongated achene less than 1.5 cm long. It is rather inconspicuous, not showy, and has a dry, hard fruit cover. Anthesis takes place during morning hours (around 10 am). Butterflies visit the flowers attracted by the brightly coloured bracts and the nectar glands aids in cross pollination (Salam, 2017). The opened flowers remain so for a day, after which the upper part of the flower tube gets twisted in a spiral. Xu *et al.*, 2009 reported the occurrence of sixteen stages during bud and flower development in one inflorescence of Bougainvillea.

6.0 Species and Cultivars

6.1.0 Species

The genus *Bougainvillea* is a very widespread genus throughout the world. It belongs to the family Nyctaginaceae and, according to the "The Plant List", contains approximately 18 species (*B. berberidifolia, B. buttiana, B. campanulata, B. glabra, B. herzogiana, B. infesta, B. lehmanniana, B. lehmannii, B. malmeana, B. modesta, B. pachyphylla, B. peruviana, B. pomacea, B. praecox, B. spectabilis, B. spinosa, B. stipitata,* and *B. trollii*). Only four species (*B.* × *buttiana, B. glabra, B. spectabilis,* and *B. peruviana*) are commercially exploited (Kobayashi, 2007). However, there are also more than 100 cultivars and three hybrid-species, the latter not yet recognized (Abarca-Vargas and Petricevich, 2018).

B. spectabilis

This species was the first member of the genus to be identified from Brazil in 1798 by German botanist Carl Ludwig Willdenow. This species is noted for its hairy leaves and stems; leaves are large and ovate, with rippling along the edges and hairs on the underside; bracts are small and can vary in colour, ranging from white, red, mauve, purple-red, or orange; fruit is a small, inconspicuous, dry, elongated achene; thorns are large; growth habit is dense, and the colourful bracts appear up and down the branches. The bloom cycle is seasonal, with plants blooming after the dry season or after a cool spell, which may trigger plants to bloom. *B. spectabilis* is Guam'territorial flower (Kobayashi *et al.*, 2007). The Yanadi tribe of Chittoor district, Andhra Pradesh, India, once used the leaves of *Bougainvillea spectabilis* to heal diabetes (Lim, 2014).

B. glabra

This climbing, evergreen member of the genus, also from Brazil, was first identified by Swiss botanis Jacques Denys Choisy in 1850. It is called as lesser bougainvillea or paper flower, and is the most common species of bougainvillea used for bonsai. The elliptical leaves are green or variegated, with a glossy sheen; glabrous (smooth, hairless), but some puberulence (presence of small hairs) may be found; bracts come in many sizes and shapes; typically they are triangular and purple

Species of Bougainvillea

B. spectabilis — ***B. glabra*** — ***B. peruviana***

B. arborea — ***B. × buttiana*** — ***B. campanulata***

B. stipitata — ***B. berberidifolia*** — ***B. spinosa***

Source: www.patagoniawildflowers.org, www.richardlyonsnursery.com, www.flickr.com

or mauve, although white bracts are also common. The bracts tend to appear along the branches and at the tips. The flowers are white to cream coloured. The thorns are small and curved at the tips. The growth habit is spreading, and the green-leaf types are fast growing. *B. glabra* is heat and drought tolerant and frost sensitive. This species blooms several times a year. *B. glabra* and *B. spectabilis* are alike in general appearance, the main differences being the bloom cycle, and *B. glabra* being hairless whereas *B. spectabilis* is hairy and velvety-felty underside of the leaves.

B. peruviana

This climbing, evergreen member of Bougainvillea from Peru was first identified by German naturalist and ex- plorer Alexander von Humboldt in 1810. This species is noted for its green bark. The growth habit is lanky. The long, thin leaves are ovate and glabrous; rounded bracts are magenta to pink in colour and may have some distortion or wrinkling, and the flowers are yellow; thorns are short and straight. This species is not as vigorous as some of the hybrids and often requires pruning to promote substantial branching. The plants may bloom several times a year if subjected to dry periods between flushes. *B. peruviana* is the most stable species compared to *B. glabra* and *B. sepctabilis*, as there is very little variation in the general shape of bracts and leaves.

B. × buttiana

Bougainvillea × buttiana is a garden hybrid of *Bougainvillea glabra* and *Bougainvillea peruviana*. (*B.× buttiana*) which stands for Mrs. Butt. It is an evergreen vine, with thorny stems and tiny trumpet shaped white flowers, usually appearing in clusters surrounded by three showy bright magenta-rose papery bracts. There are many species that are bud variations of this species such as those that are double flowered and those that are in various colours such as red, yellow and orange; leaves are ovate and dark green. This plant can be grown in a warm temperate or subtropical environment where the temperature does not fall below freezing against a south-facing wall in full sun. Numerous cultivars have been developed, of which the following have gained the Royal Horticultural Society's Award of Garden Merit: 'Miss Manila', 'Mrs Butt', 'Poulton's Special'. This species had the highest xylene removal efficiency (Sangthong, 2015). This is said to be a native species in Okinawa. However, it is limited to the coastal region where there is no frost. According to Khoshoo (1998), Bougainvillea also has three hybrid groups namely, *B. × buttiana* (glabra × peruviana), *B. × specto-peruviana* and *B. × spectoglabra*. Among these, *B. spectabilis* and *B. glabra* were reported to be more tolerant to cooler climate. Bougainvillea grows well throughout the plains of India, but only *B. glabra* thrives well at higher altitudes (Pal and Swarup, 1974).

6.2.0 Cultivars

Considering its ornamental and commercial importance in nursery trade, Research and Development (R & D) were taken up by different national research institutes (IARI, IIHR, NBRI, BARC), state agricultural universities, Agri Horticultural Societies (Kolkata and Chennai) and even by progressive nursery men. As a result, a large number of new cultivars have been developed in India.

Considering the contribution made by India, the international registration authority for new cultivars lies in the Indian Agricultural Research Institute (IARI), New Delhi, which is an international recognition conferred upon by International Society for Horticultural Science. In India, crop improvement work was started in early 20th century with the introduction of *B. spectabilis* in 1860 from Europe. However, the popularity really started with the introduction of cv. 'Mrs. Butt' from Royal Botanic Garden, Kew to The Agri Horticultural Society of India, Kolkata in 1923 (Swarup and Singh, 1995). Percy Lancaster, the renowned horticulturist, former Secretary of Agri-Horticultural Society of India, Kolkata has been credited for the development of first cultivar of Bougainvillea - 'Scarlet Queen' in 1920 and subsequently another excellent bi- coloured cultivar 'Mary Palmer' developed by him paved the way for bougainvillea cultivation in India. Some of the selected cultivars belonging to different species and some hybrids which are commonly grown in India are described below (Singh *et al.*, 2020):

- ✰ **Cultivars of *B. spectabilis:*** 'Aida', 'Boise de Rose', 'Jubilee', 'Maharaja of Mysore', 'Mrs. Lancaster', 'Rosa Catalina', 'Tomato Red', 'Lateritia' (syn. 'Dar-es-Salam', 'Brasiliensis'), 'Refulgens', and 'Speciosa'.
- ✰ **Cultivars of *B. glabra:*** 'Cypheri', 'Formosa', 'Magnifica', 'Sanderiana', 'Snow White,' 'Moonlight Madonna', 'Buddhadas', 'Dream', 'Gagarin', 'Kalyani', 'Sachidananda', 'Stanza' and 'Trinidad'.
- ✰ **Cultivars of *B. peruviana:*** 'Princess Margaret Rose' (possibly a synonym of 'Lady Hudson'), 'Mary Palmer' (patented as 'Surprise' in the U.S.A). It is a sport from 'Mrs. H.C. Buck' It is a most popular bicolored cultivar, 'Dr. B.P. Pal' 'Shubhra', 'Mrs. H.C. Buck' (Syn. 'Mahatma Gandhi'), 'Partha' A seedling of 'Princess Margaret Rose'.
- ✰ **Cultivars of *B.× buttiana* :** 'Alick Lancaster' (syn. 'Lilac Queen'), 'Enid Lancaster' also known as 'Golden Glow' and 'Gold Queen', 'Lady Mary Baring', 'Louise Wathen', 'Magenta Queen' (syn. 'Purple Queen', 'Mrs. Butt Magenta'), 'Mrs. Butt' (syn. 'Ruby', 'Crimson Lake') and 'Scarlet Queen.'
- ✰ **Some cultivars (seedlings) whose parentage are not known:** In Asia introduced by Lalbagh, Bangalore; from Kenya, 'Begum Ali Yavur Jung', 'Eclipse', 'Kayata', 'Nawab Ali Yavur Jung' and 'Pixie'.
- ✰ **Double-bracted cultivars:** In recent years, several multi- bracted or double- bracted cultivars have been introduced by Dr. J.V. Pancho of the Philippines.

Instead of the normal three bracts as found in bougainvillea, the bracts in these cultivars number between 20 and 40.

- ✰ ***B. × buttiana*:** The different cultivars in this category are listed below: 'Archana', 'Carmencita', 'Cherry Blossom' (syn. 'Bridal Bouquet'). Multi-bracted., Double Pink, Double White, 'Los Banos Beauty' (syn. Pink Champagne'), 'Marietta', 'Mahara' (syn. 'Million Dollar, Manila Magic Red, Manila Red) and 'Roseville's Delight'. A free- flowering vigorous plant with apricot-orange bracts.

Cultivars of Bougainvillea

'Mahara Magic' 'Tropical Bouquat' 'Silhouette'

'California Gold' 'Miami Pink' 'Mrs. Butt'

'Formosa' 'Pink Pixie' 'Thimma'

'Dr. P.V. Sane' 'Refulgens' 'Splendens'

Variegated Foliage of Bougainvillea

Outer Spot:
Red, White Coloured

Outer Spot:
White Coloured

Inner Spot:
Yellow Coloured

Outer Spot:
Yellow coloured

Outer Spot: Mosaic

Inner Spot: Sand

Variegated Cultivars

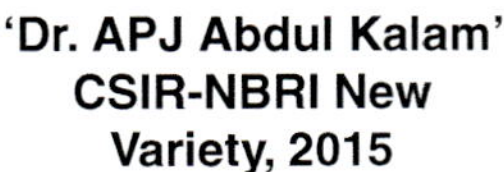

'Dr. APJ Abdul Kalam'
CSIR-NBRI New
Variety, 2015

'Los Banos Variegata'

'Abhimunya'

- **Variegated cultivars:** There are a number of cultivars with variegated leaves that are in cultivation in the garden, these are mostly bud sports. 'Archana', A double-bracted cultivar, mentioned earlier; 'Bhabha', A sport of 'Louise Wathen'; 'Glabra Variegata', 'Jawaharlal Nehru' Developed at I.I.H.R., Bangalore. The leaves are beautifully variegated yellow and green, 'L. N. Birla', 'Laxminarayan', 'Louise Wathen Variegata' (Syn. 'Louise Wathen Mediopicta'), 'Marietta', 'Rao'. A bud sport of 'Mrs. Butt', 'Scarlet Queen Variegata', 'Sharma', 'Spectabilis Variegata', 'Thimma'. A bud sport of 'Mary Palmer' with double-coloured bracts like those of 'Mary Palmer'. The young shoots are creamish- yellow and pink with some dark spots. A broad yellow variegation at the centre of the large green leaves in this cultivar makes it one of the best even when not in flower. It performs very well in Bangalore. Many types of variegated leaves of bougainvillea such as those with spots at the edge of the leaf commonly called 'Outer spots' those with spots in the inner part called 'Inner spots' and those with many blotches called 'Sand spots'. The beauty of the flower is that it can be enjoyed even when it is not blossoming.

6.3.0 Bougainvillea Cultivars and their Primary Uses

For Landscape Use

- 'Barbara Karst': this is the cultivar with large bright red bracts and a most popular standby that always performs well and blooms throughout the year.
- 'Mrs Butt': it produces red to peach bracts. This is a B. × *buttiana* cultivar.
- 'Miami Pink': cultivar with dark pink bracts suites best in the landscape.
- 'Tropical Bouquet': orange but quickly becomes pink in colour of bracts. Plant is compact and bushy in stature.
- 'California Gold': large, long-lasting gold bracts and it can grow best as vine.
- 'Royal Purple': this is the cultivar having dark purple bracts and it is a good replacement for Elizabeth Angus in humid areas.

For Containers/Hanging Baskets

- 'Helen Johnson': a dwarf red, compact and bushy grower. It is good for hanging baskets.
- 'Juanita Hatten': in this cultivar, red foilage may be slightly variegated in shades of green. Branches are freely in nature. This is also good for hanging baskets.
- 'Pink Pixie': true miniature form of Bougainvillea cultivar. It produces small pink bracts and good for small pots.
- 'Summer Snow': this produces large, pure white bracts and excellent for weddings and hanging baskets.
- 'Dr. David Barry': large lavender bract of this cultivar is good for containers.

Novelty/Special Interest Cultivars

- 'Double Red': Clusters of fluffy double bracts produces deep red bracts.

6.4.0 Commonly Produced Bougainvillea Cultivars and their Primary Uses

- 'Raspberry Ice': this cultivar is rather slow grower and produces red bracts along with variegated foliage.
- 'Double Pink': huge clusters of bright pink doubles and repeat bloomer make it good for landscaping.
- 'Surprise': this cultivar is free blooming pink in colour which may sport white flowers.
- 'Vicky': pink bracts and leaf centres creamy yellow; though some white bracts may be sported.
- 'Golden Summer': this is with white and variegated yellow-gold foliage and repeat bloomer.
- 'Sweet Dream': this is a thornless variety. Leaves are completely gold and bracts are pale lavender.

7.0 Genetics and Breeding

India is one of the major repositories of a wide range of bougainvilleas, and approximately 50 percent of the present-day cultivars have been evolved in India (Janakiram *et al.*, 2013). The work on development of bougainvillea has largely been done by the Agri-Horticultural Societies at Calcutta and Madras. The Lal Bagh Garden at Bangalore also contributed a great deal by introducing a large number of exotic cultivars, particularly the multi- bracted varieties from the Philippines and Thailand. Different breeding approaches followed in bougainvillea are hybridization, polyploidy, mutation and bud sports. A large number of varieties have been developed at the National Botanical Research Institute (NBRI) Lucknow; Bhaba Atomic Research Centre (BARC), Mumbai and Indian Institute of Horticultural Research (IIHR), Bangalore. Bougainvilleas are widely grown in Indian gardens. There are numerous varieties and cultivars that are important from floriculture point of view. Most of the present day varieties have arisen through bud sports, chance crossing and induced mutation. India is one of the major repositories of wide range of genotypes of bougainvillea and is perhaps the foremost among nations which have done appreciable good work on Bougainvillea. Extensive investigation work has been done in India on different aspects like enrichment of germplasm, agro-technology, characterization and improvement at the Indian Agricultural research institute (IARI), New Delhi; National Botanical Research Institute (NBRI), Lucknow; Indian Institute of Horticultural Research (IIHR), Bangalore; Lalbaugh Gardens, Bangalore; Bhaba Atomic Research Centre (BARC), Trombay; Agri-Horticultural Society of India, Calcutta. Different Agriculture Universities, amateur growers and nurserymen are also maintaining good amount of cultivars due to its increasing demand and popularity. Division of Floriculture and Landscaping, Indian Agricultural research Institute, New Delhi was recognized as the International Registration Authority in 1966 for bougainvillea cultivars. Since then the division is compiling, maintaining, publishing lists of names of cultivars and registering new varieties. Voluminous compilation work resulted publication of first "The International Bougainvillea Check List" in 1981 and subsequently "The New International Bougainvillea Check List" in 1999 (Singh *et al.*, 1999). These check lists are of immense importance to Bougainvillea lovers, researchers, floriculturists, landscape architects, town planners, gardeners of the world.

7.1.0 Hybridization

Since the beginning of its commercialization, producers have sought to make hybrids of *Bougainvillea* species. They have produced three recognized hybrids with striking characteristics: *B.* × *spectoperuviana, B.* × *spectoglabra,* and *B. glabra peruviana* (*B.* × *buttiana*) (Abarca-Vargas and Petricevich, 2018).

(i) ***B.* × *specto-peruviana*:** the first hybrid has no reports in the scientific literature of registration sites or scientific research works according to the databases consulted.

(ii) ***B.* × *specto-glabra*:** this plant is a hybrid of *B. spectabilis* and *B. peruviana* and it was only reported in China.

(iii) ***B. glabra-peruviana* or *B.* × *buttiana*:** this plant is a hybrid of *B. glabra* and *B. peruviana,* and it was identified and reported in Mexico, as well as in Hybridization of *Bougainvillea* enables the origin of totally new colour. The natural hybridization with polyploidy has been the single factor contributing to evolution in nature and under domestication by plant breeding (Khoshoo and Mukherjee, 1970).

Table 1: Cultivars Developed from Inter-specific Crosses

Crosses	*Cultivars*
B. peruviana × *B. glabra*	Begum Sikander, Mrs. Butt
B. peruviana × *B. spectabilis*	Wajid Ali Shah
B. buttiana × *B. peruviana*	Chitra
B. spectabilis × *B. buttiana*	Dr. R.R. Pal, Summer Time, Spring Festival
B. glabra × *B. spectabilis*	Maharaja of Mysore, Pink Beauty, Pixie, Rose Queen

Table 2: Cultivars Developed from Intra-specific Crosses*

Crosses	*Cultivars*
B. glabra (Trinida × Formosa)	Dr. H.B. Singh
B. glabra (Formosa × Trinidad)	Purple Wonder
B. spectabilis (Lalbagh × Red Glory)	Chitravati
B. peruviana (Dr. B.P. Pal × Princess Margaret Rose)	Mary Palmer Special

* Swarup and Singh (1995).

The cultivars which were evolved through hybridization at IIHR (Indian Institute of Horticulture), Bangalore are Dr H.B. Singh, Chitravati and Purple Wonder. Among these cultivars, 'Dr H.B. Singh' was patented as Krishna in Australia (Anonymous, 1996).

7.2.0 Mutation

The hybridization is not possible in multi-bracted bougainvillea due to absence of flower tubes/flowers and all varieties are not able to set seeds at all places. Hence, an alternative method, *i.e.* induced mutation breeding resulted in evolution of different new bougainvillea cultivars (Swaroop *et al.*, 2015). Mutation leads to the origin of many new forms of bougainvillea. In recent years, mutation breeding has been used as a valuable supplement to traditional methods of plant breeding which helps in the development of better cultivars (Arora and Pahuja, 2008). Most of the cultivars of bougainvillea were developed through selection of 'budsports' or by mutation breeding. Mutation breeding is one of the important methods to create variability in ornamental plants and it also reduces the time required to develop a new variety (Kannan *et al.*, 2002). Mutation breeding is the only method that can be used to improve double bracted bougainvillea, since conventional cross-breeding is not possible because of the absence of flowers (Datta, 1990). Different bract colours ranging from white to yellow, orange, magenta, red, purple and violet have arisen as a result of mutation among various forms of three basic species (Mabry and Dreiding, 1968). The relative proportion of these pigments determines the variation in bract colour (Kochhar and Ohri, 1977). Bract colour variations were due to qualitative and quantitative differences in pigment composition (Anand *et al.*, 1997). The mutation studies of bougainvillea have been carried out at National Botanical Research Institute, Lucknow and Bhabha Atomic Research Center, Bombay (Banerjee, 2010).

Table 3: Mutants Developed at National Botanical Research Institute, Lucknow

Mutant	*Parents*	*Characteristics*	*Developed/ Reported by*
'Arjuna'	Gamma ray mutant of 'Partha'	Single bract cultivar with pinkish purple colour andvariegated leaves	Gupta and Shukla (1974)
'Los Banos Variegata'	'Los Banos Beauty'	Variegated leaves with excellent pattern, attractive leaves and bract colour is mallow purple	Banerjee and Datta (1987)
'Pallavi'	'Roseville's Delight'	Variegated leaves having pale and light green combination	Banerjee *et al.* (1987)
'Mahara Variegata'	'Mahara'	Variegated leaves with creamish yellow and greenBract rhodamine purple	Datta and Banerjee (1994)
'Mahara Variegata' abnormal leaves	'Mahara'	Variegated leavesBract colour unaltered but shape and size reduced	Banerjee (2002)
'Los Banos Variegata' silver margin	'Los Banos Variegata'	Attractive variegated leaves with green and silver margin	Banerjee (2002)

Mutant	*Parents*	*Characteristics*	*Developed/ Reported by*
Los Banos Variegata'Jayanthi'	Chemical mutagen (EMS 0.02 per cent) induced mutant of cv. 'Los Banos Beauty'	Differs from the original variety in foliage colour	Jayanthi *et al.* (2000)
'Pixie Variegata'	Chemical mutagen (EMS 0.02 per cent) induced chlorophyll variegated mutant of cv. 'Pixie'.	Margin of the leaves creamish colour	Banerjee (2009)
'Los Banos Variegata' Silver Margin	Gamma ray mutant of 'Los Banos Beauty'	Leaf lamina is green and its margin is silver (1-2 mm width), variegated leaves are extremely curved	Banerjee (2012)

Bhabha Atomic Research Centre, Bombay developed mutants of Bougainvillea such as 'Lady Hudson' of 'Ceylon Variegata' (induced mutant of 'Lady Hudson'), 'Jaya' (induced mutant of 'Jayalaxmi'), 'Jayalaxmi Variegata' (induced mutant of 'Jayalaxmi') (Broertjes and Van Harten, 1998), 'Suvarna' (induced mutant of 'Lady Hudson'), 'Poultoni Variegata' (gamma ray induced mutant of 'Poultoni') and 'Silver Top' (induced mutant of 'Versicolour') (Raghava, 1999).

7.3.0 Bud Sports

Some excellent cultivars of bougainvillea were originated as a result of spontaneous bud variation namely 'Alick Lancaster', 'Bhabha', 'Cherry Blossom', 'Fantasy', 'Jawaharlal Nehru', 'Lady Mary Baring', 'L.N. Birla', 'Louis Wathen', 'Mary Palmer', 'Mrs. McClean', 'Parthasarthy,' 'Roseville's Delight', 'Shubhra' *etc.* Spontaneous bud variation resulted in three change in the cultivated Bougainvillea *i.e.* Change in bract colour, imperfect flower tube development and leaf variegation (Holttum 1955, 1957). The cv. 'Sholay' and 'Usha' are important seedling selections at IIHR from 'Red Glory' and 'Lady Hope' respectively (Anonymous, 1996). From cv. 'Arjuna' a chlorophyll variegated bud sport originated which were named as 'Abhimanyu'. It was detected by Dr. Banerjee at NBRI, Lucknow (Banerjee, 2012).

In the year 1963-1967, Dr. J.V. Pancho first reported the multi- bracted cultivars of bougainvillea in Laguna, Phillipines. The cultivar such as 'Carmentica,' 'Cherry Blossom', 'Mahara', 'Los Banos Beauty', 'Mahara Variegated', 'Pallavi', 'Rosevilles Delight', 'Archana' and 'Marietta' are multi- bracted. They are called multi- bracted cultivars as they have 20-40 bracts as compared to the normally occurring three bracted cultivars of bougainvillea. Also the flower tube is absent or rudimentary in these varieties. The multi- bracted varieties originated from the cultivars of *B* × *buttiana* (Pal and Swarup, 1974; Banerjee, 2012).

7.4.0 Polyploidy

Due to seed sterility in bougainvillea further breeding was hindered as a result of which it limits the selection of male and female parents for developing new cultivars. After detailed studies, fertility in bougainvillea was restored by colchiploidy. Thereafter, many colourful and floriferous bicolored cultivars at triploid, tetraploid and aneuploid have been raised. Some of the cultivars with induced polyploidy are 'Wajid Ali Shah', 'Mary Palmer Special', 'Dr. B.P. Pal', 'Tetra Mrs. McClean', 'Chitra' and 'Begum Sikander'.

7.5.0 Characterization

Many cultivars have been recognized based on morphological and agronomical features but environment also play a significant role in expressing varietal feature along with morphological and agronomical features. Bougainvillea cultivars are widely diverse among themselves mainly due to the colour of bracts leaf and bract size, variegation in foliage, floral tube, star, presence of pubescence *etc.* Collectively, all these factors generate a lot of confusion in the identification of the particular cultivar (Mac Daniels, 1981). Therefore, researchers are willing to seek easier and effective way to identify these cultivars.

Classification of bougainvillea cultivars based on different characters studied by Kumar *et al.* (2015) led to the characterization of a particular genotype. This quantification of existing genetic variation and identification of cultivar based on economic traits and grouping the cultivars based on genetic divergence helps in selecting appropriate cultivar and utilizes them for future bougainvillea improvement through systematic and scientific breeding approaches.

7.6.0 Molecular Characterization

In recent years for characterization of a genotype at the genomic level, several molecular approaches have been used. RAPD is most commonly used for the identification of cultivar due to its simplicity, rapidity and requirement of only a small quantity of DNA to generate numerous polymorphisms (Wight *et al.*, 1993; Cheng *et al.*, 1997). For quantifying genetic variation in plant species, morphological traits are commonly used since they provide a simple technique while simultaneously assessing genotype performance under relevant growing environments (Fufa *et al.*, 2005). Although, studies conducted by Chatterjee *et al.* (2007), Hammad, 2009 and Srivastava *et al.* (2009), where they marked RAPD as less preferred due to their less reliability and lower reproducibility. Use of more reliable and reproducible molecular markers like Simple Sequence Repeats (SSRs) becomes indispensable in Bougainvillea, like in any other plant species. SSRs or microsatellites have become genetic markers of choice in many plant species due to their multi-allelic nature, a high abundance, co-dominant inheritance, reproducibility, high degree of polymorphism, and extensive genome coverage (Varshney *et al.*, 2005). To make out the relationship between the different cultivars of bougainvillea, SSR markers were used in a study by Kumar *et al.* (2014) to characterize the cultivars. The investigation further revealed the fact that, the SSR markers are robust in detecting a high level of molecular polymorphism to characterize and in grouping the Bougainvillea cultivars besides establishing the genetic relationship and diversity which further used in crop improvement programmes.

7.7.0 Cytogenetical Studies

Basic cytogenetic studies were directed towards determination of chromosome number, mitotic and meiotic divisions, karyotypic analysis, colchiploidy *etc.* DNA content has been estimated from three basal Bougainvillea species, hybrid groups, triploid hybrid cultivars and induced tetraploids and established their relationship. Studies have solved many taxonomic problems, phylogenetic relationships and also have opened a new way to synthesize new variety through chromosomal manipulations. Pollination mechanism, breeding system and cross compatibility relationship have also been studied. 2n= 32 chromosomes were detected in two horticultural cultivars of bougainvillea (one with scarlet bract and the other with brick red bract) which morphologically resembled to *B. spectabilis* (Sen and sen, 1954). They detected a good amount of meiotic irregularities due to variable pairing of chromosomes. Forty different varieties were investigated and found 2n= 34 chromosomes, excepting three varieties where 2n= 51 chromosomes were recorded (Sharma and Bhattacharya, 1960). A large number of varieties belonging to *B. glabra*, *B. buttiana* and *B. spectabilis* were studied and 2n= 34 chromosomes were recorded. However, in some buds from red sectorial part of 'Marry Palmer' 2n= 20 chromosomes were detected (Ninan *et al.*, 1959). Cytological data was used as a tool for interpreting the evolution and affinities of different species and varieties of bougainvillea (Sharma and Bhattacharya, 1960). Natural triploidy has been reported by Banerji and Banda (1967). Aminotriazole pre-treatment was reported to be very helpful for chromosome studies in leaf and shoot tip squashes (George and Sobhana, 1976). Begum and Datta (1971) reported new 'n' number (n=8). Extensive studies were carried out on diploid and tetraploid, pollination mechanism, meiotic system and breeding system of bougainvillea (Zadoo and Khoshoo 1975; Ohri and Zadoo, 1979; Meenakshi and Bhat, 1988). They studied fifty taxa of which 48 were diploid (2n= 34, n= 17) and two were triploid (2n= 51). They attempted to examine the range of variability in various phenotypic characters within the basal species and cultivars of subsequent origin. They concluded that the basal species differ chiefly in outline, base and apex of leaf, degree of pubescence, division of main short and flowering cymes and shape of flower tube. While *B. glabra* and *B. spectabilis* are close to each other in various morphological characteristics, *B. peruviana* is distinct from both of the species. Pollination mechanism and breeding system were studied and found that 80 per cent of garden bougainvillea is male and female sterile. Fertile forms do not set seed on selfing but do so readily on crossing. Ohri and Khoshoo (1979) studied cross compatibility relationship of Bougainvillea cv. 'Mrs. H.C. Buck' and its family. They developed hybrid which was identical to 'Mrs. H.C. Buck' as tested by morphological and biochemical characters. Special attempt was made on cytological studies of variegated cultivars (Meenakshi and Bhat, 1982). They studied 13 variegated cultivars and detected gametic chromosome nmber n= 17. All variegated cultivars were grouped into three categories on the basis of degree of meiotic abnormality. DNA content of *B. peruviana* was found to be significantly different from *B. spectabilis* and *B. glabra* (Ohri and Khoshoo 1982). Their study clearly indicated that DNA content of three hybrid group *i.e. B. specto-glabra, B. specto-peruviana, B. glabra-peruviana* and synthetic hybrid were intermediate between parents and they found direct correlation of DNA content and ploidy level. 'Tetra Mrs. McClean' (1969) is such a synthesized tetraploid variety.

8.0 Growth and Flowering

It is a short day (SD) plant, with flowers forming in apical panicles on the current year wood (Ma and Gu, 2010). Bougainvillea flowers on new growth, so new growth should always be executed by pruning. Unwanted shoots, all lateral shoots are cut back to within two or three buds of the main stems. These will bear the new flowers and bracts. But pruning too frequently will suppress flowering in plant. Bougainvillea flowers profusely in high light intensities, moderate temperature along with longer nights. 8-11 hours of day length with high light intensity is best for flowering. Bougainvillea normally flowers during the short days of winter, but blooms are highly dependent on temperature. Drought stress can also stimulate flowering even with long daylight exposure. But excessive dryness leads to leaf drop and dormancy. Heavy shade is not beneficial as it inhibits flowering. Nitrogen and phosphate are critical to flowering but excessive fertilizer application may leads to make it delay. Pinching plants at 4 weeks is sometimes followed by few growers to a stake for taller plants. *Bougainvillea* × *buttiana* cultivars are mostly considered qualitative short-day plants that flower more readily and profusely under short day lengths. In sub-tropical or tropical climates like South Florida, bougainvillea flowers in late spring to early summer, and in late summer to fall, when the daylight is less than 12-h per day and night-time temperatures are above 21°C (Schoellhorn and Alvarez, 2002). Ma and Gu, (2010) built on this theory and confirmed earlier research by Steffen *et al.* (1988) that flowering in bougainvillea was controlled in some way by gibberellic acid (GA) by diverting nutrient assimilates away from the apical meristem. The GA levels are known to change in response to photoperiod, or more specifically, to the effect of far-red light on phytochrome photoreceptors (Taiz and Zeiger, 2010). The short day induction in bougainvillea 'San Diego Red', and *Bougainvillea glabra* 'Sanderiana' was thought to be the result of a complex web of interactions between GA and other hormones, as well as environmental factors such as photoperiod and light intensity (Even-Chen and Sachs, 1980).

9.0 Propagation

9.1.0 Cuttings

Propagation of bougainvillea is done commercially through cuttings. However, propagation through ground or air layering and budding are also practiced. A length of 10- 15 cm with atleast 4- 6 buds are selected for cuttings. Cuttings are planted in pot having potting media of perlite, cocopeat, sand, vermiculite and put a depth of 3-5 cm into the medium. Rooting hormone may be used before planting to promote root formation. Keep the pot in a warm place and regular spraying of water is must. In about four to six weeks the cuttings produce roots. In case of softwood cuttings night temperatures should be above 20°C and for hardwood cuttings night temperatures should be below 12-13°C. So, selection of cuttings should be done on the basis of night temperature. Before the cuttings are inserted, they may be treated with growth regulators such as IBA (indole butyric acid) IAA (indole acetic acid) and NAA (naphthalene acetic acid) in solution or talc. The concentration used is generally between 1,000 and 3,000 ppm in the 'quick-dip' method. Auxin application enhances the histological features like formation of callus tissue and differentiation of vascular tissues (Mitra and Bose, 1954). It also activates the production of roots in several evergreen ornamentals. *Bougainvillea spectabilis* when dipped for 15 second in solution of IBA at 8000 ppm, rooting percentage was found to be significantly increased and roots were found thin, fibrous and branched (Rao, 1967). Bougainvillea and other ornamental shrub species which do not show any sign of rooting in the untreated cuttings were found to produce large number of roots, weight of fresh and dry root when treated with IBA at 3000- 6000 ppm (Bose *et al.*, 1968). Singh *et al.* (2011) reported that the maximum survival and rooting percentage of cutting observed under higher concentration of IBA. Bottom heat promotes rooting. Intermittent overhead mist is required; cuttings must never be allowed to wilt. High relative humidity should be maintained around the plants at all times. Time to rooting is 6- 12 weeks. At the time of planting of cutting and transplanting, apply a broad-spectrum fungicide to prevent root rot. Some cultivars such as 'Mary Palmer', 'Thimma', 'Louis Wathen', 'Tomato Red Lateritia', *etc.*, are difficult to root from cuttings. In cultivars such as 'Thimma' and 'Cypheri' tender tip cuttings root better than hardwood cuttings. Ahmad *et al.* (2002) observed significant influence of indole- 3- butyric acid on the root production of *Bougainvillea glabra 'variegata'*. The success in rooting has been found to be affected by the position and type of cutting material selected for propagation. Tip cuttings of *B. spectabilis* and *B. glabra* cv. 'Alexandra' rooted better than those taken for lower parts of the stem (Beel and Schelhtraete, 1981). Hardwood cutting of bougainvillea when planted in open sunlight have been found better in terms of percentage of rooting and survival (Misra, 1971), while softwood cuttings rooted better under intermittent mist (Singh and Motial, 1979). Hardwood cutting of bougainvillea cv. 'Mary Palmer' showed 75 percent rooting when treated with 1500 ppm IBA, as compared to control (Kale and Bhujbal, 1972). 'Usha' cultivar of bougainvillea was also reported to root better (80 per cent) when treated with IBA at 4000 and 6000 ppm (Bhattacharjee and Balakrishna, 1983). Leaf-bud cuttings can also be used in case source material is limited. Each node of stem can be used as a cutting. The cutting, taken from shoots

Table 4: Effect of Rooting Hormones on Bougainvillea Cuttings

Cultivars	*Type of Cuttings*	*Bioregulators and Chemicals*	**Survival, **Rooting per cent*	*Reference*
'Mahara'	Hardwood cuttings	6000 ppm IBA	**100	Philip and Gopalkrishna, (1982)
'Usha'	Softwood cuttings	4000, 6000 ppm IBA, 4000 ppm NAA	**80, *100	Bhttacharjee and Balakrishna (1983)
'Torch Gory'	Hardwood cuttings	4000 ppm NAA	**70.89	Joshi *et al.* (1989)
'Akola', 'Shubra', 'Refulgence'	Hardwood cuttings	100 ppm NAA	**60	Harris and Singh (1991)
'Torch Gory'	Hardwood cuttings	2000, 2500 and 3000 ppm IBA	**100	Singh *et al.* (2011)
'Thimma'	Hardwood cuttings	2000ppm IBA	**63.33	Sahariya *et al.* (2013)

that are partially matured, consists of a leaf blade and a short piece of the stem (3- 4 cm long) with the attached axillary bud (Kobayashi, 2007). Place the bud vertically in a rooting medium and lightly cover it so that only the leaf blade can be seen.

9.2.0 Air Layering

Plants which are difficult to root from cuttings or comparatively large plant is needed for planting, layering is done. Most of the cultivars produce profuse roots in layering in about 5-6 weeks. Young, vigorously growing healthy branches are selected for layering. The leaves on selected branches should be exposed to light since these produce more food and will root faster. Branches from less than 2.54 cm in diameter are best for air layering.

9.3.0 Grafting

In cultivars that have little or no chlorophyll in their leaves are difficult to grow from cuttings and need to be grafted onto a vigorous rootstock to be propagated. Whip, tongue, approach grafting may be practiced. Cultivars with fragile root system are generally propagated through grafting. It is also used when it is desired to have multiple cultivars on one plant. The scion should be free from disease. The rootstock could be a seedling or a rooted cutting of an existing, established plant. All cut surfaces should be covered with grafting wax once the join is made to prevent moisture loss. Remove shoots from the rootstock, when the scion has firmly joined with the rootstock. Put the grafted plant in a humid environment so that the scion does not dry out (Kobayashi, 2007).

Bougainvillea can also be propagated by the inarching. The cultivars which cannot be grown either from cuttings or by layering are raised by T or Shield budding. Generally, 'Dr. R.R. Pal', a vigorous cultivar, is used as rootstock. 'Partha' is also a good rootstock and frequently used. The best budding time is February to March. Plants are also raised from seeds for obtaining new hybrids or cultivars (Singh *et al.*, 2020).

9.4.0 Micropropagation

Bougainvillea can be effectively propagated through in vitro culture. Tissue culture techniques have successfully been employed to produce lager number of difficult- to- propagate plants like *Bougainvillea*. A study conducted by Sharma, 1981 exhibited that shoot apices of *Bougainvillea glabra* cv. 'Magnifica' were induced to regenerate an average of ten shoots from their bases in response to BAP (0.5 mg/l) and IAA (1.5 mg/l). All the isolated shoots from such cultures were rooted in a medium containing 0.1 mg/l each of IBA and 2, 4, 5-T and lacking BAP. Plantlets were then successfully grown in potted soil where they flowered normally.

Best multiple shoot formation was recorded on MS medium supplemented with 1.0mg/l BAP and 250 mg/l glutamine. The regenerated shoots were successfully rooted on half strength MS medium with different concentrations and combinations of auxin. It was observed that 2.5 mg/l NAA combined with 2.5mg/l IBA gave 100 per cent root induction (Shah *et al.*, 2006). Another study by Jain, 2016 exhibited that, highest in vitro rooting (70.52 per cent) of micro-shoots was observed in the treatment where, half-strength MS medium was supplemented with IBA (1.0 mg/l) + NAA (1.0 mg/l). Hardening of in vitro propagated plants of *Bougainvillea spectabilis* rooted plantlets was done for 21 days in glass jars filled with agro peat medium {A mixture of soilrite (1) + coco peat (1) + perlite (1)} supplemented with 1/2 strength liquid inorganic MS medium and covered with polypropylene lids. The hardened plantlets were successfully transferred to the glasshouse after a short period of in vitro acclimatization.

Table 5: Effects of Plant Growth Regulators on Callus Induction and Organogenesis from Single Node of *Bougainvillea* sp.*

Genotype	*WPM*			*MS*		
	Callus Induction Per cent	*Average Callus Fresh Weight (mg)*	*Organo-genesis Induction Per cent*	*Callus Induction Per cent*	*Average Callus Fresh Weight (mg)*	*Organo-genesis Induction Per cent*
B.x buttiana	92.86a	667.29a	96.43a	85.71a	631.71a	85.71ab
B. spectabilis	82.14a	522.86a	89.29a	75.00a	510.29a	78.57ab
B. glabra	**71.00a**	**339.43b**	**78.57ab**	**71.43a**	**336.29b**	**64.29b**

* Duhoky and Al Mizory (2014).

Means followed by the same letter within each character (column) do not differ significantly ($P \leq 0.05$) according to Duncan's Multiple Range Test.

10.0 Soil and Climate

10.1.0 Soil

Bougainvillea grows well in rich, well drained, acidic (pH 5.5–6.0) soil. It does not thrive in wet soil. It has an extremely fine root system, and should be planted in well-drained soils. Excess moisture causes rotting in roots and plants. These types of media retain too much water and will rot the root system. Proper soil pH is essential because it affects the availability of mineral elements. A soil pH above 6.0 increases the possibility of micronutrient deficiencies, particularly iron. Bougainvillea is drought tolerant, salt tolerant and wind resistant plant.

10.2.0 Climate

As a tropical plant, bougainvillea requires full sun, or a light level of at least 4000 f.c. Most growers keep hanging baskets at the top of the greenhouse for maximum light exposure. Low light and shady areas are not suitable, as the plants will drop their bracts. Bougainvillea does best at elevations up to 800 meter. It can tolerate hot dry locations, at temperatures over 37°C. Plant grows well at a minimum temperature of 18°C at night and 23-35°C during the day. *B. glabra* can tolerate slightly cooler conditions (14-17°C) than *B. spectabilis* (17-20°C). Bougainvillea grows well with at least 100 cm of rainfall per year. It is better not to plant in extremely windy and unprotected areas.

11.0 Cultivation

11.1.0 Planting

The best time of planting in tropical zone is during monsoon. Planting in the winter should be avoided as the mortality rate will be high and the growth of plants will be poor. But in heavy rainfall areas planting may be done before or after the rains. In subtropical areas planting is possible almost throughout the year except during the hotter months. It can also be planted in pot, but it is advised to opt for larger size pot while planting. As it responds rapidly in suitable environments, it needs repotting time to time. Besides, root pruning should also be practiced. It helps to manage bougainvillea plant to be fit in pot. So, repotting once in every two years after and controlled root pruning is the best way to manage bougainvillea cultivation in pot. In cool climate this plant needs to be replaced from outside chilling environment. Growth hampers and ultimately, the plant may die in harsh weather. Well drained potting medium is suitable for growing bougainvillea, loamy soil, organic compost, cocopeat, *etc.* are used. If more weight is needed to stabilize the pot, sand or soil may be added. The medium needs to be well drained to promote root and plant growth. Avoid media which has high peat content and water-retention levels. These types of media retain too much water and will contribute to root rot. Media pH should be 5.5–6.0.

11.2.0 Manuring and Fertilization

Bougainvillea needs regular fertilizer with formulations having NPK ratios of 1:1:1 or 2:1:2 along with 10-12 kg of farmyard manure once a year after the June pruning. At this time, 250 gm of bone meal should also be applied to each plant. Pal and Swarup (1974) suggest to the application of 250 gm of a mixture of ammonium sulphate, superphosphate, and sulphate of potash in the proportion of 1:3:2 per fully grown bush. At the time of planting, fertilizer that is high in phosphate, and follow with a balanced liquid fertilizer at 150- 300 ppm may be applied. Nitrogen and phosphate are critical to flowering, but over-fertilization must be avoided. Too much fertilizer will promote vegetative growth instead of flowering. This is said to promote profuse flowering. Application of liquid manure by fermenting fresh cow dung or oilcake at fortnightly intervals prior to flowering has been found to be beneficial. A complete micronutrient blend or chelated spray may be scheduled to prevent leaf chlorosis and for better growth and flowering. Micronutrient applications can be used in two halves of the recommended rate at two times to avoid toxicity. To overcome this toxicity, organic fertilizer or controlled-release fertilizers, high in phosphate may also be applied to the soil. For long-term culture, topdressing with a controlled-release fertilizer is needed.

11.3.0 Irrigation

Bougainvillea needs evenly moist media during the spring, summer, and fall months, and nearly dry in winter as bougainvillea blooms better with drier winter conditions. Young plants or those which are newly planted need frequent watering. However, once established, the plants need very little or no watering. Irrigation is provided up to saturation, then the soil is left to dry out the top 5 to 7 cm before watering again. Care should be taken as soil does not get waterlogged. Too much water can lead to overly- green growth and eventually root rot; and the plant can wilt if the media is dry. During flowering watering should be restricted; otherwise, heavy watering will result in the shedding of the bracts.

11.4.0 Pruning

Bougainvillea responds well to pruning. Unless they are pruned regularly, bougainvillea grows into a tangled mass of old and new growth and overcrowding often leads to pests and diseases. Flowers are borne on new growth, so pinching back and pruning is necessary to induce new growth. Bougainvillea can be trained as a "standard," a small flowering tree with a single trunk, over arbores, into espaliers, onto walls, or to cascade down a slope (Awad *et al.*, 1988). So, a defined way of pruning and training is very much necessary for the growth, flowering as well as for the landscaping also.

Cutting of shoots should be scheduled in a regular basis. Any unwanted shoots, all lateral shoots are cut back to within two or three buds of the main stems. These will produce new shoots which will bear new flowers and bracts. Regular pruning is necessary to shape the plant and direct its growth because the shoots often grow vigorously. Bougainvilleas are prolific growers and need good pruning to force blooming and retain a pretty shape. Pruning should be done after flowering has finished, as this encourages the new growth on which the next flush of flowers will occur. To reduce the size of plants, cut them back by about a third, removing all spindly and twiggy growth. It needs to be trimmed to maintain its shape, but too much aggressive pruning of new growth will reduce bloom colour. Prune suckers from the plant's base to encourage top growth. Dead wood should be removed as it appears. The long shoots can be trained in various shapes and heights espalliers, arbours, twisted or braided trunks, or even large, fanciful animals. But the amount and the method of pruning will depend upon the cultivar and the purpose for which the plant is grown. For example, the specimen bushes growing in the lawn or on the trees and stumps do not require much pruning except pinching or tipping and removal of weak and dead branches. The side branches and erect extra growths are pruned in those plants which grow on a wall. In standards, all growths on the main stem are removed to keep the umbrella form of the plant in shape. *B. glabra* and *B. peruviana* do not need hard pruning. In northern and eastern India, the time of pruning is May to June. Tipping and pinching should also be completed by the middle of September. Manures and fertilizers are added at the time of pruning (Singh *et al.*, 2020).

12.0 Diseases and Pests

12.1.0 Diseases

12.1.1 Leaf Spot

Bougainvillea is susceptible to fungal (isolates of *Pseudomonas andropogonis*) and bacterial leaf spot (*Pseudomonas andropogonis*). The fungus causing leaf spot of bougainvillea was identified as *Passalora bougainvilleae* as described by Ellis (1976).

Diseases, Pests and Disorders of Bougainvillea

Leaf Spot

Aphid

Caterpillar Infestation

Mealy Bug Infestation

Chlorosis

Spider Mite

The reddish-brown spots that appear on leaves usually begin with new foliage and spread to give the bougainvillea a rusty looking appearance. Leaf spot lesions usually progress, enlarging and eventually distorting the plant's growth or causing the bougainvillea to lose its leaves.

Control: It can be controlled by keeping leaves and foliage dry. Pruning may be necessary to enhance air circulation and keep branches from growing together. At the first sign of infection, leaves and branches should be removed and destroyed to prevent the infection from spreading. Fungicide can minimize the spread of the infection but it cannot cure an existing infection. First report of *Burkholderia andropogonis* causing leaf spots of *Bougainvillea* sp. in Hong Kong and Clover in Canada were distinguished by Li and De Boer, 2005. Leaf spot can be effectively controlled by application of Difenconazole 25 per cent EC @ 8 ml/15 l of water.

12.1.2 Powdery Mildew

The most identifiable feature of mildew infection is the presence of white, powder-like, dusted appearance of the foliage. The infection is more apparent among the leaves than the branches. Crowded branches and dense foliage help in mildew spreading quickly. However, bougainvilleas suffering from mildew infection often develop root rot due to decreased plant vigour.

Control measure: The diseased foliage should be pruned off. During damp weather conditions, regular checking is necessary. Fungicides containing Triadimefon 25 per cent WP or Propiconazole 25 per cent EC as ingredients is helpful to prevent the spread of the disease. Potassium bicarbonate has the unique advantage of actually eliminating powdery mildew once it's there. Potassium bicarbonate is a contact fungicide which kills the powdery mildew spores quickly.

12.1.3 Root Rot

Root rot can also cause chlorosis, plant dieback, stunted growth and wilting. The disease is usually caused by fungus such as *Rhizoctonia*, *Pythium* or *Phytophthora*. The fungus infects bougainvillea roots, causing decay. Overwatering or planting bougainvillea in heavy, poorly drained soil must be avoided. Waterlogged conditions increase the likelihood that plants will develop root rot. Proper plant care and applying a broad spectrum fungicide Thiophanate-methyl 70 per cent WP @ 500-1000 gm a.i/ha or Benzimidazole group of fungicides during planting can reduce chances of infection.

12.2.0 Pests

12.2.1 Aphids

A stressed or neglected bougainvillea attracts the cowpea aphid (*Aphis craccivora*) and the melon aphid (*Aphis gossypii*). Aphids are small, soft-bodied insects with long, slender mouth parts that they use to pierce stems, leaves, and other tender plant parts and suck out plant fluids. Almost every plant has one or more aphid species that occasionally feeds on it and this is true with bougainvillea. In bougainvillea it is usually the green aphid that is found feeding on the tender new tissue. However, large populations cause curling, yellowing, and distortion of leaves and stunting of shoots. They can also produce large quantities of sticky exudates known as honeydew, which often turns black with the growth of a sooty mould fungus.

Control measures: Natural enemies can be very effective in the control of aphids. There are a number of beneficial natural enemies that attack aphids, including host-specialised parasitic wasps, as well as generalist predators such as hoverfly larvae, and adults and larvae of ladybird beetles and lacewings. Regular checking the field and removing the infected plants must be adopted to prevent the spread of aphid infestation. Insecticidal soap, neem oil, and narrow-range oil (*e.g.*, supreme or superior parafinic-type oil) provide temporary control if applied to thoroughly cover infested foliage. Timing of insecticide applications in spring is also critical. Monitoring crops from emergence and weekly afterwards for signs of aphid colonisation, particularly in crop edges are to be followed. The use of yellow sticky traps or pan traps can assist in early detection of aphids and assessment of numbers. Prevent aphid build-up in nearby weeds and other vegetation 'green bridges' to lessen the risk of nearby flights. Consider the use of seed dressings to prevent aphid building up in emerging crops early in the season. Seed treatment with imidacloprid 70 per cent WS @ 12 gm/kg of seed or Quinolphos 25 per cent EC @ 1 ml/l or Fipronil 5 per cent EC @ 1 ml/l may be useful to combat aphid.

12.2.2 Leaf Miner

Leaf miner feed by creating shallow tunnels, or mines, in young leaves of bougainvillea bushes. As the larva grows, its serpentine path of mines becomes more noticeable. Numerous mine like paths traversing the leaves of bougainvillea plants, the adults are virtually undetectable. Leaf miners can survive as a larva only in the tender, young, shiny leaf flush of bougainvillea. Older leaves that have hardened off are not susceptible unless extremely high populations are present. The larvae mine inside the lower or upper surface of newly emerging leaves cause them to curl and look distorted. The leaves never recover from this distortion and make the plant look unsightly given a heavy infestation.

Control measure: Pheromone traps should be kept about shoulder height on the affected plant. Monitoring leaf miner populations early in the season allows action to be taken early on if needed. Blue Sticky Traps attract and trap adults before they can lay eggs that will hatch and damage foliage. *Diglyphus isaea* is a beneficial wasp that parasitizes leafminer larvae, killing them before they can perpetuate. Best results are achieved when releases of *D. isaea* are made early in the season before leaf miner populations have grown to large numbers. Thoroughly spraying of Spinosad 45 per cent SC @ 3.2 ml/10 l of water to all plant surfaces is effective. Repeat applications might be needed 2-3 times throughout the growing season if damage persists.

12.2.3 Scale Insects

Armored scales, family Diaspididae, have a flattened, platelike cover that is less than 1/8 inch in diameter. The actual insect body is underneath the cover. The covers often have a differently coloured, slight protuberance. Concentric rings form as each nymphal stage (instar) secretes an enlargement to its cover. Armored scales do not excrete honeydew. Damaging species include Florida red scale and Oriental scale. Soft scales (*Coccus hesperidium*) can be smooth, cottony, or waxy and are 3-5 mm. They are usually larger and more rounded and convex than armored scales. Their surface is the actual body wall of the insect and cannot be removed. Soft scales feed in the fluid-conducting phloem tissue of the plant and excrete abundant honeydew, which is sugary water that drips from their bodies. Soft scales include black scale, brown soft scale and mealy bug. When plants are heavily infested with scales, leaves may look wilted, turn yellow, and drop prematurely. Scales sometimes curl leaves or cause deformed flowers and fruit.

Scales are often controlled by small parasitic wasps and predators including beetles, bugs, lacewings, and mites. Horticultural oils are specially refined petroleum products, often called narrow-range, superior, or supreme oils. Some botanical oils are also available. Thoroughly spraying plants at the proper time with oil alone usually provides adequate control.

12.2.4 Thrips (*Danothrips trifasciatus*)

Thrips are tiny, slender insects with fringed wings. They feed by puncturing their host and sucking out the cell contents. Certain thrips species are beneficial predators that feed only on mites and other insects. Pest species are plant feeders that scar leaf, flower, or fruit surfaces or distort plant parts. Most adult thrips are slender, minute and have long fringes on the margins of both pairs of their long, narrow wings. Immature thrips (called larvae or nymphs) are similarly shaped with a long, narrow abdomen but lack wings. Most thrips range in colour from translucent white or yellowish to dark brown or blackish, depending on the species and life stage. Thrips prefer to feed in rapidly growing tissue. Although thrips damage to leaves is unsightly, thrips activity does not usually warrant the use of insecticide sprays. Imidacloprid 17.8 per cent SL @ 3ml/10 l of water or Spinosad 45 per cent SC @ 3.2 ml/10 l of water or Oxydemeton- methyl 25 per cent EC @ 1 ml/l of water can be applied.

12.2.5 Spider Mite

Mites are common pests in landscapes and gardens and can be found feeding on many fruit trees, vines, berries, vegetables, and ornamental plants. To the naked eye, spider mites look like tiny moving dots Spider mites live in colonies, mostly on the under-surfaces of leaves; a single colony may contain hundreds of individuals. Mites cause damage by sucking cell contents from leaves. A small number of mites is not usually reason for concern, but very high population levels high enough to show visible damage to leaves can be damaging to plants, especially herbaceous ones. As feeding continues, the leaves turn yellow and drop off. Often leaves, twigs, and fruit are covered with large amounts of webbing. Damage is usually worse when compounded by water stress.

Spider mites have many natural enemies, which limit their numbers in many landscapes and gardens, especially when undisturbed by pesticide sprays. Cultural practices can have a significant impact on spider mites. Dusty conditions often lead to mite outbreaks. Apply water to pathways and other dusty areas at regular intervals. Spider mites frequently become a problem after the application of insecticides. Such outbreaks are commonly a result of the insecticide killing off the natural enemies of the mites.

12.2.6 Slugs and Snails

Snails and slugs move by gliding along on a muscular "foot." This muscle constantly secretes mucus, which later dries to form the silvery "slime trail" that signals the presence of either pest. Slugs and snails are hermaphrodites, so all have the potential to lay eggs. Adult brown garden snails lay about 80 spherical, pearly white eggs at a time into a hole in the topsoil. Snails and slugs feed on a variety of living plants as well as on decaying plant matter. A good snail and slug management program relies on a combination of methods. The first step is to eliminate, to the extent possible, all places where snails or slugs can hide during the day. Boards, stones, debris, weedy areas around tree trunks, leafy branches growing close to the ground, and dense ground covers such as ivy are ideal sheltering spots. There will be shelters that are not possible to eliminate *e.g.*, low ledges on fences, the undersides of wooden decks, and water meter boxes. Make a regular practice of trapping and removing snails and slugs in these areas.

12.2.7 Caterpillar

The bougainvillea looper (*Disclisioprocta stellata*) is a green or brown caterpillar that is roughly 3 cm long. It is also known as the inchworm or measuring worm. The looper larva mimics stems and branches very well and feeds primarily at night. *Bacillus thuringiensis* is effective against the larval stages of the looper. *Bt*, as it is commonly known, is a bacterial preparation that causes a disease in many kinds of caterpillars but does not harm beneficial insects, birds, man, or other organisms. Loopers stop feeding within hours after feeding on a sprayed leaf and die several days later. Thorough spray coverage of the tree is required for control. (*Bt* will also control other caterpillars present at the time of application). Caterpillars must ingest the pesticide to be killed. Spraying of Emamectin benzoate @ 0.5 ml/l, Spinetorum 11.7 per cent SP @ 0.5 ml/l of water or Thiomethoxam + Lambda-cyhalothrin @ 0.25ml/l of water may be helpful to control this.

12.3.0 Nutrient Deficiency

Periodically inspecting plants and foliage can reveal potential nutrient deficiencies. For example, bougainvillea with mature foliage that turns to a pale green with reddish leaves may not be getting adequate amounts of nutrients. A purplish tint to plants and leaves indicate a phosphorus deficiency. Young plant growth that develops dead areas, is likely due to a calcium deficiency. Using a balanced fertilizer, according to manufacturer instructions can prevent and treat nutrient deficiencies.

Chlorosis

Chlorosis causes yellowing on new or mature plant growth, usually due to a deficiency in iron or magnesium. Chlorosis can also be a secondary problem related to root rot. The disease may attack the leaves and/or the plant's roots. Symptoms include pale green leaves with dark veins and yellowed foliage. It can be controlled by increasing soil acidity with iron sulfate or aluminum sulfate, or by applying a micronutrient blend to plants. Application of micronutrient should be judicious as because too much may give the plant too much iron or magnesium, later causing a deficiency of another nutrient. Another treatment includes applying 1 to 2 teaspoons of Epsom salts diluted in 1 gallon of water.

Common Problems

Irregular flowering is a major concern for growers, who often use drought stress to encourage flowering. Choice of cultivar can be extremely helpful to solve this problem, however, leaf drop can be a problem as a result of over-watering, under-watering, low light levels, or cold temperatures.

13.0 Harvesting and Postharvest Management

13.1.0 Harvesting

In the autumn, bougainvillea form seed pods inside the tiny white flower in its centre. These pods are harvested and dried. From there very small seeds inside are collected. Seeds can be sown at any time of year, as long as they're kept warm. But germination may take a month or longer than that.

13.2.0 Postharvest Management

Ethylene was the major coordinator of senescence in many flowers (Nichols, 1968). Bougainvillea has a short vase life. Cameron *et al.* (1981) reported that bougainvillea bracteoles were attractive at the end of 6-day observation period and dropped 32.2 per cent treated with STS (0.5 oz/Gallon) while, 100 per cent was dropped in control plant. Ethanol is effective as ethylene inhibiting component in bougainvillea flower. Podd and Staden (2004) stated that low concentration of ethanol apparently decreased the formation of ethylene by inhibiting the action of ACC synthase.

Study has shown by Hossain *et al.* (2007) that it is possible to extend vase life of *Bougainvillea* sp. using 8 per cent and 10 per cent ethanol by causing delay senescence. Low concentration of ethanol decreased the formation of ethylene by inhibiting the action of ACC synthase as a result over all flowers (wilting, abscission, scar and colour changing) were affected physiologically (Hossain *et al.*, 2007). Another report by Moniruzzaman *et al.* (2011) suggested that the vase life of bougainvillea flower was significantly effected by ethanol treatment, and the longevity was higher in flowers treated with 8 per cent and 12 per cent ET than in those treated with water (control) and other ET concentrations.

13.3.0 Value Addition

Bougainvillea bracts are commonly used in lei making. The clusters of bracts should be clipped, preferably in the early morning, and transported them in a paper or cloth bag. After washing in cold-water soak, bracts can be wrapped in damp newspaper, placed in a plastic container or paper box. It can be stored in a

Vase Life of Bougainvillea

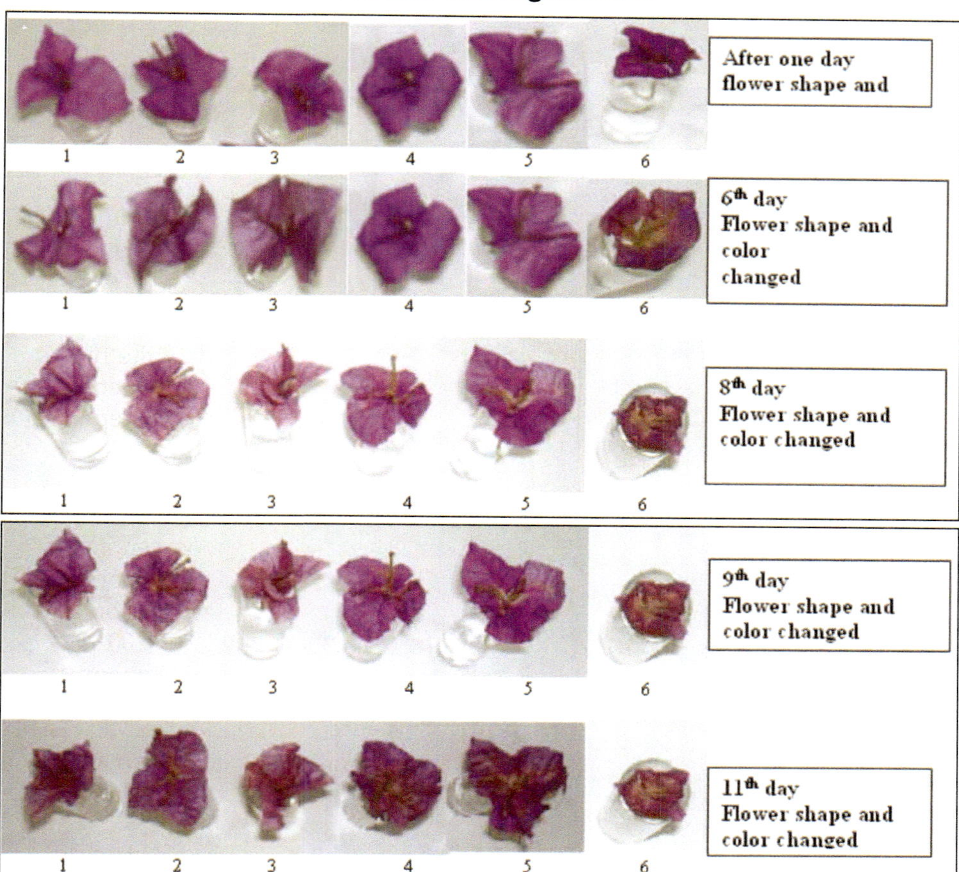

The Petal and Perianth Wilting and Abscission (Shape and colour) after Treatment. 1: Water control, 2: 2 per cent ethanol, 3: 4 per cent ethanol, 4: 8 per cent ethanol, 5: 10 per cent ethanol, 6: 70 per cent ethanol (Hossain *et al.,* 2007).

refrigerator at 4.4°C for up to 14 days. Bougainvillea bracts are also used in *haku* (in Guam, *mwar mwar*), traditional, lei-like headbands made of leaves and flowers (Kobayashi *et al.*, 2007).

Table 6: Visual Scoring of Petal Discolouration as Affected by Ethanol Concentrations (1-10 scale)*

Ethanol Concentration	*Days*							
	0	*1*	*2*	*3*	*4*	*5*	*6*	*7*
0 per cent ET	10	9	7	4	2	1	0	0
4 per cent ET	10	9	8	6	4	2	1	0
8 per cent ET	10	10	9	7	6	5	5	4
12 per cent ET	10	9	8	6	5	4	4	2
16 per cent ET	10	6	4	3	1	1	0	0
20 per cent ET	10	3	2	1	0	0	0	0
30 per cent ET	10	4	1	1	0	0	0	0
40 per cent ET	10	1	0	0	0	0	0	0
50 per cent ET	10	1	0	0	0	0	0	0

* Moniruzzaman (2011).

14.0 Biotechnology

14.1.0 Molecular Transformation

Another important aspect of bougainvillea regarding molecular transformation exhibited disease response on tomato plant. Viral diseases especially Tospovirus cause greatest economic damage to the tomato crop. Many plant species are known to synthesize proteins with antiviral properties. The genes encoding anti viral proteins (AVPs) have been isolated and cloned from many plants. The study was conducted by Wallalwar and Balasaraswathi, 2017 on synthesis of engineered Bougainvillea antiviral gene and its expression in tomato plant. Transformation of this full length BAP- cDNA showed that *E. coli* cells grew slowly and the callused of tomato tissues turned necrotic after few days. This might be due to the ribosome inactivating (RIP) property of Bougainvillea antiviral protein (BAP). To remove the ribosome inactivating (RIP) activity of BAP, primers were designed for cDNA encoding truncated/engineered Bougainvillea antiviral protein (BAP). This truncated BAP-cDNA was cloned into pGA643 vector and mobilised into Agrobacterium strain EHA105. Agrobacterium transformation was performed using hypocotyls and cotyledon as explants. The T0 putative plant with pGA643 harbouring truncated Bougainvillea antiviral protein confirmed by PCR. The crude leaf/root extract of *Bougainvillea spectabilis* Willd. was effective in inhibition of several plant viruses *viz.*, tospoviruses, tobacco mosaic virus, cowpea aphid borne mosaic virus and cucumber mosaic virus (Balasaraswathi *et al.*, 1998; Sadasivam *et al.*, 1991). The virus inhibitor in *B. spectabilis* was found to be a Bougainvillea antiviral protein (BAP). The protein was purified to homogeneity and characterized from roots of *B. spectabilis* (Balasaraswathi *et al.*, 1998).

The biosynthesis of zinc oxide nanomaterials (ZnO-NMs), with a size of ˜ 40 nm from the *Bougainvillea* paper flower aqueous extract have been done by Rauf *et al.*, 2019. The FTIR and SEM-EDX mapping analysis confirmed the size, shape and biogenic origin of ZnO-NPs. Furthermore, the purified ZnO-NMs were applied for antibacterial studies against susceptible and resistant bacterial strains and to elucidate the possible mechanism of their activity. The XTT assay and confocal imaging confirmed the ZnO-NMs materials anti-biofilm activities against medically important pathogens, *i.e. S. aureus* and *E. coli*. Moreover, the absence of cytotoxicity against healthy kidney cells (HEK-293) and erythrocytes confirmed their biocompatible nature. Furthermore, the biosynthesized ZnO-NMs showed potent anticancer activity against the breast cancer cell line (MCF-7). These biosynthesized ZnO-NMs are having excellent antimicrobial and anticancer activities and are highly biocompatible due to biogenic nature. During antimicrobial study, Zno-NMs showed excellent minimum inhibitory concentration 16 µg concentration againt *E. coli, P. aeruginosa* and *S. aureus*. While in anticancer activity, of ZnO-NMs with 15 µg/ml dose showed good response against MCF-7 cell line. Further, this killing was mechanically confirmed by ROS generation by the ZnO-NMs, which cause cell lysis by the peroxidation of membrane lipid. So, this biogenic ZnO-NMs can be used in the future for nanomaterial-based drug development (Rauf *et al.*, 2019).

15.0 References

Abdel-Salam, O.M.E., Youness, E.R., Ahmed, N.A., El- Toumy, S.A., Soulman, AM.A., Shaffie, N and Abouelfadl, D.M. (2017) *Asian Pac. J. Trop. Med.*, **10**(5) : 478-490.

Abarca-Vargas, R. and Petricevich, V.L. (2018) *Evid. Based Complement Altern Med.* Hindawi.

Anonymous. (1996) *Research programmes and progress*. IIHR, Bangalore, p. 24.

Abarca-Vargas, R., Malacara, C.F.Pe˜na. and Petricevich, V.L. (2016) *Antioxidants.*, **5**(4) : 45.

Ahmad, N., Ishtiaq, M. and Nabi, G. (2002) *Sarhad J Agri. Pakistan.*, **18**(3) : 263-270.

Anand, N., Selvaraj, Y. and Bhatt, R.N. (1997) *J. Ornam. Hort.*, **5**(1-2) : 7-11.

Arteaga Figueroa, L., Abarca-Vargas, R., Alanis, C. Garc´ýa. and Petricevich, V.L. (2017) *BioMed Research International.*, **2017** : 11.

Arora, R.N. and Pahuja, S.K. (2008) *Plant Mut. Rep.*, **2**(1) : 7-9.

Ashmawy, N.A., Behiry, S.I., Al-Huqail, A. A., Ali, H.M. and Salem, M.Z.M. (2020) *Processes.*, **8**(4) : 482.

Awad, A.E., Dawh, A.K. and Attya, M.A. (1988) *Acta Hort.*, **226**(11) : 445-454.

Balasaraswathi, R., Sadasivam, S., Ward, M. and Walker, J. (1998) *Phytochemistry.*, **47** : 1561-1565.

Banerjee, B.K., Nath, P. and Datta, S.K. (1987) *J. Nuclear Agri. Biol.*, **19**(2) : 134-136.

Banerjee, B.K. and Datta, S.K. (1987) *J. Nuclear Agri. Biol.*, **16**(1) : 48-50.

Banerjee, B.K. (2002) *National Symposium on Indian Floriculture in new millennium held at Lal Baugh Botanic Garden.*

Banerjee, B.K. (2009) National *Conference on Flori-culture for Livelihood and Profitability,* IARI, New Delhi, pp 38.

Banerjee, B.K. (2010) *Indian Bougainvillea Annual.,* **23** : 4-8.

Banerjee, B.K. (2012) *Floriculture Today.,* 34-36.

Banerji, S.K. and Banda, G.K. (1967) *Indian J. Hort.,* **24** : 106-108.

Beel, E. and Schelstraete, A. (1981) *Verbondsnievws Voor de Belgische Sierteelt.,* **25**(17) : 765-768.

Begum, J. and Datta, R.M. (1971) *Sci. and Cult.,* **37** : 295.

Bhattacharjee, S.K. and Balakrishna, M.B. (1983) *Haryana J Hort. Sci.,* **12**(1/2) : 7-12.

Bose, T.K., Singh, P.K. and Bose, S. (1968) *Indian J Hort.,* **27** : 213-217.

Broertjes, C. and Van Harten, A.M. (1988) Ornamental crops. *In*: Developments in crop science 12 " *Applied mutation breeding for vegetatively propagated crops*". Elsveir Science Publishers B.V., Amsterdam, The Netherlands, pp. 122.

Cameron, C.A., Reid, M.S. and Hickman, G.W. (1981) Cooperative Extension, University of California, pp: 6.

Chatterjee, J., Kalam, A., Mandal, A., Chakrabarty, D. and Datta, S.K. (2007) *Hort. Envi. Biot.,* **48**(1) : 43–51.

Cheng, K.T., Chang, H.C., Su, C.H. and Hsu, F.L. (1997) *Bot. Bull. Acad. Sin.,* **38** : 241-244.

Datta, S.K. (1990) Role of mutation breeding in floriculture. *In*: *Plant mutation breeding in crop improvement,* **1**: 271-281.

Datta, S.K. and Banerji, B.K. (1994) *J. Nuclear Agri. Biol.,* **23**(2) : 114-116.

Do, L.T., Aree, T., Siripong, P. Vo, N.T., Nguyen, T.T., Nguyen, P.K. and Tip-pyang, S. (2018) *PlantaMedica.,* **84**(02) : 129-134, 2018.

Duhoky, M.M.S. and Al-Mizory, L.S.M. (2014) *IOSR-JAVS.,* **6**(6) : 01-06.

Ellis, M.B. (1976) Commonwealth Mycological Institute: Kew, pp. 297.

Even-Chen, Z. and Sachs, R.M. (1980) *Plant Physiol.,* **65** : 65-68.

Ex Jussieu, C. (1789) *Gen. Pl.,* pp 91.

Fufa, H., Baenziger, P.S., Beecher, B.S., Dweikat, I., Gray-bosch, R.A. and Eskridge, K.M. (2005) *Euphytica.,* **145**: 133-146.

Ghogar A. and Jiraungkoorskul, W. (2017) *Pharmacogn Rev.,* **11**(21) : 19-22.

George, K. and Sobhana, P. (1976) *The Nucleus.,* **19**(1) : 29.

Gupta, M.N. and Sukla, R. (1974) *Indian J. Gen.,* **34**(A) : 1295-1299.

Harris, C.V. and Singh, D.B. (1991) *New Agriculturist*. 1991; **2**(I) : 19-22.

Hammad, I. (2009) *Res. J. Agri. Bio. Sci.,* **5**(1) : 63–71.

Heuer, S., Richter, S., Metzger, J.W., Wray, V., Nimtzt, M. and Strack, D. (1994) *Phytochemistry.*, **37**(3) : 761-767.

Holttum, R.E. (1938) *Gdnr's Chron.*, **103** : 164-165.

Holttum, R.E. (1955) *Agri. Hort. Assoc. Magazines.*, **12** : 2-11.

Holttum, R.E. (1957) *Malayan Agri. Hort. Assoc. Magazines.*, **14** : 13.

Hossain, A.B.M.S., Boyce, A. N. and Osman, N. (2007) *Aust J Basic Appl Sci.*, **1**(4) : 733-740.

Janakiram, T., Jain, R., Swaroop, K. and Narkar, N.D. (2013) *ICAR Newsletter.*, **19**(1) : 6-7.

Jayanthi, R., Datta, S.K. and Verma, J.P. (2000) *Indian Bougainvillea Annual.*, **15**: 9-11.

Ji, X., Lin, W., Zhang, W., Yin, F., Zhao, X., Wang, C., Liu, J., Yang, H., Chen, Y. and Liu, S. (2015) *Afr J Biotech.*, **14** : 1910-1917.

Joshi, A.R., Mahorkar, V.K. and Sadawarte, K.T.(1989) *PKV Res. J.*, **13**(2) : 166-167.

Kale, P.N. and Bhujbal, B.G. (1972) *Indian J Hort.*, **29**(3/4) : 307-309.

Kannan, M., Sathiyamurthy, V.A. and Shanker, V. (2002) *Mutagenic studies on Jasminum sambac. In Floriculture Research Trend in India.* Misra, R.L. and Misra Sanyat. (Eds)., Indian Society of Ornamental Horticulture, Divsion of Floriculture and Landscaping, IARI, New Delhi, pp 209-11.

Kobayashi, K. D., McConnell, J. and Griffis, J. (2007) *Ornamentals and Flowersk.* OF-38 by College of Tropical Agriculture and Human Resources (CTAHR).

Khoshoo, T.N. (1998) *Bougainvillea Newsletter.*, **6**(2) : 7-10.

Khoshoo, T.N. and Mukherjee, I. (1970) *Theor. Appl. Gen.*, **40** : 204-217.

Kochhar, V.K. and Ohri, D. (1977) *Z Pj1uZucht.*, **79** : 47-51.

Kochhar, V.K., Kochhar,S. and Ohri,D. (1979) *Curr. Sci.*, **48**(4) : 163-164.

Kumar, R., Prasad, A. and Bajpai, P.N. (2002) *Indian Bougainvillea Annual.*, **17** : 32-33.

Kumar, R., Mishra, R.L.S. (2002) *Floriculture Res: Trend in India*, 351-352.

Kumar, R. and Prasad, A. (2002) *Indian Bougainvillea Annual.*, 17 : 26-8.

Kumar, P.P., Janakiram, T., Bhatt, K.V., Jain, R., Prasad, K.V. and Prabhu, K.V. (2014) *Indian J. Agri. Sci.*, **84**(8) : 1024-30.

Kumar, P.P., Janakiram, T., Bhatt, K.V., Prasad, K.V. and Jain, R. (2015) *Indian J. Agri. Sci.*, **85**(5) : 661–5.

Lancaster, P. (1951) *J. Royal. Hort. Soci.*, **76** : 278.

Lim, T.K. (2014) *Springer Sci. & Buisness.*, **489**: 4.

Ma, S., and M. Gu. (2010) *In: XXVIII International Horticultural Congress on Science and Horticulture for People* (*IHC2010*)*: Inter-national Symposium on* 937. p. 237-242.

Mabry, T.J. and Dreiding, A.S. (1968) *Recent advances in phytochemistry*, Appleton-century- crofts, New York, pp. 145-150.

Mac Daniels, L.H. (1981) *Baileya.*, **21** : 77-100.

Meenakshi, S. and Bhat, R.N. (1982) *The Nucleus*, **25** (3) : 130-135.

Meenakshi, S. and Bhat, R.N. (1988) *Proc. Conf. Cyto.& Genet.*, **1** : 52-55.

Misra, A.K. (1971) *Indian J Hort.*, **28** : 68-69.

Mitra, G.C. and Bose, N. (1954) *Phytomorphology.*, **7** : 370.

Moniruzzaman, M., Islam, M.N., Hossain, M.F.B. and Rashid, M.M. (2011) *Bull. Inst. Trop. Agr.*, **34** : 69-75.

Muhammad, A.M., Huwaishli, N.A., Ali, S.M. and Jabeen, R. (2019) *Biosc.Biotech. Res.Comm.*, **12**(3).

Murugan, R. (1999) *M.Sc. Thesis*, Tamil Nadu Agricultural University, Coimbatore.

Nichols, R. (1968) *J. Hort. Sci.*, **43**: 335-349.

Ninan, T., Singh, M. and Swaminathan, M.S. (1959) *J. Indian Bot. Soc.*, **38** : 140-145.

Ohri, D. and Zadoo, S.N. (1979) *Z. Pflanzenzuchtg.*, **82** : 182-186.

Ohri, D. and Zadoo, S.N. (1986) *Plant Breed.*, **97** : 227-231.

Pal, B.P. and Swarup, V. (1974) ICAR, New Delhi.

Piattelli, M. and Imperato, F. (1970) *Phytochemistry.*, **9**(12) : 2557-2560.

Philip, J., Gopalakrishnan, P.K. (1982) *South Indian Hort.*, **30**(1) : 56-57.

Podd, L.A. and Staden, J.V. (2004) *J. Plant Growth Regul.*, **26** : 183-189.

Rauf, M.A., Oves, M., Rehman, F.U., Khan, A.R. and Husain, N. (2019) *Biomed. Pharmacother.*, **116** : 108983.

Raghava, S.P.S. (1999) *J. Ornam. Hort.*, **2**(1) : 1-6.

Rao, L.M. (1967) *J Jap. Soc. Hort. Sci.*, **36** : 445-448.

Roy, R.K. (1987) Bougainvillea in landscaping. Flower Show Bulletin, Calcutta Flower Growersâ€™ Association pp 21.

Sadasivam, S., Rajamaheswari, S. and Jeyarajan, R. (1991) *J. Ecobiol.*, **3** : 53-57.

Sahariya, K., Singh, J.N. and Singh, A. (2013) *Asian J Hort.*, **8**(1) : 140-142.

Sangthong, S., Suksabye, P. and Thiravetyan, P. (2016) *Ecotoxicol. Environ. Saf.*, **126** : 273- 280.

Schoellhorn, R. and Alvarez, E. (2002) *Florida Cooperative Extension Service Fact Sheet* ENH 874, University of Florida, Gainesville, FL.

Sen, S.N. and Sen, N.K. (1954) *Bull. Bot. Sur. Bengal.*, **8** : 140-142.

Shah, S.T., Zamir, R., Muhammad, T., Ali, H., and Pakistan, P. (2006) *Pak. J. Bot.*, **38**(4) : 953-959.

Sharma, S.C. (1986) *Bougainvillea News letter.*, **6**(2) : 5-6.

Sharma, S.C. and Roy, R.K. (2001) *Indian Bougainvillea Annual.*, **16** : 9â€"12.

Sharma, A.K., Prasad, R.N. and Chaturvedi, H.C. (1981) *Plant Cell Tiss Organ Cult.*, **1** : 33-38.

Sharma, A.K. and Bhattacharya, N.K. (1960) *The Nucleus.*, **3** : 19-76.

Singh, S.P. and Motial, V.S. (1979) *Plant Sci.*, **11** : 53-59.

Singh, S., Reddu, S.K., Sharma, S.K. and Ali, M. (2009) *Asian Journal of Chemistry.*, **21**(6) : 4744- 4748.

Singh, K.K., Rawat, J.M.S. and Tomar, Y.K. (2011) *J. Hotic. Sci. & Ornam. Pl.*, **3**(2) : 162-165.

Singh, K.K., Singh, S.P., Bairwa, B. and Chauhan, J.S. (2020) *Research Today.*, **2**(6) : 472-474.

Srivastava, R., Shukla, S., Soni, A. and Kumar, A. (2009) *Crop Breed. App. Biot.*, **9** : 154-163.

Steffen, J.D., Sachs, R.M. and Hackett, W.P. (1988) *Am. J. Bot.*, **75**: 1225-1227.

Swaroop, K., Jain, R. and Janakiram, T. (2015) *Indian J. Agri. Sci.*, **85**(9) : 1245-1247.

Swarup, V. and Singh, B. (1995) *Genetic resources of Bougainvilleas. Advances in Horticulture* vol. 12- Ornamental Plants. Eds: Chadha KL and Bhattacharya SK. Malhotra Publishing House, New Delhi- 110 064, India, pp 133-152.

Taiz, L. and Zeiger, E. (2010) *Plant physiology* (5th ed.). Sinauer Associates, Sunderland, 782 p.

The Plant List (2013) Version 1.1. Published on the Internet; http: //www. theplantlist.org/(accessed 1st January).

Tumer, N.E., Hwang, D.J. and Bonness, M. (1997) *Proceedings of the National Academy of Sciences*. **94**: 3866-3871.

Varshney, R.K., Graner, A. and Sorrells, M.E. (2005) *Trends Biot.*, **23**: 48-55.

Vukovic, N., Kacaniova, M., Hleba, L. and Sukdolak, S. (2013) *J. Essent. Oil-Bear. Plants.*, **16**(2) : 212-215.

Wallalwar, M.R. and Balasaraswathi, R. (2017) *Int J Curr Microbiol Appl Sci.*, **6**(60) : 1714-1722.

Wight, C.P., Malnar, S.J. and Fedak, G. (1993) *Genome.*, **36** : 818-820.

Xu, S., Huang, Q., Shu, Q., Chen, C. and Vick, B.A. (2009) *Sci. Horti.*, **120** : 399-405.

Zadoo, S.N. and Khoshoo, T.N. (1975) *Proceedings of Indian National Science Academy.*, **41B** : 498- 502.

Zadoo, S.N., Roy, R.P. and Khoshoo, T.N. (1976) *Indian Hort.*, **33** : 278-288.

Camellia

Debmala Mukherjee and Koushik Datta

1.0 Introduction

Camellia is a genus of flowering plants in the family Theaceae. They are found in eastern and southern Asia, from the Himalayas east to Japan and Indonesia. The Camellia genus includes over 260 species (International Camellia Society). The genus was named by Linnaeus after the Jesuit botanist Georg Joseph Kamel (Cothran, 2003), who worked in the Philippines and described a species of camellia (Sebestian, 2015). *Camellia japonica* L. is an evergreen flowering perennial with more than 3,000 named cultivars of great ornamental value (Berruti *et al.,* 2013). Camellias species have originated over 20,000 ornamental varieties, obtained mainly from *Camellia japonica,* but also from *Camellia sasanqua, Camellia saluenensis, Camellia reticulata,* and other species. The colour of camellia flowers range from white to red, and from pale to dark yellow. Because of the spectacular beauty of their flowers, ornamental camellias have spread from Asia to the whole world.

2.0 Importance and Uses

2.1.0 Aesthetic Uses

The ornamental *C. japonica, C. sasanqua* and their hybrids are the source of hundreds of garden cultivars. These ornamental species are used as specimen plants on the lawn and for colourful accents near outdoor living areas. Camellia is mostly used in the floral industry for it's foliage. Camellia arrangements are elegant and can be used for almost any kind of occasion. They have a very boheme look, so they can easily be used in wedding arrangements and bouquets. Cultivation of ornamental plants is strongly oriented to sustainable production on the one hand and on a consumer-oriented high quality product on the other hand (Berruti *et al.,* 2013).

2.2.0 Therapeutic and Other Uses

Green tea from *C. japonica* has been developed as a food source in Korea so called 'Dong BaekCha', commercialized in 2007 (Hwang *et al.,* 2004). In some parts of Japan, tea made from *C. sasanqua* leaves is popular.

Tea oil is a sweet seasoning and cooking oil made by pressing the seeds of *C. oleifera, C. japonica,* and to a lesser extent other species such as *C. crapnelliana, C. reticulata, C. sasanqua* and *C. sinensis*. Relatively little-known outside East Asia, it is the most important cooking oil for hundreds of millions of people, particularly in southern China. Camellia oil is commonly used to clean and protect the blades of cutting instruments. Camellia oil pressed from seeds of *C. japonica,* also called tsubaki oil or tsubaki- abura in Japanese, has been traditionally used in Japan for hair care. *Camellia japonica* constituted of triterpenes, saponins, glycosylated flavonoids and tannins (Yang *et al.,* 2015) also exhibit biological activity such as antioxidant, antifungal, antiviral and cytotoxic potential (Thao *et al.,* 2010).

Camellia japonica has been used in traditional medicine in Japan, China, and Korea for the treatment of stomachic illness, bleeding and inflammation (Yoshikawa *et al.,* 1996). The camellia was a symbol of Brazil's abolitionist movement during

Uses of Camellia

In Pot

As Hedge

Against Wall

As Bonsai

Designing of Wall

As Corner Decoration

Bush Camellia

At Centre of Garden

Camellia Garden

the Imperial Age. It was common practice for abolitionists to plant camellias in a show of solidarity. The following cities are nicknamed the "Camellia City" of each state: Greenville, Alabama; Sacramento, California; Fort Walton Beach, Florida; Slidell, Louisiana; McComb, Mississippi; Newberg, Oregon; and, Thomson, Georgia is nicknamed the "Camellia City of the South". The camellia is the state flower of Alabama. *C. reticulata* (Cha- hua in Yunnan) has been endowed with rich spiritual connotation (Xin *et al.*, 2015).

3.0 Origin and History

The first camellia introduced in Europe would have been in Portugal during the 16th century. The history of camellia in Europe really began in England. It was cultivated in greenhouses, at least in the beginning. The English prospected and brought back varied plants from their travels. Thus the first camellias arrived in England in the early 18th century. James Petiver, apothecary, described in 1702, a plant with a red flower named *'Thea chinensis'*. In fact, the English were trying to find the tea plant. H. Cunningham, surgeon and botanist, brought back *C. sinensis* and *C. fraterna*, the propagation of these plants to make 'tea' did not give the desired result. It is also said that the British were abused and had been given ornamental shrubs despite of tea species. Undoubtedly, it was *C. japonica*, the botanical species. In 1739, the first *Camellia japonica* was cultivated. But, it was a plant reserved to an elite, because of its rarety and cost. The plant quickly spread through the royal and princely courts of Europe. George III planted one in Pillnitz, Germany, near Dresden. It still lives there, protected in winter by a huge greenhouse, given its impressive size. It was also cultivated in Italy (Caserte), Netherlands and France. In these days *C. japonica* was the only camellia known in Europe. In 1739, the first camellias seen by Robert James and Lord Petre in a garden of England were a single red and a single white coloured flower. The gardener James Gordon was the first to introduce camellias to commerce from the nurseries he established after Lord Petre's untimely death in 1743 (Coats, 1964, 1992). With the expansion of the tea trade in the later 18th century, new varieties of camellia began to be seen in England and the first double white and red striped camellia was imported by John Slater through the British East India Company in 1792. The camellia was imported from England to America in 1797. By 1819, twenty- five camellias had bloomed in England and the first monograph on Genus *Camellia* was published by Samuel Curtis. By the 1840's, the camellia was at the height of its fashion as the luxury flower. Camellias were cultivated in the gardens of China and Japan for centuries before they were seen in Europe. The German botanist Engelbert Kaempfer called the *Camellia* as 'Japan Rose'. The camellia is native to India, China, Japan and other parts of Asia.

In the early 19th century, Henri Favre (Mayor of Nantes) discovered that the camellias could live very well out of greenhouses. The Brittany (west of France) climate is especially favourable. In the early 20th century, in the 30's, J.C. Williams, in Cornwall (England), raised the first European camellia hybrid, an hybridation between the saluenensis species, recently imported from China, and a japonica cultivar. Many hybrids of saluenensis were raised by J.C. Williams and other breeders, which were given in gratitude, the nickname 'williamsii'. One of the most famous today is 'Donation', nicknamed 'the flower power'. In all gardens of the world, their descendants can be found. New colours are searched, as well as camellias adapted to cold climates. Finally, the camellia becomes democratic.

4.0 Morphology

Camellia japonica is a flowering tree or shrub, usually 1.5- 6 metres tall, but occasionally up to 11 metres tall. The youngest branches are purplish-brown, becoming grayish- brown as they age. The alternately arranged leathery leaves are usually 5- 11 cm long, 2.5- 6 cm wide with a stalk (petiole) about 5- 10 mm long. The base of the leaf is pointed (cuneate), the margins are very finely toothed (serrulate) and the tip somewhat pointed. The flowers appear along the branches, particularly towards the ends, and have very short stems. They occur either alone or in pairs, and are 6- 10 cm across. There are about nine greenish bracteoles and sepals. Flowers of the wild species have six or seven rose or white petals, each 3- 4.5 cm long by 1.5- 2.5 cm wide; the innermost petals are joined at the base for up to a third of their length (cultivated forms often have more petals). The numerous stamens are 2.5- 3.5 cm long, the outer whorl being joined at the base for up to 2.5 cm. The three- lobed style is about 3 cm long. The fruit consists of a globe- shaped capsule with three compartments (locules), each with one or two large brown seeds with a diameter of 1- 2 cm. Fruiting occurs in September to October in the wild.

5.0 Species and Cultivars

5.1.0 Species

Camellia japonica

The most important ornamental camellia is *C. japonica*. It is believed to be native to Japan. It is also known as common camellia, Japanese camellia, or *tsubaki* in Japanese and is one of the best known species of the genus *Camellia*. Sometimes it is called the rose of winter. It is the official state flower of the US state of Alabama. In the U.S. it is sometimes called japonica, a name more often used in the UK for *Chaenomeles* (flowering quince). It is varying from small shrub to small tree; up to 30 feet or more tall. In the wild, it grows as an understory plant on hillsides. Flowers vary in size and colour, but generally are single with five to six petals, and are red. However white and pink are known. They are winter blooming, have beautiful shiny evergreen leaves, and an interesting brown trunk. Today, they have been hybridized more than any other camellia, and flowers are available in all sizes, colours, and forms. The leaf of *Camellia japonica* is rich in anti-inflammatory terpenoids such as lupeol and squalene (Majumder *et al.,* 2020).

C. japonica var. *japonica*

Named by Linnaeus, naturally occurs in forests at altitudes of 300- 1,100 metres in China, Taiwan, south Japan, and South Korea. The leaf has 1 cm long petiole, flowers between January and March.

C. japonica var. *rusticana*

Naturally occurs in forests in Zhejiang, China and Honshu in Japan. The leaf has a shorter petiole (5 mm) with fine hairs at the base. The colour of the flowers ranges from red to pink, flowers in April to May.

Species of Camellia

C. japonica *C. sinensis* *C. yunnensis*

C. petelotii *C. sasanqua* *C. pubipetala*

C. reticulata *C. nitidissima* *C. granthamiana*

C. amplexicaulis *C. oleifera* *C.tunghinensis*

Camellia sasanqua

These are the fall blooming camellias. This species is native to China and Japan, usually found growing up to an altitude of 900 metres; evergreen shrub, leaves are broad elliptic with a finely serrated margin; flowers are 5- 7 cm in diameter, white to dark pink in colour. The leaves are generally smaller than *japonicas* and flowers shatter readily, making them unsuitable for cut flowers. They bloom from September through December, and are excellent landscape plants. They will grow in either sun or shade and may not be quite as cold hardy as the *japonicas*.

Camellia reticulata (syn. *C. heterophylla*)

This is a species of flowering plant in the tea family Theaceae, native to southwestern China, in Yunnan Province. It is a loosely branched shrub or small tree, which can grow up to 10 to 15 meters. The flowers are 7- 10 cm in diameter, or larger in some cultivars, soft- pink to deep- pink and rarely almost white, with 5- 7 petals or more in some cultivars, and are produced in sub-terminal or axillary positions on the branch. This camellia is very susceptible to cold weather and has a late blooming season; August through October in the southern hemisphere and March through May in the northern hemisphere. *Camellia reticulata* is the floral emblem of Yunnan. It has a long history of cultivation, both for tea oil and for its ornamental value. Warren Hastings imported the first *C. reticulata* to England, (named 'Captain Rawes'). It remained the only known reticulata cultivated in Europe for over a century.

Camellia pubipetala

This is an endemic and endangered species with small and isolated populations occurring only in karst regions in Guangxi of south China.

Camellia oleifera

It is known as the oil camellia, and used for this purpose for years in the Orient, and has been used as a parent of hardy ornamental camellia hybrids in the USA since at least the late 1970s.

Camellia chrysantha or *C. nitidissima*

This species is best known of the yellow camellias. It came out of China around 1980. This species is endemic to the Guangxi Zhuangin southern China and northern Vietnam. It is a shrub or small tree growing up to 5 meters in height, has waxy yellow flowers. The yellow camellia has been used successfully in the hybridizing program for yellow coloured cultivars. But it is not cold hardy.

Camellia yunnanensis

C. yunnanensis is a 1.3- 7 m tall shrub or small tree. Its leaves are elliptic to broad- elliptic or ovate-elliptic, bluntly acute. They are deep green. Its flowers are white, perulate and solitary.

Camellia granthamiana

It is also a wild species discovered in Hong Kong, China, with one of the largest camellia flowers, up to 180mm in diameter.

Camellia hiemalis

This specis is a shrub or small tree with evergreen leaves and red, pink, or white flowers.

Camellia hongkongensis

It is a small evergreen tree, native to Hong Kong, bears red flowers, leaves are leathery and oblong.

Camellia vernalis

This is a hybrid between C. *japonica* and C. *sasanqua*. Size and shape are intermediate between the two species. There are very few varieties known. The cross is difficult to make.

Camellia × *williamsii*

It is a cultivar group of hybrid evergreen shrubs derived from a crossing of *C. saluenensis* with *C. japonica* (Wheeler *et al.*, 2015). It is an excellent companion to other acid- loving woodland plants such as rhododendron. It grows best in acid soil, pH.5.5- 6.5, in a sunny or partially shaded place.

Two new species of *Camellia* (Theaceae) are described by Nguyen *et al.* (2018) from Vietnam: *Camellia vuquangensis* and *Camellia hatinhensis*. The new taxa were collected from Vu Quang National Park in the centre of Vietnam. Discovery of the camellia species increased from 50- 75, making Vietnam a centre of diversity and a crucial area for more research into the diversity and distributions of *Camellia*.

5.2.0 Horticultural Forms

Camellia cultivars have been classified based on different size of flowers such as miniature, small, medium, large and forms of flowers such as anemone, peony-form, rose- form, single, semi- double, double. There are more than 3,000 cultivars developed from *C. japonica* (Cothran, 2003). Next are *C. reticulata* with over 400 named cultivars, and *C. sasanqua* with over 300 named cultivars.

Some varieties can grow to a considerable size, up to 100 m^2, though more compact cultivars are available. They are frequently planted in woodland settings, alongside other calcifuges such as rhododendrons, and are particularly associated with areas of high soil acidity. White, different shades of pink, red, carmine, purple, singly and in combination make the flowers very attractive. Improvement in *Camellia* has been achieved by hybridization among cultivars of *Camellia japonica*.

5.2.1 Single

Single flowers have five to a maximum of eight petals in one row, petals loose; flat, bowl- or cup-shaped, regular or irregular. May include petaloids; prominent display of stamens & pistils.

5.2.2 Semi-Double

Two or more rows of large regular, irregular or loose outer petals (nine or more) with an uninterrupted cluster of stamens; rows of large outer petals, with

Types of *Camellia japonica*

Single Type cv. 'Ashiya'

Single Type cv. 'Sekidotaroan'

Single Type cv. 'Kamo-honnnami'

Semi Double cv. 'Ville De Nantes'

Semi Double cv. 'Mercury Supreme'

Semi double cv. 'C.M. Wilson'

Semi Double cv. 'Royal Velvet'

Semi Double cv. 'Dr. Tinsley'

Semi Double cv. 'Songzi' Pine Cone

Semi Double Type cv. 'The Czar'

Semi Double cv. 'Dr Clifford Parks'

Types of *Camellia japonica*

Paeony Type cv.
'Frankie winn'

Paeony Type cv.
'Ann Blair Brown Variegated'

Paeony Type cv.
'Colombo'

Anemone Form cv.
'Althaeiflora'

Anemone Form cv.
'Nobilissima'

Anemone form cv.
'Chandlers Elegance'

Rose Form cv.
'Bob Hope'

Rose Form cv.
'Drama Girl'

Formal Double cv.
'Coquettii'

Formal Double cv.
'Dahlohnega'

Formal Double cv.
'Duchesse de Berry'

Formal double cv.
'White by the Gate'

the centre comprising mixed petals and stamens. It may include petaloids; petals may overlap or be set in rows for 'hose in hose' effect.

5.2.3 Double

5.2.3.1 Peony Form

A mass of raised petals with petaloids (parts of the flower that have assumed the appearance of small, narrow or twisted petals) is the specialized structure of it. Stamens may or may not be visible. Paeony form is looking like a convex mass of irregular petals and petaloids with hidden stamens; hence also called 'Informal Double'.

5.2.3.2 Anemone Form

One or more rows of large outer petals lying flat or undulating, with a mass of intermingled petaloids and stamens in the centre; it is also called 'Elegans Form'.

5.2.3.3 Rose Form

A semi-double form with one or more petaloids interrupting the cluster of stamens. Here overlapping petals show stamens in a concave centre when open. It is often referred as 'Irregular Semi-Double'.

5.2.3.4 Formal Double

It is a form of rows of overlapping petals with hidden stamens. Many rows and number of petals (sometimes more than a hundred), regularly disposed, tiered or imbricated, with no visible stamens. Usually develop with a central cone of tightly furled petals.

5.3.0 Cultivars

'Korean Fire': This is one of the best *C. japonica* cultivars and has unusually good winter hardiness. Flower of Korean Fire is single, showy, attractive red in colour and plant is evergreen, shrubby in nature. Flowers come in early spring. It is currently recommended for growing outdoors year round in USDA Zone 6.

'Flame': Flame puts forth semi-double flowers in fiery red. It is slow growing, upright to spreading shrub or small tree with age; oval, glossy, leaves with profuse winter to spring blooming flowers in milder climates. The blossoms of this cultivar stick out by their bright red colour and the blooming starts from January. This is a true heritage cultivar, and favourite of Southern gardens.

'William Barlett': It is an upright, evergreen, compact and bushy shrub with ovate, glossy, leathery, dark green leaves and, in early to mid- spring. This is late blooming camellia cultivar. Multi rows of petals in a symmetrical pattern attract attention by their double form type pink- white spotted blossoms with darker pink flecks and stripes.

'Hagoromo' : Hagoromo' is a mid- sized, spreading, evergreen shrub with dark- green, twisted, glossy leaves. It bears pale- pink, semi- double flowers in spring. Light red blossoms and bright yellow pistil are the hallmarks of this cultivar.

'Margaret Davis': This is an upright, evergreen shrub, up to 4m tall, bearing ruffled flowers from late winter to early spring with creamy- white petals edged a raspberry red. Their white blossom with pink edges, serves as an interesting touch of colour in the winter garden. This can be grown in container also.

'Juris Yellow': The medium sized anemone form camellia is composed of a row of white outer petals centred by pale yellowish petaloids interspersed with golden stamens and this was one of the earliest hybrids to display it. 'Jury's Yellow' Camellia is a long flowering hybrid that begins early in the season and continues into late season. This is a moderately vigorous *Camellia × williamsi* hybrid with an upright to rounded habit and dense branching growth habit. This large evergreen cultivar is ideal as a specimen in the garden and landscape or as a container plant.

Sl.No.	*Cultivar*	*Characteristics*
1.	Dr. Clifford Parks	An attractive, vigorous plant boasts very large, peony- shaped semi-double blossoms of scarlet red from March to April.
2.	Double Pink	Flowers double, light pink, red, stripe on the petals.
3.	Edith	A compact bushy shrub, bears large number of double, pink flowers, very attractive cultivar.
4.	Elegans	Flowers with large pink flowers which often have white streaks.
5.	GiulioNuccio	Flowers with red to pinkish petals and yellow stamens.
6.	Galaxy	Large beautiful double flowers, deep red with irregular white spots and patches.
7.	Jewel	Flowers medium-sized, double, white with pinkish tinge.
8.	Mathotiana Alba	Bear pure white flowers, and the light crimson semi-double-flowered.
9.	Margaret	Medium-sized flowers, double, in cluster, contrast colours in red and white.
10.	Margaret McCown	Large double flowers, in white, pink and purple.
11.	Raindrop	Large, well-shaped, light pink to reddish pink, double flowers.
12.	Red Cascade	A compact bushy shrub, bearing deep red, semi-double flowers.
13.	Picture	Large attractive flowers, semi-double, pinkish white with red and pink stripes.
14.	Pink Delight	Flowers large, to 6 cm across, double, pink.
15.	Spring Beauty	A compact, free-flowering shrub, red flowers turning light pink with age.
16.	Pink Prince	Medium-sized flowers, light pink, double, free-flowering cultivar.
17.	Pink Pearl	Well-shaped showy flowers, pink, double.
18.	Red Delight	Showy double flowers, deep red patches and strips of white.
19.	Red Glory	Compact bush, to 2 m high, flowers many, large, deep red in colour.
20.	Royal Red	Flowers large, to 6 cm across, glossy scarlet, double;
21.	Royal Velvet	It is graced with sizable, semi-double flowers in an intense crimson.
22.	Spring Beauty	A compact, free-flowering shrub, red flowers turning light pink with age.
23.	Triphosa	Large deep green glossy leaves and large white flowers.
24.	Tudor Baby	Large double flowers, scarlet and white in colour, very attractive when in flower.
25.	Ville de Nantes	Bloom Colour, deep red with dashes of white, petals with fine teeth. Bloom Size, medium to large.
26.	White Rose	Large attractive, double, white flowers.

6.0 Genetics and Breeding

The breeding system usually refers to all of the sexual characteristics that directly affect the genetic composition of the offspring, including the comprehensive flower characteristics, lifespan of sexual organs, flower opening patterns, degree of self-compatibility and mating systems (Wyatt, 1983). Although few species are endangered because of their breeding system (Weller, 1994), information regarding the breeding system may assist in our understanding of the causes of the endangered nature of these species (Gan *et al.*, 2013; Schemske, Husband, & Ruckelshaus, 1994). To successfully reproduce, plants need to be pollinated, which often results from mutual interactions with animals.

Camellia japonica is a widespread and morphologically diverse tree native to parts of Japan and adjacent islands. This species secretes a large quantity of nectar, its flowers are conspicuously red in colour and its petals are tough; traits consistent with 'bird pollination syndrome' (Faegri and Van der Pijl, 1971). Indeed studies have shown that *C. japonica* is visited by two bird species in Japan, white- eyes (*Zosterops japonica*) and brown- eared bulbuls (*Hypsipetes amaurotis*) (Tanaka 1980; Yumoto 1988), but the role of birds in the pollination success of *C. japonica* has not been quantitatively evaluated. Recent molecular studies have indicated that *C. japonica* has a unique genetic structure both within and among populations. First, genetic diversity within a population is very high (Wendel and Parks, 1985), even in isolated populations (Yeeh *et al.*, 1996). Second, although the overall level of population differentiation based on allelic frequency is greater than typical outbreeding trees, allelic compositions of local populations are similar throughout the species range (Wendel and Parks, 1985). Third, this species has a relatively high outcrossing rate (Wendel and Parks, 1985). *C. japonica*, long-distance gene flow by seed dispersal is unlikely because seed dispersal appears to be mediated by gravity (Ueno *et al.*, 2000) and small mammals (H. Abe, pers. comm., 2003). Consequently, the unique genetic characteristics of *C. japonica* may be mediated by pollen transfer by birds (Wendel and Parks, 1985; Oh *et al.*, 1995; Yeeh *et al.*, 1996), and this is compatible with the hypothesis that birds are the major pollen vectors of *C. japonica*. The primary pollinator of *C. pubipetala* is the sunbird *Aethopyga christinae* and its visiting frequency is quite low, whereas the honeybee *Apis cerana* is only an occasional pollinator in wild populations. Low reproductive rates in *C. pubipetala* were found to be a consequence of few species of pollinators and their low visiting frequency (Chai *et al.*, 2019). Pollen limitation may be a crucial factor that contributes to the endangered nature of this species. Artificial pollination and the release of pollinators are effective ways to increase the fruit and seed yield of this species.

Most new Camellia cultivars arise from cross pollination of existing cultivars and/wild species. This allows breeders to deliberately strive for particular breeding goals such as flower colour or size rather than the random improvements that arise unpredictably from mutations. For example the very popular *Camellia × williamsii* hybrid group resulted from a cross between various *japonica* cultivars with a species *Camellia saluensis*, originally collected from mountainous regions of China, used in breeding initially in the 1920's. This group of hybrids is especially noteworthy for its cold tolerance and 'self cleaning' flowering habit where the petals fall from the plant rapidly rather than staying put and turning an unsightly brown. Current directions with Camellia breeding include an ongoing but thus far fruitless quest for yellow and orange flower colours through a group of buttercup yellow flowered wild species such as *Camellia chrysantha*. Fragrant species such as *C. lutchuensis* and *C. oleifera* continue to be used by breeders to create hybrids with sweeter and stronger perfumes.

6.1.0 Interspecific Hybridization

There is a wide range of interspecific fertility; some species, such as *C. saluenensis*, crossing readily with a wide range of other species, and others, such as *C. sasanqua*, which will cross only with a limited number of other species and with a low rate of success. Successful intergeneric crosses are rare but the fact of their successful accomplishment should lead to further efforts in this area. Breeding goals for golden camellias are for large sized yellow flower varieties; yellow star- like flower form varieties; great tolerance in extensive management, drought, disease and pests.

List of camellia hybrids are as follows.

Sl.No.	*Interspecific Hybrid*	*Characteristics*
1.	Cornish Snow	*C. cuspidata* × *C. saluenensis;* 2.5m height, single type white flower colour.
2.	Cornish Spring	*C. cuspidata* × *C. japonica,* 2.5m height single type pink flower.
3.	Inspiration	*C. reticulatax C. saluenensis*, 4 m height with semi-double rose- pink coloured flower.
4.	Leonard Messel	*C. reticulata* × C. *williamsii,* 4 m height with semi-double rose- pink coloured flower.
5.	Royalty	*C. japonica* × *C. reticulata,* 1 m height, light red semi-double flower.
6.	Tom Knudsen	*C. japonica* × *C. reticulate,* 2.5m height, deep red, double- paeony type flower.

Sui *et al.* (2014) have successfully developed several yellow camellia varieties, such as the double yellow camellia 'Dongyue', and 'Jin bei dan xin' that has red on the surface and yellow on the back of the petal. Some varieties have been registered in the China Camellia Society. 'Dongyue' was awarded the New Variety Prize at the Camellia Exhibition held during the 2012 International Camellia Society Congress in Chuxiong. Along with these, they successfully bred double yellow camellia varieties and a series of yellow camellia hybrid varieties, such as 'Dongyue', 'Xinhuang', 'Dongmeiyingchun', 'Bi Liu', 'Jin Bei Dan Xin', 'Xin Zi', 'Shuimeiren' as well as 'Huigui' and so on.

C. tunghinensis has dense flowers, small leaves and flowers; when it was crossed with *C. nitidissima*, F_1 generations show some features much like its mother parent, *i.e.* dense flowers. When *C. nitidissima* was used as mother parent to cross with *C. hakodae* from Vietnam, F_1 offspring show their intermediate characteristics between parents, for instance, larger leaves compared to its female parent. The F_1 hybrids usually show pink or pale yellow colour of flowers, but less deep yellow in crosses

Hybrids of Camellia

'Dongyue' 'Xinhuang' 'Dongmeiyingchun'

'Jinbeidanxing' 'Xinzi' 'Shui Meiren'

'Huigui' 'H. Wirlinga Princess' 'H. Wirlinga Belle'

'Len Bray' 'R. Simpatica'

of *C. nitidissima* and red or white *C. japonica* cultivars. Deep yellow cultivars may be bred by using recurrent hybridization (backcross) or reciprocal hybridization of F_1 or F_2 as parents.

6.2.0 Self-sterility

Self- sterility in *Camellia oleifera* was explored by comparing structural and statistical characteristics following self-pollination (SP) and cross-pollination (CP). Self- sterility is a common reproductive phenomenon in Theaceae, which is mainly the result of self-incompatibility (SI) and early-acting inbreeding depression (EID) (Jiang and Wang, 1987; Sage *et al.*, 2006). SI, an initiative abortion process, can prevent reproduction after selfing and enhance heterozygosity in plants (de Nettancourt, 1997; Valtueña *et al.*, 2010).

Many hybrids will be found to be almost infertile due both to malformed pistils and to the lack of development of fertile pollen grains. Nevertheless, in many cases, these apparently totally infertile plants occasionally produce fertile pollen grains and it is often worthwhile to persevere with such difficult crosses as the results are often unusual and interesting. A proportion of such crosses produce seed with shrunken cotyledons and deformed embryos. If these are to grow, the technique of embryo culture has to be used.

Natural hybridization usually will only occur between closely related species which may have recently diverged on either side. However the main barriers to natural hybridisation are usually due to differences in ploidy level, different flowering times, and different pollination mechanisms, varying adaptability to soil and climate, or inherent incompatibility. Also amongst the camellia species hybridization between different chromosome numbers, the resultant hybrid will be infertile unless it loses its unpaired chromosomes by aneuploidity or becomes allaploid by doubling its chromosomes during the cycle of cell division. Amongst fertile camellia hybrids both mechanisms have been in evidence. Under cultivation many highly sterile individuals persist for long periods of time, through methods of vegetative propagation, without ever producing seeds. The sterile *C. reticulata* cv. 'Captain Rawes', with its unusual haploid genetic structure, but even such a highly infertile cultivar as this occasionally produces some grains of fertile pollen as evidenced by a few hybrids existing which claim it as the pollen parent (Savige, 2005). The widely cultivated varieties are mainly derived from the mutant lines and hybrid progenies of *Camellia japonica* Linn. and *Camellia reticulata* Lindl. (Wang *et al.*, 2020).

6.3.0 Mutation Breeding

Bicolored flowers are also a feature created by genetic mutations. Genetic streak variegation is an indication that the plant is a mutation (sport). In such cases the flower can become spotted or striped, usually with white on pink or red flowers. It has been said that 'Camellias are sporting fools' in as much as they have a high probability to mutate. Sometimes it is apparent in a camellia plant and sometimes it may not be seen until one day, the bloom has a streak of red or white, or the entire bloom will bloom differently from the rest of the plant. This may be just a bloom or two, an entire limb, or the entire plant all at once. The bloom may be consistently streaked with white or red, from the centre of the bloom to the edge, year to year. A difference may appear as a short stripe or elongated fleck of colour in the petals apparently beginning at the centre and extending toward the edge of the petal. Another variation of genetic variegation is a picotee around the edge of the petals, or a gradual change of colour from the centre of the bloom to the outer edge.

From the study of Li *et al.*, 2019, seven anthocyanins were detected in *C. japonica* cultivars and their bud mutation cultivars, which were cyanidin-3-*O*-β-galactoside (Cy3Ga),cyanidin-3-*O*-β-glucoside (Cy3G),cyanidin-3-*O*-(6-*O*-(*E*)-caffeoyl)-β-galactoside (Cy3Ga*E*Caf), cyanidin-3-*O*-(6-*O*-(*E*)- caffeoyl)-β-glucoside (Cy3G*E*Caf), cyanidin-3-*O*-(6-*O*-(*Z*)-*p*-coumaroyl)-β-glucoside (Cy3G*Zp*C), cyanidin-3-*O*-(6-*O*-(*E*)-*p*- coumaroyl)-β-galactoside (Cy3Ga*Ep*C) and cyanidin-3-*O*-(6-*O*-(*E*)-*p*-coumaroyl)-β-glucoside (Cy3G*Ep*C). Among the bud mutation cultivars of *C. japonica*, anthocyanins were not detected in white petals, and anthocyanins in red petals were identical with that in pink petals, but the contents of anthocyanin components and total anthocyanin in red petals were far higher than that in pink petals. The main anthocyanin components were Cy3G and Cy3G*Ep*C in red and pink petals. The proportion of Cy3G and Cy3Ga in red petals were larger than that in pink petals, while the proportion of else anthocyanins, such as Cy3G*Ep*C, were smaller than that in pink petals. Among the bud mutation cultivars of *C. japonica*, the greater the contents of various anthocyanin components and total anthocyanin were, the deeper the red of petal were. Cy3G, Cy3Ga and Cy3G*E*pC were the main anthocyanins which determined the flower colours of bud mutation in *C. japonica*, and the accumulation of their contents enhanced the red colour of petals.

'White-by-the-Gate' (*C. japonica*) is a very beautiful, medium sized, formal double, white bloom. Very occasionally, the bloom will display a narrow strip of bright red from the centre of the bloom to the outer edge. A plant may bloom perfectly white for many years, and then one year, a bloom or several blooms may display a red stripe. This stripe may appear again for several years and may be seen on that plant for several years, or it may not be seen again.

'Carters Sunburst' (*C. japonica*) is a white to blush pink flower that has a multiplicity of stripes or stripe-like flecks of red radiating from the centre of the bloom toward the outer edge. One sport of this bloom is pure white, but it did not appear until after the original seedling had been propagated for many years. This pure white camellia has held and has been named and registered as 'Chow's Han Ling'. There are several other sports of 'Carter's Sunburst'.

'Betty Sheffield' (*C. japonica*) is one of the most active sports in the camellia world. The original 'Betty Sheffield' was white, with red streaks from the centre to the outer edge.

'Lady Vansittart' (*C. Japonica*) is similar to 'Betty Sheffield'. One mutant has been named 'Yours Truly; and the other 'Lady Vansittart Red'. 'Yours Truly' is pink, streaked with red and bordered with white. 'Lady Vansittart Red' is a deep pink to red version of 'Lady Vansittart' that may exhibit a white or deeper pink streak.

'Herme' (*C. Japonica*, a.k.a. 'Jordan's Pride) is a medium, semi- double flower. It has pink petals, striped deeper pink with an irregular white edge. There are two sports of 'Herme', 'Herme Pink' and 'Herme White' 'Herme White' is white with a few deep pink stripes (usually) and 'Herme Pink' is rose pink with a few white or darker red stripes (usually).

7.0 Soil and Climate

7.1.0 Soil

The camellia species are generally well-adapted to acid soils rich in humus. Maximum species do not grow well on chalky soil or other calcium-rich soils. Camellias do not tolerate drought hence require a large amount of water, either from natural rainfall or from irrigation. However, some of the more unusual camellia species from karst soils in Vietnam, can grow without too much water. Though, wild camellias in China are found to grow in heavy soil but the plant is well accustomed with light textured composts based on peat, cotton seed waste and other coarse organic material. Camellias are admittedly found on well drained slopy land. Similar soil should be incorporated while pot planting in nurseries or garden. *C. japonica* requires a loose, well-amended, acidic (6.5 pH) soil that is moist without remaining soggy, and little competition from other plants' roots. Their roots are pretty shallow and have high oxygen demands but camellias are not heavy feeders.

7.2.0 Climate

Camellia japonica and most of its cultivars are considered to be winter hardy but sudden temperature fall up to -23°C, can damage plants permanently. Most camellias need good light to thrive and to form flower buds. But plants prefer some shade especially in hot climates where mid-day sun may scorch leaves and bleach the colour from flowers. Sometimes scorching sun even causes leaves and buds to shrivel up and fail. Where trees are not available to provide natural shade, shade structures are built. Furthermore, late frosts can damage the flower buds, resulting in misshapen flowers. Even a change in humidity can cause plants to drop buds. For container plants to thrive indoors, they must be overwintered in a temperature-controlled environment such as a greenhouse where maximum winter temperatures range from 7- 12°C.

8.0 Propagation

8.1.0 Seed Propagation

Seed propagation of camellia is rather slower propagation method than the others and it takes many years to bloom. Seeds are collected from mature seedpods and are soaked prior to planting at least for 12 hours. Seeds can be germinated faster by placing them in a plastic bag and covered with sphagnum moss, which should be misted until moist. They will usually germinate within a month. Before transplanting in pots, taproots should be snipped off. The pots are to be kept in warm location with bright but indirect sunlight.

8.2.0 Cutting

Cuttings of new growth are the easiest to root. It ensures true to type plant and blooming in short times rather than seed propagation. The ideal cutting of camellia should be healthy, disease and insect proof and it should have two or three leaves and as many growth buds. Cutting should be taken soon after the tender growth matures and hardens (May - August). The cut should preferably be made around the fifth node by removing lower leaves. A tapering cut should be made at the base of the cutting. This allows more surface for roots to develop. To reduce dehydration, the leaves left on the cutting may trim back one-half. Though not all camellias root readily from cuttings, but with good facilities, bottom heat, intermittent mist, and rooting hormones almost any camellia cutting can be rooted successfully. Cuttings should be placed in rooting media about 5 cm apart and 4- 5 cm deep. Growing media should consist of good soil, peat moss, or a combination of peat moss and sand. A covering of glass or plastic over the pot will help to retain moisture. Intermittent mist is used by nurserymen to prevent drying out. Cuttings should be kept moist, but not wet at all times.

Under optimum conditions, rooting should take place in 1-1/2 to 2 months for most cultivars. Cuttings should be ready for transplanting in six to eight months. Any box, pot, or container, or bed at least 10- 15 cm deep that provides good drainage is suitable for the medium. Although not necessary, a rooting hormone may speed up rooting if applied to cuttings before inserting into the medium. Sometimes, cuttings may delay rooting due to formation of large callus, varietal difference or temperature. Leaving the cuttings with callus in the bed would root after a certain period as it is a varietal character. Temperature should be maintained about 18-23°C for faster rooting.

8.3.0 Grafting

Grafting can be carried out during any month when the plants are not in active growth stage, however, the optimum time is during July and August just prior to the new growth starting in spring. The ideal scion is from the top growth of a plant that is healthy, free of insect pests and dormant. Two leaves with one or more buds are to be taken from where the new growth will emerge. After cutting the scion with sharp secateurs place it in a plastic bag to keep it turgid, never leaving this bag in the sun as the scion will be placed under great stress. The critical factor of grafting is the matching of the cambium layers of both the scion and rootstock. Most grafts require to be secured with grafting or budding tape around the top of the root stock to prevent the scion moving or even falling out of the cleft. Cover the soil with a layer of clean sand and enclose the graft with a large glass jar or plastic bag. The completed graft must be shaded from direct sunlight. Once the graft has formed a callus and then started to grow, the glass jar is removed gradually, preferably at night and covered again. SSR profiles can improve *C. japonica* germplasm management, and provide potential determine correlations between genetic relationship and graft compatibility among scions of *C. japonica* and genotypes of *C. oleifera* (Zhao *et al.*, 2017).

Propagation of Camellia

Seeds of *Camellia japonica*

Raising of Seedlings

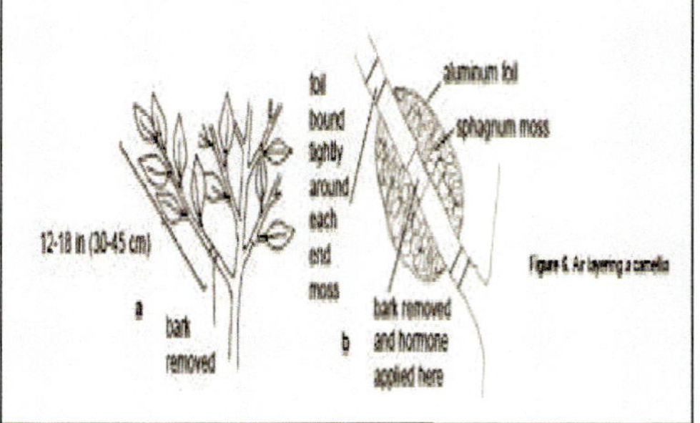

Air Layering in Camellia with Sphagnum Moss

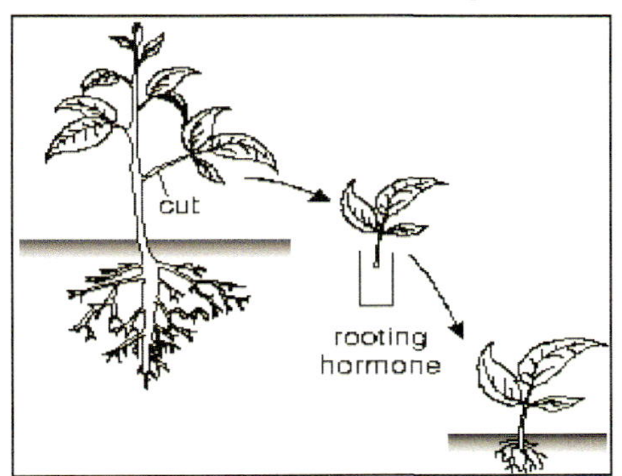

Cuttings

In vitro Culture of *C. oleifera*

Fruit of *C. japonica*

Matured Seeds

8.4.0 Air Layering

Another propagating method of camellia is air layering. It is possibly the easiest way of propagating camellias and obtaining a reasonable sized plant within six to nine months. Sphagnum moss is used to encase the cut area. The months October - December are considered to be the best time for this. At any other time root formation may be slower. At times other than October-December the bark may be more difficult to remove and scraping may be necessary.

9.0 Cultivation

9.1.0 Planting

Camellias are best planted from November to February so the roots can become established before the heat of summer. Late spring or summer planting is possible if extra care is taken. Very sandy soils should be amended by mixing 8- 15 cm of organic matter into the top 30 cm of soil. The entire planting bed, rather than individual planting holes, should be amended if possible. The planting hole should be two to three times wider and slightly shallower than the root ball. When planted, the root ball should 3-5 cm above the soil line to allow for sinking. Camellias cannot tolerate too deep planting. Add and maintain a 5-7 cm layer of mulch to insulate the root system and conserve moisture in the root zone. Avoid mulching over the root ball to allow for air exchange. Plants should be spaced according to their mature size and rate of growth, usually at least 2 m apart.

Study conducted by Berruti *et al.*, 2013 concluded that *C. japonica* is susceptible to arbuscular mycorrhizal fungus (AMF) inoculation and that a series of benefits can be achieved by inoculating pots with specific symbiotic fungi. Moreover, two newly tested materials (nutshells and rice husk) were suitable as partial peat alternatives (30 per cent v/v), and the efficacy of very low concentrations of flurprimidol to control growth and enhance flowering was underlined. Lastly, the use of cold treatment allowed the achievement of an earlier, uniform and enhanced flowering.

9.2.0 Fertilization

To harvest profuse flowering, camellia is fertilized four times in a year. However, camellias growing on a suitable site perform well with significantly less fertilizer. One or two applications a year should be adequate. Fertilizer is applied @ 12:4:12 NPK or 15:5:15 per 100 square feet of planting area in spring or early summer. Late summer fertilization may cause tender growth, which may be injured by early cold periods. Irrigation is required before and after fertilizer applications. Magnesium deficiency is the most common nutritional deficiency on camellia shrubs throughout Florida. Symptoms are most severe on the older leaves. Camellias growing on alkaline pH soils often appear chlorotic (yellow) because of deficiencies in micronutrients like iron, manganese, and zinc.

9.3.0 Irrigation

In their natural environments camellias require very little amount of water in winter when they are dormant. But in other seasons, camellia demands huge water for growth and flowering. Most flower buds form on the current year's growth from mid-summer. Camellias are fairly drought tolerant but need irrigation during extended dry periods; 2.54 cm of water should be applied every 10 to 15 days. Camellias also need 2.54 cm of water per week during flowering. They are sensitive to overwatering and succumb to root rot when watered too wet. During flowering, regular supply of water is needed. A break in water supply during these months may result in fallen flower buds and a lack of blooms later.

9.4.0 Mulching

Camellia prefers moist soil. To conserve moisture in soil for a long time, mulching is essential. Either with an organic material such as woodchips or with a plastic membrane to reduce evaporation is often used. Many enthusiasts in hot dry climates such as in California and parts of Australia, instal irrigation systems with timers set to provide water in the evening or overnight when there is low evaporation.

9.5.0 Overwintering

Despite the hibernal blooming period, camellias are sensitive to frost. Older plants can tolerate up to temperature of -8°C. Younger plants can tolerate much less. In case of indoor gardening, a warm location at the southern wall of the house and media covered with bark mulch or brushwood is practiced, when the temperatures reach above -5°C. But below that temperature, frost-protected green house or the light stairway is suitable for the overwintering of camellia. Before the beginning of the blooming period, the temperature of the environments needs to be raised to 10° to 15°C. As soon as late frost is not imminent anymore, the camellia can be resettled its location in the garden. During over wintering, the plants should not be kept under stress, water sparingly but prevent stagnant moisture.

9.6.0 Pruning

Camellias typically need minimal pruning. Necessary grooming and shaping should be done in late winter or very early spring after blooming. Pruning is done after blooming period. Selective removal of undesirable branches can be done anytime to retain a neat shape. Shearing should be avoided because it destroys the natural plant form and results in a dense outer layer of foliage that blocks light from the interior branches. Removal of dead or weak wood is followed for proper flowering. Shorten lower branches to encourage upright growth; cut back top growth to make lanky shrubs bushier.

10.0 Diseases and Pests

10.1.0 Diseases

10.1.1 Camellia Dieback and Canker

This is one of the most serious of all camellia species and is caused by the fungus *Glomerella cingulata*. It is spread by splashing water, wind or unclean pruning shears. The spores enter through wounds on the plant and can cause limbs to die. Leaves on affected branches suddenly turn yellow and wilt. Branch tips usually die. Gray

Diseases, Pests and Disorder of Camellia

Flower Blight Infection

Die Back Infection

Root Rot Infection

Algal Leaf Spot Infection

Leaf Gall Infestation

Yellow Mottle Mosaic Virus

Scale Infestation on Leaves

Scale on Leaves

Adult Stage of Scale Infestation

Aphid Infestation

Oedema Disorder

Sunscald Disorder

blotches appear on the bark and stem, and then sunken areas (cankers) develop, eventually girdling the stem. Parts of the plant above the stem canker lose vigour, wilt, and die. If infected in the lower part of the plant, entire plant death could occur. Damaged plants show more symptoms during hot, dry weather.

Maintaining micro- climate of plants as healthy as possible is of prime importance. Planting in well-drained acidic soil, avoiding wounding, and fertilizing properly may help escape of dieback. Remove diseased twigs by pruning several inches below the cankered areas. Disinfect pruning tools between all cuts, using a solution of one part household bleach to nine parts water. Fungicides, such as thiophanate-methyl or copper-based fungicides, can be applied during wet periods and normal leaf drop periods to protect fresh leaf scars from infection.

10.1.2 Camellia Flower Blight

This serious disease of camellia turns the flowers brown. Flower blight is caused by the fungus *Ciborinia camelliae* and it appears in early spring when moisture is present. Symptoms begin as small, brown, irregular spots on the flower petals. These spots quickly enlarge to cover most of the flower. The entire flower turns brown and usually drops within 24 to 48 hours of infection. Only the flowers of the plant are affected and the brown area rapidly spreads to the centre of the flower. Cold temperatures can also cause the browning of the flowers. Dark, brown veins in the petals distinguish flower blight from cold injury.

Sanitation is the best control. Pulling off and destroying all infected flowers and time to time removal of plant debris help to control spread of the disease. Replacing mulch under is to done timely as the fungus survives in the soil. Spores of the fungus can be wind-borne for up to a mile. Therefore, the best control is achieved when controls are applied to other camellia plants in the landscape. Fungicide sprays recommended for the flowers include mancozeb. Application of soil drenches, such as mancozeb or captan, around the plant every 2 weeks from late December through January may be helpful in reducing the intensity of disease.

10.1.3 Root Rot

This fungal disease is caused by *Phytophthora cinnamomi*. The initial symptoms are uniform leaf yellowing, poor growth, and wilting of the entire plant. Infected root systems lack small feeder roots and turn discoloured. Infected roots show red-brown to dark-brown colour. Death of the plant can occur rapidly, or the plant may remain in a state of decline for several years. All varieties of common Japanese camellia are susceptible, and all varieties of sasanqua camellia are resistant to this root rot.

This disease is difficult to control once plants are infected, so prevention is very much important. Cultivation in soil with good drainage is to be selected for planting or otherwise cultivation under raised bed should b followed. In areas where this disease is a problem, planting of *Camellia sasanqua* cultivars or grafting of *C. japonica* cultivars onto a sasanqua rootstock is helpful. Planting of healthy plants that show no sign of wilting or yellowing of the leaves is recommended.

10.1.4 Leaf Gall

This disease is much more common on sasanqua varieties of camellia (*Camellia sasanqua*) than on Japanese camellia (*C. japonica*). It is caused by the fungus *Exobasidium camelliae*. Leaf galls are most often observed during the spring flush of growth. New shoots and leaves become enlarged, thickened and fleshy, and appear as abnormal. The colour of the affected areas turns from light green to nearly white or pink. Later the galls rupture on the undersides of the leaves, revealing a whitish mass of spores. The galls eventually harden and become brown. Plants are seldom severely damaged.

Remove and destroy young galls before the lower leaf surfaces turn white and spores are released, or the disease will be worse the next season. Rake up and remove fallen leaves. Avoid wetting the leaves when irrigating as humid, moist, shady conditions favour gall formation. Chemical controls, such as mancozeb, are limited in effectiveness and must be applied before infection occurs. Start sprays at bud break and continue through the first of June at 7- to 14-day intervals.

10.1.5 Viruses

Camellia yellow mottle virus is transmitted by root grafts and propagation of diseased stock. This virus causes irregular, yellow, mottled, or splotchy patterns of various sizes and shapes on the leaves. Some leaves may turn entirely yellow. Irregular white blotches will appear on infected flowers. Planting of virus free plants should be followed.

10.1.6 Algal Leaf Spot

The parasitic alga *Cephaleuros virescens* is the most common causal agent of algal leaf spot on camellia. Algal leaf spots may be circular or blotchy in shape, and they are slightly raised from the plant surface. The edges of the spots may be wavy or feathered. These spots may vary in colour from a crusty gray-green to greenish-brown. However, in summer, when the alga is reproducing, the spots take on a velvety, red-brown appearance due to the production of reddish, spore-producing structures. If colonies are numerous, premature yellowing and loss of leaves can occur.

10.2.0 Pests

10.2.1 Scales

The most common insect pests of camellia are scales (*Fiorinia theae*). Scale insects feed on plants by piercing plant tissue and sucking sap. Because of their unusual appearance, populations can reach damaging levels before they are noticed. On camellia, scales usually attach to underside of leaves but some species also attach to stems. They feed and weaken the plant. Symptoms include yellowing of the upper leaf surface, fewer and smaller blossoms, leaf drop, twig die- back, and sometimes plant death. Heavy infestation causes the undersides of the leaves covered by a cottony mass.

With a light infestation, scales can be scraped off the plant and discarded. If only a few leaves are infested, handpicking and destruction of infested leaves is very effective. Spraying with horticultural oil at 2 per cent thoroughly can manage it. Soil application of dinotefuran or acephate or dimethoate systemic insecticide can be used once in the spring around the base of the infested plant to control both soft and armored scales.

10.2.2 Aphid

Aphids are small green ant-like insects that are usually visible in the spring when the new growth appears. The most common symptom of aphids is the black sooty mould which can be found covering the top of leaves. Aphids also attract ants that feed on their secretions.

10.3.0 Other Disorders

10.3.1 Bud Drop

Camellia flower buds may drop from the plant before opening, or the tips of the young buds may turn brown. One of the most common causes of bud drop is large fluctuations in temperature or moisture. Camellias perform best planted in areas with uniform moisture. Hot weather during the fall or spring as well as freezing temperature may encourage shoot growth and cause bud drop. Avoid planting varieties that bloom late in the spring. Planting in shady, cooler area may help to prevent this problem. Other plant stresses from a lack of nutrients or poor drainage can cause flower buds to drop. Camellia bud mites cause buds to develop slowly and either open late or fall off before opening. Camellias that drop their buds year after year may have a varietal problem, or it may be a problem of location that can be solved by transplanting.

10.3.2 Sunscald

Camellias planted in full sun or against a south- or west-facing direction often cause sunscald. Leaves will develop scorched or bronzed areas on the side of the plant directly exposed to the sun. Sunscald is a particular problem on camellias transplanted from shaded to sunny locations. planting in shady location or providing more shade can escspe the problem. Once the leaves have turned brown, they will not recover.

10.3.3 Oedema

Oedema (sometimes spelled edema) is a physiological disorder of camellia leaves due to excessive water uptake by the roots and a reduced ability of the foliage to transpire (or give off) this buildup of water within the leaves. The symptoms of oedema will occur primarily on the lower leaf surfaces, and at first, appear as small, water-soaked, greenish-white raised areas. Eventually, as the water pressure builds up in the lower leaf tissue, the blisters will erupt into rust-brown or yellow-brown, corky, wart-like layers of dead ruptured cells that are most characteristic of this disorder. Oedema typically occurs in late winter or early spring following wet, cool weather. With the cool temperatures, extended cloudy weather, and higher relative humidity, camellia plants will take up much more water than they can transpire. It is important to recognize that this condition, although unattractive on the foliage, does not significantly harm the health of the plant, and no spray control measures are required or effective.

Control measures include improved air movement around the plants, and an increased level of sunlight by pruning back adjacent plants and over-hanging tree limbs. Regular monitoring of soil moisture level is necessary in case of irrigation. Planting in well drained soil along with maintaining soil fertility help to reduce the problem.

10.3.4 Magnesium Deficiency

Magnesium deficiency may occur on older camellia foliage, especially if the soil contains excessive calcium or potassium. The symptom is interveinal or marginal chlorosis; leaf blades between the veins become yellow, or they may become yellow along the edges of the leaves. If the deficiency is severe enough, it can cause leaf drop. However, do not confuse this leaf drop with normal older leaf drop in late spring, which occurs on all evergreen shrubs after the new foliage emerges. Fertilize the camellias correctly during spring based upon a soil test is recommended. If magnesium deficiency is determined, irrigating camellias with Epsom salt solution of 15 ml in 5 l of water should be followed.

11.0 Harvesting and Postharvest Management

Camellia flowers attached on the plants last for seven-ten days, and flower senescence is characterized by the appearance of brown regions on the petals, followed by abscission of floral organs. In contrast, when camellia flowers are removed from the plant, they typically wilt rapidly unless held at low temperatures and high humidity (Bonner and Honda, 1950).

Woolf *et al.* (1992) reported that abscission of camellia flower buds was induced by a foliar spray with 2-4 ml 1-l ethephon [(2-chloroethyl) phosphonic acid] and that the vegetative buds and leaves were less sensitive to ethephon than flower buds. This suggests the involvement of endogenous ethylene in naturally occurring camellia flower abscission and the possibility of silver thiosulfate (STS) being effective in preventing flower abscission.

Caser *et al.*, 2015 investigated the role of ethylene in postharvest flower and leaf abscission of seven cultivars of *Camellia japonica* and evaluated the effect of inhibitors of ethylene action on their vase life. Plant materials were treated with commercial Chrysal, gaseous 1-MCP, and 1-MCP included in β-cyclodextrin-based nanosponges and compared to cut branches exposed to exogenous ethylene (1 ± 0.2 µl l-1) for 24 h and control in tap water. Visual checks for symptoms of variation in senescence, flower opening stage, flower abscission, leaf chlorophyll content (SPAD) and gloss, in addition to ethylene production levels were monitored daily. Cut branches placed in tap water lost ornamental value after ca 5.5 days of vase life and reached complete senescence after about 10 days in all the studied cultivars, regardless of supply chain, with the exception of 'Il Tramonto' with 10 and 20 days, respectively. Exogenous ethylene application reduced camellia cut branch longevity in 'Debbie' and 'R.H. Wheeler', while the other cultivars lasted as long as the control. Anti-ethylene treatments prolonged the vase life only in 'Korun Koku' and 'Il Tramonto'.

Doi and Reid, 1995 investigated abscission of flowers along with occurrence of necrotic brown spots. Open flowers of cut *Camellia japonica*. 'Kumasaka' placed in deionized water at 20°C and 60 per cent RH wilted after only two days; unopened flower buds lasted longer under these conditions, but failed to open. When open flowers were placed at high humidity, the vase life was four days, and was terminated by petal abscission. Petal wilting resulted from the failure of the cut stem to replace water lost mainly through the flower petals, and was associated with a fall in water potential, reduced fresh weight, and reduced xylem conductivity. Flower abscission was preceded by petal browning and necrosis and these symptoms occurred concomitant with the development of a sharp rise in ethylene production. Treatment of cut flowers with 0.2 mM silver thiosulfate for 24 h prevented flower abscission, but did not delay the occurrence of necrotic brown spots.

12.0 Biotechnology

12.1.0 Micropropagation

A protocol has been developed by (Hwang, 2007) for in vitro plant regeneration from seed cotyledon of *Camellia japonica* L. Adventitious buds of *Camellia japonica* were efficiently formed without callusing on the proximal region of mature seed cotyledons cultured on Murashige and Skoog (MS) medium supplemented with 0.1 mg/l 2, 4- D and 1 mg/l BA. A prolifering shoot culture was established by repeatedly sub- culturing the original cotyledon explant on fresh medium after each harvest of the newly formed shoots. Nearly 38 per cent of the shoots formed roots after being transferred to MS medium without plant growth regulators. These results suggest that cotyledon of mature seeds can be used as a source of explants for the experiments of clonal proliferation and genetic transformation in *Camellia* species.

12.2.0 Genetic Transformation

Double flower, characterized by excessive development of petals, is one of the most important traits of ornamental flowering species. Human selection over aesthetic traits is thought to play pivotal roles in the existence of vast variety of cultivated double flowers (Abbo *et al.*, 2014; Ross-Ibarra *et al.*, 2007). Recently the domestications of double flowers in some ornamental species have been recognized. In most cases, the double-flower varieties were derived from their wild ancestors bearing the single-flower (Scovel *et al.*, 1998; Liu *et al.*, 2013). Double flower domestication is of great value in ornamental plants and presents an excellent system to study the mechanism of morphological alterations by human selection (Sun *et al.*, 2014). The classic ABC model provides a genetic framework underlying the control of floral organ identity and organogenesis from which key regulators have been identified and evaluated in many plant species. Recent molecular studies have underscored the importance of C-class homeotic genes, whose functional attenuation contributed to the floral diversity in various species. Cultivated *Camellia japonica* L. possesses several types of double flowers, however the molecular mechanism underlying their floral morphological diversification

Micropropagation

Floral Visitors of ***Camellia pubipetala***
(a) Forked tailed sunbird (*Aethopyga christinae*); (b) Honeybee (*Apis cerana*).

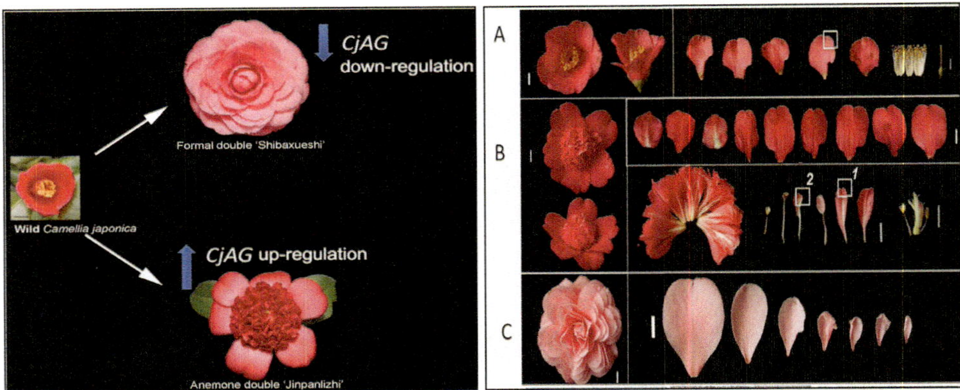

Double Flower of *C. japonica* through Biotechnology (Sun *et al.,* 2014).

Comparison of Floral Patterns in Wild and Cultivated Camellias (Sun *et al.,* 2014).

A. Wild *C. japonica* was singe-flower with canonical floral structures; B. double-flower cultivar 'Jinpanlizhi' displayed distinctive shapes between outer and inner petals; C. Double-flower cultivar 'Shibaxueshi' was a typical formal double type with gradient petals from outer layer to the inside.

remains unclear. In this study, Sun *et al.*, 2014 cloned the C-class orthologous gene *CjAG* in *C. japonica* and analyzed the expression patterns of *CjAG* in wild *C. japonica*, and performed ectopic expression in *Arabidopsis*. These results revealed that *CjAG* shared conserved C-class function that controls stamen and carpel development. Further the expression pattern of *CjAG* in two different *C. japonica* double-flower varieties, 'Shibaxueshi' and 'Jinpanlizhi' was analyzed, and showed that expression of *CjAG* was highly contracted in 'Shibaxueshi' but expanded in inner petals of 'Jinpanlizhi'. Moreover, detailed expression analyses of B- and C-class genes have uncovered differential patterns of B-class genes in the inner organs of 'Jinpanlizhi'. These results demonstrated that the contraction and expansion of *CjAG* expression were associated with the formation of different types of double flowers. This study has manifested two different trajectories of double flower domestication regarding the C-class gene expression in *C. japonica*.

The APETALA1/FRUITFUL (AP1/FUL) family genes encode MADS-box transcription factors, which are broadly involved in many aspects of floral development in higher plants. Gene duplication in the core eudicots has produced the euAP1 and euFUL clades. In this study by Sun *et al.*, 2014b isolated two homologs of the AP1/FUL family, named as CjAPL1 and CjAPL2, from *C. japonica*. Sequence and phylogenic analyses revealed that they were orthologs of FUL and AP1, respectively. Studies showed by gene expression profiling and ectopic expression in Arabidopsis that CjAPL1 and CjAPL2 potentially played different roles during floral development. Overexpression of CjAPL1/2 displayed similar phenotypes in Arabidopsis including early flowering, formation of terminal flowers, and increase in stamen and pistil numbers, but only in plants overexpressing CjAPL2 was the petal number increased. This therefore indicates that duplication of AP1- and FUL- like genes was functionally divergent in Ericales. Furthermore, higher expression levels of CjAPL1/2 were identified in four different double-flower varieties compared to wild single-flower camellias. This result provides evidence for functional diversification of AP1-like and FUL-like genes in core eudicot species and point to their roles in double flower domestication.

Chloroplast (cp) genome sequences have been used for cpDNA marker development and genetic diversity evaluation. Research conducted by Li *et al.*, 2019 that newly sequenced the chloroplast genome of *Camellia japonica* using Illumina HiSeq X Ten platform, and retrieved five other chloroplast genomes of *Camellia* previously published for comparative analyses, thereby shedding lights on a deeper understanding of the applicability of chloroplast information. The chloroplast genome sizes ranged in length from 156,607 to 157,166 bp, and their gene structure resembled those of other higher plants. There were four categories of SSRs detected in six *Camellia* cpDNA sequences, with the lengths ranging from 10 to 17bp. The *Camellia* species exhibited different evolutionary routes that lhbA and orf188, followed by orf42 and psbZ, were readily lost during evolution. Obvious codon preferences were also shown in almost all protein-coding cpDNA and amino acid sequences. Selection pressure analysis revealed the influence of different environmental pressures on different *Camellia* chloroplast genomes during long-term evolution. All *Camellia* species, except *C. crapnelliana*, presented the identical rate of amplification in the IR region. The datasets obtained from the chloroplast genomes are highly supportive in inferring the phylogenetic relationships of the *Camellia* taxa, indicating that chloroplast genome can be used for classifying interspecific relationships in this genus.

13.0 References

Abbo, S., Pinhasi, Van-Oss R., Gopher, A., Saranga, Y., Ofner, I., Peleg, Z. (2014) *Trends Plant Sci.*, **19**(6) : 351-360.

Abe, H., Ueno, S., Takahashi, T., Tsumura, Y. and Hasegawa, M. (2013) *PLoS ONE* **8**(4) : e62696.

Berruti, A., Scariot, V., Borriello, R., Bianciotto, V., Della Beffa, M.T., Christiaens, A., Van Labeke, M.C. and De Keyser, E. (2013) *Acta Hortic.*, **990** : 159-166.

Bonner, J. and Honda, S. (1950) *Camellia Res.*, **45**.

Caser, M., Seglie, L., Bizioli, R. and Scariot, V. (2015) *Adv Hortic Sci.*, **29**(2-3): 116-120.

Chai, S.F.,Chen, Z-Y., Tang, J-M., Wang, M.L., Zou, R. and Wei, X. (2019) *Plant Species Biol.*, **34** : 141-151.

Coats (1964, 1992) *The New York Botanical Garden, Curtis' Botanical Magazine*, Volume X Bronx, New York: The New York Botanical Garden, 1797.

Cothran, J.R. (2003) *Gardens and historic plants of the antebellum South. South Carolina:* University of South Carolina Press, 166-167.

de Nettancourt, D. (1997) *Sex Plant Reprod.*, **10** : 185-199.

Faegri, K. and Van der Pijl, L. (1971) *The Principles of Pollination Ecology*, 2nd edn. Pergamon Press, Oxford.

Feás, X., Estevinho, L.M., Salinero, C., Vela, P., Sainz, M.J., Vázquez-Tato, M.P. and Seijas, J.A. (2013) *Molecules.*, 18 : 4573- 4587.

Gan, X.H., Cao, L.L. and Zhang, X. (2013) *Bot. Stud.*, **54** : 50.

Hwang, E.J., Cha, Y.J., Park, M.H., Lee, J.W. and Lee, S.Y. (2004) *J. Korean Soc. Food Sci. Nutr.*, 33 : 487 493.

Hwang, S.J. (2007) *Acta Hortic.* **755** : 347-350.

Jiang, C.J. and Wang, Z.H. (1987) *J Tea Sci.*, **7** : 23-28.

Le, N.N.H., Uematsu, C., Katayama, H., Nguyen, L.T., Tran, N., Luong, D.V. and Hoang, S.T. (2017) *Korean J. Pl. Taxon.*, **47**(2) : 95-99.

Li, X-L., Yin, H-F., Fan, Z-Q. and Li, J-Y. (2019) *Sci. Agric. Sin.*, **52**(11) : 1961-1969.

Li, W., Zhang, C., Guo, X., Liu, Q., Wang, K. (2019) *PLoS One.*, **14**(5) : e0216645.

Liu, Z.X., Zhang, D.D., Liu, D., Li, F.L., Lu, H. (2013) *Plant Cell Rep.*, **32**(2) : 227-237.

Majumder, S., Ghosh, A., Bhattacharya, M. (2020) Bulletin of the National Research Centre. ***44*** *(1) : 141.*

Ming, T.L. (2000) Yunnan Science and Technology Press, Kunming, 352 pp.

Motoaki, D.I., Michael, S.R. (1996) *Postharvest Biol. Technol.*, 331-340.

Murray, J.A. (1784) Typis et impensis Jo. Christ. Dieterich, Gottingae, 987 pp.

Nguyen, L.T., Tran, N., Chiyomi, U., Hironori, K., Luong, D.V., Hoang, S.T., Nguyen, K.D., Nguyen, H.V. and Thai, T.C. (2018) *Korean J. Pl. Taxon.* **48**(2) : 115-122.

Oh, G.S., Kim, J.H., Kang, S.S., Yeeh, Y. and Chung, M.G. (1995) *Plant Species Biol.*, **10** : 155-161.

Orel, G., Wilson, P.G. (2011) *Kew Bull.*, **66** : 565-569.

Qian, J., Zhao, X., Zhao, C., Yang, H., Gou, L., Wang, W., Guo, H. (2020) *Eur J Lipid Sci Technol.*, **123** (1).

Ross-Ibarra, J., Morrell, P.L., Gaut, B.S. (2007) *Proc Natl Acad Sci U S A.*, **104**(Suppl 1) : 8641-8648.

Sage, T.L., Price, M.V., Waser, N.M. (2006) *Am J Bot.*, **93** : 254-262.

Savige, T.J. (2005) *Camellia News., No.168.*

Schemske, D.W., Husband, B.C. and Ruckelshaus, M.H. (1994) *Ecology*, **75** : 584-606.

Scovel, G., Ben-Meir, H., Ovadis, M., Itzhaki, H., Vainstein, A. (1998) *Theor Appl Genet.*, **96**(1) : 117-122.

Sebestian, K. (2015) *Centaurus*, **57** : 246-255.

Sui, N., Ji-yuan, L., and Zhong-yi, H. (2014) *Three Decades of Breeding Golden Camellia Varieties in Nanning Golden Camellia.*

Sun, Y., Fan, Z., Li, X., Li, J. and Yin, H. (2014) *Mol. Breed.*, **33** : 821-834.

Sun, Y., Fan, Z., Li, X., Liu, Z., Li, J. and Yin, H. (2014) *BMC Plant Biology.*, **14** : 288.

Tanaka, H. (1980) *Saisyu to Siiku* **42** : 382-383. (In Japanese).

Thao, N.T., Hung, T.M., Lee, M.K., Kim, J.C., Min, B.S. and Bae, K (2010) *Chem Pharm Bull* (Tokyo) 58 : 121 124.

Ueno, S., Tomaru, N., Yoshimaru, H., Manabe, T. and Yamamoto, S. (2000) *Mol. Ecol.*, **9** : 647-656.

Valtueña, F.J., Rodríguez-Riaño, T., Espinosa, F. and Ortega-Olivencia, A. (2010) *Am. J. Bot.*, **97**: 123-135.

Wang, J-H., Cai, Y-F., Li, S-F and Zhang, S-B. (2020) *Plant Divers.*, **42** (3) : 181-188.

Wang, Z.L., Wang, Y.N., Short, P., Yamaguchi, S., Shen, Y.G., Bergamini, D., Salinero, C., Corneo, A., Chen, L., Li, J.Y., Beard, D.,Powell, J. and Tsai, T.Y. (2019) *Database of International Camellia Register.*

Weller, S.G. (1994) *The relationship of rarity to plant reproductive biology*. Cambridge: Cambridge University Press.

Wendel, J.F. and Parks, C.R. (1985) *Am. J. Bot.*, **72** : 52-65.

Wheeler, L., Su, M. and Rivers, M.C. (2015) *Camellia japonica*.The IUCN Red List of Threatened Species.

Woolf, A.B., Clemens, J. and Plummer, J.A. (1992) *Hort Science.*, **27** : 32-34.

Wyatt, R. (1983) *Pollinator-plant interactions and the evolution of breeding systems*. In L. Real (Ed.), Pollination biology (pp. 51–95).Orlando: Academy Press.

Xin, T., de Riek, J., Guo, H., Jarvis, D., Ma, L. and Long, C. (2015) *J Ethnobiology Ethnomedicine.*, **11**: 74.

Yeeh, Y., Kang, S.S. and Chung, M.G. (1996) *Bot. Bull. Acad. Sinica.*, **37** : 141-146.

Yoshikawa, M., Murakami, T., Yoshizumi, S., Murakami, N., Yamahara, J. and Matsuda, H. (1996) *Chem Pharm Bull* (Tokyo) **44** : 1899 1907.

Yumoto, T. (1988) *Ecol. Res.*, **3** : 117-129.

Zhao, Y., Ruan, C., Ding, G.J. and Mopper, S. (2017) *Genet. Mol. Res.*, **16**(1).

Azalea (Rhododendron)

K. Dutta

1.0 Introduction

Azalea is a beautiful flowering shrub in the genus *Rhododendron*, generally blooms in the spring (May and June in the temperate regions). Its flowers resemble rose flowers which often lasting several weeks after flowering. It is s shade tolerant shrub, prefers living near or under trees. It is distinguished from 'true rhododendrons' by having only five anthers per flower and leaves are much smaller and pointed than leaves of the rhododendron which are much larger and leathery. Azalea is extremely grown in sub-tropical and warm temperate climate, has many cultivars varying in shape and size of the plants, colour of flowers *i.e.* single, double, semi-double, hardy and easy to grown in subtropical and temperate regions of the world.

2.0 Importance and Uses

2.1.0 Aesthetic Uses

Azalea is grown in shrubbery border, as hedge, very popular pot plants and also grown in hanging basket, shows variation in flower colour and form, planted under a deciduous tree, in sub-tropical and temperate regions of different parts of the world. The flowers are well suited for interior decoration in vases and floral arrangements.

2.2.0 Other Uses

The rhododendron is the national flower of Nepal, where the flower is considered edible and enjoyed for its sour taste. The pickled flower can last for months and the flower juice is also marketed (Regmi and Prasad, 1982). The juice of rhododendron flower is used to make squash called burans (named after the flower) in the hilly regions of Uttarakhand. It is admired for its distinctive flavor and colour. Azalea flowers have been used to make popular fruit and flower wines in Himachal Pradesh, India. The industry is promoted by the state government with tax benefits, looking to promote this industry as a full- fledged subclass of its economy (Zwitch, 2010). The herbal tea called Labrador tea (not a true tea) is made from one of three closely related species, *Rhododendron tomentosum* (Northern Labrador tea), *Rhododendron groenlandicum* (Bog Labrador tea), *Rhododendron neoglandulosum* (Western Labrador tea).

3.0 Origin and History

Azalea is native to eastern Asia (evergreen species) or North America (deciduous species). They are planted abundantly as ornamentals in the southeastern US, southern Asia, and parts of southwest Europe. According to azalea historian Fred Galle, in the United States, *Azalea indica* was first introduced to the outdoor landscape in the 1830s at the rice plantation, Magnolia- on- the- Ashley in Charleston, South Carolina. In Philadelphia, they were generally grown only in greenhouses. John Grimke Drayton (Magnolia's owner) imported the plants for use in his estate garden. In Chinese culture, the azalea is known as 'thinking of home bush' and is

Uses of Azalea

In Bed

Along a Pathway

In Pot

As Shrubbery Border

In Flower Vase

In Interior Decoration

In Flower Arragnements

immortalized in the poetry of Du Fu. The azalea is also one of the symbols of the city of Sao Paulo, Brazil.

4.0 Morphology

Rhododendron is a genus of shrubs and small to (rarely) large trees, the smallest species growing to 10-100 cm tall, and the largest, *R. protistum* var. *giganteum*, reported to 30 m tall. The leaves are spirally arranged; leaf size can range from 1- 2 cm to over 50 cm, exceptionally 100 cm in *R. sinogrande*. They may be either evergreen or deciduous. In some species, the undersides of the leaves are covered with scales (lepidote) or hairs (indumentum). They have frequently been divided based on the presence or absence of scales on the abaxial (lower) leaf surface (lepidote or elepidote). These scales, unique to subgenus *Rhododendron*, are modified hairs consisting of a polygonal scale attached by a stalk. Rhododendron are characterized by having inflorescences with scarious (dry) perulae, a chromosome number of x= 13, fruit is a septicidal capsule, an ovary that is superior (or nearly so), stamens that have no appendages, and agglutinate (clumped) pollen (Brown *et al.*, 2006).

5.0 Taxonomy

According to Linnaeus, six species of Azalea were *Azalea indica, A. pontica, A. lutea, A. viscosa, A. lapponica and A. procumbens* which he distinguished from *Rhododendron* by having five stamens. Other botanists such as Salisbury (1796) and Tate (1831) began to question the distinction between Azalea and Rhododendron, and finally in 1836, Azalea was incorporated into Rhododendron. In horticultural point of view the rhododendrons may be divided into the following groups, Evergreen rhododendrons- large group of evergreen shrubs that vary greatly in size. Most rhododendron flowers are bell-shaped and have 10 stamens; Vireya (Malesian) rhododendrons- epiphytic tender shrubs; Azaleas-group of shrubs which have smaller and thinner leaves than evergreen rhododendrons. They are generally medium-sized shrubs with smaller funnel- shaped flowers that usually have 5 stamens.

Following the cladistic analysis of Goetsch *et al.* (2005), this simplified classification was based on the discovery of three major clades (A, B, C) as follows.

Clade A

Subgenus *Rhododendron* L.: Small leaf or lepidotes (scales on the underside of the leaves). 3 sections, about 400 species, type species: *Rhododendron ferrugineum*.

Subgenus *Choniastrum* Franch. : 11 species

Clade B

Subgenus *Hymenanthes* (Blume) K.Koch: Large leaf or elepidotes (without scales), including deciduous azaleas. 2 sections, about 140–225 species, type *Rhododendron degronianum*.

Clade C

Subgenus *Azaleastrum* Planch.: Evergreen azaleas. 3 sections, about 120 species, type *Rhododendron ovatum*.

Sister Taxon

Subgenus *Therorhodion* (Maxim.) A. Gray: 2 species (*Rhododendron camtschaticum, Rhododendron redowskianun*).

According to classification given by Linnaeus, C. (1753) and Brickell (2008) azaleas are grouped into following.

Deciduous Hybrid Azaleas

Exbury hybrids – derived from the Knap Hill hybrids

Ghent (Gandavense) hybrids – Belgian raised

Knap Hill hybrids – developed by Anthony Waterer at the Knap Hill Nursery in England

Mollis hybrids – raised by Dutch and Belgian

New Zealand Ilam hybrids – derived from Knap Hill/Exbury hybrids

Occidentale hybrids – English raised

Rustica Flore Pleno hybrids – sweet-scented, double-flowered

Evergreen Hybrid Azaleas

Gable hybrids – raised by Joseph B. Gable in Pennsylvania.

Glenn Dale hybrids – US raised complex hybrids

Indian (Indica) hybrids – mostly of Belgian origin

Kaempferi hybrids – Dutch raised

Kurume hybrids – Japanese raised

Kyushu hybrids – very hardy Japanese azaleas (- 30 °C)

Oldhamii hybrids – dwarf hybrids raised at Exbury, England

Satsuki hybrids – Japanese raised, originally for bonsai

Shammarello hybrids – raised in northern Ohio

Vuyk (Vuykiana) hybrids – raised in the Netherlands

Azaleodendrons – semi-evergreen hybrids between deciduous azaleas and rhododendrons

6.0 Species and Cultivars

6.1.0 Species

The genus *Rhododendron* contains 800 to 1,100 species, vast majority of which are native to the eastern Himalaya and southeast Tibet, along with the islands of Java, Sumatra, Borneo, New Guinea, and the Philippines. The rest are broadly spread throughout the northern hemisphere in relatively small isolated populations, including Japan, northwestern North America, the Appalachian, and the Caucasus Mountains (Irving and Hebda, 1993).

Rhododendron has eight traditionally accepted subgenera based on morphology, till the consensus taxonomy used by Goetsch *et al.* (2005). *Azaleastrum; Candidastrum; Hymenanthes; Mumeazalea; Pentanthera; Rhododendron; Therorhodion; Tsutsusi. Hymenanthes*, with approximately 225 species, and subgenus *Rhododendron*, with approx. 400 species, comprise what gardeners typically describe as Rhododendrons. Two subgenera are generally known to gardeners as Azaleas, and include many fewer true species: *Pentanthera*, which comprises the deciduous azaleas, and *Tsutsusi*, which includes evergreen azaleas (Irving and Hebda, 1993).The description of some species is given below (Clarence, 2004; Alan, 2015; Charles, 2005 and Kron, 1993).

R. alabamense

This low to medium size shrub grows in Alabama and western Georgia. It can be found in moist sandy places, bears white flowers with a yellow blotch. It has a lemony-spice fragrance. The leaves are dull to slightly glossy, and dark to medium green. The flowers are produced from April to early June.

R. arborescens

Smooth azalea or sweet azalea, *R. arborescens* has white flowers, sometimes with a pinkish tinge, and red stamens. This species is native to west- central Georgia. It grows from Alabama to Pennsylvania, along the top of wet mountains and near fast moving streams. It has glossy leaves and red fall foliage. Flowers have a strong, cinnamon like fragrance. It may attain a height of 3 m. The pistil and the filaments are normally red. This characteristic makes it easy to identify the species. The flowers have white tubes. This azalea is sometimes also called the smooth azalea for its new leaf growth, has no hair, making it smooth to the touch. Its seeds are granular. Plants blooms from May to August.

R. antlanticum

It is also known as coastal or dwarf azalea, has white flowers that sometimes are flushed with pink. The plant grows 3 to 5 m tall and is smaller than most of the other native species. It is native to the Coastal Plain from Pennsylvania, blooms in the months April and May.

R. austrinum

Florida azalea, has yellow to orange flowers with a reddish tube. Some plants have a lemony fragrance. This tall growing species can grow up to 15 m, found in Florida and the Georgia- Alabama Coastal Plain. The pink colour may be a hybrid between this azalea and *R. canescens*. It blooms from March to April.

R. calendulaceum

It is commonly called the 'flame azalea' for its bright yellow, orange, or red colour. It is considered to be a very attractive species that people associate with the Appalachian Mountains. It grows from 300 to 2000 m elevation ranging from the

Species of Azalea (Rhododendron)

R. atlanticum — *R. aborescens*

R. capitatum — *R. calendulaceum*

R. canescens — *R. eastmanii*

R. luteum — *R. occidentale* — *R. vaseyi*

Piedmont to the Mountains. It can grow up to 3.5m in length. The flowers are 5- 8 cm in width. The flower tubes are glandular. It blooms in March to April.

R. canescens

This species is commonly known as Piedmont Azalea, Southern Pinxter Azalea, and Wild Azalea. It grows in areas where swamps, pocosins, and savannas are present. It has white to pinkish tubular flowers with stamens two to three times longer than the petals. This tall growing plant may attain a height of 5 m. Some have flowers with a musky- sweet, honeysuckle-like fragrance. It blooms early March to May.

R. cumberlandense

It grows most commonly in the Cumberland Mountains, called the Cumberland azalea. This species can be confused with *R. calendulaceum* and *R. flammeum* due to the colour of the flowers. They grow at higher elevations, and grow up to 4 m, despite normally growing low to the ground and twiggy. The leaves are glossy, with a dull underside, and emerge later in the season. They bloom from May to July.

R. colemanii

It is commonly called the Red Hills azalea. It grows in moist areas like near streams, hammocks, and bluffs. It starts to bloom early April to mid- May. It grows in Alabama and western Georgia along the upper Coastal Plain.

R. eastmanii

It is commonly called the May- white azalea, found only in South Carolina. It is a newer species, being named by Kron and Creel (1999), commonly found near streams. This species flowers in May and has white petals, normally with a yellow blotch.

R. flammeum

It is also known as the Oconee azalea after the river in Georgia, where it was first discovered. Previously classified as *R. speciosum*, has yellow, yellow- orange or red flowers with a large yellow blotch on the top lobe (petal). This species is native to low elevations in open woods and on slopes across central Georgia. Oconee blooms in the spring.

R. luteum

The yellow azalea or honey suckle azalea, is a species, native to southeastern Europe and southwest Asia. It is a shrub growing 3- 4 m. The leaves are deciduous, 5- 10 cm long and 2- 4 cm broad. The flowers are 3- 4 cm in diameter, bright yellow, and strongly perfumed, produced in trusses of 5- 25 together. The fruit is a dry capsule 15- 25 mm long, containing numerous small seeds.

R. occidentale

It flowers in spring and summer, with a wide range of flower colours, grows in moist areas near streams, wooded areas, bogs, bottoms of canyons, thickets, and ocean bluffs. Flower colour ranges between white, pink, and salmon, the colour blotches on the petal range from yellow to orange most commonly.

R. periclymenoides

It is a shrub, the flowers range in colour from white to pink, with no blotches on the upper petal, blooms in the spring in moist wooded areas along streams, thickets, and swamps.

R. prinophyllum

A shrub with deciduous and hairy leaves. The pink flowers are not likely to have any blotches of colour on the upper petal. It grows in swampy, wet areas like bogs and streams but can be in dry areas like rocky woods, blooming in the spring.

R. prunifolium

It is also called the plum leaf azalea. The flower colour ranges from orange to red. It has smooth, large leaves that grow early in the season, blooming from July to September. Sometimes, this species expresses pink flowers, which could be a hybrid with the sweet azalea.

R. viscosum

This is known as the swamp azalea, catch- fly azalea (due to sticky and glandular hairs), and clammy azalea. Its large land area includes growth on the coastal area of Mississippi, and along the east coast from Florida to Maine. It is sometimes confused with *R. aborescens*, but can be distinguished upon comparison of leaf texture, flowers are usually white and fragrant, bloom from May to September.

6.2.0 Cultivars

Numerous cultivars of azalea (rhododendron) have been developed as a result of hybridization among the species and cultivars. Many species and numerous cultivars of *Rhododendron* constitute a major group of flowering plants in mild to warm temperate climate in all parts of the world. Flowers are white, pink, red, purple and mauve; may be of one colour but often multicoloured in stripes, blotches, spots and edging; may be single, semi-double and double. Some popular cultivars:

'Bicolor', flowers double, pink, borders of petals purple;

'Cascade', bushy shrub, bearing red flowers in large clusters;

'Charm', flowers large, red, to 7 cm across;

'Cheers', a free-flowering bushy shrub, flowers many, purple in colour, centre red;

'Christmas Cheers', flowers appear in spring and early summer;

'Coral Bells', free-flowering cultivar, medium-sized flowers in clusters, bright pink, centre red;

'Double Delight', a shrub of medium height, bearing double red flowers;

'Fashion', flowers large, 8-10 cm across, reddish purple, centre deeper in colour;

Cultivars of Azalea (Rhododendron)

'Cascade' 'Charm' 'Coral Bells'

'Bicolor' 'Double Delight' 'Cheers'

'Formosa' 'Gumpo Whitev 'Pink Pearl'

'Golden Sunset' 'Fragrant Star' 'Purple Choice'

'Formosa', flowers attractive, purple-mauve in colour;

'George L Taber', flowers large, pink, splashed purple;

'Gumpo White', shrub of medium height, flowers large, transparent white;

'Highland', flowers semi-double, pink in colour;

'Jewel', large, single, white flowers;

'Pearl', flowers large, light pink;

'Pink Glory', large flowers, bright in colour;

'Pink Prince', attractive double flowers, pink, bordering white;

'Pink Rufle', flowers semi-double, pink, centre red, very attractive cultivar;

'Pink Shine', flowers single, pink, border of petals rose;

'Pink Show', flowers double, pink, very attractive;

'Purple Choice', flowers large, red purple, centre deep red;

'Red Face', evergreen shrub, free-flowering, bears red flowers, 5-6 cm across;

'Red Glory', very attractive shrub with purple-red flowers in clusters, blooming for several months;

'Red Ruffle', large semi-double, red flowers;

'Star', flowers pink with red centre;

'Sunset', flowers orange-red in clusters;

7.0 Genetics and Breeding

7.1.0 Genetics

Rhododendrons of subgenus *Hymenanthes* is a diploid species (2n = 2x = 26) (Jones *et al.*, 2007). The basic chromosome number of the genus *Rhododendron* is 13, most of the species and hybrids are diploids, 2n= 26. Tetraploids and hexaploids, but also octoploids and dodecaploids (2n=12x=156), are found in different species (Ammal *et al.*, 1950). There is continuous gene flow between species in nature, as they merge into each other and have subspecies and varieties. Leach, 1961 reported that since species within sub- genera cross readily with each other, introgression and polyploidization are involved in the evolution of new species.

7.2.0 Breeding

7.2.1 Hybridization

Reciprocal crosses between nine evergreen azalea species and *R. uwaense* were performed to clarify the pre- and post-fertilization barriers in this interspecific hybridization for the purpose of obtaining fragrant evergreen azaleas by Okamoto and Ureshino (2015). Unilateral incompatibility appeared in this hybridization. When evergreen azalea species were used as a seed parent, many pollen tubes stopped elongating in a style and no seed could be obtained. The reverse crosses exhibited inhibition of pollen tube penetration into ovules, chlorophyll deficiency in cotyledons, and death of young seedlings in some cross combinations. RAPD analysis revealed that 10 of the putative hybrid seedlings examined (two vigorous plants selected from respective crosses of *R. uwaense* × five evergreen azalea species) possessed specific bands derived from both parents. The plastid DNA (ptDNA) of these seedlings was inherited maternally except for seedlings of *R. uwaense* × *R. yedoense* var. *poukhanense,* and the ptDNA of hybrid ones obtained from *R. uwaense* × *R. yedoense* var. *poukhanense* was inherited paternally.

7.2.2 Cross-compatibility

The mechanism of the cross incompatibility between evergreen azalea species and yellow flowered deciduous azalea, *R. japonicum, i.e.* mechanism of unilateral cross incompatibility and albino progenies was given by Ureshino, (2008). Two strategies were useful for obtaining green progenies efficiently and many green progenies have been obtained. Among them, the progenies from three-way cross entered anthesis and set pale-yellow flowers with carotenoids. These progenies were semi-deciduous and highly sterile. To improve the semi-deciduous traits, chromosome doubling of the progenies was conducted, and the derived plants were back-crossed to evergreen azaleas.

7.2.3 Polyploidy

Polyploidy induction in *R. fortunei* with colchicine treatment was obtained when tissue-cultured plantlets treated with 0.1 per cent colchicine for 24 h showed the highest polyploid induction with a rate of 36.67 per cent. By flow cytometric analysis, 69 tetraploids and 29 octoploids were identified in the regenerated plants that were examined. Phenotypic analysis indicated that the leaves of tetraploid and octoploid plants were smaller, rounder and thicker with more abundant and longer epidermal hairs than those of diploids. Furthermore, the stomata of polyploids were larger and sparser than those of diploids. An increase in chlorophyll content was also detected in polyploids, which resulted in darker green leaves (Mo *et al.*, 2020).

7.2.4 Mutation

Six mutant plantlets from 'Alfred' and seven mutant plantlets from 'Paars' variety of azalea were selected from M1V2 generations. These mutants were selected from plantlets which were irradiated with doses 5 and 10 Gy. The dendrogram shows one big cluster and the others to be distinguished. Genetic variability induced with gamma ray and RAPD methods were used to detect mutations at the rhododendron shoot culture. Rhododendron mutants were district from controls (Atak *et al.*, 2011).

8.0 Growth and Flowering

8.1.0 Environment

Investigation was carried out by Meijon *et al.* (2011) for the promotion of flowering in azaleas by the manipulation of environmental factors, using DNA methylation levels as a marker of floral bud development and showed that the change of long-day (LD) to short- day (SD) photoperiod is the primary factor responsible for floral induction in azaleas, whereas the existence of the previous cold period as well as the physiological memory are factors which improve flower production. Furthermore, for blooming 1300 units of growing degree days under an LD were necessary. The demethylation observed after the change from LD to SD is linked to a change in cell fate which is necessary for floral transition to take place and seems to be associated with the floral signal.

8.2.0 Growth Regulators

Ranney *et al.* (1994) reported that foliar sprays and soil drenching of 50, 100, and 200 ppm of paclobutrazol on container grown rhododendron cv. 'Roseum Elegans' decreased growth with increasing rates of paclobutrazol foliar spray, whereas number of flower buds per plant increased with soil drenching of paclobutrazol. Plants of 'Catawbiense Boursault' and 'Eskimo' or 'Cannon's Double' and 'Kilian' azalea grown in containers were treated twice with chlormequat (2000, 4000 and 8000 mg/l), trinexapac-ethyl (50, 100, 200, 400 mg/l), daminozide (2500, 5000, 7500 mg/l), proxeadione calcium (75, 150, 300 mg/l) and paclobutrazol (50, 100, 200, 400 mg/l). It was observed that shoot length of subsequent growth flush following the treatments decreased with increasing rates of the growth retardants. The number of flower buds per plant increased with increasing rates of pacloburazol, chlormequat and daminozide (Marosz and Matysiak, 2005).

9.0 Propagation

Rhododendron is commonly propagated by shoot cutting and mound layering. *In vitro* micropropagation has also been standardized.

9.1.0 Seed Propagation

Seeds of *Rhododendron catawbiense* (Catawba rhododendron) were dried for 1 month, stored at 4°C (39°F) and when germinated at 25°C (77°F) in 8/16 hours photoperiod condition upto 98 per cent germination was reported by Rowe *et al.* (1994).

9.2.0 Vegetative Propagation

Jones *et al.* (2010) optimized vegetative propagation protocols for Oconee azalea, *Rhododendron flammeum* utilizing stem cuttings or in-field mound layering. Softwood stem cuttings were treated with 10,000 ppm of the potassium salt IBA (K-IBA). Fifty percent of the stems on each mound resulted in rooted layers and approximately six layers were produced per mound when treated with 5000 ppm K-IBA in June.

9.3.0 Micropropagation

Effect of growth regulators on organogenesis *in vitro* was tested in Rhododendron cv. Azuro, BohumilKavka, Catharine van Toll, Grandiflorum, Mars, Nova Zembla, Ortrud, Ovation, Prof. Scholz, Purple Splendour, Rebe and Van Werden Poelman and reported highest shoot regeneration of all cultivars on MS medium containing 6 mg/l 2iP and highest shoot multiplication on MS medium with 8- 10 mg/l 2iP in combination with 1.0 mg/l IAA (Vejsadova, H., 2008). The *in vitro* growth of explants of *Rhododendron ponticum* took place readily in Anderson's medium containing 0.072 mg/l of BA and 0.036 mg/l of NAA and rooting was

Propagation of Rhododendron

Seed Pods Production of Seedlings Growth of Seedling

Cuttings of Choots Cutting in Plugtrays Growth of Cuttings

Micropropagation of Rhododendron

***Source*: Xiangying *et al.* (2018).**

achieved by the basal dipping of the explants in hydroalcoholic solutions of 500 mg/l IAA (Cantos *et al.*, 2007). Singh *et al.* (2016) obtained multiple shoot from shoot-tip explants of *R. griffithianum* on Anderson's medium supplemented with 5 μM 2-iP and 1.0 μM IBA after twelve weeks of culture and rooting on Anderson's liquid medium supplemented with 1.5 μM IBA. The rooted plantlets were hardened in greenhouse with 84 per cent survival rate. The influence of growth regulators was studied on shoot numbers, shoot length and node numbers after 10 weeks of culture of apical shoots on ½ Anderson medium and observed that when treated with 10 mg/l 2- ip with 0.2 mg/l TDZ increased the shoot number, shoot length and node number (Rahimia *et al.*, 2013).

10.0 Soil and Climate

Deciduous azaleas, with their colourful blossoms, are popular. Some deciduous species and hybrids are very cold hardy and some tolerate warm conditions. They usually prefer subtropical and warm temperate climates. Rhododendrons grow best in light, well- drained soils with good soil aeration and an ample supply of soil moisture during the summer. Sphagnum peat moss is effective in improving soil aeration and moisture retention. Rhododendrons and azaleas, in general, require an acid soil with pH about 5.5. Soils with pH higher than 5.5 should be acidified (Reiley, 2000).

Azaleas are shallow-rooted plants that are easily damaged by excessive soil moisture. When planted in poorly drained soils, azaleas do not receive the oxygen required for healthy growth and often develop root rot diseases. They grow best in acid (4.5 to 6.0 pH), well-drained, organic soils. Good drainage is essential for azaleas. Organic matter added to the soil improves drainage in clay soils and will increase a sandy soil's ability to retain water and nutrients. Organic matter should comprise approximately 10 to 20 per cent of the total soil volume.

11.0 Cultivation

11.1.0 Planting Time

Container-grown azaleas can be planted any time of year, provided they are watered properly. Most are planted in spring when in flower. Planting in spring, however, requires special attention to watering to ensure survival through the heat of summer.

11.2.0 Manuring and Fertilization

In fertile soils, rhododendrons and azaleas can be grown well without further fertilization. For rhododendrons planted in less fertile soils, a complete fertilizer designed for acid- loving plants may be applied in late winter or early spring. In cold climates, nitrogen fertilizer should not be applied after late June, as it may promote new lush growth susceptible to winter damage.Phosphorus in fertilizer tends to favor early production of flower buds. If soil is deficient in phosphorus and since phosphorus does not readily move through the soil, it should be incorporated into the soil at planting time if needed. Magnesium is sometimes recommended for rhododendrons. Magnesium is an essential element and its deficiency symptoms

Cultivation of Rhododendron

Planting in Pit

Planting of Seedling

Seedling after Transplanting

Flowering in Beds

Flowering in Open

Cultivation in Open

Commercial Cultivation in Greenhouse

will cause yellowish areas between the leaf veins on older leaves. Lack of iron causes much the same symptoms as a lack of magnesium, but with the younger leaves showing yellowing between the veins. Iron deficiency is frequently caused by too high a soil pH. Calcium is also essential for growth. Gypsum will not raise soil pH, while lime will; therefore, lime is not generally recommended in areas with a pH above 5.5 (Reiley, 2000).

11.3.0 Irrigation

Watering newly planted azaleas is essential. This is especially true when planting in spring. Their shallow roots will dry out quickly during periods of limited rainfall. When soil appears dry, water thoroughly up to a depth of 20-30 cm. Azaleas do not tolerate water logged or completely dry soil conditions. Avoid overhead irrigation to reduce incidence of disease (Polomski and Shaughnessy, 2013). Irrigation should be scheduled to keep the soil uniformly moist. If mulch is applied and sprinkler irrigation is used, be certain to irrigate thoroughly enough to move water through the mulch and into the soil to the depth of the plants' root systems (Carter, 2006).

11.4.0 Pruning

Normally very little pruning is needed. Light maintenance pruning at the time 'dead- heading' is done which can help to keep the plant in shape. Light to moderate pruning done at the time the plant is flowering or immediately thereafter will not affect flower bud formation in the following year. Prune azaleas with frequent pinching of the tips. This will cause a compact plant with more flowers. Prune out dead wood to keep the plant healthy. Heavy pruning should be done when the plant is past the peak bloom. Avoid fall and winter pruning because this can reduce flowering (Carter, 2006).

12.0 Diseases and Pests

There are a few diseases and pests that are troublesome on azaleas. Among the diseases *Phytophthora* root rot, leaf spots, petal blight, powdery mildew and dieback are common. Other minor diseases of azalea are leaf gall, leaf folding, *Botryosphaeria* canker and dieback *etc.*

12.1.0 Diseases

12.1.1 *Phytophthora* Root Rot

It is serious fungal diseases of azalea, occur in nursery and field grown plants. It is caused by several fungal species viz, *Phytophthora cinnamomi, P. cactorum, P. citricola*. Once these pathogens are established in the soil, it can remain indefinitely. Initial symptoms consist of slow growth, off- coloured wilting of foliage especially in mid- day. Twig die- back and dropping of leaves were observed as the diseases progress. The disease can be reduced by use of healthy pathogen-free planting materials, planting in well-drained raised beds, use of slow release nitrogen fertilizers or planting of resistance cultivars of azalea like 'Alaska', 'Chimes', 'Red

Diseases and Pests of Rhododendron

Phytophthora Root Rot

Leaf Spots

Leaf Scroch

Petal Blight

Powdery Mildew

Die-back

Lace Bug

Lace Bug Infestation

Bark Scale Infestation

Rhododendron Weevil

Weevil Infestation

Borer Infestation

Wing' *etc.* Fungicide application is not curative once plants are infected, was studied by Douglas, (2011).

12.1.2 Leaf Spots

Leaf spots of azalea are caused by various fungal species *viz., Cercospora handelii, Colletotrichum* sp., *Septoria azalea* (Douglas, 2011). The symptoms usually develop on current season foliage in mid to late summer. Spots are usually visible on both upper and lower surface. Under severe conditions some defoliation occurs.The disease can be prevented by pruning and removal of infected leaves, proper plant spacing to allow good air circulation, minimizing water on the foliage from overhead irrigation, removal of fallen leaves. Spraying of some systemic fungicide is effective with the application of *Trichoderma harzianum, Streptomyces etc.*

12.1.3 Petal Blight

The fungus affects mostly azaleas and sometimes rhododendron flower petals. Southern Indica and Kurume azaleas are readily susceptible. It starts with tiny spots giving the bloom a speckled appearance. These enlarge and the flowers rot. The infection spreads from bloom to bloom. It lives in the soil during the winter. Picking and destroying affected flowers and avoidance of watering the leaves and flowers help. If it is known that the fungus is in the ground at wintertime, new mulching will help. Fungicides with captan, maneb and other chemicals can be used to control the blight.

12.1.4 Powdery Mildew

This disease causes rhododendron and azaleas to first have white powder-like spots on young leaves, which can spread to the entire upper and lower surfaces of the leaf. Cloudy, overcast weather conditions cause favorable conditions for the disease to occur. It spreads mostly by winds. Fungicides are used to treat infected plants.

12.1.5 Dieback

Dieback is characterized by dying branches. The leaves of the branch die but remain attached until late summer. It is difficult to control. Best approaches include pruning the diseased branches, using sterile equipment and avoiding any wounding of the rest of the plant. Plants in partial shade that are well- watered are more resistive to dieback

12.2.0 Pests

Insect pests affecting azalea are azalea lace bug, weevil, bark scale, borer, leaf miner, white fly *etc.* Lace bug, *Stephanitis* spp., adults and nymphs with piercing/ sucking mouthparts; Azalea bark scale, *Eriococcus azaleae* also damages by sucking plant fluids; Borer, *Synanthedon rhododendri* is a clearwing moth borer affecting rhododendron and deciduous azalea. Treating the foliage and soil surface with alternating pesticides every three weeks can be effective. If treatment with insecticide is necessary, it should be carefully timed with foliar bud break and new leaf emergence, suggested by Mills, (2017).

Harvesting and Marketing of Rhododendron

Bud Stage

Harvesting Stage

Cut Flowers

Flowers for Sale in Market

Pot Plants in Market

Loose Flowers in Market

Display of Azalea

13.0 Harvesting and Postharvest Management

13.1.0 Harvesting of Flowers

Generally, the flowers of rhododendron bloom from March and flower remain on the plant up to May, this is the best time for harvesting of rhododendron flowers.

13.2.0 Harvesting of Seed Capsules

The seeds are harvested from capsules which ripe in September-October. Hay *et al.* (2006) treated pre-dehiscent capsules of two *Rhododendron griersonianum* trees and either immediately dried in a dry-room (15 per cent relative humidity, 15°C) or placed in a high humidity room (80 per cent relative humidity, 15°C) for 30, 60, or 90 d. They observed that placing the pre-dehiscent capsules at high humidity allowed some continuation of maturation, but the longevity of these seeds was never as good as seeds collected from the plants after the same maturation period and had declined by 90 days. They suggested that 80 per cent RH and 15°C does not mimic the natural drying rate, despite the high RH irreparable ageing commenced.

14.0 Biotechnology

14.1.0 Molecular Markers

The genetic relationships of 20 Rhododendron species in Taiwan was determined by Tsai *et al.* (2003) based on the sequence of the internal transcribed spacer (ITS) region of ribosomal DNA. Sequences of the complete ITS region including ITS1, 5.8S rDNA, and ITS2, were obtained by direct sequencing of polymerase chain reaction (PCR)-amplified fragments. According to the dendrogram, six main clusters were classified among the 20 species of the genus Rhododendron in Taiwan. *Rhododendron oldhamii, R. nakaharai, R. taiwanalpinum, R. simsii, R. lasiostylum, R. rubropilosum, R. breviperulatum, R. kanehirai, and R. noriakianum* were grouped with *R. longiperulatum* in cluster I. *R.lamprophyllum* was grouped with *R. ovatum* in cluster II. *R.pseudochrysanthum, R. morri, R. hyperythrum,* and *R. rubropunctatum* were grouped with *R. formosanum* in cluster III. In addition, *R. mariesii, R. ellipticum* and *R. kawakamii* formed three independent clusters.

15.0 References

Alan S. (2015) *Flora of the southern and mid-Atlantic states*, University of North Carolina Herbarium.

Ammal, J., Enoch, I. C. and Bridgwater, M. (1950) *The Rhododendron Yearbook*. Royal Horticultural Society, pp. 78-91.

Atak, C., Celik, O. and Acik, L. (2011) *Pak. J. Bot.*, **43** : 1173-1182.

Brickell, Christopher, ed. (2008) *RHS A-Z Encyclopedia of Garden Plants*. United Kingdom: Penguin Books. pp. 1136.

Brown, Gillian K.; Craven, Lyn A.; Udovicic, Frank; Ladiges, Pauline Y. (August 2006) "Phylogenetic relationships of Rhododendron section Vireya (Ericaceae) inferred from the ITS nrDNA region" (PDF). *Australian Systematic Botany*. **19**: 329–342.

Cantos, M., Linan, J., Jose L. Garcia, Garcia-Linan, m., Miguel, A., Dominguez & Antonio T. (2007) *Central European J. Biol.*, **2** : 297-306.

Carter, K. (2006) *Botany Plant Sciences* Dept. University California Riverside, pp-1-2.

Craven, L.A., Goetsch, L.A., Hall, B.D. and Brown, G.K. (2008) *Blumea – Biodiversity, Evolution and Biogeography of Plants*, **53** : 435–442.

Daglus, S.M. (2011) *Common problems of Rhododendron and azalea*, pp. 1-8.

Don, G. (1834) *General History of Dichlamydious Plants*. iii. London: J.G. and F. Rivington. p. 843.

Goetsch, Loretta A.; Eckert, Andrew J.; Hall, Benjamin D. J. (2005) *Systematic Botany*. **30**: 616–626.

Hay, F.; Klin, J. and Probert, R. (2006) *Scientia Horticulturae*, **111**: 80-83.

Horn, Charles N. (2005) *Castanea*.**70** : 1-12.

Irving, E. and R. Hebda (1993) *Journal of the American Rhododendron Society*. American Rhododendron Society.Retrieved 4 November 2012.

Jones, J. R., Le Bude, A.V. and Ranney, T.J. (2010) *J. Environ. Hort.*, **28** : 69-73.

Jones, J.R, Ranney, T.G., Lynch, N.P., Krebs, S.L. (2007) *J. Am. Rhododend. Soc*. **61**: 220–227.

Kron, K. A. (1993) *Edinburgh J. Botany*. **50** : 249–364.

Leach, D.G. (1961) *Rhododendrons of the world*. Charles Scribner's Sons, New York.

Linnaeus, C. (1753) *"Azalea". Species Plantarum.Tomus* I. Stockholm: Laurentii Salvii. pp. 150.

Marosz, A. and Matysiak, B. (2005) *Pomologiczna*, **18** : 96-100.

Meijon, M., Feito, I., Valledor, L., Rodriguez, R., Canal, M.J. (2011) *Physiol. Plant*, **143** : 82-92.

Mills, J. (2017) *Plant Pathology & Plant-Microbe Biology*, Cornell Univ., **24**: 1-2.

Mo, L., Chen, J., Lou, X., Xu, Q., Dong, R., Tong, Z., Huang, H. and Lin, E. (2020) *Plants*, **9**: 1-13.

Okamoto, A. and Ureshino, K. (2015) *The Hortic. J*. **84** : 355-364.

Polomski, B. and Shaughnessy, D. (2013) Clemson University.

Rahimia, S., Roohangiz N., Ghaemaghami, S.A., Kalatejari, S., Farham, B. (2013) *Procedia Engg*. **59** : 240-246.

Ranney, T.G., Bir, R.E., Joseph, L. (1994) Conner and Everett P. Whitman II *Environ. Hort*. **12** : 174-178.

Regmi and Prasad, P. (1982) *An introduction to Nepalese food plants*. Royal Nepal Academy.

Reiley, H.E. (2000) *The Fundamentals of Rhododendron and Azalea Culture*, American Rhododendron Society, Canada.

Roger V. Jean; D. Barabé (1998) *Symmetry in Plants*.World Scientific.pp. 6–.ISBN 978-981-02-2621-3.Retrieved 8 November 2012.

Rowe, D.B., Frank, A. Blazich, Stuart, L. Warren, and Ranney, T.G. (1994) *Environ. Hort.*, **12** : 155-158.

Salisbury, R.A. (1796) *ProdromusStirpium in horto ad Chapel Allerton* (in Latin). London. p. 286.

Singh, K.K., Singh, M. and Chettri, A. (2016) *J. Appl. Biol. & Biotech.* **4** : 072-075.

The Strange History of 'Mad Honey' - Modern Farmer".(2018), Pliny. *Natural History*. p. 24.90.

The World Book encyclopedia. Chicago: World Book. 2004. pp. 995.

Towe, L. Clarence (2004) *American Azaleas*. Portland: Timber Press.

Tsai, C.C., Huang, S.C., Chen, C.H., Tseng, Y.H., Huang, P.L., Tsai, S.H. and Chou, C.H. J. (2003) *Hortic.Sci. & Biotech.***78** : 234-240.

Ureshino, K. (2008) *J. Japan. Soc. Hort. Sci.*, **77** : 1-6.

Vejsadova, H. (2008) *Hort. Sci.* (Prague), **35** : 90-94.

Xiangying W., Jianjun C., Chunying Z. and Zonghua W. (2018) *Ind Crops Prod.*, **126**: 459-465.

Zwitch, (2010) The-fuga-experience.com.

Poinsettia

K. Dutta

1.0 Introduction

The poinsettia (*Euphorbia pulcherrima*) is a member of the large and diverse family *Euphorbiaceae*, originated in southern Mexico and northern Guatemala. In its native habitat this species is a winter- flowering shrub that grows over 3 m, and is a common landscape plant. The sap is milky and may produce dermatitis in susceptible individuals. The umbel- like cymes are subtended by many showy, red bracts but breeders have produced many different colours including white, pink, and purple. Poinsettia is presently the most beautiful flowering shrubs and valuable pot plant.

2.0 Importance and Uses

2.1.0 Aesthetic Uses

Poinsettia (*Euphorbia pulcherrima*) is one of the most beautiful shrubs grown throughout the world grown in tropical, sub- tropical and temperate region for its showy colourful attractive bracts. It is grown both in ground and in pot; potted plants in colourful bracts are also widely used for indoor decoration; poinsettias in pots are popular in temperate countries during Christrnas for decoration when very few flowers are available for display indoors. Several species are grown as shrubs in garden and many as succulent species in rock garden.

2.2.0 Therapeutic and Other Uses

Euphorbia pulcherrima, the succulent, *E. trigona* and *E. pekinensis* are used in traditional Chinese medicine. Several *Euphorbia* species are used as food plants by the larvae of some Lepidoptera (butterflies and moths). Ingenolmebutate, a drug used to treat actinic keratosis, is a diterpenoid, found in *E.peplus*.Euphorbias are often used as hedge plants in many parts of Africa (Trejo *et al.*, 2012).

3.0 Morphology

Euphorbia pulcherrima is a shrub or small tree, typically reaching a height of 0.6- 4 metres. The plant bears dark green dentate leaves that measure 7-16 cm in length. The coloured bracts which are normally flaming red, with cultivars being orange, pale green, cream, pink, white, or marbled are often mistaken for flower petals because of their groupings and colours, but are actually leaves (Perry, 2014). The colours of the bracts are created through photoperiodism, meaning that they require darkness (at least fourteen hours at a time for 6- 8 weeks in a row) to change colour. The plants also require abundant light during the day for the brightest colour. The flowers of the poinsettia are unassuming and do not attract pollinators. They are grouped within the cyathia (small yellow structures found in the center of each leaf bunch, or false flowers) (Seltzer, 2017).

4.0 Taxonomy

The genus *Euphorbia* is one of the largest and most complex genera of flowering plants, and several botanists have made unsuccessful attempts to subdivide the

Uses of Poinsettia

In Garden | As Border | In Pot

As Specimen in Garden | Display in Glasshouse | Poinsettias in a Church

Interior Decoration | In Flower Arrangement | Along Roadways

As a Gift | Hanging Basket | Decorating as a Tree

As a Landscape | In Exhibitions | Inside Shopping Mall

genus into numerous smaller genera. According to the phylogenetic studies, *Euphorbia* can be divided into four subgenera, each containing several sections and groups, of these, subgenus *Esula* is the most basal. The subgenera *Chamaesyce* and *Euphorbia* are probably sister taxa, but very closely related to subgenus *Rhizanthium* (Steinmann and Porter, 2002; Bruyns *et al.*, 2006).

5.0 Origin and History

Indigenous to Central America, poinsettia was described as a new species in 1834. It is particularly well known for its red and green foliage and is widely used in Christmas floral displays. It derives its common English name from Joel Roberts Poinsett, the first United States Minister to Mexico, who is credited with introducing the plant to the US in the 1820s. It was known by the common name 'poinsettia' as early as 1836, derived from Joel Roberts Poinsett, a botanist and the first US Minister to Mexico, possibly as early as 1826. Prior to poinsettia, it was known as 'Mexican flame flower' or 'painted leaf' (Taylor *et al.*, 2011). For years it has been assumed that Poinsett came across the gorgeous plant in Taxco, southern Mexico, as part of a nativity scene, and sent cuttings or plants to Charleston, South Carolina in 1828. From Charleston cuttings or plants were sent to Colonel Robert Carr, a nurseryman in Philadelphia, Pennsylvania. The poinsettia was described as a new species in 1834 by German scientist Johann Friedrich Klotzsch (Klotzsch, 1834). Klotzsch credited Carl Ludwig Willdenow with the species name "pulcherrima", and the authority is given as Willd. Ex Klotzsch. The holotype had been collected in Mexico during an 1803- 1804 expedition by Alexander von Humboldt and Aimé Bonpland (Lack, 2011).

6.0 Species and Cultivars

6.1.0 Species

The present Genus *Euphorbia* has over 2000 species (Taylor *et al.*, 2011). Many species of this genus are grown as ornamental plants including large number of succulents. Some of the species are mentioned hereunder.

Euphorbia characias

It is an upright, compact evergreen shrub growing to 1.2 m tall, valued as an ornamental plant for its ability to survive drought. It is suitable for any location, in light soil and in full sun. Garden cultivars are sold under the names 'Black Pearl', 'Thelma's Giant', 'Lambrook Gold', 'Silver Swan' and 'Tasmanian Tiger'.

E. cotinifolia

It is a broadleaf red shrub native to Mexico and South America. It reaches 3.0 to 4.6 m but can be grown as a tree reaching up to 9.1 m. Small white flowers with creamy bracts bloom at the ends of the branches in summer. The purplish stems, when broken, exude a sap that is a skin irritant, grown as an ornamental plant in gardens and in pots, due to its colourful and distinctive foliage. It prefers a site with well-drained soil and full sun.

Species of Poinsettia

Euphorbia characias | *Euphorbia fulgens* | *Euphorbia milii*

Euphorbia heterophylla | *Euphorbia leucocephala* | *Euphorbia antiquorum*

Euphorbia cotinifolia | *Euphorbia pulcherrima* | *Euphorbia dendroides*

Euphorbia rigida | *Euphorbia paralias* | *Euphorbia pekinensis*

E. fulgens

The scarlet plume, is a species of flowering plant in the spurge family Euphorbiaceae, native to Mexico. It is a deciduous shrub with long, weeping branches and narrow leaves. In winter orange-red flowers growing in the leaf axils cover the length of the branches. It is not hardy, requiring temperatures above 10 °C. It must therefore be grown under glasshouse in temperate regions.

E. heterophylla

It is called Painted Euphorbia and Desert Poinsettia. It grows between 30 and 100 cm tall and has hollow stems that may be branched or simple with angular ribs. The leaves of the plant have variable shapes within and between populations. The lower leaves are alternate whereas the upper leaves are opposite and commonly have a whitish or bright red base. The stalk exudes a toxic milky white latex. The cyathia or false flowers, are located in clusters at the head of the stalk and are yellowish green. They have no petals, the red colour being part of the young leaves' colouration. The fruits are small, segmented capsules.

E. milii

The Crown of Thorns, Christ Plant, or Christ Thorn is native to Madagascar. It is a succulent subshrub or shrub growing to 1.8 m tall, with densely spiny stems. The straight, slender spines, up to 3 cm long, help it scramble over other plants. The leaves are found mainly on new growth and are up to 3.5 cm long and 1.5 cm broad. The flowers are small, subtended by a pair of conspicuous petal- like bracts, variably red, pink or white, up to 12 mm broad.

E. rigida

It is a flowering shrubs, native to southern Europe and southwest Asia. It is a bushy evergreen perennial shrub with fleshy leaves arranged in a spiral; bearing bunches of bright yellow flowers in late spring.

E. leucocephala

It is known as white poinsettia. It is native of West Indies, Mexico and Guatemala. It is a tall shrub with pinkish branches. Flowers are small yellow, bracts obovate and white. It is grown in sunny location and propagated by cuttings.

E. pulcherrima

It is known as Poinsettia or Christmas flower. It is native of Mexico and central America, one of the most beautiful shrubs grown throughout the world.

6.2.0 Cultivars

Poinsettia has many cultivars with various colour *i.e.*, pink, red, yellow, white, bicolor *etc.*, grown throughout the world. Some important cultivars of poinsettia are described as follows.

'Annette Hegg', compact bushy growth, much branched; leaves small, ovate; bracts small, broad, red;

Cultivars of Poinsettia

'Orange Spice' 'Ecke 'Ice Punch' 'White Star'

'Lemon Yellow' 'Lutea' 'Dark Red Hegg'

'Purple Shine' 'Red Cascade' 'Polly's Pink'

Ecke 'Tapestry' 'Pink Cadillac' 'Cinnamon Star'

'Annette Hegg Supreme', a sport of 'Dark Red Hegg', compact growth; bracts brilliant red in more than one row, bracts pink and cream;

'Delight', bracts pink with cream margin;

'Garden Beauty', many bright pink bracts;

'Dark Red Hegg', a beautiful plant; bracts deeper than `Annette Hegg', dark red;

Ecke 'Orange Spice', deep orange bract with dark green leaves;

'Carousel Dark Red', curled dark red bracts against dark foliage;

'Sonora White Glitter', red bracts with cream speckling;

'Whitestar', bright white bracts and medium green leaves;

'Ice Punch', bright, rosy-red bracts witha blaze of white down the middle of each bract;

'Cinnamon Star', early flowering cultivar with cinnamon-pink bracts;

'Polly's Pink', bright, fluorescent-pink bracts with dark foliage; Ecke

'Winter Rose Dark Red', incurved dark red bracts and incurved dark green foliage;

'Pink Candy', dark-pink bracts with darker pink flecks;

'Pink Cadillac',large pink bracts with very dark green foliage;

'Tapestry', red bracts and leaves withvariegated margins;

'Viking Red', large, rounded, medium-red bracts

'Lemon Yellow', bracts showy, lemon yellow in colour;

'Tutea', compact yellowish; bracts attractive, light yellow;

'Purple Shine', broadly ovate, purple bracts;

'Rosy Pink', tall growing cultivar with a row of rosy pink, elliptic bracts;

'Supreme', a cultivar with pale pink bracts;

'Yellow Star', tall, hardy cultivar with a row of elliptic, light yellow bracts, about 30 cm across.

Euphorbia pulcherrima plennissima

Cultivars

Double poinsettia: A spectacular plant with inflorescence consisting of a circle of vermilion red, large, elliptic- lanceolate bracts and a cluster of narrow bracts at the centre;

'Bicolor', large, showy clusters of double bracts, centre shades of purple, border white;

'Flaming Sphere', an unusually showy plant when in bloom; bracts many, blood red, ruffled in ball like clusters at the end of the branches;

'Starlite', bracts red, long, narrow, elliptic or lanceolate, 2- 4 rows in each cluster, up to 25 cm in diameter;

'Winter Rose', clusters of several rows of bracts, rose in colour;

'Winter White', clusters of white bract, in multiple row.

7.0 Genetics and Breeding

7.1.0 Genetics

The chromosome number of poinsettia is generally 2n= 28. Gametic chromosome numbers of 14 and 28 are reported for both *P. cyathophora* and *P. dentata*, while *P. heterophylla* shows only n = 14 (Robert, 1960). Cells at diakinesis revealed the chromosome number of *E. pulcherrima* as 2n= 26 and 2n= 28. However, the modal number is 2n= 28 with an average mean of 77.09 per cent. This confirms earlier studies on the chromosome count of *E. pulcherrima* as 2n= 28 (Moyer 1934; Ewart and Walker 1960).

7.2.0 Breeding

7.2.1 Hybridization

Euphorbia has been extensively hybridized with many cultivars available commercially. Moreover, some hybrid plants have been found growing in the wild, for instance *E.* × *martini*, a cross of *E. amygdaloides* × *E. characias* subsp. *characias*, found in southern France.

7.2.2 Mutation Breeding

Commercial cultivars of poinsettia obtained through interspecific hybridization were subject to colchicine- based mutagenesis to recover their fertility was conducted by Pan *et al.* (2019). Mutagenic treatment was conducted at different concentrations of colchicine with either lanolin or cotton serving as the matrix. The poinsettia cultivars Dulce Rosa (5 mg/l colchicine with lanolin; 10 mg/l colchicine with cotton) and Princettia- Hot Pink (15 mg/l colchicine with lanolin; 10 mg/l colchicine with cotton) gave relatively high polyploidy production efficiency and morphological mutation rate. Consequently, a total of three polyploidy mutants of 'Dulce Rosa' and 19 polyploidy mutants of 'Princettia- Hot Pink' were obtained. Both cultivars had mutants with recovered fertility, with pollen germination rate of up to 27.5 per cent. Red hues correlated with anthocyanin composition and concentration and showed common dark red colouration in cultivars 'Christmas Beauty' and 'Christmas Feeling' where cyanidin derivatives were prevalent.

In contrast, orange- red bract colour is based on the prevalent presence of pelargonidin derivatives that comprised 85 per cent of the total anthocyanin content in cv. 'Premium Red' and 96 per cent in cv. 'Harvest Orange' (synonym: 'Orange Spice'). cDNA clones of flavonoid 32 -hydroxylase (F32 H) and dihydroflavonol 4- reductase (DFR) were isolated from the four varieties, and functional activity and substrate specificity of the corresponding recombinant enzymes were studied. Kinetic studies demonstrated that poinsettia DFRs prefer dihydromyricetin and dihydroquercetin over dihydrokaempferol, and thus, favor the formation of cyanidin over pelargonidin. Heterozygosity of the F32 H was demonstrated in this cultivar, but only the mutated allele was expressed in the bracts. No correlation between F32 H- expression and the colour hue could be observed in the four cultivars (Nitarska *et al.*, 2018).

8.0 Growth and Flowering

8.1.0 Growth Regulators

Gibberellic acid is one of the growth hormones used in poinsettia (*Euphorbia pulcherrima*) to stimulate the elongation, cell division and the formation of longer internodes. Application of a high concentration of GA_3 at 50 ppm early in the season (October) produced plants of 38.4 cm in height and main branches upto 33.83 cm and Khourya *et al.* (2019) cleared that the height of the plants had a strong positive correlation with the height of the main branch and poinsettia responds to early flowering under short day condition.

A study was conducted in poinsettia (*Euphorbia pulcherrima*) cv. Single by Renu and Srivastava (2013) by dipping the cuttings in different concentrations of cycocel (1500, 3000 ppm) and alar (1000, 2000 ppm) with control (untreated). Cycocel @ 3000 ppm was found to be the best for early initiation of first vegetative bud burst, early flowering and more number of leaves, branches and maximum duration of flowering. It also reduced the internodal distance, dimension of lowest leaf and bract dimension as compared to control while alar @ 2000 ppm resulted the minimum plant height.

8.2.0 Temperature and Photoperiod

Poinsettia is responsive to temperature and photoperiod for growth and flowering. The effects of photoperiod and number of lateral shoots on vegetative growth and percentage of fully coloured bracts in five poinsettia cultivars were studied by Batelja *et al.* (2009). The cultivars were grown atthree different photoperiods; under short- day period (SD, 9 h), long-day (LD, 13 h) and natural day (ND) period. It was reported that short- day conditions generally enhanced the opening ofcentral cluster of bracts and differences among cultivars were spotted in respect of percentage of bract colouration which was greater under natural and long- day treatments than under short- day treatment.

9.0 Soil and Climate

Poinsettia is a hardy and quick- growing plant requiring sunny location and well- drained soil, rich in organic matter; show marked variation in height, size and colour of bracts. Poinsettias do well in the home and keep their colour until mid- March. The showy red, pink, white, yellow, bicolored or speckled modified leaves are called 'bracts'. With proper light and temperature, they accumulate the anthocyanin pigments that give them their colour. Poinsettias require moderately moist soil. Poinsettias thrive on indirect, natural daylight, at least six hours a day. Direct sunlight may fade the bract colour. If direct sun cannot be avoided, diffuse the light with a shade or a curtain (Eck *et al.*, 1990).

10.0 Propagation

10.1.0 Vegetative Propagation

Poinsettia is commercially propagated by cuttings. An experiment was carried out to study the effect of different concentrations of IBA and rooting media on root induction of poinsettia by Chaudhari *et al.* (2018). Among the various levels of IBA, 4000 ppm IBA was most effective for obtaining number of sprouts per cutting, number of roots per cutting, length of shoot, length of root, number of leaves per cutting, rooting percentage (76.40 per cent) and survival percentage (74.19 per cent) of rooted cuttings after 60 days of planting. Consequently, red soil + cocopeat media was found most promising for sprouting per cutting, number of roots per cutting, length of shoots, length of root, number of leaves per cutting, rooting percentage (86.84 per cent) and survival percentage (85.83 per cent) for rooted cuttings after 60 days of planting.

10.2.0 Micropropagation

Nodal segments with axillary buds of *Euphorbia pulcherrima* when cultured on MS medium supplemented with 0.5 mg/l BA and 0.1 mg/l NAA was found the best for shooting (4.8 shoots per explant). Shoots were transferred onto MS medium adding GA_3 0.3 mg/l for elongation within 4 weeks. The elongated shoots were rooted on half strength MS medium containing 2 mg/l IBA and plantlets were successfully acclimatized with 75 per cent survival rate (Tien *et al.*, 2017). A protocol was developed for multiplication of *E. pulcherrima* through somatic embryogenesis from nodal explants by Yogesh *et al.* (2003). The induction of somatic embryogenesis in red pigmented callus was achieved on MS supplemented with 2- ip (9.8 µM) and NAA (2.69 µM). Reduced level of NAA (0.54 µM) in the same medium caused maturation of somatic embryoids. Gharbia *et al.* (2016) observed 100 per cent shoot regeneration when the stem explants of *Euphorbia pulcherrima* were placed on MS medium containing 1.0 mg/l BAP combination with 0.2 mg/l of NAA. Highest number of shoots per explant (6.95) was recorded followed by (5.83) shoots per explants in MS medium containing BAP 0.5 mg/l combination with GA_3 0.1 mg/l. Shoot regeneration from leaf explants reached 77.73 per cent when the leaf explants were placed on MS medium containing BAP 1.0 mg/l and 0.2 mg/l of IAA and highest number of shoots per explant (3.00) was recorded.

11.0 Cultivation

11.1.0 Manuring and Fertilization

Arreola *et al.* (2008) observed that foliar applications of calcium improved poinsettia plant height and Ca plus B combination accelerated bract pigmentation. Foliar sprays of Ca (300, 400, and 500 mg/l), B (0.2, 0.5, and 0.8 mg/l), Mo (0.3, 0.4, and 0.5 mg/l), Ca + B (400 + 0.5 mg/l), Ca + Mo (400 + 0.4 mg/l), B + Mo (0.5 + 0.4 mg/l) and Ca + B + Mo (400 + 0.5 + 0.4 mg/l) were applied to improve the quality of *Euphorbia pulcherrima* cv. 'Supjibi Red'. Treatments were applied three times at: beginning, middle, and end of the short photoperiod. It was reported that calcium at 400 mg/l increased significantly plant height by 15.3 per cent. Leaf chlorophyll

Propagation and Cultivation of Poinsettia

Cuttings of Poinsettia

Planting of Cuttings using Polythene

Rooting of Cuttings

Commercial Production

Growth in Greenhouse

Bract Formation

Cultivation in Open

Cultivation in Greenhouse

concentration decreased by 25 per cent when bract pigmentation initiated. Calcium (300 mg) and B (0.8 mg) increased the number of transitional bracts (5.7 and 5.6, respectively) while B (0.5 mg) increased the total number of coloured bracts per shoot (8.36) compared to the rest of the treatments. Rose and White (1994) grown 'Celebrate 2' Poinsettias for 8 weeks in a controlled-environment until first signs of bract colouration was appeared. In growth stage I (GSI, weeks: 1-4) and floral induction stage (GSII, weeks: 5-8), low, medium, and high N rates (25, 75, and 125 mg N/liter, respectively) were applied by sub-irrigation (no leaching). It was observed that more than 80 per cent of shoot dry weight and more than 90 per cent of total leaf area developed during growth GSII. The higher growth of poinsettia was reported with 75 mg nitrogen at GSI and 125 mg nitrogen in GSII.

11.2.0 Pruning and Thinning

Effect of thinning to 3, 4, 5 or 6 lateral shoots on vegetative growth parameters of nine poinsettia cultivars was studied by Batelja *et al.* (2014) and found that after thinning plant height and diameter and diameter of lateral stem were affected in all the cultivars. The number of laterals affected final plant height in one experiment and lateral stem diameter. Stem diameter decreased as number of laterals increased.

11.3.0 Pinching

Effect of number and timing of pinching on reproductive growth of potted poinsettia was investigated by Abdullah *et al.* (2003). They showed that late- pinched plants were delayed in floral induction, bract colour development and finish date compared to early- pinched plants. However, the time of pinching did not significantly affect the number of bracts and bract area, once- and twice- pinched treatment delayed floral induction, bract colour development and finish dates compared to unpinched plants. The number of bracts and bract area were increased after the plants had been pinched once or twice.

12.0 Diseases and Pests

12.1.0 Diseases

Poinsettia is susceptible to some foliar problems (*Botrytis* blight, powder scab, *Alternaria* leaf spot, *Xanthomonas* leaf spot, and anthracnose) as well as many common root and stem diseases (*Pythium* and *Phytophthora* root rots, *Thielaviopsis* root rot) as well as occasional *Fusarium or Rhizoctonia* stem problems. Poinsettia mosaic virus (PMV) has been associated with minor problems, and a phytoplasmal infection has contributed free branching for more attractive plants (Daughtrey and Chase, 2016).

Botrytis Blight

The causal organism is *Botrytis cinerea.* Symptoms are produced as gray to brown blotches on bracts while target spot lesions appear on the leaves of mature plants. Damaged areas may be covered by the fuzzy gray growth (spores and mycelium) of the fungus. Because of this moldy growth, Botrytis blight is often called 'gray mold'. Botrytis may also cause stem blight on cuttings and plants, often at branch crotches (Hartman and Kaiser, 2010).

Diseases and Pests of Poinsettia

Bacterial Canker

Powdery Mildew

Leaf Blight

***Pythium* Root Rot**

Poinsettia Mosaic Virus

Leaf Spots

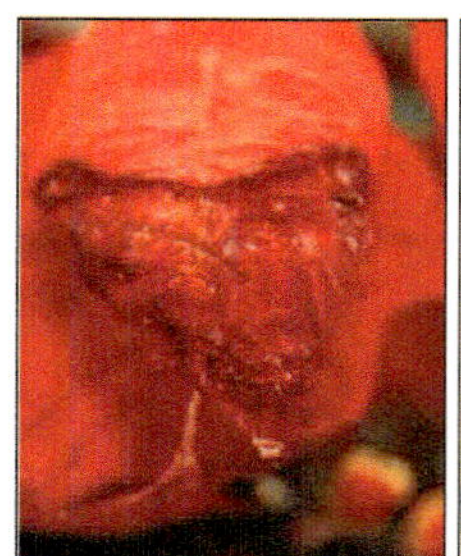

Gray Mold

Poinsettia Scab

Rhizoctonia Stem Rot

Phytophthora Root and Crown Rot

Thielaviopsis Stem and Root Rot

Mealy Bug

Diseases and Pests of Poinsettia

White Fly

Spider Mite

Thrips

Fungus Gnat

Avoid damaging plants, remove damaged tissues, space plants to provide good air circulation, heat to reduce humidity. If these practices are followed, then fungicides can help in management. Apply a systematic fungicide to protect the plants (Moorman, 2006).

Powdery Mildew

The causal organism is *Oidium* sp., the powdery mildew fungus causes yellow and brown spots to develop on susceptible leaves. Tell- tale patches of white fungal growth and spores may be evident on diseased leaves and bracts. This disease does not generally cause plant death, but it does make infected plants unmarketable. Managemnt. Apply sulphur fungicides to protect plants as soon as any mildew is found.

Scab

The causal organism is *Sphaceloma poinsettiae*. Diseased leaves develop small, raised blister- like tan or brown circular spots that may expand to dead blotches; yellowing and leaf drop follow. Gray lesions appear on stems and petioles. A purple or red border may surround these infection sites. Later, fungal growth and sporulation in the center of the lesions gives them a velvety brown appearance.

Young, diseased stems may become visibly elongated compared to healthy stems. High humidity and wet conditions favor infection and disease development.

Pythium Root Rot

Several species of *Pythium* can invade roots that have been exposed to water stress or excess soluble salts, *Pythium* root rot causes a discolouration and decay of the small absorptive roots (feeder roots). The outer layer of the root (cortex) is easily stripped off, leaving a narrow core of inner vascular tissue. Rooted cuttings may be stunted and yellow with brown, decayed roots. The loss of root function in maturing plants produces yellowing, and wilting. Plant death may occur, especially when plants are growing under stressful conditions. Use only sterile soil and clean pots and tools. Keep hose ends off the ground. Do not over- water or over- fertilize plants. Apply a fungicide to protect plants.

Rhizoctonia Root and Stem Rot

The diseases is caused by *Rhizoctonia solani*. It spreads via infested soil present on contaminated benches, pots, and tools. Conditions favoring disease development include planting too deeply and injuring the stem while planting. Cuttings and mature plants are the most vulnerable to infection. Stem lesions developing at the stem base are dry, brown, and shrunken. Resulting top symptoms include yellowing, wilting, stunting, and defoliation. A brown fungal mat may cover the infected plant tissue and nearby soil surface. Use only sterile soil and clean tools, and hang us hose ends. Do not over-fertilize. Apply a fungicide to protect plants at planting.

Bacterial Soft Rot

Bacterial soft rot is due to one or more species of *Erwinia*, most commonly *E. chrsanthemi* and or *E. carotovora*. These wound pathogens are especially aggressive in the presence of soft, succulent cuttings, along with warm temperatures and high humidity. The causal organisms are spread via contaminated tools, soil, potting media, plant debris, and water splash. Infected cuttings become soft and mushy, generally starting at the base of the stem and moving upward. Rotting tissues turn brown and often have a disagreeable odor. When present in older plants, stem decay may result in lodging.

Bacterial Canker

The diseases is caused by *Corynebacterium flaccumfaciens*, warm, moist growing conditions favor disease development. The bacterium may spread in splashing water, on hands, and on contaminated tools. Initially, longitudinal water- soaked brown streaks form on plant stems and water-soaked spots appear on the leaves. Enlarging stem lesions develop into cankers that crack and girdle the stem. As the disease progresses, cankered stems become defoliated following death of the plants.

Poinsettia Mosaic Virus

Poinsettia mosaic virus (PnMV) is mechanically transmitted and mainly spread via hands and tools. The symptoms are more evident on poinsettias grown at cooler temperatures, while plants grown at warmer temperatures may appear

Harvesting and Marketing of Poinsettia

Full Bract Stage

Packing of Poinsettia

Display of Poinsettia

In International Market

Poinsettia in Greenhouse

symptomless. Scattered yellow spots appear on infected leaves. Some spots are angular and have necrotic centers. Spots with necrotic centers could be confused with early scab infections. This virus may also cause mosaic, mottle and distortion of leaves and bracts.

12.2.0 Pests

A wide variety of insect and mite pests have been found in association with poinsettia production both in shade houses and in the greenhouse. Poinsettia is affected by whiteflies, thrips, aphids, leaf miners and spider mites (Huelsman, 2000).

White Flies

White flies are a very common pest in greenhouse production. The most common species are the greenhouse white fly (*Trialeurodes vaporariorum*) and the silverleaf white fly (*Bemisia arentifolii*). The adult silver leaf white flies are smaller, more yellow and active than the greenhouse whiteflies.

Spider Mites

Spider mites are a persistent pest problem in the greenhouse. Many species are found in the greenhouse, but the most common is the two-spotted spider mite (*Tetranychus urticae*). Spider mites are very small arthropods that develop mostly on the undersides of leaves. Spider mites cause injury to plants while feeding. Using their piercing-sucking mouth parts, they extract plant fluids. Feeding injury often give the upper leaf surface a characteristic mottled appearance.

Thrips

Thrips are also a serious insect pest in greenhouses because they are very difficult to control once a population becomes well established. Thrips feed on leaves and flowers of many host plants. The feeding injury can render a plant unmarketable and the act of feeding by thrips can transmit viruses to a susceptible host plant.

13.0 Harvesting and Postharvest Management

Poinsettia postharvest quality, *i.e.* shelf life after production in the greenhouse, is strongly influenced by both preproduction cultural practices and postproduction handling. Poinsettia is cold sensitive plant which is shipped during a cold, windy time of the year. Low temperatures are not a major issue for poinsettias, it is recommended that plants can be stored and shipped between 13 and 16 °C (Nell *et al.*, 1995). The light and temperature can influence potted poinsettia in postharvest quality retention (Scott *et al.*, 1983) by restricting air movement, inducing endogenous ethylene production, and lowering the level of light reaching the foliage.

14.0 Biotechnology

14.1.0 Molecular Markers

Randomly amplified polymorphic DNA (RAPD) techniques were used to compare the DNA from leaf tissues of nine commercial poinsettia (Euphorbia pulcherrima) cultivars. Amplification occurred in 57 out of 60 (95 per cent) tested primers. Forty- eight RAPD bands were scored from these primers, and 33 (69 per cent) were polymorphic. Sauve *et al.* (1997) placed the result of a UPGMA cluster analysis and principal components analysis placed the nine cultivars into two groups: one group consisted of 'Jingle Bells', 'Supjibi', and 'V- 17 Angelika', the other of 'V- 14 Glory', 'Red Sails', 'Jolly Red', and 'Freedom'. 'Lilo Red' and 'Pink Peppermint' belonged to the latter group, but were relatively distant from other cultivars in that group. They concluded that RAPDs are efficient for identification of poinsettia cultivars and for determination of the genetic relationships among cultivars.

15.0 References

Abdullah, T.L., Seng, O.J. and Pertanika J. (2003) *Trop. Agric. Sci.*, **26** : 15-18.

Anon. (1888, 1887), *Charleston Yearbook*. Lucas, Richardson and Co., Charleston, South Carolina. pp. 381-424.

Arreola, J.A., González, A.M.C., Luis A., Aguilar, V., León, M.T.C., Pineda. J.P. and García, E.A. (2008) *Rev. Fitotec. Mex.*, **31** : 165-172.

Batelja, K., Ban, S.G., Dumièiæ, G., Perica, S.K. (2009) *Sejemanrstvo*, **45** : 56-62.

Batelja, K., Dumièiæ, G., Perica, S., Buæan, L., Goyreta, S. (2014) Proceedings. 43rd Croatian and 3rd International Symposium on Agriculture.Opatija.Croatia, pp, 520- 524.

Bruyns, P.V., Mapaya, R.J., Hedderson, T.J. (2006) Taxon., **55** : 397-420.

Chaudhari B.B., Bhatt D., Chawla, S.L., Patel, M.A. and Parvathi, M. (2018) *J. Ornamental Hortic.*, **21**: 7-12.

Daughtrey, M. and Chase, A.R. (2016) *In book: Handbook of Florists' Crops Diseases*, pp.1-51.

Ecke, P., Matkin, O.A. and Hartley, D.E. (1990) *The Poinsettia Manual*, 3rd edition. Paul Ecke Poinsettia Ranch, Encinitas, California.

Ewart, M.S. and Walker, D.E. (1960) *J Heredity*, **51**: 203-208.

Fry, J.T. (1995) *The introduction of the poinsettia at Bartram's Garden. Bartram's Broadside* (Winter).

Gharbia, H., Danial, D. and Ibrahim, A. (2016) *IJAERS*, **3** : 132-137.

Hartman, J.R. and Kaiser, C.A. (2010) *Plant Pathology Extension*, pp. 1-6.

Huelsman, M.F. (2000) *Crop Profile for Poinsettia in Ohio.*, pp. 1-21.

Ibazez-Torres, A. (2004) *Anales de Biología*, **26** : 101-104.

Khourya, R.E., Deena, S.N., Boustanyb, N., Rizka, G.A., Geageaa, L. (2019) *Intern. J. Plant, Animal and Envir. Sci.*, **9** : 54-62.

Klotzsch, J. F. (1834) *"Beschreibungzweierneuen Euphorbienaus Mexico".Allgemeine Gartenzeitung* (in German), 2: 27.

Lack, H. Walter (2011) *Willdenowia.,* **41** : 301–309.

Moorman, G. (2006) *Plant Disease Facts*: Poinsettia Diseases.

Moyer, L.S. (1934) *Bot. Gaz.* **95**: 678- 685.

Nell, T.A., Leonard, R.T. & Barrett, J.E. (1995) *Acta Hort.* **405**: 132-137.

Nitarska, D., Stefanini, C., Haselmair-Gosch, C., Miosic, S., Walliser, B., Mikulic-Petkovsek, M., Regos, I, Slatnar, A., Debener,T., Terefe-Ayana, D., Vilperte,V., Hadersdorfer, J., Stich,K. and Halbwirth, H. (2018) *BMC Plant Biol.,* **18**: 1-12.

Pan, C., Lu, T., Pei-Jung, W.and Chen, W. (2019) *Hort. Sci.,* **54** : 1667-1672.

Perry, Leonard. (2014) *"Fun Facts About Poinsettia"*. Perry's Perennial Pages. University of Vermont Extension, Department of Plant and Soil Sciences.

RenuandSrivastava, R. (2013) *The Asian J. Hortic.,* **8** : 313-316.

Robert, D. (1960) *Annals of the Missouri Botanical Garden*. **48**. 329-341.

Rose, M.A. and White, J.W. (1994) *Hort Science,* **29** : 1309–1313.

Sauve, R., Ling, Jing-tian and Gawel, N. (1997) *Hort. Sci.,* **32**: 122-124.

Scott, L.F., Blessington, T. & Price, J. (1983) *Hort. Sci.,* **18**: 749-750.

Seltzer, Erica D., Spinner, Mary A. (2017) *"Poinsettia Facts".The Poinsettia Pages.* University of Illinois Extension. Retrieved January 30.

Steinmann, V.W. and Porter J.M. (2002) *Ann. Missouri Bot. Gard.,* **89** : 453–490.

Taylor, J.M., Lopez, R.G., Currey, C.J. and Janick, J. (2011) *Chronica Horticulturae,* **51** : 23-27.

Taylor, Judith M., Lopez, Roberto G., Currey, Christopher J., Jan, Jules (2011) *"The Poinsettia: History and Transformation". Chronica Horticulturae.,* **51** : 23–27.

Tien, L.H.T., Diep, N.T., Phong, N.V. (2017) *J. Agric. Sci. and Tech.,* **6** : 1-7.

Trejo, L., Feria Arroyo, T. P., Olsen, K. M., Eguiarte, L. E., Arroyo, B., Gruhn, J. A., Olson, M. E. (2012) *American J. Bot.* **99** : 1146-1157.

Yogesh, T., Jasrai, K. N., Thaker and D'Souza, M. C. (2003) *Plant Tissue Cult.,* **13** : 31-36.

26

Gardenia

Sk. Masudul Islam

1.0 Introduction

Gardenia is a genus of flowering plants in the coffee family, Rubiaceae, native to the tropical and subtropical regions of Africa, Asia, Madagascar and Pacific Islands, and Australia. The genus was named by Carl Linnaeus and John Ellis after Dr. Alexander Garden (1730- 1791), a Scottish- born American naturalist. There are some 140 species names applied in this genus (Govaerts *et al.*, 2010), although the number of accepted species sanctioned by modern revisions is likely to decrease. It has been in cultivation in China for at least a thousand years, and was introduced to English gardens in the mid-18th century. It is currently made up of 250 species (Judd *et al.*, 2016), of which approximately 5 per cent are listed by the International Union for Conservation of Nature under some category of extinction (IUCN, 2017). Several species of *Gardenia* are popular shrub and small trees widely grown in gardens and roadside in tropic and mild subtropics. It is also an attractive pot plant and flowering continues for a long period in favourable climate and porous soil rich in organic matter.

2.0 Importance and Uses

2.1.0 Aesthetic Uses

Gardenias can be used as screens, hedges, borders or ground covers. They also may be used as free-standing specimens or in mass plantings. These shrubs are excellent choices for fragrant flowers and handsome foliage. It is effectively used as either focus or background in informal plantings and for tropical-theme landscape plantings. Gardenia can also be an accent plant around seating areas or near windows to take advantage of their extremely fragrant white flowers.

Gardenias are prized for the strong sweet scent of their flowers, which can be very large in size in some species. Polynesian people in the Pacific islands use the fragrant blooms in their flower necklaces, which are called 'Ei' in the Cook Islands, 'Hei' in French Polynesia and 'Lei' in Hawaii. Gardenia flowers can be eaten raw, pickled, or preserved in honey.

2.2.0 Therapeutic and Other Uses

Gardenia jasminoides commonly used in traditional medicine: the roots to treat headache, dyspepsia, nervous disorders and fever, bark to treat dysentery, leaves in febriferous poultices and to treat sore eyes, flowers as emollient, and fruits to treat jaundice, dysentery, and diseases of kidneys and lungs. However, its main uses are as a dye, the fruits being used to colour food and occasionally textiles yellow, and as an ornamental.

In Thailand, seeds of *Gardenia sootepensis* are boiled with water and the solution is used as a shampoo to kill lice. Several Gardenia species are used in traditional medicine in Africa, *e.g.*, to treat fever, smallpox, leprosy, sleeping sickness, ophthalmia, after childbirth and as an aphrodisiac. Gardenia wood is sometimes used for turnery, carving and implements.

Uses of Gardenia

As Specimen Plant

As Hedge

In Pot

As Herbaceous Border

In Flower Arrangement

In Flower Vase

Gardenia jasminoides fructus (fruit) is used within traditional Chinese medicine to "drain fire" and treat certain febrile conditions, *i.e.*, have an anti-inflammatory and antipyretic effect. The fruits of *Gardenia pseudopsidium* are used to treat smallpox.

Shishihakuhito is a Chinese herbal medicine mainly composed of gardenia fruit, used to treat atopic dermatitis as it inhibits IgE-mediated histamine release (Bensky *et al.*, 2004). In the case of a woman who had developed blue-grey discolouration of the skin as a result of chronic intake of gardenia fruit extract was published (Mizawa *et al.*, 2020). Gardenic acid and gardenodic acid isolated from *Gardenia jasminoides* fruits may be used as early pregnancy-terminating agents. The lipoxygenase inhibitor 3-caffeoyl-4-sinapoylquinic acid was isolated from the fruit. In Eastern Asia, *Gardenia jasminoides* is called 'zhîzi' in China, 'chija' in Korea, and 'kuchinashi' in Japan. Its fruit is used as a yellow dye, used on fabric and food (including the Korean mung bean jelly called hwangpomuk) (Ozaki *et al.*, 2002).

3.0 Origin and History

The *Gardenia* genus is native to the tropical and subtropical regions of Africa, Asia, Madagascar, and the Pacific islands (Zongram *et al.*, 2017). Specifically, *G. thunbergia* has its origin in southern Africa and *G. jasminoides* is native to China (Mostafa *et al.*, 2013), where it has been cultivated by the inhabitants for thousands of years (Clemson-Cooperative- Extension, 2009). Two most important being *Gardenia jasminoides* (native to China), with its many cultivars, and *Gardenia thunbergia* (white gardenia, native to South Africa), which is typically grown as a rootstock. Former botanical names were *G. florida* and *G. augusta*. Evidence of *G. jasminoides* in cultivation in China dates to the Song Dynasty (960- 1279 AD), where both wild and double-flowered forms have been depicted in paintings, such as those of the Song Emperor Huizong, and the Tenth century artist Xu Xi. The Yuan Dynasty (1271- 1368) saw it on lacquerware, and the Ming Dynasty on porcelain (1368-1644). In 1794, English statesman Sir John Barrow saw gardenias in nurseries in Guangzhou, China. Daniel Solander reported that the species was brought to the United Kingdom in 1744 on the Godolphin by captain Hutchinson, who gave it to botanist Richard Warner of Woodford Row, Essex. Warner, however, was unable to propagate it until the botanist John Ellis recommended James Gordon, a gardener at Mile End. Gordon was successful in August 1757, and plants sold well thereafter; each cutting- grown plant fetching five guineas. Gardenias were first grown in the United States in 1762, in the Charleston garden of Alexander Garden, who had moved there 10 years previously (Valder, 1999; Webb *et al.*, 1761; Ellis *et al.*, 1759; Foster *et al.*, 1992; Bender, 2011).

4.0 Distribution

Gardenias are grown in Africa, Asia, Australia, Europe, North and South America, and the Pacific islands. Within China the species is found in Anhui, Fujian, Guangdong, Guangxi, Sichuan, Taiwan, Yunnan and Zhejiang provinces. Its native habitat is forest and undergrowth along streams, and on sloping and hilly terrain to an altitude of 1500 m (Chen *et al.*, 1761). Several gardenia species are endemic to Hawaii 'i, found nowhere else: *Gardenia mannii*, *G. remyi*, and *G. brighamii*; these were formerly found in dry upland areas on all of the major Hawaiian islands except Kaua'i, but they are now rare or endangered. In Greece, it is grown mainly as a pot plant for indoor use; the varieties with the greatest demand are those with double flowers, small in size and compact, which are even exported to Europe (Tsanakas *et al.*, 2013). In European countries this species is also cultivated, since protected cultivation allows its production in temperate climates (Palmateer and Chase, 2016). In the United States, gardenia is a popular pot plant (Chuenboonngarm *et al.*, 2001), which has been cultivated since the 18th century in South Carolina and more recently, in other states such as Florida (Clemson-Cooperative-Extension, 2009) and Hawaii, where it is used to make typical flower necklaces (Kobayashi and Kaufman, 2006). In Mexico there is a long tradition of its cultivation, especially in states such as Veracruz, where, sixty years ago in the municipality of Fortín it was the most representative floricultural crop (Murguía, 2007). In Argentina, it is used in commercial nurseries for cut flowers (Sosa-de-Castro *et al.*, 2003) and it is also adapted for field cultivation, mainly in the province of Tucumán, where the supply of these flowers from mid-spring is characteristic (Vidal *et al.*, 2001). In Cuba, although there are favourable climatic conditions for the adequate growth of gardenia, it is not a commercial species (Álvarez, 2008) and its cultivation is carried out only on a small scale, in gardens and domestic patios (Pérez *et al.*, 2011).

Gardenia kabaenensis is only found on Kabaena island, southeast of the main Sulawesi island, where it grows in lowland forests. *Gardenia longistipula* is endemic to South Sulawesi (vicinity of Malili, Lake Manato, and LakeTowuti). It grows in lowland forest on sandstone, limestone and ultramafic soils (Low, 2013). *Gardenia beamanii* and *Gardenia chanii* is endemic to Borneo, restricted to the northern (Sabah and East Kalimantan) and northwest parts (Brunei; possibly also adjacent parts of Sarawak). *Gardenia lanuto* is endemic to Samoa and thus far known only from Savaii and Upolu, where it appears to be uncommon. It has been noted as a tree to 10 m high and with a trunk to 15 cm in diameter, forming thickets in dense upland forest at elevations of 600- 1,000 m. It has been noted in flower in April, August, and September, and in fruit in January (Low, 2007).

5.0 Morphology

Gardenia is reliably distinguished by a number of characters, including usually shrubs or rarely trees, 30 cm to 4 m high sometimes with short shoots (*Gardenia angkorensis, G. sootepensis*), unarmed or with short shoots sometimes spinescent, with buds and young stem apices often resinous, leaves opposite or rarely ternate, sometimes clustered at stem apices, often with domatia. Inflorescences pseudo axillary and/or terminal, fascicled to cymose and several flowered or reduced to 1 flower, pedunculate to sessile, bracteate; flowers sessile to pedicellate, bisexual, monomorphic, often showy; calyx with ovary portion well developed and often longitudinally ridged; limb 5- 8-lobed or sometimes fused into a tube or conical cap that splits irregularly as corolla elongates (*i.e.*, spathaceous), often well developed; corolla white to cream, salverform or funnelform, glabrous or variously pubescent inside; lobes 5- 12, convolute in bud. Stamens 5-12, inserted in corolla throat, included or partially exserted; fruit generally yellow to orange, red- orange, or brown with pulp usually orange, baccate, leathery or fleshy, ellipsoid to sub globose.

6.0 Species and Cultivars

6.1.0 Species

Several species of *Gardenia* which are grown in gardens in tropics and subtropics are summarised hereunder.

Gardenia brighamii H. Mann

A large shrub 2- 4 m tall (although some in their natural habitat can reach up to 6 m), with light green leaves and light tan trunk. Tiny, clear to opaque yellow resin balls develop at stem tips prior to the formation of new leaves. Small flowers about 5 cm across resemble puakenikeni in petal shape. Their scent is like that of the larger bushy, non-native gardenia, with a hint of coconut oil.

G. candida A.C.Sm.

Tree to 7 m high, the young parts copiously strigillose or puberulent with hairs, the leaf blades chartaceous, elliptic, calyx copiously pilose, 4- angled or narrowly winged, the calyx limb infundibular, the hypanthium obconical- fusiform.

G. cornuta Hemsl.

Gardenia cornuta is a shrub or multi-branched small tree, 3 to 5 m tall, with erect and spreading main branches, branchlets sometimes with thick thorns, a trunk that is whitish grey, the bark is smooth, often flaking into thin peels. Leaves are smooth and crowded at the ends of twigs, usually in whorls of three, light glossy green, short- stalked. Flowers are solitary at the ends of short, rigid branches, borne in summer, November to March, white turning yellow, short- stalked, with a long, slender greenish tube. Fruits which are more attractive than the flowers, are more or less pear- shaped, woody, tipped with persistent remains of the calyx, glossy golden yellow, usually produced in large quantities between February and August. Seeds are hard and flattened.

G. hainanensis Merr.

Trees, 3- 12 m tall; branches with internodes congested to developed, resinous at apices with usually distal most several internodes coated with dried resin. Leaves opposite; obovate- oblong, adaxially shiny and glabrous. Flowers solitary; terminal or pseudo axillary, peduncle 0.4- 0.8 cm, densely puberulent, Calyx puberulent, usually covered with resin; ovary portion broadly obconical to ellipsoid, flowering in May- October. Fruiting; Berry yellow, sub globose or ovoid- ellipsoid to ellipsoid.

G. hutchinsoniana Turri.

Shrub or tree to 10 m high, evanescently sericeous on young parts; stipules firmly connate into a tubular or urceolate sheath, the leaf blades papyraceous, lanceolate- to oblong- or obovate elliptic, calyx minutely puberulent or strigillose, soon glabrate, the hypanthium fusiform- prismatic; fruits subglobose or ovoid- ellipsoid.

Species of Gardenia

Gardenia tubifera

Gardenia brighamii

Gardenia jasminoides

Gardenia psidioides

Gardenia taitensis

Gardenia volkensii

Species of Gardenia

Gardenia cornuta

Gardenia ternifolia subsp. *jovis-tonantis*

Gardenia thunbergia

Gardenia carinata

Gardenia volkensii sp. *volkensii*

Gardenia resiniflua

Species of Gardenia

Gardenia imperialis

Gardenia rutenbergiana

Gardenia remyi

Gardenia sootepensis

G. jasminoides J. Ellis

Shrubs, 0.3- 3 m tall; branches terete to flattened, with internodes developed to shortened, glabrescent or usually densely puberulent to pilosulous, becoming grey to greyish white, with buds resinous and distalmost internodes often covered with resin; leaves opposite or rarely ternate, subsessile to petiolate; petiole to 0.5 cm, densely puberulent or shortly pilosulous to glabrous; leaf blade oblong-lanceolate, obovate-oblong, obovate, oblanceolate, or elliptic, 3- 25 × 1.5- 8 cm, adaxially shiny and glabrous or sometimes puberulent on principal veins; flower solitary, terminal; peduncle 1- 10 mm, puberulent or pilosulous to glabrous; corolla white to pale yellow, simple or in cultivation sometimes doubled, outside glabrous; tube obovate or obovate-oblong to obtuse to rounded; berry yellow or orange-yellow, ovoid, sub globose, or ellipsoid, flowering in May- February.

G. jasminoides var. *fortuneana*

Shrubs are up to 1m high; flowers with doubled corolla, sterile. Flowering continues throughout year. Corolla doubled; flowers sterile, plants not setting fruit; cultivated plants.

G. *jasminoides* var. *jasminoides*

Flowers with simple corolla, fertile, flowering continues from May- February. Flowers are fertile and set fruit; wild and occasionally cultivated plants.

G. *latifolia* Aiton

Indian boxwood is a small deciduous tree or large shrub; leaves oppositely arranged, or whorled oval to obovate, smooth, with a small hairy gland in the axils of the veins on the underside; flowers appear singly at the end of branches; flowering in April- July.

G. *longistipula* Y.W. Low

Shrub to small tree, to 5 m high; leaves obovate to elliptic. Flowers are solitary, corolla hypocrateri form, cream turning light yellow, then orange yellow. Fruits; sub- globose.

G. *sootepensis* Hutch.

This species is 7- 10 m tall, often with gelatinous secretions; leaves opposite; puberulent or tomentulose. Flowers pseudo axillary usually near branch apices, yellow colour flower, solitary; flowering in June- April. Berry ellipsoid or ellipsoid-oblong.

G. *tannaens*

Tree up to 20 m high, minutely puberulent on young parts; the leaf blades papyraceous, lanceolate to oblong- elliptic, the hypanthium 5- angled, the angles continued distally into wings, the calyx limb narrowly cupuliform, becoming coriaceous in fruit, fruits sub globose.

G. *volkensii*

Small twiggy tree rarely reaching 8 m, with a relatively dense crown and a short thick and often fluted trunk; corolla large, showy white, turning yellow after a few days, then brown before dropping, tube up to 10 cm long or more.

Gardenia tubifera, Gardenia thumbergia, Gardenia ternifolia, Gardenia remifolia, Gardenia psidiodes, Gardenia imperialis are also grown in gardens.

6.2.0 Cultivars

There are many cultivars of *Gardenia*, offering considerable variation in plant size, flower shape, flowering season, and duration. Due to these characteristics, they can be used for cut flowers, as live barriers, in garden decoration, as ground cover or as isolated plants. Varieties that grow in Florida, mostly by cross- breeding or by mutations, show significant difference in the shape and size of flowers, as well as in the time of flowering, continuity and growth volume between varieties. The most important are the following according to Kobayashi and Kaufman (2006):

Cultivar	Species	Flower Colour	Flower Type	Characteristics
'Aimee Yoshioka'	*Gardenia jasminoides*	White	Large double flowers; brilliant dark green foliage.	7.62-12.7 cm in diameter; produces an abundance of flowers.
'August Beauty'	*Gardenia jasminoides*	Velvety white	Medium to large double flowers.	Flowers heavily; grows 121.92-182.88 cm high, 91.44 cm wide. with sweet fragrance Dense foliage.
'Belmont'	*Gardenia jasminoides*	White-cream	Large, double flowers 10.16-12.7 cm.	Well balanced, bushy shrub with large dark green leaves
'Chuck Hayes'	*Gardenia jasminoides*	Ivory white	Semi-double flowers.	Grows to 2 m high.
'Coral Gables'	*Gardenia jasminoides*	White	Large flowers, 7.62-12.7 cm across.	Dark green foliage on compact plants, produces many blooms.
'Daisy'	*Gardenia jasminoides*	White	Flat faced single fragrant flowers.	Grows to 2 m high.
'First Love'	*Gardenia jasminoides*	Creamy White	Very large and fragrant double blooms 10.16-12.7 cm across.	Larger than 'August Beauty'; 2 m high.
'Fortuneiana'	*Gardenia jasminoides*	Ivory white	Double, carnation-like flowers up to 10.16 cm in diameter.	Lustrous dark green leaves,moderate growth rate.
'Glazerii'	*Gardenia jasminoides*	White	Flower flat, 7.62-12.7 cm in diameter.	Compact grower; medium green foliage, bloom ideal for corsage.
'Golden Magic'	*Gardenia jasminoides*	White/yellow	Double flowers.	Flowers open pure white, turn golden yellow. This mutation was discovered on a branch of 'Mystery' that was growing in Hawaii. Grows about 1 m tall.
'Kleim's Hardy'	*Gardenia jasminoides*	Ivory white	Medium star-like single flowers withintense fragrance.	Grows to 20 cm tall, low mounding form.
'Kuchinashi'	*Gardenia jasminoides*	White	Open, 5-petaled, single-style blooms.	Compact size 20 cm high, blooms profusely.
'Miami Supreme'	*Gardenia jasminoides*	White	Large flowers, 10.16-15.24 cm in diameter.	Medium to dark green foliage.
'Mystery'	*Gardenia jasminoides*	Pure white	Double flowers 10.16-12.7 cm across.	A well-known, upright growing cultivar growing to 2 m high; needs to be pruned to keep neat.
'Radicans'	*Gardenia jasminoides*	White	Double flowers 2.54-5.08 cm across.	A dense, spreading shrub to 4 m high, it has rooting branches and narrow, dark green leaves often streaked with white. Excellent container plant or ground cover spaced 40-50 cm apart.

Cultivar	Species	Flower Colour	Flower Type	Characteristics
'Radicans Variegata'	*Gardenia jasminoides*	White	Short cultivar, 30-60 cm tall or lower	White leaf margins, coloured branches occur, very low-growing
'Veitchii Extra'	*Gardenia jasminoides*	Pure white	Double flowers, yellow center 5.08-7.62 cm in diameter.	Dense, compact, upright bushy plant to 1.5 m highw ith smaller shiny green leaves. A prolific bloomer with sweetly fragrant flowers; can be grown well in a container, excellent for cut flowers.
'White Gem'	*Gardenia jasminoides*	Ivory	Medium-fragrant, star-shaped, single flowers.	Dwarf, compact plant 30-60cm high, great in containers.
'Crown Jewel'	*Gardenia jasminoides*	Pure white	Double fragrant blooms.	Cold hardiness, a dwarf spreading habit.
'Grif's Select'	*Gardenia jasminoides*	White	Single-flowered blooms, cold hardiness upto -15 °C.	Bearing flowers in spring and early summer, to 1.5m high.

Source: Kobayashi and Kaufman (2006).

7.0 Genetics and Breeding

7.1.0 Genetics

Rubiaceae owns over 10 thousand species as reported; many species demonstrate important economic uses. With the dramatic promotion of sequencing technology, researchers are trying to solve the mysteries lying deep in the plant genomes. Therefore, aiming at studying the biodiversity and solar-related genes, acquired its complete chloroplast genome. The chloroplast genome was 155,096 bp in length with typical circle structure, it presented an 85,409 bp large single-copy (LSC) region, an 18,103 bp small single copy (SSC) region, and two 25,792 bp inverted repeat (IRs) regions. The overall GC content was 37.49 per cent, SSC and IR components showed extreme GC percentages ranging from 31.53 to 43.18. A brief online annotation reported 125 CDSs, 8 rRNAs, and 33 tRNA. To surprise, *G. jasminoides* was closer to *Coffea canephora* and *Coffea arabica* within Rubiaceae from phylogenetic analysis (Zhao *et al.*, 2020).

Inter simple sequence repeat (ISSR) markers were performed in order to analyse the genetic relationships of both taxa of *Gardenia jasminoides* var. *radicans* and *G. jasminoides* var. *grandiflora*. Over the 88 fragments, only one locus (ISSR-11-05) was specific to *G. jasminoides* var. *radicans* and only one (ISSR-09-05) *G. jasminoides* var. *grandiflora*. Although *G. jasminoides* var. *radicans* showed low levels of alleles and Shannon's information index than *G. jasminoides* for. *Grandiflora*, however, there was not significant differences ($p > 0.05$). For both taxa the mean genetic diversity of natural populations was higher than that of cultivation populations. They suggested that domestication processes via artificial selection do not have eroded the high

Cultivars of Gardenia

'Aimee Yoshioka' 'Crown Jewel' 'August Beauty'

'Belmont' 'Chuck Hayes' 'Golden Magic'

'Daisy' 'First Love' 'Double Mint'

'Sweet Tea' 'Fortuneiana' 'Glazerii'

Cultivars of Gardenia

'Kleim's Hardy' 'Ananta Dwarf' 'Miami Supreme'

'Mystery' 'Radicans' 'Veitchii Extra'

'White Gem' 'Thunberg' 'Tahitian'

'Regal' 'Amoena' 'Vietnamese'

levels of genetic diversity. ISSR markers were more effective in classifying natural populations of wild *G. fasminoides* in East Asia as well as cultivated *G. jasminoides*. The information about the phylogenetic relationship of *G. jasminoides* var. *radicans* and its closely related species is very valuable for the systematics of genus *Gardenia*, the origin of cultivated *G. jasminoides*, and future *G. jasminoides* breeding (Man Kyu Huh, 2007).

7.2.0 Breeding

Knowledge of breeding systems of endangered plants is crucial in formulating effective management plans for their conservation. Consequently, spatial distribution, flowering and fruiting patterns of *Gardenia actinocarpa* Puttock (Rubiaceae), a rare and endemic dioecious shrub of lowland tropical rainforest of Cape Tribulation, North Queensland, Australia was documented in four subpopulations over a two-year period. Sex ratio was unity in two of the sites, and significantly biased in favour of male plants in the other two sites. At most sites, analysis of spatial distribution of individuals indicated aggregated dispersion patterns irrespective of quadrat size and plant size/reproductive status. There was neither evidence of niche differentiation nor switching of the sexes. The species exhibited an extended flowering pattern of almost six months. However, male trees flowered earlier, longer, and produced more flowers than female trees. Intensity of flowering was significantly correlated with plant size in both sexes. Fruit set is very low in each of the two years of observation, but was significantly predicted by female habitat light environment, flower production and plant size. The lack of neighbourhood effects such as male tree density, male tree flower production around a focal female tree and distance from a female tree to the nearest male tree on fruit set is an indication that *G. actinocarpa* pollinators are capable of carrying pollen over long distances (Osunkoya, 1999).

8.0 Growth and Flowering

Flower induction and development are influenced by different factors (Wilkins, 1986). In nature (Kagawa, Japan) the primary apical flower of Gardenia is induced in July. Consequently, two or three vegetative shoots appear from the underlying axillary buds, which are initiated at the beginning of September. The primary flower buds are completed at the end of October and then enter into dormancy. The secondary flower buds develop very slowly during winter. Flowers appear the next year from June till August (Kosugi *et al.*, 1957; Chen, 1985).

8.1.0 Light

Light plays an important role in plant growth and development processes. The morphogenesis processes are affected by both light quality and intensity. Growth regulators have also an enormous influence (Gabarkiewicz and Rudnicki, 1995). Danuta Kozak, (2011) reported that yellow light increased the number of axillary shoots of *Gardenia jasminoides* from explants cultivated on medium containing 25 µM BA and it promoted shoot elongation most strongly on medium with the addition of 5 µM BA. Red and yellow light increased the number of roots on control medium, whereas yellow light significantly promoted root elongation growth. However, yellow and white light had the most positive effect on the average fresh weight of roots of gardenia.

Flower induction and early development are hastened by short- day treatments (SDT) (Wilkins, 1986). Conover *et al.* (1968) found that a SDT of 9 h for 4- 6 weeks was necessary to obtain highest flower numbers. A SDT given in a period of 5 weeks before natural short- day (SD) gave no better results; both treated and untreated plants showed flower buds. The same experiments in full summer advanced flowering, but the number of flowers was identical to the control (De Graaf- Van Der Zande, 1988; Beel and De Bruyn, 1992). The night temperature during induction has to be lower than 18 °C (Wilkins 1986; Morel *et al.*, 1988). Also, light intensity during induction is of great importance. Plants cultivated in full sun give more flowers than shaded plants (Rose and Dickey, 1960; Conover *et al.*, 1968). Flower development is advanced by long days (LD) (Fujisawa and Kuki, 1976; Wilkins 1986). When plants become older, flowering is less dependent on light and temperature conditions. According to these results Gardenia was catalogued as a facultative SD plant or a facultative short-long- day (SLD) plant.

Danuta Kozak, (2011) reported that yellow light increased the number of axillary shoots of *Gardenia jasminoides* from explants cultivated on medium containing 25 μM BA and it promoted shoot elongation most strongly on medium with the addition of 5 μM BA. Red and yellow light increased the number of roots on control medium, whereas yellow light significantly promoted root elongation growth. However, yellow and white light had the most positive effect on the average fresh weight of roots of gardenia.

8.2.0 Growth Regulator

Plant hormones are a naturally occurring organic material in very small quantities and affect the growth and development of plants. Growth regulator have also an enormous influence on growth and flowering in gardenia (Gabarkiewicz and Rudnicki, 1995). It has been possible to find a group of synthetic growth regulators, which resemble exactly the impact of those naturally produced. BAP showed high proliferation of *G. jasminoides* compared to 2ip and kinetin (Sayd *et al.*, 2010). Using IBA in microcuttings of *G. jasminoides*, high percentages of root *in vitro* and *ex vitro* were obtained (Pontikis, 1983; Hatzilazarou *et al.*, 2006). Dumitrescu (2002) carried out a successful combination of 0.1 mg/l IAA and 1.0 mg/l BAP for chlorophyll extract on *G.jasminoides*, and a higher range of shoot proliferation of this plant via BAP was reported by Chuenboonngarma *et al.* (2001). Pontikis (1983) revealed long and high- quality shoots in jasmine via 2iP, but Chuenboonngarm *et al.* (2001) reported 2iP produced 100 per cent chimeric plants of *G. jasminoides*. Economou and Spanoudaki (1986a) reported that BA induced axillary buds in *G. jasminoides*. However, BAP and NAA showed high proliferation compared to 2ip and kinetin on *G. jasminoides* (Sayd *et al.*, 2010). In addition, the use of IBA in micro cuttings of *G. jasminoides* produced high percentages of roots *in vitro* and *ex vitro* (Pontikis, 1983; Hatzilazarou *et al.*, 2006).

9.0 Propagation

The traditional methods of propagation of gardenia are seeds, cuttings, and grafting.

9.1.0 Seed Propagation

It is not recommended to propagate gardenia by seeds, being heterozygous, and gives plants genetically variable. The main objective of the multiplication by seed is to produce new varieties. Seeds are not used for propagation purposes (Husseini, 1993). Seeds can be sown in pots containing a 50: 50 combination of peat moss and perlite or a 50: 50 combination of peat moss and coco- peat or sand (Dirr, 1990). They should be kept in a warm place and can be covered with a transparent nylon to conserve humidity, until the first seeds germinate, which usually happens slowly and erratically. The seedlings should be transplanted when they form their second or third pair of true leaves, in this way survival rate of the seed may be increased by 20 per cent (Bradshaw, 2003).

9.2.0 Vegetative Propagation

9.2.1 Cuttings

Cuttings can be taken any time during the year, but are most successful in June, July, and August. *Gardenia thunbergia* can be propagated from seeds or cuttings (Economou, and Spanoudaki, 1985). Tip or midsection cuttings with wood 6- 8 weeks old should be cut 10- 12 cm long with at least two or three sets of leaves. Cuttings can be taken at or between nodes as they root from the cut end. Leaf removal is unnecessary and undesirable because it results in a longer rooting period. Rooting of cuttings is best under intermittent mist or in a closed-case propagating device, such as an old aquarium or clear plastic bag. Rooting media should be a 50:50 combination of clean, sharp sand and peat moss; or a 50:50 combination of peat moss and perlite or vermiculite cocopeat and sand *etc.* (Dirr, 1990). Stem cutting after flowering is the conventional method for propagation (Hutchinson, 1980). After rooting, the cutting are planted in pots and kept in semishade with high humidity level (Murguía, 2007). Although gardenia cuttings can root at any time of the year, rooting is generally most successful in June, July, and August (Park and Bradshaw, 2016). Kobayashi and Kaufman (2006) stated that gardenia takes 2 to 3 years to flower when obtained from seed, but less than a year when grown from cuttings.

9.2.2 Air-Layering

All species of *Gardenia* can be successfully propagated by tip, simple, trench, and air layering methods. In case of tip layering, pulling a branch down, inserting the tip into the ground and securing it with garden pins or weights is followed. In simple layering, bury part of a stem while leaving its tip exposed above ground. To trench layer a gardenia, an entire stem except where it joins the parent plant is buried. These methods work best for lower branches that are at or near ground level. Air layering gardenias is an effective way to propagate branches higher on the plant. A small cut is made on the stem, 10- 15 cm from the tip, and packed with moist long- fibre sphagnum peat moss around the wound and tied with plastic

Propagation of Gardenia

Seed Pod

Seedlings

Grafted Gardenia

Rooting from Cutting

Seedlings Arise from Cutting in Pots

Propagation of Gardenia

Dry Fruits

Gardenia brighamii Seed Pod

Micropropagation

Gardenia plantlet after 8 weeks on rooting medium supplemented with, (a) 3 mg/l NAA, (b) 3 mg/l NAA+ 1 mg/l $AgNo_3$, (c) 3 mg/l NAA+2 mg/l $AgNo_3$, (d) 3 mg/l NAA+3 mg/l $AgNo_3$, (e) 3 mg/l IBA, (f) 3 mg/l IBA+1 mg/l $AgNo_3$, (g) 3 mg/l IBA+2 mg/l $AgNo_3$, (h) 3 mg/l IBA+3 mg/l $AgNo_3$, (i) 2 mg/l NAA+2 mg/l IBA, (j) 2 mg/l NAA+2 mg/l IBA+1 mg/l $AgNo_3$, (k) mg/l NAA+2 mg/l IBA+2 mg/l $AgNo_3$ and (l) 2 mg/l NAA+2 mg/l IBA+3 mg/l $AgNo_3$. [Source: El-Ashry *et al.* (2018)].

wrap around the moss to maintain constant moisture and also tied with burlap or foil around the plastic wrap to block sunlight that could burn developing roots. After roots are formed on layered stems, the stems are cut below where new roots form and planted.

9.2.3 Grafting

In Central Florida and South Florida, propagation of gardenia is mostly by grafting rootstock of *Gardenia thunbergia* (Dirr, 1990). In south Florida, propagation by grafting scion from a desired cultivar to a seedling rootstock of *Gardenia thunbergia*. Rootstock seedlings, however, are difficult to obtain. Collecting and sowing the seeds from the berry in flats or pots containing a 50: 50 combination of peat moss and perlite or 50: 50 combination of peatmoss and sand or cocopeat:sand. Seeds germinate slowly and erratically. Seedlings should be removed when they form their second true leaf. By waiting until they produce their second true leaf you may increase their survival rate by 20 percent. When seedling rootstocks are approximately 15 cm or taller, and approximately a pencil thickness in stem diameter, they are ready to be grafted. Don't graft too low on the plant because the mature plant's branches may drop to the ground. As a result, the branches may root and may become infested with root knot nematodes. Pruning or grafting high prevents the problem. Where root-knot nematodes are a problem, graft scions from a desired cultivar onto seedlings of a nematode resistant rootstock such as *Gardenia thunbergia*. *G. thunbergia* is grown principally as a rootstock and imparts vigour to species grafted on it. The most successful grafts are the splice-graft and inverted saddle graft. The inverted saddle graft is easier to make and stronger than the splice-graft.

9.3.0 Micropropagation

Tissue culture technique is a useful tool, which allows rapid production of many genetically identical plants using a relatively small space, supplied at a time (Amin and Jaiswal, 1987). Tissue culture techniques play a major role in the propagation of trees and shrubs difficult to be propagated by traditional ways. Various experiments on *in vitro* propagation of ornamental plant by root culture have been reported (Kubota *et al.*, 1995). Shoot tip culture was successfully used in many of ornamental and medicinal plants. There are many reports and experiments about micropropagation of different species of ornamental. (Liu and Li, 2001).

Micropropagation of *G. jasminoides* by shoot tips (Economou and Spanoudaki, 1986a; Serret *et al.*, 1996; Sayd *et al.*, 2010) and micro shooting (Hatzilazarou *et al.*, 2006) were reported. (Chuenboonngarma *et al.*, 2001) used the young shoots of *G. jasminoides* as explant. In vitro culture of single nodes and shoot tip from *G. jasminoides* have been reported by Duhoky and Rasheed (2010). Abdullah *et al.* (2003) stated that many plantlets were obtained by culturing shoot cuttings of Gardenia in MS nutrient media with 30 g/l sucrose, 7.0 g/l agar, and different concentrations of BA and IAA. The best combination was 1mg/l BA with 0.5mg/l IAA. This treatment gave the best shoot growth suitable for rooting in primary and secondary culture by re-culturing the rootstock cutting every 6 weeks and for many times. Successful utilization of *in vitro* techniques for propagation, maintenance, and manipulation of plant germplasms has been possible for a great number of plant species. Several studies have documented that *in vitro* micropropagation can be used for clonal propagation of *G. jasminoides*, the efficiency remained low. Anyway, micropropagation of *G. jasminoides* via in vitro organogenesis using modified Murashige and Skoog (1962) medium (MS) offers higher proliferation rate of *G. jasminoides* (Suprasanna and Bapat, 2005; Wu *et al.*, 2012).

There are many reports about production of virus- free- gardenia via meristem culture method (Tyagi *et al.*, 2010). They developed micrografting technology using for ridding planting materials of diseases. The problem of low metal absorption in *G. jasminoides* can be solved at low temperatures by grafting on *G. thunbergia*; since *G. thunbergia* was able to absorb iron at low temperatures (Mastalerz, 1977). In vitro an excellent number and percentage of successful grafts on *G. thunbergia* in greenhouse were obtained in scions derived from *Gardenia jasminoides* using MS medium supplemented with 0.50 mg/l kin (Nower *et al.*, 2013).

10.0 Soil and Climate

10.1.0 Soil

Gardenia requires a rich, moist, acidic (pH 5.0- 5.5), and well-drained soil, high in organic matter and free of nematodes. Organic matter levels can be increased with soil amendments such as peat moss, ground bark, decomposed farm yard manure or compost. The soil needs to be moist at all times, Soil should be covered with a material like sawdust to make sure the soil maintains its moisture throughout the day (Kidder *et al.*, 1991). Proper soil pH is essential because it affects the availability of mineral elements. A soil pH above 6.0 increases the possibility of micronutrient deficiencies, particularly iron. Gardenia has poor to low salt tolerance and moderate wind resistance. Iron chelate can be added to the soil to lower the p^H, but care must be taken since an overdose can kill the plant, as with any other inorganic soil amendment. Continuous effort is needed to avoid micronutrient deficiencies, most notably iron (Shober and Denny, 2008).

10.2.0 Climate

Gardenias are semi-tropical plants that perform well in a mild, humid climate. It grows best in full sunlight, while it is preferred to be shaded on hot days during the summer months. High temperatures will stop gardenia flowers from blooming. The ideal temperatures for gardenia are 18°C to 21°C during the daytime and at night around 16°C. The maximum flowering of gardenia occurs under full sun or light shade in an open flower bed (Dirr, 1990).

In most low-elevation areas in Hawaii, gardenia flowers in late spring and early summer. The night temperature where the plant is grown influences flower production. Night temperature of 15.5- 16.6°C results in almost continuous bloom if plants are in healthy condition and growing well. If night temperature is lower than this, growth is reduced and the foliage is likely to become yellow-green. Gardenia is frost-hardy to about -6.6°C. Being a plant of tropical and subtropical places, it is best cultivated in intense light, although it tolerates light shading between 10,000 and 20,000 lux (Murguía, 2007).

11.0 Cultivation

As a tropical plant, gardenia thrives best in warm temperatures in humid environments. Getting those conditions is rather hard when in non- tropical latitudes, reason for which gardenias are usually cultivated indoors or in greenhouses. In warm places, though, it can be grown outdoors, either way, it prefers bright indirect sunlight or partial shade, rather than direct sunlight. Apart from the difficulties in creating the suitable conditions for the plant to live, gardenias need to be planted in an acidic soil (it is an acidophile plant). Its needs a moist soil with good drainage and desirable growing median pH between 4.5- 5.5.

11.1.0 Manuring and Fertilization

Proper fertilizer application is important for gardenia growth and flower production. A complete fertilizer with a ratio of 3:1:2 or 3:1:3 (for example, 15- 5- 10 or 15- 5- 15) of nitrogen (N), phosphorus (P_2O_5), and potassium (K_2O) is generally recommended. Apply the fertilizer two or three times per year. For continuous bloom, feed regularly with blood meal, fish emulsion, or a fertilizer formulated for "acid- loving" plants. Too much fertilizer builds up excessive levels of soluble salts in the soil, which can cause root dehydration. Therefore, it is important to follow the recommendations on the fertilizer product label. Frequently, due to a deficiency of one or more micronutrients, usually iron, gardenia leaves will become yellow (chlorotic). This deficiency can often be corrected by acidifying the soil (Yeager and Gilman, 1991).

11.2 Irrigation

Irrigation during dry periods is necessary for healthy gardenias. The soil should be kept moist at all times with regular watering, but it should not be soggy. Poorly drained, wet soils or excessive watering excludes oxygen, thereby causing root injury. If the soil is allowed to dry, shrinking and cracking also will injure roots. Irrigating with drip systems keeps water off the foliage and flowers, which helps prevent leaf spots. To help maintain adequate soil moisture, use mulch and avoid cultivation around the base of the plant. Gardenia has moderate drought tolerance. Watering is important because it affects the number of flower buds that remain to maturity on a plant. If water stress occurs in a heavily budded plant, some buds will fall before opening. Therefore, large variations in soil moisture should be avoided. Regular watering is necessary after blooming to keep plants in good condition (Kobayashi and Kaufman, 2006).

11.3.0 Pruning

Prune to keep plants compact, in proper shape, and in scale with the landscape. To maintain plants at their optimum, prune out dead wood, straggling branches, and faded flowers. Young plants growing vigorously during their first year may be pinched back at the growing points to encourage heavy branching; this prevents scraggly growth and promotes a more compact plant. The best time to prune is after flowering, because pruning earlier removes flower buds. Follow pruning with an application of fertilizer (Kobayashi *et al.*, 2006).

Cultivation of Gardenia

In Greenhouse

As a Bonsai

Growth of Seedling in Open

11.4.0 Cultivation in Pot

Gardenias do well in pots. Give the plant plenty of bright light, preferably direct sunshine for at least half a day in a sunny south or west window. Ample humidity is also important. The humidity around the plants can be increased by placing them on a pebble tray with clean pebbles, stones, or gravel partially filled with water. Running a room humidifier is an alternative. Gardenia plants prefer temperatures close to 15.5 °C at night and 21.11- 23.88 °C during the day. Maintaining uniform moisture and good drainage in the growth medium is outmost important and care should be taken not to make the soil dry. Fertilize with an acidifying fertilizer to keep the soil slightly acid, following the product label. Because gardenia is susceptible to salt burn, too much fertilizer can lead to damaging salt accumulation.

Repotting is needed when roots start to grow out from the bottom of the pot. Repot with one of the commercially prepared mixes available at garden supply stores or plant nurseries. To avoid nematodes, use a soilless or sterilized medium. A slightly acid medium high in peat moss is a good choice; the pH should be near 6.0, or slightly less. Using an acidulating fertilizer helps maintain favourable pH for gardenia (Trinklein *et al.*, 1998).

12.0 Diseases, Pests and Disorders

Gardenia is affected by pests and diseases that tend to impair its health and ornamental characteristics. The most common diseases can be of fungal or bacterial origin, and pests include nematodes, mites and insects. Mites and insects in turn can be vectors of viruses, apparently little studied in this plant species. There are also physiological disorders, whose symptoms may appear to be a disease (Yanelis Castilla- Valdés, 2018).

12.1.0 Diseases

12.1.1 Powdery Mildew

It is a fungal disease caused by *Erysiphe polygoni,* typically grow on the surface of a plant. Signs of infection include a whitish to grayish coloured, powdery, fungal growth that can occur in spots or patches, which sometimes enlarge to cover the entire leaf or other plant structure. The disease typically appears in the summer and reaches its peak in late summer. It primarily affects young leaves and shoots and causes deformed leaves and buds, leaf yellowing, and leaf drop.

Some recommended fungicides include emyclobutanil, thiophanate- methyl, chlorothalonil, propiconazole, *etc.* also applied to prevent powdery mildew. Fungicides are most effective against powdery mildew when used early in the disease development process. Some control may be obtained by sprays of sulfur or neem oil (https: / /hgic.clemson.edu, Kobayashi and Kaufman, 2006).

12.1.2 Root Rots

Various fungi, including *Phytophthora* species, *Rhizoctonia* species, and *Pythium* species, cause root rots in gardenias. Due to this disease some above-ground symptoms like leaf yellowing (oldest leaves first), leaf drop, stunting and wilting are appearing. The best management practice for root rot problems is prevention. Check the roots of nursery stock before purchasing. Also avoid overwatering, avoid planting in heavy soils. Fungicides should be used as a preventative measure and those used for managing *Pythium* include fosetyl- aluminum, phosphorous acid, fluopicolide, mefenoxam chlorothalonil *etc.* (https://hgic.clemson.edu, Kobayashi and Kaufman, 2006).

12.1.3 Canker

Canker of gardenia is typically caused by *Diaporthe gardenia*, appear as swollen, circular brownish lesions sunken with a raised edge which may develop into a gall (a swollen area resulting from infection). Cankers are most often found on stems (especially near soil level) of a gardenia; however, leaves and roots are also susceptible. The first symptom that often is noticed is the wilting of leaves on part or all of the plant. Leaves on affected branches often become a duller green colour, may yellow, and then drop. Flower buds on affected branches may drop without opening. Some preventative applications of fungicides containing chlorothalonil, thiophanate methyl, propiconazole,mancozeb are effective to control this disease (https://hgic.clemson.edu, Kobayashi and Kaufman., 2006).

12.1.4 Leaf Spots

Leaf spots of gardenia mainly caused by *Myrothecium roridum* which initially resemble as drops of water. Due to this disease leaf turns brown, and usually the center of the lesions falls, giving the appearance of a hole. It can be controlled by closely monitoring the nutrition of the plants. Application of fungicides containing azoxystrobin or pyraclostrobin, chlorothalonil, mancozeb are also effective.

12.1.5 Sooty Mould

Sooty mould is caused by a group of related fungi that grow upon sugary exudates, or honeydew, secreted by sucking insects such as aphids, scales, mealybugs, and whiteflies. Sooty mould causes charcoal- black, powdery or velvety coating on the surface of leaves and other plant parts. As a result it reduces the vigour of plants by blocking sunlight necessary for photosynthesis (food production). On the other hand, insect feeding can seriously damage a plant. Since the cause of the problem is insect activity, control measures should be directed at the insects rather than the fungus. Important first steps include identifying the plant that is infested and the insect that is causing the problem. Sooty mould usually weathers away following control of the insect infestation. Once the insects are controlled, wash the sooty mould off the leaves with soap and water (https://hgic.clemson.edu, Kobayashi and Kaufman, 2006).

12.2.0 Pests

12.2.1 Aphids (*Myzus persicae*)

Aphids are small, pear- shaped insects appear in cluster on stems, leaves, buds and flowers, suck the sap from these infected parts and developed deforms buds and discolours, curls foliage. It can be managed by applying Imidacloprid, also by

Diseases of Gardenia

Powdery Mildew

Stem Canker

Root Rots

Leaf Spots

Sooty Mould

Stem Gall

Iron Chlorosis

Bud Drop

removing with alcohol- saturated cotton swab, or wash plants with soapy water and a soft brush or cloth.

12.2.2 Citrus Mealybugs (*Planococcus citri*)

Planococcus appears as white cottony masses in the leaf axils, underside of leaves, other protected areas and suck plant juices, while heavy infestations will coat the leaves with sticky honeydew. Malathion®, Sevin®, or imidacloprid (Merit®) insecticides are effective to control mealybugs. Use of an alcohol saturated cotton swab to remove these, or wash plants with soapy water and a soft brush or cloth, or pick off with tweezers or a toothpick are also effective.

12.2.3 Scales (*Ceroplastes japonicus*)

Scale are sedentary insects that are usually green or brown hard shells attached to the stems or undersides and upper sides of leaves, often near the midrib and suck plant juices; appear as raised green or brown bumps that make stems or leaves look lumpy. It can be effectively controlled by Imidacloprid insecticide or Supreme horticultural oil. Removal of these insects with alcohol-saturated cotton swab, or wash plants with soapy water and a soft brush or cloth is recommended. Thoroughly wash undersides of leaves where pests may also reside.

12.2.4 Spider Mites (*Tetranychus urticae*)

It is a small pest primarily feed on the underside of the foliage. Spider mites suck juices from leaves causing fine stipples on foliage and silvery webs, causes colourless or whitish spots on foliage. Ultra-fine oil, insecticidal soap, neem and mitacide like Kelthane® are used to control it. Wash foliage with a strong spray of cold water and the plants with soapy water to remove insects (Kobayashi and Kaufman, 2006).

12.2.5 Thrips (*Frankliniella tritici*)

Thrips are tiny black insects that feed on flowers and leaf undersides, causes browning of the flower and petal margins, distortion of flowers or failure of buds to open. Apply Orthene insecticide to control it, but it should not be used while the flowers are still in bud, because it can burn the petals. Pyrethroid insecticides are also effectively control thrips (Kobayashi and Kaufman, 2006).

12.2.6 White Flies (*Dialeurodes citri*)

White flies are sap- feeding insects that feed on underside of leaves; develop White, cottony appearance of leaf undersides and helps in accumulation of black sooty mould. It can be controlled by successive sprays of insecticidal soaps or oils, synthetic pyrethrum, or imidacloprid insecticide.

12.2.7 Root-knot Nematode (*Meloidogyne incognita*)

Root-knot nematode are microscopic, parasitic round worms (*Meliodogyne* sp.) that live in and feed on gardenia roots, damages root and prevents normal uptake of water and nutrients which results in poor growth, stunted, chlorotic (yellow) foliage; premature wilting, low vigour, thin canopy, and leaf and/or bloom loss under relatively mild stress. Root knot nematode can be controlled by soil fumigation prior to planting. Some other useful techniques like graft onto nematode-resistant

Pests of Gardenia

Aphid

Whitefly

Thrips

Red Spider Mite

Citrus Mealybug

Root-knot Nematodes

rootstock *G.thunbergi* or incorporate wood shavings or organic matter in the soil mass to depress nematode population.

12.3.0 Disorders

12.3.1 Bud Drop

Bud drop can be seen in gardenia under stress condition. Unopened flower buds drop off. Potential causes of this include root injury, infestations of thrips or aphids, root feeding by nematodes, too much fertilizer, over- watering, under-watering, poor soil drainage, insufficient light, unusually cool weather, rapid drops in temperature, or very hot, dry weather. So it can be managed by controlling these stressors (https://hgic.clemson.edu, Kobayashi and Kaufman, 2006).

12.3.2 Chlorosis

A lack of normal green pigmentation in foliage, generally due to deficiency of one or more micronutrients (usually iron), root rot, or root- knot nematodes causes Pale green leaves with dark green veins, yellowing of foliage. But some leaf yellowing on older leaves is normal. Overall, the plant may become stunted. It can be managed by acidify the soil with aluminium sulphate, iron sulphate, or wettable sulphur. Iron chelate may be used on the soil or foliage.

13.0 Harvesting and Postharvest Management

13.1.0 Harvesting

Flowers of gardenia is harvested by cutting, not pulling, the stem below the flower and calyx (the green, leafy base). Harvest buds in the morning and open flowers in the afternoon. Use a plastic bag or container to carry the buds and flowers from the picking area. Gardenia flowers are considered very perishable and the cut flowers are primarily used in weddings, harvested as naked buds that are opened manually, and decorated with a rosette of plastic leaves. Jones (2001) reported that gardenia petals turn brown when crushed, and recommended holding buds by the stem to minimize petal bruising.

13.2.0 Postharvest Management

Cut flowers are misted with water, or sprayed with an anti-transpirant to reduce water loss during postharvest handling and transport. A lab-formulated preservative containing sucrose, $Al_2(SO_4)_3$, and citric acid was also effective in extending the vase life of cut gardenia flowers. Carbohydrates improve vase life by serving as source of energy and nutrition, especially for flower stems with developing buds (Reid and Jiang, 2012; Çelikel *et al.*, 2015). In contrast, another commonly used biocide, aluminium sulphate (AS), was very effective in extending the life of gardenia flowers.

14.0 Biotechnology

14.1.0 Embryogenesis

Young leaves of *G. jasminoides* have been studied for callus induction (Al-Juboory *et al.*, 1998; Mizukami *et al.*, 1987). There were some successful reports on development of plantlets from callus raised through leaf cultures of *Gardenia jasminoides* (Al- Juboory *et al.*, 1998). Reddy and Saritha (2012) reported that in callus induction from leaves of *in vitro* grown plantlets of *Gardenia latifolia*, high frequency was observed on MS medium supplemented with 3.0 mg/l 2,4-D. George *et al.* (1993) reported that ovary culture can be used for callus initiation via immature ovary portion of flower of *G. jasminoides*. There were some successful reports on development of plantlets of gardenia from callus raised through embryogenic callus formation on MS medium with various combinations of BA (1.0-2.0 mg/l), Kin (2.0 mg/l) and IAA (0.1 mg/l) (Reddy and Saritha, 2012).

14.2.0 Molecular Markers

RAPD- PCR was the molecular marker used to detect the genetic polymorphism within *G. jasminoides* due to the action of both sodium azide and guanidine hydrochloride. Li *et al.* (2009) performed a study to evaluate the genetic differences and mutual relationship of *G. jasminoides* at DNA level among different cultivars using ISSR molecular marker and obtained a polymorphism percentage of 55.17 per cent. Wannajindaporn *et al.* (2014) used tissue culture technique to study the effect of sodium azide on Dendrobium ornamental plant and used ISSR molecular markers to estimate the variation in mutants compared to control. This agrees with many workers which proved the genetic variation in different plants due to chemical mutagens treatments. Kishk *et al.* (2016) used ISSR molecular marker to estimate both diethyl sulphate and sodium azide mutagen on banana plant to against salt stress. guanidine hydrochloride is a more powerful mutagen and causes more genetic variations and instability than sodium azide, as was reflected in morphological parameters (*i.e.*, fresh weight, shoot length, root length, number of lateral roots, number of leaves, numbers of branches, leaf length, and leaf width) and physiological traits (*i.e.*, chlorophyll pigmentation and carotenoids) (Tawfik *et al.*, 2020).

Tsanakas *et al.* (2014) initiated a de novo transcriptome study to investigate the petal senescence in four developmental stages of cut gardenia flowers considering that the visible symptoms of senescence appear within 4 days of flower opening. De novo assembly of transcriptome sequencing resulted in 102,263 contigs with mean length of 360 nucleotides that generated 57,503 unigenes. These were further clustered into 20,970 clusters and 36,533 singletons. The comparison of the consecutive developmental stages resulted in 180 common, differentially expressed unigenes. A large number of Simple Sequence Repeats were also identified comprising a large number of dinucleotides and trinucleotides. The prevailing families of differentially expressed TFs comprise the AP2/EREBP, WRKY and the bHLH. There are 81 differentially expressed TFs when the symptoms of flower senescence become visible with the most prevailing being the WRKY family with 19 unigenes. No other WRKY TFs had been identified up to now in petal senescence of ethylene insensitive species. A large number of differentially expressed genes were identified at the initiation of visible symptoms of senescence compared to the open flower stage indicating a significant shift in the expression profiles which might be coordinated by up-regulated and/or down-regulated TFs. The expression of

16 genes that belong to the TF families of WRKY, bHLH and the ethylene sensing pathway was validated using qRT- PCR.

15.0 References

Abdullah, G.R., Al-Khateeb, A.A. and Serage, M. (2003) *J. Agricul. Mar.* Sci., **8**(1) : 35-40.

Al-Juboory, K.H., Skirvin, R.M. and Williams, D.J. (1998) *Sci. Hortic.*, **72** : 171-1788.

Al-Razzaq, S.A. and S. Younus. (2017) *Biosci. Biotechnol. Res. Asia*, **14** : 757-766.

Álvarez, A. (2008) *Rev. Jard. Bot. Nac.*, **29** : 83-100.

Amin, M.N. and Jaiswal, V.S. (1987) *Plant cell tissue Organ cult.*, **9** : 235-243.

Beel, E. and de Bruyn, P. (1992) *Verbondsnieuws Belg. Sierteelt*, **19** : 1035-1037.

Bender, S. (2011) *A Brief History of the Gardenia*. Southern Living. Birmingham, AL: Time Inc. Lifestyle Group.

Bensky, D. (2004) *Chinese Herbal Medicine Materia Medica* (3rd ed.). Seattle: Eastland Press, Inc. p. 9.

Bradshaw, J. (2003) *Gardenias*. CIR 1098. Environmental Horticulture Department, UF/IFAS Extension, USA.

Çelikel, F.G., Rybczyñski, J.J., Davey, M.R., Miku³a, A. (2015) *Post-harvest physiology of flowers from the family gentianaceae*. In: The Gentianaceae-Biotechnology and Applications. Springer Berlin Heidelberg, Berlin, Heidelberg, **2** : 287-305.

Chen, J. (1985) *Acta Hortic.*, **167** : 23-35.

Chen, T., Luo, X., Zhu, H., Taylor, C.M., Ehrendorfer, F., Lantz, H., Funston, M., Puff, C. (1761) "*Gardenia jasminoides*". Flora of China.

Chuenboongarma, N., Haroonsotea, S. and Bhamarapravatia, S. (2001) *Sci. Asia*, **27** : 137-141.

Clemson-Cooperative-Extension. (2009) *Gardenia diseases and other problems*, HGIC, USA.

Conover, C.A., Sheeman, T.J. and Poole, R.T. (1968) *Fla. Agric. Exp. Stn. Journal Series*. University of Florida, Gainesville, USA, pp. 401-408.

Danuta, K. (2011) *Acta Sci. Pol.*, **10**(4) : 65-73.

De Graaf-Van der Zande, M.T.H. (1988) Proefstation voor de Bloemisterij. Rapport 56. Aalsmeer, Nededand, p.20.

Dirr, M.A. (1990) *Manual of woody landscape plants and their identification, ornamental characteristics, culture, propagation and uses*. 4th edition. Stipes: Shampaign, Illinois.

Duhoky, M.M.S. and Rasheed, K.A. (2010) *Journal of Zankoy Sulaimani*, **13**(1): 103-120.

Dumitrescu, R. (2002) *Rom. J. Biol. -Plant Biol.*, **47**(1-2) : 19-26.

Economou, A.S. and M.J. Spanoudaki. (1985) *Hort. Sci.*, **20**(2) : 213

Economou, A.S. and Spanoudaki, M.J. (1986a) *Sci. Hortic.*, **20**(155-161) : 213-220.

Ellis, J. (1773) Directions for bringing over seeds and plants, from the East-Indies and other distant countries.

El-Ashry, A. A. E., Gabr, A. M. M., Girgis, N. D. and El-Bahr, M. K. (2018) Influence of silver nitrate on enhancing in vitro rooting of Gardenia jasminoides Ellis. *Environ. Sci. Technol.*, **11**(5) : 238-245.

Ellis, J., Webb, P.C. (1759) Philosophical Transactions of the Royal Society of London, **51** : 929-935.

Fujisawa, K. and Kuki, M. (1976) *Bull. Nara Agric. Exp. Stn. Jpn.*, **7** : 18-23

Gabarkiewicz, E. and Rudnicki, R.M. (1995) *Acta Agrobot.*, **48**(2) : 105-111.

George, P.S., Ravishankar, C.A. and Venkatarman, L.V. (1993) *Plant cell culture.*, **13**(1) : 59-62.

González, M.E. (2016) *Polianthes tuberosa L.: Review of its phylogenetic, morphological and cultural aspects*. Trop crops. **37**(3) : 120-136.

Govaerts, R., Ruhsam, M., Andersson, L., Robbrecht, E., Bridson, D., Davis, D., Schanzer, I. and Sonke´, B. (2010) World Checklist of Rubiaceae. Royal Botanic Gardens, Kew. Published on the internet at www.kew.org/wcsp/rubiaceae/.

Han, J.P., Chen, S.L., Zhang, W.S., Wang, Y.Y. and Li, X.Y. (2007) *Chin. Pharm. J.*, **42**(23) : 1774.

Hatzilazarou, S.P., Syros, T.D., Yupsanis, T.A., Bosabalidis, A.M. and Economou, A.S. (2006) *Plant Physio.*, **163**(8) : 827-836.

Huh., K.M. (2007) *J. Life Sci. Res.*, **17**(1): 24-30

Husseini, M. (1993) *Production of cut flowers and roses* (Gladiolas, Roses, and lily). Ibn Sina Library - Cairo, p.159.

Hutchinson, W.A. (1980) *Plant Propagation and cultivation*. Cincinnati, Ohio, USA: Westport, Conn.

Jones, R. (2001) *Caring for cut flowers*. Landlinks Press. Australia, p.190.

Judd, W.S., Campbell, C.S., Kellogg, E.A., Stevens, P.F., Donoghue, M.J. (2016) *Plant systematics: a phylogenetic approach*, 4th edn. Sunderland, MA: Sinauer Associates.

Kidder, G., Black, R.J. and Ruppert, K.C. (1991) *Soil pH and landscape plants*. Factsheet SL-113. Florida Cooperative Extension Service.

Kishk, A.D., Abul-Soad, A.A., Abbas, M.S., El-Shabrawi, H.M., Gaber, E.I. and Noor El-Deen, T.M. (2016) *J. Chem. Tech. Res.*, **9**(12) : 81-99.

Kobayashi, K.D. and Kaufman, A.J. (2006) *Common gardenia. Ornamentals and flowers.* Cooperative Extension Service, College of Tropical Agriculture and Human Resources, University of Hawaii at Manoa, USA.

Kosugi, K., Oyoshi, K., Sumitomo, A. and Kaneko, M. (1957) *J. Jpn. Soc. Hort Science*, **26** : 37-42.

Kozak, D. (2011) *Acta Sci. Pol.*, **10**(4) : 65-73.

Kubota, H., Sato, K., Yamada, T. and Maitani, T. (1995) *Plant Sci.*, **106**(2) : 157 -166.

Li, L., Yi, W., A-Man, Z., Pei-Lin, Z. and Shi-Liang, Z. (2009) *Chin. Tradit. Herbal Drugs*, **40**(1): 116-120.

Liu, Z. and Li, Z. (2001) *In vitro cellular and Developmental Bio.*, **37**(1) : 84-88.

Low, Y.W. (2013) *Syst. Bot.*, **38**(1) : 235-241.

Low, Y.W. and Wong, K.M. (2007) *Edinb. J. Bot.*, **64**(1) : 25.

Mastalerz, J.W. (1977) *The Greenhouse environment*, John Wiley and Sons USA. pp. 629.

Mizawa, M., Andoh, T., Shimizu, T. (2020) *JAMA Dermatology*, **156**(3) : 351.

Mizukami, H., Hirano, A. and Oashi, H. (1987) *Plant Sci.*, **48** : 11-15.

Mostafa, M.A., Alawlaqi, M.M. and Reyad, N.E.A. (2013) *J. Microbiol. Res.*, **3** : 185-196.

Mousa, G.T., Habdul-Hafeez, E.Y. and Ibrahim, O.H.M. (2015) *Pak. J. Agri. Sci.*, **52** : 651-658.

Murguía, J. (2007) *Production of orchid, anthurium, gardenia and bird of paradise.* Veracruzana University, Produce Veracruz Foundation.

Nanakorn, M., Boonmahamongkol, P., Suwanwong, S., Tantiviwat, S. and Kaveeta, L. (2007) Propagation of Gardenia obtusifolia Roxb. through tissue culture. In Proceedings of the 45th Kasetsart University Annual Conference, Kasetsart, Kasetsart University. pp. 491-497.

Nower, A.A. and Hamza. E.M. (2013) *Res. J. Appl. Sci.*, **9**(4): 3118-3128.

Osunkoya, O.O. (1999) *Biol. Conserv.*, p.347-359

Ozaki, A., Kitano, M., Furusawa, N., Yamaguchi, H., Kuroda, K., Endo, G. (2002) *Food Chem. Toxicol.*, **40**(11) : 1603-1610.

Palmateer, A.J. and Chase, A.R. (2016) *Diseases of gardenia.* In: RJ McGovern, and WH Elmer, editors, *Handbook of plant disease management: Handbook of florist's crops diseases.* Springer Int. Publishing AG, SUI. p. 2-13.

Park, S. and J. Bradshaw. (2016) *Gardenias at a glance.* Environmental Horticulture Department, UF/IFAS Extension, CIR 1098, USA.

Pérez, L., Fiallo, V.R. and González, L.R. (2011) *Jard. Bot. Nac.*, **30** : 187-201.

Pontikis, C.A. (1983) *Plant Propag.*, pp.13-14.

Reddy, Y. and Saritha, K. (2012) *Agrofor. Syst.*, **87**(3) : 591-598.

Reid, M.S., Jiang, C.Z. (2012) *Postharvest biology and technology of cut flowers and potted plants.* Chapter 1. Horticultural Reviews. Book Editor: Jules Janick, **40**: 1-54.

Rose, S.A. and Dickey, R.D. (1960) *Proc. Fla. State Hortic. Soc.*, **73** : 362-363.

Saravanakumar, K., Chelliah, R., Shanmugam, S., Varukattu, N.B., Oh, D.H., Kathiresan, K. and Wang, M. H. (2018) *J. Photochem. Photobiol. Biol.*, **185** : 126-135.

Sayd, S.S., Taie, H.A. and Taha, S.L. (2010) *Int. J. Acad. Res.*, **2**(3) : 184-191.

Serret, M.D., Trillas, M.I., Matas, J. and Araus, J.L. (1996) *Plant Cell, Tissue and Organ Cult.*, **45** : 1-16.

Shober, A. and Denny, G. (2008) *Soil pH and the home landscape or garden, Soil and Water Science* Department, Institute of food and Agricultural Sciences, University of Florida, Gainesville, FL.

Sosa-de-Castro, N.T., Álvarez, R.E. and Cabrera, M.G. (2003) Occurrence of Pestalotia sp. causing necrotic lesions in Cape Jasmine (Gardenia augusta) plants. National University of the Northeast, ARG.

Steven, F., Chongxi, Y. (1992) *Herbal emissaries*: bringing Chinese herbs to the West: A Guide to Gardening, Herbal Wisdom, and Well-Being, Healing Arts Press. pp. 185

Suprasanna, P. and Bapat, V.A. (2005) *Indian Soc. Environ. Sci. and Tech.*, p. 26.

Tawfik. E. and Fathy, M. (2020) *Plant Tissue Cult. and Biotech.*, **30**(2) : 209-218.

Tian, Y., Pu, X., Yu, H., Ji, A., Gao, R., Hu, Y., Wang, H. (2020) *Bio. Med. research international*, https: //doi.org/10.1155/2020/2903861.

Tsanakas, G.F., Polidoros, A.N. and Economou, A.S. (2013) *Sci. Hortic.*, **162** : 213-217.

Tyagi, P., Sharma, P.K. and Kothari, S.L. (2010) *Protocols for In Vitro Propagation of Ornamental Plants*, New York, USA: Humana Press, pp.39-46.

Valder, P. (1999) *Garden Plants of China*. Glebe, NSW: Florilegium, p. 289.

Vidal, M.C., Ávila, O.E. and Delgado, L.C. (2001) Advances in Plant Production of the NOA Faculty of Agronomy and Zootechnics, National University of Tucumán, ARG.

Wang, W., Shao, F., Deng, X., Liu, Y., Chen, S., Li, Y., Guo, W., Jiang, Q., Liang, H. and Zhang X. (2020) *Genet. Resour. Crop Evol.*, p.1-16.

Wannajindaporn, A., Poolsawat, O., Chaowiset, W. and Tantasawat, P.A. (2014) *Genet. Mol. Res.*, **13**(3) : 5333-5342.

Webb, E., Solander, D.C. (1761) Philosophical Transactions of the Royal Society of London, **52** : 654-661.

Wilkins, H.F. (1986) *Gardenia jasminoides*. In: A.H. Halevy (Editor), *Handbook of Flowering*. Vol. V. CRC Press, Boca Raton, FL, USA, pp. 127-131.

Wong, K.M. and Low, Y.W. (2011) *Edinb. J. Bot.*, **68** : 11-32.

Wu, Y., Wu, R., Zhang, B., Jiang, T., Li, N., Qian, K., Liu, B. and Zhang, J. (2012) *Plant Growth Regul.*, **66** : 137-143.

Yeager, T.H. and E.F. Gilman. (1991) Fertilization recommendations for trees and shrubs in home and commercial landscapes. Circular 948. Florida Cooperative Extension Service.

Zhao, K. and Zhou, Y. (2020) The chloroplast genome of Gardenia jasminoides and related phylogenetic analysis (Rubiaceae), Mitochondrial DNA Part B, **5**(2) : 1743-1745

Zongram, O., Ruangrungsi, N. and Rungsihirunrat, K. (2017) *Songklanakarin J. Sci. Technol.*, **39**(4) : 471-477.

27

Fuchsia

Debmala Mukherjee and Sk. Masudul Islam

1.0 Introduction

Fuchsia is both the common and the Latin genus name for a group of about 110 species that are found in the wild in Central and South America. The genus was named in 1703 by Father Charles Plumier in honour of a German botanist of the 16th century named Leonard Fuchs. The pronunciation of the genus in English usage is *"fyew'sha"*, or alternatively *"fyewk'see-ah"* – somewhat truer to the pronunciation of Fuchs. The native habitat of most Fuchsia species is cool, damp, densely forested areas of the Andes, although most of the cultivars are hybrids of three species: *F. magellanica, F. coccinea*, and *F. fulgens*. The crop is grown for its showy pendulous flowers which vary in colour from solid to combinations of white, red, pink, or purple. Fuchsias are produced and marketed as hanging baskets, potted plants, bedding plants or as novelties in a tree-like shape. These dainty drops of ballerina blossoms enchant us with their profusion from summer into fall.

2.0 Importance and Uses

2.1.0 Aesthetic Uses

Fuschia hybrida can be used to create a loose border of plants that forms a natural, informal barrier between areas of garden or yard. Fuchsia need to be grouped together and planted directly in the soil in order to grow large enough to become hedge-like plant. Fuchsia can grow at least 90 cm wide and it needs adequate space in this naturally large and bushy form. Hanging baskets show off the natural cascading beauty of the Fuchsia plant, livening up patio and porch spaces. *Fuchsia hybrida* needs moisture and hanging baskets made from natural materials tend to drain quickly, so it requires frequent watering to hanging Fuchsia than planted in the soil. Fuchsia can be trained to climb up a trellis to cover an unsightly garden wall or create a backdrop for flower garden. This dramatic plant pairs well with flowers such as Lavender Trumpet Vine (*Clytostoma callistegioides*) and Giant Burmese Honeysuckle (*Lonicera hildebrandiana*). Climbing together up a trellis, these vining plants will create a wall of gorgeous, showy blooms.

2.2.0 Therapeutic and Other Uses

Fresh leaves of several Fuchsia varieties are ethnomedicinally applied on wounds (Péntek and Szabó, 1985; Csepregi *et al.*, 2020), furuncles and skin inflammation as a fomentation (Papp, 2018; Boris, 2010). Anthocyanins were detected in the flowers and berries of Fuchsia species (Crowden *et al.*, 1977; Ruiz *et al.*, 2013), while in their leaves several flavonoids were present (Williams and Garnock-Jones, 1986).

Fuchsia hybrdia's blossoms and berries are edible. Often used in salads, these blooms are attractive but acidic, so use in small quantities. Fertilization in spring with organic fertilizer helps to promote flower growth when growing for culinary uses.

Humming birds are attracted to flowers that are red and pink with tubular blossoms, and *Fuchsia hybrida* is a favourite food source. Keep hummingbirds in the

Uses of Fuchsia

In Pot | Hanging Basket | Window Basket

In Decor | In Trellis | Bonsai

Garden Pots | As Hedge | As Standard

Garden Wall | In Arch | In Decoration

Source: www.green-fuse.com, www.flowerpower.com.au, www.cz.pinterest.com, www.beekenkamp.nl, www.best4hedging.co.uk, www.gardeningexpress.co.uk, www.flickr.com, www.touchofelegancellc.com

garden year-round with humming bird feeders to provide them with nourishment when the Fuchsia goes dormant in the winter.

The coloured fuchsia was first introduced as the colour of a new aniline dye called fuchsine, patented in 1859 by the French chemist François-Emmanuel Verguin. The dye was renamed magenta later in the same year, to celebrate a victory of the French army at the Battle of Magenta on June 4, 1859, near the Italian city of that name (Ball, 2003). The first recorded use of *fuchsia* as a colour name in English was in 1892 (Maerz and Paul, 1930).

3.0 Origin and History

The first fuchsia was discovered in the Dominican Republic in the late 17th Century by Father Charles Plumier, a missionary and botanist of that time. In 1703, Plumier formally published the name of his discovery as *Fuchsia triphylla, flore coccineo,* in honour of Leonhard Fuchs (1501-1566), a German physician and botanist Plumier held in high regard for his *De Historia Stirpium* (1542), the first botanical work combining very accurate description and illustrations of plants depicting flowers, fruits and seeds together on the same plant. Fuch's *De Historia Stirpium,* with its illustrations and botanical glossary, is the precursor of Botanical Illustration. Incidentally, Fuchs' name was pronounced as 'Fooks' so perhaps Fuchsias is pronounced as 'fooksias'. This original fuchsia can still be grown today under the name of *F. triphylla.*

The plant samples and seeds Charle Plumier collected in Hispaniola, among which were possibly *Fuchsia triphylla,* were lost in a shipwreck. Fortunately, the drawings and description were travelling back to Europe on another boat. So actual plants of *Fuchsia* did not reach the UK before 1788, when Captain Firth brought two species back from his South American trips, these were probably *F. coccinea* from Brazil and *F. magellanica* from Chile. Although this has been disputed by many and may even be part of the embellished story told by a Fuchsia salesman, James Lee, to sell his plants. This story was however recounted by Dickens in Household Words.

The Fuchsia has been in cultivation in Britain for roughly 200 years. The first authentic drawing of it was published in 1725 by Feuville. The actual date of the introduction into Europe of living plants is given as being between 1786 and 1789 when *F. magellanica* and *F. coccinea* were being grown. From this date the introduction of other species, doubtless by seed, continued, estimated dates are *F. lycoides* 1796, *arborescens* 1824, *microphylla* 1827, *fulgens* 1830, *corymbiflora* 1840. From then on more species seed became available and hybridisation started in earnest.

Fuchsias were at their peak of popularity in the Victorian times when the head gardeners of large houses grew pillars, standards and pyramids to line the driveways. One such gentleman was James Lye who was head gardener at Clyffe Hall, Market Lavington. He became a grower, exhibitor and hybridist of fuchsias and produced plants eight to ten feet high and four to five feet across the base and by 1866 was described as Champion Fuchsia Grower in the West of England. This popularity lasted up until the First World War when they were grown in the greenhouses and turned over to producing food. After the war the interest in fuchsias continued, particularly in the U.S.A when the American Fuchsia Society was founded in 1929.

Members travelling in Europe took back varieties which the breeders got to work on and produced a range of plants with bigger and better blooms.

From over 100 species of Fuchsias, most originate from Central and South America, but a few come from New Zealand and have the peculiarity of having blue pollen. This pollen was used by young Maori people to adorn their face, probably well before the official discovery of the Genus. In the next few years many more Fuchsia species reached the UK from South America and a plant hunting and hybridising craze soon began.

4.0 Species, Types and Cultivars

4.1.0 Species

The genus *Fuchsia* contains over 110 species placed in 12 sections that have been distinguished based on geographical, morphological, molecular and cytogenetic characters (Goulding, 1995; Berry *et al.*, 2004). The crop is animal-pollinated, mostly by humming birds (Cuevas *et al.*, 2013). Most species are hermaphroditic, although nine dimorphic species have been reported (Berry *et al.*, 2004). The history of the garden fuchsias shows that they are of hybrid origin. *Fuchsia* breeders have utilized only a small number of species, despite the rich genetic diversity of the genus (Hoshino and Berry, 1989; Goulding, 1995; Godley and Berry, 1995). The species which probably played the greatest part in the development of fuchsias are *F. coccinea, F. fulgens* and *F. magellanica.*

Fuchsia coccinea **:** It has scarlet tube and sepals with a purple corolla, the sepals are not reflexed, and the flowers are small.

Fuchsia fulgens: It has orange tube and sepals, the sepals having pale green tip, and a bright orange corolla. The tube is very long up to 2¾ inches (7cm). There are three varieties of this fuchsia which are often sold as *F. gesnariana.*

Fuchsia magellanica: It has a scarlet tube and sepals and the corolla is purple. The sepals are held horizontally. This is the hardy fuchsia seen in many gardens.

Fuchsia microphylla **:** It is characterised by the presence of male sterility (Berry *et al.*, 2004), and is one of the few subdioecious species (individuals that function as females, males and hermaphrodites at lower frequency coexist in a plant population) of the genus, making it a good system to study breeding system evolution.

Paniculate Fuchsia: Another pure species that makes a good garden plant is paniculate Fuchsia (*Fuchsia paniculata*). The sheer exuberance of this plant in bloom is captivating, as its vibrant pink flowers and ornamental berries are produced on erect shrubs that can live for 20 years in the right conditions. It demands mild, frost-free temperatures and acidic soil high in organic matter. With its slow to medium growth habit, *Fuchsia paniculata* is a good candidate for training as a tree standard.

Other important species include *Fuchsia denticulata, Fuchsia cordifolia, Fuchsia racemosa* and *Fuchsia corimbiflora.*

Fuchsia forms can be of three types *viz.* trailing, bush or vines. Trailing fuchsias have lax, cascading stems that look wonderful in a hanging basket. Bush fuchsias

Species of Fuchsia

Fuchsia coccinea

Fuchsia fulgens

Fuchsia paniculata

Fuchsia magellanica

Fuchsia microphylla

Fuchsia arborescens

Fuchsia denticulata

Fuchsia cordifolia

Fuchsia excorticata

Fuchsia boliviana

Fuchsia lycioides

Fuchsia hatschbachii

Source: www.en.wikipedia.org

have upright growth that can be used in a container or planted out in a garden. Bush and trailing fuchsias can also be trained into standards, with a long bare stem surmounted by foliage and flowers. The vine fuchsias are new on the market, and can be planted out in the garden to grow on a trellis or porch.

Hardiness of fuchsias varies, depending on the original species hardiness. 'Hardy' means the plant survives over winter in the ground, even in freezing temperatures. *F. magellanica* and its hybrids are the hardiest, surviving winters to USDA zone 6 if mulched or given some protection. 'Half-hardy' fuchsias are marginally hardy, with the top growth dying back to the ground in winter, but sending up new shoots in the spring, similar to other perennial plants. 'Tender' fuchsias will not survive below freezing temperatures.

4.2.0 Hybrids

Swingtime: The flowers of this hybrid will grace hanging baskets with blossoms from June until October in mild summer areas. Bright red sepals pop against the deeply ruffled white petals, and showy red stamens complete the red-white colour pattern. Although the heaviest flower production occurs at the end of the stems, it produces flowers up and down the length of the plant for a densely floral look.

Lady in Black: The climbing habit of this hybrid fills a new niche in the fuchsia world. The half-hardy plants climb up to 2 meter in one season, producing hundreds of flowers at any one time throughout the long blooming season. It climbs by twining tendrils, so give it a trellis or fence with thin wires or mesh to enable the vines to cling.

Dollar Princess: Blooming equally well in the sun or shade, the red and purple blooms of this hybrid blend well with warm and cool colour schemes. In spite of its exotic appearance, it is a hardy fuchsia that can survive in winter temperatures down to -15° C. The shrub produces woody branches that are part of its permanent structure, and new tender growth will emerge from these branches in the spring.

Seventh Heaven: This giant hybrid has attracted a new following of gardeners who want blooms with a greater impact than the classic dainty fuchsia flowers. Unlike some plants that produce mega flowers, like sunflowers and dahlias, these fuchsias aren't stingy with the number of blooms they yield over the growing season. Four-inch blooms are prolific on plants that trail up to 50 cm in window boxes and large containers.

Bolivian Fuchsia: It is a fast-growing plant that blooms its heart out from June to September, sporting merry clusters of red flowers that sound the dinner bell to hummingbirds. This plant even produces an edible fruit following the flowers, although the scant flesh proves to be more of a novelty than a side dish.

Alba: It is a selected cultivar of Bolivian fuchsia, bred for white flowers. The blooms are not a stark white, but a creamy pink that nevertheless look luminous under a full moon. Bi-weekly deadheading will encourage repeat blooming through fall. This is also a large shrubby form of *Fuschia*, and one that quickly succumbs to cold temperatures.

Cultivars/Hybrids of Fuchsia

'Swingtime'

'Lady in Black'

'Dollar Princess'

'Seventh Heaven'

'Bolivian Fuchsia'

'Alba'

Cultivars/Hybrids of Fuchsia

'Alice Hoffman' 'Tom Thumb' 'Riccartonii'

'Pixie' 'Mrs. W.P. Wood' 'Mrs. Popple'

'Howlett-s-hardy' 'Aurea' 'Antarctica'

Flower Colour Variation in Fuchsias

Aurea: Keeping Fuchsias in constant bloom is a tricky matter, but the golden foliage of the *F. magellanica* cultivar 'Aurea' will compensate for lack of flowers. When the fiery red flowers of 'Aurea' are taking a break, keep the spotlight on the bright leaves by pairing it with other flowers on the opposite end of the colour wheel, like red snapdragons or purple pansies.

Alice Hoffman: Dark mat-green leaves, purple-tinged when young. Sepals pinkish scarlet and Corolla is white in colour.

Chillerton Beauty: It is of spreading habit and eventually 1.2 meter high in milder parts. Leaves are ovate, acute, rather thick, pale sea-green in colour; with purplish petiole.

Corallina: Stems arching to procumbent, reddish when young. Leaves in pairs, threes or fours, mostly ovate or oblong-ovate, folded upwards along the purplish midrib; undersides tinged purple, especially on the veins. Tube and sepals scarlet; forward pointing. Petals reddish purple. This famous Fuchsia, one of the oldest of those now cultivated, was raised by Lucombe and Pince around 1844. It was not at that time highly regarded by fuchsia fanciers, the sepals not being reflexed as fashion then demanded, and the petals too infused with red, but it was found to be a vigorous climber and at one time was commonly grown against house-walls on the coasts of the south-western counties and in North Wales, where it sometimes developed a stout trunk and attained eave-height. In colder parts it is herbaceous but very effective in late summer with its brilliant flowers, graceful habit, and bronzy green leaves.

The parentage of 'Corallina' is not known. Porcher suggested it was not a hybrid but a form merely of the Brazilian *F. radicans* (*F. regia* var. *integrifolia*) and although this assertion is incorrect it is not unlikely that 'Corallina' is a hybrid from this variety or from some other Brazilian relative of *F. magellanica.*

Dr Foster: This is an erect hybrid of Fuchsia. Leaves ovate to broadly so; stems and petioles reddish. Tube and sepals scarlet; tube stout, spreading, slightly reflexed at tips. Corolla violet-mauve. The flowers are remarkably large for a hardy fuchsia.

Globosa – A dwarf, spreading shrub with glabrous, strongly toothed leaves, rounded or subcordate at the base. Flower-buds globular; half-expanded flowers balloon-shaped owing to the cohering tips of the sepals. Petals violet, about half as long as sepals. Developed by Bunney of Stratford before 1832. It is said to have been a seedling of *F. magellanica* var. *conica* but is certainly a hybrid, the other parent being in all probability the Brazilian *F. coccinea*, the influence of which is strongly suggested by the very short tube.

Madame Cornelissen: A strong-growing variety up to 91cm high, or more in mild gardens. Leaves dark green, long-tapered at the apex to an acute point; tube and sepals soft scarlet; sepals lanceolate, long-tapered, gracefully reflexed. Corolla with reduced petals, white. The flowers of this variety are beautifully formed and set off by the unusually elegant foliage.

Margaret: A very vigorous freely branching variety of vase-shaped habit, Leaves light green; tube and sepals scarlet; corolla semi-double, petals bluish violet with white flares at the base, spreading. A very showy fuchsia developed by the late W. P. Wood from 'Heritage' crossed with the white-flowered form of *F. magellanica*.

Mrs Popple: A vigorous variety eventually 120 cm high and wide, very bushy in habit. Leaves mostly ovate, folded upward along the midrib as in 'Corallina'. Tube and sepals soft crimson-scarlet; corolla violet-blue, ageing to crimson-purple; free-flowering and one of the best of the hardier sorts. It shows the influence of 'Corallina', especially in its foliage. A.M. 1934.

Mrs W. P. Wood: This fuchsia resembles the white-flowered form of *F. magellanica* but the flowers are pinker, larger and with a shorter, wider tube. It is less hardy. A seedling of 'Margaret', selfed, and like the parent developed by W. P. Wood.

Pixie: This hybrid upright, bushy habit when established. Leaves lanceolate, light green, with impressed veins, tube and sepals pinkish scarlet; corolla lavender, with red veins. This is a sport of 'Graf Witte' developed by Messrs Russell.

Riccartonii: An almost hardy deciduous shrub 1.5-1.8 meter high. Young stems and petioles reddish on the exposed side. Flowers on slender glabrous pedicels; tube and sepals scarlet; corolla violet in colour. The origin of 'Riccartonii' is believed to have been developed at Riccarton, Scotland, by Young, the gardener there, around 1830. As a greenhouse fuchsia it was soon superseded by more elaborate, larger-flowered confections, but by the end of the 1830s its hardiness had been discovered and from then on it must have become increasingly common in gardens as an open-ground plant.

Tom Thumb: Leaves medium green, ovate to lanceolate, Corolla soft violet-purple, slightly more than half as long as the sepals. A charming softly coloured fuchsia of dwarf compact habit, very free-flowering.

W. P. Wood: Leaves broad-ovate, strongly toothed, with a raised rim. Corolla reddish purple shading to scarlet at the base. This is of erect, dwarf habit.

With so many fuchsia varieties available and new hybrids released every year it is best to divide them into flower types and growth habit. In terms of flower fuchsias may be either single or double flowering and include both bell shaped flowers and the long *F. triphylla* types. Growth habit includes upright bushes, lax types and trailing types. Some are fuchsia varieties are vigorous and others are more compact and bushy.

4.3.0 Single or Double Flowering Types

Fuchsia Melanie Louise: Single flowering upright bushy habit with deep red sepals and an aubergine corolla. This is vigorous in nature and grows best in the garden or large pot.

Fuchsia Voodo: Double flowering with deep red sepals and a deep purple corolla. A trailing type well suited to a hanging basket or growing where it can tumble down; long flowering and vigorous in habit.

Fuchsia Acclamation: A double flowering trailing type with pink sepals and a clear pink corolla. Large flowers and a long flowering trailing type suited to hanging baskets.

Fuchsia Brighton Belle: Single flowering with pink sepals and corolla. A lax habit and well suited to hanging baskets, bushy growth Triphylla.

Fuchsia Priscilla Anne-Louise: Double flowers with a vigorous but slightly lax habit suited to hanging baskets, however it will outgrow them in a few years. Sepals are pink and a deep salmon corolla.

Fuchsia Jamie Evan: Single flowering lax to trailing type with pink sepals and a purple corolla with classic bell shaped flowers. Medium bushy habit, excellent in hanging baskets.

4.4.0 Trailing Types

Blush of Dawn: It has pink and light purple double flowers and can trail up to 50 cm height.

Harry Gray: It has mostly white with slight pink tinge double flowers and can trail down to 60 cm high.

Trailblazer: It has vivid pink double flowers and can trail down to 60 cm high.

Dark Eyes: It has purple and vivid red double flowers and can trail down to 60 cm high.

Indian Maid: It has purple and red double flowers and can trail up to 50 cm.

4.5.0 Upright Types

Baby Blue Eyes: It has violet and vivid red flowers and grows to 70 cm high.

Cardinal Farges: It has bright red and white single flowers and grows to 80 cm high.

Beacon: It has deep pink and purple single flowers and grows to 70 cm high.

Not all Fuchsia varieties are created equal, and some are more sun tolerant than others. Usually, red varieties with single blossoms can withstand more sun than light colours or pastels with double blooms. 'Papoose' is an example of a hardy cultivar that tolerates considerable sunlight. Other hardy varieties include 'Genii,' 'Hawkshead,' and 'Pink Fizz.'

5.0 Genetics and Breeding

Chromosome numbers of Fuchsia species (Table 1) indicated that amongst the species, only four were tetraploids whereas different ploidy levels were recorded in the cultivars (Talluri and Murray, 2009). Wanthana and Nuttha (2006) reported that chromosome number of parental varieties was 62-90 whereas chromosome number of hybrids ranged in between parental varieties, 75-90.

Table 1: Chromosome Number of different Species and Cultivars of *Fuchsia**

Species/Cultivars	*Chromosome Number (2n)*	*Section*
Fuschia arborescens Sims.	22	*Schufia*
F. boliviana Carr.	44	*Fuchsia*
F. encliandra Steud.	22	*Encliandra*
F. excorticata Linn.	22	*Skinnera*
F. fulgens Carr.	22	*Ellobium*
F. glazioviana Tamb.	44	*Quelusia*
F. hatschbachii Sekcja.	44	*Quelusia*
F. magellanica Lam.	44	*Quelusia*
F. microphylla Kunth.	22	*Encliandra*
F. minutiflora Hemsl.	22	*Encliandra*
F. procumbens Cunn.	22	*Procumbentes*
F. reflexa	22	*Encliandra*
F. splendens Zucc.	22	*Ellobium*
F. triphylla Linn.	22	*Fuchsia*
F. trumpetor	22	*Fuchsia*
'Dr Hammett'	99	Cultivar
'Gartenmeister Bonstedt'	33	Cultivar
'Timothy Hammett'	44	Cultivar

* Talluri and Murray (2009).

5.1.0 Breeding

Records show that in 1838 Galley of Leeming Bar, Yorkshire, crossed *F. fulgens* with *F. grandiflora* and Normald of Ripon, Yorkshire, raised 'Normaldii' which was an *arborescens* × *reflexa* cross. In 1841 the number of hybridists increased and many new varieties became available.

Fuchsia hybridization and selection of cultivars have been taking place since the 1830s. Much of it was carried out by early horticulturalists who were reluctant to divulge their sources or else more recently by amateur growers who were often opportunistic in their approaches and selection of material. The first hybrid was described in 1837; in 1840 the first cultivar with white sepals was obtained: 'Venus Victrix'. Popular varieties still with us today followed, like 'Tom Thumb' (1850), 'Riccartonii' (1852), 'Bland's New Striped' (1872). By contrast plants of Plumier's original *F. triphylla* didn't reach Kew before 1882.

The genus *Fuchsia* has continued to be extensively hybridised and today more than 3600 varieties have been developed. This is because fuchsia is a first class garden plant. It has got many qualities: a long flowering period, it is easily trained, has very few garden enemies and many colour combinations, which are set off in some varieties by variegated golden and purple coloured foliage. In spite of its easy nature it is a plant which is rarely grown from seed.

As an amateur fuchsia hybridist, Jack Wilson in 1967 started the first hybridization programme on Fuchsia and obtained good results from seed, but of course the resulting plants varied widely. He observed that it is rare however to have plants that are of no garden value and with luck, there are sometimes some real beauties. After gaining long experiences he decided to keep the following breeding objectives for genetic improvement of this crop.

- ☆Hardiness
- Upright growing with strong stems and close branching
- Freedom from disease
- Early flowering (late June/early July)
- Larger flowers than is usual in hardy fuchsias
- A good range of colours
- An overall good constitution.

Hundreds of named cultivars have been derived primarily from hybridizing members of section *Quelusia* (Reiter, 1944). These original artificial hybrids produced in the 19th and 20th centuries attest to the ease of hybridization and suggest a weakness of genetic barriers preventing hybridization in at least some of the species in the genus. Artificial hybridization involving *F. procumbens* has been used to study species relationships and genome homology in the south Pacific species of sections *Skinnera* and *Procumbentes*. The combinations, *F. procumbens* × *F. excorticate* and *F. procumbens* × *F. prescandens* proved to be compatible (Godley and Berry, 1995), but these were not known whether *F. procumbens* could be hybridized with species from South America. The understanding of the causes of failure to produce hybrids both among the American species and between them and the New Zealand species may allow the development and application of techniques such as stylar amputation, mentor pollen and embryo rescue, hybrid inviability or weakness.

Talluri (2007) made 49 interploidy crosses which gave variable results. The plants with different chromosome numbers were capable of forming hybrids in many diploid × tetraploid crosses such as *F. boliviana* × *F. procumbens* (16 per cent fruit set), *F. fulgens* × *F. boliviana* (60 per cent fruit set), *F. fulgens* × *F. magellanica* (38 per cent fruit set), *F. procumbens* × *F. hatschbachii* (40 per cent fruit set), *F. splendens* × *F. magellanica* (50 per cent fruit set), *F. triphylla* × *F. boliviana* (43 per cent fruit set) and *F. triphylla* × *F. glazioviana* (8 per cent fruit set). Taking the endosperm balance number (EBN) hypothesis (Johnston *et al.*, 1980) into consideration, if $2x$ (2EBN) the arbitrary EBN values for the diploid parental species involved in the production of above hybrids and $4x$ (4EBN) for polyploids were given, it confirms that at a very low-frequency crosses between species with non-matching EBNs may also be successful. Further, in some instances interspecific hybridization between the species with same EBN can also be prevented by incompatibility between the genomes of two species (Singh *et al.*, 1990). Nonetheless, the cross combination between $4x$ (4EBN) *F. boliviana* × $2x$ (2EBN) *F. procumbens*, $2x$ (2EBN) *F. procumbens* × $4x$ (4EBN) *F. hatschbachii* and $2x$ (2EBN) *F. splendens* × $4x$ (4EBN) *F. magellanica* have shown hybrid inviability as the parental genomes failed to function in unison in these EBN mismatched hybrid progeny, thereby leading to low seed germination and high seedling mortality, a postfertilization breeding barrier. In other words, only specific combinations of the parental species produce normal viable hybrids and genetically unbalanced plants are eliminated. The cross combination of *F. splendens* × *F. magellanica* and *F. triphylla* × *F. glazioviana* also showed improper embryo/endosperm development with 35 per cent and 59 per cent of aborted seeds to the normal plump seeds in the fruit, respectively (Talluri, 2007). The disharmony between parental genomes or, in some cases, improper endosperm development or endosperm failure due to imbalance in genomic imprinting between endosperm and the maternal tissue or both might lead to seed abortion (Bohs, 1991; Gutierrez-Marcos *et al.*, 2003). Talluri (2012) also recorded that crosses between species with different ploidy levels were successful in some of the cross combinations and failure of interploidy crosses was seen in many crosses irrespective of the ploidy level of the maternal parent. The imbalance between EBN of the maternal and paternal species than usually at 2 : 1 ratio for post-fertilization development of normal endosperm in these hybrid plants might have resulted in failure of these crosses and are strongly incompatible in addition to disharmony between the genomes of both species.

The first effort to study the inheritance pattern of flower colours in Fuchsia was made long back ago by Jack Wilson in 1967. He tried to ascertain the dominant and recessive characteristics in the fuchsia family. From his own experience, he found that red and purple flowers were dominant and red and white or purple and white were recessive. This was confirmed by W.P. Wood in his book A Fuchsia Survey. This in fact means that the resulting seedlings of each cross-pollination made had to be allowed to flower along with their siblings so as to grow on the resulting seedlings. In other words, the F_1 seedlings which would have the dominant characteristics must be allowed to cross-pollinate, and then the resulting seedlings would show the recessive characteristics to a greater or lesser extent. In general, fuchsia seeds if sown in March/April will flower the same year, but in some cases flowering does not occur until the second year. So, to be able to produce the F_2 generation it is necessary to grow the new seedlings for at least two years.

Later, Wanthana and Nuttha (2006) determined flower colour inheritance and observed that red colour of sepal might be controlled by dominant gene and probably, there were more than one gene that governed this character. Petal colour might be controlled with the minimal of two genes. Red petal was governed by one dominant gene and white was governed by another pair of recessive gene with possibility that recessive gene could perform epitasis on another gene.

Through DNA sequencing, it is now possible to have a fairly robust understanding of the evolution of the genus and how it can be divided into 12 sections, each containing one or more closely related species (de Cooker *et al.*, 2018). This can help guide the selection of species for future hybridization work as well as to better understand the dynamics of some of the past crosses. *Fuchsia* breeding has now shifted away from selection for cold tolerance toward utilization of a much wider array of intersectional crosses involving triphylla hybrids and smaller-flowered sections, as well as short-day-length (winter-flowering) species.

6.0 Propagation

Fuchsias are easy to propagate from cuttings. Fuchsia cuttings can be taken anytime from spring through fall, with spring being the most ideal time. Cuttings from vegetative stock plants root better than cuttings from reproductive stock plants. Maintain the stock plants vegetative by short days (less than 12 hours of light) and keep the temperature below 21°C. Cut or pinch out a young growing tip, about 5-10 cm in length, just above the second or third pair of leaves. Remove any bottom leaves and, if desired, apply rooting hormone. Insert three or four cuttings in a 7.5 cm pot or numerous cuttings in a planting tray, into a moist growing medium like sand, perlite, vermiculite, peat moss, or sterilized soil. It may help to make a hole in the growing medium with a finger or a pencil before hand for easier insertion of the cuttings. The cuttings can then be covered with ventilated plastic to retain moisture and humidity. It does speed up the rooting process. Place the cuttings in a warm location, such as a window sill or greenhouse. Within three to four weeks (or less), the cuttings should begin establishing good roots. Once these roots start, remove the plastic covering during the day to acclimate the young plants. When they have started growing well, the rooted cuttings can be removed and re-potted as needed. Once the cuttings produce some well-established roots, they can be re-potted in soil.

7.0 Soil and Climate

Fuchsias prefer humus-rich soil in the garden or use a light organic planter mix in containers with perfect drainage. Wooden containers or fibre pots keep the roots cool and allow the plant to breathe. Clay pots do the same but dry out faster.

Although fuchsias are considered shade loving plants, they need lots of light to grow and bloom. Select a location in outdoors where the plants receive direct morning sun or filtered sun all day. The warmer the climate, the more shade will be necessary. The optimum vegetative growth occurs at 20-23°C. The growth is poor below 15°C or above 26°C. Flower initiation occurs at similar rate under temperatures from 15°C to 26°C, although the most suitable temperature is 21-22°C. Fuchsias are at their best where the summer days stay below 29°C and the nights are cool. In hot and dry climates, the plants grow poorly and the flower size shrinks. Growers in those areas must provide good shade and wind protection and install automatic misting systems. Prolonged sub-freezing temperatures damage the majority of fuchsias. Hardy cultivars can survive temperatures in the teens if they are well established in the ground, have been planted deep, and have been given a heavy layer of mulch. They will freeze to the ground and re-sprout from the roots. Container plants will have to be moved under shelter. Red- and orange-flowered varieties take full sun near the ocean. White and pastel hanging basket varieties must be shaded. Some small flowered pot plant varieties will also grow in a very bright window, but usually do not do as well as outside, especially on the West Coast. If there seems to be too much stem between sets of leaves, move the fuchsia to more light. Turning the plants a quarter turn on a regular basis will also make them more symmetrical. If shade comes from trees, be careful about placing fuchsias until the trees have leafed out completely.

Propagation and Growing of Fuchsia

Berry — Seeds — Seedlings

Selection of Branch — Cuttings — Treatment (Rooting Hormone)

Planting in Plug Tray — Rooted Cutting — Ready for Transplanting

Mass Production in Polyhouse — Standard Fuchsia

Source: www.hendriksyoungplants.com, www.seasonalgardening.co.uk, www.dengarden.com, www.istockphoto.com, http://www.hoehoegrow.co.uk

8.0 Cultivation

Plant fuchsia in a container filled with any good quality commercial potting soil having porous structure and rich in organic matter. Media should retain moisture but drain well. Upright varieties may be planted with 1 cutting in 10-13 cm pots or vigorous growing varieties in 15 cm pots. Three to five cuttings are used for 25 cm hanging baskets. Place the fuchsia in bright, indirect light, as fuchsias don't do well in hot, intense sunlight. The room should be cool about 15.5 to 21.1°C during the day and a few degrees cooler during the night. The plant won't bloom in temperature above 23°C.

8.1 Irrigation

Water the plant regularly during spring and summer, providing water as needed to keep the potting mix lightly moist but not soggy. A container plant in full bloom needs water once a day or possibly twice in very warm and dry weather. Do not water a wilted plant in the midday heat if the soil is still wet. Mist the leaves to reduce surface temperatures and move the plant to a cooler location.

8.2.0 Fertilization

Fuchsias are heavy feeders that benefit from regular fertilization. Add a 50 percent dilute solution of water soluble fertilizer to the water with each irrigation. A good rule of thumb to use is to mix a balanced water soluble fertilizer half-strength and feed fuchsias each week. Switch to a "bloom" formula of fertilizer when the plants are setting buds, but do not cut out the nitrogen altogether since Fuchsias continue to grow while they are blooming. Feeding once a week with a balanced fertilizer, such as 20-20-20 (1 Tbsp per gallon of water), is recommended by the Oregon Fuchsia Society. Remember that a container plant needs regular feeding because there is little soil to hold nutrients and the frequent watering leaches them out faster. A constant liquid feed of 250 to 300 ppm nitrogen and potassium give good result in soil-less media. In cool climate, withhold fertilizer in the fall to harden plants off before they go dormant for the winter. Most fuchsias need long days to flower. Some varieties, notably the *F. triphylla* hybrids, can be in bloom all year long if the climate is mild.

8.3.0 Pruning and Pinching

Prune heavily in late winter or early spring, after the danger of frost is past. On the North Coast, it is recommended that pruning be done by mid February. Remove dead wood and most of the previous year's growth for upright fuchsias. Leave only a few strong vertical canes. The pruned branches of a hanging basket should ideally look like the spokes of a wagon wheel. Roots of container plants may also be pruned at this time. Start feeding as soon as green growth appears. Pinch out growing tips after two sets of leaves are formed. Snipping out the new leaves at the end of each branch makes two new branches grow where there is one. Repeated pinching produces bushier plants and more blooms. Flowers will open six to eight weeks after pinching is stopped. Pick off berries to prolong the blooming season. Fuchsias can be trained to grow as a bush, tree, basket, wall-pocket, espalier, or even bonsai.

It takes at least two years to train a pleasing standard shaped plant. To train a standard, start with a plant that has a good strong main stem. Pot it up with good potting soil, and start by pinching out any side shoots that appear above the leaf joints, leaving leaves on the stem itself to provide nutrients for your growing plant. Insert a stake and tie the stem to it. As needed as the plant grows, repot, re-stake and continue to pinch off side shoots. When the fuchsia is about three sets of leaves taller, pinch out the top shoot to stop upward growth. Pinch out the tips of the side shoots growing at the top of the plant until the head thickens up. When they have a nice, bushy head, remove all the leaves growing on the main stem, leaving it clear. Keep the plant staked and re-pot as required and protect in winter, and one can have an enviable, show stopping fuchsia standard.

In Europe, the most commonly used growth regulator for fuchsia height control is Daminozide (B-9) @ of 3000-4000 ppm. The first application is made at an early growth stage, when new shoots can be seen on un-pinched plants or when the shoots are 1-2 cm on pinched plants. A second treatment is required 10-14 days after the first treatment. Ancymidol (A-Rest) as a foliar spray at 25-50 ppm or a drench at 0.25-0.50 mg per 15 cm pot will also control height. In addition, more flowers may develop. The ability to produce well-shaped fuchsia through the use of growth regulators may be appealing to parks and conservatories. Robberts *et al.* (1990) tried Paclobutrazol at different concentration but they could not improve the form *of Fuchsia × hybrid.*

9.0 Diseases, Pests and Disorder

9.1.0 Diseases

9.1.1 Rust

This is a cool weather disease and usually at its worst in the fall. Provide plants with a lot of room for air movement. Pick off infected leaves and dispose of them in a sealed plastic bag and avoid overhead watering, which favours spore germination. Grow varieties that are less rust prone. Orange-flowered varieties are normally susceptible to rust disease.

In hot climates and for greenhouse growers, white fly (numerous species) can be extremely problematic. White fly populations in the early stages of population development can be held down by a vigilant program of removing infested leaves, vacuuming adults, or hosing down (syringing) with water sprays. Insecticidal soaps or oils such as neem oil may reduce but not eliminate populations.

9.2.0 Pests

9.2.1 Grapevine Moth (*Phalaenoides glycinae*)

It is the main caterpillar pest of fuchsia. Their larvae usually feed from the sheltered side of leaves, causing holes, and will eventually devour entire leaves when they reach full size. Wearing gloves, pick off minor infestations and squash them. Spray small caterpillars with Dipel. Use pyrethrum only if the infestation is severe and follow label instructions.

Diseases, Pests and Disorders of Fuchsia

Rust

Gall Mite

Whitefly

Caterpillar

Thrips

Yellowing

Source: www.gardenfocused.co.uk

9.2.2 Western Flower Thrips (*Frankiliniella occidentalis*)

This thrips pierce plant cells, causing silvering of the leaf, and feed on buds and flowers, browning the petals. Use yellow sticky traps to monitor for the presence of thrips, but to control them, prune and dispose of affected leaves and flowers. There may be repeated invasions of these bugs during hot weather, so it pays to be vigilant. Spray with insecticidal soap or horticultural oil to manage them.

9.2.3 Fuchsia Gall Mite (*Aculops fuchsiae*)

This pest is native to South America, was accidentally introduced to the West Coast in the early 1980s. Since that time, the destructive pest has created headaches for fuchsia growers across the United States and Europe. It sucks the plant juices and injects a poison that causes the fuchsia to produce gnarled and crippled growth, similar to peach leaf curl. Stems, leaves, and flowers swell and become hairy and galled. Within these galls, the mites live and breed, protected from predators. University of California's Integrated Pest Management program suggests that control may be attained by applying a spray of acaricide at two and three weeks after pruning.

9.3.0 Disorder

9.3.1 Yellowing of Leaves

The most common cause of yellowing *Fuchsia* leaves is inadequate watering. This can be due both to over and under watering. If leaves don't get enough water, they can't perform photosynthesis and they lose their healthy green colour. If they get too much water, however, their roots get clogged and can't supply the leaves with enough oxygen, resulting in yellowing fuchsia leaves.

10.0 Biotechnology

10.1.0 Micropropagation

The callus proliferation and regeneration of *Fuchsia hybrida* was successfully carried out by Chow *et al.* (1990), which was reported to occur more rapidly than that which was observed in callus derived from ovary tissues (Dabin and Vaerman, 1985; Dekeyser *et al.*, 1985). Rapid development of axillary buds from shoot-tips and nodes of 18 cultivars of *F. hybrida* has been obtained on solid Murashige and Skoog (MS) medium with 6-benzylaminopurine (BAP) and NAA as auxin (Kevers *et al.*, 2003). Tissue culture and rapid propagation of *Fuchsia alba-coccinea* Hort was also carried out using explants of stem fragments with nodes (Fugen *et al.*, 2006). Parveen and Rasheed (2013) concluded that ¾ MS media is well suited as compared to ½ and standard MS media for the *in vitro* regeneration of *F. magellanica* from stem nodules explant. Additionally, for multiplication, elongation and root development, a combination of NAA and BAP, GA_3 and NAA alone are required to be added into the ¾ MS media, respectively. Now, it can possible to grow *F. magellanica* rapidly in laboratory conditions with minimum hormonal supplementation. Induction of calluses from ovarian slices of fuchsia and somatic embryogenesis were attempted in order to produce somatic embryos, but also to precise the conditions of culture and to define the factors which can influence the propagation rate by this method of culture (Dabin and Beguin, 1987).

11.0 References

Ball, Philip (2003) In: *Bright Earth, Art and the Invention of Colour*, p. 214.

Berry, P.E., Hahn, W.J., Sytsma, K.J., Hall, J.C. and Mast, A. (2004) *American Journal of Botany*, **91**: 601–614.

Bohs, L. (1991) *Am. J. Bot.*, **78**: 1683–1693.

Boris, G. (2010) *Bachelor's Thesis*, University of Pécs, Pécs, Hungary.

Chow, Y.N., Harvey, B.M.R. and Selby, C. (1990) *Plant Cell Tissue Organ Cult.*, **22**: 17-20.

Dabin, P. and Vaerman, A.M. (1985) *Bull Soc. R. Bot. Belg.*, **118**: 172-178.

Crowden, R.K., Wright, J. and Harborne, J.B. (1977) *Phytochemistry*, **16**: 400–402.

Csepregi, Rita, Temesfoi, Viktória, Das, Sourav, Alberti, Ágnes, Tóth, Csenge Anna, Herczeg, Róbert, Papp, Nóra and Koszegi, Tamás (2020) *Antioxidants*, **9**: 166; doi: 10.3390/antiox9020166.

Cuevas, E., Jimenez, R. and Lopezaraiza-Mikel, M. (2014) *Plant Biology*, **16**: 1096–1103.

Dabin, P. and Beguin, F. (1987) *Acta Hortic.*, **212**: 725-726.

de Cooker, M.G.R.T., Goulding, E.J., Waldenmaier, J.H., Berry, P.E. (2018) In: *Ornamental Crops. Handbook of Plant Breeding* (Ed. J. Van Huylenbroec), vol 11. Springer, Cham. https: //doi.org/10.1007/978-3-319-90698-0_16

Dekeyser, A., Dabin, P. and Bouharmount, J. (1985) In: Proceedings of the International symposium (Eds. W. Horn, C.J. Jensen, W. Odenbach, O. Schieder) organized by EUCARPIA, Berlin (West), Germany, pp. 597-598.

Feuville, R.P.L. (1725) *Journal des Observations Botaniques*, **111**: 47

Fugen, G., Ruiqing, C., Zhigang, W. and Bingyao, S. (2006) *Zhiwu Ziyuan Yu Huanjing Xuebao*, **15**(3): 55-5

Godley, E.J. and Berry, P.E. (1995) *Ann. Mo. Bot. Gard.*, **82**: 473–516.

Goulding, E. (1995) In: *Fuchsias-the complete guide*, Timber Press, Portland, USA.

Gutierrez-Marcos, J.F., Pennington, P.D., Costa, L.M. and Dickinson, H.G. (2003) *Philos. Trans. R. Soc. London, Ser. B*, **358**: 1105–1111.

Hoshino, T. and Berry, P.E. (1989) *Ann.Mo. Bot. Gard.*, **76**: 585–592.

Johnston, S.A., Nijs, T.P.M., Peloquin, S.J. and Hanneman, R.E. (1980) *Theor. Appl. Genet.*, **57**: 5–9.

Kevers, C.L., Coumans-Gilles, M.F., Coumans, M. and Gaspar, T.H. (1983) *Scientia Horticulturae*, **21**: 67-71.

Maerz, A. and Paul, M.R. (1930) *A Dictionary of Colour*, New York: McGraw-Hill, p. 195.

Papp, N. (2018) Traditions and Ethnomedicine in Lueat; Lövétei Közbirtokossága és Polgármestei Hivatal: Lövéte, Romania.

Parveen, Anjum and Rasheed, Sadia (2013) *African Journal of Biotechnology,* **12**(7): 670-678.

Péntek, J. and Szabó, T.A. (1985) In: *People and Plants, Ethnobotanical Knowledge of Rural People in Tara Calatei,* Kriterion Publ. House: Bucharest, Romania.

Reiter, V. (1944) *J. Calif. Hort. Soc.,* **5**: 144–192.

Roberts, C.M., Eaton, G.W., and Seywerd, F.M. (1990) *Hort Science,* **25**(10): 1242-1243.

Ruiz, A., Hermosín-Gutiérrez, I., Vergara, C., Baer, D., Zapata, M., Hitschfeld, A., Obando, L. and Mardones, C. (2013) *Food Res. Int.,* **51**: 706–713.

Singh, A.K., Moss, J.P. and Smartt, J. (1990) *Adv. Agron.,* **43**: 199–240.

Talluri, R.S. (2007) *Ph.D. thesis,* the University of Auckland, Auckland, New Zealand.

Talluri, R.S. (2012) *Journal of Genetics,* **91**: 71-74.

Wanthana, M. and Nuttha, K. (2006) In: *Cell Genetics and Fuchsia Flower Colour Inheritance,* ARIS, pp. 10-18.

Williams, C.A. and Garnock-Jones, P.J. (1986) *Phytochemistry,* **25**: 2547–2549.

Wilson, Jack (1989) *Growing From Seed,* **3** (2).

Ixora

B. Lalramhlimi

1.0 Introduction

Ixora is a genus of flowering shrub or small tree, suitable for pot culture and flower easily under different kinds of light conditions. It is known by many names around the world as 'Rukmini', 'West Indian jasmine', 'Raktak', 'Rangan', 'Kheme', 'Viruchi', 'Santan', 'Jungle flame' and 'Jungle geranium'. The genus *Ixora* is the name of a Malabar idol, to which the flowers of some of the species are offered (Husain, 1984). It is a sun- loving and year- round flowering plant. It is grown as an ornamental plant in many regions of tropical and subtropical climates as they exhibit various forms, beautiful cluster flowers, varied flower colour, flower and leaf size, and plant height.

2.0 Importance and Uses

2.1.0 Aesthetic Uses

The shrub is evergreen with compact flowers, varied in colours which blossoms year- round or so during summer and rainy season is an important garden shrub. It is valued for the retention of the freshness of cut blossoms for a long time. It is used for hedges, landscape specimen plants, shrub border, bonsai or can be grown in large containers or boxes. The national flower of Suriname is 'Fayalobi' which is *Ixora coccinea*. It is important cut flowers for indoor decorations, also used for making garlands for idols and photoframes. The red *Ixora* flowers are very important as they are used as offerings to idols in Hindu worship, as well as in traditional Indian medicine. The petals of *I. coccinea* is a good source of bioactive ingredients and can be used in fresh-cut industry. The petals of the flower can be binded together with string or thread to make gajra/veni which is worn by women to decorate their hair. The petals of *Ixora* can be used in water and flower rangoli. The pressed dried flowers can be used for decorating greeting cards, book marks, postcards *etc.* The whole plant parts can be used for various treatments in different parts of the world.

2.2.0 Therapeutic and Other Uses

The flowers are used for amenorrhea, hypertension, whooping cough, ulcers, dysentery, haemoptysis, and cancer. They also possess estrogenic and abortifacient properties (Bachheti *et al.*, 2011; Wen and Chiu, 2011; Khare, 2007). The leaves of these plants are reported to have spasmolytic, anti-inflammatory, anti-diarrheal, antiasthmatic, hypotensive, antiulcer, antinociceptic, antiviral, antigestagenic, and anticancer activity (Faten *et al.*, 2003). The root of this plant is used in skin diseases, menorrhagia, leucorrhea, bronchial disorders, and urinary trouble. They are also used as astringent and antiseptic (Bachheti *et al.*, 2011; Khare, 2007). Barks are used as a decoction for the ailment of anemia (Ghani, 2003). The local folks and tribes especially the children eat the fruits of *Ixora coccinea* which has comparable nutritive value to sapota in their moisture content, protein, fats, total sugar, fibre, mineral matter, sodium, potassium, and energy (Nazarudeen, 2010). The potential of *I. javanica* as an antioxidant and skin lightening agent has been shown by many

Uses of Ixora

Garlands

Cut Flowers in Vase

Salad mixed with Ixora

Petals Used for Making Gajra/Veni

Ixora Flowers Mixed with other Flowers for Offering to Hindu Gods

Ixora in Pots

Uses of Ixora

Hedges

Ixora in Shrub Border

Ixora Bonsai

Ixora as Specimen Plant

Ixora in Box

studies (Oktaviyanti *et al.*, 2019). Chewing the roots of *Ixora longifolia* has been reported to ease toothache (Dontha *et al.*, 2015).

3.0 Origin and History

The centre of diversity is in south- east Asia including Southern India and Sri Lanka (Bose *et al.*, 2013). It is widely distributed in tropical Africa, America, Asia, Madagascar, and Pacific islands and 18 species (nine endemic) in China (Chen and Taylor, 2011). Today, it is widely grown as a one of the most popular ornamental shrubs. Ixora has a long history of use to treat several diseases in the Indian traditional system of medicine, the Ayurveda, and in various folk medicines. Many species of Ixora, as many as 400 species, are spread from Africa to India to Southern Asia (Dontha *et al.*, 2015) while many others species are native to India.

4.0 Morphology

Ixora is a compact, densely- branching, woody shrub or under-shrub or occasionally small trees, evergreen and can grow 1- 3 m tall depending on cultivars. It has dark green glossy leaves and colourful flower clusters that bloom year-round. Some varieties have pinkish, red, yellow, orange, scarlet, or white flowers (Keeler *et al.*, 2004). Leaves are opposite or rarely ternate; inflorescences borne on terminal principal stems (or sometimes terminal on reduced lateral stems and appearing axillary), cymose to corymbiform or paniculiform; flowers pedicellate or sessile and often fragrant. The fruit is black or red and drupaceous (Chen and Taylor, 2011).

The genera *Pavetta* and *Ixora* were traditionally treated as a single genus since Linnaeus' (1753) descriptions and until the beginning of the 20th century (Schumann, 1891; Bremekamp, 1934, 1937a). However, they exhibit distinct differences and their morphology. Salverform bifid stigmas of *Ixora* and can additionally be distinguished from *Pavetta* by a combination of the following characters: hermaphroditic flowers, articulate petioles, hypocrateriform corollas, two-carpellate ovaries containing a solitary ovule in each carpel locule, and drupaceous fruits (De Block, 1998; Lorence *et al.*, 2007). The two genera are not closely related based on combined data from *rbcL* sequences and morphological study (Andreasen and Bremer, 2000).

5.0 Taxonomy

The best modern classification was proposed by Verdcount (1958) and by Bremekamp (1966). Verdcount divided the family Rubiaceae into sub- families Rubioideae, Cinchonoideae and Guettardoideae, based on the presence or absence of raphids and of albumen, based principally upon the limited number of genera from the old world. The genus *Ixora* was placed in subfamily Cinchonoideae under the tribe Coffeeae. *Ixora* was subdivided by Bremekamp (1937a, 1937b, 1938, 1940) into three subgenera including only Malesian species: *Ixora*, *Pavettoides*, and *Sathrochlamys*, though these names are applicable to all species.

6.0 Species and Cultivars

6.1.0 Species

Ixora species are usually confined to lowland and lower montane forest up to 1700 m altitude. Some species are also found in swampy locations in the vicinity of rivers or occasionally in rice fields (*e.g. I. grandifolia*) (Ysrael and Van Valkenburg, 1999). *Ixora* comprises of about 530 species of shrubs and trees that typically grow in humid rain forest (Davis *et al.*, 2009), is the third largest genus in the family. Approximately 60 species are restricted to the neotropics and mid- Atlantic islands (Andersson, 1992), 22 species on the Fiji Islands (Smith and Darwin, 1988), ca. 34 species in continental Africa (De Block, 1998), and probably as many in Madagascar (De Block *et al.*, 2007).

The description of some of the type and common species are given below (De Block, 1998; Mouly *et al.*, 2009; Chen and Taylor, 2011).

Ixora coccinea

This species is mainly distributed across Sri Lanka, India, Burma, Laos, Cambodia, Vietnam, Malay Peninsula, Indonesia, New Guinea, and Caroline Islands. The inflorescences are erect, sessile or moderately pedunculate; the calyx tubes are short but very distinct; the corollas are white, yellow, red, or orange; tubes much longer than lobes; stamens are much shorter than corolla lobes; anthers have short cells and the styles are shortly exserted, glabrous or pubescent.

I. cuneifolia

This is mainly distributed in India, Burma, Laos, Cambodia, Vietnam, and Malay Peninsula. The inflorescences are erect or drooping, subsessile to moderately or rarely long pedunculate; narrow bracts and bracteoles; calyx tubes sometimes distinct; corollas reddish or white; stamens equal corolla lobes in length; anthers with long cells; styles glabrous, exserted portion as long as corolla lobes.

I. umbellata

I. umbellata is mainly distributed in India, Burma, Laos, Cambodia, Vietnam, and Malay Peninsula. Inflorescences are erect, subsessile or shortly pedunculate; central flowers in ultimate triads sessile and ebracteolate; well developed bracts and bracteoles; bracteoles are longer than ovaries; corollas ochre-yellow or white; tubes 3- 6 times longer than the lobes; stamens usually shorter than corolla lobes; anthers with long cells; styles glabrous, exserted portion somewhat shorter than corolla lobes.

I. cremixora

They are distributed in Madagascar and Comoro Islands. They have specific characteristic in the ovaries with descendant ovules.

I. curtisii

The species is distributed in Indonesia. The inflorescences are subsessile, subcapitate, pauciflorous; linear bracts and bracteoles, large, red; terminal flowers

Species of Ixora

Ixora odorata

Ixora finlaysoniana

Ixora coccinea

Ixora javanica

Ixora cuneifolia

Ixora parviflora

Species of Ixora

Ixora philippinensis

Ixora longibrateata

Ixora amplexicaulis

Ixora timorensis

Ixora brachiata

Ixora paludosa

Species of Ixora

Ixora malabarica

Ixora rosea

Ixora philippinensis

Ixora fulgens

Ixora grandifolia

Ixora elongata

Species of Ixora

Ixora gracilis

Ixora auricularis

Ixora pavetta

Ixora longshanensis

Ixora yunnanensis

Ixora blumei

Species of Ixora

Ixora effusa

Ixora hainanensis

Ixora williamsii

Ixora subsessilis

Ixora macrothyrsa

Ixora notoniana

Species of Ixora

Ixora polyantha

Ixora thwaitesii

Ixora undulata

Ixora casei

Ixora johnsoni

Ixora nigricans

are solitary and ebracteolate; reduced calyx tubes, lobes linear, red, (almost) as long as the corolla tube; corollas white, lobes and tube short; stamens much shorter than corolla lobes, anthers with moderately long cells; styles densely pubescent, shortly exerted.

I. drakei

Ixora drakei, Ixora humblotii, Ixora pudica and *Ixora microphylla* belong to section *Micrixora* and have small flowers with corolla tube < 1.5 cm. They are mainly distributed in Madagascar and Seychelles.

I. paludosa

It is commonly found in Tropical Asia and New Guinea, and Occidental Africa. Inflorescences are long pedunculate, multiflorous; bracts and bracteoles are narrowly triangular to filiform; corollas are white, rarely yellowish or reddish; stamens somewhat shorter than corolla lobes or of equal length, anthers with long cells; styles glabrous, exserted portion ± as long as corolla lobes.

I. filipes

This species is mainly distributed in Philippines and Celebes. The inflorescences are distinctly pedunculate, consisting of 1- 7 flowers, at the end of axillary short shoots arising on the green part of the branches; lower bracts narrowly triangular, the others wide; bracteoles ovate; corollas white; stamens shorter than corolla lobes, anthers with long cells and styles are glabrous.

I. brunnescens

The inflorescences of this species are moderately or shortly pedunculate, subpaniculate; small flowers and corolla tubes have bearded throat. They are distributed across Tropical Asia and Indonesia.

I. blumei

The main regions of distribution are Seychelles, Sri Lanka, Andaman, Nicobar Islands, India, Malay Peninsula, New Guinea, Melanesia, and Micronesia. The inflorescences are subsessiles or shortly pedunculate, trichotomously or pentachotomously corymbose; corolla tubes glabrous or pubescent inside but never bearded; the stamens as long as corolla lobes and the styles are glabrous or pilose in their middle part.

I. timorensis

The inflorescences of *I. timorensis* are subsessile or shortly pedunculate; corolla tubes bearded at throat or pubescent in its upper half inside and the styles near the middle usually densely pilose or rarely glabrous. The main regions of distribution of this species are Indonesia, Celebes, Muloccas, New Guinea, and Australia.

I. capillaris

It is found mainly in Indonesia. The inflorescences are long or very long pedunculate; corolla tubes slender or very slender, never bearded, much longer than corolla lobes; stamens frequently much shorter than corolla lobes and the styles are glabrous.

I. amplexicaulis

They are mainly distributed in Fiji and New Caledonia. The leaves are shortly petiolate to subsessile; inflorescences are multiflorous, strongly congested head-like cymes, sessile between a terminal pair of leaves (not modified); bracts linear and filiform crowded among the flowers and the calyx lobes linear or narrowly triangular.

I. paradoxalis

It is mainly distributed in regions across Philippines and Moluccas. The inflorescences are pedunculate; bracts and bracteoles are absent and the styles are glabrous.

I. dolichothyrsa

It is found mainly in New Guinea. The leaves are (sub)sessile, base rounded to cordate or subauriculate, the inflorescences are long pedunculate; bracts filiform and setaceous; calyx lobes and tube equal in length; corollas white or ochre-yellow and the styles are pubescent in their middle part.

I. whitii

This species is mainly distributed in New Guinea, Philippines, and Moluccas. The leaves are petiolate with acute base; the inflorescences are long pedunculate; basal bracts setaceous, others minute; bracteoles absent; calyces cupuliform and the styles are pubescent in their middle part.

I. philippinensis

It is mainly distributed in Philippines, Moluccas, Celebes, and Borneo. The inflorescences are shortly or moderately pedunculate, supported by a pair of sessile leaves not reduced in size; small distinct bracts and bracteoles; corollas pink or white and the styles are glabrous.

I. reducta

The species is commonly found in Madagascar and New Caledonia. The inflorescences are terminal on main and lateral branches, uniflorous, sessile but with modified inflorescence- supporting leaves present and the calyces truncate or with small triangular lobes.

I. amplifolia

It is mainly distributed from Marquesas as far westward as New Caledonia. The inflorescences are pauciflorous (often consisting of three flowers), sessile but modified inflorescence-supporting leaves present, the latter obtusely triangular to orbicular, with cordate bases.

I. chinensis

It is one of the most common native species of *Ixora* in southern China and distributed widely in Indonesia, Malaysia, Philippines, Vietnam and the tropical regions of the world. The flowers are subsessile to pedicellate; calyx glabrous; corolla red or reddish yellow, outside glabrous and tube 20- 30 mm and throat is glabrous. *Ixora chinensis* and *I. coccinea* are often confused easily mainly due to their flower colour (red to yellow), tube shorter than 5 cm and small petals, however, *I. chinensis* has cuneated leaf base and round petals while *I. coccinea* has heart shaped leaf base and pointy petals.

Husain (1984) had studied the taxonomy of the Indian species of the genus *Ixora* where he enumerated 36 species that were found across in different regions of the country.

Sl.No.	Species	Flowers and Fruits	Distribution
1	*Ixora acuminata* Roxb.	January-December	Manipur, West Bengal, Bangladesh
2	*Ixora andamensis* Brem.	May-February	Andaman and Nicobar Islands
3	*Ixora anthroantha* Brem.	April-January	West Bengal
4	*Ixora barbata* Roxb. ex Sm.	April-December	Andaman and Nicobar Islands, Uttar Pradesh
5	*Ixora brachiata* Roxb.	November-May	Maharashtra, Karnataka, Tamil Nadu, Kerala, Goa, Daman and Diu, Gujarat, Rajasthan, West Bengal
6	*Ixora brunnescens* Kurz.	January-December	Andaman and Nicobar Islands
7	*Ixora calycine* Thw.	February-September	Assam, West Bengal, Tamil Nadu, Ceylon
8	*Ixora capituliflora* Brem.	May-February	Andaman and Nicobar Islands
9	*Ixora chinensis* Lam.	January-December	Uttar Pradesh, Bangladesh, China
10	*Ixora coccinea* L.	January-December	Maharashtra, Karnataka, Tamil Nadu, Kerala, Goa, Daman and Diu, Laccadive Islands, Andaman and Nicobar Islands, Uttar Pradesh, Bangladesh, all tropical countries
11	*Ixora elongata* Heyne ex G. Don	January-December	Kerala, Karnataka, Tamil Nadu
12	*Ixora finlaysoniana* Wall.	January-December	Maharashtra, Karnataka, Tamil Nadu, Kerala, Andaman and Nicobar Islands, Uttar Pradesh, west Bengal, Assam, all tropical countries
13	*Ixora goalparensis* Brem.	May-December	Assam, West Bengal
14	*Ixora grandifolia* Zoll. & Mor.	January-December	Andaman and Nicobar Islands
15	*Ixora hymenophylla* Brem.	April-December	Andaman and Nicobar Islands
16	*Ixora javanica* (Bl.) DC.	February-November	Maharashtra, Tenasserim, Malay Peninsula, Sumatra, West Java
17	*Ixora johnsoni* Hook. f.	January-December	Kerala

Sl.No.	Species	Flowers and Fruits	Distribution
18	*Ixora leucantha* Heyne ex G. Don	January-December	Maharashtra, Karnataka, Kerala, Tamil Nadu, Assam
19	*Ixora longibrateata* Brem.	May-December	Bangladesh
20	*Ixora macrothyrsa* (Teysm. Et Binn.) T. Moore	January-December	Tamil Nadu, Karnataka
21	*Ixora malabarica* (Dennst.) Mabberley	May-December	Maharashtra, Karnataka, Tamil Nadu, West Bengal, Assam
22	*Ixora monticola* Gamble	May-December	Tamil Nadu, Kerala
23	*Ixora multibracteata* Pearson ex King & Gamble	April-December	Andaman and Nicobar Islands, South Tenasserim, Malay Peninsula, North Sumatra
24	*Ixora nigricans* R. Br.	January-December	Maharashtra, Karnataka, Kerala, Tamil Nadu, Assam, Malay Peninsula, Sumatra, Java, Bali
25	*Ixora notoniana* Wall.	January-December	Karnataka, Tamil Nadu
26	*Ixora pavetta* Andr. (non Roxb. 1820)	January-December	Maharashtra, Karnataka, Kerala, Tamil Nadu, Kerala, Andhra Pradesh, Rajasthan, Uttar Pradesh, Orissa, West Bengal, Madhya Pradesh, Gujarat, Bihar, Bangladesh, Ceylon
27	*Ixora polyantha* Wight	January-December	Maharashtra, Karnataka, Kerala
28	*Ixora pubirama* Brem.	January-December	Andaman and Nicobar Islands, West Bengal, Burma, Bangladesh
29	*Ixora rosea* Wall.	January-December	Maharashtra, Uttar Pradesh, all tropical countries
30	*Ixora roxburghii* Balakr.	January-December	Assam, Sikkim
31	*Ixora sauleirei* Gamble	May-December	Tamil Nadu
32	*Ixora subsessilis* Wall.	May-December	Assam, Tripura, Meghalaya, Arunachal Pradesh, Manipur, Nagaland, Bangladesh, Upper Burma
33	*Ixora tigriomustax* Brem.	April-January	Bangladesh
34	*Ixora thwaitesii* Hook.f.	January-May	Kerala, Ceylon
35	*Ixora undulata* Roxb.	March-September	Tamil Nadu, Andhra Pradesh, Assam, Sikkim, Arunachal Pradesh, Bihar, Orissa, West Bengal, Bhutan, Upper Burma
36	*Ixora williamsii* Sandwitch	January-December	Uttar Pradesh, Europe, N. America

6.2.0 Cultivars

Many cultivars and hybrids have been developed which are of horticultural importance through hybridization and natural crossing. The different hybrids and varieties have been arranged accordingly below:

'**Alexia**', pink and rose colour flowers;

'**Angela Bushman**', shrimp- pink to orange when first open but turns old rose as they age.;

'**Blanche**', red flowers;

'**Buttered Popcorn**', yellow flowers with little orange streak;

'**Crimson King**', bright red flowers;

'**Crimson Star**', triple flower hybrid, rare exotic variety with bright red flowers in Thailand.

'**Dainty Lady**', pink colour flowers;

'**Diora**', milky white flowers, large inflorescence, dwarf habit, heavy branching, compact and dense canopy of uniform dark green leaves;

'**Dwarf Pink**', small size about 1m tall, pink star- shaped flowers, blooms throughout the year;

'**Dwarf Red**', tiny leaves with red flowers;

'**Dwarf Yellow**', yellow flowers, tiny glossy leaves;

'**Firebrand**', red flowers, star- shaped with pointed tips;

'**Frances Perry**', yellow flowers;

'**Frankie Hipp**', large inflorescence, white flower with thin pink margin, large number of flowers, slow growing, spontaneous mutant of Nora Grant;

'**Fraseri**', bright reddish- salmon flowers;

'**Golden Spice**', bright golden yellow flowers;

'**Hedy**', red ixora with pointed petals;

'**Henry Morat**', light pink, fragrant and produces prolific blooms over a long period of time;

'**Herrera's Pink**', dark rose colour flowers;

'**Herrera's White**', white flowers;

'**Indian Rope**', red flowers and twisted waxy leaves;

'**Ixora Thai Flame**', double- tripple flowers 1/2 across that come in large clusters. Flowers change colour from yellow to red, with yellow, orange and red flowers in the same cluster, profuse blooming;

'**Jessie**', pale variety of ixora with blossoms of two colours 'orange' and 'yellow';

'**Jezebel**', more vibrant bicolor flowers of yellow and orange;

'**Kelly Gent**', orange at opening and turning salmon with age, blooms constantly;

'**Lemon Chiffon**', pale yellow colour flowers;

'**Mademoiselle**', soft pink flowers, numerous;

'**Margery**', a red orange variety, bloom continuously, all year long;

Cultivars of Ixora

'Super King' 'Nora Grant'

'Dwarf Yellow' 'Singapore Yellow'

'Frankie Hipp' 'Dwarf Pink'

Cultivars of Ixora

'Fraseri'

'Sheena'

'Maui Yellow'

'The Canary Sings'

'The Virgin'

'White Crown'

Cultivars of Ixora

'Sea Green'

'Buttered Popcorn'

'Yellow Star'

'Golden Spice'

'Yellow Delight'

'Soft Evening Glow'

Cultivars of Ixora

'Shady Lady'

'Pink Jewels'

'Mlle Nitouche'

'Hedy'

'Blanche'

'Mademoiselle'

Cultivars of Ixora

'Jessie'

'Pink Fireworks'

'Jezebel'

'Dainty Lady'

'Red Holly'

'Pretty Pumpkin'

Cultivars of Ixora

'Up and Away'

'Lemon Chiffon'

'Orange Star'

'Firebrand'

'Pink Princess'

'Pink Starlet'

Cultivars of Ixora

'Herrera's White'

'Dwarf Red'

'Maui Red'

'Strawberry Sundae'

'Maui Sunset'

'Indian Rope'

'Alexia'

'Sunkist Orange'

'Margery'

'**Maui Red**', deep red colour flowers, can tolerate partial shade as well as moderate levels of salt;

'**Maui Sunset**', 2-3 feet tall, peach-yellow flower clusters with a touch of red-orange;

'**Maui Yellow**', yellow colours flowers with glossy leaves, flower clusters spreading up to four inches;

'**Maui**', first discovered in Hawaii, Maui is thought to have a bit more cold tolerance than other varieties. The four-inch-wide flower clusters are deep orange-red when closed, but lighten a bit when open.

'**Mlle Nitouche**', red flower cluster appears as balls;

'**Nora Grant**', pinkish-red flowers, glossy leaves, can tolerate partial shade and moderate levels of salt;

'**Orange Star**', orange colour flowers;

'**Pink Fireworks**', pink to soft pink flowers;

'**Pink Jewels**', pink flowers, ball shape flower cluster;

'**Pink Pixie**', compact, upright, mounded plant habit, short internodes, dark green coloured leaves, large inflorescences with bright pink coloured flowers;

'**Pink Princess**', pink flowers;

'**Pink Starlet**', pink flower colour with pointed tip petals;

'**Pinksing**', rich pink with blooms from spring to summer and having a slight fragrance;

'**Pretty Pumpkin**', orange flowers;

'**Prince of Orange**', orange colour flowers, fast growing, evergreen;

'**Red Fairy**', red flowers;

'**Red Holly**', canopy is fully covered by florets, red flowers;

'**Sea Green**', a small to medium size shrub with unusual aqua-colour flowers, which are very fragrant. Unlike other ixoras with waxy leaves, this one has thin delicate leaves that are easy to burn with hot sun;

'**Shady Lady**', red flowers;

'**Sheena**', salmon orange flowers;

'**Singapore Yellow**', yellow flowers;

'**Soft Evening Glow**', yellow flowers;

'**Strawberry Sundae**', creamish yellow flowers with red thick borders;

'**Sunkist orange**', orange flowers, the flower petals are rounded, whereas in others they are pointed;

'**Super King**', brilliant red flowers, probably the most spectacular of the popular Ixoras;

'The Canary Sings', yellow flowers;

'The Virgin', pure white flowers;

'Up and Away', pink colour flowers, long, thin, separated petals on the flowers;

'Vista6', white flowers and large, dark green foliage, vigorous;

'White crown', white colour flowers;

'Yellow Delight', yellow flowers;

'Yellow Star', yellow flowers with pointed petals.

7.0 Genetics and Breeding

7.1.0 Genetics

Sharma and Chatterjee (1960) studied the chromosome number of most of the species of *Ixora*. He observed that *Ixora acuminata, Ixora bandhuca, Ixora chinensis, Ixora coccinea, Ixora colleri, Ixora finlaysoniana, Ixora grifithii, Ixora javanica, Ixora multibracteata, Ixora parviflora, Ixora singaporensis, Ixora stricta, Ixora subsessilis, Ixora undulata, Ixora vanduca, Ixora westii* have chromosome number, 2n= 22, except for *Ixora duffii* and *Ixora rosea* which are 2n= 20 and 2n= 33 respectively. In *Ixora finlaysoniana,* six has been taken as the basic number of the genus, from which the number eleven probably has arisen through amphidiploidy.

7.2.0 Hybridization

Hybridization work was carried out and several new species and cultivars were developed. An old English hybrid *'Ixora colei'* was apparently developed by crossing *I. coccinea* × *I. chinensis* alba parentage about 1870 by E. Cole of Manchester, England. It has massive flowers of pure white as large as 7 and half inches in diameter when fully open. It blooms intermittently during early summer and late fall and remains in flower for a considerable length of time. It prefers shade or partial shade, has compact and low spreading growth and shows more cold tolerance than most *Ixora*. It was introduced by Swedroe's Nursery in 1964 from Calcutta after being lost from England. A hybrid 'Frances Perry' was developed by crossing 'Gillete's Yellow' × 'Singapore no.1', 'Angela Busman' by crossing 'Superking × Singapore no. 1', by Swedroe in 1963. *Ixora fraseri* was probably developed by interspecific hybridization between *I. ambonensis* × *I. chinensis*. 'Henry Morat' was also developed through interspecific hybridization between *I. accuminata* × *I. coccinea* in 1943. The hybrid 'Pinksing' was developed by Swedroe in about 1960 through cross between '*I. acuminata* x Singapore no. 1'. A triple cross hybrid 'Kelly Gent' was developed by a selective cross of '*I. acuminata* x Singapore no. 1' which gave 'Pinksing' with deep pink flowers and seasonal blooming, then 'Pinksing is crossed with ''McGee's Yellow' which resulted in 'Kelly Gent'. This cross was made by Swedroe in about 1964 (Watson and Swedroe, 1965).

7.3.0 Cross-compatibility

Pollen viability (fertility) and breeding biology of three species of genus *Ixora*, namely, *I. chinensis, I. javanica* and *I. macrothyrsa* were studied by Tyagi *et al.* (1998).

Pollen germination *in vitro* showed that all three species had very high pollen fertility (> 80 per cent). None of the three species was fully self-compatible. Only partial self-compatibility was observed in *I. chinensis* and *I. javanica*. Interspecific crossing results indicated cross incompatibility among the three species studied. Causes of cross incompatibility may include cryptic chromosomal differences, chromosomal aberrations and difference in chromosome numbers. Ways to achieve cross compatibility suggested to obtain hybrids among these species for further improvement.

7.4.0 Mutation Breeding

Basiran (2000) has reported two variants or mutants through gamma rays induced mutation in shoot cuttings of *Ixora* spp. at a dose of 45Gy where he observed change in leaf size and shape, flower size and dwarfness which were in propagation stage. 'Bangkok Bow' variety is a spontaneous mutant form of diploid *Ixora coccinea*.

8.0 Growth and Flowering

8.1.0 Temperature

The effect of temperature was studied by Sakai and Higuchi (1980) on flowering of Chinese Ixora (*Ixora chinensis* Lam.). The experimental results showed that flowering was promoted at temperature between 10°C and 20°C but inhabited at 30°C, when the plants were exposed to constant temperature of 10°C, 15°C, 20°C and 30°C for 30 days in September. The most effective temperature for flower induction was approximately at 10°C temperature. In another experiment Morioka *et al.* (1989) studied the effects of minimum temperature on flowering time and uniformity of *I. chinensis* Lam., where she observed the plants grown at 7°C in greenhouse- initiated flower bud differentiation in December and subsequent transfer to higher temperatures (7, 10, 15 or 20°C) improved flower bud formation, earliness of flowering and elongation of lateral shoots. Plants grown at 6°C or 12°C until 31 December and then transferred to 18°C flowered early and uniformly.

8.2.0 Photoperiod

The effect of photoperiod was studied by Sakai and Higuchi (1980) on flowering of Chinese Ixora (*I. chinensis* Lam.) where the plants were subjected to 8 hour and 16 hour photoperiods for 60 days with high temperature (25°C) during daytime and night, and with low temperature of 25°C day and 13°C night. In budding time, there was no difference observed between the long and short photoperiods under both high and low temperature conditions.

8.3.0 Plant Growth Regulators

Jinshui *et al.* (2010) studied the different spray concentrations of GA, 6- BA, NAA and PP333 on the growth, florescence and the chlorophyll content in different periods of bud differentiation in *Ixora chinensis*. The results showed delayed florescence date 5- 6 days by GA and 2- 3 days by PP333 while NAA and 6- BA could early the florescence date, by 2- 5 days and 14- 23 days respectively. GA could promote the growth of new branches, PP333 had dwarfing effect to *Ixora chinensis* and NAA could increase the chlorophyll content of *I. chinensis*. Besides, 300 mg/L GA and 400 mg/L 6- BA had better comprehensive effects on improving plant type of *Ixora chinensis* and increasing the quality of the cutting flower. Salicylic acid (SA) and gibberellic acid (GA_3) through foliar application enhances growth and flowering of *Ixora coccinea* L. (Gad *et al.*, 2016).

9.0 Propagation

Ixora can be propagated by seeds, stem cuttings, softwood cuttings, tip cuttings, air layering and grafting.

9.1.0 Seed Propagation

The seeds of *Ixora coccinea* give 90 per cent germination with 100 ppm gibberellic acid (GA) treatment and 85 per cent germination with hot water treatment. Furthermore, without any chemicals, the plantlets of *Ixora* can be raised and domesticated (Shinde and Chavan, 2017).

9.2.0 Vegetative Propagation

9.2.1 Cutting

Reyes-Pérez *et al.* (2018) studied effect of growth regulators and substrate on the rooting of three Ixora (*I. coccinea*) varieties under greenhouse for 60 days where cuttings of three varieties with yellow, red and pink flowers were treated with different doses of NAA and IBA (1, 1.5 and 2 ml/kg) on substrates sand, burnt rice and forest soils. The results indicated that auxillary buds of *Ixora coccinea* treated with NAA and IBA and using any substrate as support, allowed reaching an optimal morphological development ensuring the survival and rooting of the plants in 100 per cent. Enhanced growth and development of *Ixora coccinea* cuttings were observed with treatment of 100 ml coconut water in combination with a polythene dome where the use of the polythene dome should be curtailed before 60 DAP (Owusu, 2020). Stem cuttings of *Ixora singaporensis* treated with IBA (2000 ppm) in softwood cuttings and IBA + NAA (3000 ppm) each in hardwood cuttings were found to be the best, respectively recording the maximum rooting percentages of 89.33 and 82.00 per cent (Prabhuling *et al.*, 2005). *Ixora* can be raised true to type only by vegetative propagation (Watson and Swedroe, 1965).

9.2.2 Air Layering

A well branched plant with stem of pencil thickness can be selected for air layering. The bark can be removed with a clean knife or cutter. A mixture of coco peat, peat moss and sand or sphagnum moss can be wrapped with enough thickness on the cut portion which is covered with plastic tied at both end with jute rope or thread. During monsoon season, roots appear as weekly as in three weeks. Cut the rooted plant from the mother plant and transfer to a pot for growing in a partial shade area. The plant will be established within two months and ready for planting in field and open areas.

Propagation of *Ixora*

Ixora Propagated by Stem Cutting

Cut 4-6 Inch Section of New Growth

Remove the Leaves Leaving 2-4 leaves at the Tip

Dip Cut Ends in Rooting Medium, however, *Ixora* can Grow without Rooting Hormones

Place the Cutting in Warm Area under Partial Shade

Rootings in *Ixora* Cuttings

Propagation of *Ixora*

Ixora Propagated by Air Layering

Cocopit, Peat Moss and sand

Remove Unwanted Leaves

Stem with Pencil Thickness

Remove the Bark Carefully without Removing the Cambium

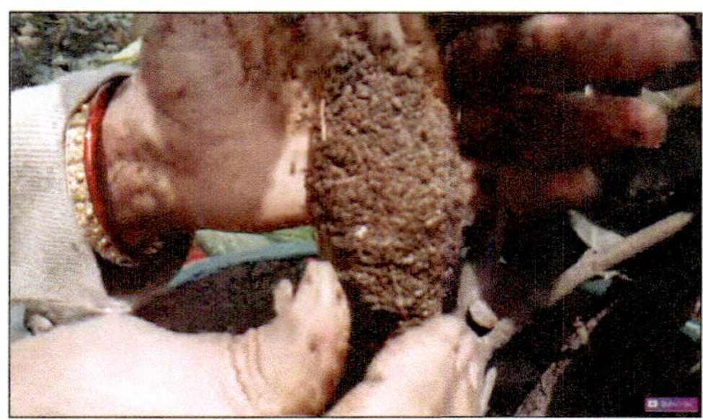

Cover with Rooting Medium with Optimum Thickness and Wrap with Plastic Polythene

Micropropagation of Ixora

A. Innoculated leaves used as explants; B. Callus induction from leaves; C. One regenerated shoot from calli; D. Many regenerated shoots from calli; E. Subculture of regenerated shoots; F. Rooting of plantlets; G. Sprouts emerging from regenerated shoots.

9.2.3 Grafting

Foliage chlorosis caused by mineral deficiencies, nematodes and other factors often limit the use of many of the better varieties in the landscape. In an effort to minimize the damage caused by the limiting factors, trials with grafting on various rootstocks were conducted. It was discovered as early as 1946 that the *Ixora parviflora* rootstock gave greater vigor, more resistance to chlorosis, and resistance to nematode damage to some of the weaker species and hybrids. This grafting operation although more expensive than growing the plant on its own root, has been very successfully performed with a cleft graft on strong seedlings or rooted cuttings of the *I. parviflora.*

Recent trials indicate that most available *Ixora* can be improved by grafting on this rootstock. The grafted plants show a greater resistance to chlorosis and cold injury. The *Ixora* 'Super King' has been observed to be incompatible when grafted on *I. parviflora* but most others have shown improvement when grafted. The use of *I. parviflora* rootstock has made it possible to produce standards in all colours. (Watson and Swedroe, 1965). Grafting was also done between *Ixora* species and *Ixora parviflora*, besides using *Ixora* hybrids as scion. *I. colei* grafted on *I. parviflora* gave best result. 'Gillette's Yellow' grafted on *I. parviflora* resulted in resistance to nematodes, chlorosis, and club-root problem. 'Singapore No.1' grafted on *I. parviflora* resulted in more resistance to chlorosis. 'Angela Bushman' hybrid grows well on own roots and slower when grafted on *I. parviflora*. *Ixora fraseri* demands shade and is very susceptible to nematodes. It must be grafted to be grown in general soil conditions successfully but grafting has not been very successful. 'Henry Morat' grafted on *I. parviflora* did not give satisfactory results.

Herrera's White has been grown successfully on its own roots but grafting on *I. parviflora* does not appear to be compatible. 'Kenny Grant' grafted on *I. parviflora* gave better results against chlorosis. *Ixora profusa* is the most susceptible to nematodes of any *Ixora* but has been very highly successful on *I. parviflora*. *Ixora rosea* grafted on *I. parviflora* roots produced slow developing prostrate growths which do not appear to be compatible in some situations. *Ixora* 'Hydrangea formis' has proven more satisfactory over a period of 12 years. The rate of growth is fairly rapid and when properly planted performs very satisfactorily on *I.* 'Hydrangea formis'. This rootstock does sucker badly but is presently in great demand. *Ixora chinensis* is very susceptible to nematodes but when grafted on *I. parviflora* the plant grows vigorously. Ixora 'Williamsii' produced larger flowers and more vigorous growth are when grafted on *I. parviflora* (Watson and Swedroe, 1965).

9.3.0 Micropropagation

Mass production of *Ixora* can be achieved through micropropagation using shoot tips, axillary buds and young leaves as explants to meet the demands in the market. Li *et al.* (2019) used young leaves of *Ixora chinensis* Lam. as explants for the first time on the Murashige and Skoog (MS) medium supplemented with different plant growth regulators. Shoot regeneration (23.7 per cent) was recorded from medium including 4.0 mg/l 6- Benzylaminopurine (6- BA), 1.0 mg/l Thidiazuron (TDZ) and 1.0 mg/l α- Naphthalene acetic acid (NAA). a survival ratio of 80.8 per cent and rooting ratio of 30.5 per cent was achieved from subculture of shoots on

MS medium fortified with 1.0 mg/l 6- BA and 0.5 mg/l NAA. Thakur and Kumar (2014) used shoot apex, leaf, node, internode and floral bud as explants which were cultured on Murashige and Skoog medium with 30g/l sucrose and 8g/l agar, KIN, NAA and 2,4- D (Hi- media, Bombay) at pH 5.6 and observed that shoot apex, leaf, internode and node explants were found suitable for callus production. Normal growth in cultured shoot apex and floral bud were observed. Internode was found most suitable for production of roots when cultured on NAA (1.5 mg/l).

10.0 Soil and Climate

Ixora requires full sunlight in moist but well- drained acid soil and can tolerate some shade (Gilman, 1999). They usually prefer tropical and subtropical regions as they are sun loving shrubs. Plants exposed to more sun will be denser and more compact, and produce more flowers.

If temperatures lowers down below 4°C, leaves and small twigs will get damaged. Although light frosts may kill the plant above ground, new growth will occur in spring.

Species and cultivars of ixora are moderately salt tolerant, but not at all tolerant of alkaline soil, which causes the leaves to become chlorotic (Floridata, 2017). A pH of around 5 is good for *Ixora;* this pH is slightly lower than for most landscape plants (Keeler *et al.,* 2004). It thrives well in high humidity and moist, friable, well-drained soil rich in organic matter.

11.0 Cultivation

11.1.0 Planting Time

Planting can be done in spring season provided they are watered properly as low temperatures can be detrimental to the plant. Planting in early rains is safe but water-logging should be avoided. *Ixora* should not be planted next to foundations or concrete walkways, as concrete surfaces have a very high pH and this can cause problems with ixora growth and flowering. Do not use pea rock as mulch. *Ixora* needs a well- drained site (Keeler *et al.,* 2004).

11.2.0 Manuring and Fertilization

Alamu (2012) studied the effect of organic manures (Sunshine; Tithonia and Cassava peel compost) @ 5 tons per hectare on *I. coccinea* where he observed better physical appearance and luxuriant look of *I. coccinea* with cassava peel compost treatment recorded as most suitable. In order to get the best growth of flowering, it is recommended to treat the plants grown in 20- cm diameter plastic pot with spraying the foliage to run-off with (2.0 gm/pot NPK combined with 5.0 ml humic acid ml/l) 6 times with 1 month interval (Arafa *et al.,* 2019).

Ixora, like other acid-loving plants, require more fertilization management than plants that are adapted to growing in alkaline soils. In alkaline soil, it is very prone to iron and manganese deficiencies. Although chelated iron and manganese can be applied to the soil as granular fertilizer, high soil pH prevents most of it from being taken up by the plant. Liquid micronutrients sprayed on the foliage work best for treating the problem. Adding organic mulches, applying soil sulfur at recommended rates, and using acidic fertilizers are all ways to help lower high pH (Keeler *et al.*, 2004). The only effective soil-applied treatments for Fe chlorosis in dwarf ixora were Fe chelates. Ferric citrate, ferrous sulphate and iron glucoheptonate are no better than the untreated controls for correcting chlorosis. The only effective foliar applied Fe source was Fe- DTPA in greening up severely chlorotic dwarf ixora (Broschat, 2003).

11.3.0 Pruning

Pruning is an essential operation, done annually in ixora where off- branches, dried twigs and diseased plant parts are removed. Repeated removal of the tips of the branches should be avoided, as it removes emerging flower buds which ultimately reduces the number of flowers. Early spring is the best time for major pruning as new growth begins to appear on plants (Keeler *et al.*, 2004).

11.4.0 Irrigation

Planting of ixora flowers may not be as easy as other ornamental plants due to its high rate of evapotranspiration which makes it require more irrigation, moisture, slightly acidic soil and humid atmospheric condition (Rahuj, 1999). Watering is much essential when planting in spring. Watering should be done thoroughly when the soil gets dried.

12.0 Harvesting and Postharvest Management

12.1.0 Harvesting

The opened flowers of Ixora are collected early morning by plucking with hands for offering to Hindu gods. The flowering branches are cut (slant cut) with clean cutting knives and wrapped with newspaper. The cut ends are dipped in water and are carried in bucket to nearby market for sale in the morning. Remove any foliage on stems that is below water which can cause decay and encourage bacterial and fungal infection. The slant cuts keep the cut ends from lying flat on bottom of bucket and also improves water uptake. Flowers should preferably be harvested in the early morning when temperature is low and plant water content is high. However, flower buds are picked usually in the late afternoon for garland making.

12.2.0 Postharvest Management

Post- harvest handling is an important operation for maintaining quality and shelf life. Improper handling can affect physical appearances and characteristic of soft tissue like petal, deteriorating the quality of the flowers. Lu *et al.* (2010) studied effects of different preservative solutions such as sucrose, KH_2PO_4, 8-hydroxyquinoline (8- HQ), and citric acid (CA) on the preservation of cut flowers of *I. chinensis*. The preservative solution (2 per cent sucrose + 150 mg/L KH_2PO_4 + 120 mg/L 8- HQ + 30 mg/L CA) was found to significantly delay the vase life of cut flowers of *I. chinensis* 6d than the control (distilled water), meanwhile, fresh weight, soluble protein content relatively decreased more slowly, and the relative electric conductivitie (REC) rose more slowly.

13.0 Diseases and Pests

There are a few diseases and pests that attack Ixora plants. It has been reported that Ixora is susceptible to several fungal diseases. Mushroom root rot can also be a problem for ixora (Gilman, 1999).

13.1.0 Diseases

Leaf Spot

This type of leaf spot is caused by *Pseudocercospora ixoricola* (Alves and Barreto, 2010). Leaf spots can also be caused by *Colletotrichum gloeosporioides* and *Pestalotiopsis ixorae* (Banerjee *et al.*, 2017). Reddy *et al.* (2016) showed that *Pestalotiopsis*- like species could occur as generalist endophytes in *Ixora* species, while Nozawa *et al.* (2017) reported a *Pseudopestalotiopsis* strain (NBRC 112267, from Kawthaung, Myanmar) from *Ixora* sp. but did not provide the morphology of the corresponding strain. Tsai *et al.* (2018) also recorded leaf spot caused by *Pseudopestalotiopsis* species (*P. ixorae* and *P. taiwanensis*) in Ixora in Taiwan. The first report of *Diaporthe* species causing leaf spots and necrosis on *Ixora chinensis* was made from Taiwan (Ariyawansa *et al.*, 2020). Spraying of fungicides, avoid overhead irrigation and field sanitation could help control these diseases.

Bacterial Leaf Spot

Lin *et al.* (2020) reported bacterial leaf spot in *Ixora*, caused by *Xanthomonas maculifoliigardeniae* (*X. axonopodis* pv. *maculifoliigardeniae*). It was first briefly reported by Miller in 1970. Copper- maneb and Agri- strep (streptomycin sulfate) provided good control and sanitation involving the use of disease free plants should help reduce or eliminate the disease (Miller, 1970).

Alternaria Blight

This type of blight is caused by *Alternaria alternata* on *Ixora coccinea* var. *bandhuca* (Devi *et al.*, 1995). The first symptoms of the infection appear during post- monsoon season as dark melanised patches on mature leaves. Fungicides can be used to treat infected plants.

Mosaic Disease

Mosaic is caused by Cucumber Mosaic Virus (CMV- Ix strain) isolated from imported *Ixora* plants (Waterworth and Povish, 1975).

13.2.0 Pests

The common pests of Ixora are aphids, scale insects, and mealy bugs. Sooty mold, a black fungus grows on the excrement of these bugs. Control of the pests will solve the sooty mold problem. Spraying of pure neem oil can be helpful. Oil spray is best applied late in the day to reduce damage to leaves by sunburn. Nematodes can also be a problem. Mulch around the plant, not coming right up against the trunk, to deter these microscopic soil worms (Keeler *et al.*, 2004).

Diseases of Ixora

Leaf Spot Caused by *C. gloeosporioides*

Leaf Spot Caused by *Pseudopestiolopsis ixorae*

Leaf Spot Caused by *Pseudopestiolopsis taiwanensis*

Bacterial Leaf Spot

Aphids

Scale under Leaf of Ixora

Mealy Bug

14.0 Biotechnology

14.1.0 Genetic Transformation

Infestation by insects such as *Ixora* leaf webber, flower webber, green horned caterpillar and looper caterpillar has become a threat to commercial exporters. Moreover, the control of these pests by means of chemical has hazardous effects on the environment and other beneficial insects and therefore, is unpopular among the buyers. As an alternative, *Ixora odorata* var. *vulcan* was transformed with the *cry* gene of *Bacillus thuringiensis* strain 6e by *Agrobacterium* mediated gene transfer. In the process, an effective plant regeneration procedure and a gene transfer system were developed. The *cry* gene of *Bacillus thuringiensis* strain 6e was isolated, purified and cloned into the *XbaI* site of the T- DNA region of the *Agrobacterium* vector pLG121Hm *via* an adapter. Recombinant pLG121Hm was transferred to *Agrobacterium* strain LBA4404 by electroporation. Positive *cry* clones were confirmed by Dot blot analysis of the plasmid DNA (extracted from the electroporated strains) with Dig labeled *cry* probe. *In vitro* grown *Ixora odorata* shoot tips (2 cm) were co-cultivated with LBA4404, harboring the *Bt* 6e *cry* cloned binary vector pLG121Hm. Co-cultivated explants were transferred to shoot multiplication medium (1/2 MS with 2mg/l BAP and 500mg/l cefotaxime) and incubated at 2500 lux with a 16h photoperiod. After 4 weeks the axillary shoots were screened on selective medium (1/2 MS supplemented with 200mg/l hygromycin). Hygromycin positive plants were used for the PCR analysis and Southern blot analysis. PCR analysis of the putative transformants, carried out with *Gus* specific primers, produced the expected 1.680kb fragment in all the hygromycin positive transformants. Southern blot analysis of these with the Dig labeled Bt 6e *cry* probe produced positive results. The positive results of both, the PCR analysis and the Southern blot confirmed genomic integration of the *Bt* 6e *cry* gene. The transformation efficiency was 20 per cent for the shoot tips whereas it was 40 per cent and 30 per cent for callus and leaf disk transformations respectively (Alahakoon *et al.*, 2007).

15.0 References

Aktar, F., Kaisar, M. A., Kabir, A. H., Hasan, C. M. and Rashid, M. A. (2009) *Dhaka University Journal of Pharmaceutical Sciences*, **8**(2): 161-166.

Alahakoon, U. I. A., Herath, H. M. V. G., Abeyrathna, P. K., Karunatileke, R. W. M. V. W. N. C., Silva, U. K. Y., Bamunuarachchige, T. C., and Bandara, J. M. R. S. (2007) *Tropical Agricultural Research*, **19**: 10-18.

Alamu, L. O. (2012) *International Journal of Academic Research in Business and Social* Sciences, **2**(6): 252.

Alves, J. L., and Barreto, R. W. (2010) *Plant disease*, **94**(2): 278-278.

Andersson, L. (1992) *Scripta Botanica Belgica*, **1**: 1-199.

Andreasen, K. and Bremer, B. (2000) *American Journal of Botany*, **87**(11): 1731-1748.

Arafa, A. M., Darwish, M. A., El-Sayed, B. A. and Ahmed, G. D. (2019) *Plant Archives*, **19**(2): 2315-2320.

Ariyawansa, H. A., Tsai, I., Withee, P., Tanjira, M., Yen, C. Y., Al-Rashed, S. and Cheewangkoon, R. (2020) *Phytotaxa*, **461**(3): 155-165.

Asolkar, L. V., Kakkar, K. K. and Chakre, O. J. (1992) Glossary of Indian Medicinal Plants with Active Principles, New Delhi: Publication and Information Directorate, pp. 72-73.

Bachheti, R. K. and Pandey, D. P. (2011) *International Journal of ChemTech Research*, **3**(3): 1028-1032.

Bamunuarachchige, T. C., Chandrananda, R. A. K. M., Damayanthi, K. P. N., and Bandara, J. M. R. S. (2000) *Tropical Agricultural Research*, **12**: 56-63.

Banerjee, A., Islam, S. and Middya, R. (2017) *Journal of Pharmacognosy and Phytochemistry*, **6**(6): 1730-1732.

Bose, S., Maji, S. and Chakraborty, P. (2013) *Journal of Pharmaceutical Sciences and Technology*, **2**: 72-74.

Bremekamp, C. E. B. (1940) *Bulletin du Jardin Botanique de Buitenzorg, s é rie 3*, **16**: 276-299.

Bremekamp, C. E. B. (1934) *Recueil des Travaux botaniques néerlandais*, **31**(1/2): 248-308.

Bremekamp, C. E. B. (1937a) *J. BOI.* **75**: 108-111, 169-175, 260-266, 295-293, 318-326.

Bremekamp, C. E. B. (1938) *J. BOI.* **76**: 330-336.

Bremekamp, C. E. B. (1966) *Acta Botanica Neerlandica*, **15**(1): 1-33.

Bremekamp, C. E. R. (1937b) *Bulletin Du Jordin Botanique Buitenzorg, Serie III*, **14**: 197-367.

Broschat, T. K. (2003) *HortTechnology*, **13**(4): 625-627.

Chatwal GR (2007) Organic Chemistry of Natural Products, Vol II, 4th Edition, Himalaya Publishing House, pp. 2.38-2.40.

Chauhan, J. S. and Vidyapati, T. J. (1993) *Indian Journal of Chemistry Section (B)*, **35**(9): 992-994.

Chauhan, J. S., Kumar, S. and Chaturvedi, R. (1984) *Phytochemistry*, **23**(10): 2404-2405.

Chen, T., and Taylor, C. M. (2011) *Ixora* Linnaeus. *Cucurbitaceae through Valerianaceae, with Annonaceae and Berberidaceae*, **19**: 177-179.

Davis, A. P., Govaerts, R., Bridson, D. M., Ruhsam, M., Moat, J. and Brummitt, N. A. (2009) *Annals of the Missouri Botanical Garden*, **96**(1): 68-78.

De Block, P. (1998) *Opera Botanica Belgica* **9**: 1- 218.

De Block, P., J. Tosh, A. Mouly, and S. Dessein. (2007) *XVIIIth AETFAT Congress, 2007.* AETFAT, Yaound é, Cameroun: 23 (Abstract).

Devi, S., Garg, P. and Dwivedi, A. K. (1995) *Archives of Phytopathology and Plant Protection*, **29**(6): 473-477.

Dontha, S., Kamurthy, H. and Mantripragada, B. (2015) *International Journal of Pharmaceutical Sciences and Research*, **7**: 567-584.

Faten, M. M. and Zedan, Z. I. (2003) *Bulletin of Pharmaceutical Science*, **26**(1): 91-96.

Floridata, 2017. *Floridata. In: Floridata Tallahassee, Florida, USA:* Floridata.http: // www.floridata.com/

Fosberg, F. R. and Sachet, H. H. (1989) *Baileya* **23**: 74-85.

Gad, M. M., Abdul-Hafeez, E. Y. and Ibrahim, O. H. M. (2016) *Journal of Plant Production*, **7**(1): 85-91.

Ghani, A. (2003) *Medicinal Plants of Bangladesh. Edn 2, Asiatic Society of Bangladesh.*

Gilman, E. F. (1999) *Ixora coccinea. University of Florida, Cooperative Extension Service. Institute of Food and Agriculture Sciences, Fact Sheet FPS-***291**: 1-3.

Hansen, J. D., Hara, A. H., Chan Jr, H. T. and Tenbrink, V. L. (1991) Efficacy of hydrogen cyanide fumigation as a treatment for pests of Hawaiian cut flowers and foliage after harvest. *Journal of economic entomology*, **84**(2): 532-536.

Husain, T. (1984) Taxonomic Studies on Indian Species of the Genus *Ixora* (Rubiaceae) (Doctoral dissertation, Aligarh Muslim University).

Jinshui, L., Luanmei, L. and Yalan, Z. (2010) *Chinese Agricultural Science Bulletin*, p. 20.

Keeler, G., Gabel, K. and Schoellhorn, R. (2004) *Ixora for South Florida. EDIS, 2004*(3).

Khare, C. P. (2007) *Indian Medicinal Plants*, Springer Private Limited, pp. 338-339.

Lee, C. L., Liao, Y. C., Hwang, T. L., Wu, C. C., Chang, F. R., & Wu, Y. C. (2010) *Bioorganic & medicinal chemistry letters*, **20**(24): 7354-7357.

Li, T., Cai, H., Wang, T., Fu, Y., Yang, W., Zhao, A. and Wang, J. (2019) *Plant Cell, Tissue and Organ Culture* (*PCTOC*), **139**(3): 605-608.

Lin, Y. R., Lee, S., Lu, C. H. and Chu, C. C. (2020) *Journal of Phytopathology*, **168**(7-8): 478-489.

Linnaeus (1753) *Species Plantarum*. Vol. I.

Lorence, D. H., Wagner, W. L., Mouly, A. and Florence, J. (2007) *Botanical Journal of the Linnean society*, **155**(4): 581-597.

Lu, L., Lin, J. and Xie, Z. (2010) *Acta Horticulturae Sinica*, **37**(8): 1351-1356.

Miller, J. W. (1970) *Plant Pathology Circular*, (101).

Mouly, A., Razafimandimbison, S. G., Khodabandeh, A. and Bremer, B. (2009) *American Journal of Botany*, **96**(3): 686-706.

Nair, S. C. and Panikkar, K. R. (1990) *Cancer letters*, **49**(2): 121-126.

Nair, S. C., Panikkar, B., Akamanchi, K. G. and Panikkar, K. R. (1991) *Cancer letters*, **60**(3): 253-258.

Nazarudeen, A. (2010) *Indian Journal of Traditional Knowledge*, **9**(2): 398-402.

Newbury, H. J. and Ford-Lloyd, B. V. (1993) *Plant Growth Regulation*, **12**: 43±51.

Nilsson, L. A., Rabakonandrianina, E., Pettersson, B. and Ranaivo, J. (1990) *Plant Systematics and Evolution*, **170**(3): 161-175.

Nozawa, S., Yamaguchi, K., Van Hop, D., Phay. N., Ando, K., Watanabe, K. (2017) *Mycoscience*, **58**(5): 328-337.

Obuzor, G. U. and Nwakanma, G. U. (2011) *International Journal of Academic Research*, **3**(2): 381-384.

Oktaviyanti, N. D. and Mun'im, A. (2019) *Heliyon*, **5**(11): e02950.

Owusu, S. E. (2020) *Journal of Agriculture and Ecology Research International*, 1-8.

Prabhuling, G., Patil, P. B., Patil, C. P. and Shanthappa, T. (2005) *Karnataka Journal of Horticulture.* **1**(4): 57-62.

Rahuj, B. (1999) *Journal of Horticultural Science and Ornamental plants.* Vol. 2(2) pp. 78.

Rastogi RP and Mehrotra BN (2002) *Compendium of Indian Medicinal Plants, Vol. IV, CDRI, Lucknow and NISC, New Delhi*, pp. 406-407.

Reddy, M. S., Murali, T. S., Suryanarayanan, T. S., Rajulu, M. G. and Thirunavukkarasu, N. (2016) *Fungal Ecology*, **24**: 70-75.

Reyes-Pérez, J. J., Luna-Murillo, R. A., Espinoza-Coronel, A., Lara-Capistrán, L., Díaz-Rivera, E. and Hernández-Montiel, L. G. (2018) *Agroproductividad*, **11**(8): 145-148.

Sakai, K. and Higuchi, H. (1980) *Research Bulletin of the Aichi-Ken Agricultural Research Center*, **12**: 109-113.

Saravanan, P. and Boopalan, E. (2011) *International Journal of Applied Biology*, **2**(2): 30-34.

Schumann, K. (1891) Rubiaceae. *Die naturlichen Pflanzenfamilien*, **4**: 1-156.

Sharma, A. K., Chatterjee, T. (1960) *Genetica*, **31**: 421-447.

Shinde, A. S., and Chavan, N. S. (2017) *Current Botany*, **8**: 155-158.

Singh, M. M., Chowdhury, S. R., Kulshreshtha, D. K., and Kamboj, V. P. (1993) *Contraception*, **48**(2): 178-191.

Smith, A. C., and S. P. Darwin (1988) *Family 168. Rubiaceae. In A. C. Smith [ed.], Flora vitiensis nova, vol. 4,* 143-362. Pacific Tropical Botanical Garden, Lawai, Kawai, Hawaii, USA.

Sommano, S., Suksathan, R., Colelli, G. and Kumpuan, W. (2016) In *VIII International Postharvest Symposium: Enhancing Supply Chain and Consumer Benefits-Ethical and Technological Issues 1194* (pp. 1187-1194).

Thakur, P. and Kumar, H. (2014) *International Journal for Exchange of Knowledge*, **1**(1): 4-7.

Tsai, I., Maharachchikumbura, S. S., Hyde, K. D. and Ariyawansa, H. A. (2018) *Mycological Progress*, **17**(8): 941-952.

Tyagi, A. P., Racule, T. and Lal, P. (1998) *South Pacific Journal of Natural Science*, **16**: 7-10.

Verdcourt, B. (1958) *Bulletin du Jardin botanique de l'Etat, Bruxelles/Bulletin van den Rijksplantentuin, Brussel*, 209-290.

Wahab, F., Subramaniam, K., Suriyamoorthy, S. and Subburaj, S. P. (2012) *Asian Pacific Journal of Tropical Biomedicine*, **2**(3): S1312-S1316.

Watson, L. E. and Swedoe, P. (1965) *Florida State Horticultural Society*, pp. 406-409.

Wen, K. C., Chiu, H. H., Fan, P. C., Chen, C. W., Wu, S. M., Chang, J. H. and Chiang, H. M. (2011) *Molecules*, **16**(7): 5735-5752.

Ysrael, M. C. and Van Valkenburg, J. L. C. H. (1999) *de Padua. Plant Resources of South-East Asia No.: Medicinal and poisonous plants 1*. Backhuys Publisher, Leiden. The Netherlands, **12**(1): 315.

Mussaenda

Sk. Masudul Islam and Arindam Bhunia

1.0 Introduction

The Mussaenda, belong to family Rubiaceae, are also known as Queen of Philippines or Flag bush group. They are genus of tropical evergreen shrubs and scrambling climbers grown for their long-lasting showy flowers. Instead of petals, which most of the common flowers are distinguished for, *Mussaendas* are known for their splendid and colourful sepal or modified leaves and continuous blooming habit. Mussaendas are native to the old-world tropics, from West and Central Africa through the Indian sub-continent, Southeast Asia and into Southern China (Mclaughlin and Garofalo, 2002). The genus *Mussaenda* comprised of 200 species of varying growth habits; many of which are wild, other are medicinal, while few are cultivated and valued as florist crops.

2.0 Importance and Uses

2.1.0 Aesthetic Uses

Ornamental Mussaendas have proved to be among the most popular, demanded and used flowering shrubs in the warm humid tropical regions (Chadha and Choudhury, 2007). They form an important focal or spot plant in the gardens whenever they are used whether singly or in group. Grown as shrubbery, it can be used to create an accent in entrance, approach to a strategic garden structure or lawn background. Flowering twigs will also make good cut flower materials for vase or wreath making (Ogbu, 2011).

2.2.0 Therapeutic and Other Uses

Traditionally Mussaenda is reported to possess number of medicinal properties. The leaves are used in the treatment of jaundice, asthma, hyperacidity, ulcers, leprosy, diuretic, wound, swells, antimicrobial, diuretic activity, hypolipidemic effect, hepatoprotective activity, fever and cough (Wesley, 2009; Sambrekar *et al.*, 2010). Gopalakrishnan and Vadivel (2011) identified twenty chemical constituents through GC-MS analysis in Mussaenda and the major constituents are quinic acid (32.87 per cent), 4-[(1E)-3-hydroxy-1-propenyl]-2-methoxyphenol (8.30 per cent), naphthalene, decahydro-2-methoxy (7.20 per cent) and 1, 2, 3-benzenetriol (7.70 per cent).

Vidyalakshmi *et al.* (2008) reviewed ethnobotanical uses of different Mussaenda species. The juice of the root of *Mussaenda frondosa* is used to treat blemishes on the tongue and the sepals are diuretic (Jayasinghe *et al.*, 2002). The infusion of the bark of *Mussaenda raiatensis* is sometimes given to an infant believed to be ill or malnourished (WHO, 1998). *Mussaenda pubescens* has been used in Chinese folk medicine as a diuretic, antichloristic and antipyretic agent. The whole plant of *Mussaenda pubescens* has been used against laryngopharyngitis, acute gastroenteritis, and dysentery and as a contraceptive agent (Dictionary of Chinese Traditional Medicine, 1986). A paste of the root of *Mussaenda roxburghii* is applied to the tongue to treat boils (Manandar and Manandar, 2002).

Uses of Mussaenda

In pot

As Specimen in Garden

As a Herbaceous Border

In Flower Arrangement

In Flower Vase

3.0 Origin and Distribution

The genus is mostly paleotropical and its centre of diversity is in tropical Asia with 100 species, followed by tropical Africa with 35 species (Bridson and Verdcourt, 1988), Madagascar with 24 species (Missouri Botanical Garden and Bergius Foundation, unpublished manuscript), and the Mascarenes with four species (Wernham, 1914). Mussaenda is characterized by a combination of valvate aestivation, fleshy or berry-like, indehiscent fruits, and numerous, small, reticulate seeds. Many species of *Mussaenda* (e.g., *M. erythrophylla, M. incana, M. parvifolia, M. philippica*) are commonly cultivated in botanical gardens throughout the world because of their beautiful, long-blooming, sturdy flowers with enlarged calyx lobes. Although recognition of *Mussaenda* as a separate genus has never been challenged, its circumscription has always been controversial (*e.g.*, Miquel, 1857; Hooker, 1880; Kurz, 1887; Schumann, 1891). Earlier, authors disagreed as to whether *Mussaenda* should include only the Asian and African species with enlarged, petaloid calyx lobes or calycophylls (also called semaphylls, sensu Leppik, 1977) and fleshy, indehiscent fruits. Miquel (1857) transferred the Asian capsular-fruited *Mussaenda uniflora* without enlarged calyx lobes, described by Don (1834), to his new genus *Aphaenandra*. Similarly, both Wernham (1916) and Li (1943) described *Pseudomussaenda* and *Schizomussaenda*, respectively, to accommodate all African and another Asian capsular-fruited *Mussaenda* with enlarged calyx lobes. On the other hand, Candolle (1830), also endorsed by Hooker (1880), Kurz (1887), and Schumann (1891), recognized a broad circumscription including *Mussaenda* species with or without semaphylls and with dehiscent or indehiscent fruits. This broad circumscription appears never to have gained acceptance. *Mussaenda* as presently circumscribed (Robbrecht, 1988; Puff *et al.*, 1993; Mabberley, 1997) includes all Afro-Asian and Malagasy *Mussaenda* species with indehiscent fruits and with or without semaphylls. The Asian Mussaenda species seem to have descended from an African progenitor that must have reached Asia via a long-distance dispersal event. The major radiations of Mussaenda seems to have occurred only after the group began to colonize Asia, where 97 species (73.48 per cent) of the 132 species Mussaenda are presently found. Despite the fact that the Afro- Asian *Mussaenda* species are shown to be closely related mainland Africa and Asia do not share in common any *Mussaenda* species. On the other hand, the most widespread African *Mussaenda* species, *M. arcuata*, is the only African *Mussaenda* species that has successfully reached the Comoro islands, Madagascar, and the Mascarenes, probably via stepping-stone dispersal. *M. erythrophylla* vars. Queen Sirikit and Light Pink were first introduced in India in 1966 at the Agri-Horticultural Society of India (Bose, 2020).

4.0 Morphology

Mussaendas are trees, shrubs, or clambering or twining lianas, rarely dioecious, unarmed; raphides absent. Leaves are opposite or occasionally in whorls of 3, with or usually without domatia; stipules persistent or caducous, interpetiolar, entire or 2- lobed. Inflorescence is terminal and sometimes also in axils of uppermost leaves, cymose, paniculate, or thyrsiform, several to many flowered, sessile to pedunculate, bracteate. Flowers are sessile to pedicellate, bisexual and usually distylous or rarely unisexual. Calyx is limb 5- lobed nearly to base, frequently some or all flowers of an inflorescence with 1(-5) white to coloured, petaloid, persistent or deciduous, membranous, stipitate calycophyll(s) with 3-7 longitudinal veins. Corolla are yellow, red, orange, white, or rarely blue (*Mussaenda multinervis*), salverform with tube usually slender then abruptly inflated around anthers, or rarely constricted at throat (*M. hirsuta*), inside variously pubescent but usually densely yellow clavate, villous in throat; lobes 5, valvate-reduplicate in bud, often long acuminate. Stamens are 5, inserted in middle to upper part of corolla tube, included; filaments short or reduced; anthers basifixed. Ovary is 2- celled, ovules numerous in each cell, inserted on oblong, fleshy, peltate, axile placentas; stigmas 2- lobed, lobes linear, included or exserted. Fruit purple to black, baccate or perhaps rarely capsular (*M. decipiens*), fleshy, globose to ellipsoid, often conspicuously lenticellate, with calyx limb persistent or caducous often leaving a conspicuous scar; seeds numerous, small, angled to flattened; testa foveolate-striate; endosperm abundant, fleshy (http://worldfloraonline.org/).

5.0 Species and Cultivars

5.1.0 Species

Mussaenda is a genus of approximately 200 species of insect-pollinated shrubs largely native to tropical Asia (Alejandro *et al.*, 2005) and exhibits considerable diversity in floral biology and sexual systems, including monomorphic species, with either approach or reverse herkogamy, stigma-height dimorphism, distyle and dioecy (Baker, 1958; Richards and Koptur, 1993; Pailler and Thompson, 1997; Naiki and Kato, 1999; Faivre, 2000; Deng, 2007; Deng and Zhang, 2008). Hence, mussaenda provides outstanding opportunities for investigating floral evolution and the origin of dioecy from stylar dimorphism. Molecular phylogenetic data indicate that a monophyletic group of *Mussaenda* species, mostly from China, has a relatively short divergence time among species (Alejandro *et al.*, 2005). The relative contribution of various isolation mechanisms during speciation is difficult to determine in deeply divergent species because of the continued accumulation of ecological and morphological differences. Therefore, studies using incipient or congeneric species can minimize these problems and may present the most relevant data for studying isolating mechanisms in plants (Kay, 2006). Two congeneric species *Mussaenda pubescens* and *M. shikokiana* that commonly occur in sympatry. Preliminary field observations suggested some natural hybridization between the two species, and therefore, *M. pubescens* and *M. shikokiana* provide a good system for examining the contributions of pre-and post-zygotic barriers to the maintenance of congeneric sympatric species.

Santapau and Henry (1973) have estimated that 14 species of this genus are endemic to Indian sub-continent. A few exotics are also grown in India, mostly introduced from Philippines and Ceylon. There are 35 species of tropical African's centre of diversity (Bridson and Verdcour, 1988; Robbreecht, 1988), out of which six are currently cultivated or of potential floricultural value (Bosch *et al.*, 2002). Among the domesticated species of primary use as florist crops, are few prized introduced species/cultivars-*M. frondosa* and *M. philippica* which are also of common sight in many Nigeria gardens, parks and commercial nurseries. They are often referred to as 'Queen of Philippines' by most Nigeria's flower merchants and gardeners, but elsewhere they are known as 'Flag bush' (Sheat and Schofield, 1995).

Two species of *Mussaenda* are distributed in Japan; *M. parviflora* grows in the Southwest islands of Japan, and *M. shikokiana* grows in Tokai, South Kinki, Shikoku and Kyushu. *Mussaenda parviflora* is dioecious (Naiki and Kato, 1999), *M. shikokiana* is self- compatible and autogamous, but these two species are closely related (Alejandro *et al.*, 2005).

A brief description of important species is given below.

Mussaenda erythrophylla (Common name: 'Ashanti blood'; Red flag bush)

It is the only West African species with a red calyx lobe and is a common forest climber in the wild, especially near streams and in secondary growth. It can grow to 10 m as a scandent shrub in its natural habitat. In cultivation however, it usually grows no more than 3 m height in a rather compact and bushy habit. It is characterized by tubular flowers, creamy white and 10 mm in diameter, with a central cushion of red felt, having a single modified sepal which is large, roundly ovate, brilliant scarlet, felt-like and borne in great profusion when in bloom. The contrast between the cream corolla and the deep red sepals is always most satisfying in the garden scenery. *M. erythrophylla* has an open, sprawling habit and requires careful pruning to maintain it as a low spreading shrub. Alternatively, it can be allowed to grow and given support, such as a trellis or an adjacent tree. It is most effective during the dry season, if adequate water supply is ensured. It is known to be tolerant to brackish conditions and will do well on wet swampy land. It grows easily from stem cutting, and sets viable seeds. This species needs full display of the day light for outdoor planting (Greensill, 1977; Purseglove, 1984; Sheat and Schofield, 1995).

M. luteola (Common name: Dwarf flag bush)

It is another species of African origin that grows to a height of 1-2m. It is more compact in habit and smaller in stature, than the other ornamental garden species. Its foliage and top canopy are scantier with rather smaller sized leaves. Flowers are yellow with an orange centre and small white sepals (flags). It thrives in the same condition as the previous species. It is also hardy and drought resistant once established (Sheat and Schofield, 1995; Alejandro *et al.*, 2005).

M. frondosa (Common name: White Queen of Philippines; White flag bush)

It is of tropical Asian origin from Indo-China to Malaysia, although it has been introduced to various parts of the world. Individual plant can grow up to 4 m height with an equal spread. The foliage is lighter green and the terminal flowers clusters have orange hollow, tubular corolla with a single white enlarge calyx lobe. When in bloom it carries masses of creamy white bracts or sepals which persist long after the tubular golden yellow flowers have fallen (Rosario, 1984; Dhillon *et al.*, 2005; Chadha and Choudhury, 2007).

Species of Mussaenda

Mussaenda erythrophylla

Mussaenda parviflora

Mussaenda frondosa

Mussaenda luteola

Mussaeade philippica

Pseudomussaenda flava

Mussaenda incana

Mussaenda arcuata

M. philippica (Common name: 'Donas'; Pink Queen of Philippines)

It is native to the Philippines. Individual plant forms a shrub or small tree, of 1-4m tall. The dark green leaves are smaller to those of *M. erythrophylla*, though less ovate and not as prominently veined. The flowers which are borne in terminal cymes consist of a yellow tubular corolla with one lobe or the calyx greatly enlarged and leaf-like. It is pink and showy, though not as striking as the bract of *M. erythrophylla*. It grows in full sun under warm spot (Brickell, 1990; Matthew and Karikari, 1990; Sheat and Schofield, 1995).

***Pseudomussaenda flava* (prev. *Mussaenda luteola, Mussaenda glabra*)**

Dwarf yellow mussaenda is native to Ethiopia, Sudan, Kenya, Uganda, Nigeria. This shrub has white floral sepals and creamy yellow flowers. While no longer in genus *Mussaenda*, it shares many traits and is the hardiest species, tolerating more dryness and cooler temperatures. This plant blooms all year round in the tropics and subtropics. The flowers also produce a faint perfume.

Other promising species worthy of gardeners' trials in warm humid tropical environment as flowering shrubs include: *M. afzelii, M. chippii* and *M. tenuiflora*. *M. chippii* is of particular importance, having some resemblance with the adapted exotic species *M. frondosa*. *M. chippii* is a common climber with an enlarged white sepal, indigenous to tropical Africa, but grows in the more humid forest region along streams or marshy areas (Greensill, 1977; Bosch *et al.*, 2002).

5.2.0 Cultivars

'Dona Evangelina': This is true tropical shrub with small yellow flowers and brightly coloured large red bracts that make the plant spectacular bloomer, flowering season from spring to fall. It cannot tolerate freeze, however planted in protected areas it is worth the care efforts.

'Dona Aurora' : This is a tropical shrub or sub-shrub that will grow to 3 m tall in tropical areas, but more likely will reach 1 m tall in containers. Clusters (corymbs) of small, tubular flowers with five spreading lobes bloom in summer, however it is the large and colourful, ovate, leaf-like sepals (to 8 cm long) that provide the real ornamental display. Some individual flowers in each cluster will develop a single enlarged sepal. Elliptic to ovate, bright green leaves (to 15 cm long). Hybrids typically feature flowers in red and/or yellow with showy sepals of white, bright red or pink.

'Queen Sirikit' : This mussaenda is a named cultivar, "Queen Sirikit", the former Queen of Thailand. It was named to commemorate her first visit to the Philipines in the 1970s. The only mussaenda ever to be given a name of a person not of Filipino extraction. It is a tropical shrub or sub-shrub that will grow to 3 m tall in tropical areas, but more likely will reach 1 m tall in containers. Clusters (corymbs) of small, tubular flowers with five spreading lobes bloom in summer, however it is the large and colourful, ovate, leaf-like sepals (to 8 cm long) that provide the real ornamental display. Some individual flowers in each cluster will develop a single enlarged sepal. Elliptic to ovate, bright green leaves (to 15 cm long). Hybrids typically feature flowers in red and/or yellow with showy sepals of white, bright red or pink.

Cultivars of Mussaenda

'Dona Evangelina' 'Dona Aurora' 'Queen Sirikit'

'Calcutta Sunset' 'Schizomussaenda' 'Marmelade'

'Dwarf Yellow Mussaenda' 'Capricorn Dream' 'Capricorn Ice'

'Aurorae' 'Bangkok Rose' 'Dona Luz'

'Calcutta Sunset' : This mussaenda is a hybrid between *M. philippicca* Dona Luz × *M. flava* and is named in resemblance a brilliant sunset of yellow peachy colours. It is a patented shrub raised from seed in 1995 in a nursery in Alipore, Calcutta, India.

This shrub is very hardy and vigorous-the result of having *P. flava* in its breeding. It is very vigorous hybrid that grows quickly to 3 m. The shrub is covered with showy yellow and salmon deepening to yellow and orange sepals. They are smaller than many other hybrids, but are very showy on the bush. In Australia, this plant has been rebranded as Mussaenda 'Calcutta Gold'.

'Schizomussaenda' : This is a rounded evergreen tropical shrub or sub-shrub that will grow to 10ft tall in tropical areas, but more likely will reach 1 m tall in containers. Clusters (corymbs) of small, tubular flowers with five spreading lobes bloom in summer, however it is the large and colourful, ovate, leaf-like sepals that provide the real ornamental display. Some individual flowers in each cluster will develop a single enlarged sepal. Elliptic to ovate, bright green leaves.

'Marmelade' : This variety is extremely rare and has orange flowers. They are tropical shrubs or sub-shrubs that will grow to 3 m tall in tropical areas, but more likely will reach 1 m tall in containers. Clusters (corymbs) of small, tubular flowers with five spreading lobes bloom in summer, however it is the large and colourful, ovate, leaf-like sepals (to 8 cm long) that provide the real ornamental display. Some individual flowers in each cluster will develop a single enlarged sepal.

'Dwarf Yellow Mussaenda' : Golden star-shaped blooms are surrounded by creamy yellow bracts. Large sepals in palest yellow-white and small yellow corollas, likes partial to full sun and moist, freely draining soil. It can be propagated by cuttings. An ever-blooming, small, evergreen shrub and is excellent in containers. It blooms especially well during the warm months. Grow it in light shade to full sun, less cold sensitive, unlike other mussaendas. It can tolerate light frost.

Mussaenda 'Capricorn Dream' has multiple scarlet red sepals and it makes quite a show in the garden.

Mussaenda 'Capricorn Ice' has multiple white sepals and deep yellow flowers.

6.0 Genetics and Breeding

Out of nearly 200 species of *Mussaenda*, chromosome numbers of only five species have been determined so far (Fagerlind, 1937; Priyadarshan and Ramachandra, 1984; Selvaraj, 1987). The chromosome number of all species studied is 2n = 22 and the basic number of the genus is x =11.

6.1.0 Reproductive Biology

Mussaenda is one of the well-known genera with flower-like inflorescences. One of the calyx lobes of some flowers in an inflorescence enlarges into a white, yellow or red petal-like lobe (ClaBen-Bockhoff, 1996).

Reproductive isolation is a fundamental requirement for speciation and includes several sequential stages. Few studies have determined the relative contributions of pre-and post-zygotic reproductive isolation in plants, especially between relative species with clear differentiation in flower form. Chen *et al.* (2014) investigated the mechanisms responsible for reproductive isolation in sympatric *Mussaenda pubescens* var. *alba* and *Mussaenda shikokiana* in Guangxi Province, China, and they made observations of flowering phenology, patterns of insect visitation, and conducted pollination experiments, including artificial hybridization. The two species had overlapping flowering times and were pollinated by overlapping pollinators; however, their relative importance differed significantly with *M. pubescens* visited more commonly by bees and *M. shikokiana* more frequently by butterflies.

They (Chen *et al.*, 2014) have not detected significant difference between interspecies and intraspecies pollen tube growth rates between *M. pubescens* var. *alba* and *M. shikokiana* in the bidirectional cross-pollination experiments, demonstrating that interspecific pollen has the ability to compete with the legitimate pollen, at least in pollen tube growth rate when they were not mixed, suggesting that the two species may lack a reproductive barrier in this aspect. Artificial pollination experiments demonstrated that interspecific cross-pollinations between *M. pubescens* var. *alba* and *M. shikokiana* produce seeds. Field investigation also detected natural hybrids in wild habitats, confirming that natural hybridization could happen between the two species.

Although *M. shikokiana* is a partly bud-selfing species with over half the ovules pollinated in buds (Chen *et al.*, 2013), the number of hybrids with *M. shikokiana* as maternal parent is not much lower than those with *M. pubescens* var. *alba* as the maternal parent.

M. pubescens is functionally dioecious with populations containing two types of individuals: a long-styled morph with short-level anthers and a short-styled form with long-level anthers. However, the long-styled morph exhibits sterile pollen and functions as a female, and the short-styled morph is female sterile and functions as male (Li *et al.*, 2010). *M. pubescens* var. *alba* possesses white flowers with stout corolla tubes and much-reduced petaloid calycophylls compared with typical *M. pubescens* (Deng and Zhang, 2004). The other species, *M. shikokiana* possesses yellow hermaphrodite flowers and is monomorphic *i.e.*, stamens are slightly shorter than a pistil. (Chen *et al.*, 2013). In Guangxi Province, China, where the geographical ranges of *M. pubescens* var. *alba* and *M. shikokiana* overlap and the two taxa are often found co-occurring in the same habitat. High rate of fruit sets by the bagging and selfing experiments shows that *M. shikokiana* is self-compatible, and even autogamous, although actual selfing rate were not yet examined (Naiki, 2008). Lower part of the stigmas and upper part of the anthers were closely situated to each other in a corolla tube of *M. shikokiana*. Upper part of the stigmas can even be touched by the anthers when the corolla tube falls. For the one- day habit of *M. shikokiana* flowers, self-compatibility and autogamy could compensate for the fruit set reduction when the number of pollinators is few, caused by bad weather or other reasons.

6.2.0 Breeding System

The breeding system of *Mussaenda* is thought to be predominantly distylous, *i.e.*, flowers of one morph have a long style and short stamens, and flowers of the other morph have a short style and long stamens (Alejandro *et al.*, 2005). Some *Mussaenda* species are morphologically distylous but functionally dioecious (Baker, 1958; Naiki and Kato, 1999). Monomorphic, self-compatible flowers of *M. shikokiana* are probably a rare case in *Mussaenda*, and thought to have evolved from a distylous ancestor by the collapse of distyly. *Mussaenda shikokiana* is closely related to dioecious *M. parviflora* (Alejandro *et al.*, 2005). The direction of the evolution in the mating system *M. shikokiana* is different from *M. parviflora*. Observation of insect visits suggests that main pollinators of *M. shikokiana* are butterflies. Several observations suggest that mussaenda is pollinated mainly by Lepidoptera. *Mussaenda parviflora* is pollinated mainly by both long-tongued butterfly and hawkmoth species (Naiki and Kato, 1999). *Mussaenda frondosa* is also thought to be pollinated by butterflies, shown by the calyx-lobe removal experiments (Borges *et al.*, 2003). Baker (1958) reported that a West African *M. elegans* was visited by butterflies.

6.3.0 Development of Cultivars

The current *Mussaenda* cultivars were developed during the 1940s and 1950s at the University of the Philippines, Los Banos. This university has been active over many decades in breeding these ornamentals and is responsible for the development of the popular cultivars 'Queen Sirikit', 'Doña Aurora' and 'Doña Luz'. Many cultivars were named after First Ladies and other notable women of the Phillipines, hence the Spanish form of address Doña in some cultivar names. 'Queen Sirikit' was named after the Queen of Thailand to commemorate a visit to the Phillipines. In the early 1900s, a mutant form of *M. philippica* was found at the base of Mt. Tuntungin. This mutant, later to be called 'Dona Aurora', exhibited five extended white sepals instead of the single, extended sepal displayed by *M. philippica*. 'Dona Aurora' was hybridized to *M. erythrophylla*, an African *Mussaenda* species which exhibits a single red sepal. The resulting progeny were self-pollinated and backcrossed to both parents (Rosario, 1984). The progeny of these crosses later became the cultivars used today.

The discovery of *M. philippica* var. aurorae (with all five calyx lobes samples were enlarged) then crossed with the African *M. erythrophylla*, which led to the development and propagation of 20 Philippine hybrids (Rosario,1998) that are known for their splendid and iridescent, enlarged calyx lobes.

6.4.0 Mutation Breeding

The genus *Mussaenda* is characterized by heterostyly, incompatibility and sterility (especially the hybrids). To expedite the development of new forms, mature fruits of *M. philippica* which has only one white petaloid calyx lobe was irradiated with the flowing doses of gamma rays: 500 r, 2 Kr, 5 Kr and 10 Kr (Rosario and Mendoza, 1999). There were no significant changes in the growth habit of the treated plants. However, they observed an improvement in the number of expanded calyx lobe of *M. philippica*. The species has one white petaloid calyx lobe, but upon treatment with 5 Kr and 10 Kr gamma rays there was an increase from one up to 6 petaloid sepals. The mutants resemble *M.* 'Dona Aurora'. This study is another proof that 'Dona Aurora' is a mutant of *M. philippica*.

7.0 Growth and Flowering

The profuse upright growth habit of some *Mussaenda* cultivars is undesirable for pot plant culture. Three growth regulators, daminozide (B-Nine), ancymidol (A-Rest), and paclobutrazol (Bonzi), were applied at two commercially recommended rates and two application methods (spray vs. drench) to determine their effects on plant height (Cramer and Bridgen, 1998). In subsequent experiments, the same growth regulators were applied at higher concentrations and in either two or three applications. The most attractive potted plants were produced with two spray applications of daminozide at 5000 mg/l or two drench applications of ancymidol at 0.5 mg/pot. Higher concentrations or additional applications excessively reduced plant height. Paclobutrazol sprays or drenches at any concentration or application number were ineffective for controlling plant height (Cramer and Bridgen, 1998).

Olosunde *et al.* (2008) evaluated that the effects of growing medium and variety on rooting characteristics of *Mussaenda philippica*. River sand medium enhanced root growth and development better than the topsoil only (control), topsoil + manure or topsoil + sawdust. The study also revealed that river sand medium gave the highest values for number of roots at 8,10 and 12 WAP and rooting percentage than topsoil, topsoil + sawdust or topsoil + manure. This could be attributed to the better aeration and good drainage of water, which could promote root growth and development. The reasons for these preferences are not clear, but Messen *et al.* (1997) found that in *Cordia alliodora* stomata conductance was higher in sand and gravel than in sawdust, suggesting that higher water content of sawdust inhibited oxygen diffusion and that anoxia caused tissue death and reduced water absorption. Presumably, therefore, species with a preference for sand and gravel are better adapted to unsaturated conditions. The stunted root growth observed form the cuttings of pink and white varieties of *M. philippica* in topsoil + sawdust medium resulted probably from high carbon-nitrogen ratio, improper decomposition and presence of growth inhibitors in the sawdust components. Stem cuttings of both pink and white varieties of *M. philippica* planted in river sand rooted earlier than those planted in other growing media, possibly due to increased root zone temperature due to proper aeration. Dawson and King (1994) observed that root zone temperature affects root initiation of stem cuttings in plants.

8.0 Propagation

8.1.0 Seed Propagation

Mussaenda can produce seed (in a small fruit), but production is poor and unreliable. Seed is significant for the purposes of hybridising but vegetative methods are used for commercial propagation.

8.2.0 Vegetative Propagation

Mussaenda is propagated mostly by hard wood and semi-hardwood cuttings. Under warm humid tropical conditions rooted cuttings are able to come to bloom in few months (2- 3) after planting. Among the cultivated species, *M. frondosa* has shown hard-to-root tendency with stem cutting (Olosunde and Fawusi, 2003;

Propagation of Mussaenda

Cryptic Dioecy in *Mussaenda pubescens*

Flowers of *Mussaenda pubescens* illustrating sex organ placement and floral tube hairs: (A) stigma protruding from the throat of the corolla in the L-morph; (B) anthers concealed in the corolla of the S-morph; (C) sterile anthers in the dense hairs of the L-morph; (D) anthers dehisced in the S-morph with the presentation of pollen grains in the dense upward-facing hairs of the corolla; (E and F) positions of the sex organs in the L-morph (E) and S-morph (F). An, Anther; St, stigma. Scale bars ¼ 1 mm (Li *et al.,* 2010).

Ogbu *et al.*, 2008). Air layering is also feasible, and is sometimes used as a means of multiplication if only a few plants are desired (Ogbu, 2011). Bose (1970) reported that tip cuttings of *M. erythrophylla* 'Queen Sirikit' with two leaves produced profuse root system in 2-3 weeks in vermiculite rooting media, under intermittent mist, throughout the year.

8.3.0 Micropropagation

Protocols were developed for propagation of *Mussaenda erythrophylla* cv. Rosea via somatic embryogenesis by manipulating growth regulators and culture conditions. Calli were induced from young leaf explants of *Mussaenda erythrophylla* cv. Rosea on Murashige, Skoog medium supplemented with 6-benzylaminopurine or kinetin along with indole-3-acetic acid or 2, 4- dichloroacetic acid within four weeks of culture. The calli were sub-cultured either in the same medium or in a modified medium for induction of embryogenic callus. In mussaenda, embryogenic calli were developed on Murashige, Skoog medium supplemented with 0.5- 1.0 mg/l 6-benzylaminopurine, 2.0-3.0 mg/l indole- 3- acetic acid, and 10 mg/l ascorbic acid. Somatic embryos were isolated and transferred to half-strength Murashige, Skoog medium supplemented with 0.25-0.5 mg/l 6-benzylaminopurine + 0.1 mg/l gibberellic acid + 5.0 mg/l adenine sulphate and 2 per cent sucrose for maturation and germination. About 70 per cent somatic embryos of mussaenda germinated. The somatic embryos of *Mussaenda erythrophylla* cv. Rosea germinated into normal plantlets with distinct shoot and well-developed root system. The somatic embryo-derived plantlets grew normally and flowered within two months of transfer to the field (Das, 2010).

9.0 Soil and Climate

9.1.0 Soil

It is grown in a completely fertile, dry and organic-rich, medium moisture soil. Plants thrive in organically enriched fertile soils that have been heavily mulched and where they are regularly fertilised. Like many other plants in the family Rubiaceae, they can be subject to nutrient deficiencies, particularly trace elements. This is all too common in high rainfall areas where nutrients are readily stripped from the soil.

9.2.0 Climate

Mussaendas perform best in humid tropical and subtropical areas, although many species and some hybrids will do well in protected areas in warm temperate areas. Plants like sheltered locations in full sun to semi-shade, and a northerly to eastern exposure (southerly in the northern hemisphere, naturebring.com).

10.0 Cultivation

Mussaenda prefers organic-rich and fertile soil which is heavily fed, these plants are regularly fertilized. Due to the lack of nutrients, the plant can be damaged. Nutrients are easily separated from the soil in high rainfall areas. Using organic fertilizers is more beneficial. Benefit from regular organic, leafy fertilizer applications. Dilute the liquid fertiliser to ½ the recommended rate and apply early in the morning or in the cool of the evening (naturebring.com).

Care should be taken to keep the bush water once or twice per week, it keeps moisture in the soil which is necessary for the plant, but do not let the mud. Water the plants during dry periods, the plants suffer from dryness and can be absorbed and die (naturebring.com).

Mussaenda bush is a stroller and benefits from annual pruning. Generally, it should not be removed more than 1/3 of the development. If more renewal is required, and then it is better to have long-term sorting, gradually reducing its size as required. Some species and varieties become climbers; they can also be trained with fences. Removal of dead flowers and bracts is also advantageous in the flowering season. As plants flower during the blooming season, faded flower clusters should be removed especially during extended periods of wet weather. This will encourage further flower production and help check source of possible fungal infection. Mussaendas can stand heavy pruning. Plants should be kept at 2 m or so, particularly if they are grown as single specimens. The lower stems tend to become bare if the plants are allowed to grown much larger. By placing mussaendas at the rear of a border containing lower-growing shrubs and herbaceous perennials, the leggy base of the plant can be concealed (Ogbu, 2011).

11.0 Diseases and Pests

Mussaendas are relatively pest and disease- free. Scale insects and variegated grasshoppers sometimes constitute minor pests problem, as well as mealy bugs and mites. The plants are especially prone to these pests's attack during period of water stress at the peak and late dry season of the warm humid tropics. Good amount of well treated organic manure and supplementary irrigation supply should be able to keep the plants healthy and strong (Ogbu, 2011).

12.0 Biotechnology

12.1.0 Molecular Markers

DNA was isolated from fresh leaves of five *Mussaenda* cultivars to study the genomic relation among them (Mukherjee *et al.*, 2012). Genomic relationship was evaluated using 21 random polymorphic DNA markers. Among the 21- decamer primers only 7 primers amplified 12 monomorphic loci. The primer OPM01 had maximum resolution power, while the primer OPQ04 had maximum RAPD primer index. Dendrogram based on the basis SHAN clustering showed two clear cut groups containing *M. leuteola* in an isolated clade. The two varieties of *M. erythrophylla* are coming under same group indicating their close association at the molecular level. Among the 193 amplified loci 181 loci were polymorphic DNA and 65 bands are exclusive amplified only in particular individual. But there were no marker band for the species *M. erythrophylla* 'Rosea'. One interesting outcome of the investigation was the closeness of *M. philipicaca* 'Aurorea' with *M. erythrophylla* 'Gining Imelda'.

Microsatellite markers were developed and characterized in *Mussaenda pubescens* for the levels of genetic diversity and changes in population genetic structure in reproductive character displacement and in shifts of sexual systems (Duan *et al.*, 2012). They amplified nineteen microsatellite loci successfully in *M. pubescens*, 17 of which were polymorphic. A maximum of eight alleles were detected per locus in 68 individuals at population level. The observed and expected heterozygosity varied from 0 to 1.000 and 0 to 0.882, respectively. These newly developed microsatellite markers will be useful in further investigations of genetic diversity and gene flow among populations of *M. pubescens* and its congeneric species.

Microsatellites have proven to be efficient genetic markers for species delimitation, phylogenetic reconstruction and detection of hybrids. Nineteen microsatellite markers for *M. pubescens* have been developed previously (Duan *et al.*, 2012), but they could not provide sufficient variation in population genetic structure analyses and in detecting interspecific hybridization in Mussaenda due to low polymorphisms. Later, Duan and Zhang (2014) reported 14 additional highly polymorphic microsatellite markers which will facilitate future studies in phylogenetic reconstruction of the genus, in gene flow analyses and hybridization detection among sympatric species, and in elucidating the population genetic differentiation patterns between the two ecological races, *viz.*, the yellow- flowered typical *M. pubescens* and the white-flowered *M. pubescens* var. *alba*.

Plastid and nuclear markers reveal high levels of genetic diversity in *M. kwangtungensis*. Indeed, haplotype diversity based on cpDNA (hT = 0.855) was higher than the mean value of angiosperms (hT = 0.68) (Petit *et al.*, 2005), while the expected heterozygosity computed on microsatellites (HE = 0.705) was higher than values reported for narrow distributed/endemic species (HE = 0.56/0.42) (Nybom, 2004). Life history traits, especially mating systems, may determine genetic structure among populations (Duminil *et al.*, 2007). Species with mixed or outcrossing mating systems was demonstrated to have significantly higher genetic diversity than selfing species (Nybom, 2004). It is clear *that M. kwangtungensis* is a typical functional dioecious species with an obligate outcrossing mating system (Li *et al.*, 2010), thus considered capable of harbouring high genetic diversity (Hamrick and Godt, 1996). The island populations displayed comparable levels of genetic diversity and differentiation with mainland populations, overall high levels of unique polymorphisms, and the greatest values of specific within population genetic diversity. No significant correlation was detected between genetic diversity of island populations and distance to mainland, as well as area of islands, except that allelic richness was significantly positively correlated with the area of islands (Shi *et al.*, 2020).

13.0 References

Alejandro, G.D., Razafimandimbison, S.G. and Liede-Schumann, S. (2005) *Amer. J. Bot.*, **92** : 544-557.

Baker, H.G. (1958) *Journal of West African Science Association*, **4** : 9-24.

Bose, T.K. (1970) *Bul. Agri. Horticultural Society of India*, **12** : 30-34

Bose, T.K. (2020) *Keynote Address, Bicentenary year, the Argi-Horticultural Society of India*, pp. 20-32

Borges, R.M., Gowda, V. and Zacharias, M. (2003) *Oecologia*, **136** : 571-573.

Bosch, C.H., Siemonsma, J.S., Lemmens, R.H.M.J. and Oyen, L.P.P. (2002) In: *Plant Resources of Tropical Africa: Basic list of species and commodity grouping*, PROTA programme, Wageningen, the Netherlands. P. III.

Brickell, C. (1990) In: *Gardeners' Encyclopedia of plants and flowers*, Dorling Kindersley, London, pp. 517.

Bridson, D. and Verdcourt, M.B. (1988) In: *Flora of tropical East Africa* (Ed. R.M. Polhill), Rubiaceae part 2. AA Balkema, Rotterdam the Netherlands, pp. 460 -475.

Candolle, A.P.DE. (1830) *Prodromus systematis naturalis regni vegetabilis Treutell and Wu¨rtz*, Paris, France.

Chadha, K.L. and Choudhury, B. (2007) In: *Ornamental Horticulture in India*, ICAR, New Delhi. pp. 172-174.

Chen, S., Luo, Z.L. and Zhang, D.X. (2013) *J. Syst. Evol.*, **51** : 731-742.

ClaBen-Bockhoff, R. (1996) In: *Second International Rubiaceae Conference Proceedings* (Eds. E. Robbrecht, C. Puff, and E. Smets), *Opera Botanica Belgica*, **7** : 329-367.

Cramer, C.S. and Bridgen, M.P. (1998) *Hort Science*, **33**: 78-81.

Das, P. (2010) *Horticultural Science*, **37**(2) : 70-78.

Deng, X.F. (2007) *Ph.D. Thesis, Chinese Academy of Sciences*, China.

Deng, X.F. and Zhang, D.X. (2004) *J. Trop. Subtrop. Bot.*, **12** : 476.

Deng, X.F. and Zhang, D.X. (2008) *Journal of Systemactics and Evolution*, **46** : 220-225.

Dhillon, B.S., Tyagi, R.K., Saxena, S. and Randhawa, G.J. (2005) In: *Plant Genetic Resources: Horticultural Crops*, Narosa Publishers, New Delhi, pp.4-7.

Dictionary of Chinese Traditional Medicine (1986) Jiangsu New Medical College, Shanghai Science and Technology Press, China, p.176.

Don, G. (1834) *A general history of the dichlamydeous plants*, vol. 3, Calyciflorae: viii 1867. London, UK.

Duan, T.T. and Zhang, D. (2014) *Journal of Genetics*, **93** : 44-47.

Duan, T.T., Gong, Wei and Zhang, Dian-Xiang. (2012) *Amer. J. Bot.*, **99**(11) : 437-439.

Duminil, J., Fineschi, S., Hampe, A., Jordano, P., Salvini, D., Vendramin, G.G. and Petit, R.J. (2007) *The American Naturalist*, **169**(5) : 662-672.

Fagerlind, F. (1937) *Acta Hortic.*, **11** : 195-470.

Faivre, A.E. (2000) *Evolution*, **54** : 847-858.

Gopalakrishnan, S. and Vadivel, E. (2011) *International Journal of Pharma and Bio Sciences*, **2** : 313-320.

Greensill, T.M. (1977) In: *Gardening in the Tropics*, Evans Publishers, London, pp.115-120.

Hamrick, J.L. and Godt, M.J.W. (1996) *Biological Sciences*, **351** : 1291-1298.

Hooker, J.D. (1880) *Flora of British India, Mussaenda*, Lovell Reeve, London, UK. **3** : 86-92.

Jayasinghe, U.L.B, Jayasooriya, C.P. and Bandara, B.M.R. (2002) *Fitoterapia*, **73**(5) : 424-427.

Kay, K.M. (2006) *Evolution*, **60** : 538-552.

Kurz, S. (1887) *Forest Flora of British Burma, Mussaenda*, M/S Bishen Singh Mahendra Pal Singh, Dehra Dun, India and M/S Periodical Experts, Delhi, India, **2** : 55-58.

Leppik, E.E. (1977) *Phytomorphology*, **27** : 161-168.

Li, A.M., Wu, X.Q., Zhang, D.X. and Barrett, S.C.H. (2010) *Ann. Bot.*, **106** : 521-531.

Li, H.L. (1943) *Journal of the Arnold Arboretum*, **24** : 99-102.

Li, Y. and Zhou, H. (2010) *Coastal Engineering*, **29**(1) : 75-82.

Mabberley, D.J. (1997) In: *The plant-book-a portable dictionary of the vascular plants*, 2nd ed. Cambridge University Press, Cambridge, UK.

Matthew, I.P. and Karikari, S.K. (1990) In: *Horticulture: Principles and Practices*, Macmillan, Lagos. p.184.

McLaughlin, J., Garofalo, J. (2002) In: *Mussaendas for South Florida Landscapes*. University of Florida, USA. http: //miamidade.ifas.ufl.edu/old/publications.

Miquel, F.A.W. (1857) *Flora van Nederlandsch-Indie¨*, Aphaenandra, Amsterdam, Netherlands. **2** : 341-342.

Mukherjee, A.K., Acharya, L., Mohapatra, T.M. and Das, P. (2012) *Indian Journal of Horticulture*, **69**(2) : 226-230.

Naiki, A. (2008) *Bulletin of the Osaka Museum of Natural History*, **62** : 21-26.

Naiki, A. and Kato, M. (1999) *Plant Species Biology*, **14** : 217-227.

Nybom, H. (2004) *Molecular Ecology*, **13** : 1143-1155.

Ogbu, J.U. (2011) In: *Proceedings of the 45th Annual Conference of the Agricultural Society of Nigeria* (ASN), pp.316-319.Sambrekar, S.N., Patil, P.A. and Kangralkar, V.A. (2010) *J. Pharmacol. Res.*, **3**(4) : 711-13.

Ogbu, J.U., Ubani, S.I. and Anaele, M.U. (2008) In: *Book of Abstract 26th national annual conference of Horticultural Society of Nigeria-HORTSON*, Adamawa State University Mubi, Nigeria, pp. 54-55.

Olosunde, O.M. and Fawusi, M.O.A. (2003) In: *Proceedings of Horticultural Society of Nigeria (HORTSON) annual conference* (Eds. A.S. Fasina, A.O. OLufolaji and V.C. Umeh), Lagos State Polytechnic, Ikorodu, pp. 121-126.

Olosunde, O.M., Olasantan, F.O., and Olubode, O.O. (2008) *Nigerian Journal of Horticultural Science*, pp.13.

Pailler, T. and Thompson, J.D. (1997) *American Journal of Botany*, **84** : 315-327.

Petit, R.J., Duminil, J., Fineschi, S., Hampe, A., Salvini, D. and Vendramin, G.G. (2005) *Molecular Ecology*, **14** : 689-701.

Priyadarshan, P.M. and Ramachandran, K. (1984) *Cytologia*, **49** : 407-413.

Puff, C., Igersheim, A. and Rohrhofer. U. (1993) *Bulletin de Jardin Botanique National de Belgique*, **62** : 35-68.

Purseglove, J.W. (1984) In: *Tropical Crops: Dicotyledons*, Longma Group Publishers, England, pp. 451-452.

Richards, J.H. and Koptur, S. (1993) *American Journal of Botany*, **80** : 31-40.

Robbrecht, E. (1988) *Opera Botanica Belgica*, **1** : 1-271.

Rosario, T.L. (1984) *Inst. Plant Breeding Bul.4, College of Agriculture*, University of the Philippines, Los Banos, College, Laguina.

Rosario, T.L. (1998) In: *Ornamental Mussaendas of the Philippines*, College of Agriculture, University of the Philippines, Los Banos, College, Laguna, The Philippines.

Rosario, T.L. and Mendoza, M.B. (1999) *Philippine Journal of Crop Science*, **24** : 58-59.

Santapau, H. and Henry, A.N. (1973) In: *A Dictionary of Flowering Plants in India*, C.S.I.R., New Delhi, pp.112.

Schumann, K. (1891) Rubiaceae. *In Die na-turlichen Pflanzenfamilien*, (Eds. A. Engler and K. Prantl) **4** : 1-156. Engelmann, Leipzig, Germany.

Selvaraj, R. (1987) *Cytologia*, **52** : 343-356.

Sheat, W.G. and Schofield, G. (1995) In: *Complete Gardening in Southern Africa*, Struik Publishers, Cape Town, p.76.

Shi, M., Wang, Y., Duan, T., Qian, X., Zeng, T. and Zhang, D. (2020) *Ecology and evolution*, **10**(20) : 11304-11321.

Vidyalakshmi, K.S., Vasanthi, H.R. and Rajamanickam, G.V. (2008) *Ethnobotanical Leaflets*, **12** : 469-475.

Wernham, H.F. (1914) *The Journal of Botany*, **52** : 64-72.

Wernham, H.F. (1916) *The Journal of Botany*, **54** : 297-301.

Wesley, J. (2009) *J. Pharmacol.Res.*, **2**(4) : 579-81.

WHO (1998) In: *Medicinal plants in the South Pacific* (Ed. Michael Doyle), WHO regional publications, Western pack series, pp.125.

Adenium

B. Lalramhlimi

1.0 Introduction

The genus *Adenium* was derived from Aden, formerly Yemen and presently the capital town of the country, while the term *'obesum'* refers to the swelling stem at its basal region. It is a deciduous plant widely used as an ornamental plant. It adapts easily in pot culture making it well suited for bonsai. During the warm moist months, the branches are adorned with glossy leaves and bright red, five cm bell shaped flowers with a white center while the branches become without blossoms and nearly leafless during the dry winter months (Dodson, 1959). It is a slow- growing shrub with no fragrance, valued for its attractive flowers and aesthetic sculptural caudex. Caudex is defined as the axis of a plant, consisting of stem and roots (Rowley, 1987) and can be above or below ground (Dimmitt *et al.*, 2009).

2.0 Importance and Uses

2.1.0 Aesthetic Uses

Desert rose is an exotic flowering succulent shrub or small tree, botanically *Adenium obesum*, known as Japanese frangipani, impala lily, mock azalea, and pink begonia that generally bear flowers on terminal corymb from June to September. The desert rose gains importance for growing as a pot plant as there was increasing demand for indoor decoration, rock garden and suitable for xeriscaping. The plants are very attractive in appearance, hardy, easy to grow, bear spectacular flowers in different colour combinations and very popular pot plants in warm climate in most parts of the world.

2.2.0 Therapeutic and Other Uses

The different parts of the plant like the sap, leaves, bark, roots, stem and latex are used in several local medicines in different parts of the world for the treatment of a wide range of diseases. The bark and root extracts are used to prepare a lotion for the treatment of different skin diseases and to kill lice (Neuwinger, 2011; Watt and Breyer- brandwijk, 2013) and bark as an abortifacient (Naji *et al.*, 2013; Wang *et al.*, 2013). Moreover, root extracts have been shown to exert cytotoxic effect against several carcinoma cell lines (Singh, 2004; Graham *et al.*, 2000). The latex is used as a medicine for the recovery of nasal drops, decaying teeth, and septic wounds (Tijjani *et al.*, 2011). Smearing the stem juice on affected area for injury, bleeding, and wounds are effective (Al-Fatima, 2019). Powder is applied externally for headache (Tounekti *et al.*, 2019). The whole plant is used for its anti-plasmodial, anti-trypanosomal and anti-leishmanial activities in Nigeria (Malebo *et al.*, 2009).The leaves of *Adenium oleifolium* are pounded to make an ointment for snakebites relief and scorpion stings by the San. The sap of *Adenium boehmianum*, *A. multiflorum*, and *A. obesum* contains toxic cardiac glycosides and is used as arrow poison throughout Africa for hunting large game (Schmelzer and Gurib-Fakim, 2008). *A. obesum* has antioxidant (Al-Ghudani and Hossain, 2015), antidiabetic (Elgebaly *et al.*, 2020), antiviral (Kiyohara *et al.*, 2012), anticancer (Ali *et al.*, 2019), antibacterial (Hossain *et al.*, 2014), organic pesticide (Abalaka *et al.*, 2013), anti-lucusticidal and antimolluscicidal (Versiani *et*

Uses of Adenium

In Pot

In Garden Pot

Hinging Pot

In Home Garden

Indoor Decoration

In Gift

Bonsai

Near Window

In Hedge

Focal Point

In Exhibition

In Landscape

Source: www.blog.nurserylive.com, www.inpinterest.com, www.patioplants.com, www.thespruce.com

al., 2014) properties. *Adenium obesum* extract is used as green inhibitor for protecting C- steel from corrosion in polluted NaCl (Fouda and Eissa, 2020). Farmers in Senegal applied Chacals Baobab (*Adenium obesum*) on cotton crops as natural insecticide for controlling cotton bollworm, spiny bollworm and Sudan bollworm.

3.0 Origin and History

Adenium obesum is originated from Africa, south of the Sahara from Senegal to Sudan and Kenya, and through Saudi Arabia, Oman, and Yemen (Plaizier, 1980, Dimmitt and Hanson, 1991). It was described as *Nerium obesum* in the Flora Aegyptiaco-Arabica by Forskâl in 1775, which was later changed to a new genus *Adenium* by Roemer and Schultes in 1819. It is distributed across countries as ornamentals in West, Central, East Africa and in Southeast Asia, while *A. multiflorum* is cultivated in southern Africa. *A. obesum, A. multiflorum* and *A. swazicum* are cultivated in the botanical garden greenhouses of Amsterdam; *A. obesum* in Brussels and in Wageningen. *A. swazicum* is endemic in Swaziland and in the adjacent areas of Mozambique and of South Africa (Plaizier, 1980). It is also naturalized in Sri Lanka and India. *Adenium obesum* is the most widespread species in the wild and in cultivation.

4.0 Morphology

Desert rose is a succulent shrub or small tree that can grow up to 0.2-5 m tall; trunk mostly up to 1 m (rarely 2 m) diameter; bark is pale green to greyish green or brown, smooth and rhizomatous. Leaves are arranged spirally, linear to obovate, entire, pubescent, or glabrous. The inflorescence is more or less dense terminal cyme; flowers are pentamerous, bisexual, showy usually appearing before the leaves. Corolla is a tube (funnel-shaped); about 2-4.5 cm × 0.9-1.7 cm; hairy to glabrous outside; lobes 1–3 cm × 0.5- 2.5 cm, spreading and pedicel 5-9 mm long. The fruit consists of two linear-oblong follicles, coherent at the base, tapering at both ends, grey to pale grey-brown, 11-22 cm long, recurved, opening by a longitudinal slit and many- seeded. Seeds are linear-oblong, slightly rough, 10- 14 mm long, pale brown, with tufts of long dirty white hairs at both ends (Plaizier, 1980; Oyen, 2006).

5.0 Taxonomy

Adenium belongs to the tribe *Nerieae* of the subfamily *Apocynoidae* (*Echitoidae*) and family Apocynaceae. It is closely related to *Pachypodium* and particularly to *Nerium*. The different morphological characters of *Nerium* and *Adenium* resemble each other. However, both genera differ in many ways; the flowers are actinomorphic in *Nerium* with laciniate scales between corolla lobes, corolla lobes longer than corolla tube, leaves 3- whorled, erect follicles, the seed has an apical tuft of hair on one side and sericeous, while flowers are slightly zygomorphic in *Adenium* with obcordate scales between corolla lobes, corolla lobes shorter than corolla tube, leaves are alternate, reflexed follicles, the seed has an apical tuft of hair at both side and glabrous. *Pachypodium* also differs from *Adenium* by its spines, which are its stipules; the stipules of *Adenium* are early caducous (Plaizier, 1980). Plaizier had classified Adenium into five species, namely, *Adenium obesum, Adenium boehmianum, Adenium oleifolium, Adenium swazicum,* and *Adenium multiflorum.* However, according to

Dimmitt *et al.* (2009), there are one species and 11 taxa, namely, *Adenium obesum, Adenium swazicum, Adenium multiflorum, Adenium boehmianum, Adenium oleifolium, Adenium somalense* 'Nova' (Tanzania), *Adenium somalense, Adenium crispum, Adenium socotranum, Adenium arabicum,* and *Adenium* sp. Oman. These are also considered by other authors to be subspecies or varieties. Rzepecky (2015) had reported a long-overlooked species from the Central-Southern Part of the Arabian Peninsula, the *Adenium dhofarense* Rzepecky sp. Nov.

6.0 Species and Cultivars

6.1.0 Species

The genus *Adenium* consists of five species in southern Africa, namely *Adenium obesum, Adenium swazicum, Adenium oleifolium, Adenium boehmianum* and *Adenium multiflorum. Adenium swazicum* was initially described as *Adenium boehmianum* Schinz var. *swazicum* (Codd, 1963) but since *Adenium boehmianum* and *Adenium swazicum* occur in completely separate areas with no intermediate specimens, Rowley reinstated *Adenium swazicum* as a species in 1974 (Plaizier, 1980). The differences among the species are highlighted below (Plaizier, 1980; Dimmitt and Hanson, 1991; Schmelzer and Gurib- Fakim, 2008; Dimmitt and Williams, 1998).

Adenium boehmianum

It is also called the Bushman poison, a rhizomatous and sturdy shrub (0.8- 2.5 m high), poisonous succulent, endemic to the mostly dry regions of southern Angola and northern Namibia. The caudex is small to large conical up to 40 cm in diameter, very long dormancy period, and deciduous. The leaves are arranged spirally and clustered near branch tips which are borne only for three months in a year. The flowers are showy, long-tubed shape, pink to mauve in colour with darker throat and hairy on outside. Flowering occurs in winter, only for few weeks. They appear from November to May (but mostly in the autumn, southern hemisphere), usually together with the leaves.

Adenium multiflorum

It is a small to large shrub, spreading to erect stems, 0.5- 3.5 m high, a succulent tree native to central and eastern Southern Africa with large roots, carrot-like (up to 1 m in diameter), shiny grey bark and poisonous white latex. The leaves follow flowers' appearance, subsessile; obovate to oblong blade. Flowers are large and showy, star-shaped, white or pale pink with a pink to dark red border, 15 prominent red nectar guides in the white throat (three per petal). The flowering time is May to August.

Adenium obesum

It is a succulent shrubby tree, branched (2 - many), tall (0.4- 4 m, rarely 5 m high), and at the extreme base up to 1 m (rarely up to 2 m in diameter); sometimes with a fleshy taproot. It is distributed in regions of Arabia, Socotra, Northeastern, Central, and rarely in Western Africa. The bark colour varies from pale greyish-green to smooth grey or smooth brown; latex fresh clear or white, dried white and

Species of Adenium

Adenium obesum — Adenium multiflorum — Adenium swazicum

Adenium crispum — Adenium boehmianum — Adenium somalense

Adenium oleifolium — Adenium socotranum — Adenium dhoforence

Adenium oman — Adenium somalense 'Nova' (Tanzania) — Adenium arabicum

sticky. The flower is cup-shaped from a short cylindrical base, full pink or somewhat darker shade towards the margins of the lobes, with darker pink stripes in the paler throat. Flowering occurs in winter. The Latin word *'obesum'* means 'fat or fleshy', refers to the habit of the stems.

Adenium oleifolium

This species is native to desert scrubland in Southern Africa, Angola, and Namibia. A dwarf succulent shrub, 40 cm tall, forming a dense mass of rather fleshy leaves and stems, with a medium- size caudex (50-80 × 15-30 cm), which is extremely bitter. The flowers are small, pale pink in colour, yellow throat with prominent red nectar guides that appear with the leaves. It has very slow growth and flowers in spring.

Adenium swazicum

This species is found in open woodland, usually on sandy brackish flats in South Africa, Swaziland, and Mozambique at altitudes of 300- 400 m. It is a succulent, small to medium shrub (0.2- 0.7 m), which has a carrot- like tuber (up to 1m diameter), which is small but with big roots and poisonous clear latex. The simple leaves are long, often folded, tapering and smooth, and borne in terminal clusters. The flowers are pink to deep purple, darker in the throat, and flowering occurs from January to April together with the leaves.

Adenium somalense

It occurs in Somalia, Kenya, Ethiopia, and extends a bit into adjacent nations. The shrub has a succulent trunk, tall with distinctly conical caudex. The flowers are small (less than 5 cm in diameter), narrower, pointed petals, vary from pink to deep red fading towards the throat with prominent nectar guides that extend onto the petal bases. This species is extremely intolerant to cold both in dry or wet soils and temperature below 10° C for a few weeks will cause the roots to die and the plant will topple. Probably for this reason and because of its large size, it is not common in cultivation.

Adenium sp. (Tanzania)

The *Adenium somalense* var. nova was discovered in the mid 1990s in Tanzania. It has a globular caudex which is buried underground, conical, with erect, barely succulent stems to about 1.5 m. In few years, rapid growth of plant develops a large caudex. Flowers are star- shaped, pink to red at the edges, fading to a white throat with prominent nectar guides. The small flowers start appearing sparsely at the end of the growing season in fall, followed by a more profuse bloom through the winter, while the plants are leafless.

Adenium socotranum

It is by far the largest species, a gigantic shrub to a tree, endemic to the island of Socotra. It has a cylindrical trunk, to 5 m in height and 2 m in diameter at the basal region. Plants rarely leaf out before mid-summer and produce stem growth for only a few weeks. The pink flowers appear for a few weeks in spring while the plant is leafless. It has become a targeted species among the adeniums due to its remote location, huge size, and recent availability in a trade. However, it has very slow growth and intolerant to the desert heat.

Adenium crispum

It occurs in a band of sandy soil near the coast of southern Somalia. The shrub is a dwarf, thin stem with fusiform caudex. The thin roots arise from near the top of this caudex below ground. In cultivation, the caudex is raised above soil level and is capable of re- rooting from the bottom. The stems are thin and hardly a foot tall. The distinctive flowers are small in size, quilled petals, larger throats and, narrower white to pinkish petals. The pink to red nectar guides in the throat may extend halfway to the tips of the petals, giving the flowers a distinctly striped aspect. It cannot tolerate cold and also intolerant to hot, humid conditions. For these reasons, it is not widely cultivated despite the beautiful flowers and compact size. Flowering in cultivation is sporadic; it seems to be most profuse during winter dormancy but may extend well into summer.

Adenium arabicum

It is native to a narrow strip near the coast on the western half of the southern Arabian Peninsula (in Yemen and Saudi Arabia). It is a dwarf to huge shrub or tree; sturdy branches, huge caudex and stem and deciduous in nature. The flowers are typically small, almost always pink, colour fades toward throat; nectar guides are strong or absent. The flowering season is mostly spring; some are summer or everblooming. It is a fast-growing species.

***Adenium* sp. Oman**

The *Adenium arabicum* in Oman is characterized by a broad, squat caudex (up to 1 m in diameter), thin stems which may root when they contact the ground. The leaves are very large, second in size only to those of *A. boehmianum*. The flowers are very small and appear in late spring while still leafless. Finally, the plants have a very slow growth and short growing season.

A. dhofarense

It is a floppy (or erect) shrub with a large caudex. It has a long dormancy period and is deciduous in nature. Flowers are small, round to star- shaped and pink colour. It has very slow growth. Leaves are large obovate, revolute margins 70- 140 × 50- 100 mm, globose to irregular caudex (but never napiform) above ground, and branches set roots when in contact with the ground. Flowering occurs from mid-March to the end of May (Rzepecky, 2015).

6.2.0 Cultivars

There have been numerous cultivars developed through hybridization among species and cultivars. These cultivars yield wide range of colours from white to dazzling reds, single to multi-whorled, nearly ever-blooming to ever-blooming, fast growing and resistant to abiotic stresses.

Notable varieties are as follows:

'**Amiability**', Taiwanese variety, flowers usually have a narrow black edge, and are borne profusely several months a year;

'**Arrogant**', red flowers;

'**Asha**', an Arizona back-crossed to *A. obesum*, huge, bright pink flowers, nearly everblooming, fast growing;

'**Black Pine**', resembles '**Daeng Nabanant**' in its 100+ mm size, but it has a longer blooming season;

'**Black Queen**', Taiwanese variety, flowers are dark during moderately hot weather;

'**Black Raspberry**', red flowers, striped throat, blooms over long season;

'**Black Ruby**', profuse flowering, tall conical caudex, col tolerant;

'**Boyce Thompson**', *A. swazicum*, deep purple;

'**Crimson Picotee**', *A. obesum* × *A. multiflorum*, large leaves, evergreen, thick and sturdy stems, broad red border on the white petals.

'**Crimson Star**', sturdy stem, erect to ascending, large flower, solid, deep blood-red and are borne nearly year-round;

'**Crown**', red cultivar, sturdy plant, flowers almost year-round;

'**Daeng Nabanant**', huge flowers are borne sporadically, red flowers, 100 mm size;

'**Deang Udam Sap**', dark red, superior flower diameter, good flower longevity;

'**Double Sweet Heart**', pink, good flower longevity, double type flower form, 10 petals per flower;

'**Emperor Star**', waterfall types, strong plant, good shaped flowers and colour much of the year, variation of flower streaking at different seasons;

'**Endless Sunset**', solid deep pink, everblooming;

'**Grumbley White**', extremely floriferous, pure white;

'**Gujarat Adenium- 1 (Gad-1)**', multi- petalous, red colour, good flower longevity;

'**Gujarat Adenium- 2 (Gad-2)**', multi- petalous, novel purple colour, profuse flowering habit;

'**Hansoti Dwarf**', water use is minimal, flowering during dormancy and deciduous;

'**Harry Potter**', pink, bigger flower size diameter;

'**Heart of gold**', white with pink border petals, yellow throats;

'**Incandescent**', offspring of 'Black Ruby' × 'MAD 218'

'**King of Purple**', double flowers that open purple with red stripes on every petal, blooms from spring until fall;

Cultivars of Adenium

'Hansoti Dwarf' 'Crown' 'Amiability'

'Black Queen' 'Vigil' 'Singapore'

'Daeng Nabanant' 'Heart of Gold' 'Emperor Star'

'Noble Queen' 'Soft' 'Black Pine'

Cultivars of Adenium

'Taiwan Home Run' 'MAD #286' 'Grumbley White'

'Incandescent' 'Gujarat Adenium-1' 'Gujarat Adenium-2'

'Black Raspberry' 'Black Ruby' 'King of Purple'

'Red Everbloomer' 'Red Cloud' 'Crimson Star'

'**MAD #286**', a pink picotee with yellow throat;

'**Mor Lok Dok**', white flowers;

'**Mung Siam**', purple flowers, single whorl;

'**Noble queen**', Taiwan hybrid, A. *obesum* and not *A. multiflorum* because it flowers year-round and is evergreen;

'**Perpetual Pink**', *Adenium swazicum*, sturdy stems, large flower size, long blooming, pale pink to deep raspberry red flower colour;

'**Picotee**', pink flowers;

'**Red Cloud**', large, shiny green leaves, red flowers;

'**Red Ever-bloomer**', *Adenium obesum*, upright stems, red flower colour, long blooming season;

'**Red Ribbons**', thin weak stem, very large flower, deep red with only slight fading toward center, petals hang down like ribbons;

'**Singapore**', *A. arabicum*, very large flowers, deep pink flowers (10 cm or over 4 inches in diameter), flowers up to 10 months per year, massive bloom;

'**Soft**', a giant (100 mm), Taiwan introduction;

'**Sudarshan**', dark pink flowers;

'**Taiwan dwarf**', pink flowers;

'**Taiwan Home Run**', red petals, flower colour is stable in hot weather, blooms nearly year-round;

'**Vigil**', brilliant orange- red colour, white of the throat extends onto the petal bases;

'**Vithoons White**', white flowers;

'**Volcanic Sunset**', sturdy and erect stem, crowded leaves, short internodes, large flowers, deep red, fading slightly toward the throat.

7.0 Genetics and Breeding

7.1.0 Hybridization

Singh *et al.* (2020) performed crosses involving ten parents (7 females and 3 males) where he studied different morphological characters. The germplasm *viz.*, MSDSH1 (Mung Siam × Double Sweet Heart) showed superior performances in number of flowers per cluster (10.23) and flower longevity on plant (11.91 days) as compared to parent (Mung Siam and Double Sweet Heart) and Local Pink with novel flower colour having attractive purplish coloured ten petals. It was renamed as G. Ad.2 (Gujarat Adenium- 2) and released in SVRC, Gujarat.

The traits of some hybrids given by Dimmitt, a famous *Adenium* breeder are presented in table form:

Taxon	Plant Form, Size	Caudex Size Relative to Plant	Dormancy Require-ment	Cold Tolerance	Flowers	Season	Notes
A.obesum modern cultivars	Medium shrubs; often weak stems	Small to medium	None	Poor	Round, large; wide range of colours; doubles	Best cultivars flower year-round	Fast growth
A.obesum × A.swazicum	Medium shrubs; floppy (or erect) stems	Small to medium; often huge roots	None	Good	Large, round; intense solid colours	Often year-round	Fast growth
A.obesum × A.crispum	Medium shrubs; erect to floppy stems	Small to large	Short to none	Poor	Medium; often strongly striped & spotted	Often year-round	Slow growth
A.arabicum × A.obesum	Large to huge; usually erect	Large to huge	Short to none	Very good	Large, round; deep pink to red	Spring to year-round	Fast growth
A.arabicum × A.crispum	Medium; erect	Very large	Short to none	Good	Small, narrow petals; tend to be striped	Spring to year-round	-
A.arabicum × A.obesum × A.crispum	Medium to large; erect	Large	Short to none	Good	Medium to large; pink to red; often striped	Spring to year-round	Fast growth

7.2.0 Cross-compatibility

There was cross incompatibility between African *Adenium obesum* and close relative *Adenium socotranum*. However, *Adenium socotranum* is cross compatible with the Arabian ones. There was no success in experiment between *A. dhofarense* and the African *A. obesum* crosses as reported by Mark Dimmitt (1998). By cross pollination of *A. obesum* with *A. swazicum* and *A. arabicum* some amazing varieties have been produced by breeders across the globe. Similarly, *A. crispum* cross pollinated with *A. somalense*, *A. swazicum* and *A. obesum* have produced exotic hybrids which are collectors delights and breeders' pride (Kumar, 2015).

7.3.0 Mutation Breeding

An experiment was carried out where the seeds of *Adenium somalense* var. *somalense* were acutely irradiated (200- 300 Gy dosage) as well as growing plants of *Adenium obesum* with different total doses at the dose rate of 0.96 Gy/hr. Several mutants were obtained; among the outstanding mutants from gamma rays treated adenium, three of them were registered as new varieties (Jompuk *et al.*, 2008).

8.0 Growth and Flowering

8.1.0 Light

Investigation was carried out by McBride *et al.* (2014b) for studying the effect of light intensity with three treatments, i.e, full sun [with a measured maximum photo-synthetically active radiation (PAR) of 1850 μmol/m^2/s], 30 per cent shade (1255 μmol/m^2/s), or 50 per cent shade (943 μmol/m^2/s) and nutrition level on growth and flowering of *A. obesum* 'Red' and 'Ice Pink'. Different nutritional level included 1.25- L pots top dressed with controlled-release fertilizer Nutricote Plus (18N-2.6P-6.6K) at different rates 0.4, 0.9, or 1.4 g of nitrogen (N) per pot. Highest flower numbers and quality ratings were recorded in 30 per cent shade. Greatest canopy heights and widths but poor flower numbers and quality ratings were recorded in 50 per cent shade level. The highest root dry weights were measured at 1.4 g of N under both full sun and 30 per cent shade. Caudex burn can occur under extreme heat or full sun exposure if leaf canopy does not shade it.

8.2.0 Temperature

Adenium plants can thrive beautifully under hot and warm regions. However, they will not tolerate cold freezing temperatures. Stem tips can be damaged at temperature below 4°C, and may even cause plant death if low temperature persists (McLaughlin and Garofalo, 2002). Plants can be grown in pots or containers to ease shifting to protected structures to protect them from cold spells. In contrast, when temperatures exceed 37°C, most Adenium plants slow their growth and stop flowering (Dimmitt and Hanson, in press). Dormancy is induced in all taxa when nights regularly fall below about 10°C. When completely dry and dormant, plants can tolerate near-freezing temperatures, though there is high incidence of root- rot (Dimmitt, 1998).

8.3.0 Plant Growth Regulators

The vegetal growth retardant paclobutrazol (PBZ) was applied in six treatments 5; 10; 20; 40 and 80 ml of the active ingredient, and water as control in studies focused on reducing the size of *Adenium obesum* plant. The dose of 80 ml of PBZ reduced the length of aerial part in 37.29 per cent in relation to the control. Despite the decrease observed in the plant final size, the phytotoxic effects caused by PBZ prevented its utilization in the *Adenium obesum* commercial production. The vegetative growth inhibition associated with PBZ has been linked to the application method (Mauk *et al.*, 1990), which is the application via soil up to reaching field capacity, it proved to be more efficient than the application via aerial part spraying, when it comes to plant size reduction (Pardos *et al.*, 2005; Al-khassawneh *et al.*, 2006).

9.0 Propagation

Desert rose can be propagated by seed, stem cuttings, grafting, air layering, and micropropagation. Seeds are most commonly used for propagation and are the least expensive method.

9.1.0 Seed Propagation

The seeds germinate within one to two weeks under soil temperature of 29°C in soil mixture and ready for transplanting after a month. Seed-grown plants can flower in 12 months under proper care and optimum conditions. The plants originating from seeds exhibit more swollen caudex and primary roots than those propagated from cuttings (Colombo *et al.*, 2015).

9.2.0 Vegetative Propagation

Adenium multiflorum, A. oleifolium and *A. swazicum* can be easily propagated through stem cuttings. Cuttings can be taken from the tip of the stem, 15 cm or longer, treated with auxins and fungicide before planting in pots. McBride (2012) had performed rooting of cuttings in two varieties of *A. obesum*, 'Ice Pink' and 'Red', and reported the use of 8000 mg/l of indole- butyric acid (IBA) was most suitable for rooting of apical cuttings. Grafting was practiced in Wageningen and Amsterdam where *Adenium* spp. has been successfully grafted on a *Nerium oleander* stock which resulted in much richer flowering material (Plaizier, 1980). Grafting is mainly done for commercial purposes. The root- stocks are raised from seeds for grafting. Air-layering of *Adenium* is more likely to be successful during hot, humid weather and used to produce few plants of relatively large size for special purposes (Dimmitt *et al.*, 2009, McLaughlin and Garofalo, 2002; Hartmann *et al.*, 2002).

9.3.0 Micropropagation

Kanchanapoom *et al.* (2010) cultivated *Adenium obesum* shoot tips using a combination of the growth regulators BA and NAA to induce indirect organogenesis and a high frequency of 5.20 ± 1.10 shoots per explant using 22.2 µM BA. High rooting and survival were achieved using MS medium supplemented with 0.3 per cent activated charcoal and without any growth regulators. Experiments in micropropagation of *Adenium obesum* showed that the addition of 0.5 µmol/L IBA improved the frequency of rooting significantly under *in vitro* conditions (Xiaomei *et al.*, 2003). Another study by (Liu *et al.*, 2004) found that using NAA (0.2 to 0.4 mg/l) + BA (1 to 2mg/l) induced callus in *Adenium obesum* cultures. The medium with 2, 4-D @1~3 mg/l or 2, 4-D @1~ mg/l + BA @ 0.5~1.0 mg/l was the optimal mixture on which the callus was yellow and loose, and suitable for suspension culture.

10.0 Soil and Climate

10.1.0 Soil

Different types of potting media are used for growing adeniums in containers. The recommended pH for potting media containing predominantly organic matter is between 5.0 and 5.8 (Kampf, 2000). The plants exhibited higher growth rates, caudex growth and greater nutrients build-up in dry matter by using sand + Amafibra® 47 coconut fiber and vermiculite + Amafibra® 47 coconut fiber as potting media and are recommended for cultivating this species (Colombo *et al.*, 2016). In seedling production, the substrates should have high aeration and low density to avoid rotting of seeds. Among the recommended materials, substrates based on semi-composted pine bark, pure or in combination with coconut fiber,

Propagation and Growing of Adenium

Pod

Seeds

In Plug Tray

Seedlings

Propagation by Cutting

Potted Plant Production of Adenium

Copper Deficiency

Zinc Deficiency

Boron Deficiency

Propagation and Growing of Adenium

Micropropagation

(a) Multiple shoot formation from shoot tip culture; (b) Callus formation; (c) Rooted plantlet, (d) Three-month-old plantlets [Kanchanapoom *et al.* (2010)].

provide a high emergence percentage of seedlings, *i.e.* 97 per cent and 98 per cent, respectively (Colombo *et al.*, 2017).

10.2.0 Climate

Desert roses are mainly native to tropical regions and require warm and hot climates. Most adeniums will tolerate near freezing nights in winter if medium is dry. They perform very well under sub-tropical and greenhouse conditions. The most vigorous vegetative growth occurs with moderate to high humidity.

11.0 Cultivation

11.1.0 Planting Time

Desert rose can be planted in containers any time throughout the year under green house. Adeniums grow best under warm, moist and sunny conditions. However, direct extreme heat may burn new growth on the plant.

11.2.0 Manuring and Fertilization

Adenium if fertilized with heavy nitrogen gives slender, rapid growth, big leaves, and increased internode spaces. On the other hand, excessively fast growth is soft and susceptible to rot as well as aesthetically unappealing. To produce high-quality plants in 1.25- L pots, *Adenium* 'Red' should be fertilized with 1.08 g N per liter of the potting mix. For plants grown in 3.0-L pots, N rates of 0.36 g or 0.72 g per liter of potting mix would be recommended with a preference for 0.36 g (McBride *et al.*, 2014a). Phosphorus is also an essential element and its deficiency will give stunted plants with small, deep green leaves and poor flowering. However, plants require a little amount of phosphorus and in excess will give long and stretched growth and also renders other micronutrients insoluble. A dosage of 1-2 grams per one liter water of Calcium nitrate + Potassium nitrate in the liquid feed is used to provide the macronutrients calcium, potassium, and nitrogen; compare this to 1- 2 grams per 100 liters water of Copper sulphate to provide the micronutrient copper (Hosco India, 2015).

11.3.0 Pruning

Adenium responds well to pruning which make them suitable for pot culture besides their slow growth, salt tolerance, and drought tolerance. They are valued for their beautiful structure, multiple flowers and thickened caudex. So, pruning has become an important operation to get desired well balanced aesthetic structure. Pinching is done on seedling plants to create branching during period of vigorous growth in late winter or early spring. Old leafless stems, long and dry can be pruned back in spring to improve and rejuvenate the plant and promote new growth which will bear flowers. Repotting is ideal with fresh media after pruning. Only a little pruning is required in general.

11.4.0 Repotting

Adeniums respond well to annual or biannual repotting. The old media may be thrown or washed off if they are soggy or degraded or it may be retained and fresh media may be used to fill the rest of the pot. Root pruning can be done while repotting massive adenium roots with thickened secondary roots, by cutting off these roots at the junction with the caudex to give some space for more media to enter the pots. Turmeric powder (100 g) mixed with cinnamon powder (20 g) in clean drinking water (5L) as an organic treatment of cuttings have been found to yield good results on adeniums for repotting (Kumar, 2015). After repotting, the plants can be watered just enough and kept in partial shade for one week.

11.5.0 Irrigation

Timely watering with optimum amount is very crucial in adenium cultivation. A combination of wet and cold or over- watering in a sticky, poorly draining medium can kill a well- established plant. A night temperature of below 10°C can cause root rot so the plant needs to be kept dry. The amount of water to be given to plant is based on the kind of media used. Substrates such as vermiculite, carbonized rice husk and semi-composted pine bark, among others, can be used to facilitate proper irrigation. However, attention should be given to irrigation management, since adenium cuttings are susceptible to rot, due to their high-water content (Colombo *et al.*, 2018). During hot and dry weather, daily watering may be required while under cool winter conditions, watering may be every 7- 10 days. Plants do not require frequent watering when they attain few years and slow growth.

12.0 Diseases and Pests

There are some diseases and pests that are affecting adenium cultivation. Most of the diseases can be traced back to excessive water in media or leaves. Among the diseases leaf spots, tomato spotted wilt, cucumber mosaic and little leaf are common are major diseases. Other minor diseases of adenium include damping off, root rot, anthracnose and dieback *etc.*

12.1.0 Diseases

12.1.1 Aristastoma Leaf Spots

This disease is caused by *Aristastoma* spp. (McMillan Jr. *et al.*, 1997). Symptoms include an irregular, oval to circular, rusty brown, necrotic lesions 5 to 15 mm in diameter on plant foliage. Large leaf spots developed tan centers. Fungicides can be used to treat infected plants.

12.1.2 Leaf Spots

This disease is caused by *Nigrospora sphaerica* (de Jesus *et al.*, 2020). The symptoms appear as chlorotic spots initially on the edge of the leaf blade to brown lesions that subsequently increased in size. Spraying of fungicides will help to treat the infected plants. Fungicides should not be sprayed on flowers during hot sunny day to prevent from sunburn.

12.1.3 Spotted Wilt

This disease is caused by tomato spotted wilt virus and is transmitted by thrips (Adkins and Baker, 2005). Virus-like foliar symptoms, including chlorotic ring and

Diseases and Pests of Adenium

Leaf Blight | Stem Rot | Dry Rot

Caudex Rot | Mealy Bug | Aphid

Stink Bug | Mealy Bug | Spider Mite

Soft Scale | Caterpillar | Phytoplasma

Source: www.tropicanursery.com, www.link.springer.com

line patterns were observed on the plant. This disease can be prevented by choosing healthy pathogen-free planting materials. Infected plants can be uprooted and burnt. Spraying of systemic insecticides can be recommended to control the vectors.

12.1.4 Mosaic

This disease is caused by cucumber mosaic virus and is transmitted by aphids (Baker *et al.*, 2003). Later, Chen *et al.* (2010) observed CMV infection in *A. obesum* plants occurring in Taiwan. Symptoms included chlorosis on leaf margins, localized lesions, systemic mosaic and necrotic ringspots. No chemical control is available that can effectively control this disease. To achieve this, it is crucial to remove weeds and diseased plants from the field, as well as use clean and sanitized tools. Another option consists of the use of resistant varieties or the so-called trap crops.

12.1.5 Little Leaf Disease

This disease is caused by *Candidatus* phytoplasma asteris (Raj *et al.*, 2006). Symptoms included phyllody and excessive development of internodes resulting in little leaves. Infected plants can be uprooted and burnt.

12.2.0 Pests

Insect pests affecting adenium are mealy bug, aphids, spider mites, weevils, whitefly, cochineals, caterpillar, seed bug, stink bugs, soft scale, lepidopteran larvae *etc.* Mealy bug is one of the most serious pests that infect the above ground and the roots of the plant. It causes distortion of shoots and flowers. While repotting, the roots can be checked for presence of root mealy bug. Milk weed bugs (*Spilostethus pandurus* or *Caenocoris nerii*) are orange coloured beetles and they destroy the seed pods by feeding on them (Kumar, 2015). Lepidopteran larvae attack in spring when the new shoots appear and feed on the tender fresh leaves and flower buds. Stink bugs appear at the end of monsoon and suck on the juices of seed pods. The thrips and aphids are also vectors of the viral diseases which spread easily if not controlled. Insecticidal soap and micro-encapsulated diazinon (Knox.Out) are safe for the plants and systemic Dimethoate 267EC is not phytotoxic if used as labelled at temperatures below 32°C. However, Cygon has the same active ingredient but the "inert" solvent kills foliage (Dimmitt, 1998). Soft scale can be easily controlled by pruning off the infested leaves and shoots.

13.0 Biotechnology

13.1.1 Molecular Markers

The genetic diversity of 10 genotypes of *Adenium obesum* was determined in India by Chavan *et al.* (2018) based on DNA based molecular marker RAPD. The genotype 'Deang Udam Sap' and 'Picotte' were found to be more genetically divergent genotypes followed by genotypes 'Deang Udam Sap' and 'Mor Lok Dok'. Genotypes 'Double Sweet heart' and 'Vithoon's White' were found to be genetically closest genotypes followed by genotypes 'Mor Lok Dok' and 'Picotte'. Genetic similarity between genotype 'Arrogant' and 'Mung Siam Purple' was also found to be considerable.

14.0 References

Abalaka, S.E., Fatihu, M.Y., Ibrahim, N.D.G. and Ambali, S.F. (2014) *Journal of Pharmaceutical Research International*, **4**(9): 1041-1052.

Adkins, S. and Baker, C. A. (2005) *Plant disease*, **89**(5): 526-526.

Al-Ghudani, M.K.N. and Hossain, M.A. (2015) *Asian Pacific Journal of Tropical Disease*, **5**: S155–S158.

Ali, A. Q., Farah, M.A., Abou-Tarboush, F.M., Al-Anazi, K.M., Ali, M. A., Lee, J. and Mahmoud, A.H. (2019) *Saudi Journal of Biological Sciences*, **26**(3): 547-553.

Al-khassawneh, N.M, Karam, N.S and Shibli, R.A. (2006) *Scientia Horticulturae*, **107**(2): 187-193.

Baker, C.A., Achor, D. and Adkins, S. (2003) *Plant disease*, **87**(8): 1007-1007.

Chavan, S.K., Singh, A, and Barkule, S.R. (2017) *Ecology, Environment and Conservation*, **24**: S403.

Chen, Y.K., Chang, Y.S., Lin, Y.W. and Wu, M.Y. (2012) *Plant disease*, **96**(4): 593-593.

Codd, L.E. (1963) In: *Apocynaceae. Flora of Southern Africa*, Vol. 26, Botanical Research Institute, Pretoria, P. 244.

Colombo, R.C., Cruz, M.A.D., Carvalho, D.U.D., Hoshino, R.T., Alves, G.A.C. and Faria, R.T.D. (2018) *Ornamental Horticulture*, **24(**3): 197-205.

Colombo, R. C., Favetta, V., de Carvalho, D.U., da Cruz, M.A., Roberto, S.R. and de Faria, R.T. (2017) *Ornamental Horticulture*, **23**(3): 250-256.

Colombo, R.C., Favetta, V., de Melo, T.R., de Faria, R.T. and de Aguiar e Silva, M.A. (2016) *Australian Journal of Crop Science*, **10**(2): 258-263.

Colombo, R.C., Favetta, V., Yamamoto, L.Y., Alves, G.A.C., Abati, J., Takahashi, L.S. A. and Faria, R.T.D. (2015) *Journal of Seed Science*, **37**(4): 206-213.

Das, A.B., Mohanty, S. and Das, P. (1999) *Cytobios*, **98**: 95-104.

de Jesus, J.M.I., de Paula, K.L.M., Rodrigues, A.B.L. and da Cunha, M.G. (2020) *Journal of Plant Pathology*, **102**(4): 1379-1379.

Dimmitt, M.A. (1998) *Cactus and Succulent Journal*, **70**(2): 59-64.

Dimmitt, M.A. and Williams, A. (1998) In: *The Adenium pages, Tucson Cactus and Succulent Society*, accessed from Adenium Species page 1 (tucsoncactus.org) on 28 February, 2021.

Dimmitt, M.A. and Hanson, C. (1991) *Cactus and Succulent Journal*, **63**: 223–225.

Dimmitt, M.G., Joseph, G. and Palzkill, D. (2009) In: *Adenium: Sculptural elegance, floral extravagance*. Scathingly Brilliant Idea, Tucson, AZ.

Dodson, J.W. (1959) *Cactus and Succulent Journal*, **31**: 55.

Elgebaly, H., Germoush, M., Mosa, N., Zahou, F., Soffar, A., Alotaibi, N. and Mahmoud, A. M. (2020) *International Journal of Pharmacology*, **16**(4): 310-318.

Fouda, A.E.A.S. and Eissa, M. (2020) *Journal of Bio-and Tribo-Corrosion*, **6**(4): 1-11.

Graham, J. G., Quinn, M. L., Fabricant, D. S. and Farnsworth, N. R. (2000) *Journal of Ethnopharmacology*, **73**(3): 347-377.

Hartmann, H.T., Kester, D.E., Davies, F.T. and Geneve, R.L. (2002) In: *Plant propagation: principles and practices*, Prentice Hall. 7: iii-880.

Hosco India (2015) *Fertilizing pot plants. Retrieved from Plastic Pots from Horticultural Supplies Co. India* (*hoscoindia.com*)

Hossain, M.A., Al-Abri, T.H.A., Al-Musalami, A.H.S., Akhtar, M.S. and Said, S. (2014) *Asian Pacific Journal of Tropical Disease*, **4**: S985-S989.

Jompuk, P., Wongpiyasatid, A., Lamseejan, S. and Chusreeaeom, K. (2008) *Mutation induction using acute and chronic gamma irradiation on some vegetatively propagated ornamental crops in Thailand.*

Kampf, A.N. (2000) In: *Selection of materials for use as substrate, Substrates for plants: the basis of plant production in containers. Porto Alegre: Genesis*, pp. 139-145.

Kanchanapoom, K., Sunheem, S. and Kanchanapoom, K. (2010) *Notulae Botanicae Horti Agrobotanici Cluj-Napoca*, **38**(3): 209-213.

Kiyohara, H., Ichino, C., Kawamura, Y., Nagai, T., Sato, N., Yamada, H. and Abdel-Sattar, E. (2012) *Phytomedicine*, **19**(2): 111-114.

Kumar, A. (2015) *Adorable adeniums. Accessed from adorable_adeniums_book.pdf* (ramanink.com) on 2nd March, 2021.

Liu, T., Chen, B. and Chan, X. (2004) *The effect of hormone on callus culture of Desert Rose Adenium obesum*. Economic Forest Researches.

Malebo, H. M., Tanja, W., Cal, M., Swaleh, S.A.M., Omolo, M. O., Ali, A. H., Se´quin, U., Hamburger, M., Brun, R., Ndiege, I. O. (2009) *Tanzania Journal of Health Research*, **11**: 226–234.

Mauk, C.S., Unrath, C.R, Blankenship, S.M. and Lehman, L.J (1990) *Plant Growth Regulation*, **9**(1): 27-35.

McBride, K.M. (2012) *Ph.D. Thesis*, University of Florida, USA.

McBride, K.M., Henny, R.J., Mellich, T.A. and Chen, J. (2014a) *Hort Science*, **49**(12): 1518-1522.

McBride, K., Henny, R.J., Chen, J. and Mellich, T.A. (2014b) *Hort Science*, **49**(4): 430-433.

McLaughlin, J. and Garofalo, J. (2002) *Desert Rose, Adenium obesum: nursery production.* University of Florida Cooperative Extension Service. Miami-Dade County Extension. 66: 1-2.

McMillan Jr, R.T., Graves, W.R. and Leahy, R.M. (1997) *Plant Disease*, **81**(8): 960-960.

Naji, E., Rershi, S.M. and Luca, R. (2013) *Journal of Pharmacy and Phytotheraputics*, **1**: 5–7.

Neuwinger, H.D. (2011) *African Traditional Medicine: A Dictionary of Plant Use and Applications*. Medicinal Pharmaceutical Sciences, Stuttgart, Germany, pp. 589–594.

Oyen, L.P.A. (2006) *Adenium obesum* (Forssk.) Roem. and Schult. [Internet] *Record from PROTA4U*. (Eds. G.H. Schmelzer, A. Prota Gurib-Fakim), Plant Resources of Tropical Africa/Ressources végétales de l'Afrique tropicale), Wageningen, Netherlands.

Pardos, M., Calama, R., Montero, G. and Pardos, J.A (2005) *Hort Science*, **40**(6): 1773-1776.

Plaizier, A.C. (1980) *A revision of Adenium Roem. and Schult. and of Diplorhynchus Welw. ex Foc. & Hiern* (*Apocynaceae*) (No. 80-12). Veenman.

Raj, S.K., Khan, M.S., Kumar, S. and Snehi, S.K. (2006) *New Disease Reports*, **14**: 5-5.

Rowley, G.D. (1987) *Caudiceform and Pachycaul Succulents*. Strawberry Press. Mill Valley, CA.

Rzepecky, A. (2015) *Cactus and Succulent Journal*, **87**(3): 129-135.

Schmelzer, G.H. and Gurib-Fakim, A. (2008) In: *Medicinal Plants. Plant Resources of Tropical Africa*. pp. 43–49.

Singh, A., Chavan, S.K., Bhandari, A.J., Parekh, V., Shah, H.P., Patel, A.I. and Patel, B. N. (2020) *Electronic Journal of Plant Breeding*, **11**(02): 346-350.

Singh, C. (2004) *U.S. Patent Application No. 10/404,902*.

Stoffel Petrus Bester (June 2004) *South African National Biodiversity Institute's plant information website*.

Tijjani, M.S. and Sallau, S.I. (2011) *Bayero Journal of Pure and Applied Sciences*, **4**(1): 79–82.

Tounekti, T., Mahdhi, M. and Khemira, H. (2019) *Evidence-Based Complementary and Alternative Medicine*, doi.org/10.1155/2019/3190670.

Versiani, M.A., Ahmed, S.K., Ikram, A., Ali, S.T., Yasmeen, K. and Faizi, S. (2014) *Chemistry and Biodiversity*, **11**(2): 171-180.

Wang, H., Nair, M.G., Strasburg, G.M., Chang, Y.C., Booren, A.M., Gray, J.I. and DeWitt, D.L. (2013) *Journal of Natural Products*, **62**: 294–296.

Watt, J.M. and Breyer-brandwijk, G.M. (2013) *The Medicinal and Poisonous Plants of Southern and Eastern Africa*, Second ed. E and S Livingstone Ltd, Edinburgh and London, pp. 941–946.

Xiaomei, L., Chen, X. and Chen, B. (2003) *Journal of Hubei Agricultural College*, **23**: 91-94.

31

Calliandra

Subhashis Kundu and Debmala Mukherjee

1.0 Introduction

Calliandra, is an attractive shrub with finely divided leaves and clusters of colourful powder-puff like flowers. It belongs to the family Fabaceae and is native to tropical and subtropical region of America. It is large, multiple branched, evergreen shrub having silky leaf of glossy copper coloured when new and turn dark metallic green when old. This genus contains about140 species (de Souza, 2013). *Calliandra* cultivars are diploid. This plant is commonly known as powder puff because of its bright puffy shape flowers. It mainly produces flowers of watermelon pink, deep red, white milk colour generally in warm months. These plants flower all year round, but the best blooming is in spring and summer. They can be easily pruned and ideal for designing topiary. For its beautiful flowers and shrubby nature, it is popular as hedges and boundary plants.

2.0 Importance and Uses

2.1.0 Aesthetic Uses

Calliandra is a popular ornamental plant for its colourful attractive flowers. It is popular among parks, homes and institutional gardens. In its native area, it was unused and it was not until it was introduced into Indonesia that its potential value and multipurpose uses became apparent. Coloured species of calliandra are widely used as informal hedges, buffer strips around parking lot, divider along road, single yard specimen, along fence as well as container or pot plant.

2.2.0 Therapeutic and Other Uses

The species, *C. calothyrsus* are mainly used as firewood, fodder, reforestation, soil stabilisation, soil improvement and minor secondary production (pulp and paper, honey and shellac, leaf meal). *Calliandra calothyrsus* as a nitrogen fixing tree has been used as a green manure to improve the fertility status of soil in rotation with sugar cane and in alley cropping with corn (Gichuru and Kang, 1989). The plant produces high yields of high protein leaf material on less fertile soils and is therefore very suitable as a green manure. However, the relatively high levels of polyphenols (tannins) present in the leaves of calliandra slow the rate of microbial. In Nigeria, roots are used in the treatment of haemorrhoids. Elsewhere, decoction of the flowers used as blood purifier and tonic.The ethanolic root extract of *C. portoricensis* exhibited the highest activity for inhibitory and reversal activities, hence possessed antisickling properties *in vitro* with *C. portoricensis* being the more active plant (Amujoyegbe, 2014).

3.0 Origin and History

Callindra is originated in tropical and subtropical region of America, specifically Bolivia, is native to the humid and sub- humid regions of Central America and Mexico. It was introduced into Indonesia in 1936 for green manure and as a shade tree, the planting being financially encouraged during the 1970s. It has since been

Uses of Calliandra

As Hedge | At Centre of Garden | Road Divider

As Pergola | Against Wall | Screening

Along Fence | As Topiary | In Pot

Calliandra Bonsai

introduced into many other tropical countries, particularly in South- East Asia and Africa (Ethiopia, Uganda, Kenya, Tanzania, Rwanda, Zimbabwe), and also in Australia, Brazil, Bolivia, Colombia and Hawaii (Palmer *et al.*, 1994; Hess *et al.*, 2006). In Uganda, calliandra is the most cultivated fodder tree, and is preferred to leucaena, mulberry (*Morus alba*), tagasaste (*Chamaecytisus palmensis*) and gliricidia (Wambugu *et al.*, 2006). Calliandra was considered a promising shrub for high- land areas (Place *et al.*, 2009) and was, therefore, encouraged for agroforestry projects in East Africa (Franzel *et al.*, 2014).

4.0 Distribution

The plant is native to the humid and sub- humid regions of Central America and Mexico and is found from southern Mexico to Central Panama, approximately between 8° and 19° N (Macqueen, 1992). It was introduced from Guatemala to Java, Indonesia in 1936 (Verhoef, 1939) as a possible source of green manure and shade for coffee plantations and is now found throughout the Indonesian archipelago. During the 1970s, sponsored by the Indonesian State Forest Corporation, *C. calothyrsus* was planted in many areas of Java and elsewhere in Indonesia in what has been termed 'a true self-perpetuating greening movement'. More recently it has been introduced to other areas of southeast Asia and is also under experimental evaluation in Africa, Australia, Brazil, Bolivia and Hawaii.

5.0 Morphology

Calliandra is an almost evergreen, thornless small legume tree, usually about 5-6 m high, but it can reach a height of 12 m. It has a straight trunk up to 30 cm in diameter and many branches that form a dense canopy. The bark is very variable in colour, from white to red brown or blackish brown (Palmer *et al.*, 1994). Calliandra has a fast growing, vigorous root system that develops down to a depth of 1.5-2 m within 4-5 months. The root system encompasses both superficial adventitious roots and deep growing roots, sometimes developing a taproot (Orwa *et al.*, 2009). Calliandra form root nodules with *Rhizobium*. New sprouts are readily formed from the root system and facilitate coppicing. Under annual coppicing of stems of 3-5 cm diameter, the tree can survive for many years (FAO, 2016).

Calliandra leaves are alternate and bi- pinnately compound. The inflorescence, borne at the apex, is a showy spike-like raceme 10-30 cm in length. It bears several clusters of purplish-red flowers that are 4-6 cm long with very conspicuous stamens, hence the name 'calliandra' (means 'beautiful male') was given. Flowering continues throughout the year under sufficient soil moisture, but in its native environment is concentrated between October and January (late wet season). Flowering can cease during dry seasons of more than 4-5 months. Flowers are andromonoecious, bearing both hermaphrodite (bisexual) and staminate (male) flowers.The fruits are broadly linear, flattened, pubescent, dehiscent pods, 8-11 cm long × 1 cm broad, and brown in colour. They contain 3-15 seeds. The seeds are ellipsoid, flattened, 5-7 mm long, and mottled dark brown in colour (Palmer *et al.*, 1994).

6.0 Species and Cultivars

6.1.0 Species

About 140 species of evergreen shrubs belong to this genus (de Souza, 2013); natives of tropical and subtropical America and also of India.

Calliandra brevipes

A bushy shrub with drooping and spreading branches, to 2 m high; leavespinnately compound, each about 2.5-5.0 cm long; leaflets very small, 3- 6 mm long, light green, linear- oblong; flowersin roundish clusters, with numerous fine pink stamens, 5 cm across. This is a beautiful shrub with a fine feathery foliage, thrives well in a semi-shady location and well drained soil in mild climate; flowers almost throughout the year, abundantly during the winter months; multiplied from seeds or air-layering.

Calliandra marginate

One of the best-known flowering shrubs in cultivation, erect, up to 2.5 m high with an umbrella-shaped appearance; leaves bipinnate with 3 leaflets on each pinna, leaflets obliquely-elliptic, 2-3 cm long, upper surface green glabrous, tomentose beneath; flowers appear in axillary clusters, hemispherical, about 3.5 cm across, numerous bright scarlet stamens.Flowering continues for almost throughout the year and more abundantly in summer and rains; it prefers fertile soil and full sun; usually grown from seedsor propagated by cuttings or air layering. But this species does not respond to pruning.

Calliandra haematocephala

A tall, spreading shrub, about 2.5 m high; leavesalternate, bipinnately compound, each of the 2 pinnae bears 8- 10 pairs of leaflets, 1.5- 3.0 cm long, obliquely oblong or lanceolate, bright green; flowersmany in a head, becoming hemispherical when open, about 6 cm across, with numerous bright crimson stamens.A well-shaped, evergreen shrub with dark green foliage; bears numerous crimson flowers during autumn and winter; pruned in early rains for maintaining the height and shape, if necessary.

Calliandra houstoniana

A perpetual flowering, straggling plant with terete branches; leaves alternate, bipinnate, almost as long as a pinna and each pinna has 60- 80 pairs of small green leaflets; flowers appear in clusters, buds dark brown; each flower has 5 fleshy, dark brown petals; stamens numerous, long, purple- white, 3.8 cm long. It grows better in a shady place; flowers appear the year round, less showy than most of the species.

Calliandra tweedii

A tall shrub, about 3 m high; branches spreading almost horizontally; leaves bipinnate with 10- 14 pairs of pinnae, 7.5- 10.0 cm long, many leaflets, narrow, oblong, 1.5- 2.5 cm long and dull green; flowersmany, appear in hemispherical heads, 5- 6 cm across, axillary, upper part of stamens pink and lower part white.

Species of Calliandra

C. brevipes | *C. haematocephala* | *C. houstoniana*

C. calothyrsus | *C. eriophylla* | *C. tergemina*

C. surinamensis | *C. grandiflora* | *C. riparia*

C. tergemina var *emerginata* | *C. tweedii* | *C. californica*

Source: www.en.wikipedia.org

Calliandra physocalyx

It is a shrub with pink-and-white or pink-and-red flowers. Like other members of the genus *Calliandra,* the filaments of the stamens are long and colourful, about 7.5 centimetres long. The species was first scientifically described in 1988 (Hernandez and Sousa, 1988).

Calliandra hybrida

A hybrid of *Calliandra hematocephala* × *Calliandra tweedii* raised at the Agri-Horticultural Society of India, Kolkata. A tall, much-branched shrub, reaching a height of 2.5 m; leaves pinnately compound, leaflets 6- 9 pairs, oblique-elliptic or linear, 2- 3 cm long; flowers many in axillary powder- puff like clusters, about 6- 7 cm across; filaments numerous, deep pink, lower portion white. 'Alipore', another selection from hybrid seedlings, leaflets of this cultivar are smaller than those of *Calliandra hybrid* and flowers do not open fully.Unlike the parents, the hybrid flowers all the year round; an ideal plant in a shrubbery or as a specimen shrub.

Calliandra protoricensis

A large, tall shrub is about 2.5 m high; leaves bipinnate with 2- 5 pairs of pinnae, each of three leaflets, obliquely- obovate, 2.5- 4.0 cm long; flowerswhite, 12- 20, in almost spherical heads, about 5 cm across. It bears large number of white flowers in winter and spring; fairly quick- growing plant needs pruning almost every year.

Calliandra houstoniana var. *calothyrsus*

It is a small to medium-sized tree, which can be either single or multi-stemmed depending upon the seed source. It can tolerate only partial shade and grows best when in open situations. However, it is tolerant of repeated lopping and also browsing at ground level. The flowering period of *C. houstoniana var. calothyrsus* can extend all over the year if sufficient moisture is available, but it peaks between October and January within its native range in Mesoamerica and during September to December in Sri Lanka (Chamberlain, 2001). Flowering will cease over the dry season where greater than 4 months of dry conditions occurs (Cook *et al.,* 2005). Leaves drop in areas which experience dry periods greater than four months.

6.2.0 Cultivars

No cultivars of *C. calothyrsus* have been released. Only few accessions are found.

'Patulul' provenance (CIAT 22316): selected in Guatemala, collected by OFI, UK,produced high levels of both leaf and wood in multi-location trials, and also shown to have superior nutritive value.

'Bandung' and '**Maduin**' landraces: Produced high levels of both leaf and wood in multi-location trials. Widely distributed to exotic locations(Indonesia, collected by OFI, UK).

'Embu' landrace: KARI/ICRAF, Kenya. Closely related to Indonesian landraces.

7.0 Genetics and Breeding

The haploid chromosome number reported for *Calliandra* is *n* =11 (Chamberlain, 2001).Predominantly, calliandra possesses out- crossing with a weak, possibly late-acting, self- incompatibility system. Calliandra is sparsely seedy due to its nocturnal flowering habit and cross-pollination largely by bats and moths (Brewbaker, 2004). Pollination is achieved by hawkmoths (Sphingidae), bats of the genus *Glossophaga* and other less specialised fruit bats. Pollination in *C. calothyrsus* is effected predominantly by nectivorous bats that possess long, extensible tongues covered in bristle- like papillae. In the native range, these bats are members of subfamily Glossophagina, *e.g. Glossophaga soricina*. Less specified fruit bats and large moths will also visit and pollinate *C. calothyrsus*. The nectivores bats often forage over small distances, but longer range pollination, either by these bats or by fruit bats cannot be discounted (Chamberlain, 2001).

Calliandra species has heteromorphic pollen grains shed in ellipse-shaped.There is evidence that *C. calothyrsus* exhibits a late-acting self-incompatibility system and can tolerate selfing at a low level. Fruit set will therefore be limited by the behaviour of potential pollinators visiting the plant. *Callindra calothyrsus* appears to be predominantly outcrossing, with a weak, gametophytic, possibly late acting, self incompatibility system (Chamberlain, 2001). It has been assumed that, *C. calothyrsus* was out- crossing and self- incompatible on the basis of its andromonoecious sexual system, the slightly protandrous nature of the flowers, 10 mm difference in length between the style and stamens, and its pollination by bats and moths (Owour, 1996; Macqueen, 1992). In natural population, a variety of factors, *e.g.*, floral phenology, pollinator behaviour and population size may influence the out crossing versus selfing rate, which may in turn affect the genetic structure of populations.

The interplant movements of visitors to *C. calothyrsus* may also affect fruit set. Movements between flowers of the same plant will result in geitonogamy and a low level of fruit set, whereas movements between flowers of different plants will increase the chance of xenogamy and high fruit set (Arroyo, 1976). If, however, the correct pollen lands, the plant may be unable to mature every fruit because of resource, spatial, or physical restrictions.

7.1.0 Hybridization

7.1.1 Interspecific Hybridization

Species within calliandra series Racemosae typically display allotropic distribution, with geographic isolation rather than reproductive isolation, suggested as the reason for speciation within the group of taxa (Macquen and Hermandez, 1997). Occasionally, symmetric populations (different species occupying the same geographic area) has been observed, which have arisen as a result of habitat disturbance, followed by colonisation by one or other of the species. Sympatry primarily exists between *C. calothyrsus* and *C. houstoniana*, *C.houstoniana* and *C. grandiflora*, and *C. houstoniana*and *C. juzepczukii* and putative hybrids have been documented between *C. houstoniana* and *C. houstoniana* and *C. juzepczukii* (Macquen and Hermandez, 1997). Although, interspecific hybridization is widely used to incorporate new combination of characters in putative hybrids.

7.1.2 Natural Hybridization

Interspecific hybridization between calliandra series Racemosa was suspected to occur naturally between *C. calothyrsus* and *C. houstoniana* in Hondurus (near Meambar, Comayagua and Santa-Rosa, Copan) and in Mexico (Colotlipa, Guerrero and Yajalon, Chiepas) and between *C. houstoniana* and *C. juzepczukii* in Mexico (Ocozocautla, Chiapas) (Macqueen and Hernandez, 1997). If, speciation has occurred in the past through a process of geographic rather than reproductive isolation, habitat disturbance following human activity (*e.g.* agriculture and road construction) in combination with the pioneer nature of many calliandra species, may have led to hybridization between previously separated species. An intermediate morphological form observed in Flores, Guatemala has been identified as a hybrid between *C. clothyrsus* and *C. houstoniana*subsp. *houstoniana* and exhibited reduced fertility (Chamberlain, 2001).

Artificial hybrids have been produced between *C. calothyrsus* and *C. houstoniana* and *C. houstoniana* and *C. juzepczukii* (Chamberlain, 2001). Preliminary trials in Honduras suggest that there may be hybrid vigour for growth. Precocious flowering may be associated with hybrids, which produced structurally perfect flowers. Although, interspecific hybridization of calliandra is at very primary stage and hence needs more attention.

8.0 Soil and Climate

8.1.0 Soil

Calliandra is an aggressive pioneer species that can be found in disturbed areas such as roadsides, river banks and shifting cultivation plots (Palmer *et al.*, 1994). Calliandra does well under a wide range of soils and is outstanding in those of low fertility (Wiersum *et al.*, 1997). It cannot survive more than two weeks of water-logging (Wambugu*et al.*, 2006). Calliandra grows well on a wide range of light textured, low fertility soils from acidic sandy soils to deep volcanic loams. It does not withstand compact, poorly drained, alkaline calcareous soils (Orwa *et al.*, 2009; Palmer *et al.*, 1994). Calliandra has an extremely fine root system, and should be planted in well-drained soils. Excess moisture causes rotting in roots and plants. These types of media retain too much water and will rot the root system. Proper soil pH is essential because it affects the availability of mineral elements. A soil pH above 6.0 increases the possibility of micronutrient deficiencies, particularly iron. Calliandra is drought tolerant, salt tolerant, and wind resistant.

8.2.0 Climate

Calliandra prefers tropical climate, grows from sea level up to an altitude of 1800- 2200 m, but does better up to 1300 m. It grows best with annual rainfall between 700 and 3000 mm, but will tolerate 4000 mm in the wet tropics. It likes annual temperatures ranging between 22 and 28°C, with the mean temperature of the hottest months being within 24- 30°C and of the coldest months within 18-24°C. Calliandra does not withstand frost but is tolerant of dry spells lasting from 1 to 7 months. But temperature less than 16°C, decreases flowering of calliandra. For potted plants moderate to high humidity (50 per cent RH) is needed for better flowering.

9.0 Propagation

9.1.0 Seed Propagation

This is an easy method of calliandra propagation. The seed production season is during the dry season from June to September. After collecting, drying and storing of the seeds these can germinate without treatment, but a quicker germination is achieved by soaking the seeds in cold water for 48 hours or 24 hours in warm water.The seeds could be stored for 1-2 years. Seeds will germinate in about 2 to 3 weeks. Seedlings are raised in nursery bed and when they attain a height of 15-20cm, transplanting is done. (Orwa *et al.*, 2009).

9.2.0 Vegetative Propagation

9.2.1 Stump

Another method is of propagation is done through stumps. In this case the seedlings are grown for four months until they reach a height of 75 to 100 cm, subsequently they are top- and root-pruned to about 30 to 20 cm respectively. Stumps are very useful for interplanting among other trees or for a better establishment on steep slopes.

9.2.2 Cuttings

Propagation of calliandra is also done through cuttings. A length of 8- 10 cm stem tip cuttings are mainly used for propagation. Cuttings are dipped in root promoting hormone and planted in pot with potting media made up of perlite and peat in 1:1 ratio with a depth of 2.5- 5 cm. Keep the pot warm (21°C) but regular spraying of water is a must. In twomonths, it should take root and start to grow into a new plant.

Calliandra are also propagated by air or ground layering and the layered shoots root within 40 days, if rooting hormone is applied.

10.0 Cultivation

10.1.0 Planting

Planting of *Calliandra calothyrsus* can be either done by direct seeding or by seedlings. For direct seeding, the rows should be free of weeds and the ground should be roughly cultivated. The seeds should be sown at 1 to 3 cm depth.Plantation of seedlings is done at the beginning of the rainy season and an initial treatment of fertilizer is given. The early growth is slow but after the mycorrhizal infection the plant grows up to 3.5m in 6 months. Because of the slow early growth weeding is necessary in the first year but afterwards the canopy closes and weed is no longer problematic. Incase of row, planting distance is kept at least 1.2 to 1.8 m. For planting

Propagation and Growing of Calliandra

Pods and Seeds of Calliandra

Seeds

Seedlings in Polybags

Seedling in Pot

Ready for Transplanting

Alley Cropping with Maize

Established for Fodder

Contour Hedge

Flowering in Calliandra

Pollinator, *Glossophaga soricina*

Source: www.tropicalforages.info, www.jollykabirizi.com, www.flickr.com

along fence 0.9 to 1.2 m or more from house or walkway for planting is preferred. The dwarf cultivar of powderpuff can also be planted in pot, but it is advised to opt for larger size pot while planting.Within the first year calliandra flowers and bears fruits and after 12 months the plants can be 3 to 5 m tall and 5 cm in diameter at stump height. At this stage the plant has the optimal size for the use of firewood. Harvesting is done at the end of the dry season and for a faster re- sprouting, the cutting is done 20 to 50 cm above the ground. In Java, calliandra was successfully included in a 10- year rotation based on 4 years calliandra, 4 years sugarcane and 2 years maize (Wiersum *et al.*, 1997). However, in Samoa alley cropping of calliandra with taro over 4 years did not sustain taro yields (Wiersum *et al.*, 1997).

10.2.0 Manuring and Fertilization

Calliandra needs proper nutrition for proper growth and better flowering. It needs fertilizer atleast three times in a year; spring, summer and autumn with a quality granular fertilizer. *C. calothyrsus* species has a symbiotic relationship with certain soil bacteria, these bacteria form nodules on the roots and fix atmospheric nitrogen. Some of this nitrogen is utilized by the growing plant but some can also be used by other plants growing nearby. Supplements like bone meal and liquid fertilizer can also be used to promote higher numbers of florets. For potted plants balanced water soluble fertilizer diluted 2 times and used within every 2 weeks.

10.3.0 Irrigation

Calliandra can tolerate drought once it is established, but it needs evenly watering during the dry spells in summer. Calliandra does best in regular irrigation; care should be taken as soil does not get waterlogged. Too much water can lead to overly-green growth and eventually root rot; in drought, the plant can wilt. Use humid tray or humidifier to maintain relative humidity for potted plants.

10.4.0 Pruning

Overtime, this shrubby plant will grow into a tree, reaching several meters high and willspread about as wide. Pruning will help it to branch out, giving powderpuff plant a fuller, bushier look, as well as more blooms. Mainly hard pruning done due to spring month to keep it up to 90 cm. Hard pruning doesn't affect the plant very much, in fact this plant makes a charmy bonsai. Regular pollarding of calliandra promotes coppicing for more than 10 years. In alley cropping, pollarding should be carried out at 4 month intervals to limit shading of the companion crop (Orwa *et al.*, 2009).

11.0 Diseases and Pests

11.1.0 Diseases

As calliandra is widely spread around the tropics, a wide range of pests and diseases are proving adapted to the species and adversely affecting its productivity, even causing death. Fungal attacks can cause serious damages on *Calliandra calothyrsus* when the plants are cut back too low prior to rainfall or during the wet season. Another potentially devastating disease of calliandra is dieback. The primary cause of this disease was notorious due to a complex of fungal species isolated from diseased specimens. The isolation of *Fusarium oxysporum* and a *Phomopsis* species is consistent with earlier diagnosis of similar specimens from Uganda (Simons, 2000), suggesting that one or both of the pathogens are the primary causes of the disease. Roothaert *et al.* (1998) reported *Nectria ochroleuca* as the cause of similar dieback symptoms on calliandra, and application of well- rotted cow dung around infected trees, uprooting and burning infected trees to be effective against dieback.

11.2.0 Pests

While usually pest-free, powderpuff bush can beoccasionally infested by mites, caterpillars, or otherchewing insects. A beetle pest, *Tragocephala guerini* was observed damaging the branches of *Calliandra calothyrsus* in Cameroon (Gauhl, 1998), Central Africa. Wilting of *C. calothyrsus* plants was found to be associatedwith rodents girdling the stem base partly or entirely, with insect larvae feeding in branches. The larvaebore into and feed on branch wood, hollowing the centre and leaving onlythe bark. In each attacked branch, only one larva was observed.At regular distances of 2–4 cm, they construct aeration holes through the bark. Further, *Pachnoda ephippiata* might feed on flowers, fruits and foliage, causing floral abortion and poor seed production. Bruchid beetles can destroy a high proportion of stored seed if seed is not kept under cold conditions.

12.0 Harvesting

Flowering may start in the first year, but good fruit set starts from second year onwards. Fruits ripen 3 months after anthesis. The plants will often flower all year round. When grown for fuel, wood yields have been reported in the order of 15-40 tonnes per hectare just one year after planting, with annual coppice harvests, continuing for 10- 20 years. Annual wood production potential is 5- 65 m^3/ha. *Calliandra* forms seed inside pods. After harvesting, pods are dried and seeds are collected. Annual forage yields of 7- 10 t and up to 20 t dry matter/ha have been obtained under variable growing conditions (Kabi *et al.*, 2008; Palmer *et al.*, 1994). Seed viability can be maintained for several years in hermetic storage at 3°C with 6- 10 per cent moisture content.

13.0 References

Amujoyegbe, O.O., Agbedahunsi, J. M. and Akanmu, M. A. (2014) *Eur. J. Med. Plants.*,**4**(2): 206-219.

Arroyo, M.T.K. (1976) *Taxon.*, **25**: 543-548.

Brewbaker, J.L. (2004) *Encyclopedia of Forest Sciences.*

Capucho, L.C. and Teixeira, S.P. (2013) *S. Afr. J. Bot.*, **89**: 227-233.

Chamberlain, J.R. (2001) *Tropical Forestry Papers*, No. 40 : vi + 100 pp.Oxford Forestry Institute, University of Oxford, UK.

Cook, B.G. Pengelly, B.C. Brown, S.D. Donnelly, J.L., Eagles, D.A., Franco, M.A. Hanson, J., Partridge, I.J., Peter, M. and Schultze-Kraft, R. (2005) *Tropical Forages: an interactive selection tool.* Brisbane, Australia: CSIRO, DPI&F, CIAT, ILRI.

de Souza, E.R., de Queiroz, L.P., Lewis, G.P., Forest, F., Schnadelbach, A.S., van den Berg, C. (2013) *Taxon.*, **62** (6): 1200-1219.

FAO, 2016. *Grassland Index. A searchable catalogue of grass and forage legumes*. FAO, Rome, Italy.

Gauhl, F., Pasberg-Gauhland, C. and Goergen, G. (1998) *Agrofor. Syst.*,**41**: 213-218.

Gichuru, M.P. and Kang, B.T. (1989) *Agrofor. Syst.*,**9**: 191-203.

Macqueen, D.J. and Hernandez, H.M. (1997) *Kew Bulletin.*, **52**: 1-50.

Franzel, S., Carsan, S., Lukuyu, B., Sinja, J. and Wambugu, C. (2014) *Curr. Opin. Environ. Sustain.*, **6**: 98-103.

Hernandez, H.M. and Mario Sousa, S. (1988) *Syst. Bot.*, **13** (4): 519-524.

Hess, H.D., Tiemann, T.T., Noto, F., Franzel, S., Lascano, C. E. and Kreuzer, M. (2006) *Agroforest. Syst.*, **68** (3): 209-220.

Kabi, F., Bareeba, F.B. (2008) *Anim. Feed Sci. Technol.*, **140** (1-2): 178-190.

Macqueen, D.J. (1992) *Commonwealth Forestry Review*, **71**: 20-34.

Orwa, C., Mutua, A., Kindt, R., Jamnadass, R. and Anthony, S. (2009) *Agroforestree Database: a tree reference and selection guide version* 4.0. World Agroforestry Centre, Kenya.

Palmer, B., Macqueen, D.J. and Gutteridge, R.C. (1994) *Trop.Grassl. Soc. Austr. Inc.*, Queensland, Australia.

Place, F., Roothaert, R., Maina, L., Franzel, S., Sinja, J. and Wanjiku, J. (2009) *ICRAF Occasional Paper* No. 12. Nairobi: World Agroforestry Centre.

Roothaert, R., Karanja, G.M., Kariuki, I.W., Paterson,R., Tuwei, P., Kiruiro, E., Mugwe, E. and Franzel, S.(1998) *'Calliandra for livestock'*. Technical BulletinNo. 1. Embu, Kenya.

Simons, S.A. (2001) *Report on a new pest/disease problem threatening Calliandra species and other agroforestry species in Uganda*, AFRENA, Uganda Project.

Verhoef L. (1939) *Calliandra calothyrsus Meissn. Korte Mededeelingen 79. Boschbouwproefstation, Bogo*r, Java, Indonesia.

Wambugu, C., Franzel, S., Cordero, J. and Stewart, J. (2006) *World Agroforestry Centre*, Nairobi, Kenya, Oxford Forestry Institute, Oxford, U.K., pp.172.

Wiersum, K.F. and Rika, I.K. (1997) *Calliandracalothyrsus* Meisner. Record from Proseabase. (Eds. I. FaridahHanum, L.J.G. van der Maesen), *PROSEA* (*Plant Resources of South-East Asia*) Foundation, Bogor, Indonesia.

32

Crossandra

S. Sangeetha Priya, C. Aswath,
S. Sambandamurthi and Swati Saha

1.0 Introduction

Crossandra is an important commercial flower, which is commonly called the firecracker flower, is a species of flowering plant in the family Acanthaceae. It is mainly grown in India, tropical Africa and Madagascar (Bailey, 1963). In India, its cultivation is restricted in South India, and produced flowers in the open field. In temperate areas, they are grown under glass or olyethylene cover. It is usually grown in containers, but can be attractive in beds as well. The flowers have no perfume but stay fresh for several days on the bush. A common bedding plant in subtropical regions, crossandra is equally suitable as good indoor plant. Its foliage can reach a maximum height of 90 cm, but often stays around 60 cm feet indoors. The flowers are most commonly salmon to apricot in colouration. Some cultivars have been bred to produce yellow, pink, or red flowers as well.

2.0 Importance and Uses

2.1.0 Aesthetic Uses

The flowers are commonly used for hair adornment. Though not fragrant, flowers are very popular because of its attractive bright colour, light weight and good keeping quality. Plants are suitable for beautification of garden as hedge, pot plant, cottage garden, hanging basket, ground cover and also excellent for bedding. These are used for making garland, either alone or in combination with jasmine flowers. Using crossandra flowers in combination with jasmine is becoming increasingly popular in India, particularly in southern parts, for hair adornments, because the jasmine flowers provide colour contrast and the desired fragrance.

2.2.0 Therapeutic and Other Uses

Crossandra is well known medicinal plant in various region of India. This plant is one of the most chosen varieties for folk medicine. Flower extract used in various conditions like fever, headache, aperitif, pain *etc.* (Gundamaraju and Verma, 2012). The aqueous fresh flower extract of crossandra exhibited anti- solar activity, therefore, Crossandra flowers can be used in various sunscreen formulations (Pulipati *et al.*, 20014).

3.0 Origin, Species and Cultivars

3.1.0 Origin

All the species of crossandra are natives of the Arabian Peninsula, tropical Africa, Madagascar, India and Sri Lanka (Brickell, 1996).

3.2.0 Species

Crossandra belongs to the family Acanthaceae. There are 50 species, but only few like *Crossandra infundibuliformis*, *C. guineensis*, *C. mucronata*, *C. nilotica* and *C. subacaulis* are cultivated. The species grown for commercial flower production is *C. infundibuliformis*.

Uses of Crossandra

In Pot

In Hanging Basket

In Home Décor

In Container

Floral Crown

Floral Strings

Loose Flower

In Garland

Hair Decor

Floral Ornaments

In Floral Rangoli

Along Pathway

Uses of Crossandra

Ground Cover

In Mixed Border

Under a Tree

In Home Garden

Source: www.calyx.com.au, www.m.mediaamazon.com, www.getflowersdaily.com, www.englishgardens.com, www.terrepolicycentre.com, www.ingrini.com, www.thespruce.com, www.imagesvc.meredithcorp.io, www.toptropicals.com, www.5.imimg.com, www.live.staticflickr.com, www.1.bp.blogspot.com

C. *infundibuliformis* syn. C. *undulaefolia*

Plants 100 cm tall, erect, evergreen shrubs; leaves ovate, margin- wavy; glossy; flowers fan-shaped, orange yellow to salmon-pink; born axillary; spikes cone-shaped, 10 cm long. Cultivar Mona Walhead- Compact; flowers deep salmon-pink. 'Orange', 'Delhi', 'Lutea Yellow' and 'Sebaculis Red' are also important cultivars.

C. nilotica

Plants 60 cm tall, evergreen, sub- shrubs; leaves elliptic, hairy; flowers fan-shaped, vivid red to orange, dense; axillary or terminal spikes, 7 cm long, bracts hairy.

3.3.0 Cultivars

ICAR-IIHR, Bengaluru released six varieties of Crossandra namely 'Arka Ambara', 'Arka Kanaka', 'Arka Shreeya', 'Arka Shravya' and 'Arka Chenna'. The salient features of these varieties are given here.

'Arka Ambara': The F_1 hybrid selection, a cross between Local x Lakshmi crossandra, has its unique character of bigger size flowers, which is 69 per cent higher than the local cultivar. The plant height resembles that of local cultivar, however, inflorescence length, hairy structures on inflorescence resembles that of Lakshmi. The weight per 100 flowers was 13.01 g compared to its parents 6.24 and 6.47 g, respectively. Arka Ambara was found to be suitable as potted plant with shorter and compact growth habit (Bharathi *et al.*, 2018).

'Arka Kanaka': The F_1 hybrid selection is a cross between Local yellow × Delhi crossandra showed a unique character of bigger size flowers, which is 67 per cent higher than the local cultivar and also for its novel colour of Orange (Orange group 21 A). The weight per 100 flowers was 11.9 g compared to its parents 5.05 and 6.45 g, respectively.

'Arka Shreeya': It is a hybrid between 'Local' × 'Arka Ambara'. The plant and flower morphology of this hybrid resembles that of 'Arka Ambara', however, with a change in flower colour. The flower size and yield is superior with the hybrid 'Arka Ambara' and got a unique colour (Orange Red group 32A). 'Arka Shreeya' found to be suitable as potted plant with shorter and compact growth habit (Bharathi *et al.*, 2018).

'Arka Shravya': It is a hybrid between *Crossandra Nilotica* × 'Mangalore Local'. The size and the yield of the flower are bigger than it's both the parents. Interestingly, the long rachis length of the inflorescence and opening of 4 flowers at a time can be used as cut flower, in addition to the loose flower for garlands. The cut inflorescence along with rachis kept in a small jar with oasis, the opened flowers will stay for 4 days in inflorescence. The small jars with crossandra flowers can be used for decorating small table space.

'Arka Chenna': This is developed through mutation. There is no seed set in this mutant. The flowers can be harvested four months after planting. The flower is medium sized unlike the other hybrids. It is slightly bigger than the local (20 per cent). The yield is 4 times higher than the local variety. It yields 40 kg loose flowers

Species of Crossandra

Crossandra infundibuliformis

Crossandra guineensis

Crossandra nilotica

Crossandra undulaefolia

Crossandra flava

Crossandra pungens

Crossandra mucronata

Crossandra greenstockii

Crossandra subacaulis

Crossandra zuluensis

Crossandra puberula

Crossandra johanninae

Source: www.toptropicals.com, www.plantbook.co.za, www.static.inaturalist.org, www.commons.wikimedia.org, www.mozambiqueflora.com, www.bushtrucker.ch/assets, www.lifestyleseeds.co.za, www.toptropicals.com, www.plantsguru.com, www.image.slidesharecdn.com, www.giardinaggio.it

Species of Crossandra

Crossandra fruticulosa | *Crossandra albolineata* | *Crossandra grandidieri*

Crossandra quadridentata | *Crossandra stenandrium* | *Crossandra nobilis*

Crossandra longipes | *Crossandra greenstockii* | *Crossandra longispica*

Crossandra rupestris | *Crossandra multidentata* | *Crossandra isaloensis*

Source: http://portugal.inaturalist.org

Cultivars of Crossandra

'Tropic Yellow Splash' 'Tropic Flame' 'Orange Marmalade'

'Arka Ambara' 'Arka Kanaka' 'Arka Chenna'

'Lutea Yellow' 'Sebaculis Red' 'Delhi Crossandra'

'Orange Crossandra' 'Shamrock Crossandra' 'Sundance'

Source: www.iihr.res.in, www.panamseed.com, www.image.slidesharecdn.com, www.mybageecha.com, www.slideshare.net, www.in.pinterest.com

per week per 1000 plants. The shelf life is slightly higher than the local. The stalk strength is 20 per cent more than the local variety. The colour is like local which is orange.

'Mona Walled' is a Swedish clone with compact form of 38- 45 cm height, dark green, shinny leaves, and salmon-rose flowers (Anon., 2021a).

'Florida Summer' and 'Florida Sunset' (also known as 'Dania') are two different crossandra species than 'Mona Wallhed'. They have narrower leaves, are more cold tolerant and grow best in full sun or semi-shade (Anon., 2021a).

Other cultivars of crossandra are 'Laxmi', 'Dr. A.P.J. Abdul Kalam', 'Delhi Crossandra'.

Ramachu and Thangam (2010) reported that 'CG- 4', 'CG- 2' and 'CG- 7' accessions were tall and 'CG- 6' was dwarf in stature. Moreover, accessions 'CG- 4' was quite early in flowering while 'CG- 2', 'CG- 5' and 'CG- 6' were late flowering. Furthermore, maximum spike length, corolla tube length and flower diameter were recorded with 'CG- 2.' Accession 'CG-5' produced highest number of spikes per plant. The best performance with respect to number of flowers/spike, weight of 100 flowers and flower yield per plant was noticed with 'CG- 2.'

'Arka Shravya' was found superior for rachis length, number of spikes/plant, number of flower/spike, number of flowers/picking, weight of flowers/picking, yield/plant, least days for opening of first flower (Priyanka *et al.*, 2017; Bhosale *et al.*, 2018; Tejaswi *et al.*, 2019; Gowsika *et al.*, 2019; Tejaswi *et al.*, 2020).

The genotype 'ACC- 1' recorded more number of spikes/plant and more number of flowers/spike, while, 'Arka Ambara' recorded maximum 100 flowers weight (9.68 g) (Priyanka *et al.*, 2017); 'Arka Shreeya' showed earliest spike initiation and 50 per cent flowering, while, 'Kadiyam Local 2' found to be earliest in complete flowering (79.67 days) (Tejaswi *et al.*, 2019); 'Arka Kanaka' had maximum stem girth, and 'Arka Shreeya' recorded the maximum flower diameter and 100 flower weight (Gowsika *et al.*, 2019). However, 'Arka Ambara' and 'Nilakottai' local recorded more weight of 100 florets and more days for flower opening in spike, respectively, whereas, 'ACS-2' recorded more number of flowers per 10 g weight, 'Arka Kanaka' showed the more number of primary and secondary branches and longest shelf life (Tejaswi *et al.*, 2020).

4.0 Genetics and Breeding

4.1.0 Genetics

Orange crossandra is tetraploid (2n= 40), sets seeds profusely, breeds true and produces bright orange-coloured flowers. The cultivar Delhi is triploid (2n= 30) and produces more attractive flower of bright deep orange colour. Lutea Yellow is also tetraploid (2n= 40) and the flowers are of orange- yellow colour. Sebaculis Red is another tetraploid (2n= 40) and hardy cultivar which possesses high degree of tolerance to nematodes.

Lalitha and Kumar (1993) recorded high estimates of heritability and genetic advance for flower yield/plant and number of flowers/plant. Path analysis indicated that the number of flowers/plant and spikes/plant were major factors contributing to flower yield.

4.2.0 Breeding

The breeding work done at the Tamil Nadu Agricultural University (India) resulted in a hybrid between two local types collected from Palni and Marudhamalai hills. This hybrid produces larger flowers and has high degree of field tolerance to nematode-fungus complex. The flowers are very attractive with yellowish orange petals. Another hybrid, evolved by crossing the orange- flowered cultivar and the Sebaculis Red, produces very attractive dark pink flowers. On the basis of mean values and variability, 8 improved strains were selected for desirable floral traits, including high 100- flowers weight, long corolla tubes and high flower yield/plant from gamma-rays and ethyl methane sulfonate (EMS) treated crossandra plants (Lalitha *et al.*, 1993). EMS 30 mM produces higher intermodal length, number of leaves, length and breadth of leaf, number of branches and length of branch (Dhivya *et al.*, 2015). However, Balakrishnan *et al.* (2015) found beneficial effect of gamma radiation at 10 and 29 kR on growth and yield traits.

Dhivya *et al.* (2015) reported that the LD_{50} value for gamma rays for seed germination and seedling survival was ranged from 20-30 kR. Days for first flowering delayed upon exposure to increase in dose of gamma rays. The number of spikes per plant, flowers/spike, length of spike, length of corolla tube, flowers and per per plant was increased at lower radiation dose of 10 and 20 kR. Leaf abnormalities were also observed at higher doses. Stimulating effect of gamma radiation was observed at 10 and 20 kR.

The breeding programme at ICAR- IIHR, Bangalore involved hybridization between the Local cultivars of Karnataka, Tamil Nadu, Andhra Pradesh, Pudichery and improved cultivars of 'Lakshmi' and 'Delhi Crossandra' yielded many F_1 hybrids with varying morphological characters. Two F_1 hybrids 'IIHR2004- 09' and 'IIHR2004- 11' were selected among the large number of hybrids, for their unique larger size, which is 67 and 69 per cent higher than the Local cultivars, respectively. These hybrids were released as 'Arka Ambara' and 'Arka Kanaka'. These hybrids may be recommended for commercial cultivation of loose flowers and potted plants. Since the flowers are novel in colour and size they may fetch fancy price in the market (Aswath *et al.*, 2007).

The other two hybrids 'IIHR2005- 1' and 'IIHR2005- 2' performed well and were superior over local cultivars and on par with 'Arka Ambara' and 'Arka Kanaka'. 'IIHR2005- 1' is ahybrid between Local × 'Arka Ambara'. This resembles arka Ambara with change in flower colour. 'IIHR2005- 2' is a hybrid bwtween *Crossandra nilotica* × Mangalore Local. The size and yield of flowers was higher than both the paresnts. Interestingly the long rachis length of the inflorescence and opening of four flowers at a time can be exploited to use this as cut flower, in addition to loose flower for garlands (Aswath *et al.*, 2009).

Guidelines for the conduct of test for distinctiveness, uniformity and stability for crossandra have been developed by the nodal centre ICAR-IIHR, Bengaluru for protection of crossandra varieties under PPV & FRA, New Delhi. The examples varieties for 17 traits (8 vegetative and 9 flower traits) have been given enumerated in the guideline (Anon., 2021).

5.0 Growth and Flowering

Crossandra commences flowering 2- 3 months after planting and continues to bear flowers throughout the year with a drop in production during the rainy season. The flowers open in sequence from the base of the spike. Two flowers which are diagonally opposite in the spike open at the same time. It takes about two days for complete opening of the flower. Depending on the length of spike, it takes nearly 15-25 days to complete flowering on a spike.

5.1.0 Environmental Factors

5.1.1 Photoperiod

Gibbs *et al.* (1989) applied 4 post-production light levels (10, 20, 30 or 40 $\mu mol/s/m^2$ from cool- white fluorescent lamps) and 4 light duration (6, 12, 18 and 24 h daily) to *C. infundibuliformis* cv. Mona Walhed when the plants had an average raceme length of 2-3 cm but had not started to flower. Good quality plants with enhanced flowering duration were obtained with light durations of >12 h and/or light intensities of >20 $\mu mol/s/m^2$.

5.2.0 Growth Regulators

Rooted cuttings of *C. infundibuliformis* cv. Delhi treated with gibberellic and or BA (100 or 200 ppm), with or without 1 per cent urea, 1 month after planting and subsequently at bimonthly intervals resulted in maximum leaf area and number of branches, yield/ha, number of spikes/month and spike length with 200 ppm BA + 1 per cent urea, and total floret number/spike was highest with 200 ppm GA_3 (Sreekala *et al.*, 1999b).

GA_3 at 150 ppm was found to be more effective in growth and flowering response of triploid crossandra cv. Delhi (Bini Sundar *et al.*, 2008). Plants sprayed with gibberellic acid (200 ppm) resulted in maximum plant height, leaf area, dry matter, number of branches, plant spread, days taken to flower spike initiation, days taken to first harvest, duration of flowering and flower yield per plant in crossandra genotype ACC- 1 (Pujeri *et al.*, 2019).

Paclobutrazol caused retardation of growth and significant reduction in yield, whereas, maximum plant height, number of branches and spikes per plant were recorded with chloremquat 500 ppm (Sreekala *et al.*, 2000). Combination of maleic hydrazide (MH) 500 ppm + Ferrous sulphate 0.5 per cent was found best in improving growth, yield and quality attributes followed by maleic hydrazide 500 ppm (Karuppaiah *et al.*, 2019).

6.0 Propagation

6.1.0 Seed Propagation

Crossandra is propagated by seeds. Seeds are sown mainly in seed beds during mid February. Tetraploids are commonly propagated through seeds. Seed rate is

Propagation and Growing of Crossandra

Seeds   Seedling in Pots   Transplanting in Beds

By Cuttings

In vitro Shoot Regeneration, Multiplication and Rooting (*Source*: Kalaivani *et al.* 2019)

In Drip System   Field Cultivation   Open Field Cultivation

Source: www.ccari.res.in, www.gumtree.com.au, www.slideshare.net, www.i.ytimg.com, www.hoeandshovel.com, www.iihr.res.in, www.upload.wikimedia.org

5kg/ha. 60 days old seedlings are transplanted in the main field (TNAU Agritech Portal).

6.2.0 Vegetative Propagation

6.2.1 Cuttings

Both soft wood and semi- hard wood cuttings are used for raising planting materials. Generally triploids are propagated through terminal cuttings. Cuttings are taken in February- March or mid July-August and it take about 4-6 weeks to form roots. Aruna and Mangaiyarkarasi (2020) had taken three types of cuttings *viz.*, terminal cuttings (10-15 cm length with 2 nodes), middle cuttings (10-15 cm length with 2 nodes) and bottom cuttings (10-15 cm length with 2 nodes) taken leaving 30 cm from the terminal region and cuttings were dipped in 100, 200, 300 ppm of IBA along with control prior to planting in pro-trays. IBA 200 ppm with middle cuttings prior to planting proved superior in respect of root length, root volume, shoot girth. Rehana *et al.* (2020) found that combination of terminal cuttings and IBA quick dip at 3000 ppm for 10 seconds performed best to different root and shoot characters as well as field establishment of rooted cuttings in var. Arka Shravya. For planting in 1 ha, 41700 terminal cuttings having 10- 15 cm length are required (TNAU Agritech Portal).

6.3.0 Micropropagation

Girija *et al.* (1999) reported that in *C. infundibuliformis* the maximum number of multiple shoots was obtained from axillary and apical buds on MS medium containing 1 mg BA/l. At the end of the second subculture, development of 45- 49 shoots per explant was noticed from the axillary bud culture, compared with 40-43 shoots per explant from apical buds.

Kalaivani *et al.* (2019) carried out *in vitro* propagation of crossandra var. 'Arka Shravya' and 'Pondicherry local' through direct organogenesis. Nodal segments found to be the most suitable explants with respect to response for *in vitro* regeneration and multiplication. Nodal segments pre-treated with 0.2 per cent bavistin (30 min) + 0.5 per cent streptomycin (45 min) + 70 per cent ethanol (30 sec.) + 0.5 per cent NaOCl (5 min) registered the highest survival *per cent*. However, shoot tips pre-treated with 0.2 per cent bavistin (30 min) + 0.3 per cent streptomycin (45 min) + 70 per cent ethanol (30 sec) + 0.5 per cent NaOCl (5 min) registered the highest survival per cent. MS medium + BAP (2 mg/l) was found to be best for nodal segment cultures with respect to the maximum percentage response to shoot proliferation, days to shoot emergence and shoots/explant. MS medium + BAP (2 mg/l) + GA_3 (1 mg/l) was found to be best with response to highest shoot elongation percentage of shoots and average length of shoots. Efficient rooting was achieved with ½ MS medium + IBA (1 mg/l) gave maximum *in vitro* rooting percentage. Almeida *et al.* (2021) revealed that BAP at 0.5 mg/l induced multiple buds on explants and IBA 0.5 mg/l promoted rooting in crossandra.

7.0 Soil and Climate

7.1.0 Soil

Crossandra grows in fertile, humus rich and well drained soil. In an experiment, Velu (1988a) reported that increasing levels of calcium carbonate adversely affected plant growth and proportionally reduced the availability of iron, regardless of the form in which it was supplied. Alkaline or saline soils are not suitable because plant develop deficiency symptoms showing chlorosis. Bharathi *et al.* (2018) found that potting media comprising soil + sand + FYM (2:1:1) recorded maximum internodal length of branches and number of spikes per plant.

7.2.0 Climate

Crossandra is a tropical plant. It grows well in a sunny location and the temperature around 30°C. Plants are prone to low temperature and frost. It is shade tolerant to some extent, but susceptible to low temperature and frost (TNAU Agritech Portal).

8.0 Cultivation

8.1.0 Planting

Seedlings are ready for transplanting when they have 4-5 pairs of leaves. Similarly, cutting should be transplanted when sufficient roots are developed. Seedlings or rooted cuttings, treated with fungicide and nematicide, should be planted on the side of ridges at a spacing of 30 cm between plant and planting should be done in moist soil. The distance between 2 ridges should be 60 cm. Providing partial shade to the plants in summer is beneficial to maintain the health of plants and obtain higher yield of flowers.

8.2.0 Irrigation

Crossandra needs adequate moisture in soil, immediately after planting. Irrigation schedule mainly depend on the soil and weather conditions. Plant is drought tolerant, but requires free watering for good yield. Klock-moore and Broschat (2001) revealed that the maximum final shoot dry weight of crossandra was noticed when plants watered every day compared to every 3 days. Also, shoot dry weight was maximum in sub- irrigated pots over overhead watered pots.

8.3.0 Manuring and Fertilization

High fertility in the soil is essential for good yield. In addition to the basal application of 25 t/ha farmyard manure or compost, periodical top dressing with fertilizers is essential. The first application of fertilizers should be done in 50-60 days after planting, at the rate of 50 kg of urea, 100 kg of superphosphate and 60 kg of muriate of potash per hectare twice at an interval of 6 months, except in the rainy season.

Gowthami *et al.* (2019) reported that nitrogen and potassium at 150 kg + 60 kg/ ha improved number of spikes/plant, spike length, number of florets/spike, floret length and flower yield/plant. Priyanka *et al.* (2018) observed that minimum days to spike initiation, days to first harvest, higher number of flowers/spike, spikes/ plant, maximum flower yield was recorded with 75:45:45 kg NPK + 18.75 t FYM + 5 t vermicompost/ha and maximum 100 flower weight with 50:30:30 kg NPK + 18.75 t FYM + 5 t vermicompost/ha at a closer spacing of 60 cm × 30 cm in crossandra genotype ACC-1 (Table 1).

Table 1: Effect of different Level of Nutrients on Flowering Attributes of Crossandra*

Treatments	*Days taken to Spike Initiation*	*Days Taken to First Harvest*	*Duration of Flowering*	*No. of Flowers/ Spike*	*No. of Spikes/ Plant*	*100 Flower Weight (g)*	*Flower Yield (t/ha)*
100:60:60 kg NPK + 25 t FYM/ha	107.76	118.20	48.83	24.76	107.96	3.70	2.82
75:45:45 kg NPK + 18.75 t FYM + 2 t vermicompost/ ha	97.16	108.00	50.08	30.91	110.76	3.54	4.71
50:30:30 kg NPK + 12.5 t FYM + 2 t vermicompost/ ha	104.83	116.51	46.11	24.88	99.30	3.75	3.05
75:45:45 kg NPK + 18.75 t FYM + 5 t vermicompost/ ha	92.65	103.61	51.90	35.80	111.48	3.20	4.82
50:30:30 kg NPK + 12.5 t FYM + 5 t vermicompost/ ha	106.48	117.93	43.08	22.73	81.93	3.46	2.66
SEm (±)	0.41	0.33	0.47	0.81	1.62	0.06	0.11
CD @ 5 per cent	1.25	1.03	1.43	2.48	4.94	0.17	0.33

* Priyanka *et al.* (2018).

Gajalakshmi and Abbasi (2002) reported that the compost obtained from water hyacinth had positive impact which would help in dispelling the apprehension of farmers that compost obtained from pernicious weeds like water hyacinth may have deleterious effect on other plants (Table 2).

Swaminathan and Sambandamurthi (2000) reported that in triploid crossandra plant height, number of branches, number of leaves, leaf area per plant, number of spikes per plant, spike length, number of flowers per spike and flower yield per plant were the highest in N 120 kg + K_2O 70 kg/ha + *Azospirillum brasilense* 2 kg/ha + FYM 30 t/ha. Azospirillum + N + K_2O proved to be more beneficial than Azospirillum or FYM alone. Bhavanisanker and Vanangamudi (1999) reported that combined application of 100 per cent recommended N as urea + azospirillum and 75 per cent recommended P as superphosphate + phosphobacteria gave the highest flower yield (1.106 and 1.474 kg/plant) at 3 and 17 months after planting, respectively. Prabudoss (2011) revealed that crossandra plants receiving 75 per cent NPK along with *Azospirillum* and phosphobacteria registered maximum growth and yield.

Table 2: Effect on the Application of Water Hyacinth Compost/ Vermicompost on Growth of Crossandra*

Attributes	*Time of Observation*	*Control*	*Water Hyacinth Compost*	*Water Hyacinth Vermicompost*
Plant height (cm)	90th day	23.14 ± 2.19	27.34 ± 2.68	28.86 ± 2.7
	180th day	28.5 ± 2.82	33.76 ± 3.78	35.34 ± 3.78
No. of leaves	90th day	72 ± 6.42	81.8 ± 4.71	83.4 ± 4.03
	180th day	78 ± 6.07	89.2 ± 5.49	93.8 ± 3.87
Root length (cm)	90th day	14.78 ± 0.89	14.3 ± 1.15	16.2 ± 0.89
	180th day	15.52 ± 1.03	15.9 ± 1.01	17.46 ± 0.70
Shoot-root ratio	90th day	2.48 ± 0.39	3.44 ± 0.33	3.4 ± 0.23
	180th day	2.76 ± 0.36	3.18 ± 0.27	3.5 ± 0.30
Total dry weight	90th day	17.76 ± 0.58	18.7 ± 0.64	18.96 ± 0.75
	180th day	19.34 ± 0.59	20.9 ± 1.03	21.64 ± 1.42
Days to first flowering	-	45 ± 5.1	39.8 ± 4.75	36.4 ± 3.44
No. of flowers per plant	90th day	11.4 ± 4.03	16.6 ± 4.84	17.2 ± 2.71
	180th day	19.4 ± 3.01	24 ± 4.38	26.8 ± 3.43
Inflorescence length (cm)	90th day	8.94 ± 0.29	9.38 ± 0.36	9.78 ± 0.53
	180th day	9.16 ± 0.27	9.54 ± 0.22	10.02 ± 0.48
Harvest index	90th day	6.79 ± 1.19	7.91 ± 1.05	9.29 ± 1.29
	180th day	7.15 ± 1.11	7.89 ± 1.38	9.21 ± 1.17

* Gajalakshmi and Abbasi (2002).

The use of oil cakes (at 15 g/plant) with *Glomus mosseae* (1000 chlamydospores/ plant) was found to promote better plant growth, reduced nematode multiplication rate, and increased root colonization by the mycorrhiza and the number of mycorrhizal propagules (Nagesh and Reddy, 1997). Mycorrhizal inoculation showed positive effect on plant height, early flowering, increased flower number, increased flower weight, reduced weight loss over 24 h, maximum leaf P concentration and increased total dry weight compared to uninoculated control (Vaingankar and Rodrigues, 2015). Velu (1988b) obtained maximum number of flowers (101/plant) in pot grown crossandra plants receiving basal FYM + 1 per cent $FeSO_4$ + 2 per cent urea applied as a foliar spray once every 30 days. The control plants produced 55 flowers per plant.

Aruna *et al.* (2007) found that N at 180 kg/ha along with foliar spray of $ZnSO_4$ (0.5 per cent) + $FeSO_4$ (1.0 per cent) recorded increased plant height, number of branches, number of leaves, leaf area, number of spikes/plant, number of flowers/spike and flower yield. Application of increased levels of nitrogen and foliar spray of micronutrients recorded highest yield of 1.54 g, 60.43 g, 99.12 g and 171.59 g at 3rd, 6th, 9th and 12th month after planting.

Prabudoss (2011) reported that biofertilizers *Azospirillum* and phosphobacteria induces or enhances the growth, development and yield of crop plant through 'N' fixation, 'P' solubilization and also by synthesizing growth promoting substances like IAA, GA and cytokinines. The *Crossandra* plants receiving 75 per cent recommended dose of NPK fertilizers along with *Azospirillum* and phosphobacteria registered maximum growth and yield, however, 100 per cent NPK + *Azospirillum* and phosphobacteria was on par with many growth parameters. Hence, it reduces 25 per cent N and P fertilizers by intending *Azospirillum* and phophobacteria.

Rajarajan *et al.* (2014) investigated the effect of Biostimulator *i.e.* seaweed extract of *Gracilaria corticata* on the growth and chemical composition of *Crossandra infundibuliformis*. Spraying of plants with seaweed extract at higher rate increased plant height, photosynthetic pigments, P and K content of leaves, total sugar content and NO_3 content.

Ambika *et al.* (2019) found that application of 100 per cent recommended dose of nitrogen, phosphorus and potassium + *Azotobacter* + PSB + KSB improved the growth, flowering and yield attributes in *cv.* 'Arka Ambara'. Increase in plant height, number of branches, plant spread, flower and yield traits, the early first spike initiation, first flower harvest, number of flowers/spikes and flower yield/ha and it was on par with 75 per cent RDF + *Azotobacter* + PSB + KSB.

Biju *et al.* (2020) studied the effect of micronutrients on growth and flower yield in 'Delhi crossandra'. Foliar application of 0.5 per cent $MnSO_4$ + 0.5 per cent $FeSO_4$ recorded better growth and yield. Increase in plant height, number of branches, number of leaves and spikes/plant, number of flowers/spike, spike length, stalk length, weight of 100 flowers and yield per plant was also observed.

8.4.0 Weed Control

Weeds not only draw water and nutrients but also act as host plant for a number of pests and diseases. Manual weeding is done at early stage of growth.

8.5.0 Pinching

Removal of terminal growing portion of stem, *i.e.*, pinching is practised when young to encourage bushy growth and axillary branches. Maturity of flowers bud is delayed by this practice. Bharathi *et al.* (2018) reported that double pinching was found to be better in increasing number of primary branches, intermodal length of branches, plant spread, number of spikes/plant and spike length (Table 3).

Table 3: Effect of Pinching on Performance of Crossandra under Pot Culture*

Parameter	*Single Pinching*	*Double Pinching*	*No Pinching*	*S.E.m±*	*CD at 5 per cent*
Plant height (cm)	47.40	42.87	48.84	3.37	NS
No. of leaves/plant	154.92	155.05	170.99	2.80	7.81
No. of primary branches/plant	3.56	4.11	3.28	0.12	0.35
No. of secondary branches/plant	3.34	3.46	3.52	0.13	NS
Internodal length of branches (cm)	6.27	6.43	5.86	0.16	0.44
Plant spread NS (cm)	32.08	28.26	29.37	0.96	2.67
Plant spread EW (cm)	12.19	21.13	20.12	0.66	1.85
Days taken for flowering	50.85	63.51	43.29	0.56	1.55
No. of spikes/plant	14.35	15.74	13.46	0.43	1.19
Flower diameter (cm)	4.34	3.72	3.95	0.27	NS
Spike length (cm)	15.97	17.65	17.08	0.20	0.57
Shelf life (days)	19.99	20.36	20.54	0.25	NS

* Bharathi *et al.* (2018).

8.6.0 Pruning

Pruning of crossandra plant improve flower quality and plant vigour. Light pruning is followed in late winter *i.e.*, after flowering. Removal of dead spikes are also an important practices followed in crossandra. Once the seeds are formed and reach the advanced stage of development, growth and flowering are severely affected. After the flowering is over, the spent spikes are removed regularly.

9.0 Diseases and Pests

9.1.0 Diseases

Fungal Diseases

The most damaging disease is wilt caused by *Rhizoctonia* sp. and *Fusarium solani*. Lokesh *et al.* (2008) found that flower rot in crossandra caused by *Fusarium pallidoroseum* which produced purplish flowers, shortened internodes and decayed flowers colonized with cottony colonies of the fungus. Qiu *et al.* (2020) first reported basal stem and root rot caused by *F. solani* on *Crossandra infundibuliformis*. Indiresh *et al.* (1989) reported that *cvs.* Yellow and Bright Orange were susceptible (50 and 85 per cent disease incidence respectively) to *F. oxysporum* f. sp. *solani* while cv. Orange was moderately susceptible (38 per cent). The cultivar Green was resistant with 3 per cent incidence of disease. The characteristic symptoms of the disease are dropping of leaves, marginal yellowing and rotting of roots and rootlets. The invasion by nematode (*Pratylenchus* sp.) will predispose the plant for the fungus to penetrate and cause several wilting. The affected plants should be removed. If the symptoms are noticed at the early stage, drenching the soil with 1 per cent bordeaux or 0.3 per cent thiram will keep the pathogen under control. Jeyarajan

Diseases and Pests of Crossandra

Leaf Spot, Stem and Root Rot, Wilt

Stem Rot, Nematode, Damage by Nematode

Damage by White Fly, Damage by Thrips, Aphid

Whitefly, Phytoplasma (Leaf Yellowing), Alternanthera Mosaic Virus

Source: www.krishi.icar.gov.in, www.ndrs.org.uk, www.gd.eppo.int, www.slideshare.net, www.agritech.tnau.ac.in, www.link.springer.com, www.researchgate.net; Vadiveli *et al.* (1976)

et al. (1987) reported that addition of neem cake to soil reduced wilt (*Fusarium solani*) of crossandra flowers. Treatment with *Trichoderma viride* isolates effectively reduced the wilt incidence in crossandra at field level (Jegathambigai *et al.*, 2009). Application of *T. viride* isolate showed strongest antagonistic activity towards *F. solani* followed by *Aspergillus niger* and *A. terreus*. Both the antagonists gave better disease control when used individually than in combinations. Disease severity of 33 and 40 per cent was recorded when *T. viride* and *A. niger* were applied at 15 g/5 kg soil (Rani and Sivam, 2009).

Ramachandan *et al.* (2000) reported that foot and root rot (*Phytopthora nicotianae*) is a serious problem in *C. undulaefolia* (*C. infundibuliformis*), disease incidence being high in low lying areas. Naidu (2000) reported a serious disease of *C. infundibuliformis* caused by *Sclerotium rolfsii* (*Corticium rolfsii*) in Andhra Pradesh, India.

9.2.0 Pests

Scales (*Saissetia nigra* and *S. viridis*), white fly (*Lipaleyrodes* sp.) and bug (*Gymnencia affinis*) are important insect pests (Sivagami Vadivelu *et al.*, 1976; Sundara Babu and Sivagami Vadivelu, 1980). Spray of methyl preparation at 0.01 per cent or 0.7 per cent phosalone twice at fortnightly interval will control these pests. Hussain *et al.* (1996) recommended the use of neem (*A. indica*) and pongamia (*P. pinnata*) oils, both at 4 per cent, for controlling *Orthezia insignis* on crossandra. They were also more effective than 2 insecticides tested (quinalphos and monocrotophos). *Contarinia* sp. was recorded for the first time as a pest of crossandra near Bangalore, India. The larvae attacked the buds, seriously affecting flower production (Rani and Mohan, 1999). Citrus mealy bud (*Planococcus citri*) found crossandra as its new host plant in India and elsewhere. Since, the application of conventional insecticides failed to control it, biological control were tried in which release of 500 larvae of Australian ladybird beetle (*Cryptolaemus montrouzeri*) in an area of 100 m^2 cleared almost all the mealy bugs after 3 months (Mani and Krishnamoorthy, 2007).

Several species of nematodes have been found to affect crossandras (Sivagami-Vadivelu and Muthukrishnan, 1980). The common ones are *Meloidogyne incognita* (root-knot nematode), *Pratylenchus delattrei* (lesion nematode) and *Longidorus africanus* (needle nematode). Plants affected by these nematodes exhibit stunted growth, curled leaves with purple discolouration, chlorotic symptoms and discolouration of the flowers. The size of leaf and flower is also reduced. The roots affected by lesion nematode show brownish necrotic streaks initially and later the whole roots decay. In case of infestation by root-knot nematode, root galls of various sizes are formed on the roots. The needle nematode causes forking and clubbing of root tips. In case of severe infection, the plants decline and die.

Crossandra should not be raised in nematode infested field. In the nursery, application of nematicides, *viz.*, temik 10 or phorate 10 G @ 25 g per m^2 will control the nematodes. In general, Nagesh *et al.* (1999) reported that mycorrhizal (*Glomus fasciculatum*) inoculation (especially in combination with neem cake) not only reduced nematode infection but also significantly enhanced plant growth parameters and flower yield in pot. Nagesh and Reddy (1995) studied the management of the root-knot nematodes of crossandra, *Paecilomyces lilacinus* and *Verticillium lecanii* with leaf extracts of castor (*Ricinus communis*) and neem (*Azadirachtaindica*) (5 per cent each) and used as a bare root dip and soil drench. Combinations of *V. Lecanii* and *P. lilacinus* (each at 2 × 104 spores/ml) with 5 per cent neem leaf extract resulted in significantly higher plant growth parameters and flower yield. The root gall index was lowest when plants were treated with *V. lecanii* plus neem leaf extract (5 per cent), followed by *V. lecanii* plus castor leaf extract (5 per cent).

Jothi *et al.* (2004) found that neem oil cake was effective in reducing the infestation of *Pratylenchus delatterei* either alone or when coupled with carbofuran 3G with better establishment of seedlings, increased plant biomass and flower yield.

Rao (2007) reported that treatment of nursery bed with the formulation of *Pochonia chlamydosporia* and *Pseudomonas fluorescens* each at 50 g/m^2 and seed treatment with *P. fluorescens* significantly reduced the nematode population in roots and soil with increased per cent of parasitisation and flower yield of crossandra. However, integrated used of both the biocontrol agents did not have pronounced effect on root colonization. Ramakrishnan *et al.* (2014) revealed that talc formation of *T. viride* could effectively manage nematode fungal disease complex (*M. incognita* and *F. oxysporum*) and enhance flower production by 28 per cent.

10.0 Harvesting and Postharvest Management

10.1.0 Harvesting

Flowering starts a month after transplanting. Fully opened flowers are picked once in two days. Flowers should be picked early in the morning by pulling the corolla out of the calyx. But in some places it is the practice to pick the flowers along with ovary so that the weight of the flower is increased.

10.2.0 Yield

Crossandra flowers are very light and on an average about 1500 flowers make on kilogram. Flower yield increases as the plants grow producing more number of laterals. Bhavanisanker and Vanangamudi (1994) recorded 1.474 kg flowers/plant from 17 month old plants.

10.3.0 Postharvest Management

For local markets, the flowers are packed in cloth or polythene bags. Plucked flowers remain fresh for 36 h. Salvi *et al.* (1998) reported that pre-harvest sprays of 5 or 10 ppm CCC and all concentrations of kinetin and triadimefon (5, 10, 15, 20 ppm) increased the longevity by at least one day compared with untreated controls (from 4-5 or 6 days). Of the postharvest sprays, 5 and 10 ppm CCC, 15 ppm kinetin and 15 ppm triadimefon increased longevity by 1 or 2 days. Sreekala *et al.* (1999a) observed that spikes from plants treated with 100 ppm BA without urea had the longest vase life (8.65 days). Highest spike fresh weight was recorded from the 200 ppm GA treatment. BA at 200 ppm + urea had the highest stem length (14.1 cm). Treatment with paclobutrazol (PBZ) at 500 ppm resulted in the lowest postharvest weight loss (0.21 g). Spikes treated with sucrose + $CaCl_2$ had the longest vase life (8.2 days) and the lowest postharvest weight loss (0.30 g).

Harvesting, Value Addition and Marketing of Crossandra

Freshly Harvested Flowers

Garland Making with Flowers

Packaging to Transport from Field (Left, *Source*: IIHR) to Market

Selling in Local Market as Loose Flower

11.0 References

Almeida, J.L., Diniz, J.D.N. and Hernandez, F.F.F. (2021) *Rev. Bras. Hortic. Ornam.*, **14**(2) : 115-122.

Ambika, K., Kumari, V., Naik, R., Akshatha, M.V. and Khalid (2019) *Int. J. Chem. Sci.*, **3**(1): 38-40.

Anonymous (2021) http: //www.plantauthority.gov.in/pdf/DUS per cent 20Guidelines per cent 20Crossandra.pdf

Anonymous (2021a) https: //www.plantgrower.org/uploads/6/5/5/4/65545169/ crossandra_chapter.pdf

Aruna, P. and Mangaiyarkarasi, R. (2020) *Int. J. Chem. Stud.*, **8**(1): 1765-1767.

Aruna, P., Rajangam, J., Rani P.G. and Manivannan, M.I. (2007) *Asian J. Hortic.*, **2**(2) : 169-177.

Aswath, C. Rao, T.M. and Ramachandran, N. (2007) *J. Ornam. Hortic.*, **10**(1): 64-66.

Aswath, C., Rao T.M. and Ramachandran, N. (2009) *J. Ornam. Hortic.*, **12**(3): 211-212.

Bailey, L.H. (1963) *The Standard Cyclopedia of Horticulture*, The Macmillan Company, New York.

Balakrishnan, S., Juliet, H.S. and Dhivya, M. (2015) *Int. J. Agric. Sci. Res.*, **5**(3): 267-272.

Bharathi, T.U., Rao, T.M. and Aswath, C. (2018) *Int. J. Curr. Microbiol. App. Sci.*, **7**(12): 1028-1035.

Bhavanisanker, K. and Vanangamudi, K. (1999) *South Indian Hort.*, **47**: 125-129.

Bhosale, P.B., Kadam, M.B., Katwate S.M. and Pawar, B.G. (2018) *Int. J. Appl. Res.*, **4**(3): 204-206.

Biju, A., Bargavi, G., Bhavani, K., Chinjumol, K., Divyadharshini, V., Indu Rani, C., Jeeva, S. and Masilamani, P. (2020) *J. Sci. Agric. Eng.*, X(XXXIV).

Bini Sundar, S.T., Sadasakthi, A., Kumar, G.A. and Priya, M.V. (2008) *Res. Crops*, **9**(2): 335-337.

Brickell (1996) *A-Z Encyclopedia of Garden Plant*, Dorling Kindersley, London, p. 319.

Dhivya, M., Balakrishanan, S. and Hephziba, S.J. (2015) *Int. J. Agric. Sci. Res.*, **5**(3): 267-272.

Dhivya, M., Balakrishnan, S. and Thangaselvabai, T. (2015) *Int. J. Agric. Sci. Res.*, **5**(3): 199-204.

Gajalakshmi, S. and Abbasi, S.A. (2002) *Bioresour. Tech.*, **85**(2): 197-199.

Gibbs, M.M., Blessington, T.M., Price, J.A. and Wang, Y.T. (1989) *Hort Science*, **24**: 705.

Girija, S., Ganapathi, A. and Vengadesan, G. (1999) *Scientia Hort.*, **82**: 331-337.

Gowsika, S., Kavitha, M., Sankari, A. and Vetrivelkalai, P. (2019) *Int. J. Chem. Stud.*, **7**(3): 3394-3396.

Gowthami, L., Nageswararao, M.B., Umajyothi, K. and Umakrishna, K. (2019) *J. Plant Dev. Sci.*, **11**(2): 85-88.

Gundamaraju, R. and Verma, T.M. (2012) *Int. J. Pharm. Sci. Res.*, **3**(11): 4545-4548.

Hussain, M.A., Puttaswamy and Virakatamoth, C.A. (1996) *Insect Environ.*, **2**: 85-86.

Indiresh, K.M., Uthaiah, B.C. and Setty, T.A.S. (1989) *Curr. Res. Univ. Agri. Sci.*, Bangalore, **18**: 6-7.

Jegathambigai, V., Wijeratnam, R.S.W. and Wijesundera, R.L.C. (2009) *Asian J. Plant Pathol.*, **3**(3): 50-60.

Jeyarajan, R., Bhaskaran, R., Jayaraj, S. and Doraiswamy, S. (1987) *Natural Pesticides from the neem tree Azadirachtaindica A. Juss and other tropical plants*, pp. 635-644.

Jothi, G., Babu, R.S., Ramakrishnan, S. and Rajendran, G. (2004) *Bioresour. Tech.*, **93**(3) : 257-259.

Kalaivani, K., Rajadurai, K.R., Hemaprabha, K. and Kannan, M. (2019) *Int. J. Chem. Stud.*, **7**(2): 1606-1609.

Karuppaiah (2019) *Hortic. Int. J.*, **3**(1): 1-5.

Klock-moore, K.A. and Broschat, T.K. (2001) *HortTech.*, **11**(3): 456-460.

Lalitha, S. and Kumar, N. (1993) *South Indian Hort.*, **41**: 287-292.

Lalitha, S., Kumar, N. and Khader, M.A. (1993) *South Indian Hort.*, **41**: 355-359.

Lokesh, S., Govindappa, M., Raghavendra, V.B. and Sudisha, J. (2008) *J. Gen. Plant. Pathol.*, **74**(4): 338-340.

Mani. M. and Krishnamoorthy, A. (2007) *J. Biol. Control*, **21**(2): 283-285.

Nagesh, M. and Reddy, P.P. (1997) *Mycorrhiza News*, **9**: 12-14.

Nagesh, M., Reddy, P.P. and Rao, M.S. (1999) *Mycorrhiza News*, **11**: 11-13.

Naidu, P.H. (2000) *Indian Phytopath.*, **53**: 496-497.

Prabudoss, V. (2011) *Int. J. Curr. Res.*, **3**(7): 314-319.

Priyanka, T.K., Kamble, B.S., Anuradha, R.W., Naveenakumar and Shirol, A.M. (2017) *J. Pharmacogn. Phytochem.*, **6**(6): 252-254.

Priyanka, T.K., Kamble, B.S., Subiya, R.K., Anuradha, R.W. and Gasti, V.D. (2018) *J. Pharmacogn. Phytochem.*, **7**(1): 487-490.

Pujeri, A., Kamble, B., Tirakannanavar, S. and Gasti, V.D. (2019) *Int. J. Chem. Stud.*, **7**(4): 635-638.

Pulipati, S., Vamsi, A.K.Ch. and Srinivasa Babu P. (2014) *Asian Pac. J. Health Sci.*, **1**(2): 104-106.

Qiu, F., Xu, G., Xie, C.P., Li, X. and Zheng, F.Q. (2020) *Australas. Plant Pathol.*, **49**(6). doi: 10.1007/s13313-020-00735-3.

Rajarajan, R., Selvaraju, S. and Vasanth, M. (2014) *Int. J. Adv. Res. Biol. Sci.*, **1**(8): 29-33.

Ramachandran, N., Purushothaman, S.M. and Veena (2000) *Plant Disease Res.*, **15**: 234-236.

Ramachu, K. and Thangam, M. (2010) *J. Ornam. Hortic.*, **13**(2): 138-141.

Ramakrishnan, S., Devrajan, K., Poornima, K., Deepa, S. and Jain, R. (2014) *Indian J. Nematol.*, **44**(2): 123-125.

Rani, B.J. and Mohan, N.J. (1999) *Insect Environ.*, **5** : 81.

Rao, M.S. (2007) *J. Ornam. Hortic.*, **10**(2) : 110-114.

Rehana, S.K., Madhavi, M., Dorajee Rao, A.V.D. and Subbaramamma, P. (2020) *J. Pharmacogn. Phytochem.*, **9**(2): 1469-1475.

Salvi, B.R., Valsalakumari, P.K., Rajeevan, P.K. and Geetha, C.K. (1998) *J. Orna. Hort.*, **1**: 75-77.

Sivagami-Vadivelu B. and Muthukrishnan, T.S. (1980) *Nat. Sem. Proc. Tech. Comm. Flower Crops,* Tamil Nadu Agriculture University, Coimbatore, pp. 29 & 30.

SivagamiVadivelu, B., Vasantharaj David and Rajamohan, N. (1976) *Pesticides*, **10**: 39-40.

Sreekala, C., Mathew, K.L., George, T.S. and Krishnan, S. (1999a) *South Indian Hort.*, **47**: 351-356.

Sreekala, C., Mathew, K.L., George, T.S. and Rajeevan, P.K. (1999b) *J. Ornam. Hortic.*, **2**: 124-126.

Sreekala, C., Mathew, K.L., George, T.S. and Rajeevan, P.K. (2000) *Hortic. J.*, **13**(2): 65-70.

Sundara Babu, P.C. and SivagamiVadivelu B. (1980) *Pestology*, **4**: 15.

Swaminathan, V. and Sambandamurthi, S. (2000) *Crop Res. Hisar*, **20**: 334-337.

Tejaswi, R., Salma, Z. and Aditya, G. (2020) *J. Pharmacogn. Phytochem.*, **SP6** : 262-266.

Tejaswi, R., Salma, Z., Dorajee Rao, A.V.D. and Naresh, S. (2019) *Bull. Env. Pharmacol. Life Sci.*, **8**(4): 100-104.

Vadivelu, S., David, B.V. and Rajamohan, N. (1976) *Pesticides*, **9**(12): 39-40.

Vaingankar, J.D. and Rodrigues, B.F. (2015) *J. Plant Nutr.*, **38**: 1478-1488.

Valu, G. (1988b) *Res. Dev. Report.*, **5**: 105-106.

Velu, G. (1988a) *Res. Dev. Report.*, **5**: 89-90.

Solidago

K. Dutta and Debmala Mukherjee

1.0 Introduction

Solidago is a beautiful flowering plant of the family *Asteraceae*. It is commonly called 'goldenrod'. The genus name stems from the Latin *'solido'*, which means 'to make whole', and refers to the medicinal properties of the genus. Most of the species are herbaceous perennials, few are semi- woody, found wild in open areas such as meadows, prairies, and savannas. They are widespread throughout the world most of them originating from North America or confined to this part of the world (Beck *et al.*, 2014), a few species are native to South America and Eurasia (WFO, 2020). Some species are found growing wild in Western Ghats of South India.

2.0 Importance and Uses

2.1.0 Aesthetic Uses

Solidago is used as cut flowers, in flower vase for interior decoration and also suitable for dried flower arrangement. Most of the *Solidago* species occur in the spontaneous flora and cultivated as decorative plants, with beautiful yellow flowers, blooming almost year round. In India, goldenrod is grown in tropics, sub- tropics and mild temperate climate as perennial edge and hedge plant; in herbaceous border and also grown in pots. The cut branches with flowers are commonly used in floral arrangement with other flowers, and the display is attractive. *S. canadensis* is cultivated as an ornamental plant in Europe and is used as a bee plant. In north-eastern Croatia, it is still valued by beekeepers (Stefanic *et al.*, 2003). In Europe, *S. canadensis* is considered an environmental weed (Weber, 2003) as this is found in forest edges, dry meadows, wetland edges, rail sides, roadsides and wasteland. *S. altissima* is a species typically found in abandoned fields (Yamato *et al.*, 1999).

2.2.0 Therapeutic and Other Uses

The dry shoot extracts of *S. canadensis* is used in European for urological and antiphlogistical remedy (Apati *et al.*, 2003). The aerial parts of European goldenrod is used for centuries as anti-inflammatory, spasmolytic, and diuretic remedies in the traditional medicine as a urological agent in kidney and bladder inflammation, urolithiasis, and cystitis (Cai *et al.*, 2014). All the species of *Solidago* contain carotenoid- derived yellowish flowers and act as a good source of antioxidant (Horvath *et al.*, 2010).

3.0 Morphology

Solidago is a perennial plant mainly growing from woody rhizomes. The stems are decumbent or erect. Most of the species are unbranched, but some display branching habit in the terminal part. In some species, the basal leaves drop off before flowering. The leaf margins, commonly show serration, some leaves may display trinerved venation. The flower heads have distinct ray and disc florets. Only ray florets are female, others are male, hermaphroditic or entire sterile. Floret corollas are usually yellow in colour, but white ray florets are found in *Solidago bicolor*.

Uses of *Solidago*

As Specimen Plant

Along Road Side

Beside Water Bodies

At Sea Shore

In Flower Vase

In Floral Arrangement

It is short- day plant and bloom in late summer or in early spring. Some species produce abundant nectar when moisture is plentiful, or when the weather is warm and sunny (Semple and Cook, 2003). It is a perennial plant which can attain three waves of flowering per year. The flower heads are usually grouped in complex inflorescences.

4.0 Origin and Distribution

Solidago canadensis is an erect rhizomatous perennial plant native to Mexico, eastern and southern USA and Canada, between the latitudes 26°N and 65°N. It has been introduced in Europe and part of Asia, especially China and Australasia (Webber, 2003). Moron *et al.* (2009) reported that this species is closely related to *S. gigantea*, most invasive species in Europe which possesses large size rhizomes, vigorous growth and allelopathic effect.

5.0 Species and Cultivars

5.1.0 Species

The genus *Solidago* consists of about 190 species. Few examples of cultivated goldenrods include *S. bicolor*, *S. caesia*, *S. canadensis*, *S. cutleri*, *S. riddellii*, *S. rigida*, *S. shortii*, and *S. virgaurea* (Jelitto and Schacht, 1995). The description of some species is given below:

Solidago aestivalis

It is known as swamp wrinkle- leaved goldenrod. This species tends to flower 20- 35 days earlier than *Solidago rugosa* (Blake, 1931).

S. auriculata

It is commonly called clasping goldenrod or eared goldenrod. It is mostly found in rocky forests over calcareous rocks, although it is found along streams. It is a perennial plant, grow up to 20 cm tall, spreading by means of underground rhizomes. Small yellow flower heads are produced in late summer.

S. buckleyi

It is known as Buckley's goldenrod, native to central North America. It is an uncommon species found mainly in the Ozark Mountains of Arkansas and Missouri. *Solidago buckleyi* is a perennial herb with large woody taproots, basal leaves wilt before flowering time (Gleason, 1968).

S. canadensis

It is known as Canadian goldenrod, a herbaceous perennial plant, native to north- eastern and north- central North America. It has upright growing habit with small yellow flowers produced in branched pyramidal shaped inflorescence, flowering generally occurs from July to October (Donald, 2005).

S. gigantea

It is a North American species, commonly known as giant goldenrod. It is a

Species of *Solidago*

S. canadensis  *S. gigantea*  *S. rugosa*

S. lepida  *S. multiradiata*  *S. chilensis*

Cultivars of *Solidago*

Golden Fleece 'Little Lemon' 'Sweety' 'Straehlen Krone'

perennial herb, reaches up to 2 m, spreading by means of rhizomes. The flower heads are yellow, containing both disc florets and ray florets (Semple and Cook, 2006).

S. lepida

It is known as western Canada goldenrod, a perennial herb produce small flower heads, each with ray florets surrounding the disc florets.

S. multiradiata

It is known as rocky Mountain goldenrod or alpine goldenrod, native to North America. It is an erect plant having linear or lance- shaped leaves, sometimes spoon- or spatula-shaped, mostly hairless but serrated edges. The inflorescence is a cluster of several flower heads, each flower head contains many yellow disk florets and ray florets.

S. ptarmicoides

It is known as white flat- top goldenrod or upland white aster, native to the central and eastern Canada. It has distinctive white to cream- coloured flowers. Leaves are narrow and linear. This species prefers dry sandy soils and grassy meadows (Reznicek, 2011).

S. nemoralis

It is known as gray goldenrod or field goldenrod or dwarf goldenrod, is native to North America. Unlike other goldenrods, it is smaller, grows up to 50 cm. The stems are reddish to gray- green and have white hairs. Flowering occurs in late summer (Belt, 2009).

S. velutina

It has a basal rosette of leaves, present up to flowering. One plant can produce as many as 500 small yellow flower heads in a large, branched inflorescence at the top of the plant (Nesom, 1989).

S. virgaurea

It is known as European goldenrod, widespread across most of Europe as well as North Africa and northern, central, and south-western Asia. It is grown as a garden flower with many different cultivars. It flowers profusely in late summer.

5.2.0 Cultivars

The goldenrod cultivars listed below, have unique plant size and blossoms characteristics (Hayes, 2021).

Bluestem: This cultivar grows up to one meter. It blooms from August with clusters of daisy-like flowers on blue- green stems. Bluestem grows well in shady areas and in dry soil.

Golden Fleece: It blooms from August to October with large sprays of golden-yellow flowers. When it's not in bloom, produces heart- shaped foliage.

Seaside: It blooms from August to October, thrives in sandy soil. It grows well in seaside gardens, thriving and loving saline soil.

Little Lemon: Plant height is small, blooms from July to August, grow well in a container. The size of plant also makes it ideal for edging or border plant.

Showy Goldenrod: It blooms from July to September. It has densely packed bright yellow flowers on red stems.

The other important cultivars are Ballardi, Golden Gate, Golden Wing, Monte D'oro, Monte Solo, Peter Pan, Straehlen Krone, Super and Tara Gold.

6.0 Genetics

The basic chromosome number of *Solidago* is n= 9 and there are diploid (2n= 18) (Beaudry and Chabot, 1957), triploid, tetraploid and hexaploid forms (2n= 54) (Kabuce and Priede, 2010). In Europe, only diploid plants have been reported (Musial, 1989).

Chromosome numbers were analysed by Morton *et al.* (2018) for 133 individuals of 41 taxa and cultivars and four hybrids in the goldenrod genus *Solidago* from Canada, the United States and botanical gardens in Europe. They found the chromosome number of *S. juliae*, *S. radula*, S. *mexicana*, and *S. velutina* ssp. *californica*, all having somatic chromosome number, 2*n*= 36. Beaudry and Chabot (1957) proposed a working hypothesis for the origin of the hexaploid *S. altissima* (2n= 54), produced as a result of hybridization between diploid *S. canadensis* or *S. lepida* (2n= 18) and the tetraploid *S. gigantea* (2n = 36) with subsequent doubling of chromosomes. Hybrids between *S. altissima* and *S. virgaurea* were reported from Denmark, Norway and Sweden (Sunding, 1989).

Solidago is generally self-sterile and insect-pollinated plant. The major pollinating agents of *S. canadensis* are honeybees (*Apis mellifera*), bumblebees (species of *Bombus*), soldier beetles (*Chauliognathus pennsylvanicus*) and syrphid flies (Werner *et al.*, 1980).

7.0 Growth and Flowering

Solidago virgaurea complex is thought to produce seeds during a short growing season at high elevations by starting to flower earlier, and the large seed size is advantageous for seedling establishment. Therefore, elevational changes in flowering phenology and reproductive traits are thought to be an adaptation to the short growing season at high elevations (Sakurai and Takahashi, 2016).

Vegetative growth and asexual reproduction are especially vigorous during summer due to high temperature and soil drought stress. But seed germination requires rainy season. Most *S. canadensis* produces flowers between September to January and fruits in late October (Huang *et al.*, 2007).

8.0 Propagation

8.1.0 Seed Propagation

Solidago is commonly propagated by wind- disseminated seeds and spreading underground rhizomes, depending upon the species. *S. canadensis* is propagated

Propagation of Solidago

Seeds of Solidago

Vegetative Propagation through Rhizome

Seedling

Vegetative Propagation in Large Scale

In vitro Propagation of *Solidago virgaurea* (John et al., 2012).

A. Multiple Shootlets formation from the nodal segment of *S. virgaurea* cultured on MS + Kin 4.5 mg/l + IAA 0.5 mg/l;B. Shoot Elongation of *S. virgaurea* cultured on MS + Kin 4.5 mg/l + IAA 0.5 mg/l;C. *In vitro* rooltets formation from the in vitro derived shootlet of S. virgaurea cultured on MS + BAP 0.5 mg/l + IAA 4.5 mg/l;D. Hardened Plantlet of *S. virgaurea*

by both seed and rhizomes. It was reported that seeds required cold treatment for germination. Seeds from North American plants germinated more easily after cold-stratification for 1- 12 weeks. The larger the achene, the higher is the probability of germination. Seeds from European species do not show dormancy and do not need scarification or stratification for germination. The optimum temperature for seed germination is 25- 30°C (Werner *et al.*, 1980).

8.2.0 Vegetative Propagation

Solidago is multiplied by vegetative means with the help of rhizomes. Rhizomes break dormancy at the beginning of April, each rhizome produce only one shoot from the apical tip. After emergence of new leaves, shoot elongation is started. The inflorescence first appears in late July to early August. The new rhizome buds are normally visible in August but growth and development occurs in autumn after fruiting is completed.

8.3.0 Micropropagation

An *in- vitro* propagation protocol has been established from nodal s*et al.* (2012). Highest frequency of shoot proliferation (91.8 ± 0.64) and maximum negments of *S. virgaurea* by John umber (12.1 ± 0.20) per node was observed in Murashige and Skoog's (MS) medium augmented with 4.5 mg/l of Kinetin in combination with 0.5 mg/l of indole- 3- acetic acid. Highest frequency (91.7 ± 1.43) of rooting and maximum number (3.8 ± 0.28) of rootlet per shootlet was achieved on MS medium fortified with 4.5 mg/l of indole- 3- acetic acid in combination with 0.5 mg/l of 6-benzylaminopurine (BAP). Regenerated plants were successfully transferred to field (78 per cent).

In another study on *S. canadensis* by Li *et al.* (2012), the highest callus production from leaf segments are obtained on Murashige and Skoog's medium (MS medium) supplemented with 1.0 mg/l naphthalene acetic acid (NAA) and 1.0 mg/l 6-benzylaminopurine (BAP). Adventitious shoots regenerated directly from leaf explants without an intermediate callus phase with the highest shoot induction percentage of 87.2 per cent. The largest number of adventitious shoots per leaf explant (3.2) was obtained on MS medium supplemented with 0.4 mg/l NAA and 2.0 mg/l BAP. MS medium supplemented with 0.1 mg/l NAA and 1.0 mg/l BAP was the best medium for axillary shoot regeneration from nodal segments.

9.0 Soil and Climate

Generally, solidago prefers sandy loam and red loam soils with proper drainage. It also performs well in poor and dry soil. It can be grown in varied climatic conditions; cooler climate gives high quality cut flowers. *S. canadensis* can tolerate a fairly wide range of soil fertility and texture conditions (Werner *et al.*, 1980).

10.0 Cultivation

10.1.0 Planting

Solidago can be cultivated throughout the year. Soil is ploughed to fine tilth. The bed should be prepared by turning the soil under to a depth of 20- 30 cm.The rhizomes are planted in flat beds at 45 × 45 cm spacing. It prefers full sun, can also tolerate shady condition.

10.2.0 Irrigation

Watering is done normally once in three days to keep the soil always wet and to induce proper growth and development. Goldenrod needs plenty of water in summer months but they can withstand drought conditions at maturity.

11.0 Diseases and Pests

Solidago is infested with a number of diseases and pests. Some common diseases and pests are as follows (Burpee, 2021).

11.1.0 Diseases

11.1.1 Aster Yellows

The common symptoms of aster yellows are stunted growth or excessive growth; petals are deformed and turn green. The disease spreads by leafhoppers. Spray any systemic insecticide to manage leafhoppers.

11.1.2 Powdery Mildew

Powdery mildew occurs on leaves in humid weather conditions. White powdery masses appear on leaf surface. Spraying of wettable Sulphur at 2g/l or Azoxystrobin @ 1m/l is recommended to manage the disease.

11.1.3 Rust

The pathogen causes rusty coloured spots on foliage and stems of the plant. Removal of infected parts of plants is recommended.

11.1.4 Bacterial Leaf Spot

The symptoms appear as small translucent spots with a broad yellowish edge, slowly enlarge and become angular or irregularly circular with a reddish center. The pathogen may also affect the flower heads.

11.1.5 Solidago Virus

Dark green mottling of the leaves noticed on young leaves, turn yellowish and the plant becomes stunted.

11.2.0 Pests

11.2.1 Aphids

Greenish, red or black coloured sucking insects, generally feed on the undersides of leaves. They leave a sticky residue on foliage that attracts ants. Rinse affected plant under running water to remove the pests and spray any systemic insecticide or horticultural oil.

11.2.2 Leafhoppers

Leafhoppers can cause injury to leaves and spread diseases. Insecticides like Pyrethrins or Azadirachtin may be applied to suppress leafhopper populations.

11.2.3 Spider Mites

These are red, black, brown or yellow in colour, suck the plant juices removing chlorophyll and injecting toxins which cause white dots on the foliage. The foliage turns yellow and become dry. They multiply quickly and thrive in dry conditions. It can be controlled by periodically wash down of plants with a strong spray of water and be sure to get good coverage of the lower leaf surfaces. Use insecticidal soap for house plants or broad- spectrum miticide can give good control.

11.2.4 Slugs

These pests feed at night and are mostly a problem in damp weather. They make leave large holes in the foliage or eat leaves entirely.

11.2.5 Spittlebugs

The young insects feed on the leaves and stems. The adults have large frog- like eyes. They have generally one generation each year.

12.0 Harvesting and Yield

Inflorescences appear from June onwards and flowering occurs at the end of July, depending up on the species. Peak flowering time is between mid- August and the end of September, but flowering can continue through October. The small seeds are produced in large numbers. Rhizome is formed and grows in late autumn (Weber, 2000). The first harvesting of flowers is done 75 days after planting and when 25 per cent of the flowers will open, they are harvested for marketing. On an average about 3 lakhs stems per hectare is obtained.

13.0 Biotechnology

13.1.0 Genetic Transformation

Goldenrod has monochromatic yellow carotenoid- containing flowers that cannot be modified using classical breeding approaches due to a limited gene pool. To generate solidago with novel colours through metabolic engineering, Skaliter *et al.* (2019) developed a procedure for its regeneration and transformation. In an attempt to promote anthocyanin biosynthesis, they generated transgenic solidago expressing snapdragon (*Antirrhinum majus*) Rosea1 and Delila, as well as *Arabidopsis thaliana* Production of Anthocyanin Pigment 1 (PAP1) transcription factors. Transgenic cv. Golden Glory expressing cauliflower mosaic virus 35S-driven *PAP1* generated red flowers that accumulated delphinidin and its methylated derivatives, as compared to control yellow flowers in the GUS- expressing plants. This protocol was efficiently used in genetic engineering for novel colouration in solidago.

14.0 References

Beaudry, J.R. and Chabot, D.L. (1957) *Contributions de l'Institut Botanique de l'Université de Montréal*, **70**: 65-72.

Beck, J.B., Semple, J.C., Brull, J.M., Lance, S.L., Phillips, M.M., Hoot, S.B. and Meyer, G.A. (2014) *Appl. Plant Sci.*, **2**: 1300093.

Belt, S. (2009) *USDA NRCS, Norman A. Berg National Plant Materials Center*, Beltsville, MD.

Blake, S.F. (1931) *Flora of North America, Solidago auriculata* Shuttleworth: Eared or clasping goldenrod.

Burpee, W.A (2021) *How to Plant Solidago Perennials,* https: //www.burpee.com/gardenadvicecenter/encyclopedia/perennials/learn-about solidago/encyclopedia__solidago-article.html.

Cai, T., Caola, I., Tessarolo, F., Piccoli, F., D'Elia, C., Caciagli, P., Nollo, G., Malossini, G. Nesi, G. and Mazzoli, S., Palmieri, A., Verze, P., Malossini, G., Mirone, V. and Johansn, T.E.B. (2014) *World J. Urol.*, **32**: 1007-1014.

Donald D.C. (2005) *A Naturalist's Guide to Field Plants: An Ecology for Eastern North America*. Syracuse University Press, pp, 9.

Gleason, H. A. (1968) *The Sympetalous Dicotyledoneae*. **3**: 596.

Hayes, B. (2021) https: //www.gardeningchannel.com/grow-goldenrod-flowers-solidago/

Horvath, G., Molnar, P., Farkas, A., Szabo, L., Turcsi, E. and Deli, J. (2010) *Chromatographia,* **71**: S103–S108. doi: 10.1365/s10337-010-1510-4.

Huang, H., Guo, S. and Chen, G. (2007) *Frontiers of Biology in China,* **2**(2): 196-204.

Jelitto, L. and Schacht, W. (1995) *Hardy Herbaceous Perennials:* A-K; Vol. 2, L-Z. Timber Press. pp. 629.

John, P.P.J., Revathy, I and Johnson, M. (2012) *Res Plant Biol.,* **2**(4): 08- 15.

Kabuce N. and Priede, N., (2010) *NOBANIS - Invasive Alien Species Fact Sheet - Solidago canadensis.*, USA: NOBANIS, 10 pp. https: //www.nobanis.org/globalassets/speciesinfo/s/solidago-canadensis/solidago-canadensis.pdf

Li, J., Kang Y., Qiang, S. and Peng, G. (2012) *Acta Soc. Bot. Pol.,* **81**(1): 53-60.

Morton, J.K., Venn, J. and Semple, J.C. (2018) *Rhodora,* **119**: 341-348.

Musial, K. (1989) *Acta Biol. Cracoviensa, Series Bot.,* **31**: 73-84.

Nesom, G.L. (1989) Taxonomy of *Solidago velutina* (Asteraceae: Astereae) with a new, related species from México.

Reznicek, A. A., Voss, E. G. and Walters, B. S. eds. (2011) *"Solidago ptarmicoides".* Michigan Flora Online. University of Michigan Herbarium.

Sakurai, A. and Takahashi, K. (2016) *Plant Species Biol.,* DOI: 10.1111/1442-1984.12153

Semple, J.C. and Cook, R.E. (2006) *"Solidago gigantea". In Flora of North America Editorial Committee (ed.). Flora of North America North of Mexico* (FNA). 20.

Skaliter, O., Ravid, J., Shklarman, E., Ketrarou, N., Shpayer, N., Ben Ari, J., Dvir, G., Farhi, M., Yue Y and Vainstein, A. (2019) *Front. Plant Sci.* **10**: 1561. doi: 10.3389/fpls.2019.01561.

Sunding, P., (1989) Naturalized Solidago (golden rod) species in Norway. *Blyttia,* **47**(1): 23-27

Toiu, A., Vlase, L., Vodnar, D.C., Gheldiu, A.M. and Oniga, I. (2019) *Molecules*, **24**: 2666.

Werner, P.A., Bradbury, I.K. and Gross, R.S. (1980) *Can. J. Plant Sci.*, **60** (4): 1393-1409.

World Flora Online. Consortium WFO (2020) *World Flora Online*. Available online: http: //www. worldfloraonline.org (accessed on 3 August, 2020).

Genesis of Education, Research and Development of Floriculture and Ornamental Horticulture in India: A Retrospective

Garden of The Agri-Horticultural Society of India, established in 1872, Calcutta- A Pilgrimage

*T.K. Bose**

**Trained in Ornamental Horticulture and Landscaping at the Royal Botanic Garden, Kew, U.K. in 1961; Horticulturist and Secretary, the Agri-Horticultural Society of India, 1960-72. Lecturer, Landscape Gardening (part-time); Former Bengal Engineering College, Shibpur, West Bengal, 1962-1965; Project Co-ordinator, All India Co-ordinated Floriculture Improvement Project, Indian Council of Agricultural Research (ICAR); Professor and Head, Department of Horticulture, Dean, Faculty of Agriculture and Director of Research, (1976- 1996), Bidhan Chandra Krishi Viswavidyalaya, West Bengal, India.*

1.0 Introduction

Floriculture is now a remunerative farming system, particularly for small and marginal farmers supplying flowers and plants to domestic markets, providing employment to thousands of families in flower cultivation, value addition and marketing. The areas under flower crops in India are 3,03,210 ha with a production of 22,63,200 tonnes loose flower and 6,46,530 tonnes cut flower stems during 2018-19 (Table 1), the area being the largest in open cultivation of flowers in the world. In spite of the fact that export market of flowers and plants is growing fast and many developing countries have utilized the opportunities for their economic growth through export of cut-flowers and pot plants, India is yet to create an impact on export trade. Therefore, the research priorities in floriculture in India have been intensified for production of commercially important flowers for domestic market, and also for export, from selection of suitable cultivars, standardization of protected cultivation technology for yield and quality improvement, and application of effective postharvest management to ensure regular supply of consignments of fresh flowers round the year.

Table 1: State-wise Production of Flowers in India during 2018-19*

State	*Area ('000 ha)*	*Production ('000 MT)*	
		Loose Flower	*Cut Flower*
Andhra Pradesh	28.34	530.35	0.00
Arunachal Pradesh	0.00	0.00	0.00
Assam	5.10	32.37	56.16
Bihar	1.40	10.10	0.11
Chhatisgarh	13.75	60.07	59.05
Gujarat	20.50	195.86	0.00
Haryana	5.96	72.80	3.33
Himachal Pradesh	0.71	5.67	12.36
Jammu & Kashmir	0.28	27.00	4.45
Jharkhand	0.98	4.47	4.14
Karnataka	24.76	178.03	75.21
Kerala	53.26	0.08	44.84
Madhya Pradesh	29.83	355.09	0.00
Maharashtra	11.45	59.93	0.03
Manipur	0.03	0.01	0.25
Meghalaya	0.01	0.00	0.02
Mizoram	0.18	0.00	2.25
Nagaland	0.04	0.00	24.40
Odisha	6.61	24.90	28.54
Punjab	2.09	13.07	0.00
Rajasthan	3.48	4.85	0.00

State	Area ('000 ha)	Production ('000 MT)	
		Loose Flower	Cut Flower
Sikkim	0.24	16.50	0.09
Tamil Nadu	37.26	510.21	23.91
Telangana	4.70	22.22	11.72
Tripura	0.00	0.00	0.00
Uttar Pradesh	21.36	46.73	65.81
Uttarakhand	1.56	3.02	14.92
West Bengal	28.49	77.27	214.37
Others	0.83	12.61	0.59
All India Total	303.21	2263.20	646.53

* NHB Horticulture Database (2020).

2.0 Education

Education in horticulture in India was started in 1950's, in few colleges of agriculture in Pune, Banaras, Coimbatore, Agra and Kanpur, where floriculture was a minor part. As a specialized branch of horticulture, course on floriculture and ornamental horticulture was first introduced at the Collage of Agriculture, Calcutta University in 1961, in collaboration with The Agri-Horticultural Society of India, established in 1820, where apart from teaching, intensive research on floriculture was also taken up in M. Sc. (Ag.) and Ph.D. programmes.

In 1960s, teaching and research in floriculture were also started at the Division of Horticulture, Indian Agricultural Research Institute (IARI), New Delhi. Later, a Division of Floriculture and Landscaping was established at IARI. Presently, the scientists of Divisions of Floriculture and Landscaping are actively engaged in teaching and research in floriculture and have been contributing to the development of floriculture.

Education, research and training in floriculture and ornamental horticulture were introduced as one of the courses in the Department of Horticulture after the establishment of Agricultural Universities starting from 1960 onwards and continued for many years. Subsequently, in many Agricultural/Horticultural Universities, Faculty/College of Horticulture was established and Department of Floriculture and Ornamental Horticulture was introduced as a major course in M.Sc. (Ag.) and Ph.D. programmes. At present, there are 70 Agricultural Universities in the country where floriculture is a major subject of teaching and research in horticulture.

Apart from the facilities of education and research in floriculture created by the Universities, the Indian Council of Agricultural Research (ICAR), realizing the importance of education, research and training in floriculture for the economic development of the country, sanctioned a Coordinated Project on Floriculture in 1971, for carrying out intensive research for improvement of quality and production of commercially important flower crops, and increase the demand in domestic and export market. Funds, facilities and staff were provided to the Agricultural

Uses of Flowers and Ornamental Plants

Lotus in Pond

Garland to Worship

God Rose Garden

Flower Arrangement

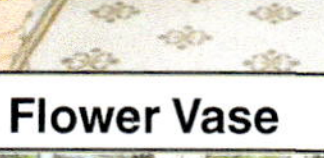

Flower Vase

Design with Ornamental Plants

Lalbagh Botanical Garden

Botanical Garden in Ooty

Uses of Flowers and Ornamental Plants

Pot Plants

Herbaceous Border

Shrubbery

Dry Flower Arrangement

Perfume from Flowers

Plants for Environment Protection

Road Side Plantation

Flowers in Hanging Basckets

Flowers in Window Box

Commercial Flowers in India

Rose Gladiolus Gerbera

Chrysanthemum Tuberose Marigold Jasmine

Carnation Anthurium Orchid

Lilium China Aster Antirrhinum Bird of Paradise

Important Ornamental Plants

Gustavia *Spathodia* *Monodora*

Pyrostegia *Wisteria* *Allamanda*

Euphorbia pulcherrima *Hibiscus* *Bougainvillea*

Lagerstroemia indica *Hydrangea* in Pot *Gardenia*

Important Ornamental Plants

Cymbidium | Dendrobium | Cattleya

Aglaonema | Dieffenbachia | Calathea

Syngonium | Spathiphyllum | Tulip

Dahlia | Amaryllis | Hemerocallis

Universities and also to the research institutes like Indian Agricultural Research Institute (IARI), National Botanical Research Institute (NBRI), Indian Institute of Horticulture Research (IIHR) working for the improvement of flowers.

Teaching and research on breeding and improvement of ornamental plant, protected cultivation (green house) of flowers and pot plants, post-harvest management of cut blooms and loose flowers, biotechnological investigations of important ornamentals are major courses in post-graduate syllabus in floriculture. Apart from improvement in quality and production of flowers, investigations on postharvest management of flowers including grading, packaging, transport and increase in vase life of flowers have been standardised in several institutes.

During the last 50 years, the Agricultural Universities and the Institute have significantly contributed to research, education and development in floriculture and ornamental horticulture.

Expansion of domestic markets of flowers and ornamental plants and simultaneous improvement of quality and production for various uses has contributed immensely to the economic development of the growers in most part of the country.

Education in floriculture and ornamental horticulture has also proved effective in planning and development of landscape garden in suitable locations including community garden, parks in cities and town and in regional planning, throughout the country and their management. Expertise in ornamental horticulture has also an important role in the protection of environment by using plants in landscape planning and its implementation.

In cities and towns nurseries have been established as a source of annual and perennial ornamental plants including trees, shrubs, climbers, bulbous plants *etc.* and application of propagation technology and nursery management practices has improved the quality of planting materials and availability of healthy ornamental plants of enormous types which greatly helped in developing garden and landscaping throughout the country.

2.1.0 Publication of Text and Reference Books in Floriculture

Prior to 1960, almost no authentic and comprehensive book on ornamental horticulture and gardening was published in India. Later, after introduction of floriculture and ornamental horticulture as a subject in M.Sc. (Ag.) and Ph.D courses in Agriculture Universities, several text and reference books were published which proved as valuable publications on the subject. The first authentic publication on ornamental plants and gardening in India was written by Mr. Kasturi Rangan in 1961; Late Mr. L.N. Birla, eminent industrialist and former President of The Agri-Horticultural Society of India, wrote a book on landscape gardening.

The book on roses by Dr. B.P. Pal, published in 1968 was an excellent publication from India. 'Trees in Calcutta' by Mr. A.P. Benthal and 'Beautiful Flowering Trees' by Dr. M.S. Randhawa were popular books on trees.

Apart from publication of text and reference books on fruits, vegetables, medicinal plants, spices and plantation crops, propagation and nutrition of

Books on Floriculture and Ornamental Horticulture Edited by T.K. Bose and Associates

Revised 3rd Editions of 'Commercial Flowers' in Five Volumes

horticultural crops, a series of books on floriculture, ornamental horticulture, and gardening was written/edited by T.K. Bose and his associates which included 'Gardening in India' (Bose and Mukherjee, 1970), Orchids of India, a comprehensive publication of Indian Orchids and their cultivation by Bose and Bhattacharjee, 1980 was accepted as a valuable publication on Indian Orchids, 'Tropical Garden Plants in Colour' with about 1500 colour illustrations, 'Floriculture and Landscaping' (Bose *et al.*, 1999). The 2nd edition of the book entitled 'Commercial Flowers' in two volumes edited by Bose *et al.* (2003), based on updated research review on major commercial flowers proved to be valuable text and reference books, and were accepted worldwide. Third edition of 'Commercial Flowers' is being revised for publication in five volumes with adequate illustrations. A book entitled 'Ornamental Plants and Garden Design in Tropics and Sub-tropics', edited by Bose *et al.* (2015) is an illustrated publication with more than 7000 illustrations, in two volumes. Illustrated publication on 'Trees and House Plants' was accepted as valuable guide on important ornamental plants (Bose *et al.*, 2002). 'Trees of the World' Volume I with text and colour illustrations of about 1500 trees species is a pioneer publication on such large number of trees (Bose *et al.*, 1999). Dictionary of Horticulture in ten volumes with about 7000 pages text and 6000 colour illustrations provides important and comprehensive information on horticulture including flowers and ornamental plants.

The publication on 'Biotechnology of Horticultural Crops' edited by V.A. Parthasarathy and others (2002) in three volumes including important commercial flowers and ornamental plants proved very useful as reference book in biotechnological research. Dr. V.A. Parthasarathy and Dr. C. Aswath have also contributed in the publication of reference book on Floriculture and Ornamental Horticulture.

All these publications mentioned above have been widely used as text and reference books by the students, researchers, teachers in floriculture in the Agricultural Universities and Institutions. The illustrated publications on flowers and ornamental plants edited by T.K. Bose and his associates since 1970 have also proved very useful to professional and amateur growers and landscape architects for planning and implementation of landscape gardens and protection of environment. All these publication had worldwide demand.

In the meanwhile, valuable publication on floriculture and ornamental horticulture has been edited by Dr. S.K. Bhattarcharjee and Dr. R.L. Mishra and their associates.

3.0 Research

3.1.0 The Agri-Horticultural Society of India

In India, scientific research on floriculture and ornamental horticulture was first taken up at The Agri-Horticultural Society of India (AHSI), in the garden in Calcutta, established in 1872, the first landscape garden in India with thousand of species and cultivars of ornamental plants. Hybridization and selection in *Hibiscus, Ixora, Cassia, Plumeria* and bulbous plants like *Amaryllis, Hemerocallis* and *Crinum* by S. Percy Lancaster, the then Secretary of the Society, between 1930 and 1954 produced many improved cultivars. The Society is also the pioneer institution in India to carry out systematic and intensive research on the improvement of floriculture and ornamental horticulture from 1962-1980 at the Birla Laboratory, and in the garden of the society. Facilities for research on floriculture including temperature controlled glasshouse, mist propagation chamber and tissue culture laboratory, first of its kind in India were provided for research.

Major research investigations carried out at the society included:

i) Regulation of growth and flowering of ornamental plants (shrubs and annuals) by using chemicals. Growth regulating chemicals like Cycocel, B-Nine, Phosfon, Paclobutrazol proved effective in developing compact and attractive pot plants with improvement in flowering, depending on plant species and chemicals used.

ii) Investigation on nutritional requirements of commercial flowers-carnation, aster, antirrhinum, petunia, dianthus *etc.*

iii) Propagation of hard-to-root cuttings from about 200 species and cultivars of perennial ornamentals under mist, their survival and growth.

The Agri-Horticultural Society of India, Kolkata

The AHSI Garden

Dr. Rev. William Carey

Laboratory of AHSI

Mussaenda 'Queen Sirikit'

Bougainvillea 'Million Dollar'

Alpinia zerumbet 'Variegata'

Introduced by T.K.Bose at AHSI

Royal Botanic Gardens, Kew, UK

A View of Royal Botanic Gardens, Kew

Victoria Lily under Greenhouse, Royal Botanic Gardens

Physiology of adventitious root formation in cuttings was also investigated intensively.

iv) Micropropagation of orchids, bulbous plants and house plants *in vitro* and selection of elite cultivars. Protocols of micropropagation were standardized and published in scientific journals.

v) Effect of growth, flowering and seed formation in ornamental plants (shrubs, bulbous plants and annuals) in temperature controlled glasshouse in Birla Laboratory.

vi) Improvement of cultivars and cultivation practices in tropical bulbous plants like *Hippeastrum, Hemerocallis,* and tuberose.

vii) Standardization of cultivation practices and management of house plants (indoor) in different environments (light and temperature), growing media and nutrition.

viii) Apart from the submission of thesis for M.Sc. (Ag.) and Ph. D. degree results of the investigations were published in scientific journals in India and abroad.

ix) The research projects were carried out in collaboration with College of Agriculture, Calcutta University, and Bidhan Chandra Krishi Viswa Vidyalaya (BCKV, after 1974), and financed by ICAR, CSIR, UGC and Industry. More than 20 research associates received Ph.D. degree from Calcutta University and BCKV after submission of thesis, based on the result of research on the above mentioned projects on the improvement of ornamental plants between 1965 and 1976. Many M.Sc. (Ag.) students in Horticulture with specialization in Floriculture successfully completed research at the Society, as a part of the course in horticulture.

The research associates who worked under the guidance of T.K. Bose, former secretary, The Agri-Horticulture Society of India, on different research projects became pioneer and eminent scientists in floriculture and ornamental horticulture in India.

3.2.0 ICAR-Indian Agricultural Research Institute (IARI), New Delhi

Investigations on the improvement of roses through breeding and selection by Dr. B.P. Pal, Former Director General (ICAR) in collaboration with the Division of Vegetable crops and Floriculture, IARI, New Delhi evolved several outstanding cultivars of roses *viz.,* 'Dr M.S. Randhawa', 'Dulhan', 'Lalima', 'Maharani', 'Delhi Princess', 'Rose Sherbet', 'Raja Surendra Singh of Nalagarh', 'Belle of Punjab', 'Chamber di Kali'.

Interspecific and intervarietal hybridization in gladiolus at IARI also produced several outstanding cultivars with beautiful flowers. Significant research on chrysanthemum, hollyhock, amaryllis, marigold and other ornamental plants have been also carried out. *(see chapter 1 in revised Commercial Flowers, volume 3 – Breeding and Improvement of Ornamental Plants).*

Dr. Vishnu Swaroop was a pioneer, and initiated research on rose, marigold, gladiolus *etc.* in 1960s at the Division of Vegetable Crops and Floriculture at IARI. A centre of research of the All India Coordinated Research Project in floriculture was also located at IARI.

3.3.0 ICAR-Indian Institute of Horticultural Research (IIHR), Bengaluru

A Division of ornamental crops at IIHR started in 1968 with the objectives on development of improved varieties of ornamental plants along with appropriate production technology including protected cultivation. Over the years, a good number of varieties and hybrids for open field as well as for protected cultivation have been released for cultivation among the farmers throughout the country. It had made significant contributions in the improvement of rose, hibiscus, china aster, gladiolus, tuberose, crossandra, carnation, gerbera, jasmine, chrysanthemum and annual flowering plants. The institute also carries out work on post harvest management and dehydration of ornamental crops. Research on floriculture at IIHR was intensified after a centre of all India coordinated research project on floriculture was sanctioned in 1972. In accordance with the technical programme, investigations on the improvement of flower production in chrysanthemum, aster, gladiolus, were undertaken. During the last five decades IIHR has contributed immensely in the development of improved cultivars of important commercial flowers and also standardized the cultural practices for increasing flower production including greenhouse cultivation.

IIHR has developed several high yielding and attractive rose varieties *viz.*, 'Arka Pride', 'Arka Ivory', 'Arka Sukanya' *etc.* Rose variety 'Arka Red' has bright red colour with double flower, best suitable for growing in outside beds and as a cut flower, yields about 40 flowers per plant per year. A large number of fragrant genotypes belonging to *Rosa hybrida, R. centifolia, R. damascena, R. chinensis* and bourbon roses have been evaluated for yield and quality. Three rose varieties *viz.*, 'Arka Sinchana', 'Arka Sharmeeli' and 'Arka Kinnari' have been identified as potential garden rose varieties. Different rose genotypes have been evaluated against black spot of rose caused by *Diplocarpon rosae* under natural infection in 2019. Rose variety 'Arka Pink' (Hybrid IIHR 16-8) was developed which have semi double flowers and a shelf life of more than 7 days.

Three varieties of China aster *viz.*, 'Arka Aadya', 'Arka Archana', 'Arka Kamini' have been evaluated for landscape use. Chrysanthemum variety 'Arka Pink Star' has been evaluated for day neutral trait and pot culture. Some other varieties of chrysanthemum released from IIHR are 'Chandrika', 'Indira', 'Kirti', 'Pankaj', 'Rakhi', 'Red Gold', 'Yellow Gold' *etc.* In marigold, development of a petaloid male sterile line has been identified in both orange and yellow colour group. French marigold, 'Arka Honey', a dwarf photo insensitive variety with double coloured flower, bears flowers throughout the year with a yield of 5- 6 tons per acre and variety 'Arka Pari', also photo-insensitive and orange in colour has been released for cultivation. Many of the gladiolus varieties like 'Arka Pratham', 'Arka Gold', 'Arka Aayush', 'Arka Amar'; tube rose cultivars 'Shringar', 'Suvasini', 'Prajwal'

Research Institutes and State Agricultural Universities Engaged in Teaching and Research on Floriculture and Ornamenal Horticulture

ICAR-IARI Campus, New Delhi

Dr. B.P. Pal, FRS
Ex-DG, ICAR,
Pioneer Rose Breeder

TNAU, Coimbatore

Gerbera Cultivation in Greenhouse, BCKV, W.B.

ICAR-Directorate of Floriculture Research, Pune

ICAR-NRC Orchids, Sikkim

ICAR-IIHR, Bengaluru

CSIR-NBRI, Lucknow

PAU, Ludhiana

ICAR-NBPGR, New Delhi

Tissue Culture Lab, KAU, Kerala

and 'Vibhav'; chrysanthemum cultivars 'Arka Kirti', 'Arka Chandrika'; Crossandra cultivars 'Arka Ambara', 'Arka Kanaka' have become very popular among the farmers. Some bougainvillea cultivars like 'Chitravati', 'Dr. H.B. Singh', 'Jawaharlal Nehru', 'Purple Wonder', 'Sholay', 'Usha' *etc.* and some Hibiscus varieties like 'Anuradha', 'Basant', 'Bharat Sundari', 'Geetanjali', 'Priya', 'Smt. Indira Gandhi', 'Dilruba' *etc.* have also been released by IIHR. Extensive investigation has also been carried out to standardize cultural practices for improving the production of commercial flowers in the field and under protected condition.

3.4.0 National Botanical Research Institute, Lucknow (Uttar Pradesh)

Extensive investigations at NBRI on mutation breeding in chrysanthemum and tuberose resulted in the production of several outstanding cultivars. Breeding and selection of bougainvillea and gladiolus was also major programme in the improvement of ornamental plants.

NBRI has also developed many improved varieties of several ornamental plants. The important cultivars among them released are 'Amar Kiran', 'Amar Parvarti', 'Summer Queen', 'Amar Tarang' in *Amaranthus tricolour*; 'Archana', 'Begam Sikander', 'Mary Palmer Special', 'Mahara Variegata', 'Chitra', 'Dr. B.P.Pal', 'Parthasarathy', 'Wazid Ali Shah' *etc.* in Bougainvillea; 'Archana', 'Basant Bahar', 'Gazel', 'Manmohan', 'Priyadarshini', 'Rim Jhim', 'Sada Bahar', 'Pitamber' *etc.* in gladiolus; 'Angara', 'Salmon Beauty', 'Sharada', 'Sukumari' in rose; 'Rajat Rekha', 'Swarna Rekha' in tuberose. Some large flower mutant of chrysanthemum *viz.*, 'Aruna', 'Asha', 'Kanak', 'Pitamber' and some small flowered mutants like 'Agnishikha', 'Alankar', 'Basanti', 'Manbhavan', 'Surekha Yellow', some bud sports like 'Kasturba Gandhi', 'Sonar Bangla', 'White Cloud' of chrysanthemum have also been developed at NBRI.

3.5.0 ICAR-National Research Centre for Orchid (NRC Orchids), Sikkim

The Sikkim state authorities handed over 22.19 acres of land belonging to Regional Agricultural Centre along with all other assets to ICAR for establishment of NRC Orchids. In 1997, a campus for research on temperate orchids was established in Pakyong, Sikkim with a major focus of research on collection, characterization, evaluation, conservation and utilization of available germplasm in north eastern region for orchids. Now-a-days the centre involved in the development of attractive marketable varieties/hybrids, molecular characterization, standardization of agro-techniques, post- harvest management, and production of quality planting materials through tissue culture and creation of repository of information related to all aspects of orchids in the country. However, the climate of the centre is not suitable for successful cultivation of tropical and mild sub-tropical orchid of commercial importance like *Vanda, Dendrovbium, Cattleya, Phalaenopsis etc.*

3.6.0 ICAR-Directorate of Floricultural Research (DFR), ICAR

In view of the importance of floriculture in India, ICAR has established Directorate of Floricultural Research in 2009 during the XIX Group Meeting of All

Improved Cultivars of Ornamental Plants

Pusa Priya | Arka Honey | Pink Dona

R.R. Pal | Intense | Elite

Priscilla | Rose 'Madhosh' | *Bougainvillea* 'Thima'

Him Pushkar | Apeldoorn | Trumpet

India Coordinated Research Project on Floriculture, held at IARI, New Delhi to promote and strengthen floricultural research and enhance the technological base in floriculture. Initially it was located at IARI campus. In 2014, the Directorate was shifted to its new location *i.e.* College of Agriculture Campus, Shivajinagar, Pune to intensify research and development of floriculture and ornamental horticulture.

3.7.0 All India Coordinated Floriculture Improvement Project

In order to improve the cultivars of commercially important flowers and their production, a co-ordinated project on floriculture improvement was sponsored and sanctioned by the Indian Council of Agricultural Research, Govt of India in 1971, when Dr. B.P. Pal** was the Director General, ICAR. The centres of the coordinated project were located in different agricultural universities and institutes for carrying out investigations on the improvement in flower production of elite cultivars in different States and locations of various agro-climatic regions. Successful implementation of the research programmes in most of the centres of the project has contributed significantly in the improvement and development of floriculture in the country. There has been a steady increase in the demand of flowers and ornamental plants in India for which the area under flower cultivation production per unit area and nursery industry have also increased enormously during the last five decades.

The scientists in the project located in the agricultural universities also participate in teaching, research and training as members of the faculty of Horticulture and also in the Department of Floriculture and Landscaping now existing in many universities. Successful implementation of coordinated project has enormous impact on floriculture improvement. Facilities for research on biotechnology, postharvest technology of flowers, and establishment of greenhouse for improving the production of flowers were also provided through the project.

Results of research on improvement in production of commercial flowers in the coordinated projects are analysed critically in the workshop of the project and improved technology of production is recommended for adoption by the growers. Similarly, improved varieties identified from research are also released for commercial cultivation. As a result of intensive research and trials made at the centres of the coordinated projects significant development has taken place in floriculture and ornamental horticulture in India.

Greenhouse Cultivation

As a result of extensive investigation on the management of greenhouse and production technology in many agricultural universities and institutes in India, successful cultivation of commercial flowers like rose, gerbera, chrysanthemum, carnation, anthurium, gladiolus, lilium and foliage plants are now being grown under protection (greenhouse) for the improvement in production of high value flowers and plants of superior quality which are now available in most of the cities

---

** *I take this opportunity to express my respectful gratitude to Dr. B.P. Pal, F.R.S. Former Director General, Indian Council of Agricultural Research, Govt. of India, for his love towards Nature, particularly flowers and ornamental plants. He was also an eminent rose breeder of international repute.*

Greenhouse and Open Field Cultivation

Rose in Greenhouse

Carnation

Anthurium

Seedling Production

Mist Unit in Greenhouse

Gerbera Cultivation

Marigold Cultivation in Open

Tuberose Cultivation in Open

Gladiolus Cultivation in Open

and towns at a reasonable price has been standardized. Improvement of quality and production of cut flowers and pot plants by growing under protection was also a major objective of AICRP on Floriculture.

3.8.0 Research at the Agricultural Universities in India

3.8.1 Bidhan Chandra Krishi Viswavidyalaya (BCKV), West Bengal

Extensive investigations on major commercial flower crops were carried out successfully at the experimental station of the Department of Horticulture, BCKV through the implementation of a centre of co-ordinated project on floriculture. Selection of high yielding cultivars with superior flower quality and standardization of cultivation practices in many important commercial flowers in India, widely used as cut flower and loose flowers including aster, antirrhinum, gladiolus, marigold, tuberose, chrysanthemum (for cut and loose flower), rose, carnation and dahlia (pompon for cut flower) were carried out from 1976 to 1996 under the guidance of T.K. Bose. Later, the Department of Horticulture at BCKV became a Faculty of Horticulture with five departments including Floriculture and Landscaping. The results of the investigations which were communicated to the growers through demonstration, workshop and bulletins proved to be of immense importance in increasing flower production in the State and also in the country. As a result, commercial production of flowers became highly remunerative and the area under flower cultivation extended from about 8,000 to 28,490 hectares in West Bengal, one of the largest flower growing states in the country, in about four decades. Intensive investigations on the post harvest management of cut flowers and loose flowers have also been taken up. Hybridization, selection and mutation breeding have resulted in the development of cultivars of superior quality and higher flower production of marigold, chrysanthemum and tuberose which have been accepted by the commercial growers in the State and elsewhere. Many students and research associates have successfully completed M.Sc.(Ag.) and Ph.D. courses in floriculture and landscaping of the faculty.

3.8.2 Other Agricultural Universities

Significant contributions through research and development in flowers of commercial importance have been made by the Punjab Agricultural University, Ludhiana, Tamil Nadu Agricultural University, Coimbatore; University of Agricultural Sciences, Bangalore; Agricultural College, Pune; Kerala Agricultural University; Sher-e-Kashmir Agricultural University, mainly because of traditional demand of flowers for daily use and favourable climate for flower and seed production, greenhouse cultivation and nursery management. Improvement in floriculture was possible due to research in both temperate and sub-tropical climates in Himachal Pradesh from the facilities provided in the coordinated project to the Dr. Y.S. Parmar University of Horticulture and Forestry, Solan; Navsari Agricultural University, Navsari, Gujarat also made significant contribution in floriculture development in the state. Many other Agricultural Universities in the country are also intensively working for the improvement of Floriculture in different parts of India.

4.0 Development in Floriculture and Ornamental Horticulture

4.1.0 Introduction of Ornamental Plants in India

Introduction of ornamental plants in India was first made by the Moghul Emperors for their formal garden decoration in various parts of the country, mainly conifers and roses. Later, the British rulers in India and the Portuguese introduced many plant species and cultivars of economic and ornamental importance.

4.1.1 The Agri-Horticultural Society of India

During the last 200 years, The (Royal) Agri-Horticultural Society of India founded by Rev. Dr. William Carey in 1820, and The (Royal) Indian Botanic Garden (established in 1787) introduced hundreds of species and cultivars of ornamental plants which included trees, shrubs, climbers, annuals and biennial flowering plants, house plants, palms, cacti and other succulents, orchids, bulbous plants, roses, chrysanthemums, dahlias and many others, mainly from tropical America, Europe, South and tropical Africa, China, Japan, South-East Asia *etc.*, from early nineteenth century and thereafter. The plants introduced through The Agri-Horticultural Society of India, Indian Botanic Garden, Indian Forest Research Institute proved to be of immense importance as the source of planting materials of ornamental importance in the country for developing landscape gardens and plantations for their aesthetic value, improving environment by reducing pollution level, contributing to the economic development of people from commercial floriculture including export of flowers and plants.

The Agri-Horticultural Society of India, also developed the first informal landscape garden in Calcutta, 1872 which became a popular garden design in India and elsewhere, using ornamental plants of many species and cultivars including those introduced by the Society.

Apart from the ornamental plants which constituted a major group, the Agri-Horticultural Society of India introduced many types of vegetables, temperate fruits which were grown in different agro-climatic regions in the country for testing their performance. Subsequently, seeds and planting materials of introduced plants were distributed for extending their cultivation in suitable areas. The society also contributed to enormous agricultural development in India from 19th to early 20th century by introducing improved cultivars of many field crops from different parts of the world. Introduction of various horticultural and plantation crops, hundreds of species of ornamental plants and their performance in different parts of India had been reported in the Journal of the Agri-Horticultural Society, first published in 1827, Transactions and Bulletins of the society.

Among the ornamental trees, species of *Acer, Acacia, Adansonia, Bauhinia, Brownea, Cryptomeria, Cotoneaster, Cupressus, Colvillea, Erythrina, Eucalyptus, Lagerstoemia, Monodora, Plumeria, Pachira, Spathodea, Tabebuia* and many others were introduced by the Agri-Horticultural Society and the Indian Botanic Garden. Among ornamental shrubs, species and cultivars of *Abelia, Bauhinia, Bougainvillea, Brunfelsia, Calliandra, Dombeya, Eranthenmum, Euphorbia, Fuchsia, Gardenia, Hibiscus, Hydrangea, Ixora, Mussaenda, Nerium, Petrea arborea, Polyscias, Portlandia, Rhododendron*

(Azalea), *etc*. were introduced which are important garden plants in tropical and subtropical climate of India. *Mussaenda erythrophylla* 'Queen Sirikit' and 'Light Pink', *Bougainvillea* 'Million Dollar' and *Alpinia zerumbet* 'Variegata', introduced by T.K. Bose from Thailand in 1966 through the Society proved to be very popular garden plants in tropical parts of the country. Many species of climbers including *Allamanda cathartica, Beaumontia grandiflora, Echites caryophyllata*, species of *Lonicera, Ipomoea, Passiflora, Pyrostegia venusta, Wisteria sinensis, etc*. were introduced in India. Many species of bulbous plants including *Achemenes, Amaryllis, Caladium, Dahlia, Eucharis, Hippeastrum, Haemanthus, Hemerocalis* and seeds of annual flowering plants now widely grown throughout the country like *Antirrhinum, Dianthus, Phlox, Petunia, Tagetes* (marigold) *Verbena* and others were also introduced by the Society in the nineteenth and early twentieth century and grown in its gardens for distribution of seeds and planting material.

Many attractive cultivars of rose, chrysanthemum, dahlia, bougainvillea, species and cultivars of *Solenostemon* (coleus) and *Begonia*, first introduced through the Society are popular garden plants for decades.

Many species and cultivars of orchids of commercial importance as cut flowers and pot plants including *Cattleya, Dendrobium, Phalaenopsis, Vanda, etc*., were first introduced in the country by the Society.

Species and cultivars of *Aglaonema, Calathea, Dieffenbachia, Monstera, Neoregelia, Philodendron; etc*, introduced in Society's garden are now popular house plants in pot and also grown in ground in semi-shade in tropics and subtropics in the country.

Among hundreds of ornamental plants introduced by the Agri-Horticultural Society of India and through other sources, a few important species and cultivars are illustrated herein. The introductions made by the society used to be exhibited by the growers to the public for creating awareness on the scope of agricultural and horticultural development. The first exhibition in agriculture and horticulture was organized by the society in 1828.

4.1.2 ICAR-National Bureau of Plant Genetic Resources (NBPGR), New Delhi

This Institute has been introducing plants of economic importance including ornamentals plants from different countries for testing their performance, research and commercial cultivation of elite types during the last 6 decades. Many plants of ornamental importance brought to India through NBPGR are summarized hereunder.

Some notable introduction of rose cultivar are 'Spotless Pink', Altissimo Deep Red', 'Crimson Scarlet', 'Doris Tysteman', 'Dutch Gold', 'Fragrant Delite', 'Swan Lake,' 'Topekar', 'White Cocode'. About thirty three species of *Rosa* were introduced from Australia, Canada, Denmark, Germany, Holland, U.K. and U.S.A. and amongst them, *Rosa canina, Rosa laxa, Rosa hugonis, Rosa pomifera* are important, A large number of exotic genotypes of *Anemone, Begonia, Chrysanthemum, Dahlia, Freesia, Gladiolus, Gloxinia, Hyacinthus, Iris, Narcissus, Lilium* and Tulips were introduced. In gladiolus, 158 cultivars were introduced from USA, UK, Canada, Egypt, Israel and USSR.

Gladiolus cardinalis, Gladiolus byzantinus were introduced from Canada, Israel; *Gladiolus communis* from Denmark, Hybrids of *Freesia* 'Olympaid Jones' and 'Olympiad Gloe' from Denmark. NBPGR introduced 74 collections of carnation from Australia, Canada, Cyprus, France, Germany, UK and USA. Among species, *Dianthus spiculiformis, Dianthus fragrans, Dianthus pinifolius, Dianthus balbisii, Dianthus barbatus* and *Dianthus plumarius* were also introduced by NBPGR. Two species of cyclamen, *Cyclamen persicum* and *Cyclamen herifolium,* were introduced from Denmark, Germany and U.K.

In flowering annuals, notable exotic introductions include species and cultivars of *Aster, Antirrhinum, Calendula, Celosia, Delphinium, Escholtzia, Impatiens, Lathyrus, Petunia, Stock, Salpiglossis, Tagetes, Tropaeolum, Viola* and *Zinnia; Viola tricolour* and *Viola wittrockina* were introduced from Canada, France, UK and USA. *Zinnia angustifolia* is an introduction from Australia and France. Other notable introductions of annuals are *Antirrhinum majus* from Australia, France and China, *Calendula officinalis* from France and cultivars of *Tagetes erecta, Tagetes patula and Tagetes tenuifolia* from Australia, Holland, Slovakia and USA. A large number of exotic cultivars of orchids belonging to the genera *Cymbidium, Cattleya, Dendrobium* and *Paphiopedilum* were also introduced.

Many ornamental plants have also been introduced by amateurs and professional growers, which have proved to be important garden plants. During the last three decades, many cultivars of rose, chrysanthemum, dahlia, gerbera, anthurium, gladiolus, carnation, annual flowering plants, orchids and foliage plant have been introduced in India through various sources, which have enriched the germplasm of ornamental plants in the country for further improvement and commercial cultivation.

4.1.3 The Indian Botanic Garden, Kolkata (West Bengal)

The garden established in 1787, introduced a large number of species of trees, shrubs, climbers and other perennial plants many of which are now being grown as ornamental plants in the gardens of our country.

4.1.4 Botanic Garden in Lalbagh and Ooty

Botanic garden at Lalbagh has introduced large number of species and cultivars of ornamental plants suitable for tropical and sub-tropical regions many of which are now popular garden plants. In the Botanic garden at Ooty, many species of ornamental plants suitable for cultivation in temperate regions have been successfully introduced.

5.0 Advances in Floriculture and Ornamental Horticulture in India

Advances in floriculture and ornamental horticulture in India started about 5 decades ago which have been expanding and flourishing with the production of good quality flowers and plants and simultaneous increase in domestic and export trade. In order to cope with technological advances, various research institutes and agricultural universities are providing training facilities to students and potential entrepreneurs for promoting establishment of modern hi-tech production centres in floriculture and ornamental horticulture. Both Govt. of India and the States are also providing financial and technical assistance to make projects viable and profitable.

It is relevant to mention few commercial organisations which have contributed in the development of floriculture in India.

In the development of commercial floriculture including production of elite cultivars, standardization of protocol for protected cultivation (green house), hybrid seed production, and facilities of micropropagation for exports of seeds, cut flowers, plants and plantlets are considered as achievements in floriculture in India. The advancement in floriculture and ornamental horticulture also significantly promote in development of the country.

5.1.0 The Indian Society of Ornamental Horticulture

In order to promote development of Floriculture and Ornamental horticulture, a national seminar was organized in 1965 in Kolkata, in which about 90 horticulturists from all over India participated and the Society of Ornamental Horticulture was established. T. K. Bose was the founder secretary of the society. Later the office of the society was shifted at the Division of Floriculture and Landscaping, ICAR-IARI, New Delhi. The society organises seminars and publishes a journal 'Journal of Ornamental Horticulture' at regular interval.

5.2.0 Indo-American Hybrid Seeds (IAHS), Bengaluru, Karnataka

The Indo-American Hybrid Seeds (IASH), a pioneering Indian company was started by Dr. Manmohan Attavar in the year 1965 to produce hand pollinated hybrid petunia seeds. F_1 hybrid seeds of *Antirrhinum, Geranium, Impatiens, Nicotiana,* and ornamental chilli were also produced in climate controlled greenhouses under close supervision of specialists. In addition, open pollinated seeds of several important flowering plants are produced and supplied to domestic and overseas customers.

IAHS Pvt. Ltd. has perfected the art of growing healthy and attractive indoor potted plants. These plants are produced in greenhouse/net house and acclimatized to indoor environment prior to sale. Product mix is continually updated to satisfy customer's preference. There is a separate tissue culture laboratory for the multiplication of elite clones of different plant species. It also imparts laboratory technique courses to graduate and post-graduate students of biotechnology from educational institutions. IAHS has marketing centres in cities and towns in India and it also provides extension support to the farmers and growers to promote the use of quality seeds and plants for the improvement in the production of vegetables and flowers.

5.3.0 KF (Kumar Florist) Bio-plants, Pune, Maharashtra

KF Bio-plants is the largest plant Biotech Company in India. A joint venture of Kumar properties and Florist Holland B.V., is the leading producer and supplier of quality plants of various floricultural and ornamental like orchids, gerbera, anthurium, carnation *etc.*, forestry plants, fruit plants, high-value vegetable seeds and seedlings. The company is a global producer with the annual production to the

Activities of KF (Kumar Florist) Bioplants and Indo-American Hybrid Seeds (IAHS)

Flower Show at KF Bioplants

KF Bioplants Greenhouses

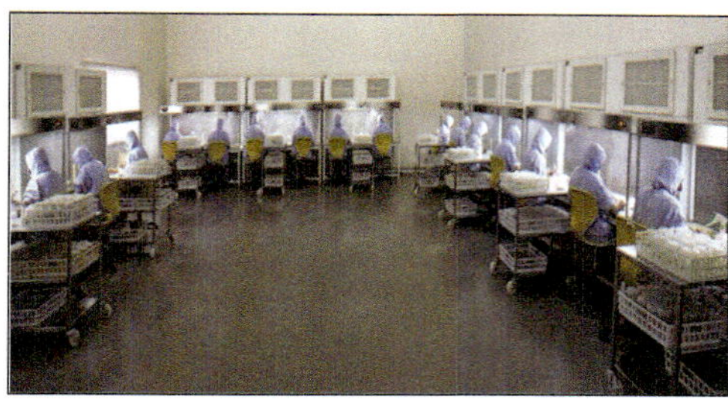

KF Bioplants Tissue Culture Laboratory

KF Bioplants Nursery Area

Anthurium at IAHS

Nursery Area of IAHS

Gardening Centre of IAHS

tune of 60 million plantlets. Driven by innovation, KF Bio-plants, an ISO 9001: 2008 certified company has been a catalyst of success for more than 22 years in the field of horticulture. KF Bio-plants contributes 90 per cent of the domestic floriculture market making it the largest plants biotech company in India and exports its world-class products to more than 30 countries. It exports planting materials to various countries mainly, The Netherlands, Australia, Japan, South Korea, Indonesia, Pakistan, Kenya, Sri Lanka, Mauritius, Nepal, Italy, New Zealand, Mexico, USA, Columbia and Other South American countries. Technical tie-ups with global giants allow the company to bring the best products and processes that match stringent international standards. Advantages of KF Bio-plants are its collaboration with global leaders in floriculture and horticulture. It is a treasure house of *in vitro* or hardened plants, ornamentals and also other high value horticultural crops. Constantly seeking to bring the best in the world to the growers, the company has tie-ups with the commercially successful companies.

6.0 Lapses in Floriculture Development in India

Apart from cultivation of field crops and vegetables, flowers are also intensively grown as commercial crops by small and marginal farmers with small holding throughout the country. Small farmers living in remote villages, earning their livelihood from the cultivation of flowers and value addition and not often aware of cultivars of commercial flowers of better quality, higher yield and improved method of cultivation which have adversely affected their efforts for generation of more income and ensure livelihood. The growers of flower in open field like those of other perishable crops are often compelled to sell their produce (cut blooms and loose flowers) at throwaway prices, before they lose their freshness and they are often deprived of fair price.

Though post-harvest management practices of some high value cut blooms like rose gladiolus, lilium, chrysanthemum *etc.* are adopted by the florists for keeping the freshness of flowers of longer periods, most of the cut and loose flowers from annuals which constitute the largest bulks of flowers produced in the country are not treated except to keep the stem in cut flowers in fresh water. Existing methods of packaging and transportation to wholesale and retail market damage the flowers and reduce their freshness within very short periods. About 30 per cent of cut and loose flowers brought to market by the growers and sellers are damaged at the end of the day and often remain unsold. In the majority of flower markets in India, adequate storage arrangements have not been provided to reduce the post-harvest loss of flowers and increase income of the growers.

6.1.0 Suggestions for Further Development and Generation of Income of the Growers

The lapses jeopardizing sustainable livelihood due to low income of thousands of small and marginal growers of flowers in different parts of India are surmountable. The department of horticulture in the different States in the country should form a separate sell for development of floriculture and solve the problems including production of quality flowers, formation of Farmers Producers Organization (FPO)

Marketing of Flowers in India and the Netherlands

Selling of Flowers in Shop in India

Selling of Flowers in Supermarket in India

Selling of Loose Flowers in Roadside Market in India

Selling of Flowers in Wholesale Market in India

Aalsmeer Flower Auction, the Netherlands

Quality Planting Material Production and Vase Life Extension of Ornamentals

Micropropagation Protocols of *Dieffenbachia* (Left) and Cacti (Right)

Vase Life Extension of Eustoma (Left) and Rose (Right)

Vase Life Extension of different Ornamentals

of flower growers, creation of adequate facilities for post-harvest management of cut and loose flowers including grading, packaging, prolonging of vase life, cool chain system and storage to keep the flowers fresh for a long time and increase the income of growers.

7.0 Important Suggestions for the Improvement of Teaching Floriculture and Ornamental Horticulture in India

7.1.0 Establishment of Model Landscape Garden with Major Features

From my experience in teaching, research and development of Floriculture and Ornamental Horticulture, as a landscape consultant and editor/author of books on the subject including those with thousands of colour illustrations, which have created a significant impact on the knowledge of floriculture in India and abroad, for the last 60 years, I have realised that the students, specializing in Floriculture and Ornamental Horticulture/Landscaping in most of the Agricultural Universities in India (about 70) are not provided necessary scope and facilities for identifying hundreds of species and thousands of cultivars of ornamental plants in the major groups – trees, shrubs, climbers, house plants herbaceous foliage plants, orchids, cacti and other succulents, bulbous plants, aquatic ornamentals, annual and biennial flowering plants and many others. Hence, this students specializing in the subject are not aware of the importance and uses including landscaping, growth and flowering habit, methods of propagation, cultivation in open and greenhouse, training and pruning, control of diseases and pests of many species and cultivars of ornamentals, which are of immense importance to develop a concept in Ornamental Horticulture in totality and ideas for further improvement.

In view of the facts mentioned above, it is suggested that a model landscape garden should be established in the agricultural university campus before introducing the course of floriculture and ornamental horticulture.

Research on breeding and improvement of ornamental plants would prove futile without knowing the existing germplasm, in the absence of a garden for teaching and research.

7.2.0 Facilities for Protected Cultivation of Ornamental Plants for Teaching and Research

Protected cultivation of ornament plants for various uses – greenhouse cultivation of flowers and pot plants for improving the quality and production, raising of seedling, installation of mist propagation unit in polyhouse for large scale multiplication of perennial ornamentals from cuttings and hardening of propagated plants are of major importance to promote production and cultivation of ornamental plants. Suitable protection is also needed for the establishment and management of nursery of ornamental plants. Environment controlled greenhouse with automation systems are often used in temperate climate for the production of cut flowers and pot plants throughout the year.

7.3.0 Development of Facilities for Application of Biotechnology for Teaching and Research

During the last four decades, application of biotechnology has developed important methods for research investigations on the improvement of horticultural crops in general, and ornamental plants in particular, including protoplast fusion, identification of molecular markers, genetic transformation, cryopreservation of germplasm *etc*. Successful micropropagation of many species and cultivars of ornamental plants *in vitro* has proved very effective in large scale multiplication of disease- free ornamental plants throughout the year and in a short time. It is therefore suggested that necessary facilities should be made available in the Department of Floriculture and Ornamental horticulture for teaching and research.

7.4.0 Development of Facilities for Teaching and Research on Harvesting and Postharvest Management of Flowers and Plants

In order to reduce post harvest loss of perishable ornamentals including cut blooms, loose flowers and pot plants and maintain their freshness for a longer period, apart from determining the optimum stage of harvest, technology for effective postharvest management including grading, packaging, storage and methods of vase life enhancement needs to be standardized and necessary facilities provided for education and research to the students and researchers.

7.5.0 Integrated Pest Management: Theory and Practice

In floriculture, integrated pest management and avoiding the use of toxic chemical pesticides, which is often ignored should be a major subject of instruction and application not only to reduce cultivation cost but also to protect the growers and users from health hazard. Flowers and ornamental plants are often used for indoor decoration and should therefore be free from toxic pesticides.

Non-pesticide management (NPM) practice developed in India has been accepted as a technology to eliminate the use of toxic pesticides in agriculture by FAO.

7.6.0 Teaching, Research and Development of Orchids for Domestic and Export Markets

Although India is one of the most important centres of origin (genera and species) in warm, sub-tropical and temperate regions, including commercially important *Vanda, Dendrobium, Phalaenopsis, Cymbidium, Paphiopedilum,* species, hybrids and cultivars of which are commercially important as cut flowers and pot plants throughout the world. In spite of favourable climate in many locations in India, suitable for commercial orchid cultivation in large scale for supplying to domestic and export market, no serious effort has so far been made in the country. The research centre of orchids in Sikkim is suitable for growing temperate orchids only. However, the demand for warm and subtropical orchids is much higher. It is therefore suggested that the department of floriculture and ornamental horticulture in Agriculture Universities located in suitable climate for cultivation of warm and sub- tropical orchids, should provide emphasis on teaching, research and development for their commercial exploitation. It is relevant to mentioned that Thailand, mainly tropical in climate exports orchids (cut flowers and pot plants) amounting to millions of US dollar per year.

7.7.0 Application of Irrigation Systems in the Cultivation of Ornamental Plants

Systems for irrigating ornamentals in greenhouse, polytunnel or in the field or containers, to promote growth, yield and quality of plants and save water sources need to be standardised through demonstration and practice.

7.8.0 Organic Cultivation of Flowers for Commercial Purpose in Open Field

Area under cultivation of flowers for commercial purpose in India in open field is the largest in the world and the growers are mostly poor, small and marginal farmers. In the present flower cultivation practices, use of chemical fertilizers and toxic pesticides not only increases the cost of cultivation but also adversely affects the yield and quality of flowers. The flowers thus produced lose freshness in a short time, compared to those grown using organic matter only. In order to improve the yield and quality of flowers at a lower cost of cultivation and fertility of soil, it is recommended that organic cultivation practices of flowers under field condition should be standardized through teaching and research for the development of floriculture in the country.

7.9.0 Production of F_1 Seeds and Postharvest Management of Seeds of Ornamental Plants for Maintaining Quality and Viability

In developed countries, improvement of plant growth, yield and quality of flowers of large number of annual and biennial flowering plants have been achieved by using F_1 seeds during the last many decades. In order to improve the yield and quality of annuals and biennials grown in India, it is also suggested to provide a thorough knowledge on technology of F_1 seed production and postharvest management of seeds to the students and researchers and scope of testing and evaluating the performance of F_1 seed.

7.10.0 Production of Pot Plants and their Management

In view of the shortage of garden space in ground, lapses in establishment and management in garden, use of pot plants (flowering and foliage), suitable for decorating spaces in domestic and corporate houses, in celebrations and public functions, has been increasing steadily, particularly in towns and cities and nurseries are supplying container grown plants adequately to meet the demand of the customers.

Technology of production of different groups of ornamental plants in containers and their management to keep the plants attractive for long period needs to be standardized, which should be an important subject for development of floriculture.

8.0 Conclusion

Education, research and training in floriculture and ornamental horticulture have also enhanced capacity building of the floriculturists and increased success rate in the implementation of projects relating to flower production, nursery development and planning and establishment of garden for higher employment and income generation through enormous floricultural development in India. The current scenario of the world floriculture research reveals a paradigm shift from traditional methods to modern technologies. However, floriculture research in India requires structural change and strategic orientation in R&D towards motivation for diversification and modernization. In order to ensure supply of high quality uniform plant materials in large quantity, development of low cost, environmental friendly, and economically viable production technology is essential. Post-harvest management should be fine tuned and storage capacity should be ensured with the studies on in-depth physiological and biochemical changes related to flower senescence process. The adverse effects of climate change and variability affecting crop productivity require intervention to produce climate resilient varieties which can perform more efficiently under adverse weather conditions.

In Appreciation

It is my pleasure to acknowledge the dedicated and enormous contributions of Dr. Vishnu Swarup, Dr. S. K. Bhattacharjee and Dr. R.L. Mishra (Indian Agricultural Research Institute), Dr. C.R. Muthukrishnan (Tamil Nadu Agricultural University), Dr. J. S. Arora (Punjab Agricultural University), Dr. L.P. Jadav and Dr. R.G. Maity (Bidhan Chandra Krishi Viswavidyalaya), Dr. P. Das (Regional Plant Resource Centre, Bhubaneswar), Dr. P.N. Khusoo and Dr. R. K. Roy (National Botanical Research Institute) and Dr. Manmohan Attavar (Indo-American Hybrid Seeds, Bengaluru), in teaching, research and development of Floriculture and Ornamental Horticulture in India since 1960s.

9.0 References

Bidhan Chandra Krishi Viswavidyalaya, Mohanpur, Nadia, West Bengal, https://www.bckv.edu.in/news_details.php.

Bose, T.K. (1965) *Handbook of Shrubs*, Oxford IBH Publishing Co., Kolkata.

Bose, T.K. (2019) *Key Note Address, Bicentinary Celebration of The Agri-Horticultural Society of India*, Kolkata.

Bose, T.K. and Bhattacharya, S.K. (1980) *Garden Plants*, Oxford IBH Publishing Co., Kolkata.

Bose, T. K. and Bhattacharjee, S.K. (1980) *Orchids in India*, Naya Prakash, Kolkata.

Bose, T.K., Chowdhury, B. and Sharma, S.P. (1991) *Tropical Garden Plants in Colour*, Horticulture and allied publishers, Kolkata.

Bose, T.K., Choudhury, B., Sharma, S.P. and Pal, P. (2004) *Garden Plants in Colour-* House Plants, Naya Udyug, Kolkata.

Bose, T. K., Das, P. and Maity, G.G. (1999) *Trees of the World*, Regional Plant Research Centre, Bhubaneswar. India, Vol 1.

Bose, T.K., Maity, G.G., Sanyal, D. and Sadhu, M.K. (2007) *Dictionary of Horticulture*, Vols. **1-10**, Naya Udyug, Kolkata.

Bose, T.K., Maity, R.G. and Dhua, R.S. (1999) *Floriculture and Landscaping*, Naya Prakash, Kolkata.

Bose, T.K. and Mukherjee, D. (1970) *Gardening in India*, Oxford IBH Publishing Co., Kolkata.

Bose, T. K., Singh, L.J., Sadhu, M.K. and Maity, T.K. (2015) *Ornamental Plants and Garden Design in Tropics and Subtropics*, Astral International, New Delhi. Vols **1& 2**.

Bose, T.K., Yadav, L.P., Pal, P., Das, P. and Parthasarathy, V.A. (2002) *Commercial Flowers*, Naya Prokash, Vols **1&2**.

CSIR-National Botanical Research Institute, Pratap Marg, Lucknow, https://nbri.res.in/

Das T.M. (2012) *Indian Biologist*, **20**, 205.

Division of Floriculture and Landscaping, Krishi Anusandhan Bhawan, New Delhi, India, https://www.icar.org.in/content/horticultural_division.

Division of Floriculture and Medicinal Crops, IIHR, Hessaraghatta, Bengaluru, https://www.iihr.res.in/division-flower-and-medicinal-crops.

ICAR-NRC for Orchids, Pakyong, Sikkim, https://www.nrcorchids.nic.in/index.php/en/

ICAR-National Bureau of Plant Genetic Resources, New Delhi, http://www.nbpgr.ernet.in/

NHB Horticulture Database (2020) National Horticulture Board, Gurugram, Haryana.

Pal, B.P. (1961) *The Rose in India*. ICAR, New Delhi.

Randhawa, M.S. (1961) *Beautiful trees and Gardens.* ICAR, New Delhi.

Randhawa, M.S. (1983) *Flowering Trees*. National Book Trust of India, New Delhi.

Index

E

F

G

H

I

J

K

L

Q

R

U

V

W

X

Y